Apress®

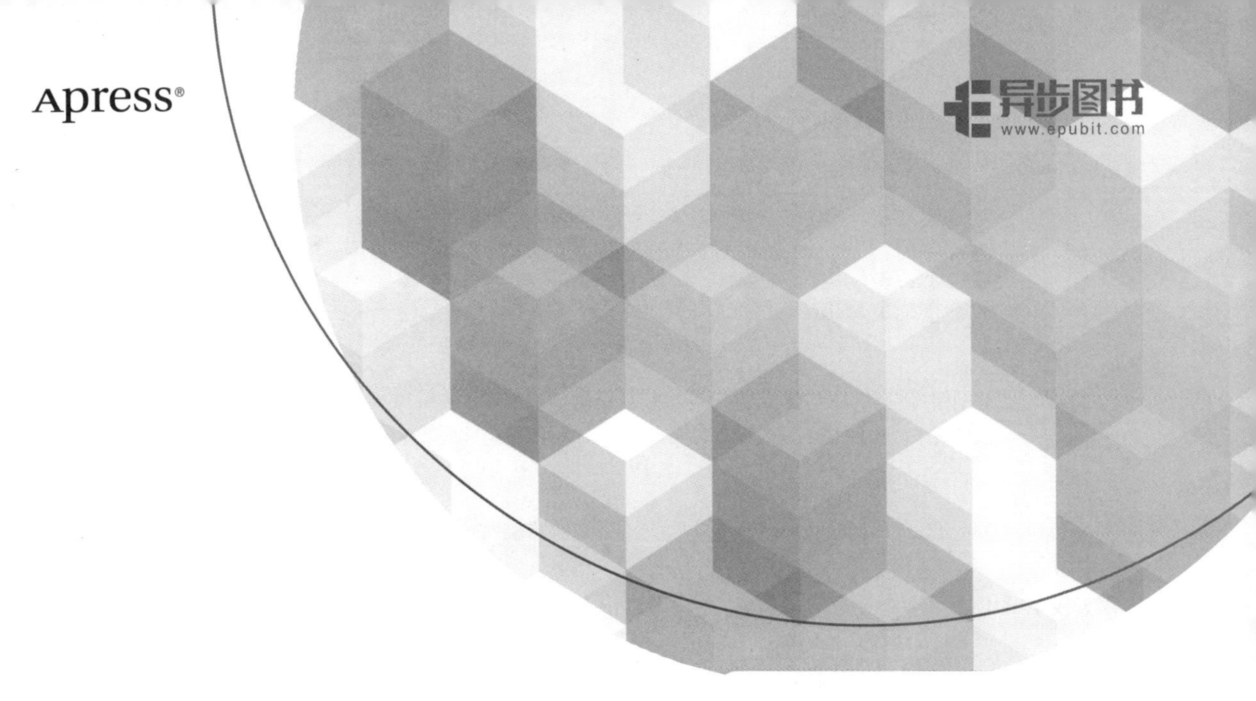

精通ASP.NET Core MVC（第7版）

Pro ASP.NET Core MVC 2

Seventh Edition

[美] 亚当 · 弗里曼（Adam Freeman） 著

郝冠军 孙臻 闫小迪 张渧易 译

人 民 邮 电 出 版 社

北 京

图书在版编目（CIP）数据

精通ASP.NET Core MVC : 第7版 / (美) 亚当·弗里曼 (Adam Freeman) 著 ; 郝冠军等译. -- 北京 : 人民邮电出版社, 2021.6
ISBN 978-7-115-55961-6

Ⅰ. ①精… Ⅱ. ①亚… ②郝… Ⅲ. ①网页制作工具一程序设计 Ⅳ. ①TP393.092.2

中国版本图书馆CIP数据核字(2021)第021780号

版权声明

◆ 著 [美] 亚当·弗里曼（Adam Freeman）
译 郝冠军 孙 臻 闫小迪 张淯易
责任编辑 谢晓芳
责任印制 王 郁 焦志炜
◆ 人民邮电出版社出版发行 北京市丰台区成寿寺路 11 号
邮编 100164 电子邮件 315@ptpress.com.cn
网址 https://www.ptpress.com.cn
北京市艺辉印刷有限公司印刷
◆ 开本：787×1092 1/16
印张：44.25
字数：1437 千字 2021 年 6 月第 1 版
印数：1 – 2 000 册 2021 年 6 月北京第 1 次印刷
著作权合同登记号 图字：01-2019-4802 号

定价：169.90 元
读者服务热线：(010) 81055410 印装质量热线：(010) 81055316
反盗版热线：(010) 81055315
广告经营许可证：京东市监广登字 20170147 号

内 容 提 要

本书详细解释 ASP.NET Core MVC 的架构、功能和应用，深入阐述构建现代可扩展的 Web 应用程序的工具、技术和方法，揭示如何为.NET Core 平台创建轻型的移动端应用程序。本书主要内容包括 MVC 模式、C#基本特性、Razor、Visual Studio、MVC 应用程序的单元测试、实际应用程序的创建、URL 路由、高级路由特性、控制器、依赖注入、过滤器、API 控制器、视图、视图组件、标签助手、模型绑定、模型验证、ASP.NET Core Identity、模型约定和操作约束等。

本书适合.NET 开发人员和 Web 开发人员阅读，也可供计算机相关专业的师生阅读。

作 者 简 介

Adam Freeman 是一位经验丰富的 IT 专家，曾在一系列公司担任高级职务，最后的职务是一家全球性银行的首席技术官（Chief Technology Officer，CTO）和首席运营官（Chief Operating Officer，COO）。退休以后，他喜欢写作和长跑。

技术审校者简介

Fabio Claudio Ferracchiati 是一名高级咨询师、高级分析师/开发人员，擅长使用微软的技术。他目前就职于 BluArancio 公司。他是微软认证的.NET 解决方案开发人员、微软认证的.NET 应用程序开发人员、微软认证的专业人士，还是一位笔耕不辍的作者。在过去十年中，他为意大利和国际杂志撰文，在计算机领域与他人合著了十几本图书。

推　荐　序

虽然 ASP.NET Core MVC 3 已经发布了，但是本书的翻译和出版非常及时，内容非常不错，满足了成千上万希望学习.NET 的开发人员对更好的学习资料的强烈需求，感谢本书的译者花时间翻译了一本好书，帮助中国的开发者利用 ASP.NET Core MVC 构建应用。

本书是 ASP.NET 方面的畅销书，作者 Adam Freeman 在本书里对 ASP.NET Core MVC 进行了详细介绍，说明了如何从 ASP.NET Core MVC 中获得最大收益。结合具体示例和代码，本书展示了 ASP.NET Core MVC 的方方面面，解释了如何使用 ASP.NET Core MVC 构建 Web 应用程序。

本书围绕 ASP.NET Core MVC 深入探讨构建可扩展的 Web 应用程序所需要的工具和技术，有助于开发人员为.NET 平台生成更精简的应用程序。

张善友
微软最有价值专家、腾讯云最有价值专家、华为云最有价值专家
深圳市友浩达科技有限公司 CEO

服务与支持

本书由异步社区出品，社区（https://www.epubit.com/）为您提供相关服务和支持。

提交勘误

作者和编辑尽最大努力来确保书中内容的准确性，但难免会存在疏漏。欢迎您将发现的问题反馈给我们，帮助我们提升图书的质量。

当您发现错误时，请登录异步社区，按书名搜索，进入本书页面，单击“提交勘误”，输入勘误信息，单击“提交”按钮即可（见下图）。本书的作者和编辑会对您提交的勘误进行审核，确认并接受后，您将获赠异步社区的 100 积分。积分可用于在异步社区兑换优惠券、样书或奖品。

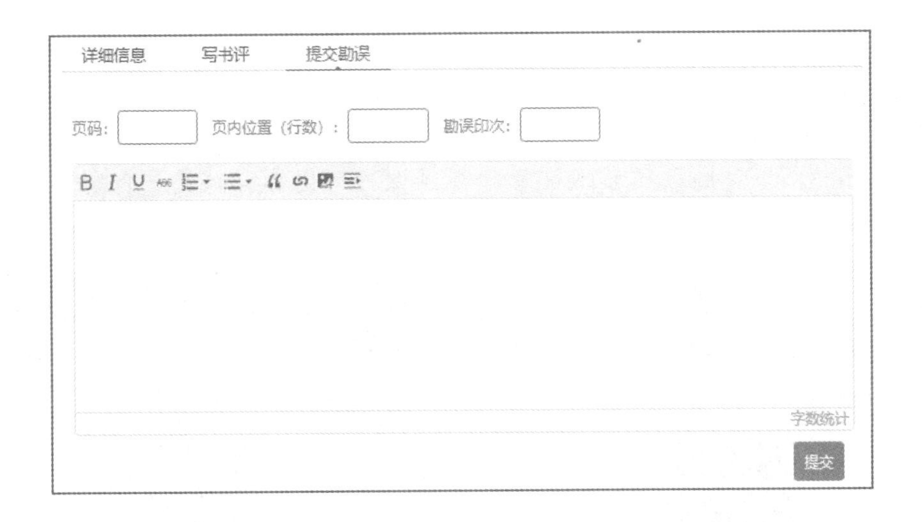

扫码关注本书

扫描下方二维码，您将会在异步社区的微信服务号中看到本书信息及相关的服务提示。

与我们联系

我们的联系邮箱是 contact@epubit.com.cn。

如果您对本书有任何疑问或建议，请您发邮件给我们，并请在邮件标题中注明本书书名，以便我们更高效地做出反馈。

如果您有兴趣出版图书、录制教学视频，或者参与图书翻译、技术审校等工作，可以发邮件给我们；有意出版图书的作者也可以到异步社区在线投稿（直接访问 www.epubit.com/contribute 即可）。

如果您所在学校、培训机构或企业想批量购买本书或异步社区出版的其他图书，也可以发邮件给我们。

如果您在网上发现有针对异步社区出品图书的各种形式的盗版行为，包括对图书全部或部分内容的非授权传播，请您将怀疑有侵权行为的链接通过邮件发送给我们。您的这一举动是对作者权益的保护，也是我们持续为您提供有价值的内容的动力之源。

关于异步社区和异步图书

“异步社区”是人民邮电出版社旗下 IT 专业图书社区，致力于出版精品 IT 图书和相关学习产品，为作译者提供优质出版服务。异步社区创办于 2015 年 8 月，提供大量精品 IT 图书和电子书，以及高品质技术文章和视频课程。更多详情请访问异步社区官网 https://www.epubit.com。

“异步图书”是由异步社区编辑团队策划出版的精品 IT 专业图书的品牌，依托于人民邮电出版社近 30 年的计算机图书出版积累和专业编辑团队，相关图书在封面上印有异步图书的 LOGO。异步图书的出版领域包括软件开发、大数据、人工智能、测试、前端、网络技术等。

异步社区

微信服务号

目　　录

第一部分　ASP.NET Core MVC

第二部分 ASP.NET Core MVC 详解

第一部分　ASP.NET Core MVC

ASP.NET Core MVC 对于使用微软平台的 Web 开发人员来说是一次彻底的转变。它强调清晰的架构、设计模式和可测试性，并且不会试图隐藏 Web 的工作方式。

本书第一部分旨在介绍 MVC 开发的基本概念，包括 ASP.NET Core MVC 中的新功能，并在实践中体验框架的使用方式。

第 1 章　ASP.NET Core MVC 背景

ASP.NET Core MVC 是一个来自微软的 Web 应用程序开发框架，它结合了模型-视图-控制器（MVC）体系结构的有效性和整洁性、敏捷开发的想法和技术，以及.NET 平台的最佳部分。在本章中，你将了解微软创建 ASP.NET Core MVC 的原因，看看它如何与其前身和替代品进行比较。最后，本章将概述 ASP.NET Core MVC 中的新特性以及本书所涵盖的内容。

1.1　ASP.NET Core MVC 的历史

最早的 ASP.NET 是在 2002 年推出的，当时微软热衷于保护其在传统桌面应用程序开发中的主导地位，并将互联网视为威胁。图 1-1 说明了当时出现的 ASP.NET Web Forms 技术栈。

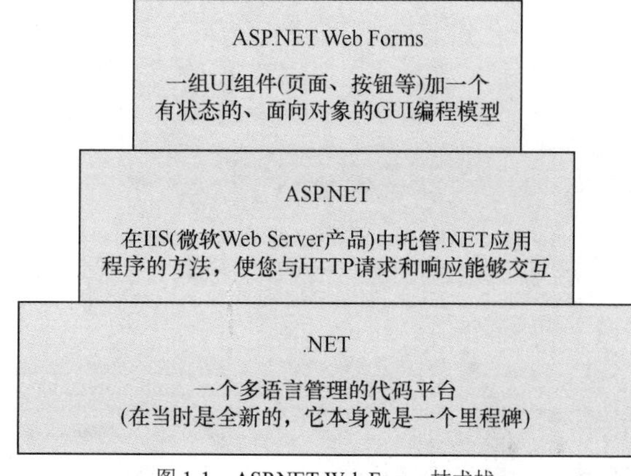

图 1-1　ASP.NET Web Forms 技术栈

1.1.1　ASP.NET Web Forms

微软试图使用 ASP.NET Web Forms 将用户界面（User Interface，UI）模拟为服务器端控件对象层，隐藏超文本传输协议 HTTP（本身是无状态的）和超文本标记语言 HTML（当时许多开发人员对此还不熟悉）。每个控件在请求之间保持自己的状态，在需要时自动渲染为 HTML，并将客户端事件（如按钮单击）与服务器端相应的事件处理程序代码相关联。实际上，Web 窗体是一个巨大的抽象层，旨在通过 Web 传递经典的事件驱动的图形用户界面（Graphics User Interface，GUI）。

其思想是使 Web 开发的体验与开发桌面应用程序一致。开发人员可以基于有状态的用户界面进行考虑，而无须使用一系列独立的 HTTP 请求和响应。微软可以使 Windows 桌面开发人员向新型 Web 应用程序开发领域实现无缝转型。

ASP.NET Web Forms 存在的问题

传统的 ASP.NET Web Forms 开发从原则上来说曾经很好，但事实证明它更加复杂。

- View State 权重：跨请求维护状态的实际机制（称为“View State”）导致在客户端和服务器之间传输大量数据。即使是大小适中的 Web 应用程序，这些数据也可能会达到几百千字节，并且每次请求都会来回传递数据，从而导致响应时间更慢，并增加了服务器的带宽需求。
- 页面生命周期：连接客户端事件与服务器端事件处理程序代码（页面生命周期的一部分）的机制

可能会变得复杂和棘手。很少有开发人员在不产生 View State 错误或发现某些事件处理程序莫名失败的情况下，能在运行时成功操纵控件层。

- 关注点分离的错误理念：ASP.NET Web Forms 的代码隐藏模型提供了将应用程序代码从 HTML 标记中移除到单独的代码隐藏类中的方法。这是为了分离逻辑层和表现层，但实际上，又鼓励开发人员将表现层代码（如操纵服务器端控件树）与其应用程序逻辑（如操纵数据库数据）混在这些怪异的后台代码类中。最终的结果可能是难以理解的。
- 对 HTML 的有限控制：服务器控件将其自身呈现为 HTML，但不一定是你想要的 HTML。在 ASP.NET 的早期版本中，HTML 输出无法符合 Web 标准，或者不能很好地利用层叠样式表（CSS），并且服务器控件生成难以预测且复杂的 ID 属性，这些属性难以使用 JavaScript 访问。这些问题在最近的 ASP.NET Web Forms 版本中有所改进，但是获取你期望的 HTML 仍然是比较困难的。
- 有漏洞的抽象：ASP.NET Web Forms 尽可能隐藏 HTML 和 HTTP。当你尝试实现自定义行为时，你经常会放弃抽象，这迫使你对回发事件机制进行逆向工程，或执行笨拙的操作以使其生成所需的 HTML。
- 低可测试性：ASP.NET Web Forms 的设计人员无法预料到自动测试将成为软件开发的重要组成部分。他们设计的紧密耦合架构不适合单元测试。集成测试也可能是一个挑战。

ASP.NET Web Forms 并非一无是处，实际上，微软为提高标准合规性和简化开发流程付出了巨大的努力，甚至从原始的 ASP.NET MVC 框架中获取了一些功能，并将其应用于 ASP.NET Web Forms。当你需要快速的结果时，ASP.NET Web Forms 表现优异，你可以在一天内拥有一个相当复杂的 Web 应用程序。但除非你在开发过程中足够小心，否则你会发现你创建的应用程序难以测试和维护。

1.1.2 起初的 MVC 框架

2007 年 10 月，微软发布了一个基于现有 ASP.NET 平台的新开发平台，旨在直接回应对 ASP.NET Web Forms 的批评和竞争平台（如 Ruby on Rails）的普及。新平台称为 ASP.NET MVC 框架，并反映了 Web 应用程序开发的新兴趋势，如 HTML 和 CSS 标准化、REST Web 服务、有效的单元测试以及开发人员应该接受 HTTP 的无状态特性的想法。

支持最初 MVC 框架的概念现在看起来很自然而且显而易见，但是它们在 2007 年的.NET Web 开发世界中是缺乏的。ASP.NET MVC 框架的引入使微软的 Web 开发平台重新回到了现代。

MVC 框架还表明微软的态度发生了重大变化，微软以前曾试图控制 Web 应用程序工具链中的每个组件。现在微软基于 jQuery 等开源工具构建了 MVC 框架，从竞争（并且更为成功的）平台中获得了设计约定和最佳实践，并将源代码发布到 MVC 框架，供开发人员审查。

起初的 MVC 框架存在的问题

微软在创建 MVC 框架时，基于现有的 ASP.NET 平台，这是有道理的，因为该平台具有很多固有的底层特性，ASP.NET 开发人员都熟知和理解这些特性。

但是，将 MVC 框架移植到最初为 ASP.NET Web Forms 设计的平台上是需要妥协的。MVC 框架开发人员逐渐喜欢使用配置设置和代码调整，来禁用或重新配置对 Web 应用程序没有任何影响但对整个程序正常工作来说必需的特性。

随着 MVC 框架的普及，微软开始将一些核心功能添加到 ASP.NET Web Forms 中。结果越来越不相匹配，其中需要用来支持 MVC 框架的独特设计特性被扩展到支持 ASP.NET Web Forms，却为了让所有的东西融合在一起而让设计变得更加不相匹配。同时，微软开始使用创建 Web 服务（Web API）和实时通信（SignalR）的新框架来扩展 ASP.NET。新的框架添加了自己的配置和开发约定，每个都有自己的优点和特异之处，结果导致了零乱的混乱状态。

1.2 ASP.NET Core

2015 年，微软公布了 ASP.NET 和 MVC 框架的新方向，也就是本书要讨论的主题 ASP.NET Core MVC。

ASP.NET Core 基于.NET Core 构建，它是.NET Framework 的跨平台版本，没有 Windows 平台特定的应用程序编程接口（Application Programming Interface，API）。虽然 Windows 仍然是主要的操作系统，但 Web 应用程序越来越多地托管在云平台的小型而简单的容器中，并且通过采用跨平台方法，微软扩展了.NET 的覆盖面，使得 ASP.NET Core 应用程序能够部署到更广泛的托管环境中。另外，ASP.NET Core 还有一个额外的优点——开发人员可以在 Linux 系统和 macOS 上创建 ASP.NET Core Web 应用程序。

ASP.NET Core 是一个全新的框架。它更简单、更容易处理，并且没有来自 ASP.NET Web Forms 的遗留问题。另外，由于它基于.NET Core，因此它支持在一系列平台和容器上进行 Web 应用程序开发。

ASP.NET Core MVC 提供了基于新的 ASP.NET Core 平台构建的初始 ASP.NET MVC 框架的功能。它集成了以前由 Web API 提供的功能，能以更自然的方式生成复杂内容，并且使关键开发任务（如单元测试）更简单。

1.2.1 ASP.NET Core MVC 2 的新特性

ASP.NET Core MVC 2 的发布版本重点关注整合，它在早期版本中引入了一些工具和平台变更。ASP.NET Core MVC 2 需要.NET Core 2，它具有很多扩展的 API，现已在其他 Linux 发行版上受到支持。有用的变化包括一个新的元数据包系统（它简化了 NuGet 包的管理），一个新的 ASP.NET Core 配置系统，以及对 Entity Framework Core 2 的支持。最重要的新功能是 Razor Pages，它尝试重新构建应用，并使用更现代的平台创建与 Web Pages 相关的开发风格，但 MVC 开发人员对 Razor Pages 并不感兴趣（因而在本书中没有描述）。

1.2.2 ASP.NET Core MVC 的主要优点

本节将简要介绍新的 MVC 平台如何克服 ASP.NET Web Forms 和初始 MVC 框架的遗留问题，以及如何完善 ASP.NET。

1. MVC 架构

ASP.NET Core MVC 遵循称为模型-视图-控制器（MVC）的模式，指导 ASP.NET Web 应用程序及其包含的组件之间的交互。

区分 MVC 架构模式和 ASP.NET Core MVC 实现很重要。MVC 模式并不新颖，它可以追溯到 1978 年 Xerox PARC 的 Smalltalk 项目，但是它作为 Web 应用程序的一种开发模式，已经得到了广泛普及，原因如下。

- 用户与遵守 MVC 模式的应用程序进行交互是遵循自然循环的。用户执行一个动作，应用程序进行响应，更改其数据模型并向用户传递更新的视图。然后重复这一循环。这对作为一系列 HTTP 请求和响应来传输的 Web 应用程序来说非常合适。
- Web 应用程序需要组合几种技术（例如数据库、HTML 和可执行代码等），通常分为一系列层次。这些组合产生的模式会自然地映射到 MVC 模式的概念上。

ASP.NET Core MVC 实现了 MVC 模式，与 ASP.NET Web Forms 相比，ASP.NET Core MVC 极大改善了关注点的分离。实际上，ASP.NET Core MVC 实现了 MVC 模式的变体，特别适用于 Web 应用程序。你将在第 3 章中更多地了解此架构的理论和实践。

2. 可扩展性

ASP.NET Core 和 ASP.NET Core MVC 已构建为一系列具有明确特征的独立组件，能满足.NET 接口的需求，也可构建在抽象基类上。你可以轻松地用自己的实现替换关键组件。一般来说，ASP.NET Core MVC 为每个组件提供了以下 3 个选项。

- 使用组件的默认实现（对于大多数应用程序来说应该是足够的）。
- 从默认实现派生一个子类来调整其行为。
- 使用接口或抽象基类的新实现完全替换组件。

你将从第 14 章开始，了解各种组件、如何替换以及为什么要调整或替换每个组件。

3. 严格控制 HTML 和 HTTP

ASP.NET Core MVC 能够生成清晰、符合标准的标签。它的内置标签帮助器能产生符合标准的输出，但与 ASP.NET Web Forms 相比，有更重要的理念上的变化。ASP.NET Core MVC 并不会生成一些难以控制的 HTML 控件，而是鼓励你创建简单而优雅的标签，并使用 CSS 进行样式化。

当然，如果你想要为诸如日期选择器或级联菜单之类的复杂 UI 元素使用一些现成的小部件，那么 ASP.NET Core MVC 采用的“无特定要求”方法可以很轻松地使用各种最佳组合的客户端库，如 jQuery、Angular 或 Bootstrap CSS 库。ASP.NET Core MVC 与这些库相互配合得很好，微软已包含这些模板以启动新的开发项目。

ASP.NET Core MVC 与 HTTP 协调工作。你可以控制在浏览器和服务器之间传递的请求，因此你可以根据需要调整用户体验。使用 Ajax 更加容易，创建 Web 服务来接收浏览器 HTTP 请求是一个简单的过程。

4. 可测试性

ASP.NET Core MVC 架构在使应用程序变得可维护和可测试方面提供了良好的开端，因为你可以将不同的应用程序关注点自然地分离成独立的部分。此外，ASP.NET Core 平台和 ASP.NET Core MVC 框架的每个部分都可以为单元测试进行隔离和替换，可以使用任何流行的开源测试框架（如 xUnit）。

在本书中，你将看到如何为 ASP.NET MVC 控制器编写整洁、简单的单元测试示例。为了模拟各种场景，这些示例使用各种测试和模拟策略来支持框架组件的虚构或模拟实现。即使你以前从来没有写过单元测试，这也是一个很好的开始。

可测试性不仅仅是单元测试的问题。ASP.NET Core MVC 应用程序也可以与 UI 自动化测试工具一起使用。你可以编写模拟用户交互的测试脚本，而不需要猜测框架将生成哪些 HTML 元素结构、CSS 类或 ID，你不必担心页面结构发生意外的变化。

5. 强大的路由系统

统一资源定位器（URL）的风格随着 Web 应用技术的发展而发展，比如，以下 URL 越来越少见。

```
/App_v2/User/Page.aspx?action=show%20prop&prop_id=82742
```

取而代之的是一种更简单、更干净的格式：

```
/to-rent/chicago/2303-silver-street
```

采用这种 URL 结构有一些很好的理由。第一，搜索引擎会对 URL 中找到的关键字加权。搜索“rent in Chicago”（芝加哥租房）更有可能找到更简单的网址。第二，许多网络用户现在已经足够了解 URL，并乐于通过在浏览器的地址栏中键入导航选项。第三，当人们理解 URL 的结构时，才更有可能链接它，与朋友分享，甚至通过手机朗读。第四，它不会向公共 Internet 暴露你的应用程序的技术细节、文件夹和文件名结构，因此你可以自由地更改底层的实现，而不会破坏所有的传入链接。

整洁的 URL 在早期的框架中很难实现，但 ASP.NET Core MVC 默认使用称为 URL 路由的功能来提供整洁的 URL。这样可以控制你的 URL 模式及其与应用程序之间的关系，自由地为用户创建有意义和有用的 URL 模式，而无须遵守预定义模式。当然，这意味着你可以轻松地定义一种现代 REST 风格的 URL 模式。

6. 现代 API

微软的.NET 平台随每个主版本的发展而发展，支持甚至定义了现代编程的最新方向。ASP.NET Core MVC 是为.NET Core 构建的，因此其 API 可以充分利用 C#程序员熟悉的语言和运行时创新，包括 await 关键字、扩展方法、lambda 表达式、匿名和动态类型以及语言集成查询（LINQ）。

许多 ASP.NET Core MVC API 方法和编码模式与早期平台相比遵循更整洁、更具表现力的方式。不要担心，如果你尚不了解最新的 C#语言特性，第 4 章会提供 MVC 开发中最重要的 C#特性总结。

7. 跨平台

以前的ASP.NET版本特定于Windows系统，需要使用Windows桌面编写Web应用程序，使用Windows服务器才能部署和运行它们。微软使ASP.NET Core支持跨平台，包括开发和部署。ASP.NET Core可用于不同的平台，包括macOS和一系列流行的Linux发行版。跨平台支持使得部署ASP.NET Core MVC应用程序变得更加容易，并且可以很好地支持应用程序容器平台，如Docker等。

当前大多数ASP.NET Core MVC开发很可能会使用Visual Studio完成，但微软也创建了一个名为Visual Studio Code的跨平台开发工具，这意味着ASP.NET Core MVC开发不再局限于Windows平台。

8. ASP.NET Core MVC是开源的

与以前的Microsoft Web开发平台不同，你可以免费下载ASP.NET Core和ASP.NET Core MVC的源代码，甚至修改和编译自己的版本。当你要调试跟踪进入一个系统组件并希望进入其代码内部（甚至阅读原始的程序员注释）时，这是非常有价值的。如果你正在构建一个高级组件，并希望了解进一步开发的可能性，或想了解内置组件如何进行实际工作，这也是非常有用的。

你可以从GitHub下载ASP.NET Core和ASP.NET Core MVC的源代码。

1.3 预备知识

为了充分利用本书，你应该熟悉Web开发的基础知识，了解HTML和CSS的工作原理，并掌握C#的相关知识。如果你对客户端细节（如JavaScript）有些模糊，请不要担心。重点是本书中的服务器端开发，你可以通过示例获取所需的内容。第4章将总结MVC开发中最有用的C#语言特性，如果你正在从早期版本转到最新的.NET版本，你将发现它们非常有用。

1.4 本书的结构

本书分为两部分，每一部分都涵盖了一系列相关主题。

本书第一部分将从ASP.NET Core MVC的背景开始，解释MVC模式的优点和实际影响，介绍ASP.NET Core MVC的功能，并描述每个ASP.NET Core MVC程序员需要学习的工具和C#语言功能。

在第2章中，你将通过创建一个简单的Web应用程序，深入了解主要组件、构建块以及它们如何组合在一起。然而，本书第一部分主要介绍如何开发一个名为SportsStore的项目，通过该项目，展示从开始到部署的实际开发流程，并介绍ASP.NET Core MVC的主要特性。

本书第二部分将解释用于构建SportsStore应用程序的ASP.NET Core MVC功能的内部工作原理。该部分将展示每个功能如何工作，解释它所扮演的角色，并展示可用的配置和自定义选项。本书第一部分介绍广泛的背景基础，第二部分则深入讨论细节。

1.5 如何获取本书的示例代码

在GitHub网站上搜索“pro-asp.net-core-mvc-2”，即可下载本书所有章节的示例代码。下载的内容不需要修改，并包含所有必要的资源。

1.6 联系作者

如果你在使用本书的示例代码时遇到问题，或者如果你在本书中发现问题，你可以通过电子邮件adam@adam-freeman.com 联系作者，作者会尽力帮助你。

1.7 小结

本章介绍了 ASP.NET Core MVC 的背景以及它是如何从 ASP.NET Web Forms 和起始的 ASP.NET MVC 框架发展而来的，阐述了使用 ASP.NET Core MVC 的好处以及本书的结构。在下一章中，你将在一个简单的示例程序中看到 ASP.NET Core MVC 先进的特性。

第2章　首个 MVC 应用程序

鉴赏一款软件开发框架的最佳方式就是深入其中并使用它。在本章中，你将使用 ASP.NET Core MVC 创建一个简单的数据输入应用程序。整个开发过程可分解为多个小的步骤，你将了解到清晰的 MVC 应用程序的构造。当然为了简化，也会跳过一些技术细节。但别担心，如果你是 MVC 新人，你会发现许多感兴趣的内容。对于使用但未解释的部分，本章也会提供参考引用，以便于你发现所有的细节。

> **本书的内容更新**
>
> 微软对.NET Core 和 ASP.NET Core MVC 有一项活跃的开发计划，这意味着在你阅读本书时可能会有更新的版本可用。要求读者每隔几个月就买一本新书是不现实的，尤其是在大多数改动相对较小的情况下。因此，作者将免费更新本书，并将内容放在 GitHub 上，以应对由次要版本引起的更改。这种更新对于作者来说是一个实验（对于 Apress 来说也是），作者还不知道这些更新可能采取什么形式，尤其是因为作者不知道未来主要版本的 ASP.NET Core MVC 将包含什么，但目标是通过补充包含的示例来延长本书的寿命。作者无法承诺这些更新会是什么样的，它们会采取什么形式，或者在将它们添加到本书的新版本之前，作者会花费多长时间。当新的 ASP.NET Core MVC 版本发布时，请保持开放的心态并检查本书的 GitHub 主页。如果你对如何改进更新有想法，请发电子邮件至 adam@adam-freeman.com。

2.1　安装 Visual Studio

本书使用的集成开发环境是 Visual Studio 2017，它提供了使用 ASP.NET Core MVC 时所需的各种功能。本书中的示例使用的是免费的 Visual Studio 2017 社区版本，有需要的读者可以从 Visual Studio 官网下载。当你安装 Visual Studio 时，你必须选择.NET Core cross-platform development 选项，如图 2-1 所示。

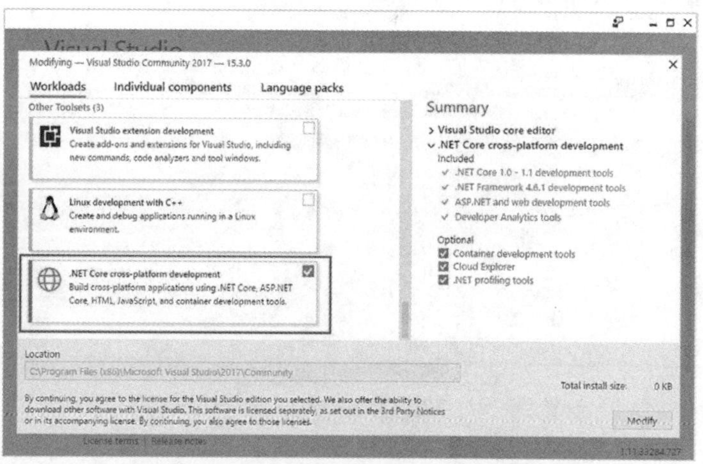

图 2-1　选择.NET Core cross-platform development 选项

注　意

Visual Studio 2017 早于 ASP.NET Core MVC 2 发布。如果你已为早期版本的 ASP.NET Core MVC 安装了 Visual Studio，则必须使用最新的 Visual Studio 更新。你可以运行 Visual Studio 安装程序，并为正在使用的 Visual Studio 版本选择 Update 选项。

提　示

Visual Studio 只支持 Windows。你可以在其他平台上创建 ASP.NET Core MVC 应用程序并使用本书 Visual Studio 示例中的代码，但是其他平台可能不包含本书示例中使用的工具。

详情可参见第 13 章。

2.2　安装.NET Core 2.0 SDK

Visual Studio 安装包含了 ASP.NET Core MVC 开发所需的所有功能，但不包括必须单独下载和安装的.NET Core 2.0。

访问微软官网，下载并运行适用于 Windows 的.NET Core SDK 安装程序。执行完安装程序后，打开新的命令提示符或 PowerShell 窗口并运行以下命令以显示已安装的.NET 版本：

```
dotnet --version
```

2.3　创建新的 ASP.NET Core MVC 项目

本节从创建一个新的 ASP.NET Core MVC 项目开始讲述。在左侧窗格中从 File 菜单中选择 New→Project 菜单来打开 New Project 对话框。如果你选择 Installed→Visual C#→Web，你将看到 ASP.NET Core Web Application（.NET Core）项目模板，按照图 2-2 选择项目类型。

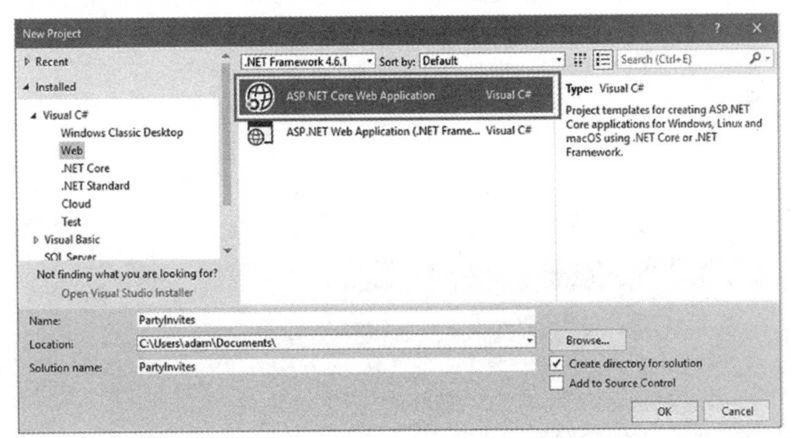

图 2-2　选择 ASP.NET Core Web Application 项目模板

提　示

项目模板的选择可能会令人困惑，因为它们的名称非常相似。ASP.NET Web Application（.NET Framework）模板会使用 ASP.NET 和 MVC 框架的遗留版本创建项目，这一框架早于 ASP.NET Core。另外两个模板用于创建 ASP.NET Core 应用程序，它们在运行时会有不同，原则上你可以任意选择.NET Framework 或.NET Core 选项，第 6 章会解释它们之间的区别。但是请注意，本书使用了.NET Core 选项，所以请你尽量也选择这个选项以确保能获得和示例代码相同的结果。

将新项目的名称设置为 PartyInvites，单击 OK 按钮继续，你将看到另一个对话框，要求你为项目设置初始内容。确保从下拉菜单中选择.NET Core 和 ASP.NET Core 2.0 以配置初始项目，如图 2-3 所示。

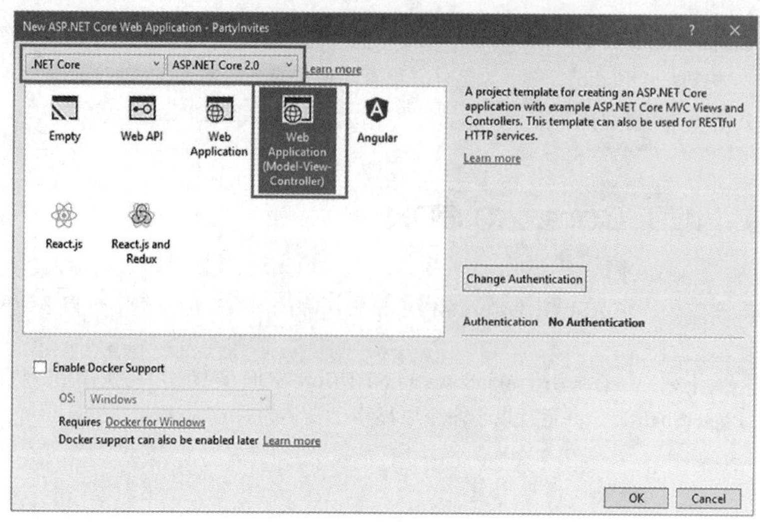

图 2-3 进行初始项目配置

这里有几个模板选项，每个选项都会创建一个具有不同起始内容的项目。在本章中，选择 Web Application（Model-View-Controller）选项，该选项会设置 MVC 应用程序使用预定义的内容启动开发。

注 意

只有本章使用 Web Application（Model-View-Controller）项目模板。作者不喜欢使用预定义的项目模板，因为它们鼓励开发人员将一些重要的特性（如身份验证）视为黑盒。本书的目标是让你了解和管理 MVC 应用程序的各个方面，所以本书的其余部分均使用了空模板。这一章是关于快速入门的，所以使用 Web Application（Model-View-Controller）模板非常合适。

单击 Change Authentication 按钮，并确保选中了 No Authentication 单选按钮，进行身份验证设置，如图 2-4 所示。这个项目不需要任何身份验证，但是第 28～30 章解释了如何对 ASP.NET 应用程序进行安全保护。

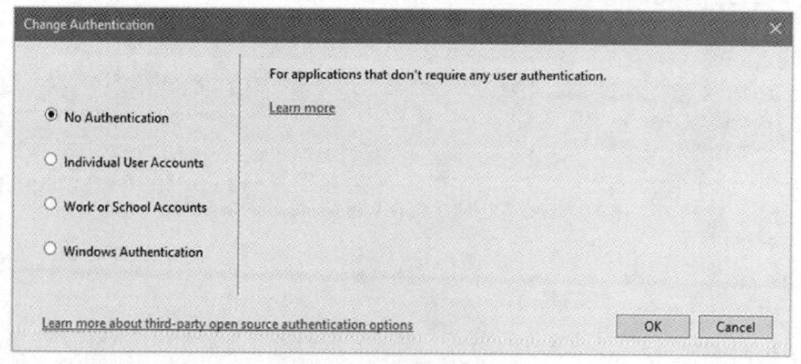

图 2-4 进行身份验证设置

单击 OK 按钮，然后关闭 Change Authentication 对话框。确保没有勾选 Host in the Cloud 复选框，然后单击 OK 按钮创建 PartyInvites 项目。

在 Visual Studio 中创建了这个项目后，你将看到 Solution Explorer 窗格中显示了许多文件和文件夹，如图 2-5 所示。这是使用 Web Application 模板创建的新 MVC 项目的默认项目目录，之后你将快速了解 Visual Studio 创建的每个文件和文件夹的用途。

提 示

如果你看到的是 Pages 文件夹而不是 Controllers、Models 和 Views 文件夹，那么说明你选择的是 Web Application 模板而不是（相似而令人困惑的）Web Application（Model-View-Controller）模板。你必须删除你创建的项目并重新开始。

你可以通过从 Debug 菜单中选择 Start Debugging 来调试运行应用程序（如果提示你启用调试，只需要单击 OK 按钮）。当你这样做时，Visual Studio 会编译应用程序，使用名为 IIS Express 的应用程序服务来运行它，并打开 Web 浏览器来请求应用程序的内容。你可以在图 2-6 中看到结果。

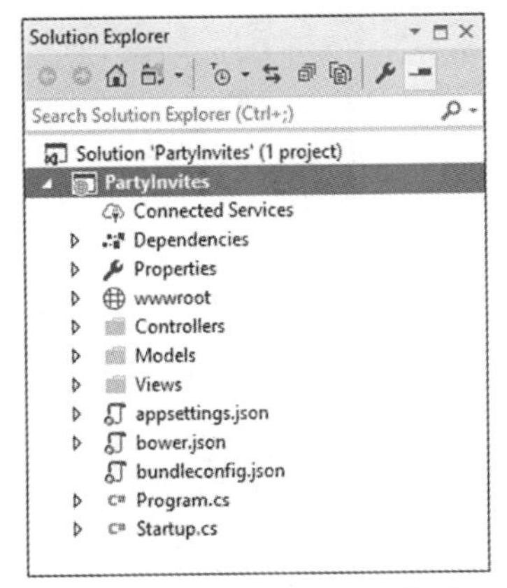

图 2-5 ASP.NET Core MVC 项目的初始文件与文件夹结构

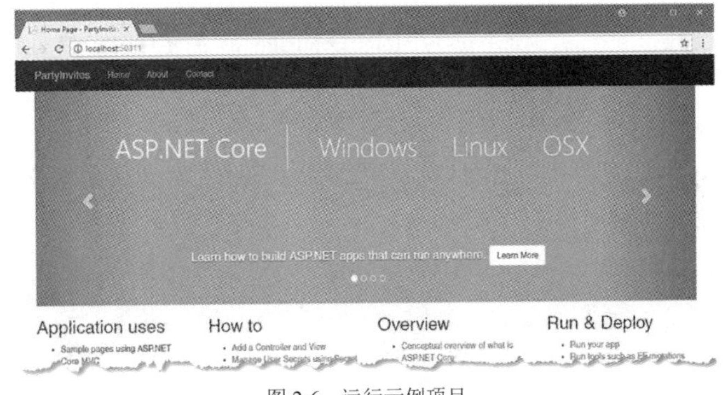

图 2-6 运行示例项目

当 Visual Studio 使用 Web Application（Model-View-Controller）模板创建项目时，它会添加一些基本的代码和内容，这就是你在运行应用程序时所看到的内容。在本章的其他部分，将替换这些内容，以创建一个简单的 MVC 应用程序。

当你完成的时候，请通过关闭浏览器窗口停止调试，或者返回 Visual Studio，并从 Debug 菜单中选择 Stop Debugging 以停止调试。

Visual Studio 通过打开浏览器来显示项目。选择你安装的任何浏览器，单击 IIS Express 工具栏按钮右边的箭头，从 Web Browser 菜单的选项列表中选择要使用的浏览器，如图 2-7 所示。

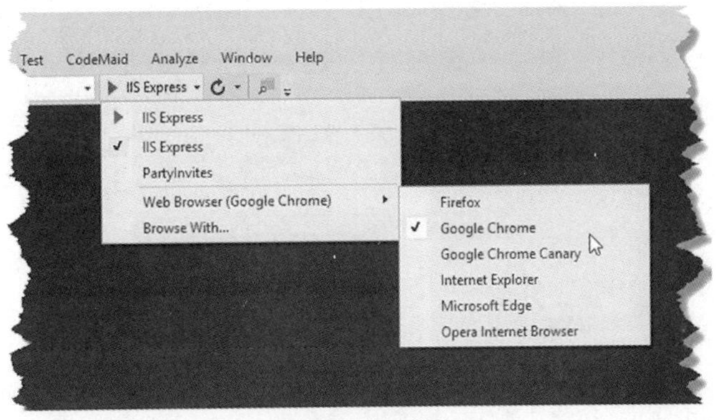

图 2-7 选择浏览器

本书的所有截图将使用谷歌 Chrome 或谷歌 Chrome 金丝雀版浏览器，但你也可以使用任何别的浏览器（包括 Microsoft Edge 和最新版本的 Internet Explorer）来显示书中的示例。

2.3.1 添加控制器

在 MVC 模式中，传入的请求由控制器来处理。在 ASP.NET Core MVC 中，控制器是一个 C#文件（通常从 Microsoft.AspNetCore.Mvc.Controller 类继承，这个类已内置于 MVC 控制器的基本类中）。

控制器中的每个公共方法都称为操作（action）方法，这意味着你可以通过 Web 基于某种 URL 来调用执行操作方法。MVC 约定通常会将控制器放置在 Controllers 文件夹中，当创建项目时，Visual Studio 会自动创建这个文件夹。

提 示

你不需要一定遵循这个或大多数其他的 MVC 约定，但是建议你仍然遵守——因为它将帮助你理解本书中的示例。

Visual Studio 为项目添加了一个默认的控制器类，如果你在 Solution Explorer 中展开 Controllers 文件夹，就可以看到它。这个文件被命名为 HomeController.cs。控制器类的名称的后半部分包含 Controller 这个单词，这意味着当你看到一个名为 HomeController.cs 的文件时，就知道它包含一个名为 Home 的控制器，Home 是 MVC 应用中使用的默认控制器。单击 Solution Explorer 中的 HomeController.cs 文件，通过 Visual Studio 打开它并进行编辑，你将看到代码清单 2-1 所示的 C#代码。

代码清单 2-1 Controllers 文件夹中 HomeController.cs 文件的初始内容

```
using System;
using System.Collections.Generic;
using System.Diagnostics;
using System.Linq;
using System.Threading.Tasks;
using Microsoft.AspNetCore.Mvc;
using PartyInvites.Models;

namespace PartyInvites.Controllers {
    public class HomeController : Controller {
        public IActionResult Index() {
            return View();
        }

        public IActionResult About() {
            ViewData["Message"] = "Your application description page.";

            return View();
        }

        public IActionResult Contact() {
            ViewData["Message"] = "Your contact page.";

            return View();
        }

        public IActionResult Error() {
            return View(new ErrorViewModel { RequestId = Activity.Current?.Id
                ?? HttpContext.TraceIdentifier });
        }
    }
}
```

将 HomeController.cs 中的代码改为代码清单 2-2 所示的样子：只保留一个方法并且删除其他的方法，更改结果类型及其实现，同时删除没有使用的命名空间所在的 using 语句。

代码清单 2-2 更改 Controller 文件夹中的 HomeController.cs

```
using Microsoft.AspNetCore.Mvc;

namespace PartyInvites.Controllers {

    public class HomeController : Controller {
```

```
        public string Index() {
            return "Hello World";
        }
    }
}
```

这些修改没有什么特别的效果，但可以作为很好的演示。这里更改了名为 Index 的方法，以返回字符串“Hello World”。接下来，从 Visual Studio 的 Debug 菜单中选择 Start Debugging 以再次运行这个项目。

提 示

如果之前的应用程序还在运行，请从 Debug 菜单中选择 Restart，或者选择 Stop Debugging 后再选择 Start Debugging，开始这个新的应用程序的调试。

浏览器将向服务器发出一个 HTTP 请求。默认的 MVC 配置意味着这个请求将使用 Index 方法（这是一个操作方法）来处理，并且方法的返回结果将被发送回浏览器，如图 2-8 所示。

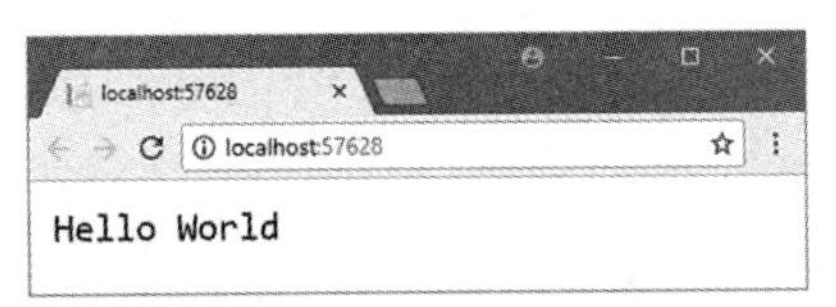

图 2-8 操作方法的输出

提 示

请注意，Visual Studio 已将浏览器定向到 57628 端口。你的浏览器在请求这一 URL 时，返回的很可能是一个不同的端口号，因为 Visual Studio 在创建项目时会分配一个随机的端口号。如果查看 Windows 任务栏的通知区域，你将找到 IIS Express 图标。这是完整的 IIS 应用服务器的一个简化版本，它包含在 Visual Studio 中，用于在开发过程中发布 ASP.NET Core 的内容和服务。第 12 章将展示如何将 MVC 项目部署到生产环境中。

2.3.2 理解路由

除了模型、视图和控制器之外，MVC 应用程序还使用了 ASP.NET 路由系统，它决定了如何把 URL 映射到控制器和 action 上。路由是用来决定如何处理请求规则的。当使用 Visual Studio 创建 MVC 项目时会自动添加一些默认的路由。你可以请求以下任何一个 URL，它们将直接指向 HomeController 的 Index 操作方法。

- /。
- /Home。
- /Home/Index。

因此，当浏览器请求 http://yoursite/或 http://yoursite/Home 时，它将从 HomeController 的 Index 方法返回输出。你可以通过在浏览器中更改 URL 来测试一下效果。此时 URL 会是 http://localhost:57628/（你的端口号部分可能与此处的不同），如果你将/Home 或/Home/Index 添加到 URL 并按下 Enter 键，你看到的应该是 MVC 应用程序输出“Hello World”。

这是关于 ASP.NET Core MVC 约定的一个很好的示例。在本例中，我们约定项目中存在一个名为 HomeController 的控制器，并且该控制器为渲染首页的默认控制器。在此假设 Visual Studio 为新项目创建的默认路由配置也遵循这个约定。由于确实遵循了这个约定，因此这里自动得到了前面列表中 URL 的支持。如果没有遵循该约定，则需要修改配置，以指向此处创建的其他控制器。由于这只是一个简单的小示例，因此采用默认配置即可。

2.4　渲染页面

上一个示例的输出并不是HTML，而只是字符串“Hello World”。要为浏览器请求生成HTML响应，我们需要视图（view），视图将告诉MVC如何为浏览器生成请求的响应。

2.4.1　创建并渲染视图

我们需要做的第一件事是修改Index的操作方法，如代码清单2-3所示。更改部分以粗体显示，这是本书遵循的惯例，以使示例更容易理解。

代码清单2-3　修改控制器以在HomeController.cs中呈现视图

```
using Microsoft.AspNetCore.Mvc;

namespace PartyInvites.Controllers {

    public class HomeController : Controller {

        public ViewResult Index() {
            return View("MyView");
        }
    }
}
```

当从一个操作方法返回一个ViewResult对象时，就是在指示MVC去渲染一个视图。可通过调用View方法来创建ViewResult，并指定要使用的视图，比如MyView。如果运行该应用程序，你可以看到MVC会试图查找所要使用的视图，参见图2-9所示的错误消息。

图2-9　错误消息

错误消息对你是很有帮助的，它不仅解释了MVC无法找到上面为操作方法指定的视图，而且显示了在哪些地方进行了查找。视图存储在Views文件夹中，并将其组织为子文件夹。例如，与Home控制器相关联的视图存储在一个名为Views/Home的文件夹中。不是特定于单个控制器的视图则被存储在一个名为Views/Shared的文件夹中。当使用Web Application（Model-View-Controller）模板时，Visual Studio会自动创建Home和Shared文件夹，并放入一些占位用的视图。

下面开始创建视图，在Solution Explorer中右击Views→Home文件夹并从弹出的菜单中选择Add→New Item。Visual Studio将为你提供一个项目模板列表。使用左侧窗格选择ASP.NET的类别，然后在中央窗格中选择MVC View Page选项，创建视图，如图2-10所示。不要使用RazorPage模板，它与MVC框架无关。

提　示

你将在Views文件夹中看到一些已经存在的文件，这些文件被Visual Studio添加到项目中以提供一些初始内容，其中一些你已经在图2-7中看到。你可以忽略这些文件。

在Views/Home文件夹中创建一个名为MyView.cshtml的视图文件。Visual Studio将自动创建Views/Home/MyView.cshtml文件并打开它，进入编辑页面。视图文件的初始内容只是一些注释和占位符。请将它们替换为代码清单2-4所示的内容。

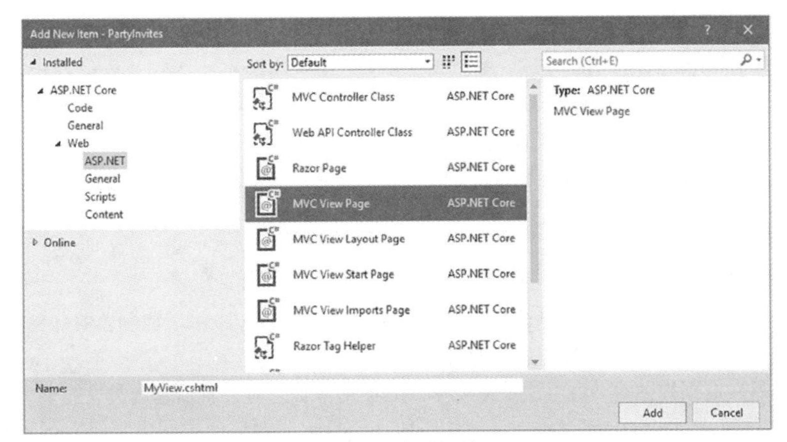

图 2-10 创建视图

代码清单 2-4 用如下代码替换 Views/Home 文件夹下 MyView.cshtml 文件的内容

```
@{
    Layout = null;
}

<!DOCTYPE html>

<html>
<head>
    <meta name="viewport" content="width=device-width" />
    <title>Index</title>
</head>
<body>
    <div>
        Hello World (from the view)
    </div>
</body>
</html>
The new contents of the view file are mostly HTML. The exception is the part that looks like this:
...
@{
    Layout = null;
}
...
```

提 示

初学者很容易在错误的文件夹中创建视图文件。如果你没有在 Views/Home 文件夹中创建一个名为 MyView.cshtml 的文件，请删除你创建的文件并再次尝试。

这是一个通过 Razor 视图引擎进行解释的表达式（expression），它处理视图的内容并生成发送到浏览器的 HTML。这是一个简单的 Razor 表达式，它告诉 Razor 选择不使用布局，它就像 HTML 模板，将被发送到浏览器（将在第 5 章中详述）。现在我们先暂时不讨论 Razor 的问题，稍后回过头来再讨论。我们现在想要查看视图的创建效果，请从 Debug 菜单中选择 Start Debugging 以运行应用程序。你应该会看到图 2-11 所示的结果。

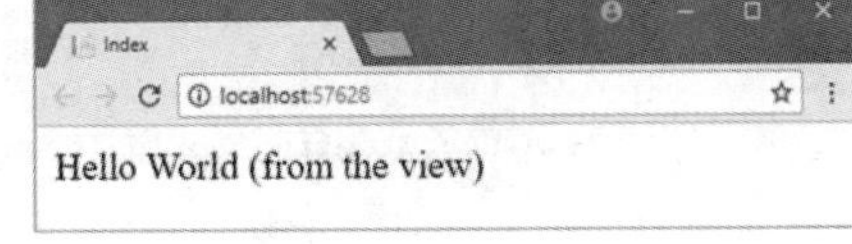

图 2-11 测试视图的结果

在最初编辑 Index 的操作方法时，返回的是一个字符串值。这意味着 MVC 除了将字符串值传递给浏览器之外什么都没做。现在，Index 方法返回了一个 ViewResult，MVC 渲染了一个视图并返回它所生成的 HTML。因为已经告诉 MVC 应该使用哪个视图，所以 MVC 会使用命名约定来自动找到它。我们约定视图具有操作方法的名称，并且包含在一个以控制器命名的文件夹/Views/Home/MyView.cshtml中。

除了字符串和ViewResult对象外，还可以从操作方法返回其他结果。例如，如果返回一个RedirectResult，浏览器将被重定向到另一个URL。如果返回一个HttpUnauthorizedResult，则会强制用户登录。这些对象统称为操作结果。Action Result系统允许你封装和重用操作中的常见响应。第17章将对其做更多的介绍并演示一些不同的方法。

2.4.2 添加动态输出

Web应用程序平台的核心要点就是构造和显示动态输出。在MVC中，控制器的任务是构造一些数据并将其传递给视图，视图再负责将其渲染成HTML。

从控制器传递数据到视图的一种方法是使用ViewBag对象，它是控制器基本类型的一个成员。ViewBag是一个动态对象，可以为它分配任意属性，使这些值在随后渲染的任何视图中都保持可用。代码清单2-5演示了在HomeController.cs文件中如何通过这种方式传递一些简单的动态数据。

代码清单2-5 在Controllers文件夹下的HomeController.cs文件中传递动态数据

```
using System;
using Microsoft.AspNetCore.Mvc;

namespace PartyInvites.Controllers {

    public class HomeController : Controller {

        public ViewResult Index() {
            int hour = DateTime.Now.Hour;
            ViewBag.Greeting = hour < 12 ? "Good Morning" : "Good Afternoon";
            return View("MyView");
        }
    }
}
```

为ViewBag.Greeting属性赋值，便是为视图提供数据。ViewBag.Greeting属性直到被赋值时才会形成，你不需要提前定义类，这使得你可以更加自如方便地将数据从控制器传递给视图。在视图中再次检索ViewBag.Greeting属性可以获取其数据值，如代码清单2-6所示，它显示了对MyView.cshtml文件所做的相应更改。

代码清单2-6 在Views/Home文件夹下的MyView.cshtml文件中对ViewBag.Greeting的数据值进行检索

```
@{
    Layout = null;
}

<!DOCTYPE html>

<html>
<head>
    <meta name="viewport" content="width=device-width" />
    <title>Index</title>
</head>
<body>
    <div>
        @ViewBag.Greeting World (from the view)
    </div>
</body>
</html>
```

我们在代码清单2-6中添加的是一个Razor表达式，当MVC使用视图生成响应时，将对其进行求值。当在控制器的Index方法中调用View方法时，MVC定位了MyView.cshtml视图文件并要求Razor视图引擎解析文件的内容。Razor 会寻找类似于本例中添加的这种表达式并处理它们。在本例中，只要处理这个 Razor 表达式，就会在视图中插入你在操作方法里赋给 ViewBag.Greeting 的值。

属性名 ViewBag.Greeting 只是一个普通的代号，你可以用任何属性名替换它而不会改变其工作方式，只需要保证在控制器中使用的名称与在视图中使用的名称匹配即可。你可以将值分配给多个属性来实现多个数据值从控制器到视图的传递。请通过启动项目来看看这些更改的效果，来自 MVC 的动态响应如图 2-12 所示。

图 2-12 来自 MVC 的动态响应

2.5 创建一个简单的数据录入程序

本章剩下的部分将通过构建一个简单的数据录入程序探索 MVC 更多的基本特性。本节将加快进度以演示 MVC 的实际操作，因此将跳过一些关于背后工作原理的解释。但别担心，后面的章节将深入讨论这些主题。

2.5.1 设置场景

假设有个朋友决定举行一场新年夜派对，她邀请你为此创建一个 Web 应用程序，以便受邀人能够进行电子回复。要求如下。

- 显示有关聚会信息的主页。
- 可用于回复的表单。
- 实现电子回复表单（RSVP）的验证，将显示感谢页面。
- 总结页面，显示谁会来参加聚会。

下面将介绍如何修改完善在本章开始时创建的 MVC 项目，并添加上述几个要求。通过前面介绍的内容可以实现第一个要求，并添加一些 HTML 到现有视图中，以提供有关聚会的详细信息。代码清单 2-7 显示了对 Views/Home/MyView.cshtml 文件所做的修改。

代码清单 2-7 在 MyView.cshtml 文件中显示聚会的详细信息

```
@{
    Layout = null;
}

<!DOCTYPE html>

<html>
<head>
    <meta name="viewport" content="width=device-width" />
    <title>Index</title>
</head>
<body>
    <div>
       @ViewBag.Greeting World (from the view)
       <p>We're going to have an exciting party.<br />
       (To do: sell it better. Add pictures or something.)
       </p>
    </div>
</body>
</html>
```

运行应用程序，通过从 Debug 菜单中选择 Start Debugging，你将看到聚会的详细信息（详细信息用一个占位符表示，你只要懂得操作方式就好），如图 2-13 所示。

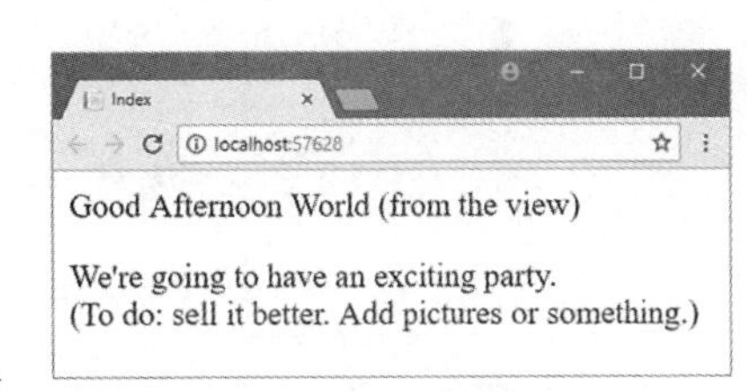

图 2-13 添加到视图中的 HTML 表示聚会的详细信息

2.5.2 设计数据模型

在MVC中，M代表模型，它是应用程序中最重要的部分。模型是现实世界中实际对象、过程和规则的表示。模型通常称为领域模型（Domain Model），包含应用程序域中要建立的C#领域对象（Domain Object）和操作它们的方法。视图和控制器以一致的方式向客户端暴露领域对象，而设计良好的MVC应用程序往往从设计良好的模型开始，这是作为控制器和视图添加的关键点。

这里只是朋友间使用的一个简单程序，并不需要复杂的PartyInvites项目模型。接下来，将会创建一个名为GuestResponse的领域类，它负责存储、验证和确认电子回复。

MVC约定将组成模型的类放置在名为Models的文件夹中。要创建Models文件夹，可右击PartyInvites项目（项目中已包含Controllers和Views文件夹），从弹出菜单中选择Add→New Folder，并设置文件夹的名称为Models。

为了创建类文件，在Solution Explorer中右击Models文件夹并从弹出菜单中选择Add→Class。将新类的名称设置为GuestResponse.cs并单击Add按钮。编辑这个新的类文件的内容，如代码清单2-8所示。

代码清单2-8 在Models文件夹下的GuestResponse.cs文件中定义GuestResponse领域类

```
namespace PartyInvites.Models {

    public class GuestResponse {

        public string Name { get; set; }
        public string Email { get; set; }
        public string Phone { get; set; }
        public bool? WillAttend { get; set; }
    }
}
```

提 示

你可能已经注意到，WillAttend属性是一个可以为空的bool值，这意味着可以是true、false或null。2.5.8节会解释这一点。

2.5.3 创建第二个操作和强类型视图

这个应用程序的目标之一是包含一个电子回复表单，因此需要定义一个可以接收请求的操作方法。单个控制器类可以定义多个操作方法，而约定是将相关的操作集中在同一个控制器中。代码清单2-9显示了如何向主控制器中添加一个新的操作方法。

代码清单2-9 在Controllers文件夹下的HomeController.cs文件中添加一个操作方法

```
using System;
using Microsoft.AspNetCore.Mvc;

namespace PartyInvites.Controllers {

    public class HomeController : Controller {

        public ViewResult Index() {
            int hour = DateTime.Now.Hour;
            ViewBag.Greeting = hour < 12 ? "Good Morning" : "Good Afternoon";
            return View("MyView");
        }

        public ViewResult RsvpForm() {
            return View();
        }
    }
}
```

电子回复表单的操作方法是一种不带参数调用的视图方法，它告诉 MVC 渲染与操作方法关联的默认视图，默认视图与操作方法的名称需要相同，在本例中为 RsvpForm.cshtml。

右击 Views/Home 文件夹并从弹出菜单选择 Add→New Item。从 ASP.NET 类别中选择 MVC View Page 模板，将新文件命名为 RsvpForm.cshtml。然后单击 Add 按钮创建文件。更改文件的内容，如代码清单 2-10 所示。

代码清单 2-10 更改 Views/Home 文件夹中 RsvpForm.cshtml 文件的内容

```
@model PartyInvites.Models.GuestResponse

@{
    Layout = null;
}

<!DOCTYPE html>

<html>
<head>
    <meta name="viewport" content="width=device-width" />
    <title>RsvpForm</title>
</head>
<body>
    <div>
        This is the RsvpForm.cshtml View
    </div>
</body>
</html>
```

虽然内容主要是 HTML 代码，但添加了@model Razor 表达式，用于创建强类型视图。强类型视图用于渲染特定类型的模型，如果指定要处理的类型（比如 PartyInvites.Models 中的 GuestResponse 类），MVC 可以创建一些便于使用这种类型的方法（MVC 本身提供了一些便捷的方法，可以在视图中方便地使用这种类型）。

要测试新的操作方法及其视图，可以从 Debug 菜单中选择 Start Debugging 来启动应用程序，并使用浏览器导航到/Home/RsvpForm 这个 URL。

MVC 将使用前面描述的命名约定来将请求引导到由主控制器定义的 RsvpForm 这个操作方法。RsvpForm 方法会告诉 MVC 渲染默认视图，默认视图使用另一个应用程序命名约定来渲染 Views/Home 文件夹中的 RsvpForm.cshtml 文件。图 2-14 显示了结果。

图 2-14 渲染第二个视图的结果

2.5.4 链接操作方法

如果从 MyView 视图中创建一个链接，这样访客就可以看到 RsvpForm 视图，而不必知道针对特定操作方法的 URL，具体操作如代码清单 2-11 所示。

代码清单 2-11 在 MyView.cshtml 文件中向 RSVP Form 视图添加一个链接

```
@{
    Layout = null;
}

<!DOCTYPE html>

<html>
<head>
    <meta name="viewport" content="width=device-width" />
```

```
    <title>Index</title>
</head>
<body>
    <div>
        @ViewBag.Greeting World (from the view)
        <p>We're going to have an exciting party.<br />
        (To do: sell it better. Add pictures or something.)
        </p>
        <a asp-action="RsvpForm">RSVP Now</a>
    </div>
</body>
</html>
```

代码清单2-11添加了一个拥有asp-action属性的元素，asp-action属性是标签助手，它会在渲染视图时执行Razor指令。asp-action属性是将href属性添加到包含指向操作方法的URL的a元素中的指令。第24～26章将详述标签助手的工作方式，asp-action属性是最简单的一种标签助手，它告诉Razor插入一个URL，这个操作方法由正在渲染当前视图的同一控制器定义。你可以通过启动项目看到标签助手创建的链接，如图2-15所示。

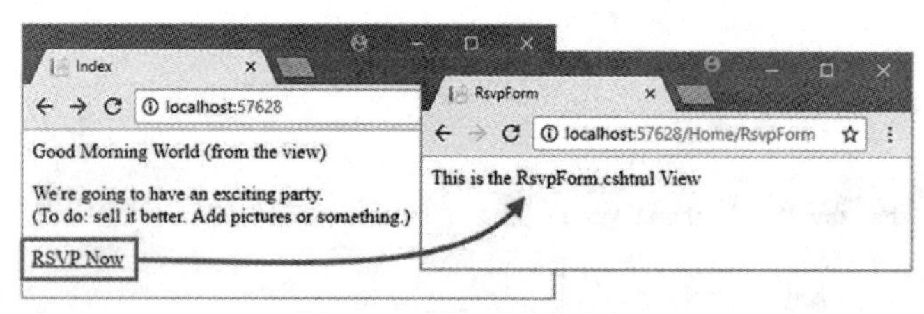

图2-15 链接两个操作方法

启动应用程序，在浏览器中移动光标到RSVP Now链接上，你将看到链接指向以下URL（允许Visual Studio给你的项目分配不同的端口号）：

```
http://localhost:57628/Home/RsvpForm
```

这里有一个重要的原则，你应该使用MVC提供的特性（feature）来生成URL，而不是将它们硬编码到你的视图中。当使用标签助手为a元素创建href属性时，它会检查应用程序的配置以确定URL。这让应用程序的配置可以支持不同的URL格式，而不需要再去更新任何视图。第15章将解释原理。

2.5.5 建立表单

既然已经创建了强类型视图，并且可以从Index视图中访问它，接下来就构建RsvpForm.cshtml文件的内容并转换成用于编辑GuestResponse对象的HTML表单，如代码清单2-12所示。

代码清单2-12 在Views/Home文件夹下的RsvpForm.cshtml文件中创建一个表单视图

```
@model PartyInvites.Models.GuestResponse

@{
    Layout = null;
}

<!DOCTYPE html>

<html>
<head>
    <meta name="viewport" content="width=device-width" />
    <title>RsvpForm</title>
</head>
<body>
```

```
    <form asp-action="RsvpForm" method="post">
        <p>
            <label asp-for="Name">Your name:</label>
            <input asp-for="Name" />
        </p>
        <p>
            <label asp-for="Email">Your email:</label>
            <input asp-for="Email" />
        </p>
        <p>
            <label asp-for="Phone">Your phone:</label>
            <input asp-for="Phone" /></p>
        <p>
            <label>Will you attend?</label>
            <select asp-for="WillAttend">
                <option value="">Choose an option</option>
                <option value="true">Yes, I'll be there</option>
                <option value="false">No, I can't come</option>
            </select>
        </p>
        <button type="submit">Submit RSVP</button>
    </form>
</body>
</html>
```

这里为 GuestResponse 模型类的每个属性都定义了一个 label 元素和一个 input 元素（或为 WillAttend 属性定义了一个 select 元素）。标签助手的另一个功能就是把每个元素都使用 asp-for 属性与模型属性关联起来。标签助手属性可通过配置将元素绑定到模型对象。下面是使用标签助手生成 HTML 并将其发送到浏览器的示例：

```
<p>
    <label for="Name">Your name:</label>
    <input type="text" id="Name" name="Name" value="">
</p>
```

这里使用 label 元素的 asp-for 属性设置了 for 属性的值，使用 input 元素的 asp-for 属性设置了 id 和 name 元素的值。这在目前看起来并不是特别有用，但是你以后会发现将元素与模型属性相关联会提供额外的优势。

更直接的是应用于 form 元素的 asp-action 属性，它通过应用程序的 URL 路由配置将 action 属性设置为一个 URL，该 URL 将针对特定的操作方法，如下所示：

```
<form method="post" action="/Home/RsvpForm">
```

与应用于元素的 helper 属性一样，这种方法的好处是可以更改应用程序使用的系统 URL，而由标签助手生成的内容也将自适应这些更改。

你可以通过运行应用程序并单击 RSVP Now 链接（见图 2-15）来查看表单，如图 2-16 所示。

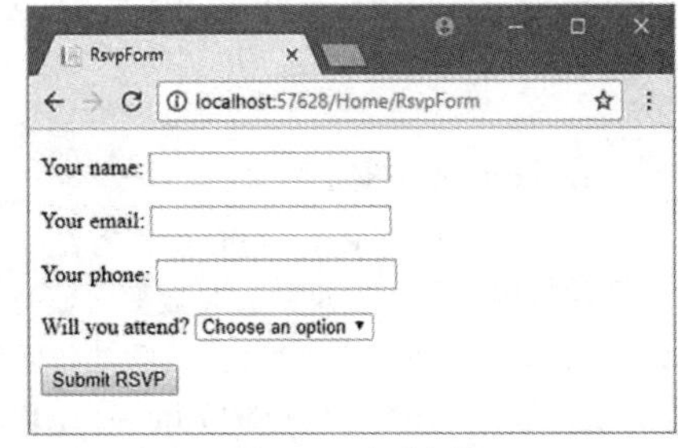

图 2-16 查看表单

2.5.6 接收表单数据

当把表单发布到服务器时，我们还没有告诉 MVC 我们要做什么。现在单击 Submit RSVP 按钮只会清除我们已输入表单的所有值。这是因为在 Home 控制器中表单仅仅返回了同一个 RsvpForm 的操作方法，它只告诉 MVC 再次渲染视图（所以表单内容没有保存而被刷新掉）。

为了接收和处理提交的表单数据，这里将使用一个核心的控制器特性。这里将添加第二个 RsvpForm 操作方法来创建以下内容：

● 响应 HTTP GET 请求的方法。GET 请求是当用户每次单击链接时浏览器发出的请求。在该例中，这个操作方法将负责在第一次访问/Home/RsvpForm 时显示初始的空白表单。

- 响应 HTTP POST 请求的方法。默认情况下，使用 Html.BeginForm()渲染的表单在被浏览器提交时，将发出 POST 请求。在该例中，这个操作方法将负责接收提交的数据并决定如何处理它们。

在独立的 C#方法中处理 GET 和 POST 请求有助于保持控制器代码整洁，因为这两种方法有不同的职责。虽然这两个操作方法都调用同一个 URL，但是 MVC 会基于使用的是 GET 还是 POST 请求来采取不同的方法处理请求。代码清单 2-13 显示了对 HomeController 类所做的更改。

代码清单 2-13 在 HomeController.cs 文件中添加一个操作方法来支持 POST 请求

```
using System;
using Microsoft.AspNetCore.Mvc;
using PartyInvites.Models;

namespace PartyInvites.Controllers {
    public class HomeController : Controller {

        public ViewResult Index() {
            int hour = DateTime.Now.Hour;
            ViewBag.Greeting = hour < 12 ? "Good Morning" : "Good Afternoon";
            return View("MyView");
        }

        [HttpGet]
        public ViewResult RsvpForm() {
            return View();
        }

        [HttpPost]
        public ViewResult RsvpForm(GuestResponse guestResponse) {
            // TODO: store response from guest
            return View();
        }
    }
}
```

这里已经将 HttpGet 特性添加到现有的 RsvpForm 操作方法中，MVC 知道该方法只可用于 GET 请求。然后，添加了一个重载版本的 RsvpForm 操作方法，该方法接收一个 GuestResponse 对象。如果将 HttpPost 特性应用于此方法，MVC 会使用这个方法处理 POST 请求。随后几节将解释这些附加的操作是如何工作的。这里还导入了 PartyInvites.Models 命名空间，这样就可以引用 GuestResponse 模型类型，而不需要类名。

1. 使用模型绑定

第一个重载的 RsvpForm 操作方法与前面的 RsvpForm.cshtml 文件渲染的是同一个视图，参见图 2-16 所示的表单。第二个重载的参数变得较有趣，考虑到操作方法是在响应 HTTP POST 请求时被调用的，而 GuestResponse 是一个 C#类，二者是如何连接的呢?

答案是模型绑定（Model Binding），这是一个很有用的 MVC 特性，通过解析传入的数据，并使用 HTTP 请求中的键/值对填充领域模型类型的属性。

模型绑定是一种强大且可定制的特性，它消除了直接处理 HTTP 请求带来的烦琐工作，让你直接操作 C#对象，而不是处理浏览器发送的单个数据值。GuestResponse 对象将作为参数传递给操作方法，该对象会自动填充来自表单字段的数据。第 26 章将详细介绍模型绑定的细节，包括如何定制它。

这个应用程序的最后一个目标是提供总结页面，其中包含参加聚会的人员信息，这意味着需要跟踪收到的响应。通过创建对象的内存集合来实现这一点。虽然在实际应用中这可能不太好用，因为当应用程序停止或重新启动时，响应数据将丢失，但是这种方法将允许开发人员保持对 MVC 的持续关注，并创建一个很容易被重置为初始状态的应用程序。

提 示

第 8 章将演示 MVC 的持久存储和数据访问，这是更贴合实际使用的一个示例。

添加一个文件到项目的 Models 文件夹并右击，选择 Add→Class，将文件命名为 Repository.cs 并使用它来定义类，如代码清单 2-14 所示。

代码清单 2-14 Models 文件夹下的 Repository.cs 文件的内容

```
using System.Collections.Generic;

namespace PartyInvites.Models {
    public static class Repository {
        private static List<GuestResponse> responses = new List<GuestResponse>();

        public static IEnumerable<GuestResponse> Responses {
            get {
                return responses;
            }
        }

        public static void AddResponse(GuestResponse response) {
            responses.Add(response);
        }
    }
}
```

Repository 类及其成员是静态的，这将使开发人员能够轻松地从应用程序的不同位置存储和检索数据。MVC 提供了一种更复杂的方法来定义通用的功能，称为依赖注入（dependency injection），这将在第 18 章中详述，但静态类是一种很好的方法，可以用于简单的程序。

2. 存储响应

既然可以存储数据了，接下来就更新接收 HTTP POST 请求的操作方法，如代码清单 2-15 所示。

代码清单 2-15 更新 HomeController.cs 文件中的操作方法

```
using System;
using Microsoft.AspNetCore.Mvc;
using PartyInvites.Models;

namespace PartyInvites.Controllers {

    public class HomeController : Controller {

        public ViewResult Index() {
            int hour = DateTime.Now.Hour;
            ViewBag.Greeting = hour < 12 ? "Good Morning" : "Good Afternoon";
            return View("MyView");
        }

        [HttpGet]
        public ViewResult RsvpForm() {
            return View();
        }

        [HttpPost]
        public ViewResult RsvpForm(GuestResponse guestResponse) {
            Repository.AddResponse(guestResponse);
            return View("Thanks", guestResponse);
        }
    }
}
```

所要做的就是处理发送到请求中的表单数据，比如处理传递给操作方法的 GuestResponse 对象，将其作为参数传递给 Repository.AddResponse 方法，以存储响应。

为什么模型绑定不像 Web 表单绑定？

第 1 章解释了传统 ASP.NET Web Forms 的一个缺点是将 HTTP 和 HTML 的细节对开发人员隐藏。你可能想知道，在代码清单 2-15 中用于从 HTTP POST 请求创建 GuestResponse 对象的 MVC 模型绑定是否也在做同样的事情。

答案为不是。模型绑定去除了烦琐和易导致错误的任务，使开发人员不需要检查 HTTP 请求和提取每一个数值，但（这是最重要的部分）如果开发人员想手动处理一个请求，仍然可以使用模型绑定，因为 MVC 提供了对所有请求数据的轻松访问方式。MVC 没有对开发人员隐藏任何东西（所有数据和操作都是可见的），而且其中还有许多特性可以更简易地使用 HTTP 和 HTML。当然，为了使用这些特性，你需要手动选择。

这似乎是一个细微的差别，但是随着你对 MVC 的了解，你将感受到这种开发体验与传统 ASP.NET Web Forms 是完全不同的，并且你总是可以知道如何处理应用程序接收的请求。

在 RsvpForm 操作方法中调用 View 方法，使得 MVC 渲染一个名为 Thanks 的视图，并将 GuestResponse 对象传递给视图。要创建视图，可在 Solution Explorer 中右击 View/Home 文件夹并从弹出菜单中选择 Add→New Item。在 ASP.NET 类别中选择 MVC View Page 模板，命名为 Thanks.cshtml，单击 Add 按钮。Visual Studio 将创建 Views/Home/Thanks.cshtml 文件并打开它进行编辑。更改该文件的内容，如代码清单 2-16 所示。

代码清单 2-16 Views/Home 文件夹下的 Thanks.cshtml 文件的内容

```
@model PartyInvites.Models.GuestResponse

@{
    Layout = null;
}

<!DOCTYPE html>

<html>
<head>
    <meta name="viewport" content="width=device-width" />
    <title>Thanks</title>
</head>
<body>
    <p>
        <h1>Thank you, @Model.Name!</h1>
        @if (Model.WillAttend == true) {
            @:It's great that you're coming. The drinks are already in the fridge!
        } else {
            @:Sorry to hear that you can't make it, but thanks for letting us know.
        }
    </p>
    <p>Click <a asp-action="ListResponses">here</a> to see who is coming.</p>
</body>
</html>
```

Thanks.cshtml 视图文件基于在 RsvpForm 操作方法中传递给 View 方法的 GuestResponse 属性的值并使用 Razor 来显示内容。Razor 的@model 表达式指定了强类型视图的领域模型类型。

使用 Model.PropertyName 来访问领域对象中的属性值。举一个示例，要获取 Name 属性的值，可调用 Model.Name。不用担心 Razor 的语法是没有意义的——第 5 章将更详细地解释它。

现在 Thanks 视图已经创建了，这是一个使用 MVC 处理表单的基本示例。在 Visual Studio 中，从 Debug 菜单中选择 Start Debugging 以启动应用程序，单击 RSVP Now 链接，向表单添加一些数据，然后单击 Submit RSVP 按钮。你将看到图 2-17 所示的结果（当名字不是 Joe 或者你不能参加聚会时，会显示不同的感谢信息）。

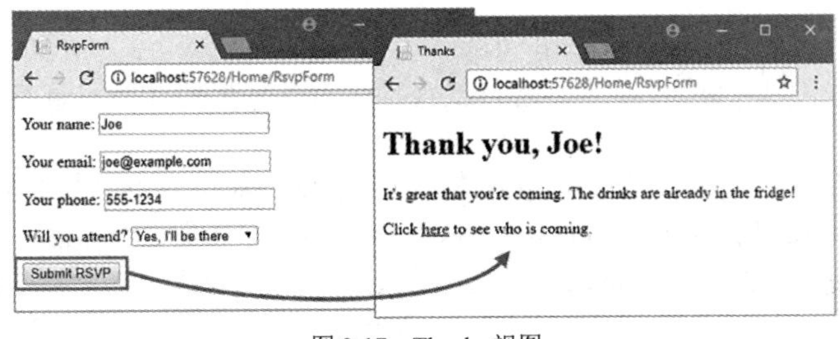

图 2-17 Thanks 视图

2.5.7 显示响应

在 Thanks.cshtml 文件的最后，添加一个 a 元素来创建一个链接以显示即将参加聚会的人的列表。下面使用 asp-action tag helper 属性创建一个 URL，该 URL 指向一个名为 ListResponse 的操作方法，如下所示：

```
...
<p>Click <a asp-action="ListResponses">here</a> to see who is coming.</p>
...
```

如果将鼠标指针悬停在浏览器的链接上，你将看到它指向/Home/ListResponses 这个 URL。这与 Home 控制器中的任何操作方法都不对应，如果单击链接，就会看到一个空页面。打开浏览器的开发者工具并查看服务器发送的响应，将显示 404 Not Found（Chrome 浏览器有点奇怪，它不向用户显示错误消息，但是第 14 章会解释如何生成有意义的错误消息）。

这里将通过创建使 URL 对应的目标为 Home 控制器的操作方法来解决这个问题，如代码清单 2-17 所示。

代码清单 2-17 在 Controllers 文件夹下的 HomeController.cs 文件中添加一个操作方法

```
using System;
using Microsoft.AspNetCore.Mvc;
using PartyInvites.Models;
using System.Linq;

namespace PartyInvites.Controllers {

    public class HomeController : Controller {

        public ViewResult Index() {
            int hour = DateTime.Now.Hour;
            ViewBag.Greeting = hour < 12 ? "Good Morning" : "Good Afternoon";
            return View("MyView");
        }

        [HttpGet]
        public ViewResult RsvpForm() {
            return View();
        }

        [HttpPost]
        public ViewResult RsvpForm(GuestResponse guestResponse) {
            Repository.AddResponse(guestResponse);
            return View("Thanks", guestResponse);
```

```
        }

        public ViewResult ListResponses() {
            return View(Repository.Responses.Where(r => r.WillAttend == true));
        }
    }
}
```

新的操作方法名为 ListResponse，它使用 Repository.Responses 属性作为参数调用 View 方法。这就是操作方法为强类型视图提供数据的方式。对 GuestResponse 对象的集合使用 LINQ 进行过滤，使得只有将要参加聚会的响应被使用。

ListResponse 这个操作方法没有指定应该用于显示 GuestResponse 对象集合的视图名称，这意味着将使用默认的命名约定，MVC 将在 Views/Home 和 Views/Shared 文件夹中查找名为 ListResponses.cshtml 的视图。可在 Solution Explorer 中右击 View/ Home 文件夹，并从弹出菜单中选择 Add→New Item 以创建视图。在 ASP.NET 类别中选择 MVC View Page 模板，命名为 ListResponses.cshtml 并单击 Add 按钮。编辑新视图的内容，如代码清单 2-18 所示。

代码清单 2-18 Views/Home 文件夹下的 ListResponses.cshtml 文件的内容：显示邀请函

```
@model IEnumerable<PartyInvites.Models.GuestResponse>

@{

    Layout = null;
}

<!DOCTYPE html>

<html>
<head>
    <meta name="viewport" content="width=device-width" />
    <title>Responses</title>
</head>
<body>
    <h2>Here is the list of people attending the party</h2>
    <table>
        <thead>
            <tr>
                <th>Name</th>
                <th>Email</th>
                <th>Phone</th>
            </tr>
        </thead>
        <tbody>
            @foreach (PartyInvites.Models.GuestResponse r in Model) {
                <tr>
                    <td>@r.Name</td>
                    <td>@r.Email</td>
                    <td>@r.Phone</td>
                </tr>
            }
        </tbody>
    </table>
</body>
</html>
```

Razor 视图文件有.cshtml 文件扩展名，因为它们是 C#代码和 HTML 元素的混合体。你可以在代码清单 2-18 中看到这一点，这里使用 foreach 循环来处理每个 GuestResponse 对象，它在操作方法中使用 View 方法传递给视图。与原生的 C#中的 foreach 循环不同，Razor 的 foreach 循环的主体包含添加到响应中的 HTML 元素，这些元素将被发送回浏览器。在这个视图中，每个 GuestResponse 对象都会生成一个 tr 元素，tr 元素中又会包含一个 td 元素，td 元素中是对象属性的值。

要查看工作中的列表，可以通过从 Debug 菜单中选择 Start Debugging 来运行应用程序，提交一些表单数据，然后单击链接查看响应列表。你将看到自应用程序启动以来输入的所有数据，如图 2-18 所示。视图呈现数据的方式有些简陋，但是现在已经足够完成我们的功能了，本章后面将介绍应用程序的其他样式。

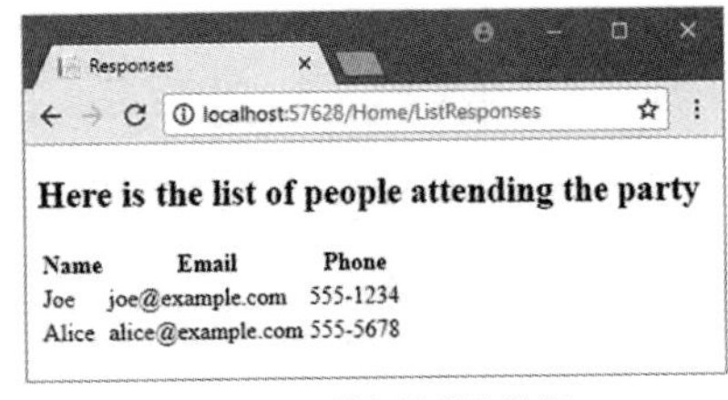

图 2-18 输入的所有数据

2.5.8 添加验证

现在该向应用程序添加数据验证功能了。如果程序缺乏验证，用户就可以输入无意义的数据，甚至提交一个空表单。在 MVC 应用程序中，验证通常应用于领域模型而不是用户界面。这意味着你只在一个地方定义了验证，但能够在模型类使用的应用程序的任何地方生效。MVC 支持声明式验证规则（Declarative Validation Rule），这是通过 System.ComponentModel.DataAnnotations 命名空间中的 Attribute 进行定义的。这意味着验证约束是使用标准 C#注解属性来表示的。代码清单 2-19 展示了如何将这些属性应用到 GuestResponse 模型类。

代码清单 2-19 在 Models 文件夹下的 GuestResponse.cs 文件中添加验证以应用相关属性

```
using System.ComponentModel.DataAnnotations;

namespace PartyInvites.Models {

    public class GuestResponse {

        [Required(ErrorMessage = "Please enter your name")]
        public string Name { get; set; }
        [Required(ErrorMessage = "Please enter your email address")]
        [RegularExpression(".+\\@.+\\..+",
            ErrorMessage = "Please enter a valid email address")]
        public string Email { get; set; }

        [Required(ErrorMessage = "Please enter your phone number")]
        public string Phone { get; set; }

        [Required(ErrorMessage = "Please specify whether you'll attend")]
        public bool? WillAttend { get; set; }
    }
}
```

MVC 会自动检测属性并在模型绑定过程中使用它们来验证数据。由于已经导入包含验证属性的命名空间，因此不再需要限定它们的名称就可以引用它们。

提 示

正如前面提到的，WillAttend 属性使用可为 null 的 bool 型。这样做是为了能够应用 Required 注解属性。如果使用常规的 bool 型，那么通过模型绑定接收到的值只有 true 或 false，你将无法判断用户是否选择了值。可空的 bool 型有三个可能的值——true、false 和 null。如果用户没有选择值，浏览器将发送 null，这让 Required 注解属性能够报告验证错误。这是 MVC 将 C#特性与 HTML 和 HTTP 混合在一起的一个很好的示例。

你可以在控制器类中使用 ModelState.IsValid 属性来检查是否存在验证问题。代码清单 2-20 显示了如何在 Home 控制器类里启用 POST 的 RsvpForm 操作方法中检查表单验证错误。

代码清单 2-20 在 Controllers 文件夹下的 HomeController.cs 中检查表单验证错误

```
using System;
using Microsoft.AspNetCore.Mvc;
using PartyInvites.Models;
using System.Linq;

namespace PartyInvites.Controllers {

    public class HomeController : Controller {

        public ViewResult Index() {
            int hour = DateTime.Now.Hour;
            ViewBag.Greeting = hour < 12 ? "Good Morning" : "Good Afternoon";
            return View("MyView");
        }

        [HttpGet]
        public ViewResult RsvpForm() {
            return View();
        }
        [HttpPost]
        public ViewResult RsvpForm(GuestResponse guestResponse) {
            if (ModelState.IsValid) {
                Repository.AddResponse(guestResponse);
                return View("Thanks", guestResponse);
            } else {
                // there is a validation error
                return View();
            }
        }

        public ViewResult ListResponses() {
            return View(Repository.Responses.Where(r => r.WillAttend == true));
        }
    }
}
```

Controller 基类提供了一个名为 ModelState 的属性，该属性提供关于将 HTTP 请求数据转换成 C#对象的信息。如果 ModelState.IsValue 属性返回 true，那么可以知道 MVC 能够满足通过 GuestResponse 类的属性指定的验证约束。因此验证通过，将展现 Thanks 视图。

如果 ModelState.IsValue 属性返回 false，那么可以知道发生验证错误了。ModelState 属性返回的对象提供了每个问题的详细信息，但开发人员并不需要知道细节，因为通过一个特别有用的特性，无须任何参数就能调用 View 方法，并自动地请求用户解决任何问题。

当 MVC 渲染视图时，Razor 可以访问与请求相关的所有验证错误的细节，而标签助手可以访问细节并向用户显示验证错误。代码清单 2-21 显示了如何为 RsvpForm 视图添加验证用的标签助手属性。

代码清单 2-21 在位于 Views/Home 文件夹下的 RsvpForm.cshtml 文件中添加验证总结

```
@model PartyInvites.Models.GuestResponse

@{

    Layout = null;
}

<!DOCTYPE html>
```

```
<html>
<head>
    <meta name="viewport" content="width=device-width" />
    <title>RsvpForm</title>
</head>
<body>
    <form asp-action="RsvpForm" method="post">
        <div asp-validation-summary="All"></div>
        <p>
            <label asp-for="Name">Your name:</label>
            <input asp-for="Name" />
        </p>
        <p>
            <label asp-for="Email">Your email:</label>
            <input asp-for="Email" />
        </p>
        <p>
            <label asp-for="Phone">Your phone:</label>
            <input asp-for="Phone" /></p>
        <p>
            <label>Will you attend?</label>
            <select asp-for="WillAttend">
                <option value="">Choose an option</option>
                <option value="true">Yes, I'll be there</option>
                <option value="false">No, I can't come</option>
            </select>
        </p>
        <button type="submit">Submit RSVP</button>
    </form>
</body>
</html>
```

将 asp-validation-summary 属性应用于 div 元素，并在渲染视图时显示验证错误列表。asp-validation-summary 属性的值来自 ValidationSummary 枚举，这个枚举指定了验证总结中将包含哪些类型的验证错误。这里指定包含所有类型的验证错误，这是大多数应用程序开始编写时的常用方式，第 27 章将描述其他的取值并解释它们的工作原理。

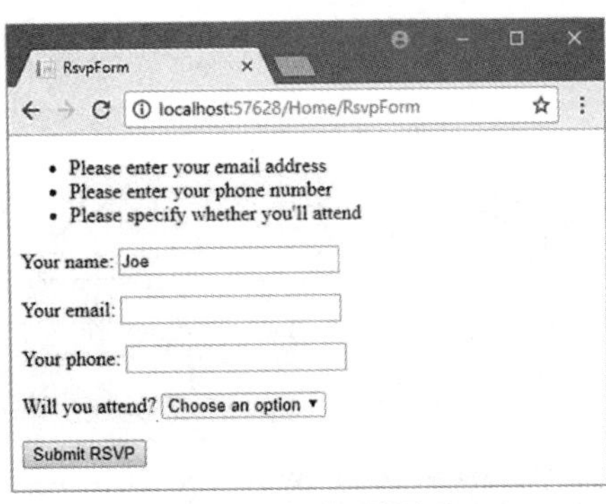

图 2-19 验证错误

要查看验证总结是如何工作的，运行应用程序，只填写 Name 字段，然后不输入任何其他数据就提交表单。你将看到的验证错误如图 2-19 所示。

RsvpForm 这个操作方法将不会渲染 Thanks 视图，直到满足应用到 GuestResponse 类的所有验证约束为止。注意，当 Razor 使用验证总结显示视图时，输入 Name 字段中的数据（合法数据）将被保留并再次显示。这是模型绑定的另一个好处——简化了表单数据的工作。

注 意

使用过 ASP.NET Web Forms 的读者知道服务器控件（server control）的概念，通过将序列化的值保存到一个名为__VIEWSTATE 的隐藏字段中，服务器控件可保持状态。MVC 模型绑定与服务器控件、回发或视图状态等 Web 表单概念无关。MVC 不会将隐藏的__VIEWSTATE 字段注入所渲染的 HTML 页面。相反，它通过设置输入元素的 value 属性来包含数据。

高亮显示无效字段

标签助手属性与模型属性通过元素关联了一个方便的特性，可以用来和模型绑定结合。当模型属性验证失败时，标签助手属性将生成略有不同的 HTML。这里是在没有验证错误时为“手机号”字段生成的 HTML：

```
<input type="text" data-val="true" data-val-required="Please enter your phone number"
id="Phone" name="Phone" value="">
```

相比之下，以下是在用户提交表单而不将任何数据输入文本字段后（这是验证错误，因为将 Required 验证属性应用到了 GuestResponse 类的 Phone 属性）生成的 HTML：

```
<input type="text" class="input-validation-error" data-val="true"
    data-val-required="Please enter your phone number" id="Phone"
    name="Phone" value="">
```

这里用粗体高亮显示了不同之处——标签助手属性将输入元素添加到一个名为 input-validation-error 的类中。我们可以利用这个特性创建一个样式表，其中包含这个类的一些 CSS 样式，以及其他不同的 HTML 标签助手属性以丰富显示效果。

MVC 项目中的约定是将交付给客户的静态内容（如 CSS 样式表等）放置到 wwwroot 文件夹中，根据内容的类型进行组织排序，所以 CSS 样式表被放入 wwwroot/css 文件夹，JavaScript 文件则被放入 wwwroot/js 文件夹，等等。

要创建一个 CSS 样式表，可以右击 wwwroot/css 文件夹，在 Visual Studio 的 Solution Explorer 中，选择 Add→New Item，导航到客户端部分，并从模板列表中选择 Style Sheet，如图 2-20 所示。

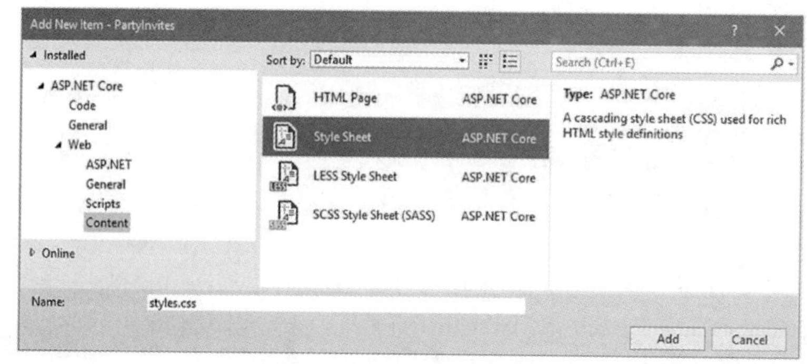

图 2-20　创建一个 CSS 样式表

提　示

当使用 Web Application(Model-View-Controller)模板创建项目时，Visual Studio 会在 wwwroot/css 文件夹中创建 site.css 文件。可以忽略这个文件，你在本章中不会用到它。

将文件的名称设置为 styles.css，单击 Add 按钮创建 CSS 样式表，编辑这个文件，如代码清单 2-22 所示。

代码清单 2-22　wwwroot/css 文件夹下的 styles.css 文件的内容

```
.field-validation-error    {color: #f00;}
.field-validation-valid    { display: none;}
.input-validation-error    { border: 1px solid #f00; background-color: #fee; }
.validation-summary-errors { font-weight: bold; color: #f00;}
.validation-summary-valid  { display: none;}
To apply this stylesheet, I have added a link element to the head section of the RsvpForm
view, as shown in Listing 2-23.
```

为了应用这个样式表，需要向 RsvpForm 视图的头部添加一个 link 元素，如代码清单 2-23 所示。

代码清单 2-23　在 Views/Home 文件夹下的 RsvpForm.cshtml 文件中应用样式表

```
...
<head>
    <meta name="viewport" content="width=device-width" />
```

```
    <title>RsvpForm</title>
    <link rel="stylesheet" href="/css/styles.css" />
</head>
...
```

link 元素使用 href 属性指定样式表的位置。注意，URL 中省略了 wwwroot 文件夹。ASP.NET 的默认配置包括对服务静态内容（例如图像、CSS 样式表和 JavaScript 文件）的支持，并且自动将请求映射到 wwwroot 文件夹。第 14 章将详述 ASP.NET 和 MVC 配置过程。

提　示

在处理样式表时，有一个特殊的标签助手，如果有许多文件可以管理的话，使用这个标签助手将十分方便。详见第 25 章。

随着样式表的应用，当提交数据时会突出显示验证错误，如图 2-21 所示。

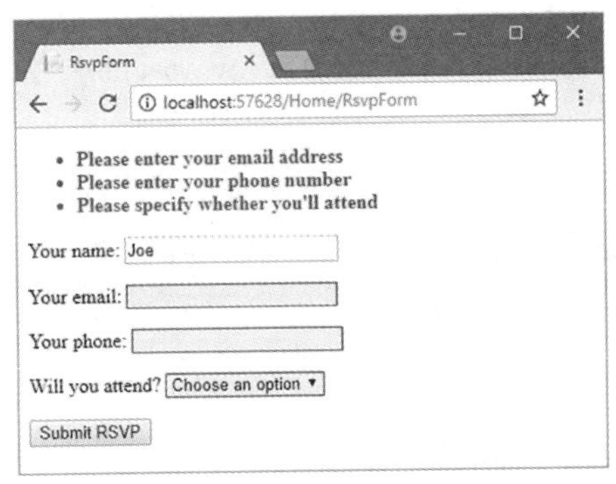

图 2-21　自动突出显示的验证错误

2.5.9　设置内容样式

至此，应用程序的所有功能目标都已完成，但是应用程序的整体外观很差。当你使用 Web Application（Model-View-Controller）模板创建项目时，Visual Studio 已经内置安装了一些常见的开源库。虽然作者不喜欢使用模板项目，但作者确实喜欢微软选择的库。其中一个叫作 Bootstrap，它是一个很好的 CSS 框架，最初是由 Twitter 开发的，目前已经成为一个主要的开源项目，并且已经成为 Web 应用程序开发的主流框架之一。

注　意

在撰写本书时，Bootstrap 3 是当前最新版本，并且第 4 代版本已在开发。微软可能会选择在 Visual Studio 的后期版本中更新 Web Application（Model-View-Controller）模板所使用的 Bootstrap 版本，这可能会导致内容以不同的方式显示。这对于本书的其他章节来说并不是问题，因为本书将展示如何显式地指定某个库的版本，以便得到预期的结果。

1. 设置欢迎视图

Bootstrap 的基本特性是通过运用于 HTML 元素的类起作用的，这些类与定义在 wwwroot/lib/bootstrap 文件夹中的 CSS 选择器相关联。Bootstrap 的相关细节可以从 getbootstrap 网站获取。你也可以通过代码清单 2-24 中的 MyView.cshtml 视图文件看到如何将一些基本样式应用到视图文件。

代码清单 2-24　添加 Bootstrap 到 Views/Home 文件夹下的 MyView.cshtml 文件中

```
@{
    Layout = null;
}

<!DOCTYPE html>
```

```
<html>
<head>
    <meta name="viewport" content="width=device-width" />
    <title>Index</title>
    <link rel="stylesheet" href="/lib/bootstrap/dist/css/bootstrap.css" />
</head>
<body>
    <div class="text-center">
        <h3>We're going to have an exciting party!</h3>
        <h4>And you are invited</h4>
        <a class="btn btn-primary" asp-action="RsvpForm">RSVP Now</a>
    </div>
</body>
</html>
```

这里添加了 link 元素，它的 href 属性可以用于从 wwwroot/lib/bootstrap/dist/css 文件夹加载 bootstrap.css 文件。通常约定将第三方的 CSS 和 JavaScript 包安装到 wwwroot/lib 文件夹中，第 6 章将详述用于管理这些包的工具。

导入 Bootstrap 样式表后，需要对元素进行样式化。这只是一个简单的示例，所以只使用少量的 Bootstrap CSS 类——text-center、btn 以及 btn-primary。

text-center 类可以使元素及子元素的内容居中。btn 类用于修饰 button、input 或其他作为按钮使用的元素的样式，而 btn-primary 类用于指定按钮的颜色范围。你可以通过运行应用程序看到效果，如图 2-22 所示。

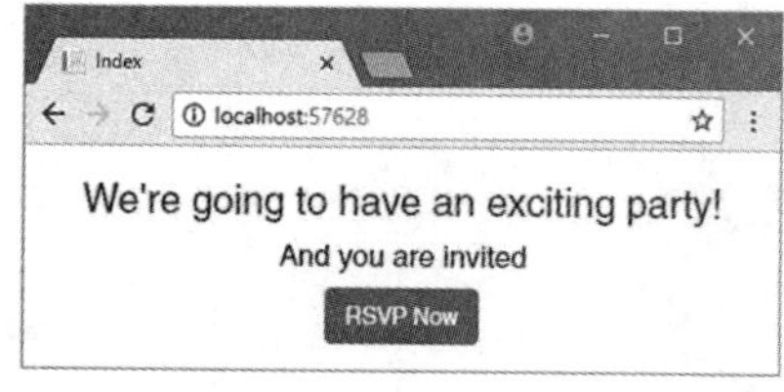

图 2-22 样式化视图的效果

2. 设置 RsvpForm（电子回复）表单视图

Bootstrap 定义了可以用于样式表的类。这里不打算详细介绍，但是会展示如何将这些类应用到代码清单 2-25 中。

代码清单 2-25 在 Views/Home 文件夹下的 RsvpForm.cshtml 文件中添加 Bootstrap

```
@model PartyInvites.Models.GuestResponse

@{
    Layout = null;
}

<!DOCTYPE html>

<html>
<head>
    <meta name="viewport" content="width=device-width" />
    <title>RsvpForm</title>
    <link rel="stylesheet" href="/css/styles.css" />
    <link rel="stylesheet" href="/lib/bootstrap/dist/css/bootstrap.css" />
</head>
<body>
    <div class="panel panel-success">
        <div class="panel-heading text-center"><h4>RSVP</h4></div>
        <div class="panel-body">
            <form class="p-a-1" asp-action="RsvpForm" method="post">
                <div asp-validation-summary="All"></div>
                <div class="form-group">
                    <label asp-for="Name">Your name:</label>
                    <input class="form-control" asp-for="Name" />
                </div>
                <div class="form-group">
```

```
                <label asp-for="Email">Your email:</label>
                <input class="form-control" asp-for="Email" />
            </div>
            <div class="form-group">
                <label asp-for="Phone">Your phone:</label>
                <input class="form-control" asp-for="Phone" />
            </div>
            <div class="form-group">
                <label>Will you attend?</label>
                <select class="form-control" asp-for="WillAttend">
                    <option value="">Choose an option</option>
                    <option value="true">Yes, I'll be there</option>
                    <option value="false">No, I can't come</option>
                </select>
            </div>
            <div class="text-center">
                <button class="btn btn-primary" type="submit">
                    Submit RSVP
                </button>
            </div>
        </form>
    </div>
</div>
</body>
</html>
```

这个示例中的Bootstrap类创建了一个标题，只是为了提供一下结构布局。为了对表单进行样式化，使用了form-group类，该类用于对包含标签和相关输入或选择的元素进行样式化。你可以在图2-23中看到样式的效果。

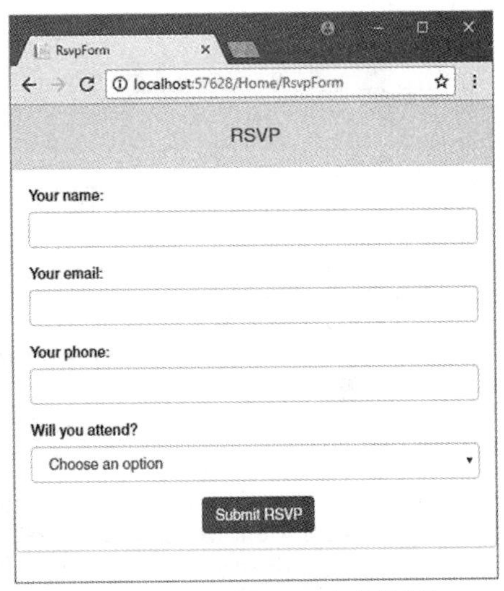

图2-23　样式化RsvpForm视图的效果

3. 样式化Thanks视图

下一个要样式化的视图文件是Thanks.cshtml，你可以在代码清单2-26中看到具体是如何完成的，使用的CSS类与在其他视图中使用的相似。为了使应用程序更容易管理，尽可能避免重复代码和标记是一条很好的原则。MVC提供了几个特性来帮助减少重复，后面的章节将对此进行描述。这些特性包括Razor布局、分部视图和视图组件。

代码清单2-26　在Thanks.cshtml文件中使用Bootstrap

```
@model PartyInvites.Models.GuestResponse

@{
    Layout = null;
}

<!DOCTYPE html>

<html>
<head>
    <meta name="viewport" content="width=device-width" />
    <title>Thanks</title>
    <link rel="stylesheet" href="/lib/bootstrap/dist/css/bootstrap.css" />
</head>
<body class="text-center">
    <p>
```

```
        <h1>Thank you, @Model.Name!</h1>
        @if (Model.WillAttend == true) {
            @:It's great that you're coming. The drinks are already in the fridge!
        } else {
            @:Sorry to hear that you can't make it, but thanks for letting us know.
        }
    </p>
    Click <a class="nav-link" asp-action="ListResponses">here</a>
    to see who is coming.
</body>
</html>
```

图 2-24 显示了样式化 Thanks 视图的效果。

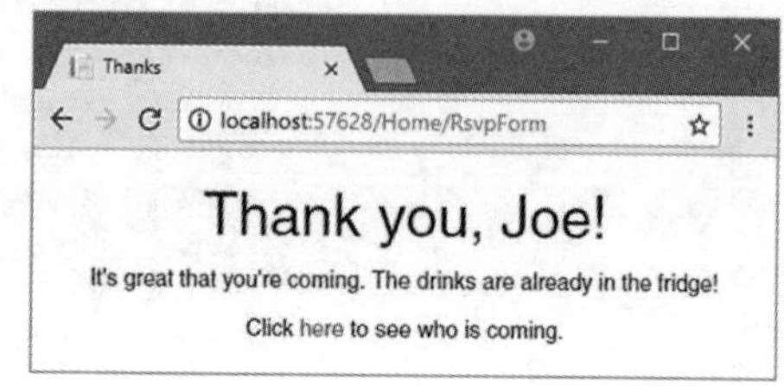

图 2-24　样式化 Thanks 视图的效果

4. 样式化列表视图

最后一个需要样式化的视图是 ListResponses，它是一个用于展示参加派对人员信息的列表。在样式化内容时遵循与所有 Bootstrap 样式相同的基本操作，如代码清单 2-27 所示。

代码清单 2-27　在 Views/Home 文件夹下的 ListResponses.cshtml 文件中添加 Bootstrap

```
@model IEnumerable<PartyInvites.Models.GuestResponse>

@{
    Layout = null;
}

<!DOCTYPE html>

<html>
<head>
    <meta name="viewport" content="width=device-width" />
    <link rel="stylesheet" href="/lib/bootstrap/dist/css/bootstrap.css" />
    <title>Responses</title>
</head>
<body>
    <div class="panel-body">
        <h2>Here is the list of people attending the party</h2>
        <table class="table table-sm table-striped table-bordered">
            <thead>
                <tr>
                    <th>Name</th>
                    <th>Email</th>
                    <th>Phone</th>
                </tr>
            </thead>
            <tbody>
                @foreach (PartyInvites.Models.GuestResponse r in Model) {
                    <tr>
                        <td>@r.Name</td>
                        <td>@r.Email</td>
                        <td>@r.Phone</td>
                    </tr>
                }
            </tbody>
        </table>
    </div>
</body>
</html>
```

图 2-25 展示了样式化 ListResponses 视图的效果。将这些样式添加到视图中，即可完成这个应用程序。

应用程序现在已满足所有的开发目标，并且外观上也有了很大的改进。

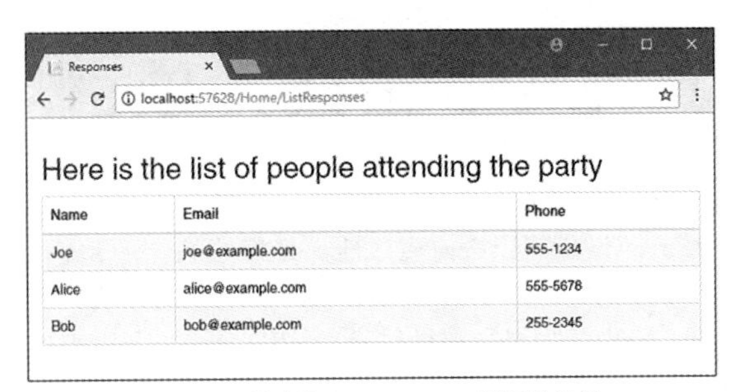

图 2-25 样式化 ListResponses 视图的效果

2.6 小结

在本章中，我们创建了一个新的 MVC 项目，并使用它构造了一个简单的数据录入程序，你看到了 ASP.NET Core MVC 的大体架构和方法。本章跳过了一些关键特性（包括 Razor 语法、路由和测试），但是后面的章节将深入讨论这些主题。下一章将描述 MVC 的设计模式，这是使用 ASP.NET Core MVC 实现有效开发的基础。

第3章　MVC模式、项目与约定

在深入探讨 ASP.NET Core MVC 的核心内容之前，最好熟悉一下 MVC 的设计模式、隐藏在后面的核心思想，以及将 MVC 项目转换为 ASP.NET Core MVC 项目的方法。你可能已经对本章将要讨论的一些想法和约定有所了解，特别是如果之前做过 ASP.NET 或 C#开发工作的话。如果没有，建议仔细研读一下本章，以加深对 MVC 底层知识的理解，从而帮助你更好地投入到本书接下来的框架特性学习中。

3.1　MVC 简史

MVC（Model-View-Controller，模型-视图-控制器）在 20 世纪 70 年代后期就已经出现，产生于 Xerox PARC（施乐公司的帕洛阿尔托研究中心）的 Smalltalk 项目，当时被构思为早期 GUI 应用程序的一种组织方式。最初的 MVC 模式的某些细节依赖于 Smalltalk 特有的概念，如屏幕和工具等，但是其更广泛的概念仍然适用于现在的应用程序，而且特别适用于 Web 应用程序。

3.2　MVC 模式

从高级术语角度看，MVC 模式意味着一个 MVC 应用程序将被分成至少 3 部分。

- 模型，它包含或代表用户使用的数据。
- 视图，它用来呈现模型中的一些用户界面。
- 控制器，它用来处理输入请求，执行模型操作，并选择渲染给用户的视图。

MVC 架构的每一部分都是被明确界定和自包含的，我们称之为关注点分离。模型中操纵数据的逻辑仅仅在模型内部，展示数据的逻辑仅仅在视图中，处理用户请求和输入的代码仅仅包含在控制器中，每一部分都有明确的分工。这样，无论应用程序有多大，在整个生命周期中它都会变得易于维护和扩充。

3.2.1　模型

模型中包含了用户使用的数据。有两种类型的模型：一种是视图模型，只代表从控制器传递到视图的数据；另一种是领域模型，包含业务域中的数据，以及创建、存储和操纵这些数据的操作、转换和规则，这些统称为模型逻辑。

模型是应用程序工作的定义域。例如，在银行应用程序中，模型代表了应用程序支持的任何事情，比如账户、总账、客户的信用额度以及模型中可以操控数据的操作，如存款和提款。模型也负责维护整体的状态和数据的一致性，比如，确保所有的交易都被添加到分类账中并且客户没有从账户中取走超出账户余额的钱。

对于 MVC 的每个组件，接下来分别描述哪些内容应该包含在内，哪些内容不应该包含在内。在应用程序中使用 MVC 构建模型时，注意以下几点：

- 包含领域数据；
- 包含用户创建、管理和修改领域数据的逻辑；
- 提供一个可以公开模型数据和操作的干净的 API。

以下做法是错误的：

- 暴露模型数据是怎样被获取或组织的（换句话说，数据存储机制的细节不能暴露给控制器和视图）；
- 包含基于用户交互改变模型的逻辑（因为那是控制器的工作）；
- 包含向用户显示数据的逻辑（因为那是视图的工作）。

保证模型与控制器和视图逻辑分离的好处是可以使逻辑的测试工作更加简单（第 7 章将会介绍单元测试）。不仅如此，这样做还可以使整个应用程序的优化和维护工作变得简单。

提　示

许多不熟悉 MVC 模式的开发者对数据模型中包含逻辑处理的事情感到困惑，请相信，MVC 模式的目标是将数据和逻辑分离。有些人认为 MVC 模式的目标是将一个应用划分成 3 个功能区域，每个功能区域都包含逻辑和数据模块，其实这是误解。MVC 模式的目标并不是要从模型中消除逻辑，相反，而是确保模型中只包含用于创建和管理模型数据的逻辑。

3.2.2 控制器

控制器在 MVC 模型中是数据模型和视图之间联系的纽带。控制器定义了对数据模型进行操作的业务逻辑的动作，以及视图给用户提供视图数据的动作。

使用 MVC 模式构建控制器时应该包含基于用户交互更新模型的动作；不应该包含管理数据外观的逻辑（那是视图的工作），同时不应该包含管理数据持久化的逻辑（那是模型的工作）。

3.2.3 视图

视图包含了向用户展示数据或者从用户那里获取数据以便控制器使用的逻辑。视图应该包含需要展示给用户的逻辑和标注，不应该包含复杂的逻辑（这些最好放在控制器中）和创建、存储或操控域模型的逻辑。

视图可以包含逻辑，但是包含的逻辑应该比较简单或者很少使用。在视图中放置的任何事情，即使是最简单的方法调用或表达式，也会使整个应用变得难以测试和维护。

3.2.4 MVC 的 ASP.NET 实现

顾名思义，ASP.NET Core MVC 使 MVC 模式适用于 ASP.NET 和 C#开发。在 ASP.NET Core MVC 中，控制器是 C#类，通常派生于微软的 AspNetCore.Mvc.Controller 类。从控制器派生而来的类中的每个公有方法都是一种操作方法，与一个 URL 相关联。当一个请求被发送给一个与操作方法相关联的 URL 时，便会执行操作方法中的语句，以执行领域模型上的一些操作，然后选择一个视图显示给客户端。图 3-1 显示了控制器、模型和视图之间的这种交互。

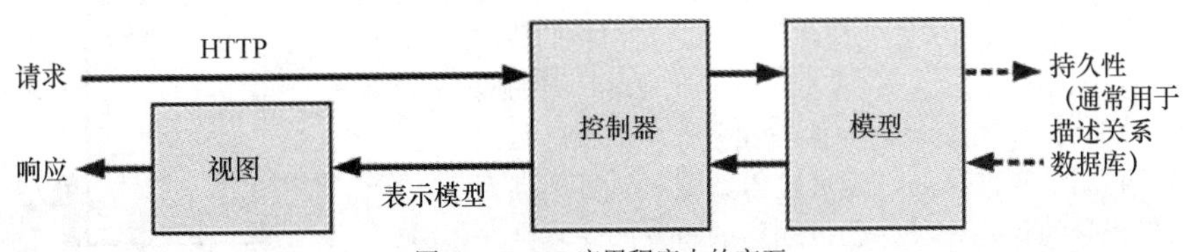

图 3-1　MVC 应用程序中的交互

ASP.NET Core MVC 使用名为 Razor 的视图引擎组件来负责处理视图，以生成浏览器需要的结果。Razor 视图是一些包含 C#逻辑的 HTML 模板，用于处理模型数据以产生随着模型变化而生成的动态内容。第 5 章将会介绍 Razor 的工作原理。

ASP.NET Core MVC 对领域模型的实现没有任何约束。你可以使用常规的 C#对象创建模型，并且可以使用.NET支持的任何数据库、对象关系映射（ORM）框架或者.NET支持的其他数据工具来实现持久化。

单页应用程序

Web应用程序开发历史倾向于将浏览器视为呈现HTML和响应鼠标单击的简单显示设备，这称为Web应用程序的往返样式。每次用户单击链接时，都会向ASP.NET Core MVC应用程序发送一个HTTP请求，在这种应用程序中，控制器选择一个由Razor呈现并发送回浏览器的视图，以便向用户显示一个新的HTML页面。所有的逻辑、数据和状态都存在于ASP.NET Core MVC服务器中，这简化了开发过程，意味着除了确保浏览器能够处理Razor视图中包含的HTML功能外，不必太关注浏览器。

相比之下，单页应用程序将浏览器合并到应用程序平台中。服务器负责管理应用程序的数据，而运行浏览器的JavaScript代码则请求数据，将其显示给用户，并响应用户交互。在单页应用程序中，模型、视图和控制器共享浏览器和服务器。

应用程序的ASP.NET Core MVC服务器部分提供对应用程序数据的访问，而不是将完整的HTML页面发送到浏览器，这些数据是由JavaScript框架（如Angular或React）查询和显示的。

单页应用程序可能比往返应用程序更具响应性，但它们的创建更复杂，需要同时具备C#和JavaScript技能才能进行有效的开发，难度不可低估。第20章将演示如何使用ASP.NET Core MVC来提供此类应用程序中的数据，但不会演示单页应用程序开发。作者最喜欢的JavaScript框架是Angular，如果要在ASP.NET Core MVC中使用Angular，请参阅作者编写的另一本书——*Angular for ASP.NET Core MVC*。

3.3 MVC与其他模式的比较

当然，MVC并不是唯一的软件架构模式。还有很多其他的模式，并且其中的一些至少曾经是非常流行的。通过比较，你也可以了解关于MVC的更多知识。接下来的章节将简要介绍几个创建应用的不同方法，并将它们与MVC进行对比。一些与MVC大同小异，其他一些则完全不同。

并不是说MVC对所有场景都是最佳模式。作者主张选取最佳方法来解决手头的问题。你会看到，在某些场景下，有些其他模式与MVC的表现相当，甚至要优于MVC。在选择模式时，建议做出明智的选择。

3.3.1 “智能UI”模式

最常用的设计模式之一称为“智能UI”模式。大多数程序员在其职业生涯中有过创建智能UI程序的经历。如果使用过Windows Forms或ASP.NET Web Forms，那你一定也有这样的经历。

要创建一个UI应用程序，通常，开发人员会通过把一组组件或控件拖放到设计界面或画布的方式来构建用户界面。控件可以创建用户的按压按钮事件、单击事件、鼠标移动事件以与用户交互。开发人员为控件的这些事件编写一系列事件处理代码。当特定组件的特定事件被触发时便会调用这些代码。这就创建了一种单片式的应用程序，如图3-2所示。处理用户界面与业务逻辑的代码是混合在一起的，它们没有做分离。定义输入数据、执行数据查询或修改用户账号的代码片段都已按期望的事件顺序耦合在一起。

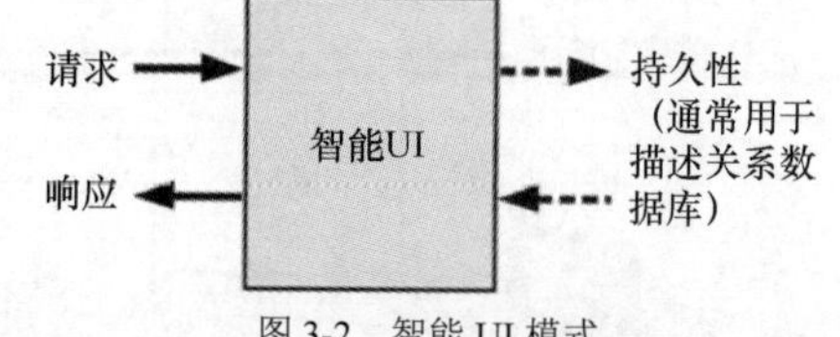

图3-2 智能UI模式

智能UI项目非常适合简单的项目，因为你很快就可得到一些很好的结果（你将会在第8章看到，MVC开发在交付结果之前需要做一些精心的准备工作和前期投入）。智能UI项目也适用于用户界面的原型开发。这些界面设计工具确实是很不错的。如果你正在与客户沟通并且正需要捕获客户对外观和接口流程的需求，那么智能UI系统可以帮助你以快速响应的方式生成和测试一些不同的方案。

智能 UI 模式最大的缺点就是难以维护和扩展。将领域模型与带有用户界面代码的业务逻辑混搭在一起会导致工作的重复。换言之，为了支持新增加的组件，需要复制并粘贴相同的业务逻辑代码片段。查找所有重复的部分并修改错误是非常难的事情。要添加新的功能而不影响已有的功能是几乎不可能的，对智能 UI 应用程序进行测试也是一件难事。唯一的办法就是模拟用户的交互行为，这项工作不仅效果不理想，而且很难保证有一个全面的测试覆盖率。

在 MVC 中，智能 UI 模式通常是一种反模式（anti-pattern）。这种反模式的出现，至少在某些地方，是因为大多数人是在经历了长期的智能 UI 应用程序开发与维护导致的失控之后，才来到 MVC 世界以寻求另一种开发方案。

另外，拒绝难以控制的智能 UI 模式是错误的。并不是智能 UI 模式中的所有东西都那么不堪，其中也有很多积极的方面。智能 UI 模式是非常易于开发的，并且开发起来速度比较快。组件和设计工具方面的开发商投入很大的精力，使得用户的开发经历尽量愉悦，即使几乎没有任何开发经验的人在几小时内也可以开发出具有专业外观和合理功能的应用。

智能 UI 应用最大的弱点是可维护性差，但不会在小型开发项目中出现。如果要为小客户开发一款简单的工具，那么智能 UI 应用是很好的选择，MVC 应用程序的额外复杂性则根本无法保证。

3.3.2 理解模型-视图架构

智能 UI 应用中可能会出现维护问题的地方是业务逻辑，业务逻辑横跨多个应用以至于改变或增加功能都是十分艰难的。模型-视图（model-view）架构则为此做了改进，它将业务逻辑抽取出来形成单独的域模型。这样做之后，数据、处理过程和规则都集中在应用程序中的同一个部件中，如图 3-3 所示。

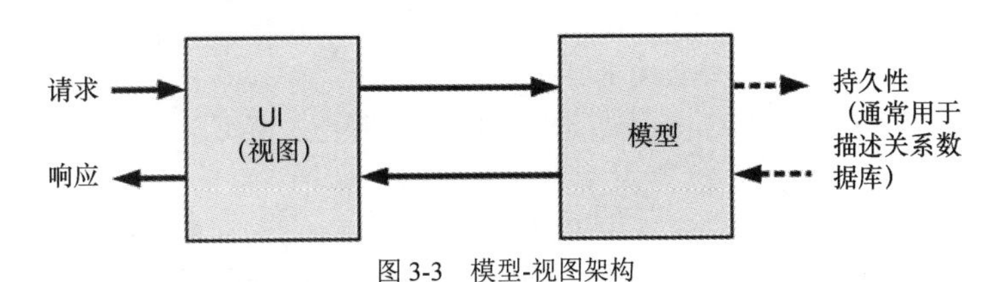

图 3-3 模型-视图架构

模型-视图架构可以说是对智能 UI 模式的一种整体上的改进，比方说更加易于维护了，但同时也来带来两个问题。第一个问题，因为 UI 和域模型是紧密结合的，所以对其中任何一个进行单元测试都比较困难。第二个问题来自实践而不是模式的定义。模型通常会包含大量的数据访问代码，其实并不需要这样，但通常是这样的，这就意味着数据模型不仅仅包含业务数据、操作和规则。

3.3.3 经典的 3 层架构

为了解决“模型-视图”架构存在的问题，可使用 3 层模式（见图 3-4）将持久性代码从域模型分离出来并放置在一个新的组件中，这个组件称作数据访问层（DAL）。

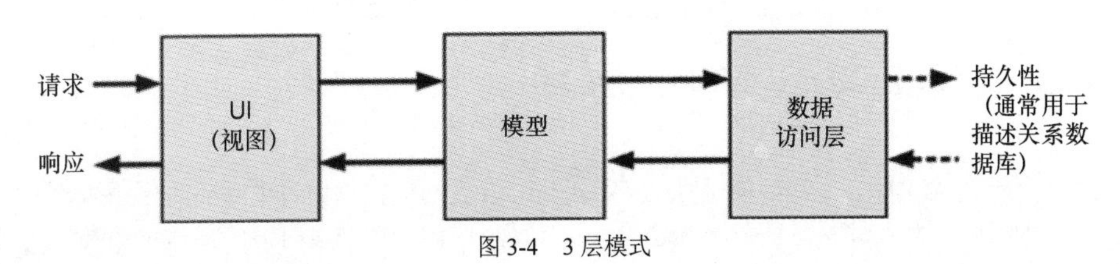

图 3-4 3 层模式

3 层架构是应用程序中使用最广泛的模式。它对 UI 的实现方式没有限制，并且不需要很复杂的操作就可以实现很好的关注点分离（SoC）。另外，我们注意到，有了数据访问层（DAL），单元测试也会变得相对更容易一些。我们可以很容易地观察到经典的 3 层架构应用程序和 MVC 模式的相似之处。不同的是，当 UI 层直接耦合到 GUI 框架（例如 Windows 窗体或 ASP.NET Web 窗体）的 click 事件时，要自动地执行单元测试几乎是不可能的。由于3层架构应用程序的UI部分可以很复杂，因此有很多代码无法严格测试。

最坏情况下，3层架构在UI层严重缺乏约束，也就是说，许多此类没有真正实现关注点分离（SoC）的应用程序，都被错误地认为是智能UI应用程序。随之而来的就是最坏结果——无法测试、难以维护且极其复杂的应用程序诞生了。

3.3.4　MVC的多样性

前面已经介绍了MVC应用程序的核心设计原则，尤其是ASP.NET Core MVC方面的内容，还解释了MVC模式与其他模式不同的方面以及为适应项目范围和目的对MVC所做的添加、调整和调节工作。下面简要介绍MVC最普遍的两种变化。了解这些变化对于使用ASP.NET Core MVC并没有太大的帮助，这里只是为了保证信息的完整，因为你将听到许多在讨论软件模式时用到的术语。

1. MVP（Model-View-Presenter，模型-视图-呈现器）模式

MVP是MVC的一种变体，以便更容易地适应状态的变化，例如Windows窗体或ASP.NET Web窗体。为了得到最好的智能UI模式而不引起通常会伴随而来的错误，这值得尝试。

在这种模式中，呈现器具有与MVC控制器相同的职责，但它与状态化视图具有更直接的关系，根据用户的输入和动作，直接管理着UI组件中显示的数据。MVP模式有以下两种实现。

- **被动式视图**（passive view）实现，在这种实现中，视图不包含逻辑。视图是UI控件的容器，由呈现器直接进行操纵。
- **监管控制器**（supervising controller）实现，在这种实现中，视图可能要负责一些具有表现逻辑的元素，如数据绑定，并被给予对领域模型数据源的引用。

这两种实现方式之间的差别涉及视图如何智能化。任何一种方式下，呈现器与GUI框架都是解耦的，这使得呈现器的逻辑更简单且更适合单元测试。

2. MVVM（Model-View-View Model，模型-视图-视图模型）模式

MVVM模式是MVC的最新变体，源于微软，并应用于WPF。在MVVM模式中，模型和视图具有与MVC相同的作用。不同的是MVVM中关于视图模型的概念，视图模型是用户界面的一种抽象表示，它既暴露了视图中待显示的数据，也暴露了能够通过UI进行的操作。与MVC控制器不同，MVVM视图模型没有视图（或者任何特定UI技术）的概念。MVVM视图使用WPF的绑定（binding）特性，将视图控件暴露的属性（下拉菜单中的条目或按钮的单击效果）与视图模型暴露的属性双向地关联在一起。

提　示

MVC也使用术语视图模型（view model），但指的是一种简单的模型类，只用于从控制器向视图传递数据。而领域模型（domain model）则相反，指的是数据、操作以及规则的完整表示。

3.4　ASP.NET Core MVC项目

在创建一个新的ASP.NET MVC项目时，Visual Studio会为你提供项目中需要的一些初始内容选项。这样不仅可以为新手开发人员简化整个学习过程，并且可以应用一些节省时间的最佳实践来完成常见的功能和任务。作者对使用项目或代码模板的方法并不热衷。提供这种模板的初衷是好的，但是只拿来执行总觉得平庸。ASP.NET和MVC的特色之一就是在定制个性化的平台时有非常大的灵活性。Visual Studio创建和填充的这些项目、类和视图会让用户难以摆脱他人的样式，并且这些内容和配置太过普通，一般可用性也不大。可能微软也不知道用户需要什么样的应用程序，所以他们做的是涵盖所有的基础内容。但是这种笼统的方式反而让作者摒弃这些默认的内容。

建议从一个空的项目开始，添加需要的文件夹、文件和软件包，这样你不仅可以了解 MVC 的工作方式，还可以完全操控应用程序中包含的所有内容。

当然，作者的偏好不应该完全代表你的开发经验，可能你会发现模板更好用，特别是对于 ASP.NET 新手以及那些还没有开发适合自己的开发样式的开发者来说。你可能还会觉得项目模板是十分有用的资源并且是想法的来源，但是在你完全了解它的工作方式之前，你应该谨慎地向应用程序中添加任何功能。

3.4.1　创建项目

当首次创建新的 ASP.NET 项目时，可以从以下 3 个基本的起点——Empty 模板、Web API 模板和 Web Application（Model-View-Controller）模板之一开始，如图 3-5 所示。

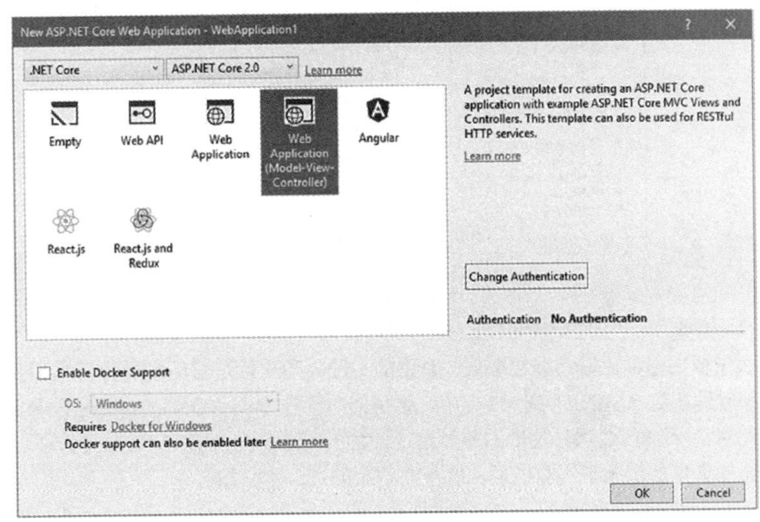

图 3-5　ASP.NET 项目模板

Empty 模板包含 ASP.NET 的内核管道，但不包括 MVC 项目所需要的库或配置。Web API 模板包含 ASP.NET 内核和 MVC，其中的示例应用程序演示了如何接收和处理来自客户端的 Ajax 请求，第 20 章将对此进行描述。

Web Application（Model-View-Controller）模板包括 ASP.NET Core 和 MVC，其中包含的示例应用程序演示了如何生成 HTML 内容。Web API 和 Web Application（Model-View-Controller）模板可以提供不用的方案来为用户配置进行身份验证和授权访问功能。

其他模板提供了适合使用单页应用程序框架（Angular 和 React）与 Razor 页面的初始内容（允许代码和标记混合在一个文件中，合并控制器和视图的角色，并权衡 MVC 模型的一些优点以获得简单性）。

项目模板给人的印象可能是你需要遵循特定的路径来创建特定类型的 ASP.NET 应用程序，但情况并非如此。这些模板只是在完成相同功能时不同的起点。在使用任何模板创建的项目中都可以随时添加需要的功能。例如，处理 Ajax 请求，以及身份验证和授权功能都是从空的项目模板开始的。

因此，项目模板之间的真正区别是最初的库集、配置文件、代码以及 Visual Studio 在创建项目时添加的内容。最简单的模板和最复杂的模板之间存在很多差异，图 3-6 展示了每种模板在创建新的项目之后的解决方案管理器（Solution Explorer）的样子。对于 Web Application（Model-View-Controller）模板，因为单个列表太长了，所以必须将解决方案管理器放在不同的文件夹中。

使用 Web Application（Model-View-Controller）模板添加到项目中的额外文件看起来令人望而生畏，但其中一些只是占位符或常见功能的示例实现。一些其他文件设置了 MVC 或配置了 ASP.NET。其他的是客户端库，你将在程序生成 HTML 时添加这些库。现在的文件列表看起来可能比较长，但在你读完整本书后，你就会了解到每一个文件所要完成的功能。

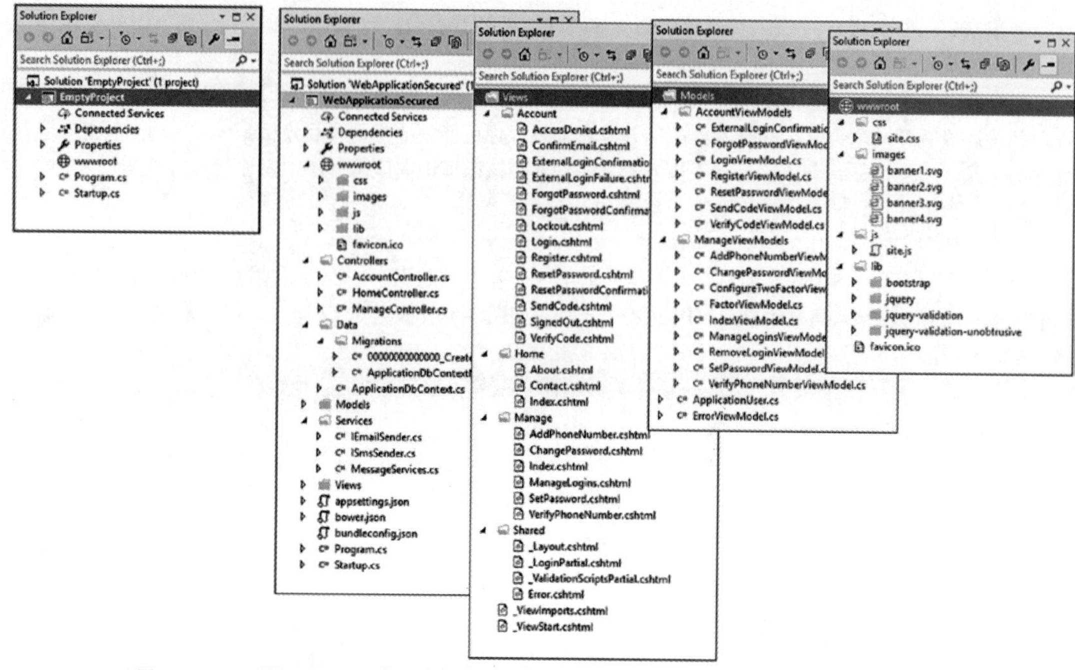

图 3-6 Empty 模板和 Web Application（Model-View-Controller）模板添加到新项目中的默认内容

无论使用什么模板创建项目，都会发现一些公用的文件夹和文件是必然出现的。项目中的某些项具有特殊的功能，它们被硬编码到 ASP.NET 或 MVC 中，也可能包含在 Visual Studio 提供的支持工具中。其他的则取决于大多数 ASP.NET 或 MVC 项目使用的命名约定。表 3-1 描述了 ASP.NET Core MVC 项目中一些重要文件夹，其中一些文件夹默认情况下不会出现，这会在稍后的章节中介绍。

表 3-1　重要的文件夹

文件夹	描述
/Areas	area 是一种分解大型应用程序的方法
/Dependencies	提供对项目依赖的所有包的细节描述
/Components	定义视图组件类的地方，这些类用来显示自包含特征（例如购物车）
/Controllers	用来放置控制器类的地方，也可以将控制器类放在任何其他地方，它们都会被编译到同一个程序集中
/Data	这是定义数据库相关类的地方，但作者更习惯在 Models 文件中对它们进行定义
/Data/Migrations	这里存储了数据库架构的详细信息，以便存储应用程序的数据，在第 8～11 章中，作为 SportsStore 项目的一部分，使用了迁移
/Models	用来存放视图模型和域模型类，也可以在项目的任何地方或另一个单独的项目中定义模型类
/Views	保存视图和分部视图，通常与使用控制器命名的相关文件夹放在一起
/Views/Shared	这里包含非特定控制器的布局和视图
/Views/_ViewImports.cshtml	这个文件用于指定 Razor 视图中的命名空间，也可用来设置标签助手
/Views/_ViewStart.cshtml	这个文件用于指定 Razor 视图引擎的默认布局
/wwwroot	这是放置静态内容（如 CSS 文件和图像）的地方，同时也是 Bower 包管理器安装 JavaScript 和 CSS 软件包的地方

下面列出一些重要的文件。

- appsettings.json：包含可适应于不同环境（如开发、测试、生产环境）的配置信息。该文件常用于数据库服务器连接字符串和日志/调试配置。
- /bower.json：默认情况下，这个文件是隐藏的，其中包含了由 Bower 软件包管理器管理的文件列表。
- /<project>.csproj：用于指定项目的一些基本配置选项，包括使用的 NuGet 属性。
- /Program.cs：负责配置应用程序的宿主平台。
- /Startup.cs：用来配置应用程序。

3.4.2 关于 MVC 的约定

MVC 项目中有两种约定。其中一种是关于如何构造项目的建议。例如，按常规将依赖的第三方 JavaScript 和 CSS 包放在 wwwroot/lib 文件夹中，其他 MVC 开发人员也希望在这里能找到这些包，软件包管理器也会默认将这些包安装在这里。当然，你也可以随意重命名或删除 lib 文件夹，或者把软件包放到其他地方。只要视图中的脚本和链接元素指向的位置正确，MVC 就可以正常地运行应用程序。

另一种约定源于配置方面的原则，这也是 Ruby on Rails 如此受欢迎的原因之一。通过配置，不需要显式地配置控制器与视图之间的关联。例如，只需要遵循特定的文件命名规则就可以使程序正常执行。在此约定下，更改项目结构的灵活性就变得很小了。

提　示

如果将标准的 MVC 组件替换为自己的实现，所有这些约定就可以改变了。本书将介绍不同的方法，以帮助我们解释 MVC 应用程序是如何工作的，这些是你在大多数项目中会碰到的约定。

1. 关于控制器类的约定

控制器类的名称都是以 Controller 结尾的，例如 ProductController、AdminController 和 HomeController。当从项目的其他位置引用控制器时，如使用 HTML 辅助方法时，只需要指定名称的第一部分（如 Product），MVC 就会自动将 Controller 追加到名称中，并开始查找控制器类。

提　示

可以通过创建模型约定来改变这一点，详见第 31 章。

2. 关于视图的约定

视图保存在/Views/Controllername 文件夹中。例如，与 ProductController 类关联的视图将保存在/Views/Product 文件夹中。

提　示

请注意，这里在 Views 文件夹中省略了控制器类的 Controller 部分，比如使用的是/Views/Product 而不是 Views/ProductController。

MVC 希望操作方法的默认视图应以操作方法命名。例如，与列表（List）操作方法相关联的默认视图应名为 List.cshtml。因此，对于 ProductController 类中的列表（List）操作方法，默认视图应该是/Views/Product/List.cshtml。当你在操作方法中返回调用 View 方法的结果时，将使用默认视图，如下所示：

```
...
return View();
...
```

要按名称指定不同的视图，如下所示：

```
...
return View("MyOtherView");
...
```

请注意，上面所列的视图没有包括文件扩展名或视图路径。在查找视图时，MVC 首先会在以控制器命名的文件夹中查找，然后在/Views/Shared 文件夹中查找。这意味着可以将多个控制器使用的视图放在/Views/Shared 文件夹中，MVC 可以找到它们。

3. 关于布局的约定

布局的命名约定是在文件中加上下划线（_）字符，布局文件放在/Views/Shared 文件夹中。默认情况下，所有视图都会使用/Views/_ViewStart.cshtml文件。如果不想将默认版式应用于视图，可以更改_ViewStart.cshtml中的设置（或完全删除该文件）并在视图中指定其他版式，如下所示：

```
@{
    Layout = "~/_MyLayout.cshtml";
}
```

也可以对给定视图禁用任何布局，如下所示：

```
@{
    Layout = null;
}
```

3.5 小结

本章首先介绍了MVC架构模式，然后讨论了领域模型的意义，并讲述了依赖注入。依赖注入能够消除组件耦合，以强制应用程序各部件之间严格分离。下一章将介绍C#语言特性，它们已广泛应用于MVC Web应用开发中。

第 4 章 C#基本特性

本章主要描述在 Web 应用开发中难理解或经常引起混淆的 C#特性。因为这不是专门介绍 C#的书，所以对于每个 C#特性，本书只提供简单的示例，以便读者在后续的章节中能够更清楚地了解示例，并在自己的项目中充分利用这些特性。

表 4-1 列出了本章要解决的问题。

表 4-1 本章要解决的问题

问题	解决方法	代码清单
避免访问空的属性引用	使用 null 条件控制运算符	代码清单 4-5～代码清单 4-8
简化 C#属性	使用自动实现的属性	代码清单 4-9～代码清单 4-11
简化字符串组合	使用字符串插值	代码清单 4-12
测试对象的类型或特征	使用模式匹配	代码清单 4-13～代码清单 4-16
一次性创建对象并设置其属性	使用对象或集合初始化器	代码清单 4-17 和代码清单 4-18
为不能修改的类添加功能	使用扩展方法	代码清单 4-19～代码清单 4-26
简化委托和声明方法的使用	使用 Lambda 表达式	代码清单 4-27～代码清单 4-34
使用隐式类型	使用 var 关键字	代码清单 4-35
创建对象而不定义其类型	使用匿名方法	代码清单 4-36 和代码清单 4-37
简化异步方法的使用	使用 async 和 await 关键字	代码清单 4-38～代码清单 4-41
在不定义静态字符串的情况下获取类方法或属性的名称	使用 nameof 表达式	代码清单 4-42 和代码清单 4-43

4.1 准备示例项目

在本章中，为了演示语言特性，使用 ASP.NET MVC Core Web Application（.NET Core）模板创建了一个新的 Visual Studio 项目，并命名为 LanguageFeatures，如图 4-1 所示，取消勾选 Add to Source Control 复选框，单击 OK 按钮。

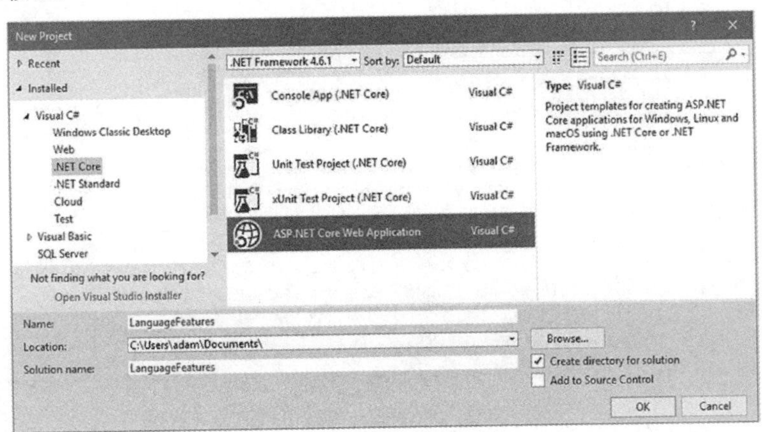

图 4-1 选择项目类型

当需要呈现不同的项目配置时，可以选择 Empty 模板，如图 4-2 所示。在对话框顶部的列表中选择.NET Core 和 ASP.NET Core 2.0，确保将 Authentication 选项设置为 No Authentication，并在单击 OK 按钮创建项目之前，取消选中 Enable Docker Support 复选框。

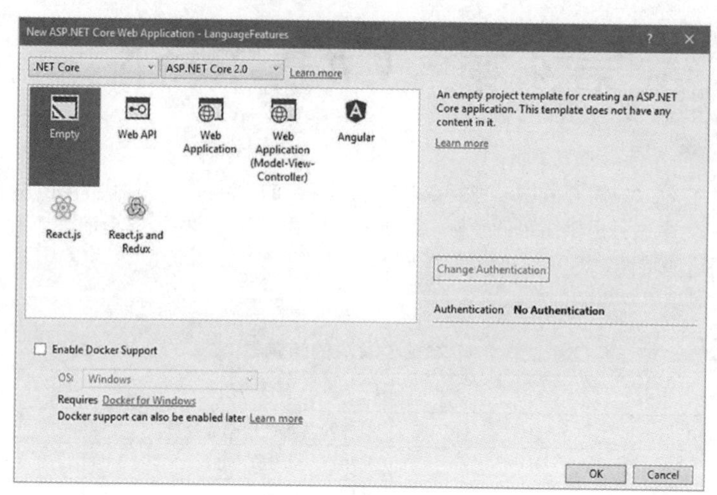

图 4-2 选择项目模板

4.1.1 启用 ASP.NET Core MVC

使用 Empty 项目模板创建的项目中包含最小的 ASP.NET 核心配置，并且没有提供任何 MVC 支持，这就意味着 Web Application（Model-View-Controller）模板添加的占位内容尚不存在，要使控制器和视图起作用，还需要执行一些额外的操作步骤。本节虽然介绍了启用 MVC 设置所需要的步骤，但不会深入剖析每一个步骤的细节。

要启用 MVC 框架，请对 Startup 类进行代码清单 4-1 所示的更改。

代码清单 4-1 在 LanguageFeatures 文件夹下的 Startup.cs 文件中启用 MVC 框架

```
using System;
using System.Collections.Generic;
using System.Linq;
using System.Threading.Tasks;
using Microsoft.AspNetCore.Builder;
using Microsoft.AspNetCore.Hosting;
using Microsoft.AspNetCore.Http;
using Microsoft.Extensions.DependencyInjection;

namespace LanguageFeatures {

    public class Startup {

        public void ConfigureServices(IServiceCollection services) {
            services.AddMvc();
        }

        public void Configure(IApplicationBuilder app, IHostingEnvironment env) {
            if (env.IsDevelopment()) {
                app.UseDeveloperExceptionPage();
            }

            //app.Run(async (context) => {
            //    await context.Response.WriteAsync("Hello World!");
            //});

            app.UseMvcWithDefaultRoute();
```

```
        }
    }
}
```

本书第 14 章将介绍怎样配置 ASP.NET Core MVC 应用程序，但代码清单 4-1 中额外添加的两个语句使得应用程序具有基本的 MVC 设置并使用默认的配置和约定。

4.1.2 创建 MVC 应用程序组件

MVC 设置好了之后，我们便可以添加 MVC 应用程序组件来演示重要的 C#语言特性。

1. 创建模型

我们从创建一个简单的模型类开始，接下来就可以使用一些模型数据了。添加一个名为 Models 的文件夹并在其中创建一个文件，命名为 Product.cs 以定义代码清单 4-2 中的类。

代码清单 4-2 Models 文件夹下的 Product.cs 文件的内容

```
namespace LanguageFeatures.Models {
    public class Product {

        public string Name { get; set; }
        public decimal? Price { get; set; }

        public static Product[] GetProducts() {

            Product kayak = new Product {
                Name = "Kayak", Price = 275M
            };

            Product lifejacket = new Product {
                Name = "Lifejacket", Price = 48.95M
            };

            return new Product[] { kayak, lifejacket, null };
        }
    }
}
```

Products 类定义了 Name 和 Price 属性，还有一个名为 GetProducts 的静态方法用于返回 Products 数组。这个数组中的一个元素已设置为 null，用来展示一些有用的 C#特性。

2. 创建控制器和视图

对于本章中的示例，我们使用一个简单的控制器来演示不同的 C#语言特性。创建 Controllers 文件夹并在其中添加 HomeController.cs 类文件，其中的内容如代码清单 4-3 所示。如果使用默认的 MVC 配置，MVC 会默认把 HTTP 请求发送到 Home 控制器。

代码清单 4-3 Controllers 文件夹下的 HomeController.cs 文件的内容

```
using Microsoft.AspNetCore.Mvc;

namespace LanguageFeatures.Controllers {
    public class HomeController : Controller {

        public ViewResult Index() {
            return View(new string[] { "C#", "Language", "Features" });
        }
    }
}
```

Index 操作方法使得 MVC 呈现默认视图并且在 HTML 中传递一个字符串数组给客户端。为了创建相应的视图，添加 Views/Home 文件夹（首先创建 Views 文件夹，然后在其中创建 Home 子文件夹），再在其中创建 Index.cshtml 文件，这个文件的内容如代码清单 4-4 所示。

代码清单 4-4 Views/Home 文件夹下的 Index.cshtml 文件的内容

```
@model IEnumerable<string>
@{ Layout = null; }

<!DOCTYPE html>
<html>
<head>
    <meta name="viewport" content="width=device-width" />
    <title>Language Features</title>
</head>
<body>
    <ul>
        @foreach (string s in Model) {
            <li>@s</li>
        }
    </ul>
</body>
</html>
```

从 Debug 菜单中选择 Start Debugging 以运行示例应用程序，你会看到图 4-3 所示的输出。

由于本章中所有示例的输出都是文本，因此浏览器中显示的信息可如下呈现：

```
C#
Language
Features
```

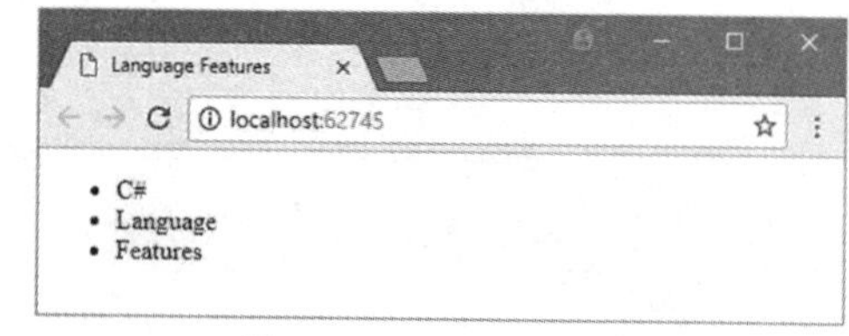

图 4-3 运行示例应用程序

4.2 运用 null 条件运算符

null 条件运算符可以使我们更优雅地判断 null 值。在 MVC 开发中，我们需要判断请求是否包含头部或值，还需要判断模型中是否包含特殊的数据项，因此会有大量的 null 值需要检测。处理 null 值的传统方式是进行显式检测，但是当对象及其属性都需要检查时，这种方式就变得冗长且容易出错，null 条件运算符使这一过程变得更加简洁明了，如代码清单 4-5 所示。

代码清单 4-5 在 Controllers 文件夹下的 HomeController.cs 文件中检测 null 值

```
using Microsoft.AspNetCore.Mvc;
using System.Collections.Generic;
using LanguageFeatures.Models;

namespace LanguageFeatures.Controllers {
    public class HomeController : Controller {

        public ViewResult Index() {

            List<string> results = new List<string>();

            foreach (Product p in Product.GetProducts()) {
                string name = p?.Name;
                decimal? price = p?.Price;
                results.Add(string.Format("Name: {0}, Price: {1}", name, price));
            }

            return View(results);
        }
```

```
    }
}
```

Product 类定义的 GetProducts 静态方法会返回一个对象数组。在控制器的 Index 操作方法中对这些值进行检查以获得 Name 和 Price 值的列表。问题是返回的对象以及对象中属性的值都有可能为 null。也就是说，不能在 for 循环中仅仅通过 p.Name 或 p.Price 来引用值，否则会导致 NullReferenceException。为了避免这一点，可使用 null 条件运算符，如下所示：

```
...
string name = p?.Name;
decimal? price = p?.Price;
...
```

null 条件运算符是一个问号（符号？）。如果 p 为 null，name 也将设置为 null；如果 p 不为 null，name 将设置为 Person.Name 属性的值。同样，Price 属性也是如此。请注意，在使用 null 条件运算符时，指派的变量必须能够为 null，这就是为什么将 price 变量声明为可以为 null 的十进制数值。

4.2.1 null 条件运算符的连接运算

null 条件运算符可以连在一起以遍历整个对象的层次结构，这使它成为简化代码以及进行安全导航的有效工具。在代码清单 4-6 中，在嵌套引用的 Product 类中添加一个属性，创建更加复杂的对象层次结构。

代码清单 4-6 在 Models 文件夹下的 Product.cs 文件中添加一个属性

```
namespace LanguageFeatures.Models {
    public class Product {

        public string Name { get; set; }
        public decimal? Price { get; set; }
        public Product Related { get; set; }

        public static Product[] GetProducts() {

            Product kayak = new Product {
                Name = "Kayak", Price = 275M
            };
            Product lifejacket = new Product {
                Name = "Lifejacket", Price = 48.95M
            };

            kayak.Related = lifejacket;

            return new Product[] { kayak, lifejacket, null };
        }
    }
}
```

每一个 Product 对象都有一个可以引用其他 Product 对象的 Related 属性。在 GetProducts 方法中，设置代表皮艇的 Product 对象的 Related 属性。代码清单 4-7 展示了如何利用 null 条件运算符检验对象属性但不引起错误。

代码清单 4-7 检测在 Controllers 文件夹下的 HomeController.cs 文件中嵌套的 null 值

```
using Microsoft.AspNetCore.Mvc;
using System.Collections.Generic;
using LanguageFeatures.Models;

namespace LanguageFeatures.Controllers {
    public class HomeController : Controller {
```

```
        public ViewResult Index() {

            List<string> results = new List<string>();

            foreach (Product p in Product.GetProducts()) {
                string name = p?.Name;
                decimal? price = p?.Price;
                string relatedName = p?.Related?.Name;
                results.Add(string.Format("Name: {0}, Price: {1}, Related: {2}",
                    name, price, relatedName));
            }

            return View(results);
        }
    }
}
```

null 条件运算符可以被应用于属性链中的任意部分，就像下面这样：

```
...
string relatedName = p?.Related?.Name;
...
```

赋值逻辑是这样的：当 p 为 null 或者 p.Related 为 null 时，relatedName 为 null；否则，relatedName 为 p.Related.Name 属性的值。运行程序，你将在浏览器窗口中看到如下输出：

```
Name: Kayak, Price: 275, Related: Lifejacket
Name: Lifejacket, Price: 48.95, Related:
Name: , Price: , Related:
```

4.2.2 联合使用 null 条件运算符和 null 合并运算符

通过将 null 条件运算符（单个问号）和 null 合并运算符（两个问号）联合起来使用，可以有效地将 null 值反馈给应用程序，如代码清单 4-8 所示。

代码清单 4-8 在 Controllers 文件夹下的 HomeController.cs 文件中结合使用 null 条件运算符和 null 合并运算符

```
using Microsoft.AspNetCore.Mvc;
using System.Collections.Generic;
using LanguageFeatures.Models;

namespace LanguageFeatures.Controllers {
    public class HomeController : Controller {

        public ViewResult Index() {

            List<string> results = new List<string>();

            foreach (Product p in Product.GetProducts()) {
                string name = p?.Name ?? "<No Name>";
                decimal? price = p?.Price ?? 0;
                string relatedName = p?.Related?.Name ?? "<None>";
                results.Add(string.Format("Name: {0}, Price: {1}, Related: {2}",
                    name, price, relatedName));
            }
                return View(results);
        }
    }
}
```

null 条件运算符可以保证在使用对象属性的过程中不会抛出 NullReferenceException 异常，而 null 合并运算符可以保证在浏览器的显示结果中不包含 null 值。运行程序，你将看到浏览器窗口中显示如下结果：

```
Name: Kayak, Price: 275, Related: Lifejacket
Name: Lifejacket, Price: 48.95, Related: <None>
Name: <No Name>, Price: 0, Related: <None>
```

4.3　使用自动实现属性

C#支持自动实现属性，我们之前在定义 Person 类的属性时使用过这项功能，如下所示：

```
namespace LanguageFeatures.Models {
    public class Product {

        public string Name { get; set; }
        public decimal? Price { get; set; }
        public Product Related { get; set; }

        public static Product[] GetProducts() {

            Product kayak = new Product {
                Name = "Kayak", Price = 275M
            };
            Product lifejacket = new Product {
                Name = "Lifejacket", Price = 48.95M
            };

            kayak.Related = lifejacket;

            return new Product[] { kayak, lifejacket, null };
        }
    }
}
```

这项功能使我们可以在定义属性的时候不必实现 get 和 set 主体。使用自动实现属性让我们可以像如下示例这样定义属性：

```
...
public string Name { get; set; }
...
```

以上代码等效于以下代码：

```
...
public string Name {
    get { return name; }
    set { name = value; }
}
...
```

这种语法特征称为语法糖（syntactic sugar），这使得 C#的使用变得更加友好。在本例中，你可以通过为赋值属性消除冗余代码而无须大规模地修改代码的运行逻辑。任何可以使代码变得更加易于编写和维护的优化都是有益的，特别是在大型项目中。

4.3.1　初始化自动实现的属性

自动实现的属性在 C# 3.0 以上的版本中都支持，可以在不需要构造函数的情况下设置初始值，如代码清单 4-9 所示。

代码清单 4-9 在 Models 文件夹下的 Product.cs 文件中初始化自动实现的属性

```
namespace LanguageFeatures.Models {
    public class Product {

        public string Name { get; set; }
        public string Category { get; set; } = "Watersports";
        public decimal? Price { get; set; }
        public Product Related { get; set; }

        public static Product[] GetProducts() {
            Product kayak = new Product {
                Name = "Kayak",
                Category = "Water Craft",
                Price = 275M
            };
            Product lifejacket = new Product {
                Name = "Lifejacket", Price = 48.95M
            };

            kayak.Related = lifejacket;

            return new Product[] { kayak, lifejacket, null };
        }
    }
}
```

为自动实现的属性赋值并不会阻止后续的 setter 方法对属性值进行改变，但是在需要使用构造函数为大量默认属性赋值的情况下，可以让代码更加简洁。代码清单 4-9 已将 Water Craft 赋给 Category 属性，并且初始值是可以更改的。

4.3.2 创建只读的自动实现属性

我们还可以创建只读属性，在初始化自动实现的属性时忽略 set 关键字即可实现这一点，如代码清单 4-10 所示。

代码清单 4-10 在 Models 文件夹下的 Product.cs 文件中创建只读的自动实现属性

```
namespace LanguageFeatures.Models {
    public class Product {

        public string Name { get; set; }
        public string Category { get; set; } = "Watersports";
        public decimal? Price { get; set; }
        public Product Related { get; set; }
        public bool InStock { get; } = true;

        public static Product[] GetProducts() {

            Product kayak = new Product {
                Name = "Kayak",
                Category = "Water Craft",
                Price = 275M
            };
            Product lifejacket = new Product {
                Name = "Lifejacket", Price = 48.95M
            };

            kayak.Related = lifejacket;

            return new Product[] { kayak, lifejacket, null };
```

```
        }
    }
}
```

InStock 属性被初始化为 true 并且不能更改，但是可以在类的构造函数中进行赋值，如代码清单 4-11 所示。

代码清单 4-11 在 Models 文件夹下的 Product.cs 文件中为只读的自动实现属性赋值

```
namespace LanguageFeatures.Models {
    public class Product {

        public Product(bool stock = true) {
            InStock = stock;
        }

        public string Name { get; set; }
        public string Category { get; set; } = "Watersports";
        public decimal? Price { get; set; }
        public Product Related { get; set; }
        public bool InStock { get; }
        public static Product[] GetProducts() {
            Product kayak = new Product {
                Name = "Kayak",
                Category = "Water Craft",
                Price = 275M
            };

            Product lifejacket = new Product(false) {
                Name = "Lifejacket",
                Price = 48.95M
            };

            kayak.Related = lifejacket;

            return new Product[] { kayak, lifejacket, null };
        }
    }
}
```

构造函数允许将只读属性的值指定为参数，如果没有提供值，默认为 true。在使用构造函数设置之后，属性值就无法更改了。

4.4 使用字符串插值

string.Format 方法是构造包含数据值的字符串的传统 C#工具。如下是 Home 控制器中的一个示例：

```
...
results.Add(string.Format("Name: {0}, Price: {1}, Related: {2}",
                    name, price, relatedName));
...
```

C# 6.0 增加了对另外一种方式（称为字符串插值）的支持，这样就避免了字符串模板中的{0}引用必须与参数变量匹配。字符串插值直接使用变量名就可以了，如代码清单 4-12 所示。

代码清单 4-12 在 Controller 文件夹下的 HomeController.cs 文件中使用字符串插值

```
using Microsoft.AspNetCore.Mvc;
using System.Collections.Generic;
using LanguageFeatures.Models;
```

```
namespace LanguageFeatures.Controllers {
    public class HomeController : Controller {

        public ViewResult Index() {

            List<string> results = new List<string>();

            foreach (Product p in Product.GetProducts()) {
                string name = p?.Name ?? "<No Name>";
                decimal? price = p?.Price ?? 0;
                string relatedName = p?.Related?.Name ?? "<None>";
                results.Add($"Name: {name}, Price: {price}, Related: {relatedName}");
            }

            return View(results);
        }
    }
}
```

插值字符串以$作为前缀并且包含槽（hole），这是对包含在{and}字符内的值的引用。计算字符串时，可利用指定的变量或常量来填充这些槽。

Visual Studio 提供了用于创建插值字符串的智能感知支持，当键入{符号时，提供了可用方法的列表；这就最大限度减少了人为的输入错误，并且结果的字符串格式更易理解。

提 示

字符串插值支持 string.Format 方法可用的所有指定格式。格式说明符是槽的一部分，因此 $"Price:{price:C2}将把 price 值格式化为带两位小数的货币值。

4.5 使用对象和集合初始化器

当在 Product 类的静态方法 GetProducts 中创建一个对象时，使用对象初始化器可以实现创建对象和初始化属性值一步完成，如下所示：

```
...
Product kayak = new Product {
    Name = "Kayak",
    Category = "Water Craft",
    Price = 275M
};
...
```

这是可以使 C#语言易于使用的另一个语法糖特征。没有这个功能，我们就不得不调用 Product 构造函数，然后使用新创建的对象来设置每个属性的值，像下面这样：

```
...
Product kayak = new Product();
kayak.Name = "Kayak";
kayak.Category = "Water Craft";
kayak.Price = 275M;
...
```

类似的还有集合初始化器，利用它可以实现创建集合和初始化集合内容一步完成。例如，创建一个字符数组，如果没有初始化器，将需要指定数组的大小以及每一个数组元素的值，如代码清单 4-13 所示。

代码清单 4-13 在 Controllers 文件夹下的 HomeController.cs 文件中初始化一个对象

```
using Microsoft.AspNetCore.Mvc;
using System.Collections.Generic;
```

```
using LanguageFeatures.Models;

namespace LanguageFeatures.Controllers {
    public class HomeController : Controller {

        public ViewResult Index() {
            string[] names = new string[3];
            names[0] = "Bob";
            names[1] = "Joe";
            names[2] = "Alice";
            return View("Index", names);
        }
    }
}
```

使用初始化器可以使设定数组元素的值成为所构造集合的一部分，这种结构隐式地为编译器提供了数组的大小，如代码清单 4-14 所示。

代码清单 4-14 在 Controllers 文件夹下的 HomeController.cs 文件中使用初始化器

```
using Microsoft.AspNetCore.Mvc;
using System.Collections.Generic;
using LanguageFeatures.Models;

namespace LanguageFeatures.Controllers {
    public class HomeController : Controller {

        public ViewResult Index() {
            return View("Index", new string[] { "Bob", "Joe", "Alice" });
        }
    }
}
```

数组元素被包含在“{”和“}”之间，这使得集合的定义更加简明并且可以实现集合的嵌套定义。代码清单 4-14 中的代码与代码清单 4-13 中的代码有相同的效果，如果运行示例应用程序，你将在浏览器窗口中看到如下内容：

```
Bob
Joe
Alice
```

使用索引初始化器

C#梳理了集合初始化器的用法，可使用索引来创建字典这样的集合。代码清单 4-15 展示了如何通过传统的 C#方式重写 Index 操作方法来初始化一个字典。

代码清单 4-15 在 Controllers 文件夹下的 HomeController.cs 文件中初始化一个字典

```
using Microsoft.AspNetCore.Mvc;
using System.Collections.Generic;
using LanguageFeatures.Models;

namespace LanguageFeatures.Controllers {
    public class HomeController : Controller {

        public ViewResult Index() {
            Dictionary<string, Product> products = new Dictionary<string, Product> {
                { "Kayak", new Product { Name = "Kayak", Price = 275M } },
                { "Lifejacket", new Product{ Name = "Lifejacket", Price = 48.95M } }
            };
            return View("Index", products.Keys);
```

```
        }
    }
}
```

初始化此类集合的语法过于依赖“{”和“}”了，尤其是在利用对象初始化器来定义集合值的时候。C#编译器支持使用一种更加自然的方法来初始化索引集合，这种方法与在初始化集合后检索或更改值是一致的，如代码清单4-16所示。

代码清单4-16　在Controllers文件夹下的HomeController.cs文件中使用C#集合初始化器语法

```
using Microsoft.AspNetCore.Mvc;
using System.Collections.Generic;
using LanguageFeatures.Models;

namespace LanguageFeatures.Controllers {
    public class HomeController : Controller {

        public ViewResult Index() {
            Dictionary<string, Product> products = new Dictionary<string, Product> {
                ["Kayak"] = new Product { Name = "Kayak", Price = 275M },
                ["Lifejacket"] = new Product { Name = "Lifejacket", Price = 48.95M }
            };

            return View("Index", products.Keys);
        }
    }
}
```

以上操作的效果等同于创建一个字典，键为Kayak和Lifejacket，值为Product对象，但元素是通过用于其他集合操作的索引表示法创建的。如果运行示例应用程序，在浏览器窗口中将可以看到如下结果：

```
Kayak
Lifejacket
```

4.6　模式匹配

最近对C#最有用的补充之一是对模式匹配的支持，模式匹配可用于测试对象是特定类型的还是具有特定特征。另一种形式是语法糖，它可以极大地简化复杂的条件语句块。is关键字用于执行类型测试，如代码清单4-17所示。

代码清单4-17　对Controllers文件夹下的HomeController.cs文件执行类型测试

```
using Microsoft.AspNetCore.Mvc;
using System.Collections.Generic;
using LanguageFeatures.Models;

namespace LanguageFeatures.Controllers {
    public class HomeController : Controller {

        public ViewResult Index() {

            object[] data = new object[] { 275M, 29.95M,
            "apple", "orange", 100, 10 };
            decimal total = 0;
            for (int i = 0; i < data.Length; i++) {
                if (data[i] is decimal d) {
                    total += d;
                }
            }
```

```
            return View("Index", new string[] { $"Total: {total:C2}" });
        }
    }
}
```

is 关键字执行类型检查，如果值是指定的类型，就将值赋给新变量，如下所示：

```
...
if (data[i] is decimal d) {
...
```

如果 data[*i*]中存储的是十进制值，那么上述表达式的计算结果将为 true。data[*i*]的值将被分配给变量 *d*，这样就可以在随后的语句中使用该值，而无须执行任何类型转换。is 关键字将只匹配指定的类型，这意味着只处理 data 数组中的两个值（数组中的其他元素是 string 和 int 值）。

如果运行应用程序，浏览器窗口会显示以下输出：

```
Total: $304.95
```

switch 语句中的模式匹配

模式匹配也可以在 switch 语句中使用，switch 语句支持 when 关键字，用于在 case 语句匹配值时进行限制，如代码清单 4-18 所示。

代码清单 4-18 Controllers 文件夹下的 HomeController.cs 文件中的模式匹配

```
using Microsoft.AspNetCore.Mvc;
using System.Collections.Generic;
using LanguageFeatures.Models;

namespace LanguageFeatures.Controllers {
    public class HomeController : Controller {

        public ViewResult Index() {

            object[] data = new object[] { 275M, 29.95M,
                "apple", "orange", 100, 10 };
            decimal total = 0;
            for (int i = 0; i < data.Length; i++) {
                switch (data[i]) {
                    case decimal decimalValue:
                        total += decimalValue;
                        break;
                    case int intValue when intValue > 50:
                        total += intValue;
                        break;
                }
            }

            return View("Index", new string[] { $"Total: {total:C2}" });
        }
    }
}
```

要匹配特定类型的任何值，请在 case 语句中使用类型和变量名，如下所示：

```
...
case decimal decimalValue:
...
```

以上 case 语句能匹配任何十进制值，并将其赋给名为 decimalValue 的新变量。

为了更有选择性，可以包含 when 关键字，如下所示：

```
...
case int intValue when intValue > 50:
...
```

以上 case 语句能匹配 int 值，并将它们赋给名为 intValue 的变量，但仅当值大于 50 时才这么做。如果运行应用程序，浏览器窗口会显示以下输出：

```
Total: $404.95
```

4.7 使用扩展方法

扩展方法是在类中添加不属于用户且不能直接修改的方法的简便方式。代码清单 4-19 展示了 ShoppingCart 类的定义，将它添加到 Models 文件夹的 ShoppingCart.cs 文件中，用于表示 Product 对象的集合。

代码清单 4-19 Models 文件夹下的 ShoppingCart.cs 文件的内容

```
using System.Collections.Generic;

namespace LanguageFeatures.Models {

    public class ShoppingCart {
        public IEnumerable<Product> Products { get; set; }
    }
}
```

这个简单的类可作为 Product 对象列表的封装器（对于本例我们只需要一个基础类）。假设我们需要定义 ShoppingCart 类中 Product 对象的总值，但是不能改变类本身，由于它来自第三方，因此我们可能无法获取源代码。这时候我们就可以使用扩展方法来添加我们需要的功能。代码清单 4-20 展示了添加到 Models 文件夹中的 MyExtensionMethods 类。

代码清单 4-20 Models 文件夹下的 MyExtensionMethods.cs 文件的内容

```
namespace LanguageFeatures.Models {

    public static class MyExtensionMethods {

        public static decimal TotalPrices(this ShoppingCart cartParam) {
            decimal total = 0;
            foreach (Product prod in cartParam.Products) {
                total += prod?.Price ?? 0;
            }
            return total;
        }
    }
}
```

第一个参数前面的 this 关键字表明 TotalPrices 是一个扩展方法，用于告诉.NET 在这种情况下可以对 ShoppingCart 类使用扩展方法。我们可以使用 cartParam 参数引用扩展方法作用的 ShoppingCart 实例。这里的方法枚举了 ShoppingCart 中的 Product 并且返回 Product.Price 属性的总值。代码清单 4-21 显示了如何在 Home 控制器的操作方法中应用扩展方法。

注 意

扩展方法不能打破类为自身的方法、字段和属性定义的访问规则。可以利用扩展方法扩展类的功能，但是只能使用允许访问的类成员。

代码清单 4-21 在 Controllers 文件夹下的 HomeController.cs 文件中使用扩展方法

```
using Microsoft.AspNetCore.Mvc;
using System.Collections.Generic;
using LanguageFeatures.Models;

namespace LanguageFeatures.Controllers {
    public class HomeController : Controller {

        public ViewResult Index() {
            ShoppingCart cart
            = new ShoppingCart { Products = Product.GetProducts() };
            decimal cartTotal = cart.TotalPrices();
            return View("Index", new string[] { $"Total: {cartTotal:C2}" });
        }
    }
}
```

关键的声明语句如下：

```
...
decimal cartTotal = cart.TotalPrices();
...
```

在 ShoppingCart 对象中调用 TotalPriccs 方法时就像这个方法是 ShoppingCart 类的成员一样，尽管它是一个由不同类定义的扩展方法。如果扩展类在当前类的作用域内，.NET 将可以查找到它们，因为它们在同一个命名空间中，或者在 using 语句指定的命名空间中。如果运行示例程序，浏览器窗口会显示以下输出：

```
Total: $323.95
```

4.7.1 将扩展方法应用于接口

我们也可以创建应用于接口的扩展方法，这允许我们在实现接口的所有类中调用它们。代码清单 4-22 展示了更新的 ShoppingCart 类，它实现了 IEnumerable<Product>接口。

代码清单 4-22 在 Models 文件夹下的 ShoppingCart.cs 文件中实现一个接口

```
using System.Collections;
using System.Collections.Generic;

namespace LanguageFeatures.Models {

    public class ShoppingCart : IEnumerable<Product> {
        public IEnumerable<Product> Products { get; set; }

        public IEnumerator<Product> GetEnumerator() {
            return Products.GetEnumerator();
        }

        IEnumerator IEnumerable.GetEnumerator() {
            return GetEnumerator();
        }
    }
}
```

现在可以更新扩展方法，以便处理 IEnumerable<Product>，如代码清单 4-23 所示。

代码清单 4-23 在 Models 文件夹下的 MyExtensionMethods.cs 文件中更新一个扩展方法

```
using System.Collections.Generic;

namespace LanguageFeatures.Models {

    public static class MyExtensionMethods {
```

```
        public static decimal TotalPrices(this IEnumerable<Product> products) {
            decimal total = 0;
            foreach (Product prod in products) {
                total += prod?.Price ?? 0;
            }
            return total;
        }
    }
}
```

第一个参数的类型被更改为IEnumerable<Product>，也就是说，方法体中的每一次for循环都直接作用在Product对象上。使用接口意味着可以计算由任何IEnumerable<Product>枚举的Product对象的总值，其中既包括ShoppingCart实例，也包含Product对象数组，如代码清单4-24所示。

代码清单4-24 在Controllers文件夹下的HomeController.cs文件中对数组使用扩展方法

```
using Microsoft.AspNetCore.Mvc;
using System.Collections.Generic;
using LanguageFeatures.Models;

namespace LanguageFeatures.Controllers {
    public class HomeController : Controller {
        public ViewResult Index() {

            ShoppingCart cart
                = new ShoppingCart { Products = Product.GetProducts() };

            Product[] productArray = {
                new Product {Name = "Kayak", Price = 275M},
                new Product {Name = "Lifejacket", Price = 48.95M}
            };

            decimal cartTotal = cart.TotalPrices();
            decimal arrayTotal = productArray.TotalPrices();

            return View("Index", new string[] {
                $"Cart Total: {cartTotal:C2}",
                $"Array Total: {arrayTotal:C2}" });
        }
    }
}
```

如果运行示例程序，你将看到如下结果，结果表明不管Product对象是如何收集的，我们都能从扩展方法中获得相同的结果：

```
Cart Total: $323.95
Array Total: $323.95
```

4.7.2 创建过滤扩展方法

扩展方法可以用于筛选对象的集合。作用在IEnumerable<T>上并且返回IEnumerable<T>的扩展方法可以使用yield关键字将选择条件应用于源数据中的项，以生成结果集。代码清单4-25演示了这样一种扩展方法，已将其添加到MyExtensionMethods类中。

代码清单4-25 在Controller文件夹下的MyExtensionMethods.cs文件中添加过滤扩展方法

```
using System.Collections.Generic;

namespace LanguageFeatures.Models {

    public static class MyExtensionMethods {

        public static decimal TotalPrices(this IEnumerable<Product> products) {
```

```
            decimal total = 0;
            foreach (Product prod in products) {
                total += prod?.Price ?? 0;
            }
            return total;
        }
        public static IEnumerable<Product> FilterByPrice(
                this IEnumerable<Product> productEnum, decimal minimumPrice) {

            foreach (Product prod in productEnum) {
                if ((prod?.Price ?? 0) >= minimumPrice) {
                    yield return prod;
                }
            }
        }
    }
}
```

FilterByPrice 扩展方法采用一个附加参数来帮助筛选那些 Price 属性与参数相匹配或高于参数的 Product 对象，代码清单 4-26 展示了具体用法。

代码清单 4-26 在 Controllers 文件夹下的 HomeController.cs 文件中使用过滤扩展方法

```
using Microsoft.AspNetCore.Mvc;
using System.Collections.Generic;
using LanguageFeatures.Models;

namespace LanguageFeatures.Controllers {
    public class HomeController : Controller {

        public ViewResult Index() {

            Product[] productArray = {
                new Product {Name = "Kayak", Price = 275M},
                new Product {Name = "Lifejacket", Price = 48.95M},
                new Product {Name = "Soccer ball", Price = 19.50M},
                new Product {Name = "Corner flag", Price = 34.95M}
            };

            decimal arrayTotal = productArray.FilterByPrice(20).TotalPrices();

            return View("Index", new string[] { $"Array Total: {arrayTotal:C2}" });
        }
    }
}
```

当我们对 Product 对象数组调用 FilterByPrice 方法时，TotalPrices 方法只返回那些价格高于 20 美元的值，用于计算总值。如果运行示例程序，浏览器窗口会显示如下输出：

```
Total: $358.90
```

4.8 使用 Lambda 表达式

Lambda 表达式会导致很多混淆，Lambda 表达式本身极具简化的特性就会产生一种迷惑。为了理解要解决的问题，我们沿用之前定义的 FilterByPrice 扩展方法。此扩展方法用来根据价格筛选 Product 对象，也就是说，如果还要通过姓名进行筛选，那么需要再定义一个过滤扩展方法，如代码清单 4-27 所示。

代码清单 4-27 在 Models 文件夹下的 MyExtensionMethods.cs 文件中添加一个过滤扩展方法

```
using System.Collections.Generic;

namespace LanguageFeatures.Models {
```

```
    public static class MyExtensionMethods {

        public static decimal TotalPrices(this IEnumerable<Product> products) {
            decimal total = 0;
            foreach (Product prod in products) {
                total += prod?.Price ?? 0;
            }
            return total;
        }

        public static IEnumerable<Product> FilterByPrice(
                this IEnumerable<Product> productEnum, decimal minimumPrice) {

            foreach (Product prod in productEnum) {
                if ((prod?.Price ?? 0) >= minimumPrice) {
                    yield return prod;
                }
            }
        }

        public static IEnumerable<Product> FilterByName(
                this IEnumerable<Product> productEnum, char firstLetter) {

            foreach (Product prod in productEnum) {
                if (prod?.Name?[0] == firstLetter) {
                    yield return prod;
                }
            }
        }
    }
}
```

代码清单4-28展示了如何在控制器中使用两种过滤扩展方法来创建两个不同的总量。

代码清单4-28 在Controllers文件夹下的HomeController.cs文件中使用两种过滤扩展方法

```
using Microsoft.AspNetCore.Mvc;
using System.Collections.Generic;
using LanguageFeatures.Models;

namespace LanguageFeatures.Controllers {
    public class HomeController : Controller {

        public ViewResult Index() {

            Product[] productArray = {
                new Product {Name = "Kayak", Price = 275M},
                new Product {Name = "Lifejacket", Price = 48.95M},
                new Product {Name = "Soccer ball", Price = 19.50M},
                new Product {Name = "Corner flag", Price = 34.95M}
            };

            decimal priceFilterTotal = productArray.FilterByPrice(20).TotalPrices();
            decimal nameFilterTotal = productArray.FilterByName('S').TotalPrices();

            return View("Index", new string[] {
                $"Price Total: {priceFilterTotal:C2}",
                $"Name Total: {nameFilterTotal:C2}" });
        }
    }
}
```

第一个过滤扩展方法选择价格为20美元以上的所有商品，第二个过滤扩展方法选择名称以字母S开头的商品。如果运行示例程序，你将在浏览器窗口中看到以下输出：

```
Price Total: $358.90
Name Total: $19.50
```

4.8.1 定义函数

可以不断重复以上过程，并为感兴趣的每一个属性和属性组合创建不同的过滤扩展方法。更优雅的方式是将处理枚举的代码从选择条件中分离出来。用 C#处理这个比较简单，C#通过允许函数作为对象传递实现了这一操作。代码清单 4-29 展示了一个扩展方法，用于过滤 Product 对象的枚举，但将选择结果中元素的决策操作委托给了一个单独的函数。

代码清单 4-29 在 Models 文件夹下的 MyExtensionMethods.cs 文件中创建常规的 Filter 方法

```
using System.Collections.Generic;
using System;

namespace LanguageFeatures.Models {

    public static class MyExtensionMethods {

        public static decimal TotalPrices(this IEnumerable<Product> products) {
            decimal total = 0;
            foreach (Product prod in products) {
                total += prod?.Price ?? 0;
            }
            return total;
        }

        public static IEnumerable<Product> Filter(
                this IEnumerable<Product> productEnum,
                Func<Product, bool> selector) {

            foreach (Product prod in productEnum) {
                if (selector(prod)) {
                    yield return prod;
                }
            }
        }
    }
}
```

Filter 方法的第二个参数是一个接收 Product 对象并返回 bool 值的函数。Filter 方法为每个 Product 对象调用该函数，如果返回 true，就将 Product 对象添加到结果中。要使用 Filter 方法，可以指定另一个方法或创建独立的函数，如代码清单 4-30 所示。

代码清单 4-30 在 Controllers 文件夹下的 HomeController.cs 文件中使用函数来过滤 Product 对象

```
using Microsoft.AspNetCore.Mvc;
using System.Collections.Generic;
using LanguageFeatures.Models;
using System;

namespace LanguageFeatures.Controllers {
    public class HomeController : Controller {

        bool FilterByPrice(Product p) {
            return (p?.Price ?? 0) >= 20;
        }

        public ViewResult Index() {
```

```
            Product[] productArray = {
                new Product {Name = "Kayak", Price = 275M},
                new Product {Name = "Lifejacket", Price = 48.95M},
                new Product {Name = "Soccer ball", Price = 19.50M},
                new Product {Name = "Corner flag", Price = 34.95M}
            };

            Func<Product, bool> nameFilter = delegate (Product prod) {
                return prod?.Name?[0] == 'S';
            };

            decimal priceFilterTotal = productArray
                .Filter(FilterByPrice)
                .TotalPrices();
            decimal nameFilterTotal = productArray
                .Filter(nameFilter)
                .TotalPrices();

            return View("Index", new string[] {
                $"Price Total: {priceFilterTotal:C2}",
                $"Name Total: {nameFilterTotal:C2}" });
        }
    }
}
```

但这两种方式都不理想。FilterByPrice 这样的方法会扰乱类的定义。创建 Func<Product,bool>对象虽然可以避免这个问题，但其笨拙的语法难以阅读和维护。Lambda 表达式可以解决这个问题，从而以一种更加优雅且更具表现力的方式来定义函数，如代码清单 4-31 所示。

代码清单 4-31 在 Controllers 文件夹下的 HomeController.cs 文件中使用 Lambda 表达式

```
using Microsoft.AspNetCore.Mvc;
using System.Collections.Generic;
using LanguageFeatures.Models;
using System;

namespace LanguageFeatures.Controllers {
    public class HomeController : Controller {

        public ViewResult Index() {

            Product[] productArray = {
                new Product {Name = "Kayak", Price = 275M},
                new Product {Name = "Lifejacket", Price = 48.95M},
                new Product {Name = "Soccer ball", Price = 19.50M},
                new Product {Name = "Corner flag", Price = 34.95M}
            };

            decimal priceFilterTotal = productArray
                .Filter(p => (p?.Price ?? 0) >= 20)
                .TotalPrices();
            decimal nameFilterTotal = productArray
                .Filter(p => p?.Name?[0] == 'S')
                .TotalPrices();

            return View("Index", new string[] {
                $"Price Total: {priceFilterTotal:C2}",
                $"Name Total: {nameFilterTotal:C2}" });
        }
    }
}
```

在代码清单 4-31 中，Lambda 表达式以粗体显示。参数的类型不需要指定，可自动推断。符号“=>”

连接了参数和 Lambda 表达式的结果。在示例程序中，名为 p 的 Product 参数被连接到 bool 型结果。如果 Price 属性的值大于或等于 20 美元（第一个表达式），或者 Name 属性以 S 开头（第二个表达式），那么参数 p 为 true。虽然代码的工作方式与单独的方法和函数代理相同，但它更简洁，对于大多数人来说易读性更强。

Lambda 表达式的其他形式

不需要在 Lambda 表达式中表达委托的逻辑。我们可以很容易地调用方法，就像下面这样：

```
prod => EvaluateProduct(prod)
```

如果在委托函数中需要具有多个参数的 Lambda 表达式，就必须将这些参数包含在括号中，如下所示：

```
(prod, count) => prod.Price > 20 && count > 0
```

最后，如果需要在具有多个语句的 Lambda 表达式中使用逻辑，那么可以换用大括号（{}），并使用 return 语句返回结果，如下所示：

```
(prod, count) => {
    // ...multiple code statements...
    return result;
}
```

不需要在代码中使用 Lambda 表达式，但它们确实是表达复杂函数的一种简单清晰的方式。

4.8.2 使用 Lambda 表达式实现方法和属性

在 C# 6.0 中，对 Lambda 进行了扩展以便它们可以实现方法和属性。在 MVC 开发中，特别是在编写控制器时，通常会使用包含单个语句的方法来选择要显示的数据和要呈现的视图。在代码清单 4-32 中，重写了 Index 操作方法。

代码清单 4-32 在 Controllers 文件夹下的 HomeController.cs 文件中创建公共 action 模式

```
using Microsoft.AspNetCore.Mvc;
using System.Collections.Generic;
using LanguageFeatures.Models;
using System;
using System.Linq;

namespace LanguageFeatures.Controllers {
    public class HomeController : Controller {

        public ViewResult Index() {
            return View(Product.GetProducts().Select(p => p?.Name));
        }
    }
}
```

操作方法 Index 将从静态的 Product.GetProducts 方法中获取 Product 对象的集合，并使用 LINQ 生成 Name 属性的值，然后用作默认视图的视图模型。如果运行示例程序，你将在浏览器窗口中看到以下输出：

```
Kayak
Lifejacket
```

浏览器窗口中也会有一个空的列表项，因为 GetProducts 方法在结果中包含了 null 引用，但这对本章的这一部分并不重要。

当一个方法的主体由单个语句组成时，可以将其重写为 Lambda 表达式，如代码清单 4-33 所示。

代码清单 4-33 在 Controllers 文件夹下的 HomeController.cs 文件中将操作方法重写为 Lambda 表达式

```
using Microsoft.AspNetCore.Mvc;
using System.Collections.Generic;
using LanguageFeatures.Models;
using System;
using System.Linq;

namespace LanguageFeatures.Controllers {
    public class HomeController : Controller {

        public ViewResult Index() =>
        View(Product.GetProducts().Select(p => p?.Name));
    }
}
```

方法的 Lambda 表达式省略了 return 关键字，并使用=>将方法签名（包括参数）与其实现相关联。代码清单 4-33 所示的 Index 操作方法的工作方式与代码清单 4-32 中的相同，但表达式更简洁。这种基本方法也可以用来定义属性。代码清单 4-34 显示了如何在 Product 类中使用 Lambda 表达式来添加属性。

代码清单 4-34 在 Models 文件夹下的 Product.cs 文件中将属性表示为 Lambda 表达式

```
namespace LanguageFeatures.Models {
    public class Product {

        public Product(bool stock = true) {
            InStock = stock;
        }

        public string Name { get; set; }
        public string Category { get; set; } = "Watersports";
        public decimal? Price { get; set; }
        public Product Related { get; set; }
        public bool InStock { get; }
        public bool NameBeginsWithS => Name?[0] == 'S';

        public static Product[] GetProducts() {

            Product kayak = new Product {
                Name = "Kayak",
                Category = "Water Craft",
                Price = 275M
            };

            Product lifejacket = new Product(false) {
                Name = "Lifejacket",
                Price = 48.95M
            };

            kayak.Related = lifejacket;

            return new Product[] { kayak, lifejacket, null };
        }
    }
}
```

4.9 使用类型推断和匿名类型

C#的 var 关键字允许你在不显式指定变量类型的情况下定义局部变量，如代码清单 4-35 所示，这称为类型推断（type inference）或隐式类型。

代码清单 4-35 在 Controller 文件夹下的 HomeController.cs 文件中使用类型推断

```
using Microsoft.AspNetCore.Mvc;
using System.Collections.Generic;
using LanguageFeatures.Models;
using System;
using System.Linq;

namespace LanguageFeatures.Controllers {
    public class HomeController : Controller {
        public ViewResult Index() {
            var names = new [] { "Kayak", "Lifejacket", "Soccer ball" };
            return View(names);
        }
    }
}
```

names 变量并不是没有类型，相反，要求编译器从代码中推断出类型。编译器检查数组声明，并推断出 names 是一个字符串数组。运行示例程序，将生成以下输出：

```
Kayak
Lifejacket
Soccer ball
```

使用匿名类型

通过组合对象的初始化器和类型推断，可以创建简单的视图模型对象，它们对于在控制器和视图之间传输数据很有用，而不必定义类或结构，如代码清单 4-36 所示。

代码清单 4-36 在 Controller 文件夹下的 HomeController.cs 文件中创建匿名类型

```
using Microsoft.AspNetCore.Mvc;
using System.Collections.Generic;
using LanguageFeatures.Models;
using System;
using System.Linq;

namespace LanguageFeatures.Controllers {
    public class HomeController : Controller {

        public ViewResult Index() {
            var products = new [] {
                new { Name = "Kayak", Price = 275M },
                new { Name = "Lifejacket", Price = 48.95M },
                new { Name = "Soccer ball", Price = 19.50M },
                new { Name = "Corner flag", Price = 34.95M }
            };

            return View(products.Select(p => p.Name));
        }
    }
}
```

products 数组中的每个对象都是匿名类型的对象，但这并不是说在这种情况下 JavaScript 变量是动态的，而只是意味着编译器将自动创建类型定义。强制类型转换仍在执行。例如，可以获取并设置在初始化器中定义的属性。如果运行示例程序，你将在浏览器窗口中看到以下输出：

```
Kayak
Lifejacket
```

```
Soccer ball
Corner flag
```

C#编译器根据初始化器中参数的名称和类型来生成类。两个具有相同属性名和类型的匿名类型对象将被分配给同一个自动生成器类。这意味着 Product 数组中的所有对象都将具有相同的类型，因为它们拥有相同的属性。

提 示

必须使用 var 关键字定义匿名类型对象的数组，由于在代码被编译之前不会指定类型，因此不知道要使用哪种类型。匿名类型对象数组中的元素必须拥有相同的属性，否则，编译器无法确定数组类型应该是什么。

为了证明这一点，更改代码清单 4-37 的输出，以便显示类型名称而不是 Name 属性的值。

代码清单 4-37 在 Controllers 文件夹下的 HomeController.cs 文件中显示 Name 属性的类型

```
using Microsoft.AspNetCore.Mvc;
using System.Collections.Generic;
using LanguageFeatures.Models;
using System;
using System.Linq;

namespace LanguageFeatures.Controllers {
    public class HomeController : Controller {

        public ViewResult Index() {
            var products = new [] {
                new { Name = "Kayak", Price = 275M },
                new { Name = "Lifejacket", Price = 48.95M },
                new { Name = "Soccer ball", Price = 19.50M },
                new { Name = "Corner flag", Price = 34.95M }
            };

            return View(products.Select(p => p.GetType().Name));
        }
    }
}
```

products 数组中的所有对象都被分配相同的类型，类型名称并不友好并且不能直接使用，你可能会看到与以下输出不同的名字：

```
<>f__AnonymousType0`2
<>f__AnonymousType0`2
<>f__AnonymousType0`2
<>f__AnonymousType0`2
```

4.10 使用异步方法

C#中最近添加的一项重要内容是对异步方法（asynchronous method）的处理方式的改进。异步方法会在后台进行，并在完成后通知你，这样在执行后台工作时代码就能够处理其他事务。异步方法是从代码中消除瓶颈，并允许应用程序利用多个处理器和处理器内核并行执行工作的重要工具。

在 MVC 中，异步方法可以提高应用程序的总体性能，这是因为服务器在调度和执行请求的方式上有了更大的灵活性。C#关键字 async 和 await 用于异步工作的执行。

为此，需要在示例项目中添加一个新的.NET 程序集，以便可以进行异步 HTTP 请求。在 Solution Explorer 窗格中右击 LanguageFeatures 文件夹，在弹出的菜单中选择 Edit LanguageFeatures.csproj，添加一些内容，如代码清单 4-38 所示。

代码清单 4-38 在 LanguageFeatures 文件夹下的 LanguageFeatures.csproj 文件中添加程序集引用

```
<Project Sdk="Microsoft.NET.Sdk.Web">

  <PropertyGroup>
    <TargetFramework>netcoreapp2.0</TargetFramework>
  </PropertyGroup>

  <ItemGroup>
    <Folder Include="wwwroot\" />
  </ItemGroup>

  <ItemGroup>
    <PackageReference Include="Microsoft.AspNetCore.All" Version="2.0.0" />
    <PackageReference Include="System.Net.Http" Version="4.3.2" />
  </ItemGroup>

</Project>
```

在保存 LanguageFeatures.csproj 文件后，Visual Studio 将下载 System.Net.Http 程序集并将其添加到项目中。第 6 章将更详细地描述这一过程。

4.10.1 直接使用任务

C#和.NET 对异步方法有很好的支持，但代码往往十分冗长，并且不习惯并行编程的开发人员通常会因为不同的语法而感到困惑。例如，代码清单 4-39 显示了 MyAsyncMethods 类中定义的 GetPageLength 异步方法，可将其添加到 MyAsyncMethods.cs 类文件。

代码清单 4-39 Models 文件夹下的 MyAsyncMethods.cs 文件的内容

```
using System.Net.Http;
using System.Threading.Tasks;

namespace LanguageFeatures.Models {

    public class MyAsyncMethods {

        public static Task<long?> GetPageLength() {

            HttpClient client = new HttpClient();

            var httpTask = client.GetAsync("http://apress.com");

            return httpTask.ContinueWith((Task<HttpResponseMessage> antecedent) => {
                return antecedent.Result.Content.Headers.ContentLength;
            });
        }
    }
}
```

上述代码使用一个 System.Net.Http.HttpClient 对象请求 Apress 主页的内容并返回内容的长度。.NET 代表将被异步完成的任务。Task 对象根据后台工作产生的结果强制进行类型化。所以，当调用 HttpClient.GetAsync 方法时，得到的是一个 Task<HttpResponseMessage>对象。这说明请求将在后台执行，并且返回的结果是一个 HttpResponseMessage 对象。

提 示

当使用“后台运行”这样的字眼时，表示跳过很多细节，只聚焦 MVC 相关的关键点。.NET 对异步方法和并行编程的支持是非常好的，如果要创建能够利用多核和多处理器硬件的真正高性能的应用程序，建议你进一步了解它。随着更多功能被介绍，你将看到 MVC 如何使创建异步 Web 应用程序变得容易。

在本例中，使用 ContinueWith 方法处理从 HttpClient 返回的对象。在 GetAsync 方法中，使用一个 Lambda 表达式在 HttpResponseMessage 中返回一个属性的值，其中包含从 Apress Web 服务器获取的内容的长度。

```
...
return httpTask.ContinueWith((Task<HttpResponseMessage> antecedent) => {
    return antecedent.Result.Content.Headers.ContentLength;
});
...
```

请注意，以上代码使用了两次 return 关键字。第一次使用 return 关键字是为了指定返回一个 Task <HttpResponseMessage>对象，当任务完成时，将返回 ContentLength 头部的长度。ContentLength 头部会返回一个可为 null 的 long 值，这意味着 GetPageLength 方法的返回结果是 Task <long?>，如下所示：

```
...
public static Task<long?> GetPageLength() {
...
```

不要介意这种方式是否有道理，也不要担心，因为很多人也会遇到同样的问题，为此微软在 C#中添加了两个关键字来简化异步方法的使用。

4.10.2 使用 async 和 await 关键字

C#引入了两个关键字，专门用于简化异步方法（如 HttpClient.GetAsync）的使用。这两个关键字是 async 和 await，代码清单 4-40 展示了怎样应用它们来简化异步方法的使用。

代码清单 4-40 在 Models 文件夹下的 MyAsyncMethods.cs 文件中使用 async 和 await 关键字

```
using System.Net.Http;
using System.Threading.Tasks;

namespace LanguageFeatures.Models {

    public class MyAsyncMethods {

        public async static Task<long?> GetPageLength() {

            HttpClient client = new HttpClient();

            var httpMessage = await client.GetAsync("http://apress.com");

            return httpMessage.Content.Headers.ContentLength;
        }
    }
}
```

在调用异步方法时使用了 await 关键字，从而告诉 C#编译器要等待 GetAsync 方法返回 Task 结果，然后继续执行同一方法中的其他语句。

应用 await 关键字意味着可以处理从 GetAsync 方法返回的 Task 结果，尽管 GetAsync 只是常规方法并且只返回给 HttpResponseMessage 对象一个值。最重要的是，可以通过正常的方式使用其他方法的 return 关键字返回结果，在本例中也就是 ContentLength 属性的值。这是一种顺理成章的技术，意味着我们不必担心 ContinueWith 方法和 return 关键字的多次使用。

使用 await 关键字时，还必须将 async 关键字添加到方法签名中，就像在示例中所做的那样。方法的返回结果类型不变。示例中的 GetPageLength 方法仍然返回 Task <long? >，这是因为 await 和 async 关键字是使用一些巧妙的编译技巧实现的：允许使用一种更自然的语法，但并不会改变应用它们的方法要执行的操作。在调用 GetPageLength 方法时仍然要处理 Task <long? >结果，因为仍有后台操作会生成可为 null 的 long 值，当然，程序员也可以选择使用 await 和 async 关键字。

以上模式将贯穿于整个 MVC 控制器，这使得编写异步操作方法变得简单，如代码清单 4-41 所示。

代码清单 4-41 在 Controllers 文件夹下的 HomeController.cs 文件中定义异步操作方法

```
using Microsoft.AspNetCore.Mvc;
using System.Collections.Generic;
using LanguageFeatures.Models;
using System;
using System.Linq;
using System.Threading.Tasks;

namespace LanguageFeatures.Controllers {
    public class HomeController : Controller {

        public async Task<ViewResult> Index() {
            long? length = await MyAsyncMethods.GetPageLength();
            return View(new string[] { $"Length: {length}" });
        }
    }
}
```

这里已将 Index 操作方法的结果修改为 Task<ViewResult>，从而告诉 MVC 该操作方法将返回一个 Task 对象并在完成时生成一个 ViewResult 对象，提供渲染视图所需的详细信息和数据。这里为方法定义添加了 async 关键字，这样就可以在调用 MyAsyncMethods.GetPathLength 方法时使用 await 关键字了。MVC 和.NET 负责处理后续事项，从而得到易于编写、易于阅读、易于维护的异步代码。如果运行该应用程序，你将看到类似于下面的输出内容（可能长度是不同的，因为 Apress 网站的内容经常更改）：

```
Length: 54576
```

4.11 获取名称

在 Web 应用开发中有许多任务需要引用参数、变量、方法或类的名称，比如处理用户输入时引发的异常或者创建验证错误。传统的方法是使用硬编码的方式获得名称，如代码清单 4-42 所示。

代码清单 4-42 在 Controller 文件夹下的 HomeController.cs 文件中对名称进行硬编码

```
using Microsoft.AspNetCore.Mvc;
using System.Collections.Generic;
using LanguageFeatures.Models;
using System;
using System.Linq;

namespace LanguageFeatures.Controllers {
    public class HomeController : Controller {

        public ViewResult Index() {

            var products = new [] {
                new { Name = "Kayak", Price = 275M },
                new { Name = "Lifejacket", Price = 48.95M },
                new { Name = "Soccer ball", Price = 19.50M },
                new { Name = "Corner flag", Price = 34.95M }
            };

            return View(products.Select(p => $"Name: {p.Name}, Price: {p.Price}"));
        }
    }
}
```

调用 LINQ Select 方法将生成一个字符串序列，其中的每一个字符串都包含对 Name 和 Price 属性的硬编码引用。运行应用程序会在浏览器窗口中生成以下输出：

```
Name: Kayak, Price: 275
Name: Lifejacket, Price: 48.95
Name: Soccer ball, Price: 19.50
Name: Corner flag, Price: 34.95
```

这种方法存在的问题是容易出现错误，要么因为名称键入有误，要么因为代码被重构，而字符串中的名称却没有正确更新。结果容易产生误导，特别容易出现问题。C#引入了 nameof 表达式，由编译器负责生成名称字符串，如代码清单 4-43 所示。

代码清单 4-43 在 Controllers 文件夹下的 HomeController.cs 文件中使用 nameof 表达式

```
using Microsoft.AspNetCore.Mvc;
using System.Collections.Generic;
using LanguageFeatures.Models;
using System;
using System.Linq;

namespace LanguageFeatures.Controllers {
    public class HomeController : Controller {
        public ViewResult Index() {

            var products = new [] {
                new { Name = "Kayak", Price = 275M },
                new { Name = "Lifejacket", Price = 48.95M },
                new { Name = "Soccer ball", Price = 19.50M },
                new { Name = "Corner flag", Price = 34.95M }
            };

            return View(products.Select(p =>
                $"{nameof(p.Name)}: {p.Name}, {nameof(p.Price)}: {p.Price}"));
        }
    }
}
```

编译器通过形如 p.Name 的方式处理引用，以便仅将最后一部分包含在字符串中，从而生成与前面示例中相同的输出。Visual Studio 提供了 nameof 表达式的智能感知支持，可提示你选择引用，并且在重构代码时正确地更新表达式。由于编译器负责处理 nameof 表达式，因此使用无效引用会导致编译错误，从而可以收到错误引用或过期引用的通知。

4.12 小结

本章概述了一名合格的 MVC 程序员需要了解的关键 C#特征。C#是一种非常灵活的语言，提供各种不同的方法来处理各种问题，但这些问题又是 Web 应用开发过程中经常会遇到的，这在本书的很多示例中都有体现。下一章将介绍 Razor 视图引擎，并解释如何将之应用于 MVC Web 应用程序中以生成动态内容。

第 5 章　使用 Razor

在 ASP.NET Core MVC 应用程序中，视图引擎（view engine）负责处理发送给客户端的内容。MVC 框架中默认的视图引擎称为 Razor，用来为 HTML 文件添加注释说明并将这些动态内容插入发送给浏览器的输出中。

本章将介绍一个用于快速解读 Razor 语法的工具，这样当你看到它的时候，就可以马上将其识别出来。本章不会非常详细地介绍 Razor 的内容，把本章看作 Razor 语法的速成指南即可。随着本书其他章节介绍 MVC 的其他功能，我们再慢慢深入介绍。关于在具体语境中如何理解 Razor，参见表 5-1。

表 5-1　在具体语境中理解 Razor

问题	答案
它是什么？	Razor 是负责将数据合并到 HTML 文档中的视图引擎
它有什么作用？	动态生成内容的能力对编写 Web 应用程序是必不可少的，Razor 提供了可以使 C#语句轻松地与 ASP.NET Core MVC 的其他部分协同工作的能力
怎样使用？	将 Razor 表达式添加到视图文件的静态 HTML 中，Razor 表达式可协助生成客户端请求的响应
有哪些容易出现的问题或限制？	Razor 表达式可以包含几乎所有的 C#语句，并且很难决定逻辑应该属于视图还是控制器，从而削弱 MVC 关注点分离的核心理念
有其他的做法吗？	可以编写自己的视图引擎。也有一些第三方的视图引擎可以使用，但它们往往在特定情况下可行，不提供长期支持

表 5-2 列出了本章要完成的操作。

表 5-2　本章要完成的操作

操作	解决方法	代码清单
访问视图模型	使用@model 表达式定义模型类型，使用@Model 表达式访问模型对象	代码清单 5-5、代码清单 5-14、代码清单 5-17
使用类型名称而不限定它们的命名空间	创建视图导入文件	代码清单 5-6 和代码清单 5-7
定义多视图共享的内容	使用布局	代码清单 5-8～代码清单 5-10
定义默认布局	使用视图启动文件	代码清单 5-11～代码清单 5-13
将数据从控制器传递到视图模型之外的其他视图	使用视图包	代码清单 5-15 和代码清单 5-16
有选择地生成内容	使用 Razor 条件表达式	代码清单 5-18 和代码清单 5-19
为数组和集合中的每个元素生成内容	使用 Razor 的 foreach 表达式	代码清单 5-20 和代码清单 5-21

5.1　准备示例项目

为了演示 Razor 是怎样工作的，首先创建一个 ASP.NET Core Web Application（.NET Core）项目并取名为 Razor，就像之前的章节描述的那样。接下来，在 Startup.cs 文件中启用了 MVC 的默认配置，如代码清单 5-1 所示。

代码清单 5-1　在 Razor 文件夹下的 Startup.cs 文件中启用 MVC 的默认配置

```
using System;
using System.Collections.Generic;
using System.Linq;
```

```
using System.Threading.Tasks;
using Microsoft.AspNetCore.Builder;
using Microsoft.AspNetCore.Hosting;
using Microsoft.AspNetCore.Http;
using Microsoft.Extensions.DependencyInjection;

namespace Razor {
    public class Startup {

        public void ConfigureServices(IServiceCollection services) {
            services.AddMvc();
        }

        public void Configure(IApplicationBuilder app, IHostingEnvironment env) {
            if (env.IsDevelopment()) {
                app.UseDeveloperExceptionPage();
            }
            //app.Run(async (context) => {
            //    await context.Response.WriteAsync("Hello World!");
            //});
            app.UseMvcWithDefaultRoute();
        }
    }
}
```

5.1.1 定义模型

接下来，创建模型文件夹 Models，并在其中添加名为 Product.cs 的类文件，用于定义代码清单 5-2 中的简单模型类。

代码清单 5-2 Models 文件夹下的 Product.cs 文件的内容

```
namespace Razor.Models {

    public class Product {

        public int ProductID { get; set; }
        public string Name { get; set; }
        public string Description { get; set; }
        public decimal Price { get; set; }
        public string Category { set; get; }
    }
}
```

5.1.2 创建控制器

Startup.cs 文件中的配置如下：默认情况下，MVC 将请求发送到名为 Home 的控制器。创建 Controllers 文件夹，并在其中添加名为 HomeController.cs 的类文件，用于定义代码清单 5-3 中的简单控制器。

代码清单 5-3 Controllers 文件夹下的 HomeController.cs 文件的内容

```
using Microsoft.AspNetCore.Mvc;
using Razor.Models;

namespace Razor.Controllers {
    public class HomeController : Controller {

        public ViewResult Index() {
            Product myProduct = new Product {
```

```
                ProductID = 1,
                Name = "Kayak",
                Description = "A boat for one person",
                Category = "Watersports",
                Price = 275M
            };

            return View(myProduct);
        }
    }
}
```

以上控制器定义了一个名为 Index 的操作方法来创建和填充 Product 对象的属性。将 Product 对象传递给 View 方法，以便渲染视图时将其用作模型。在调用 View 方法时，并不指定视图文件的名称，这样操作方法便会使用默认视图。

5.1.3　创建视图

为了给 Index 操作方法创建默认视图，创建 Views/Home 文件夹，在其中添加 MVC 视图页面文件 Index.cshtml，其中的内容如代码清单 5-4 所示。

代码清单 5-4　Views/Home 文件夹下的 Index.cshtml 文件的内容

```
@model Razor.Models.Product

@{
    Layout = null;
}

<!DOCTYPE html>
<html>
<head>
    <meta name="viewport" content="width=device-width" />
    <title>Index</title>
</head>
<body>
    Content will go here
</body>
</html>
```

后面将介绍 Razor 视图的其他部分，并演示其中一些内容的不同作用。在学习 Razor 的时候，请记住一点，视图的存在是为了向用户表达模型的一个或多个方面。也就是说，可利用从一个或多个对象中检索到的数据来生成 HTML 页面文件。如果你仍记得我们始终试图建立可以发送到客户端的 HTML 页面，那么就会觉得 Razor 所做的一切都是有意义的。如果运行示例应用程序，你将看到图 5-1 所示的输出。

图 5-1　示例应用程序的输出

5.2　使用模型对象

让我们从 Index.cshtml 视图文件的第一行开始。

```
...
@model Razor.Models.Product
...
```

Razor 表达式以@字符开头。在本例中，@model 表达式声明了将从操作方法传递到视图的模型对象的类型。这样就可以通过@model 来访问视图模型对象的方法、字段和属性，代码清单 5-5 显示了对 Index 视图所做的简单补充。

代码清单 5-5 在 Views/Home 文件夹下的 Index.cshtml 文件中引用视图模型对象的属性

```
@model Razor.Models.Product

@{
    Layout = null;
}

<!DOCTYPE html>
<html>
<head>
    <meta name="viewport" content="width=device-width" />
    <title>Index</title>
</head>
<body>
    @Model.Name
</body>
</html>
```

注 意

这里使用@model 声明视图模型对象的类型，而使用@Model 访问 Name 属性。

如果运行应用程序，你将看到图 5-2 所示的输出。

使用@model 表达式指定类型的视图称为强制类型视图（strongly typed view）。当键入@model 和句点时，Visual Studio 便会弹出@model 表达式成员可以使用的名称建议，如图 5-3 所示。

图 5-2 在视图中读取属性的输出

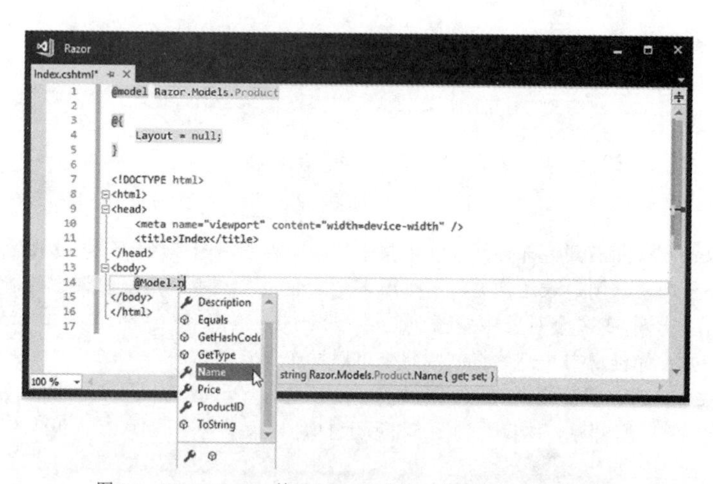

图 5-3 Visual Studio 基于@model 表达式为成员名称提供建议

Visual Studio 对成员名称的可视化建议有助于避免在 Razor 中出现错误。根据个人意愿，可以忽略这些建议，Visual Studio 将高亮显示成员名称有问题，以便进行更正，就像使用常规的 C#类文件一样。图 5-4 列举了一个示例，开发人员在其中试图引用@Model.NotARealProperty。Visual Studio 已意识到开发人员在模型类型中指定的 Product 类没有这样的属性，因而在编辑器中高亮显示错误。

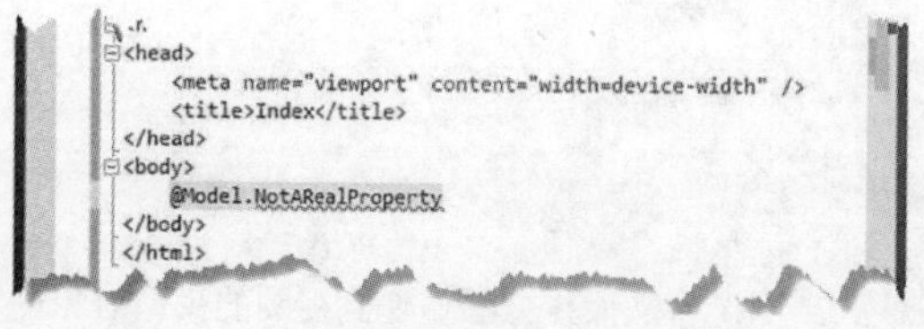

图 5-4 Visual Studio 为@Model 表达式报告错误

使用视图导入

当在 Index.cshtml 文件的开头定义模型对象时，必须导入包含模型类的命名空间，如下所示：

```
...
@model Razor.Models.Product
...
```

默认情况下，在强制类型的 Razor 视图中引用的所有类型都必须使用命名空间进行限定。当模型对象有唯一的引用类型时，这不是什么大问题；但是当需要编写更复杂的 Razor 表达式时，就会使视图的可读性变差。

可以通过在项目中添加视图导入文件来指定要搜索类型的命名空间集合。视图导入文件放在 Views 文件夹中，并命名为_ViewImports.cshtml。

注 意

Views 文件夹中名称以下画线（_）开头的文件不会返回给用户，从而将想要呈现的视图文件和支持它们的文件区分开。视图导入文件和布局模板（稍后介绍）是以下画线为前缀的。

要创建视图导入文件，请在解决方案资源管理器（Solution Explorer）中右击 Views 文件夹，从弹出菜单中选择 Add→New Item，然后从 ASP.NET 类别中选择 MVC View Imports Page 模板，如图 5-5 所示。

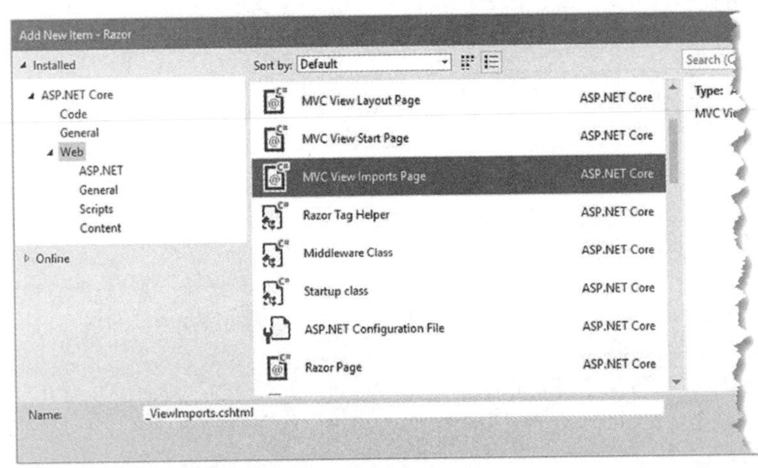

图 5-5 创建视图导入文件

Visual Studio 会自动将文件的名称设置为_ViewImports. cshtml，单击 Add 按钮以创建该文件。代码清单 5-6 展示了添加完表达式之后的视图。

代码清单 5-6 Views 文件夹下的_ViewImports.cshtml 文件的内容

```
@using Razor.Models
```

Razor 视图中用来搜索类的命名空间由@ using 表达式指定，后面跟着相应的命名空间。在代码清单 5-6 中，已为 Razor.Models 命名空间添加了一项，其中包含了示例应用程序的模型类。

现在，Razor.Models 命名空间就包含在视图导入文件中了，可以从 Index.cshtml 文件中移除命名空间，如代码清单 5-7 所示。

代码清单 5-7 在 Views/Home 文件夹下的 Index.cshtml 文件中不使用命名空间引用模型类

```
@model Product

@{
```

```
    Layout = null;
}

<!DOCTYPE html>
<html>
<head>
    <meta name="viewport" content="width=device-width" />
    <title>Index</title>
</head>
<body>
    @Model.Name
</body>
</html>
```

提 示

还可以将@using 表达式添加到单个视图文件中，这将允许在单个视图中使用没有命名空间的类型。

5.3 使用布局

Index.cshtml 视图文件中还有另外一个重要的 Razor 表达式：

```
...
@{
    Layout = null;
}
...
```

这是一个 Razor 代码块，允许在视图中包含 C#语句。Razor 代码块以@“{”开始，以“}”结束，里面包含的语句将在呈现视图时执行。

以上 Razor 代码块将 Layout 属性的值设置为 null。Razor 视图被编译为 MVC 应用程序中的 C#类，并且用基类定义了 Layout 属性。第 21 章将介绍具体的工作方式。将 Layout 属性设置为 null 的效果就是告诉 MVC 视图是自包含的，并将渲染客户端所需的所有内容。

自包含的视图对于简单的示例应用程序表现比较好，但是一个真正的项目可以有几十个视图，有些视图将共享内容。在视图中复制共享内容又很难管理，尤其是当需要进行更改并且必须跟踪所有需要更改的视图时。

当一个模板包含公共的内容并且可以应用于一个或多个视图时，比较好的方法就是使用 Razor 布局。对布局进行更改时，更改将自动影响使用布局的所有视图。

5.3.1 创建布局

布局通常由多个控制器使用的视图共享，并存储在 Views/Shared 文件夹中，这是在查找文件时 Razor 会自动查看的位置之一。为了创建布局，先创建 Views/Shared 文件夹，右击后从弹出菜单中选择 Add→New Item。从 ASP.NET 类别中选择 MVC View Layout Page 模板，并将文件名设置为_BasicLayout.cshtml，如图 5-6 所示。单击 Add 按钮创建这个文件（与视图导入文件一样，布局文件的名称也以下画线开头）。

代码清单 5-8 显示了由 Visual Studio 创建的_BasicLayout.cshtml 文件的初始内容。

代码清单 5-8 Views/Shared 文件夹下的_BasicLayout.cshtml 文件中的初始内容

```
<!DOCTYPE html>
<html>
<head>
```

```
    <meta name="viewport" content="width=device-width" />
    <title>@ViewBag.Title</title>
</head>
<body>
    <div>
        @RenderBody()
    </div>
</body>
</html>
```

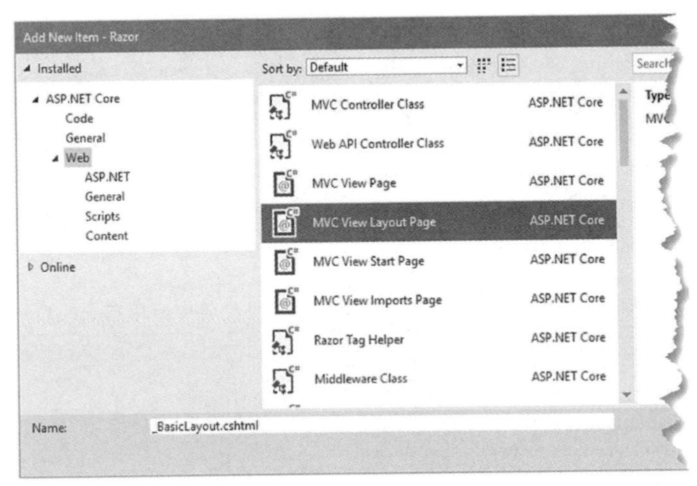

图 5-6 创建布局

布局是一种特殊的视图形式，代码中已高亮显示@表达式。@RenderBody 方法的调用会将操作方法定义的视图内容插入布局标记中。

```
...
<div>
    @RenderBody()
</div>
...
```

布局中的另一个 Razor 表达式会查找名为 ViewBag.Title 的属性，以便设置 title 元素的内容。

```
...
<title>@ViewBag.Title</title>
...
```

ViewBag 允许在应用程序中传递数据值，本例中是指在视图和布局之间传递。在将布局应用于视图时，你将看到这是如何工作的。

布局中的 HTML 元素将应用于任何使用它们的视图，并且提供了一个用于定义公共内容的模板。在代码清单 5-9 中，在布局中添加了一些简单的标记，这样模板效果就很明显了。

代码清单 5-9 在 Views/Shared 文件夹下的_BasicLayout.cshtml 文件中添加内容

```
<!DOCTYPE html>
<html>
<head>
    <meta name="viewport" content="width=device-width" />
    <title>@ViewBag.Title</title>
    <style>
        #mainDiv {
            padding: 20px;
```

```
            border: solid medium black;
            font-size: 20pt
        }
    </style>
</head>
<body>
    <h1>Product Information</h1>
    <div id="mainDiv">
        @RenderBody()
    </div>
</body>
</html>
```

这里添加了标头元素以及一些 CSS 来为包含@RenderBody 表达式的 div 元素的内容设置样式，这样就可以清楚地知道哪些内容来自布局，哪些内容来自视图。

5.3.2 使用布局

要将布局应用于视图，我们需要设置 Layout 属性的值，并移除现在由布局提供的 HTML，如代码清单 5-10 所示的 html、head 和 body 元素。

代码清单 5-10 在 Views/Home 文件夹下的 Index.cshtml 文件中应用布局

```
@model Product

@{
    Layout = "_BasicLayout";
    ViewBag.Title = "Product Name";
}

Product Name: @Model.Name
```

Layout 属性将指定用于视图的布局文件，但不使用.cshtml 文件扩展名。Razor 将在/Views/Home 和 Views/Shared 文件夹中查找指定的布局文件。

上面还设置了 ViewBag.Title 属性。在渲染视图的时候，布局会使用该属性设置 title 元素的内容。

即使是非常简单的应用，视图的这种转换也是非常激动人心的。布局包含了任何 HTML 响应所需的所有结构，这使得视图只需要关注向用户呈现数据的动态内容就可以了。当 MVC 处理 Index.cshtml 文件时，将应用布局来创建统一的 HTML 响应，如图 5-7 所示。

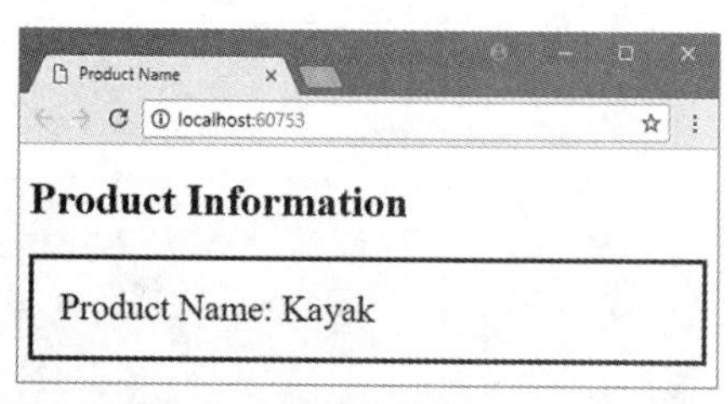

图 5-7 为视图使用布局后的效果

5.3.3 应用视图启动文件

另一个需要交代的地方，就是必须为将要使用的每个视图指定布局文件。因此，如果需要重命名布局文件，就必须找到引用布局文件的每个视图并进行更改，这非常容易出错。最重要的是，这与 MVC 应用程序的易维护宗旨背道而驰。

这个问题可以通过使用视图启动文件（view start file）来解决。在渲染视图的时候，MVC 将查找一个名为_ViewStart.cshtml 的文件。这个文件的内容将被视为包含在视图文件本身中，我们可以使用这个功能自动设置 Layout 属性的值。

要创建视图启动文件，请右击 Views 文件夹，从弹出菜单中选择 Add→New Item，然后从 ASP.NET 类别中选择 MVC View Start Page 模板，如图 5-8 所示。

Visual Studio 会自动将文件的名称设置为_ViewStart.cshtml，单击 Add 按钮，Visual Studio 为这个文件创建的初始内容如代码清单 5-11 所示。

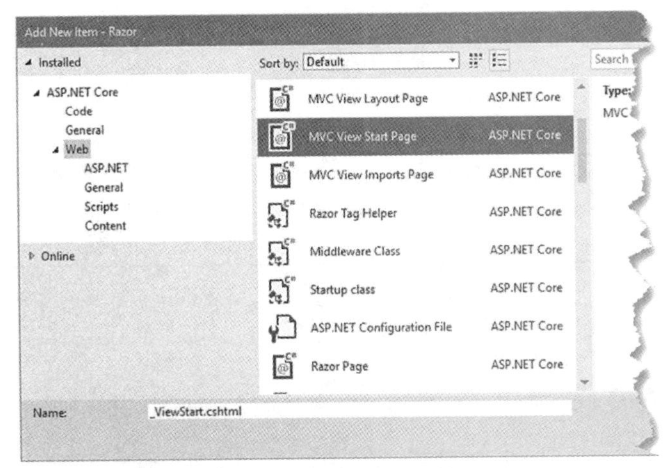

图 5-8 创建视图启动文件

代码清单 5-11 Views 文件夹下的_ViewStart.cshtml 文件的初始内容

```
@{
    Layout = "_Layout";
}
```

为了将布局应用于应用程序中的所有视图，更改分配给 Layout 属性的值，如代码清单 5-12 所示。

代码清单 5-12 在 Views 文件夹下的_ViewStart.cshtml 文件中应用默认视图

```
@{
    Layout = "_BasicLayout";
}
```

因为视图启动文件包含 Layout 属性的值，所以可以从 Index.cshtml 文件中删除相应的表达式，如代码清单 5-13 所示。

代码清单 5-13 更新 Views/Home 文件夹下的 Index.cshtml 文件以体现视图启动文件的使用

```
@model Product

@{
    ViewBag.Title = "Product Name";
}

Product Name: @Model.Name
```

不必指定要使用的视图启动文件。MVC 将自动定位视图启动文件并使用里面的内容。视图启动文件中定义的值优先，这样就可以方便地重写视图启动文件。

还可以使用多个视图启动文件为应用程序的不同部分设置默认值。Razor 会查找离正在处理的视图最近的视图启动文件，这意味着可以通过在 Views/Home 或 Views/Shared 文件夹中添加视图启动文件来覆盖默认设置。

警 告

务必了解从视图启动文件中省略 Layout 属性与将 Layout 属性设置为 null 之间的区别。如果视图是自包含的，并且不希望使用布局，可将 Layout 属性设置为 nulll。如果省略 Layout 属性，MVC 将认为需要布局，并且应该使用已在视图启动文件中找到的值。

5.4 使用 Razor 表达式

至此，我们已经了解了视图和布局的基本知识，接下来介绍 Razor 支持的各种不同类型的表达式，以及如何使用它们来创建视图内容。在优秀的 MVC 应用程序中，操作方法和视图的角色之间存在着明确的差异，如表 5-3 所示。

表 5-3 操作方法和视图的角色差异

内容	可以做什么	不能做什么
操作方法	将视图模型对象传递到视图	将格式化数据传递到视图
视图	使用视图模型对象向用户显示内容	修改视图模型对象

为了充分发挥 MVC 的优势，我们需要尊重并保证应用程序不同部分之间的分离。正如你将看到的，Razor 可以为我们做很多工作，包括使用 C#语句，但是不能使用 Razor 执行业务逻辑或以任何方式操作域模型对象。

作为一个简单的示例，代码清单 5-14 显示了一种在 Index.cshtml 文件中添加表达式的方式。

代码清单 5-14 在 Views/Home 文件夹下的 Index.cshtml 文件中添加一个表达式

```
@model Product

@{
    ViewBag.Title = "Product Name";
}

<p>Product Name: @Model.Name</p>
<p>Product Price: @($"{Model.Price:C2}")</p>
```

你可以在操作方法中设置 Price 属性的值，并将其传递给视图。虽然这样可以奏效，但是采用这种方法会破坏 MVC 模式自身的优点，并降低对未来变化的响应能力。虽然 ASP.NET Core MVC 不强制正确使用 MVC 模式，但你应该了解不这么做可能带来的影响。

处理数据与格式化数据

区分处理数据和格式化数据是非常重要的。视图会格式化数据而不是将对象的属性格式化为字符串，这就是之前将 Product 对象传递给视图的原因。处理数据（包括选择要显示的数据对象）是控制器的职责，控制器将调用模型来获取和修改所需的数据。有时很难弄清楚处理和格式化之间的界限是什么，但作为一条经验法则，谨慎起见，建议把除了最简单的表达式以外的任何东西从视图中拿出来移到控制器中。

5.4.1 插入数据

使用 Razor 表达式可以轻松将数据插入标记中。最常见的方法是使用@Model 表达式，Index 视图已包含具体示例，如下所示：

```
...
<p>Product Name: @Model.Name</p>
...
```

还可以使用 ViewBag 功能插入数据，也就是之前在布局中用来设置 title 元素内容的那项功能。ViewBag 可用于将数据从控制器传递到视图以补充模型，如代码清单 5-15 所示。

代码清单 5-15 在 Controller 文件夹下的 HomeController.cs 文件中使用 ViewBag 功能

```
using Microsoft.AspNetCore.Mvc;
using Razor.Models;

namespace Razor.Controllers {
    public class HomeController : Controller {
```

```
        public ViewResult Index() {
            Product myProduct = new Product {
                ProductID = 1,
                Name = "Kayak",
                Description = "A boat for one person",
                Category = "Watersports",
                Price = 275M
            };

            ViewBag.StockLevel = 2;

            return View(myProduct);
        }
    }
}
```

ViewBag 属性会返回一个可用于定义任意属性的 dynamic 对象。在代码清单 5-15 中，定义了一个 StockLevel 属性，并将属性值指定为 2。由于 ViewBag 是动态的，因此不必事先声明属性名称，但这意味着 Visual Studio 无法为 ViewBag 属性提供自动完成的建议。

何时使用 ViewBag 以及何时扩展模型是经验和个人偏好问题。作者喜欢只使用 ViewBag 来提供说明如何呈现数据的视图提示，而非用于显示给用户的数据。如果确实要将视图包用于显示给用户的数据，可以使用@ViewBag 表达式来访问它们，如代码清单 5-16 所示。

代码清单 5-16 在 Views/Home 文件夹下的 Index.cshtml 文件中显示视图包

```
@model Product

@{
    ViewBag.Title = "Product Name";
}

<p>Product Name: @Model.Name</p>
<p>Product Price: @($"{Model.Price:C2}")</p>
<p>Stock Level: @ViewBag.StockLevel</p>
```

图 5-9 显示了运行结果。

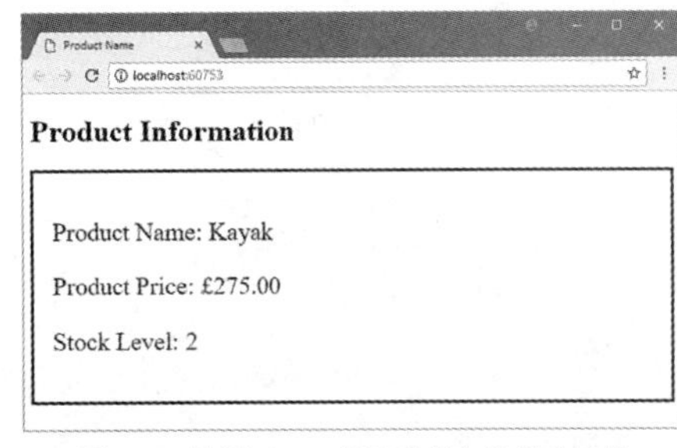

图 5-9 使用 Razor 表达式插入数据的结果

5.4.2 设置属性值

到目前为止，已为所有的示例设置了元素的内容，但也可以使用 Razor 表达式来设置元素的属性值。代码清单 5-17 显示了如何使用@Model 和@ViewBag 表达式来设置 Index 视图中元素的属性值。

代码清单 5-17 在 Views/Home 文件夹下的 Index.cshtml 文件中使用 Razor 表达式来设置属性值

```
@model Product

@{
    ViewBag.Title = "Product Name";
}

<div data-productid="@Model.ProductID" data-stocklevel="@ViewBag.StockLevel">
    <p>Product Name: @Model.Name</p>
    <p>Product Price: @($"{Model.Price:C2}")</p>
    <p>Stock Level: @ViewBag.StockLevel</p>
</div>
```

这里使用 Razor 表达式设置了 div 元素的 data 属性值。

提 示

名称由 data-作为前缀的属性，多年来一直是一种创建自定义属性的非正式方法，而今天作为 HTML5 的一部分已成为正式标准的一部分。它们经常被用到，所以 JavaScript 代码可以定位特定的元素，并且 CSS 样式可以得到更精准的应用。

如果运行示例程序并查看发送到浏览器的 HTML 源代码，你将看到 Razor 已经设置了属性的值：

```
<div data-productid="1" data-stocklevel="2">
    <p>Product Name: Kayak</p>
    <p>Product Price: £275.00</p>
    <p>Stock Level: 2</p>
</div>
```

5.4.3 使用条件语句

Razor 能够处理条件语句，也就是说，我们可以根据视图数据中的值调整视图的输出。这种技术是 Razor 的核心，使得使用者可以创建复杂但流畅的布局，同时又易读、易于维护。在代码清单 5-18 中，修改 Index 视图以包含条件语句。

代码清单 5-18 在 Views/Home 文件夹下的 Index.cshtml 文件中使用 Razor 条件语句

```
@model Product

@{
    ViewBag.Title = "Product Name";
}

<div data-productid="@Model.ProductID" data-stocklevel="@ViewBag.StockLevel">
    <p>Product Name: @Model.Name</p>
    <p>Product Price: @($"{Model.Price:C2}")</p>
    <p>Stock Level:
        @switch (ViewBag.StockLevel) {
            case 0:
                @:Out of Stock
                break;
            case 1:
            case 2:
            case 3:
                <b>Low Stock (@ViewBag.StockLevel)</b>
                break;
            default:
                @: @ViewBag.StockLevel in Stock
                break;
        }
    </p>
</div>
```

为了使用条件语句，请在 C#条件关键字的前面放置@字符，如示例中的@switch。就像使用常规的 C#代码块一样，代码块的终止符号为右大括号“}”。

在 Razor 代码块内，可以通过定义 HTML 和 Razor 表达式将 HTML 元素和数据包含在视图输出中，如下所示：

```
...
<b>Low Stock (@ViewBag.StockLevel)</b>
...
```

元素或表达式不需要放在引号中或者以特殊的方式表示它们——Razor 引擎会将它们解释为要处理的输出。但是，如果要在视图中不包含 HTML 元素时向其中插入文本，就需要对 Razor 进行特殊标注，并将所在行的前缀写成如下形式：

```
...
@: Out of Stock
...
```

@:字符是为了防止 Razor 被解释为 C#语句，这是遇到文本时的默认行为。你可以在图 5-10 中看到条件语句的结果。

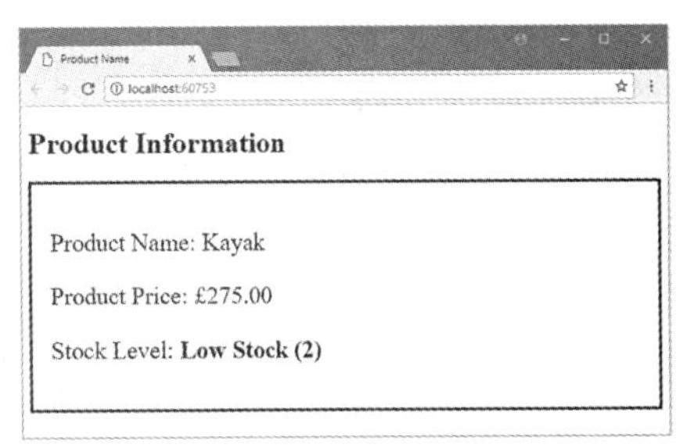

图 5-10 在 Razor 视图中使用开关表达式

条件语句在 Razor 视图中很重要，因为它们允许根据视图从操作方法接收的数据来对内容进行更改。在这里我们添加一些额外的演示，代码清单 5-19 演示了如何在 Index.cshtml 文件中添加 if 语句。

代码清单 5-19 在 Views/Home 文件夹下的 Index.cshtml 文件的 Razor 视图中使用 if 语句

```
@model Product

@{
    ViewBag.Title = "Product Name";
}

<div data-productid="@Model.ProductID" data-stocklevel-"@ViewBag.StockLevel">
    <p>Product Name: @Model.Name</p>
    <p>Product Price: @($"{Model.Price:C2}")</p>
    <p>Stock Level:
        @if (ViewBag.StockLevel == 0) {
            @:Out of Stock
        } else if (ViewBag.StockLevel > 0 && ViewBag.StockLevel <= 3) {
            <b>Low Stock (@ViewBag.StockLevel)</b>
        } else {
            @: @ViewBag.StockLevel in Stock
        }
    </p>
</div>
```

这个条件语句的结果与 switch 语句相同，这里演示的是如何让 Razor 视图与 C#条件语句相匹配。第 21 章将会解释这里的条件语句是怎样工作的。

5.4.4 枚举数组和集合

在编写 MVC 应用程序时，通常需要枚举数组的内容或某些其他类型对象的集合，并针对每个元素内容执行相应的操作。为了演示这是如何完成的，在代码清单 5-20 中，修改 Home 控制器中的 Index 操作方法，以便将一个 Product 对象数组传递给视图。

代码清单 5-20 在 Controllers 文件夹下的 HomeController.cs 文件中使用数组

```
using Microsoft.AspNetCore.Mvc;
using Razor.Models;

namespace Razor.Controllers {
    public class HomeController : Controller {

        public IActionResult Index() {
            Product[] array = {
                new Product {Name = "Kayak", Price = 275M},
                new Product {Name = "Lifejacket", Price = 48.95M},
                new Product {Name = "Soccer ball", Price = 19.50M},
                new Product {Name = "Corner flag", Price = 34.95M}
            };
            return View(array);
```

```
            }
        }
    }
```

Index 操作方法创建了一个包含简单数据的 Product[]对象，并将它们传递给 View 方法，以便使用默认视图呈现数据。在代码清单 5-21 中，更改 Index 视图的模型类型，并使用 foreach 循环枚举数组中的每个对象。

提　示

代码清单 5-21 中的 Model 不需要以@字符作为前缀，因为它是 C#表达式的一部分。我们很难弄清楚何时需要使用@字符，但是当不需要使用时，Visual Studio 智能感知功能会通过下画线提示有错误。

代码清单 5-21　在 Views/Home 文件夹下的 Index.cshtml 文件中枚举数组

```
@model Product[]

@{
    ViewBag.Title = "Product Name";
}

<table>
    <thead>
        <tr><th>Name</th><th>Price</th></tr>
    </thead>
    <tbody>
        @foreach (Product p in Model) {
            <tr>
                <td>@p.Name</td>
                <td>@($"{p.Price:C2}")</td>
            </tr>
        }
    </tbody>
</table>
```

@foreach 语句用于枚举 Model 数组的内容，并为每个数组元素生成一行数据。以上代码在 foreach 循环中创建了名为 p 的局部变量，然后使用 Razor 表达式@p.Name 与@p.Price 引用属性名称和价格，结果如图 5-11 所示。

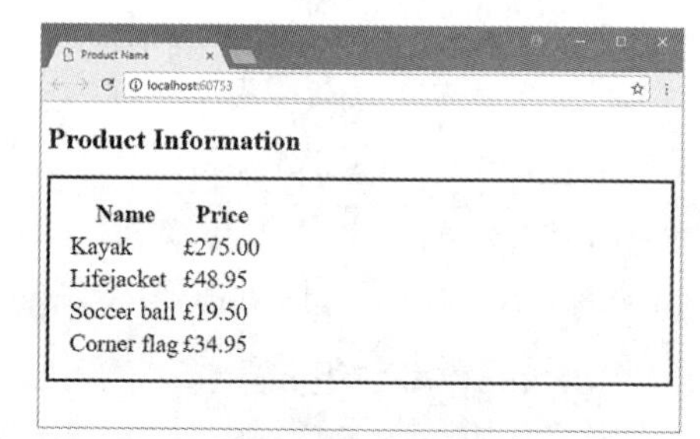

图 5-11　使用 Rzaor 枚举数组的结果

5.5　小结

本章概述了 Razor 视图引擎以及如何用它生成 HTML，介绍了如何通过视图模型对象和 ViewBag 来引用从控制器传来的数据，以及如何使用 Razor 表达式根据数据定制对用户的响应。接下来的章节将介绍更多关于 Razor 用法的示例，第 21 章将详细介绍 MVC 视图的机制。下一章将介绍 Visual Studio 为使用 ASP.NET Core MVC 项目而提供的一些功能。

第 6 章　使用 Visual Studio

本章主要介绍 Visual Studio 为开发 ASP.NET Core MVC 项目提供的关键特性。表 6-1 列出了本章要完成的操作。

表 6-1　本章要完成的操作

操作	方法	代码清单
添加包到项目中	使用 NuGet 工具管理.NET 包，使用 Bower 管理客户端包	代码清单 6-6～代码清单 6-8
查看视图或类更改的效果	使用迭代开发模式	代码清单 6-9～代码清单 6-11
在浏览器中显示详细消息	使用开发人员调试页	代码清单 6-12
获取有关程序执行的详细信息和控制信息	使用调试器	代码清单 6-13
使用 Visual Studio 重新加载一个或多个浏览器	使用浏览器链接	代码清单 6-14 和代码清单 6-15
减少 HTTP 请求的数量以及 JavaScript 和 CSS 文件所需的带宽大小	使用 Bundler & Minifier 扩展	代码清单 6-16～代码清单 6-23

6.1　准备示例项目

在本章中，将使用 Empty 模板创建一个新的名为 WorkingWithVisualStudio 的 ASP.NET Core Web Application（.NET Core）项目。在 Startup.cs 文件中用默认配置启用 MVC，如代码清单 6-1 所示。

代码清单 6-1　在 WorkingWithVisualStudio 文件夹下的 Startup.cs 文件中启用 MVC

```
using System;
using System.Collections.Generic;
using System.Linq;
using System.Threading.Tasks;
using Microsoft.AspNetCore.Builder;
using Microsoft.AspNetCore.Hosting;
using Microsoft.AspNetCore.Http;
using Microsoft.Extensions.DependencyInjection;
namespace WorkingWithVisualStudio {
    public class Startup {

        public void ConfigureServices(IServiceCollection services) {
            services.AddMvc();
        }

        public void Configure(IApplicationBuilder app, IHostingEnvironment env) {
            app.UseMvcWithDefaultRoute();
        }
    }
}
```

6.1.1　创建模型

创建 Models 文件夹，并添加名为 Product.cs 的类文件，用来定义代码清单 6-2 中的类。

代码清单 6-2 Models 文件夹下的 Product.cs 文件的内容

```
namespace WorkingWithVisualStudio.Models {

    public class Product {
        public string Name { get; set; }
        public decimal Price { get; set; }
    }
}
```

为了创建一个简单的 Product 存储对象，在 Models 文件夹中添加名为 SimpleRepository.cs 的类文件，用来定义代码清单 6-3 中的类。

代码清单 6-3 Models 文件夹下的 SimpleRepository.cs 文件的内容

```
using System.Collections.Generic;

namespace WorkingWithVisualStudio.Models {
    public class SimpleRepository {
        private static SimpleRepository sharedRepository = new SimpleRepository();
        private Dictionary<string, Product> products
            = new Dictionary<string, Product>();

        public static SimpleRepository SharedRepository => sharedRepository;

        public SimpleRepository() {
            var initialItems = new[] {
                new Product { Name = "Kayak", Price = 275M },
                new Product { Name = "Lifejacket", Price = 48.95M },
                new Product { Name = "Soccer ball", Price = 19.50M },
                new Product { Name = "Corner flag", Price = 34.95M }
            };
            foreach (var p in initialItems) {
                AddProduct(p);
            }
        }

        public IEnumerable<Product> Products => products.Values;

        public void AddProduct(Product p) => products.Add(p.Name, p);
    }
}
```

这个类会将模型对象存储在内存中，这意味着当应用程序停止或重新启动时，对模型所做的任何更改都将丢失。非持久化存储对于本章中的示例来说已经足够，但对于许多实际项目来说并不可取。关于创建持久化存储模型对象的示例，请参见 8 章。

注 意

代码清单 6-3 定义了一个名为 SharedRepository 的静态属性，使用它就可以在整个应用程序中对 SimpleRepository 对象进行访问。这虽然不是最好的做法，但可以通过它展示一下在 MVC 开发过程中经常会遇到的一个问题。第 18 章将介绍如何使用共享组件来更好地解决这个问题。

6.1.2 创建控制器和视图

在项目中创建 Controllers 文件夹，并添加名为 HomeController.cs 的类文件，用来定义代码清单 6-4 中的控制器。

代码清单 6-4 Controllers 文件夹下的 HomeController.cs 文件的内容

```
using Microsoft.AspNetCore.Mvc;
using WorkingWithVisualStudio.Models;

namespace WorkingWithVisualStudio.Controllers {
    public class HomeController : Controller {

        public IActionResult Index()
            => View(SimpleRepository.SharedRepository.Products);
    }
}
```

Index 操作方法会获取所有模型对象并将它们传递给 View 方法以呈现默认视图。为了添加默认视图，创建 Views/Home 文件夹，在其中添加名为 Index.cshtml 的视图文件，里面的内容如代码清单 6-5 所示。

代码清单 6-5 Views/Home 文件夹下的 Index.cshtml 文件的内容

```
@model IEnumerable<WorkingWithVisualStudio.Models.Product>
@{ Layout = null; }

<!DOCTYPE html>
<html>
<head>
    <meta name="viewport" content="width=device-width" />
    <title>Working with Visual Studio</title>
</head>
<body>
    <table>
        <thead>
            <tr><td>Name</td><td>Price</td></tr>
        </thead>
        <tbody>
            @foreach (var p in Model) {
                <tr>
                    <td>@p.Name</td>
                    <td>@p.Price</td>
                </tr>
            }
        </tbody>
    </table>
</body>
</html>
```

Index 视图中包含一个表格，使用 Razor foreach 循环为每个模型对象创建行，其中每一行都包含 Name 和 Price 属性。如果运行示例应用程序，你将看到图 6-1 所示的结果。

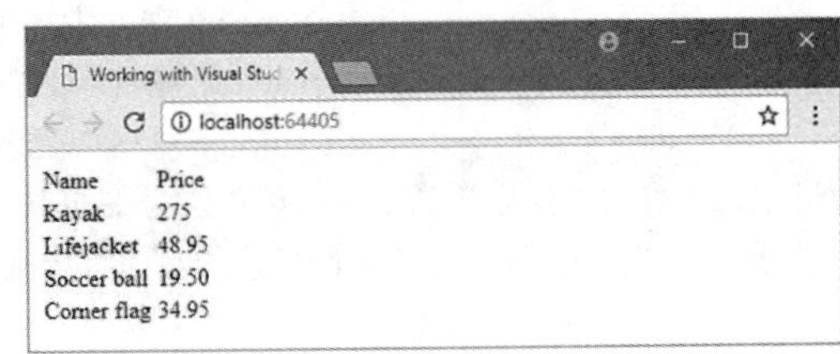

图 6-1 运行示例应用程序的结果

6.2 管理软件包

ASP.NET Core MVC 项目需要两种不同类型的软件包，后面将描述其中的每种类型以及 Visual Studio 为管理它们而提供的工具。

6.2.1 NuGet

为了管理包含在项目中的.NET 包，Visual Studio 提供了一个图形工具——NuGet。要打开该图形工具，请选择 Tools→NuGet Package Manager→Manage NuGet Packages for Solution，弹出的界面如图 6-2 所示。

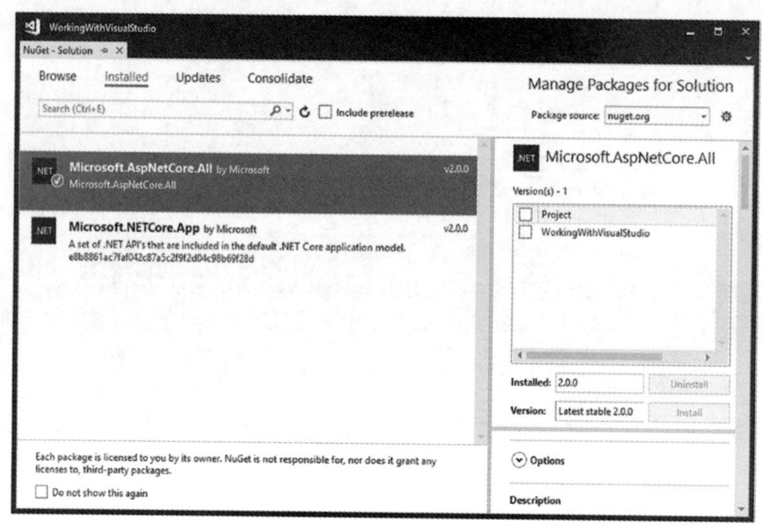

图 6-2 弹出的界面

Installed 选项卡提供了项目中已安装的包的摘要信息，Browse 选项卡可用于定位和安装新的包，Updates 选项卡可用于列出已发布的最新版本的软件包。

Microsoft.AspNetCore.All 包

如果使用了早期版本的 ASP.NET Core，你将明白向新项目添加大量 NuGet 包的必要性。ASP.NET Core 2 采用不同的方法，依赖一个名为 Microsoft.AspNetCore.All 的包。

Microsoft.AspNetCore.All 包是元包（meta-package），里面包含了 ASP.NET Core 和 MVC 框架所需的所有单个 NuGet 包，这意味着不需要逐个添加包。在发布应用程序时，将删除元包中未被应用程序使用的单个包，以确保不部署其他没有用到的包。

NuGet 包列表和位置

NuGet 工具在<projectname>.csproj 文件中提供了包的有关信息，其中<projectname>可由项目名称替换。对于示例应用程序，这意味着 NuGet 包的详细信息存储在名为 WorkingWithVisualStudio.csproj 的文件中。Visual Studio 不会在 Solution Explorer 窗格中显示.csproj 文件。要编辑该文件，请在 Solution Explorer 窗格中右击项目，然后在弹出的菜单中选择 Edit WorkingWithVisualStudio.csproj。Visual Studio 将打开该文件进行编辑。.csproj 文件是 XML 格式的，你将看到类似下面这样的元素，从而将 ASP.NET Core 元包添加到项目中：

```
...
<ItemGroup>
  <PackageReference Include="Microsoft.AspNetCore.All" Version="2.0.0" />
</ItemGroup>
...
```

为每个包指定名称和所需的版本号。尽管元包包括 ASP.NET Core MVC 所需的所有功能，但你仍必须将包添加到项目中，以便可以使用附加功能。可以使用图 6-2 所示的界面或使用命令行工具来添加软件包。也可以直接编辑.csproj 文件，Visual Studio 会检测到更改并下载和安装相应的软件包。

当使用 NuGet 将一本包添加到项目中时，这个包将与它所依赖的所有包一起自动安装。可以通过在 Solution Explorer 窗格中选择 Dependencies→NuGet 来研究 NuGet 包及其依赖项。ASP.NET Core 元包具有大量的依赖项，其中一些依赖项如图 6-3 所示。

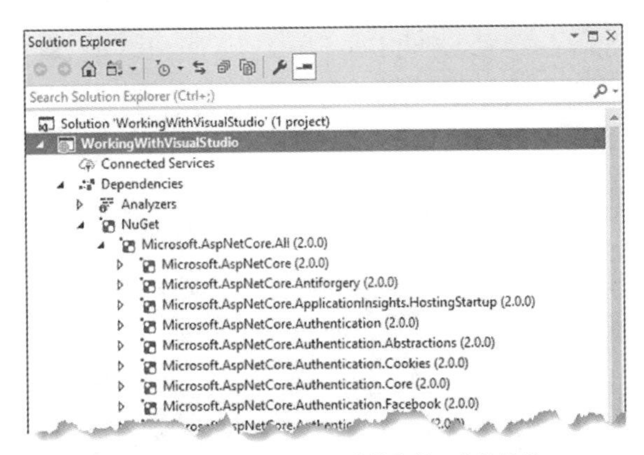

图 6-3 Solution Explorer 窗格中的一些依赖项

6.2.2 Bower

客户端包（client-side package）是包含发送到客户端的内容的包，如 JavaScript 文件、层叠样式表或图像。NuGet 过去也用来管理这些项目，但是 ASP.NET Core MVC 依赖于 Bower 工具。Bower 是一个开源工具，已经独立于 Microsoft 和.NET 开发完成，并被广泛应用于非 ASP.NET Web 应用程序的开发中。

注 意

Bower 最近已弃用。但是，Bower 仍在积极维护中，并且对 Bower 的支持已集成到 Visual Studio 中。

在某些时候，可以期望 Microsoft 支持其他用于管理客户端包的工具，但是在这种情况发生之前，应该继续使用 Bower。

1. Bower 包列表

Bower 包可通过 bower.json 文件指定。要创建 Bower 配置文件，请在 Solution Explorer 窗格中右击 Working WithVisualStudio 项目，从弹出的菜单中选择 Add→New Item，然后在左侧窗格中选择 ASP.NET Core→Web→General，在右侧窗格中选择 Bower Configuration File 项目模板，如图 6-4 所示。

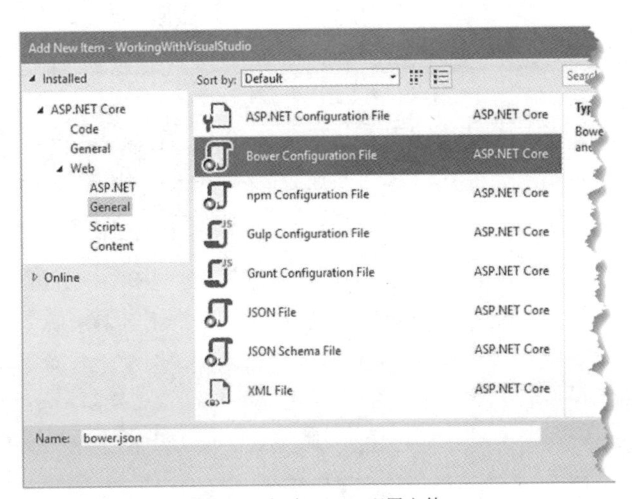

图 6-4 创建 Bower 配置文件

Visual Studio 已将名称设置为 bower.json，单击 Add 按钮将该文件添加到项目中，默认内容如代码清单 6-6 所示。

代码清单 6-6 bower.json 文件的默认内容

```
{
  "name": "asp.net",
  "private": true,
  "dependencies": {
  }
}
```

代码清单 6-7 显示了将客户端包添加到 bower.json 文件中的方法，这是通过添加与 project.json 文件相同格式的项到 dependencies 部分来实现的。

提 示

Bower 软件包的存储库参见 Bower 网站，可以在其中搜索要添加到项目中的软件包。

代码清单 6-7 在 bower.json 文件中添加包

```
{
  "name": "asp.net",
  "private": true,
  "dependencies": {
    "bootstrap": "3.3.7"
  }
}
```

以上代码将 Bootstrap CSS 包添加到了示例项目中。当编辑 bower.json 文件时，Visual Studio 将为你提供一个包名列表，并列出可用包的版本，如图 6-5 所示。

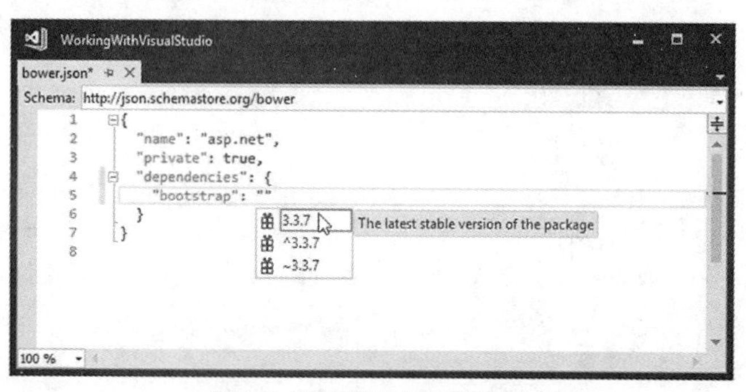

图 6-5 列出客户端包的可用版本

在编写本书时，Bootstrap 包的最新版本是 3.3.7。但是请注意，Visual Studio 提供了 3 个版本号，分别是 3.3.7、^3.3.7 和~3.3.7。在 bower.json 文件中，版本号可以不同的方式指定，一些常用模式如表 6-2 所示。指定包的最安全方法是使用显式的版本号。这样就能确保始终使用相同的版本，除非特意更新 bower.json 文件来更换其他版本。

提 示

对于本书中的示例，直接创建和编辑.json 文件。该文件易于编辑，有助于你得到期望的结果。Visual Studio 为管理 Bower 软件包提供了一个图形工具，右击 bower.json 文件，从弹出的菜单中选择 Manage Bower Package 即可打开该工具。

表 6-2 bower.json 文件中版本号的一些常用模式

模式	描述
3.3.7	安装与给定的版本号完全匹配的软件包，例如 3.3.7
*	使用星号将允许 Bower 下载和安装任何版本的软件包
>3.3.7 >=3.3.7	使用>或>=作为版本号的前缀将允许 Bower 安装大于以及大于或等于给定版本号的软件包的任何版本

续表

模式	描述
<3.3.7 <=3.3.7	使用<或<=作为版本号的前缀将允许 Bower 安装小于以及小于或等于给定版本号的软件包的任何版本
~3.3.7	使用波浪符（~字符）作为版本号的前缀将允许 Bower 安装即使版本号的最后一位不匹配的版本。例如，指定~3.3.7 将允许 Bower 安装 3.3.8 或 3.3.9 版本，但不能安装 3.4.0 版本
^3.3.7	使用插入符（^字符）作为版本号的前缀将允许 Bower 安装即使版本号的第二位或补丁号不匹配的版本。例如，指定^ 3.3.0 将允许 Bower 安装 3.3.1、3.4.0 和 3.5.0 版本，但不可以安装 4.0.0 版本

Visual Studio 会监控 bower.json 文件是否更改，并自动使用 Bower 工具下载和安装软件包。当把更改保存到代码清单 6-7 所示的文件时，Visual Studio 将下载 Bootstrap 包并将其安装到 wwwroot/lib 文件夹中，如图 6-6 所示。

就像 NuGet 一样，Bower 可管理添加到项目中的包的依赖。对于一些高级功能，Bootstrap 依赖 jQuery JavaScript 库，这就是为什么在图 6-6 中有两个包。展开 Solution Explorer 窗格中的依赖项，就可以查看包及其依赖项的列表，如图 6-7 所示。

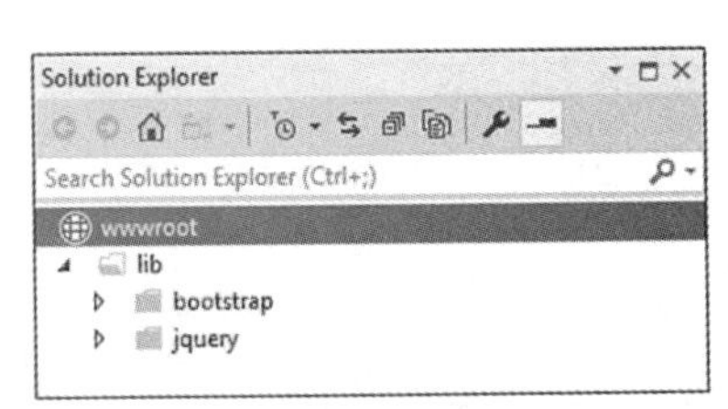

图 6-6 为项目添加客户端包

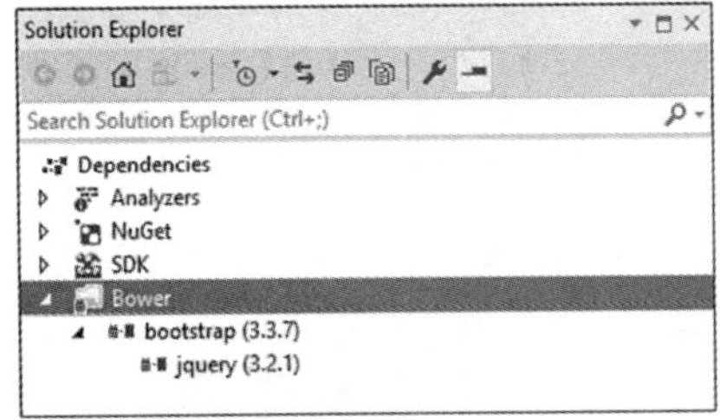

图 6-7 检查客户端包及其依赖项

2. 更新 Bootstrap 包

在本书的其余部分，使用了 Bootstrap CSS 框架的预发布版本。在撰写本书时，Bootstrap 团队正在开发第 4 版，并且已经发布了几个早期版本。这些版本被标记为 alpha，但是质量很高，并且它们足够稳定，可以在本书的示例中使用。当考虑是选择使用即将过时的 Bootstrap 3 还是使用 Bootstrap 4 的预发布版本来编写这本书时，最终作者决定使用新版本，尽管在最终发布之前，用于设置 HTML 元素样式的一些类名可能会更改。这意味着你必须使用相同版本的 Bootstrap 包才能从示例中获得预期的结果。

要更新 Bootstrap 包，请更改 bower.json 文件中的版本号，如代码清单 6-8 所示。

代码清单 6-8 在 WorkingWithVisualStudio 文件夹下的 bower.json 文件中更改包的版本

```
{
  "name": "asp.net",
  "private": true,
  "dependencies": {
    "bootstrap": "4.0.0-alpha.6"
  }
}
```

在保存对 bower.json 文件所做的更改时，Visual Studio 将下载新版的 Bootstrap 包。

6.3 迭代开发

Web 应用开发通常是一个迭代过程，可以对视图或类进行细微的更改，并运行应用程序来测试它们的效果。让 MVC 和 Visual Studio 协同工作便可以实现这种迭代并方便快捷地看到更改效果。

6.3.1 修改 Razor 视图

在开发过程中，一旦收到来自浏览器的 HTTP 请求，对 Razor 视图所做的更改就会立即生效。为了演示这是如何工作的，请从 Debug 菜单中选择 Start Debugging，启动应用程序，打开浏览器并显示数据后，对 Index.cshtml 文件所做的更改如代码清单 6-9 所示。

代码清单 6-9 更改 Index.cshtml 文件

```
@model IEnumerable<WorkingWithVisualStudio.Models.Product>
@{ Layout = null; }

<!DOCTYPE html>
<html>
<head>
    <meta name="viewport" content="width=device-width" />
    <title>Working with Visual Studio</title>
</head>
<body>
    <h3>Products</h3>
    <table>
        <thead>
            <tr><td>Name</td><td>Price</td></tr>
        </thead>
        <tbody>
            @foreach (var p in Model) {
                <tr>
                    <td>@p.Name</td>
                    <td>@($"{p.Price:C2}")</td>
                </tr>
            }
        </tbody>
    </table>
</body>
</html>
```

将更改保存到 Index 视图中，然后单击浏览器中的 Reload 按钮以重新加载当前网页。对视图的更改（添加页眉以及将 Price 属性设置为美元格式）将生效并显示在浏览器中，如图 6-8 所示。

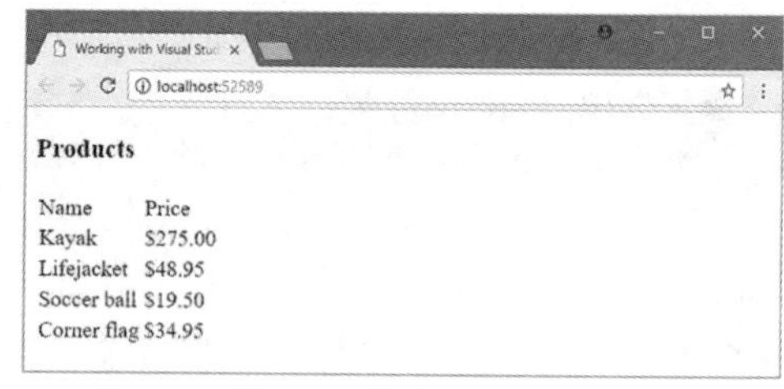

图 6-8 对视图进行更改

提 示

第 21 章将介绍 Razor 视图的使用过程。

6.3.2 对 C#类进行更改

对于 C#类（包括控制器和模型），处理更改的方式取决于如何启动应用程序。后面会介绍两种可用方式，可以通过 Debug 菜单中的不同菜单项进行选择，如表 6-3 所示。

表 6-3 Debug 菜单项

菜单项	描述
Start Without Debugging	当接收到 HTTP 请求时，项目中的类将自动编译，这是一种更动态的开发体验。应用程序在没有调试器的情况下运行，所以不能控制代码的执行
Start Debugging	在这种开发模式下，必须显式地编译项目并重新启动应用程序以使更改生效。调试器在程序运行时添加到应用程序中，这样可以检查运行状态并分析任何问题

1. 自动编译类

在正常开发过程中，快速迭代可以使你立即看到更改的效果，无论是添加新操作还是更改视图模型选择的数据都可以。对于此类开发，当浏览器接收到 HTTP 请求后，Visual Studio 便能立即检测到更改并自动重新编译类。要查看此操作的工作方式，请从 Visual Studio 的 Debug 菜单中选择 Start Without Debugging 菜单项。在浏览器显示了应用数据后，对 Home 控制器的更改如代码清单 6-10 所示。

代码清单 6-10 在 HomeController.cs 文件中筛选模型数据

```
using Microsoft.AspNetCore.Mvc;
using WorkingWithVisualStudio.Models;
using System.Linq;
namespace WorkingWithVisualStudio.Controllers {
    public class HomeController : Controller {

        public IActionResult Index()
            => View(SimpleRepository.SharedRepository.Products
                        .Where(p => p.Price < 50));
    }
}
```

以上代码使用 LINQ 来筛选 Product 对象，只有 Price 属性小于 50 的才能被传递给视图。将更改保存到控制器类文件并重新加载浏览器窗口，而无须停止或重新启动 Visual Studio 中的应用程序。浏览器的 HTTP 请求将触发编译过程，应用程序将使用修改后的控制器类重新启动，从而产生图 6-9 所示的结果，其中省略了表格中的 Kayak 产品。

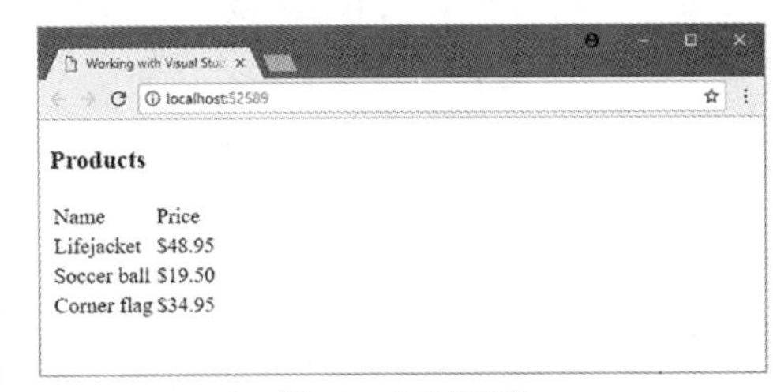

图 6-9 自动编译类

当一切都按计划进行的时候，自动编译功能是非常好用的。但是当编译或运行过程中出现错误时，错误会直接显示在浏览器而不是 Visual Studio 中，此时我们便很难找出哪里出了状况。例如，代码清单 6-11 显示了如何为存储库中的模型对象集合添加 null 引用。

代码清单 6-11 在 SimpleRepository.cs 文件中添加 null 引用

```
using System.Collections.Generic;

namespace WorkingWithVisualStudio.Models {
    public class SimpleRepository {
        private static SimpleRepository sharedRepository = new SimpleRepository();
        private Dictionary<string, Product> products
            = new Dictionary<string, Product>();

        public static SimpleRepository SharedRepository => sharedRepository;
        public SimpleRepository() {
            var initialItems = new[] {
                new Product { Name = "Kayak", Price = 275M },
                new Product { Name = "Lifejacket", Price = 48.95M },
                new Product { Name = "Soccer ball", Price = 19.50M },
                new Product { Name = "Corner flag", Price = 34.95M }
            };
            foreach (var p in initialItems) {
                AddProduct(p);
            }
            products.Add("Error", null);
        }

        public IEnumerable<Product> Products => products.Values;

        public void AddProduct(Product p) => products.Add(p.Name, p);
    }
}
```

Visual Studio 的智能感知功能将突出显示语法问题，但在应用程序运行之前，不会显示 null 引用这样的问题。重新加载浏览器中的页面时才会编译 SimpleRepository 类，并且应用程序会被重启。当 MVC 创建控制器类的实例以处理来自浏览器的 HTTP 请求时，HomeController 构造函数才会实例化 SimpleRepository 类，这将反过来尝试处理在列表中添加的 null 引用。null 值会引发一个问题，但此时还不清楚这个问题是什么，因为浏览器没有显示有用的消息。

2. 启用开发人员异常页面

在开发过程中，当出现错误时，在浏览器窗口中显示更多有用的信息会大有帮助。可以通过启用开发人员异常页面来完成，我们需要对 Startup 类进行配置更改，如代码清单 6-12 所示。

代码清单 6-12 在 Startup.cs 文件中启用开发人员异常页面

```
using System;
using System.Collections.Generic;
using System.Linq;
using System.Threading.Tasks;
using Microsoft.AspNetCore.Builder;
using Microsoft.AspNetCore.Hosting;
using Microsoft.AspNetCore.Http;
using Microsoft.Extensions.DependencyInjection;

namespace WorkingWithVisualStudio {
    public class Startup {

        public void ConfigureServices(IServiceCollection services) {
            services.AddMvc();
        }
        public void Configure(IApplicationBuilder app, IHostingEnvironment env) {
            app.UseDeveloperExceptionPage();
            app.UseMvcWithDefaultRoute();
        }
    }
}
```

第 14 章将详细介绍 Startup 类的作用，但是现在只需要知道调用 UseDeveloperExceptionPage 扩展方法以设置错误描述页面就可以了。

重新加载浏览器窗口，自动编译过程将重新生成应用程序，并在浏览器中生成更有用的错误消息，如图 6-10 所示。

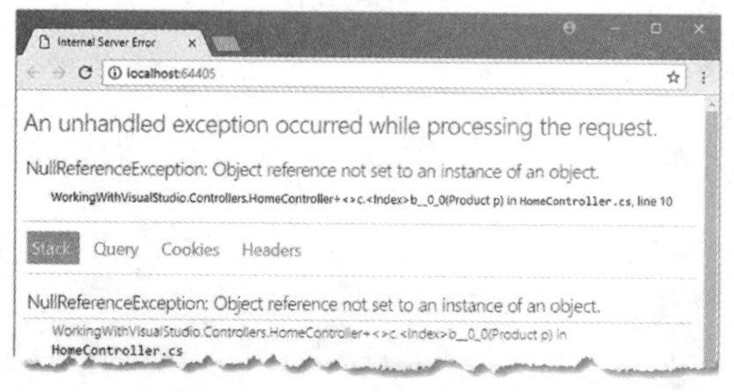

图 6-10 一个开发人员异常页面中的错误消息

浏览器显示的错误消息足以解决简单的问题，比如由于迭代开发，最近新更改的内容可能是引起问题的原因。但是，对于复杂的问题或者那些不能立即显现的问题，需要使用 Visual Studio 调试器。

3. 使用调试器

Visual Studio 还支持使用调试器运行 MVC 应用程序，可以暂停执行以检查应用程序的状态和代码逻辑规定的请求应遵循的路径。这需要使用不同的开发模式，因为在重新启动应用程序之前对 C#类所做的修改不会起作用（尽管对 Razor 视图所做的更改会自动生效）。

这种开发方式不像使用自动编译那样具有动态性，但是 Visual Studio 调试器非常好用，可以在浏览器窗口中显示信息，从而使开发人员对应用程序的工作方式有更深层次的了解。

要使用调试器运行应用程序，请从 Visual Studio 的 Debug 菜单中选择 Start Debugging。Visual Studio 将在启动应用程序之前编译项目中的 C#类，也可以使用 Build 菜单手动编译代码。

示例应用程序中仍然包含 null 引用，也就是说，由 SimpleRepository 类抛出的未处理的 NullReferenceException 将中断应用程序并将执行控制传递给开发人员，如图 6-11 所示。

提 示

如果调试器没有截获异常，请从 Visual Studio 的菜单栏中选择 Debug→Windows→Exception Settings 并确保 Common Language Runtime Exceptions 列表中所有的异常类型都被检测到。

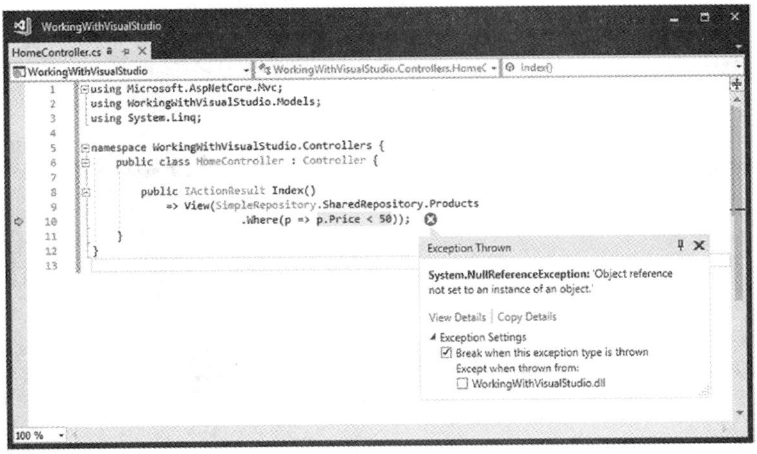

图 6-11 未处理的异常

1）设置断点

调试器并不会指出导致问题的原因，只会指出问题出现的地方。Visual Studio 高亮显示的语句表明在使用 LINQ 筛选对象时出了问题，但只需要做少量的工作就可以了解详细信息并找到根本原因。

断点（breakpoint）用于告诉调试器停止应用程序的执行而转为手动执行。可以检查应用程序的状态，查看正在发生的情况，还可以选择再次恢复执行。

要创建断点，请右击代码语句，然后从弹出的菜单中选择 Breakpoint→Insert Breakpoint。作为演示，我们将断点应用于 SimpleRepository 类中的 AddProduct 方法，如图 6-12 所示。

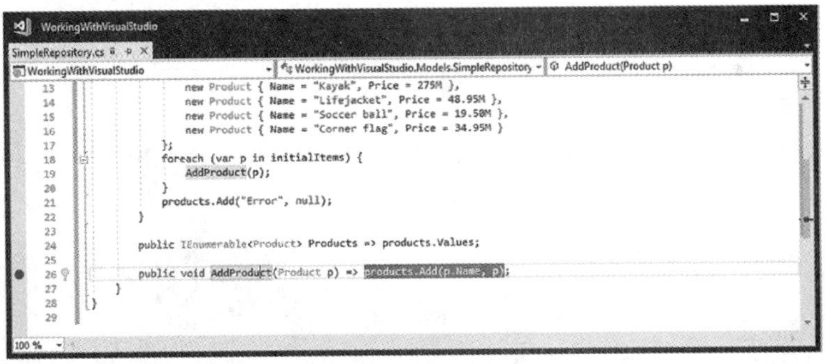

图 6-12 创建断点

选择 Debug→Start Debugging，启动应用程序，如果应用程序已经处于运行状态，就使用调试器或者选择 Debug→Restart 以重启应用程序。在浏览器初始化 HTTP 请求时，也会初始化 SimpleRepository 类。当程序执行到断点时，就会停止执行。

此时，可以使用 Visual Studio 的 Debug 菜单项或窗口顶部的控件来控制应用程序的执行；或者选择 Debug→Windows，启用调试视图来检查应用程序状态。

2）在代码编辑器中查看数据值

断点最常见的用途是在代码中跟踪 bug。在修复 bug 之前，我们必须弄清楚究竟发生了什么，而 Visual Studio 提供的最有用的功能之一就是能够在代码编辑器中查看和监控变量的值。

如果将鼠标指针移动到变量 p 上，并移到调试器高亮显示的 AddProduct 方法上，就会弹出一个窗口，显示 p 的值，如图 6-13 所示。

这个例子可能不那么确切，因为数据对象与断点是在相同的构造函数中定义的，但这个功能适用于任何变量。可以浏览各个变量以查看其属性和字段值。每个变量的右侧都有一个小的图钉按钮，在代码继续执行时可以用它监控变量的值。

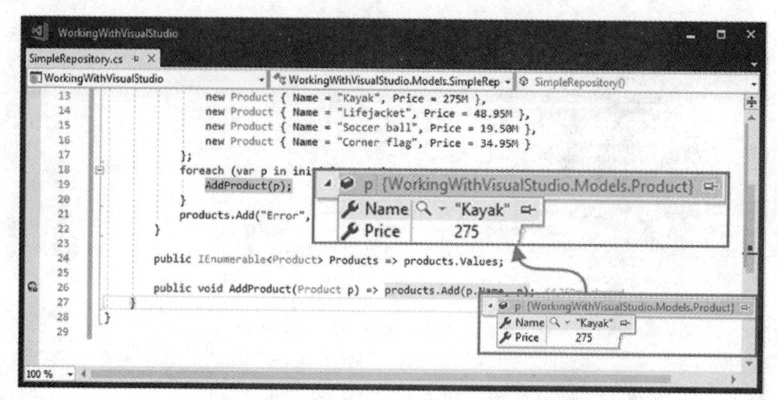

图 6-13 检查数据值

将鼠标指针悬停在变量 p 上并固定到 Product 引用上，展开引用后，还可以锁定 Name 和 Price 属性，从而实现图 6-14 所示的效果。

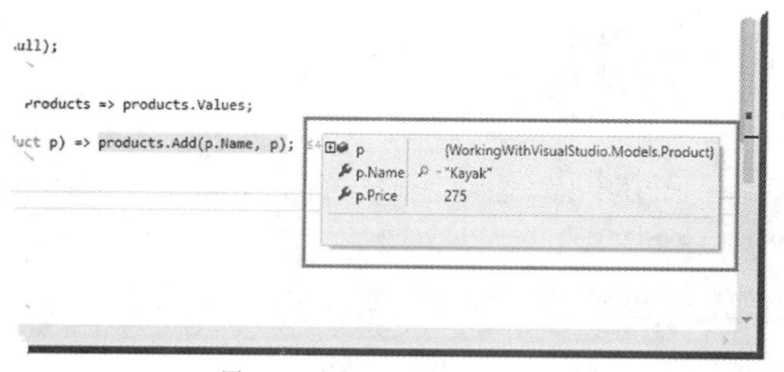

图 6-14 在代码编辑器中锁定值

从 Visual Studio 的 Debug 菜单中选择 Continue 以继续执行应用程序。由于应用程序正在执行 foreach 循环，因此当再次遇到断点时将停止执行。被锁定的值将显示 p 变量及其属性值的变化，如图 6-15 所示。

图 6-15 监控锁定值的变化

3）使用 Locals 窗口

Locals 窗口可通过选择 Debug→Windows→Locals 菜单项来打开。Locals 窗口以类似于锁定值的方式显示数据值，但显示的是所有与断点相关的本地对象，如图 6-16 所示。

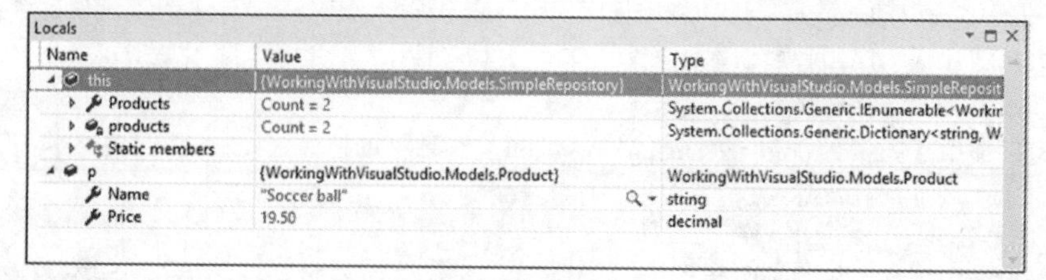

图 6-16 Locals 窗口

每次选择 Debug→Continue 时，应用程序就会继续执行，下一个对象将由 foreach 循环处理。如果一直继续下去，就会看到 null 引用出现，在 Locals 窗口中也可以看到。窗口和锁定的值显示在代码编辑器中。通过使用调试器控制应用程序的执行，可以跟踪代码的流程并了解正在发生的事情。

我们可以通过清理 Product 对象的集合来修复 null 引用问题，还可以选择另一种方法来使控制器更加健壮，如代码清单 6-13 所示，这里使用了 null 条件运算符来检查 null 值（如第 4 章所述）。

代码清单 6-13 在 HomeController.cs 文件中解决 null 引用问题

```
using Microsoft.AspNetCore.Mvc;
using WorkingWithVisualStudio.Models;
using System.Linq;

namespace WorkingWithVisualStudio.Controllers {
    public class HomeController : Controller {
        public IActionResult Index()
            => View(SimpleRepository.SharedRepository.Products
                        .Where(p => p?.Price < 50));
    }
}
```

要禁用断点，可右击已应用断点的代码语句，并从弹出的菜单中选择 Delete Breakpoint。重新启动应用程序，你将看到图 6-17 所示的简单数据表。

与那些需要使用错误扫描软件才能解决的问题相比，这个问题比较简单，但是 Visual Studio 调试器是非常好用的，通过使用应用程序的各种可视化视图和控制执行操作，我们可以真正挖掘到错误的细节。

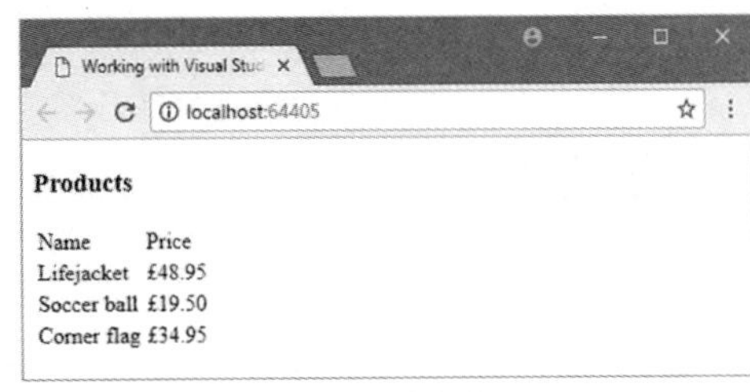

图 6-17 修复错误

6.3.3 使用浏览器链接

浏览器链接（browser link）功能可以通过将一个或多个浏览器置于 Visual Studio 的控制之下来简化开发过程。尤其是在需要查看一系列浏览器的更改效果时，这项功能最有用。浏览器链接在使用类的自动编译功能时非常有效，因为它可以用来修改项目中的任何文件，并显示更改的效果，而不必切换到浏览器并手动重新加载页面。

1. 设置浏览器链接

启用浏览器链接需要对 Startup 类进行配置更改，如代码清单 6-14 所示。

代码清单 6-14 在 WorkingWithVisualStudio 文件夹下的 Startup.cs 文件中启用浏览器链接

```
using System;
using System.Collections.Generic;
using System.Linq;
using System.Threading.Tasks;
using Microsoft.AspNetCore.Builder;
using Microsoft.AspNetCore.Hosting;
using Microsoft.AspNetCore.Http;
using Microsoft.Extensions.DependencyInjection;

namespace WorkingWithVisualStudio {
    public class Startup {

        public void ConfigureServices(IServiceCollection services) {
            services.AddMvc();
        }

        public void Configure(IApplicationBuilder app, IHostingEnvironment env) {
            app.UseDeveloperExceptionPage();
            app.UseBrowserLink();
            app.UseMvcWithDefaultRoute();
        }
    }
}
```

2. 使用浏览器链接

要了解浏览器链接的工作方式，请从 Visual Studio 的 Debug 菜单中选择 Start Without Debugging。Visual Studio 将启动应用程序并打开新的浏览器选项卡以显示结果。查看发送到浏览器的 HTML，你将看到其中包含一块附加的内容，如下所示：

```
<!DOCTYPE html>
<html>
<head>
    <meta name="viewport" content="width=device-width" />
    <title>Working with Visual Studio</title>
</head>
<body>
    <h3>Products</h3>
    <table>
        <thead>
            <tr><td>Name</td><td>Price</td></tr>
        </thead>
        <tbody>
                <tr><td>Lifejacket</td><td>&#xA3;48.95</td></tr>
                <tr><td>Soccer ball</td><td>&#xA3;19.50</td></tr>
                <tr><td>Corner flag</td><td>&#xA3;34.95</td></tr>
        </tbody>
    </table>
<!-- Visual Studio Browser Link -->
<script type="application/json" id="__browserLink_initializationData">
    {"requestId":"968949d8affc47c4a9c6326de21dfa03","requestMappingFromServer":false}
</script>
<script type="text/javascript"
    src="http://localhost:55356/d1a038413c804e178ef009a3be07b262/browserLink"
    async="async">
</script>
<!-- End Browser Link -->
</body>
</html>
```

提 示

如果没有看到附加部分，请从图 6-18 所示的菜单中选择 Enable Browser Link 并重新加载浏览器。

Visual Studio 会将一对 script 元素添加到发送到浏览器的 HTML 中，以打开一个具有较长生命周期的 HTTP 连接并返回给应用程序服务器，以便 Visual Studio 可以要求浏览器重新加载页面。如果 script 元素不显示，请检查以确保在图 6-18 所示的菜单中选择了 Enable Browser Link 选项。代码清单 6-15 显示了对 Index 视图所做的更改，并演示了使用浏览器链接的效果。

代码清单 6-15 向 Index.cshtml 文件添加一个时间戳

```
@model IEnumerable<WorkingWithVisualStudio.Models.Product>
@{ Layout = null; }

<!DOCTYPE html>
<html>
<head>
    <meta name="viewport" content="width=device-width" />
    <title>Working with Visual Studio</title>
```

```
</head>
<body>
    <h3>Products</h3>
    <p>Request Time: @DateTime.Now.ToString("HH:mm:ss")</p>
    <table>
        <thead>
            <tr><td>Name</td><td>Price</td></tr>
        </thead>
        <tbody>
            @foreach (var p in Model) {
                <tr>
                    <td>@p.Name</td>
                    <td>@($"{p.Price:C2}")</td>
                </tr>
            }
        </tbody>
    </table>
</body>
</html>
```

保存对视图文件所做的更改，并从 Visual Studio 工具栏的 Browser Link 菜单中选择 Refresh Linked Browsers，如图 6-18 所示。如果浏览器链接不起作用，请尝试重新加载浏览器或重新启动 Visual Studio 并重试。

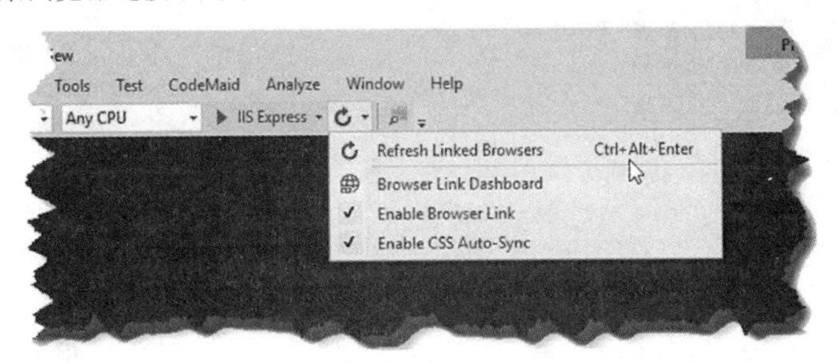

图 6-18 使用浏览器链接重新加载浏览器

发送到浏览器的嵌入 HTML 中的 JavaScript 代码将重新加载页面，显示最终的效果——添加了一个简单的时间戳。每次选择 Visual Studio 菜单项时，浏览器都会对服务器发出一个新的请求。这个请求会使 Index 视图生成一个新的具有更新时间戳的 HTML 页面。

注 意

浏览器链接的 script 元素仅在成功的响应中才会嵌入，也就是说，如果在编译类文件、呈现 Razor 视图或处理请求时发生异常，浏览器和 Visual Studio 之间的连接就会丢失，并且一旦解决了问题，就必须使用浏览器重新加载页面。

3. 使用多浏览器

浏览器链接可用于在多个浏览器中同时显示应用程序，这项功能在你希望解除浏览器之间的实现差异（特别是在实现自定义 CSS 样式表时）或查看应用程序如何在桌面浏览器和移动浏览器的组合中呈现时很有用。

要选取浏览器，请从 Visual Studio 工具栏的 IIS Express 下拉菜单中选择 Browse With 选项，如图 6-19 所示。

Visual Studio 会显示它所知道的浏览器的列表。图 6-20 显示了作者系统中安装的浏览器，其中一些是 Windows 操作系统自带的（比如 Internet Explorer 和 Edge），另一些是作者自己安装的常用浏览器。

Visual Studio 会在安装过程中查找常用的浏览器，你也可以使用 Add 按钮来设置未自动发现的浏览器。你还可以设置浏览器栈这样的第三方测试工具，通过将浏览器运行在云托管的虚拟机上，实现了不必人工管理大型的操作系统和浏览器矩阵就可以进行测试。

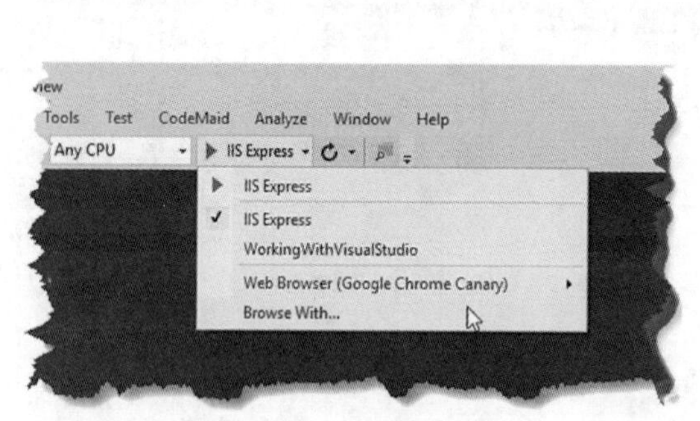

图 6-19 选择多浏览器

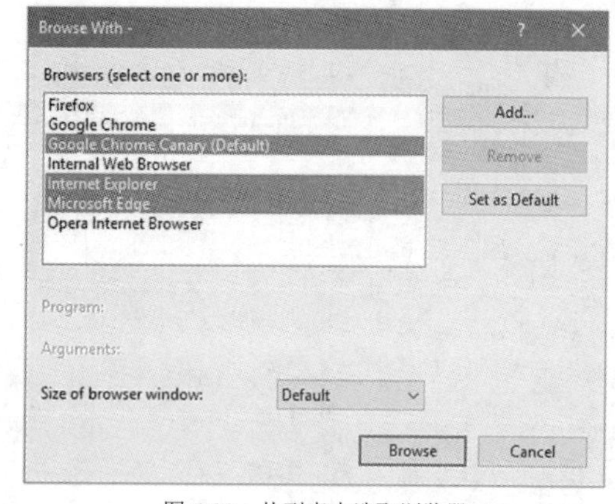

图 6-20 从列表中选取浏览器

这里选择了 Google Chrome Canary（Default）、Internet Explorer 和 Microsoft Edge 浏览器。单击 Browse 按钮将启动这 3 种浏览器，并使它们加载示例应用程序的 URL，如图 6-21 所示。

可以通过选择 Browser Link Dashboard 菜单项来查看浏览器链接的管理情况，这将打开图 6-22 所示的窗口，其中显示了每个浏览器输出的是哪个 URL 的内容，每个浏览器都可以单独刷新。

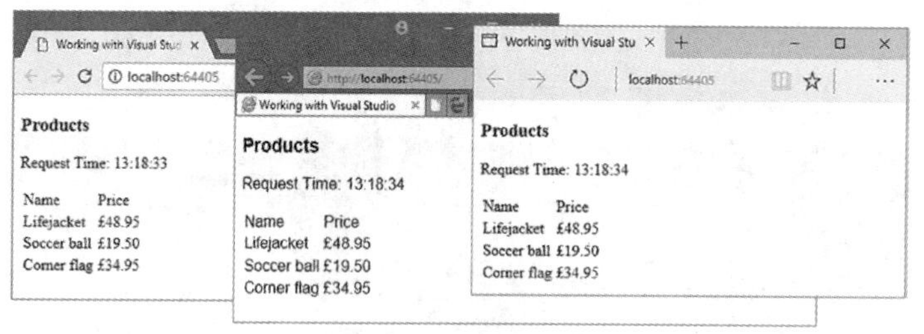

图 6-21 使用多浏览器

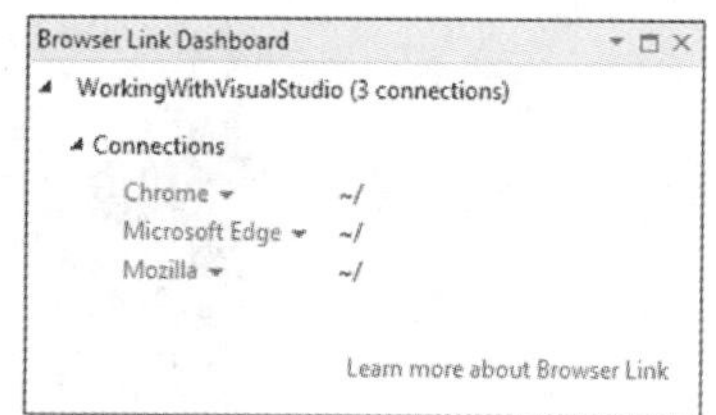

图 6-22 Browser Link Dashboard 窗口

6.4 部署 JavaScript 和 CSS

在创建 Web 应用程序的客户端部分时，通常会创建许多自定义的 JavaScript 和 CSS 文件，它们用于补充由 Bower 安装的包。这些文件需要进行处理以优化它们在生产环境中的传递速率，从而最大限度地减少 HTTP 请求的数量以及将它们传送到客户端所需的网络带宽。本节将介绍微软为执行这项任务而提供的 VisualStudio 扩展。

6.4.1 启用静态内容传递

ASP.NET Core 允许从 wwwroot 文件夹向客户端提供静态文件，但是当使用 Empty 模板创建项目时，默认情况下这是不允许的。可通过将代码清单 6-16 所示的语句添加到 Startup 类中来启用对静态文件的支持。

代码清单 6-16 在 WorkingWithVisualStudio 文件夹下的 Startup.cs 文件中启用对静态文件的支持

```
using System;
using System.Collections.Generic;
using System.Linq;
using System.Threading.Tasks;
using Microsoft.AspNetCore.Builder;
using Microsoft.AspNetCore.Hosting;
using Microsoft.AspNetCore.Http;
using Microsoft.Extensions.DependencyInjection;
```

```
namespace WorkingWithVisualStudio {
    public class Startup {

        public void ConfigureServices(IServiceCollection services) {
            services.AddMvc();
        }

        public void Configure(IApplicationBuilder app, IHostingEnvironment env) {
            app.UseDeveloperExceptionPage();
            app.UseBrowserLink();
            app.UseStaticFiles();
            app.UseMvcWithDefaultRoute();
        }
    }
}
```

6.4.2　为项目添加静态内容

为了演示打包和缩小化处理过程，我们需要在项目中添加一些静态内容，并使其具备交付给客户端的能力。首先，创建 wwwroot/css 文件夹，这是存储自定义 CSS 文件的地方。然后，使用 Style Sheet 模板添加名为 first.css 的文件，如图 6-23 所示。Style Sheet 模板位于 ASP.NET Core→Web→Content 部分。

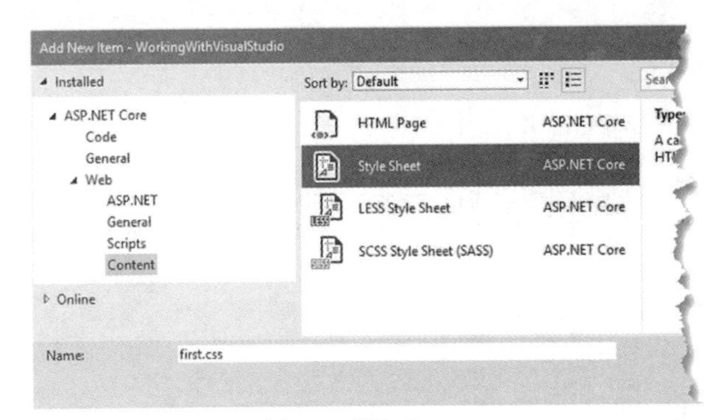

图 6-23　添加 first.css

编辑 first.css 文件以添加代码清单 6-17 所示的 CSS 样式。

代码清单 6-17　wwwroot/css 文件夹下的 first.css 文件的内容

```
h3 {
    font-size: 18pt;
    font-family: sans-serif;
}
table, td {
    border: 2px solid black;
    border-collapse:collapse;
    padding: 5px;
    font-family: sans-serif;
}
```

重复上述过程，在 wwwroot/css 文件夹中创建另一个样式表，名为 second.css，如代码清单 6-18 所示。

代码清单 6-18　wwwroot/css 文件夹下的 second.css 文件的内容

```
p {
    font-family: sans-serif;
    font-size: 10pt;
    color: darkgreen;
```

```
    background-color:antiquewhite;
    border: 1px solid black;
    padding: 2px;
}
```

自定义的 JavaScript 文件存储在 wwwroot/js 文件夹中。使用 JavaScript File 模板创建名为 third.js 的文件，如图 6-24 所示。

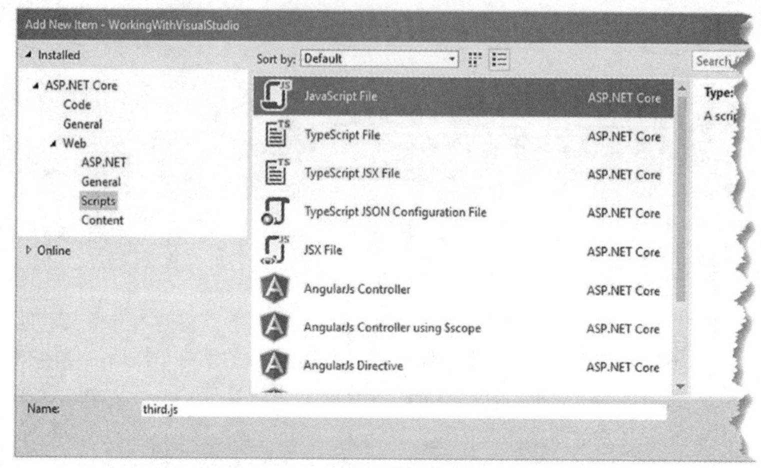

图 6-24 创建一个 JavaScript 文件

在 third.js 文件中添加一些简单的 JavaScript 代码，如代码清单 6-19 所示。

代码清单 6-19 wwwroot/js 文件夹下的 third.js 文件的内容

```
document.addEventListener("DOMContentLoaded", function () {
    var element = document.createElement("p");
    element.textContent = "This is the element from the third.js file";
    document.querySelector("body").appendChild(element);
});
```

这里还需要另外一个 JavaScript 文件。在 wwwroot/js 文件夹中创建 fourth.js 文件，在其中添加代码清单 6-20 所示的代码。

代码清单 6-20 wwwroot/js 文件夹下的 fourth.js 文件的内容

```
document.addEventListener("DOMContentLoaded", function () {
    var element = document.createElement("p");
    element.textContent = "This is the element from the fourth.js file";
    document.querySelector("body").appendChild(element);
});
```

6.4.3 更新视图

最后的准备步骤是更新 Index.cshtml 文件以使用新的 CSS 样式表和 JavaScript 文件，如代码清单 6-21 所示。

代码清单 6-21 为 Views/Home 文件夹下的 Index.cshtml 文件添加 script 和 link 元素

```
@model IEnumerable<WorkingWithVisualStudio.Models.Product>
@{ Layout = null; }

<!DOCTYPE html>
<html>
<head>
    <meta name="viewport" content="width=device-width" />
    <title>Working with Visual Studio</title>
```

```
    <link rel="stylesheet" href="css/first.css" />
    <link rel="stylesheet" href="css/second.css" />
    <script src="js/third.js"></script>
    <script src="js/fourth.js"></script>
</head>
<body>
    <h3>Products</h3>
    <p>Request Time: @DateTime.Now.ToString("HH:mm:ss")</p>
    <table>
        <thead>
            <tr><td>Name</td><td>Price</td></tr>
        </thead>
        <tbody>
            @foreach (var p in Model) {
                <tr>
                    <td>@p.Name</td>
                    <td>@($"{p.Price:C2}")</td>
                </tr>
            }
        </tbody>
    </table>
</body>
</html>
```

如果运行示例应用程序，你将看到图 6-25 所示的内容。现有内容中已添加了 CSS 样式，JavaScript 代码中也添加了新内容。

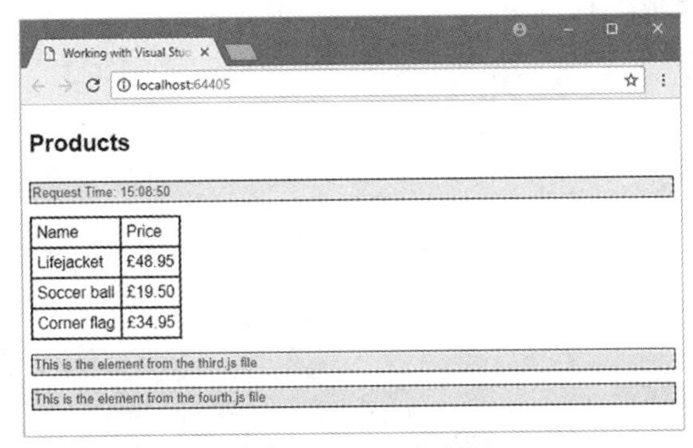

图 6-25　运行示例应用程序

6.4.4　MVC 应用程序中的打包和缩小

目前共有 4 个静态文件，所以浏览器必须发出 4 次请求才能获得这些静态文件，而且请求这些文件中的任何一个都比向客户端传递信息占用的带宽要大，因为它们包含了许多只对开发人员有意义但对浏览器没有意义的空白字符和变量名。

合并相同类型的文件称为打包。使文件变小称为缩小。这两个任务都是由 Visual Studio 的 Bundle & Minifier 扩展在 ASP.NET Core MVC 应用程序中执行的。

1. 下载 Visual Studio 扩展

第一步是安装扩展。选择 Tools→Extensions and Updates 菜单项，然后单击 Online 类别以显示可用的 Visual Studio 扩展库。在窗口右上角的搜索框中输入 Bundler & Minifier，如图 6-26 所示。找到 Bundler & Minifier 扩展，然后单击 Download 按钮将其添加到 Visual Studio 中。完成安装并重新启动 Visual Studio。

2. 打包和缩小文件

一旦安装 Bundler & Minifier 扩展并重新启动 Visual Studio，就可以选择同一类型的多个文件，将它们打包在一起并缩小它们的内容。例如，在 Solution Explorer 窗格中选择 first.css 和 second.css 文件，右击后，从弹出的菜单中选择 Bundler & Minifier→Bundle and Minify Files，如图 6-27 所示。

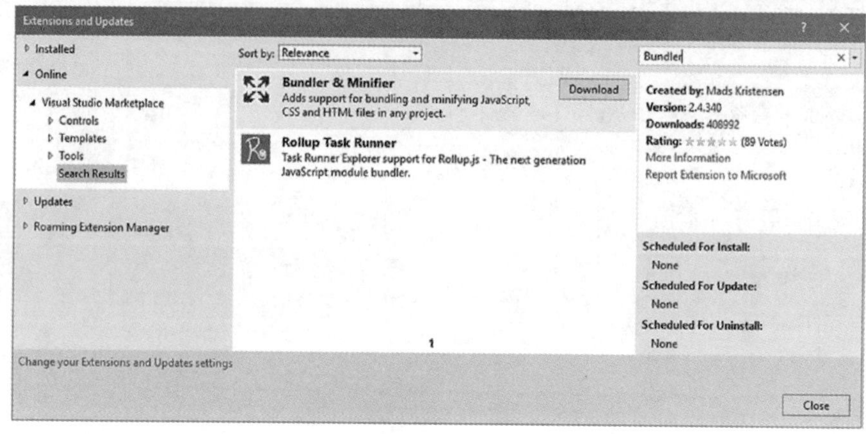

图 6-26 找到 Visual Studio 扩展

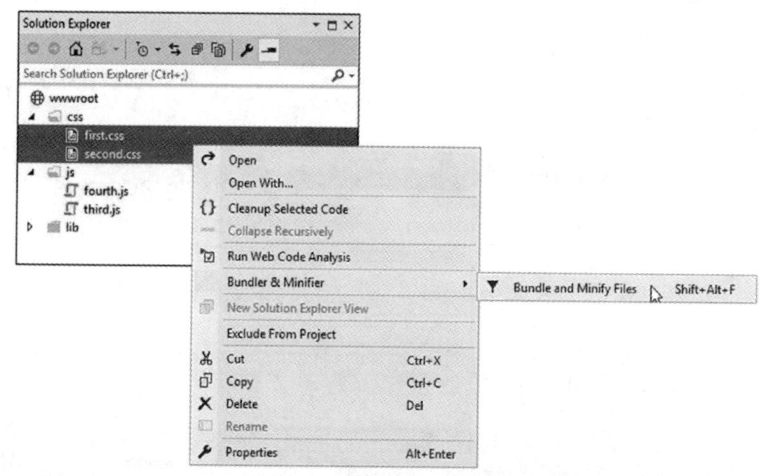

图 6-27 打包和缩小 CSS 文件

将输出文件另存为 bundle.css，Bundler & Minifier 扩展将处理这个 CSS 文件。Solution Explorer 窗格中将显示新的 bundle.css 文件，可以展开它以显示 bundle.min.css 压缩文件。打开这个压缩文件后，你将看到两个独立的 CSS 文件的内容已被合并，并且所有的空白符都已被删除。你可能不愿意直接使用这个文件，但它非常小，只需要一个 HTTP 连接就能将 CSS 样式传递给客户端。

对 third.js 和 fourth.js 文件重复这个过程，在 wwwroot/js 文件夹中创建 bundle.js 和 bundle.min.js 文件。

警 告

一定要按照浏览器加载它们的顺序选择文件，以保留输出文件中的样式或代码语句的顺序。例如，在选择 fourth.js 文件之前，请确保选择了 third.js 文件，以确保代码按正确的顺序执行。

在代码清单 6-22 中，已将单独文件的 link 元素替换为 Index.cshtml 文件中请求打包和缩小文件的元素。

代码清单 6-22 在 Index.cshtml 文件中使用打包和压缩的文件

```
@model IEnumerable<WorkingWithVisualStudio.Models.Product>
@{ Layout = null; }
```

```
<!DOCTYPE html>
<html>
<head>
    <meta name="viewport" content="width=device-width" />
    <title>Working with Visual Studio</title>
    <link rel="stylesheet" href="css/bundle.min.css" />
    <script src="js/bundle.min.js"></script>
</head>
<body>
    <h3>Products</h3>
    <p>Request Time: @DateTime.Now.ToString("HH:mm:ss")</p>
    <table>
        <thead>
            <tr><td>Name</td><td>Price</td></tr>
        </thead>
        <tbody>
            @foreach (var p in Model) {
                <tr>
                    <td>@p.Name</td>
                    <td>@($"{p.Price:C2}")</td>
                </tr>
            }
        </tbody>
    </table>
</body>
</html>
```

运行应用程序，虽然不会有任何可视化的更改，但打包和压缩的文件已经向浏览器提供了在单独的文件中定义的所有样式和代码。

在执行打包和缩小操作时，Bundler & Minifier 扩展会将文件的处理记录在 bundleconfig.json 文件中并保存在项目的根目录下。下面是为示例应用程序中的文件生成的配置：

```
[
  {
    "outputFileName": "wwwroot/css/bundle.css",
    "inputFiles": [
      "wwwroot/css/first.css",
      "wwwroot/css/second.css"
    ]
  },
  {
    "outputFileName": "wwwroot/js/bundle.js",
    "inputFiles": [
      "wwwroot/js/third.js",
      "wwwroot/js/fourth.js"
    ]
  }
]
```

Bundler & Minifier 扩展将自动监控输入文件的更改，并在发生更改时重新生成输出文件，以保证所做的任何编辑都反映在打包和缩小的文件中。作为演示，代码清单 6-23 显示了对 third.js 文件所做的更改。

代码清单 6-23 third.js 文件中的改动

```
document.addEventListener("DOMContentLoaded", function () {
    var element = document.createElement("p");
    element.textContent = "This is the element from the (modified) third.js file";
    document.querySelector("body").appendChild(element);
});
```

保存文件后，Bundler & Minifier 扩展将重新生成 bundle.min.js 文件。如果重新加载浏览器，你将看到图 6-28 所示的更改效果。

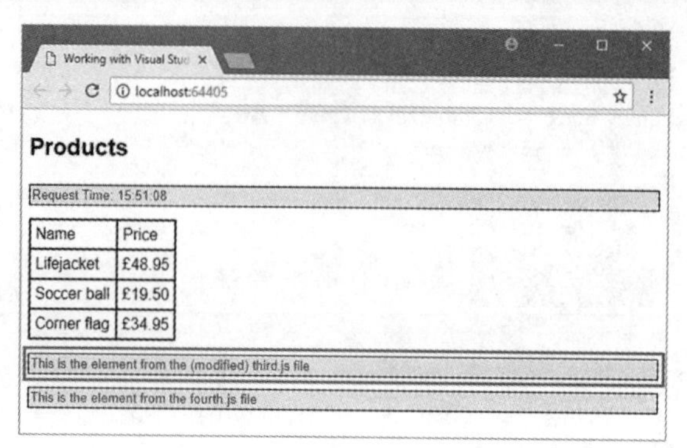

图 6-28 打包和缩小的文件中的更改检测

6.5 小结

本章介绍了 MVC 项目的结构，描述了两个可用的.NET 运行时功能，以及 Visual Studio 为 Web 应用开发提供的一些功能，包括自动类编译、浏览器链接以及打包和缩小。下一章将介绍 ASP.NET Core MVC 项目是如何进行单元测试的。

第 7 章　对 MVC 应用程序进行单元测试

本章演示如何对 MVC 应用程序进行单元测试。单元测试是一种测试形式，由于其中的单个组件已与应用程序的其他部分隔离，因此可以彻底验证其行为。ASP.NET Core MVC 已设计为便于创建单元测试，Visual Studio 为各种单元测试框架提供了支持。本章展示如何设置单元测试项目，说明如何安装最流行的测试框架之一，并描述编写和运行测试的过程。表 7-1 总结了本章要完成的操作。

表 7-1　　本章要完成的操作

操作	方法	代码清单
创建单元测试	创建单元测试项目，安装一个测试包，添加包含测试的类	代码清单 7-5 和代码清单 7-6
为单元测试隔离组件	使用接口分离应用程序组件，在单元测试中使用带有受限测试数据的伪实现	代码清单 7-7～代码清单 7-14
对不同的数据值运行同样的单元测试	使用参数化的单元测试，也可通过方法或属性获取测试数据	代码清单 7-15～代码清单 7-17
创建伪测试对象	使用 mocking 框架	代码清单 7-18 和代码清单 7-19

决定是否进行单元测试

能够简单地进行单元测试是使用 ASP.NET Core MVC 的优势之一，但这不适用于所有人。

作者喜欢单元测试，经常在自己的项目中使用单元测试，但不是所有的项目，并且也不像你所预期的那样。作者会专门为特性和功能编写单元测试，它们难以编写，而且很可能是部署中 bug 的来源。在这种情况下，单元测试有助于确立那些重要的想法。仅仅测试需要实现的东西，可以帮助处理关于潜在问题的想法，这应该发生在开始处理实际的 bug 之前。

也就是说，单元测试是一种工具，只有你才知道需要执行多少测试。如果没有发现单元测试的好处，或者有更适合的方法论，那就不要因为单元测试时髦而使用它（但是，如果没有更好的方法论，并且根本没有执行测试，那么你可能会让用户发现 bug，这非常不理想。你不必做单元测试，但是你真的需要考虑做一些其他测试）。

如果以前没有接触过单元测试，那么鼓励你尝试一下，看看它是如何工作的。如果你不是单元测试的粉丝，那么可以跳过本章，继续进入第 8 章，在那里构建更现实的 MVC 应用程序。

7.1　准备示例项目

本章将继续使用你在第 6 章中创建的项目。对于本章，将在仓储中添加新的 Product 对象以添加支持。

7.1.1　启用内置的标签助手

本章使用内置的标签助手为链接元素设置 href 属性。本书将在第 23～25 章说明标签助手是如何工作的。为了简单地启用它们，可右击 Views 文件夹，从弹出的菜单中选择 Add New Item，然后从 ASP.NET 类别中选择 MVC View Imports Page 项目模板，创建一个视图导入文件，Visual Studio 将自动设置文件名为 _ViewImports.cshtml，然后单击 Add 按钮，添加代码清单 7-1 所示的代码。

代码清单 7-1　Views 文件夹下的_ViewImports.cshtml 文件的内容

```
@addTagHelper *, Microsoft.AspNetCore.Mvc.TagHelpers
```

以上语句将启用内置的标签助手，包括你很快要在Index视图中使用的标签助手。可以添加using语句，从项目中导入命名空间，但是Index视图不是本章中示例应用程序的重要部分，并且使用命名空间引用模型类型也不是问题。

7.1.2 为控制器添加操作方法

第一步是为Home控制器添加操作方法，Home控制器用来渲染通过浏览器输入和接收数据的视图，如代码清单7-2所示。这些操作方法遵循你第2章使用过的模式，第17章将详细说明这些模式。

代码清单7-2 在Controller文件夹下的HomeController.cs文件中添加操作方法

```
using Microsoft.AspNetCore.Mvc;
using WorkingWithVisualStudio.Models;
using System.Linq;
namespace WorkingWithVisualStudio.Controllers {
    public class HomeController : Controller {

        SimpleRepository Repository = SimpleRepository.SharedRepository;

        public IActionResult Index() => View(Repository.Products
                 .Where(p => p?.Price < 50));

        [HttpGet]
        public IActionResult AddProduct() => View(new Product());

        [HttpPost]
        public IActionResult AddProduct(Product p) {
            Repository.AddProduct(p);
            return RedirectToAction("Index");
        }
    }
}
```

7.1.3 创建数据输入表单

为允许用户创建新的产品，在Views/Home文件夹中创建名为AddProduct.cshtml的Razor视图，并且与Home控制器中的AddProduct方法所呈现的视图一样，遵循默认的文件名和位置约定。代码清单7-3展示了新视图的内容，它基于你在第6章使用Bower加入项目的Bootstrap CSS包。

代码清单7-3 Views/Home文件夹下的AddProduct.cshtml文件的内容

```
@model WorkingWithVisualStudio.Models.Product
@{ Layout = null; }

<!DOCTYPE html>
<html>
<head>
    <meta name="viewport" content="width=device-width" />
    <title>Working with Visual Studio</title>
    <link rel="stylesheet" href="/lib/bootstrap/dist/css/bootstrap.min.css" />
</head>
<body class="p-2">
    <h3 class="text-center">Create Product</h3>
    <form asp-action="AddProduct" method="post">
        <div class="form-group">
            <label asp-for="Name">Name:</label>
            <input asp-for="Name" class="form-control" />
        </div>
        <div class="form-group">
            <label asp-for="Price">Price:</label>
```

```
            <input asp-for="Price" class="form-control" />
        </div>
        <div class="text-center">
            <button type="submit" class="btn btn-primary">Add</button>
            <a asp-action="Index" class="btn btn-secondary">Cancel</a>
        </div>
    </form>
</body>
</html>
```

该视图包含一个 HTML 表单，其使用 HTTP POST 请求发送 Name 和 Price 值到 Home 控制器的 AddProduct 操作方法。内容已使用 Bootstrap CSS 包进行装饰。

7.1.4　更新 Index 视图

最后的准备步骤是更新 Index 视图，以便包含到新表单的链接，如代码清单 7-4 所示。这里还借此机会删除了你在上一章中使用的 JavaScript 文件，使用 Bootstrap 替换了自定义的 CSS 样式表，并应用于视图中的 HTML 元素。

代码清单 7-4　更新 Views/Home 文件夹下的 Index.cshtml 文件的内容

```
@model IEnumerable<WorkingWithVisualStudio.Models.Product>
@{ Layout = null; }

<!DOCTYPE html>
<html>
<head>
    <meta name="viewport" content="width=device-width" />
    <title>Working with Visual Studio</title>
    <link rel="stylesheet" href="/lib/bootstrap/dist/css/bootstrap.min.css" />
</head>
<body class="p-1">
    <h3 class="text-center">Products</h3>
    <table class="table table-bordered table-striped">
        <thead>
            <tr><td>Name</td><td>Price</td></tr>
        </thead>
        <tbody>
            @foreach (var p in Model) {
                <tr>
                    <td>@p.Name</td>
                    <td>@($"{p.Price:C2}")</td>
                </tr>
            }
        </tbody>
    </table>
    <div class="text-center">
        <a class="btn btn-primary" asp-action="AddProduct">
            Add New Product
        </a>
    </div>
</body>
</html>
```

如果运行示例应用程序，你将会看到新装饰的内容和 Add New Product 按钮，用于引导到数据输入表单。如果提交表单，将添加新的 Product 对象到存储库中，并重定向浏览器以便初始的应用程序视图被展示出来，如图 7-1 所示。

提　示

切记此例中的存储库仅仅在内存中存储对象，这意味着在应用程序重新启动之时，创建的任何新产品都将被丢弃。

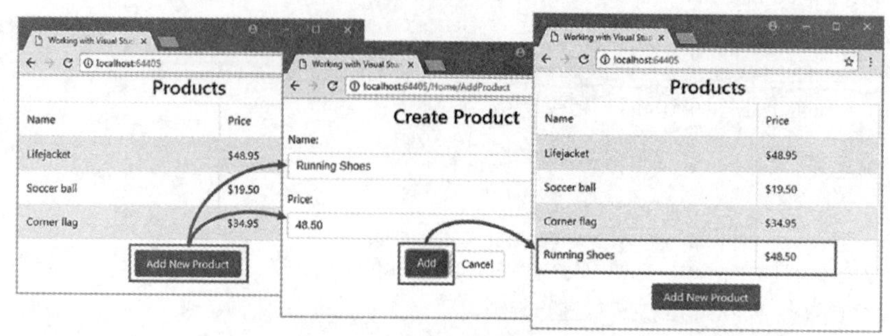

图 7-1　运行示例应用程序

7.2　测试 MVC 应用程序

单元测试用于验证应用程序中独立组件的行为和特性，ASP.NET Core 和 ASP.NET Core MVC 被设计为尽可能为 Web 应用程序设置和运行单元测试。本节将说明如何在 Visual Studio 中设置单元测试，并演示如何为 MVC 应用程序编写单元测试，还将介绍一些使得单元测试更为简单和可靠的有用工具。

有多种单元测试包可用，本书中使用的名为 xUnit.net，它与 Visual Studio 能够良好集成，可用来为 ASP.NET 编写单元测试。表 7-2 介绍了 xUnit.net 的背景。

表 7-2　　xUnit.net 的背景

问题	答案
xUnit.net 是什么？	xUnit.net 是单元测试框架，可以用来测试 ASP.NET Core MVC 应用程序
为何有用？	xUnit.net 是编写良好的测试框架，可以很容易集成到 Visual Studio 中
如何使用？	测试被定义为带有 Fact 或 Theory 特性注解的方法。在方法体中，使用 Assert 类定义的方法对测试期望的结果与实际的值进行比较
有何缺陷或限制？	单元测试的主要缺陷是不能有效隔离测试中的组件。xUnit.net 特有的最大问题是缺失文档。一些基本信息请参见 GitHub 网站，但是高级用法需要做一些试验
有何替代品？	有许多测试框架可用，两个流行的选择是 MSTest（来自微软）和 NUnit

注　意

单元测试中的任何问题都与个人偏好有关，并且每个人的意见并不相同。类似地有些开发者不喜欢将单元测试的代码从应用程序代码中分离出来，倾向于将测试定义在同一个项目中，甚至定义在同一个类文件中。这里介绍的是常用方式，也是本书遵循的方式，但是，如果不适合你，那么你应该尝试其他的测试风格，直至找到自己喜欢的。

7.2.1　创建单元测试项目

对于 ASP.NET Core 应用程序，通常创建单独的 Visual Studio 项目来保存单元测试，每个测试定义为 C#类中的一个方法。使用单独项目意味着可以部署应用程序而不需要部署测试。

为了创建测试项目，在 Solution Explorer 窗格中，右击 WorkingWithVisualStudio 解决方案，从弹出的菜单中选择 Add→New Project。在弹出的 Add New Project 对话框中，在左侧选择 Visual C#→.NET Core，在中间选择 xUnit Test Project (.NET Core)模板，如图 7-2 所示。

警　告

确保选择了正确的项目模板。Visual Studio 提供了多种单元测试的项目模板，并且它们有类似的名称。

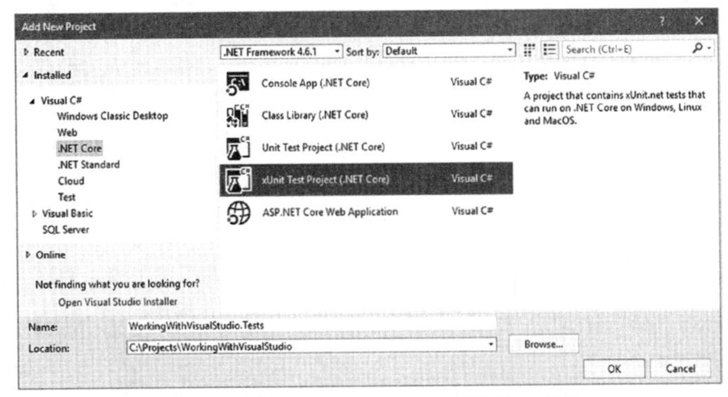

图 7-2 选择 xUnit Test Project (.NET Core)模板

约定的命名方式是将单元测试项目命名为<ApplicationName>.Tests。将新项目的名称设置为 WorkingWithVisualStudio.Tests，然后单击 OK 按钮以创建新项目。Visual Studio 将会创建项目并安装 xUnit 对应的 NuGet 包以及依赖的包。

删除默认测试类

Visual Studio 添加了一个 C#类到该测试项目中，它会导致后面示例的测试结果变得难以理解。右击 WorkingWithVisualStudio.Tests 项目中的 UnitTest1.cs 文件，然后从弹出菜单中选择 Delete。在提示对话框中单击 OK 按钮，Visual Studio 将会删除该类文件。

7.2.2 创建项目引用

为了使主项目中的类对于测试项目可用，右击 Solution Explorer 窗格中的 WorkingWithVisualStudio.Tests 项，然后从弹出菜单中选择 Add→Reference。选中 Solution 部分的 WorkingWithVisualStudio 项，如图 7-3 所示。

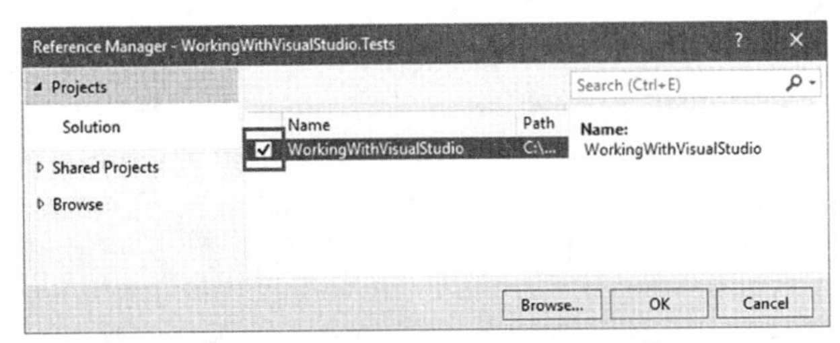

图 7-3 选择 WorkingWithVisualStudio 项

单击 OK 按钮以创建对应用程序项目的引用。你可能在 Solution Explorer 窗格的测试项目的 Dependencies 中看到一个处理中的图标，但是一旦你构建该项目这个图标将会消失。

编写和运行单元测试

现在所有的准备工作已经完成，可以编写一些测试。为了开始，如代码清单 7-5 所示，在 WorkingWithVisualStudio.Tests 项目中添加名为 ProductTests.cs 的类文件，这是一个简单的类文件，但是其中包含了从单元测试开始所需的一切。

代码清单 7-5 WorkingWithVisualStudio.Tests 文件夹下的 ProductTests.cs 文件的内容

```
using WorkingWithVisualStudio.Models;
using Xunit;

namespace WorkingWithVisualStudio.Tests {

    public class ProductTests {
```

```
        [Fact]
        public void CanChangeProductName() {

            // Arrange
            var p = new Product { Name = "Test", Price = 100M };

            // Act
            p.Name = "New Name";

            //Assert
            Assert.Equal("New Name", p.Name);
        }

        [Fact]
        public void CanChangeProductPrice() {

            // Arrange
            var p = new Product { Name = "Test", Price = 100M };

            // Act
            p.Price = 200M;

            //Assert
            Assert.Equal(100M, p.Price);
        }
    }
}
```

注 意

CanChangeProductPrice方法包含故意设置的错误，这将在本节稍后部分解决。

ProductTests类中有两个单元测试，其中的每个来自WorkingWithVisualStudio项目中Product模型类的不同行为。测试项目包含许多类，其中的每个可以包含多个单元测试。

通常，测试方法的名称描述了测试的内容，类的名称描述了正在测试的内容。这使得在项目中容易组织测试，在通过Visual Studio运行测试的时候，容易理解所有测试的结果。例如，名称ProductTests指出测试的是Product对象。

应用于每个方法的Fact特性指出这是一个测试。在方法体中，单元测试遵循称为组织（Arrange）、行动（Act）、断言（Assert）的A/A/A模式。组织是指为测试设置条件，行动是指执行测试，断言是指验证结果是否像预期的那样。

测试的组织和行动部分都是常规的C#代码，但是断言部分由xUnit.net处理，xUnit.net提供了一个名为Assert的类，其方法用于检查行动的结果是否如预期的那样。

提 示

Fact特性和Assert类都定义在命名空间Xunit中，因此所有的测试类中都必须有using语句。

Assert类的方法是静态的，用于比较不同种类的期望值和实际值，表7-3展示了Assert方法的常见用法。

表7-3 xUnit.net提供的Assert方法的常见用法

名称	描述
Equal(expected,result)	断言结果与期望值相等。该方法的重载版本用于比较不同的类型和集合。该方法的另一个版本接收一个额外的参数对象，该对象实现了用于比较对象的IEqualityComparer<T>接口
NotEqual(expected,result)	断言结果不等于期望值
True(result)	断言结果为true

续表

名称	描述
False(result)	断言结果为 false
IsType(expected,result)	断言结果为指定类型
IsNotType(expected,result)	断言结果不为指定类型
IsNull(result)	断言结果为 null
IsNotNull(result)	断言结果不为 null
InRange(result,low,high)	断言结果介于 low 和 high
NotInRange(result,low,high)	断言结果不介于 low 和 high
Throws(exception,expression)	断言指定的表达式抛出指定的异常类型

每个 Assert 方法都允许进行不同类型的比较，如果得不到预期的结果，就引发异常。异常用于标志测试失败。在代码清单 7-5 所示的测试中，使用 Equal 方法来检测属性的值是否被正确修改：

```
...
Assert.Equal("New Name", p.Name);
...
```

7.2.3 编写并运行单元测试

Visual Studio 支持通过测试管理器（Test Explorer）查找和运行测试。可以通过选择 Test→Windows→Test Explorer 找到测试管理器，如图 7-4 所示。

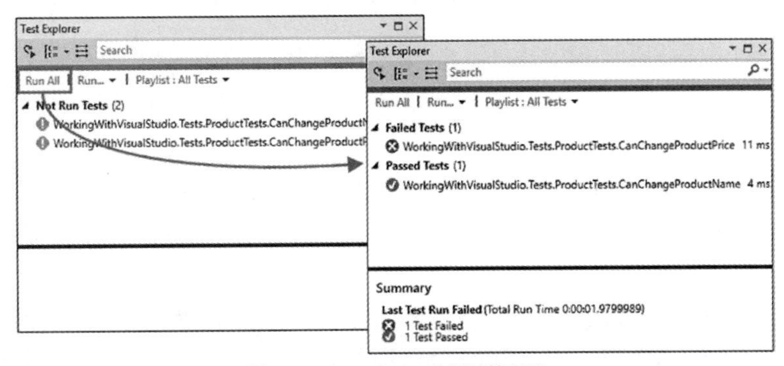

图 7-4 Visual Studio 的测试管理器

提 示

如果在测试管理器中没有看到单元测试，就请重新构建解决方案，编译触发单元测试的发现过程。

可通过单击 Test Explorer 窗口中的 Run All 按钮来运行全部测试。Visual Studio 将使用 xUnit.net 来运行项目中的测试并显示结果。如上所述，CanChangeProductPrice 测试包含导致测试失败的错误。问题源自 AssertEqual 方法的参数，其使用测试结果来比较源 Price 属性值而不是修改的值。代码清单 7-6 纠正了这个错误。

代码清单 7-6 更正 ProductTests.cs 文件中的测试

```
using WorkingWithVisualStudio.Models;
using Xunit;

namespace WorkingWithVisualStudio.Tests {

    public class ProductTests {
```

```
        [Fact]
        public void CanChangeProductName() {
            // Arrange
            var p = new Product { Name = "Test", Price = 100M };

            // Act
            p.Name = "New Name";

            //Assert
            Assert.Equal("New Name", p.Name);
        }

        [Fact]
        public void CanChangeProductPrice() {

            // Arrange
            var p = new Product { Name = "Test", Price = 100M };

            // Act
            p.Price = 200M;

            //Assert
            Assert.Equal(200M, p.Price);
        }
    }
}
```

提 示

当测试失败时，最好先检查测试的准确性，再检查测试的组件，特别是测试新的或刚刚修改的组件。

如果拥有大量测试，那么全部运行测试需要花费一些时间。为了快速迭代，测试管理器提供了不同的选项来测试子集。最有用的子集是失败的测试集，如图7-5所示。重新运行不正确的测试，测试管理器将显示没有测试失败。

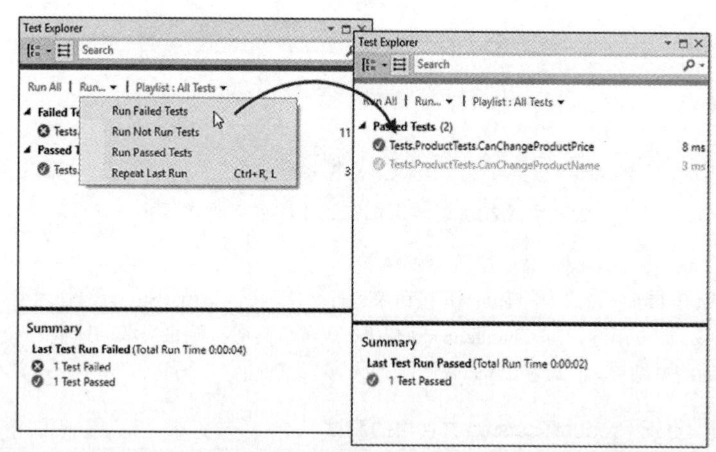

图7-5 有选择地运行测试

7.2.4 隔离组件以进行单元测试

为类似于Product这样的类编写单元测试是很容易的。不是因为Product类简单，而是因为Product类是自包含的，这意味着当在Product对象上执行操作时，可以确认是在测试由Product类提供的功能。

对于 MVC 应用程序中的其他组件来说，情况会变得复杂，因为它们之间存在依赖。你所定义的下一个测试集将处理控制器，检查在控制器和视图之间传递的 Product 对象的顺序。

当比较通过自定义的类实例化的对象时，需要使用 xUnit.net 中的 Assert.Equal 方法来接收一个实现了 IEqualityComparer<T>接口的参数，以便对象可以比较。第一步是在测试项目中添加名为 Comparer.cs 的文件，用它定义代码清单 7-7 中的类。

代码清单 7-7 WorkingWithVisualStudio.Tests 项目中 Comparer.cs 文件的内容

```
using System;
using System.Collections.Generic;

namespace WorkingWithVisualStudio.Tests {

    public class Comparer {

        public static Comparer<U> Get<U>(Func<U, U, bool> func) {
            return new Comparer<U>(func);
        }
    }

    public class Comparer<T> : Comparer, IEqualityComparer<T> {
        private Func<T, T, bool> comparisonFunction;

        public Comparer(Func<T, T, bool> func) {
            comparisonFunction = func;
        }

        public bool Equals(T x, T y) {
            return comparisonFunction(x, y);
        }

        public int GetHashCode(T obj) {
            return obj.GetHashCode();
        }
    }
}
```

这些类允许使用 Lambda 表达式，而不是针对每种组件类型定义新的类以创建 IEqualityComparer<T>对象。这不是必需的，但能简化单元测试类中的代码，使其易读、易于维护。

现在很容易进行比较，可以说明应用程序组件之间的依赖问题。添加名为 HomeControllerTests.cs 的文件到 WorkingWithVisualStudio.Tests 项目中，定义代码清单 7-8 所示的单元测试。

代码清单 7-8 WorkingWithVisualStudio.Tests 项目中 HomeControllerTests.cs 文件的内容

```
using Microsoft.AspNetCore.Mvc;
using System.Collections.Generic;
using WorkingWithVisualStudio.Controllers;
using WorkingWithVisualStudio.Models;
using Xunit;

namespace WorkingWithVisualStudio.Tests {
    public class HomeControllerTests {

        [Fact]
        public void IndexActionModelIsComplete() {
            // Arrange
            var controller = new HomeController();

            // Act
```

```
            var model = (controller.Index() as ViewResult)?.ViewData.Model
                as IEnumerable<Product>;

            // Assert
            Assert.Equal(SimpleRepository.SharedRepository.Products, model,
                Comparer.Get<Product>((p1, p2) => p1.Name == p2.Name
                    && p1.Price == p2.Price));
        }
    }
}
```

以上代码清单中的单元测试会检查Index操作方法传递给视图的所有存储库中的对象（此刻忽略行动部分，第17章将说明ViewResult类及其在MVC应用程序中扮演的角色）。目前，知道Index操作方法返回的模型数据就足够了。

如果运行测试，你将看到测试失败了，指出存储库中的对象集不同于Index操作方法返回的对象集。为了弄清楚失败的原因，有个问题需要解决：假设测试是在Home控制器上执行，但是控制器类依赖于SimpleRepository类，这使得难以揭示错误来自测试目标还是应用程序的另一部分。

这个示例应用程序简单到可以通过查看HomeController和SimpleRepository类的代码，就能轻松指出问题所在。在实际应用程序中，目视检查并不容易，依赖链使得代码难以理解导致测试失败。通常，存储库将依赖于某种持久化存储系统（比如数据库）以及提供访问的库，单元测试可作用于整个复杂的组件链，其中任何一个都可能导致错误。

当目标只针对应用程序的一小部分时，单元测试是有效的，比如独立的方法或类。你需要做的是能够从应用程序的其他部分隔离Home控制器，以便可以限制测试的范围，并排除任何存储库导致的影响。

隔离组件

隔离组件的关键是使用C#接口。为了将控制器和存储库分离，在Models文件夹中添加名为IRepository.cs的类文件，用它定义代码清单7-9所示的接口。

代码清单7-9 Models文件夹下的IRepository.cs文件的内容

```
using System.Collections.Generic;

namespace WorkingWithVisualStudio.Models {

    public interface IRepository {

        IEnumerable<Product> Products { get; }
        void AddProduct(Product p);
    }
}
```

这个接口没有什么特别之处（除没有定义常见Web应用程序需要的全部操作之外。但是，添加这样的接口可以使你能够轻松隔离组件进行测试。第一步是更新SimpleRepository类以便实现新的接口，如代码清单7-10所示。

代码清单7-10 在Model文件夹下的SimpleRepository.cs文件中实现接口

```
using System.Collections.Generic;

namespace WorkingWithVisualStudio.Models {
    public class SimpleRepository : IRepository {
        private static SimpleRepository sharedRepository = new SimpleRepository();
        private Dictionary<string, Product> products
            = new Dictionary<string, Product>();

        public static SimpleRepository SharedRepository => sharedRepository;
```

```
        public SimpleRepository() {
            var initialItems = new[] {
                new Product { Name = "Kayak", Price = 275M },
                new Product { Name = "Lifejacket", Price = 48.95M },
                new Product { Name = "Soccer ball", Price = 19.50M },
                new Product { Name = "Corner flag", Price = 34.95M }
            };
            foreach (var p in initialItems) {
                AddProduct(p);
            }
            products.Add("Error", null);
        }

        public IEnumerable<Product> Products => products.Values;

        public void AddProduct(Product p) => products.Add(p.Name, p);
    }
}
```

下一步是修改控制器，以便引用存储库的属性使用接口而不是类，如代码清单 7-11 所示。

代码清单 7-11 在 Controller 文件夹下的 HomeController.cs 文件中添加存储库属性

```
using Microsoft.AspNetCore.Mvc;
using WorkingWithVisualStudio.Models;
using System.Linq;

namespace WorkingWithVisualStudio.Controllers {
    public class HomeController : Controller {
        public IRepository Repository = SimpleRepository.SharedRepository;

        public IActionResult Index() => View(Repository.Products
                    .Where(p => p?.Price < 50));
        [HttpGet]
        public IActionResult AddProduct() => View();

        [HttpPost]
        public IActionResult AddProduct(Product p) {
            Repository.AddProduct(p);
            return RedirectToAction("Index");
        }
    }
}
```

提 示

ASP.NET Core MVC 支持使用更优雅的方式来解决这个问题，这称为依赖注入，将在第 18 章说明。依赖注入经常导致混乱，因此本章将以更简单、更手动的方式来隔离组件。

这可能不是一个重大变化，但是它允许在测试期间改变控制器使用的存储库。代码清单 7-12 更新了控制器的单元测试，以便它们使用特殊版本的存储库。

代码清单 7-12 在 HomeControllerTests.cs 文件的单元测试中隔离控制器

```
using Microsoft.AspNetCore.Mvc;
using System.Collections.Generic;
using WorkingWithVisualStudio.Controllers;
using WorkingWithVisualStudio.Models;
using Xunit;

namespace WorkingWithVisualStudio.Tests {
```

```
    public class HomeControllerTests {

        class ModelCompleteFakeRepository : IRepository {

            public IEnumerable<Product> Products { get; } = new Product[] {
                new Product { Name = "P1", Price = 275M },
                new Product { Name = "P2", Price = 48.95M },
                new Product { Name = "P3", Price = 19.50M },
                new Product { Name = "P3", Price = 34.95M }};
            public void AddProduct(Product p) {
                // do nothing - not required for test
            }
        }

        [Fact]
        public void IndexActionModelIsComplete() {
            // Arrange
            var controller = new HomeController();
            controller.Repository = new ModelCompleteFakeRepository();

            // Act
            var model = (controller.Index() as ViewResult)?.ViewData.Model
                as IEnumerable<Product>;

            // Assert
            Assert.Equal(controller.Repository.Products, model,
                Comparer.Get<Product>((p1, p2) => p1.Name == p2.Name
                    && p1.Price == p2.Price));
        }
    }
}
```

这里定义了一个假的IRepository接口实现，它仅仅实现了测试需要的属性，并使用始终一致的测试数据（使用真实数据库可能不是这种情况，特别是在你与其他开发人员共享，并且他们正在进行自己的修改时）。

修订之后的单元测试还是失败了，表明问题来自HomeController类的Index操作方法，而不是依赖的组件。通过单元测试进行的操作方法是非常简单的，检查问题也是显而易见的。

```
...
public IActionResult Index() => View(Repository.Products.Where(p => p.Price < 50));
...
```

问题来自LINQ表达式的Where方法，Where方法用于过滤Price属性大于或等于50的Product对象。在这一点上，可以确信产生问题的原因。但是在做修改之前，最佳实践是创建测试来确认问题，如代码清单7-13所示。

代码清单7-13 在WorkingWithVisualStudio.Tests项目的HomeControllerTests.cs文件中添加测试

```
using Microsoft.AspNetCore.Mvc;
using System.Collections.Generic;
using WorkingWithVisualStudio.Controllers;
using WorkingWithVisualStudio.Models;
using Xunit;
namespace WorkingWithVisualStudio.Tests {
    public class HomeControllerTests {

        class ModelCompleteFakeRepository : IRepository {

            public IEnumerable<Product> Products { get; } = new Product[] {
                new Product { Name = "P1", Price = 275M },
                new Product { Name = "P2", Price = 48.95M },
                new Product { Name = "P3", Price = 19.50M },
                new Product { Name = "P3", Price = 34.95M }};
```

```
        public void AddProduct(Product p) {
            // do nothing - not required for test
        }
    }

    [Fact]
    public void IndexActionModelIsComplete() {
        // Arrange
        var controller = new HomeController();
        controller.Repository = new ModelCompleteFakeRepository();

        // Act
        var model = (controller.Index() as ViewResult)?.ViewData.Model
            as IEnumerable<Product>;

        // Assert
        Assert.Equal(controller.Repository.Products, model,
            Comparer.Get<Product>((p1, p2) => p1.Name == p2.Name
                && p1.Price == p2.Price));
    }

    class ModelCompleteFakeRepositoryPricesUnder50 : IRepository {

        public IEnumerable<Product> Products { get; } = new Product[] {
            new Product { Name = "P1", Price = 5M },
            new Product { Name = "P2", Price = 48.95M },
            new Product { Name = "P3", Price = 19.50M },
            new Product { Name = "P3", Price = 34.95M }};

        public void AddProduct(Product p) {
            // do nothing - not required for test
        }
    }

    [Fact]
    public void IndexActionModelIsCompletePricesUnder50() {
        // Arrange
        var controller = new HomeController();
        controller.Repository = new ModelCompleteFakeRepositoryPricesUnder50();

        // Act
        var model = (controller.Index() as ViewResult)?.ViewData.Model
            as IEnumerable<Product>;

        // Assert
        Assert.Equal(controller.Repository.Products, model,
            Comparer.Get<Product>((p1, p2) => p1.Name == p2.Name
                && p1.Price == p2.Price));
    }
  }
}
```

提　示

这些测试中含有大量的重复代码。7.3 节将描述如何简化测试。

上面定义了一个新的仅含有 Price 属性值小于 50 的 Product 对象的存储库，并且你将在新的测试中仅使用这个存储库。如果运行这个测试，你将看到测试成功了，这增加了问题是由 Index 操作方法中 Where 方法的用法导致的权重。

在真实项目中，了解测试失败的原因时，需要将测试的目的与应用程序的规范相协调。可能的情况是，Index 操作方法应该通过 Price 来过滤 Product 对象，在这种情况下，将需要修改测试。这是常见的结果，失败的测试并不总是表明应用程序中真正存在问题。从另一个角度说，如果 Index 操作方法不应该过滤模型对象，则需要更正修改，如代码清单 7-14 所示。

代码清单 7-14 移除 Controller 文件夹下的 HomeController.cs 文件中的 LINQ 过滤器

```
using Microsoft.AspNetCore.Mvc;
using WorkingWithVisualStudio.Models;
using System.Linq;

namespace WorkingWithVisualStudio.Controllers {
    public class HomeController : Controller {
        public IRepository Repository = SimpleRepository.SharedRepository;

        public IActionResult Index() => View(Repository.Products);

        [HttpGet]
        public IActionResult AddProduct() => View(new Product());

        [HttpPost]
        public IActionResult AddProduct(Product p) {
            Repository.AddProduct(p);
            return RedirectToAction("Index");
        }
    }
}
```

测试驱动开发

本章遵循最常见的单元测试风格，先开发应用程序的功能，再进行测试以确保按需工作。因为大多数开发者首先考虑应用程序代码，然后进行测试，所以这种风格很受欢迎。

这种开发方式的问题在于倾向于只关注部分应用程序代码，这些代码要么难以编写，要么需要重度调试，从而使功能的某些方面仅仅得到部分测试或未经测试。

另一种开发方式是测试驱动开发（TDD）。TDD 中有很多变化，但核心思想是在实现功能本身之前为功能编写测试。首先编写测试可以使你更仔细地考虑正在实现的规范，以及如何知道功能已经正确实现。TDD 不是深入实现细节，而是提前考虑成功或失败的度量。

你编写的测试一开始都将失败，因为你的新功能还没有实现。但是随着你为应用程序添加代码，你的测试将逐渐从红色变为绿色，并且在所有功能都完成的时候，测试将全部通过。TDD 需要训练，它确实产生了更全面的测试集，并且生成更可靠的代码。

如果重新运行测试，你将看到它们全部通过，如图 7-6 所示。

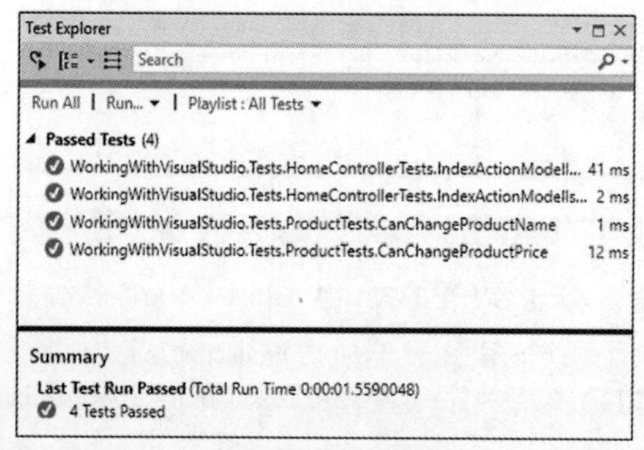

图 7-6 所有测试已通过

对于这样一个简单的问题，看起来需要做很多工作，但是在实际的应用程序中，测试特定组件的能力是至关重要的。若已经识别出问题，只有在有效隔离组件的情况下，才能验证和修复程序。

7.3 改进单元测试

7.2 节介绍了在 Visual Studio 中编写和运行单元测试的基本方法，并强调了隔离正在测试的组件的重要性。本节将介绍一些更加高级的工具和特性，你可以使用它们更简洁、更有表现力地编写测试。如果已浸入单元测试文化，那么你最终可以得到很多测试代码，并且代码的清晰性也变得重要起来，特别是当需要修改测试以反映开发过程中应用程序的变化并进入维护阶段时。

7.3.1 参数化单元测试

为 HomeController 类编写的测试显现出仅针对某些数据值的问题。为了测试这种情况，最终创建了两个类似的测试，每个都有自己的假的存储。这是一种重复的方式，特别是因为这些测试之间唯一的区别是存储中 Product 对象的 Price 属性的 decimal 值。

xUnit.net 提供对参数化测试的支持。其中的测试数据已从测试中删除，以便单个方法可以用于多个测试。在代码清单 7-15 中，使用参数化测试功能来删除用于 HomeController 类的测试中的重复数据。

代码清单 7-15 在 HomeControllerTests.cs 文件中参数化单元测试

```
using Microsoft.AspNetCore.Mvc;
using System.Collections.Generic;
using WorkingWithVisualStudio.Controllers;
using WorkingWithVisualStudio.Models;
using Xunit;

namespace WorkingWithVisualStudio.Tests {
    public class HomeControllerTests {

        class ModelCompleteFakeRepository : IRepository {

            public IEnumerable<Product> Products { get; set; }

            public void AddProduct(Product p) {
                // do nothing - not required for test
            }
        }

        [Theory]
        [InlineData(275, 48.95, 19.50, 24.95)]
        [InlineData(5, 48.95, 19.50, 24.95)]
        public void IndexActionModelIsComplete(decimal price1, decimal price2,
                decimal price3, decimal price4) {
            // Arrange
            var controller = new HomeController();
            controller.Repository = new ModelCompleteFakeRepository {
                Products = new Product[] {
                    new Product {Name = "P1", Price = price1 },
                    new Product {Name = "P2", Price = price2 },
                    new Product {Name = "P3", Price = price3 },
                    new Product {Name = "P4", Price = price4 },
                }
            };

            // Act
            var model = (controller.Index() as ViewResult)?.ViewData.Model
                as IEnumerable<Product>;

            // Assert
            Assert.Equal(controller.Repository.Products, model,
```

```
                Comparer.Get<Product>((p1, p2) => p1.Name == p2.Name
                    && p1.Price == p2.Price));
        }
    }
}
```

参数化单元测试使用 Theory 特性而不是标准的 Fact 特性，还使用了 InLineData 特性，以允许为单元测试方法定义参数的指定值。由于 C#限制了特性中数据值的表达方式，因此在测试方法上定义了 4 个 decimal 参数用于为 InlineData 特性提供值。可在测试方法中使用 decimal 值来生成 Product 对象的数组，用于设置假的存储库对象的 Products 属性。

每个 Inline 特性定义了一个单独的单元测试，在 Visual Studio 测试管理器中显示为单独的项，如图 7-7 所示。测试管理器中的条目显示了将用于单元测试方法中参数的值。

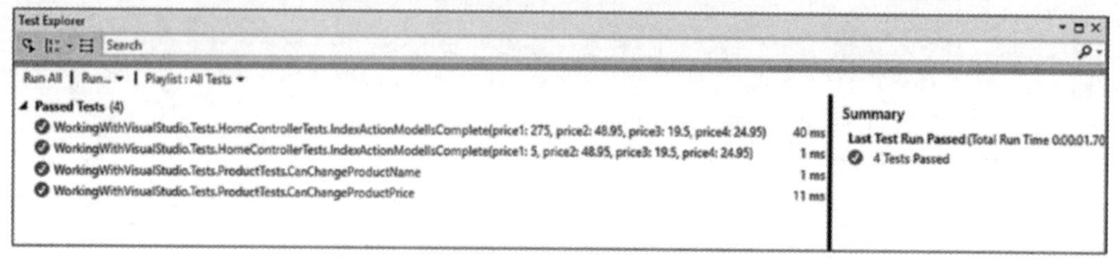

图 7-7 Visual Studio 测试管理器中的参数化测试

从方法或属性中获取测试数据

在特性中表达数据时会受到限制，这限制了 InlineData 特性的有效性，但另一种替代方法是创建静态方法或属性来返回测试需要的对象。在这种情况下，对数据的定义方式没有限制，可以创建更广泛的测试值。为演示如何工作，在单元测试项目中添加名为 ProductTestData.cs 的文件，用它定义代码清单 7-16 所示的类。

代码清单 7-16 WorkingWithVisualStudio.Tests 项目中 ProductTestData.cs 文件的内容

```
using System.Collections;
using System.Collections.Generic;
using WorkingWithVisualStudio.Models;

namespace WorkingWithVisualStudio.Tests {

    public class ProductTestData : IEnumerable<object[]> {

        public IEnumerator<object[]> GetEnumerator() {
            yield return new object[] { GetPricesUnder50() };
            yield return new object[] { GetPricesOver50 };
        }

        IEnumerator IEnumerable.GetEnumerator() {
            return this.GetEnumerator();
        }

        private IEnumerable<Product> GetPricesUnder50() {
            decimal[] prices = new decimal[] { 275, 49.95M, 19.50M, 24.95M };
            for (int i = 0; i < prices.Length; i++) {
                yield return new Product { Name = $"P{i + 1}", Price = prices[i] };
            }
        }

        private Product[] GetPricesOver50 => new Product[] {
            new Product { Name = "P1", Price = 5 },
            new Product { Name = "P2", Price = 48.95M },
            new Product { Name = "P3", Price = 19.50M },
            new Product { Name = "P4", Price = 24.95M }};
    }
}
```

测试数据由实现了 IEnumerable<object[]>接口的类提供，并返回一个对象数组序列。序列中的每个对象数组包含一个将要传递给测试方法的参数集。改进测试方法以便接收 Product 对象数组，Product 对象数组为测试数据添加了另外一层。该层是对象数组的枚举，其中的每个枚举元素都包含一个 Product 对象数组。测试数据的这种深度结构可能令人困惑，但保证正确是最重要的，因为在 Xunit.net 试图传递给测试方法的参数个数不匹配方法签名时，测试将不会工作。

构建测试数据类，以便使用私有方法或属性定义单独的测试数据集，然后通过 GetEnumerator 方法组合到对象数组序列中。为了演示不同的技术，这里使用方法和属性创建了 Product 对象数组，但是还有另一种方法（选择由正在测试的数据类型来驱动）。代码清单 7-17 展示了如何通过 Theory 特性的测试数据类来设置测试。

代码清单 7-17 在 HomeControllerTests.cs 文件中使用测试数据类

```
using Microsoft.AspNetCore.Mvc;
using System.Collections.Generic;
using WorkingWithVisualStudio.Controllers;
using WorkingWithVisualStudio.Models;
using Xunit;

namespace WorkingWithVisualStudio.Tests {
    public class HomeControllerTests {

        class ModelCompleteFakeRepository : IRepository {

            public IEnumerable<Product> Products { get; set; }

            public void AddProduct(Product p) {
                // do nothing - not required for test
            }
        }

        [Theory]
        [ClassData(typeof(ProductTestData))]
        public void IndexActionModelIsComplete(Product[] products ) {
            // Arrange
            var controller = new HomeController();
            controller.Repository = new ModelCompleteFakeRepository {
                Products = products
            };

            // Act
            var model = (controller.Index() as ViewResult)?.ViewData.Model
                as IEnumerable<Product>;

            // Assert
            Assert.Equal(controller.Repository.Products, model,
                Comparer.Get<Product>((p1, p2) => p1.Name == p2.Name
                    && p1.Price == p2.Price));
        }
    }
}
```

提 示

如果希望在与单元测试相同的类中包含测试数据，那么可以使用 MemberData 特性代替 ClassData。MemberData 特性使用字符串来配置将提供 IEnumerable<object[]>的静态方法名称，序列中的每个数组可作为测试方法的参数集。

将特性 ClassData 配置为测试数据类的类型，在本例中是 ProductTestData。在运行测试的时候，Xunit.net

将创建 ProductTestData 类的实例，并用来为测试获取测试数据序列。

注　意

在测试管理器中查看测试代码，你将看到 IndexActionModelIsComplete 测试有单个入口，尽管 ProductTestData 类提供了两个测试数据集。这发生在测试数据对象不能被序列化的情况下，可以通过在测试对象上应用 Serializable 特性来解决。

7.3.2　改进假的实现

隔离组件的有效性需要类的假的实现以提供测试数据，或者检查组件应有的行为。可以创建实现了 IRepository 接口的类，这是一种有效的方式，但会导致为期望运行的每种测试创建实现类型。作为示例，代码清单 7-18 展示了为检查在 Index 操作方法中是否只调用了存储库中 Products 方法一次所需要做的额外工作（当担心组件对存储库进行重复查询时，这种测试十分常见，会导致多个数据库查询）。

代码清单 7-18　为 HomeControllerTests.cs 文件添加单元测试

```
using Microsoft.AspNetCore.Mvc;
using System.Collections.Generic;
using WorkingWithVisualStudio.Controllers;
using WorkingWithVisualStudio.Models;
using Xunit;
using System;

namespace WorkingWithVisualStudio.Tests {
    public class HomeControllerTests {

        class ModelCompleteFakeRepository : IRepository {

            public IEnumerable<Product> Products { get; set; }

            public void AddProduct(Product p) {
                // do nothing - not required for test
            }
        }

        [Theory]
        [ClassData(typeof(ProductTestData))]
        public void IndexActionModelIsComplete(Product[] products ) {
            // Arrange
            var controller = new HomeController();
            controller.Repository = new ModelCompleteFakeRepository {
                Products = products
            };
            // Act
            var model = (controller.Index() as ViewResult)?.ViewData.Model
                as IEnumerable<Product>;

            // Assert
            Assert.Equal(controller.Repository.Products, model,
                Comparer.Get<Product>((p1, p2) => p1.Name == p2.Name
                    && p1.Price == p2.Price));
        }

        class PropertyOnceFakeRepository : IRepository {
            public int PropertyCounter { get; set; } = 0;

            public IEnumerable<Product> Products {
                get {
```

```
                PropertyCounter++;
                return new[] { new Product { Name = "P1", Price = 100 } };
            }
        }

        public void AddProduct(Product p) {
            // do nothing - not required for test
        }
    }

    [Fact]
    public void RepositoryPropertyCalledOnce() {
        // Arrange
        var repo = new PropertyOnceFakeRepository();
        var controller = new HomeController { Repository = repo };

        // Act
        var result = controller.Index();

        // Assert
        Assert.Equal(1, repo.PropertyCounter);
    }
  }
}
```

假的实现并不总是简单的数据源，它们也可以用于断言组件执行工作的方式。在这个示例中，添加了一个简单的计数器属性，它在每次读取存储库的 Products 属性时都会递增，可使用 Assert.Equal 方法来确保属性仅被调用一次。

1. 添加 mocking 框架

创建假的对象会导致失控，最好的方式是使用假的框架，也就是所谓的 mocking 框架（假的和仿冒的对象之间有一点技术区别，但是现在，测试工具为易于使用而模糊了它们之间的区别，所以可以交替使用这两个术语）。本章使用的 mocking 框架名为 Moq，其背景如表 7-4 所示。

表 7-4 Moq 框架的背景

问题	答案
是什么？	Moq 是一个用于在应用程序中创建组件的仿冒实现的软件包
有何用？	Moq 框架使得为单元测试创建仿冒组件以隔离应用程序的部件变得容易
如何使用？	Moq 使用 Lambda 表达式来定义仿冒组件的功能，仅仅需要定义测试可能使用的功能
有何缺陷或限制？	习惯语法可能需要一点努力，请从 GitHub 网站查看文档和示例
有何替代品？	有多种替代框架可用，包括 NSubstitue 和 FakeItEasy。所有这些框架都提供类似的功能，可在它们之间进行选择

要安装 Moq，在 Solution Explorer 窗格中右击 WorkingWithVisualStudio.Tests 项目，从弹出的菜单中选择 Manage NuGet Packages。单击 Browse 按钮，在搜索框中输入 moq。从包列表中选择 Moq，如图 7-8 所示，单击 Install 按钮，在项目中添加包。

注 意

把 Moq 添加到单元测试项目中，而不是应用程序项目中。

2. 创建 mock 对象

创建 mock 对象意味着告诉 Moq 期望何种类型的对象，然后配置其行为，将 mock 对象用于测试的目标。在代码清单 7-19 中，使用 Moq 来替换用于测试 HomeController 的两个假的存储库。

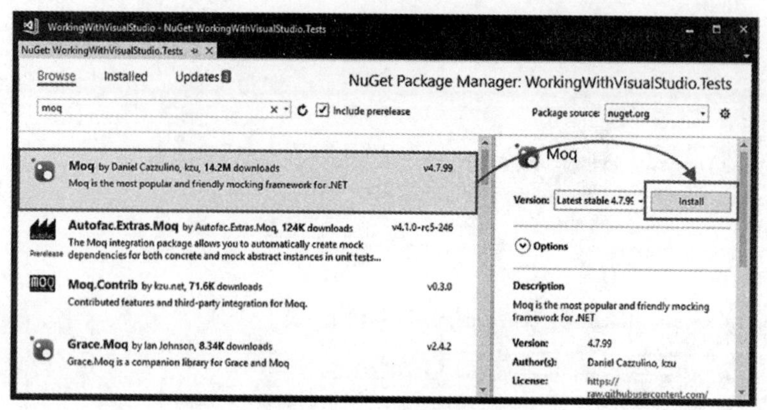

图 7-8　为单元测试项目添加包

代码清单 7-19　在 HomeControllerTests.cs 文件中使用 mock 对象

```
using Microsoft.AspNetCore.Mvc;
using System.Collections.Generic;
using WorkingWithVisualStudio.Controllers;
using WorkingWithVisualStudio.Models;
using Xunit;
using System;
using Moq;

namespace WorkingWithVisualStudio.Tests {
    public class HomeControllerTests {

        [Theory]
        [ClassData(typeof(ProductTestData))]
        public void IndexActionModelIsComplete(Product[] products ) {

            // Arrange
            var mock = new Mock<IRepository>();
            mock.SetupGet(m => m.Products).Returns(products);
            var controller = new HomeController { Repository = mock.Object };

            // Act
            var model = (controller.Index() as ViewResult)?.ViewData.Model
                as IEnumerable<Product>;

            // Assert
            Assert.Equal(controller.Repository.Products, model,
                Comparer.Get<Product>((p1, p2) => p1.Name == p2.Name
                    && p1.Price == p2.Price));
        }

        [Fact]
        public void RepositoryPropertyCalledOnce() {

            // Arrange
            var mock = new Mock<IRepository>();
            mock.SetupGet(m => m.Products)
                .Returns(new[] { new Product { Name = "P1", Price = 100 } });
            var controller = new HomeController { Repository = mock.Object };
            // Act
            var result = controller.Index();

            // Assert
            mock.VerifyGet(m => m.Products, Times.Once);
        }
```

```
        }
    }
```

使用 Moq 框架使你可以删除 IRepository 接口的假的实现，并使用很少几行代码替换它们。本节不会详细介绍 Moq 框架支持的不同功能，但是会在示例中介绍 Moq 框架的使用方式（有关 Moq 框架的示例和文档，参见 GitHub 网站）。本书的其余内容及示例将说明如何测试不同类型的 MVC 组件。

第一步是创建一个新的 mock 对象，指定将要实现的接口，如下所示：

```
...
var mock = new Mock<IRepository>();
...
```

创建的 mock 对象仿冒了 IRepository 接口。下一步是定义测试所需的功能。不像常规实现接口的类，mock 对象仅仅配置测试所需的行为。对于第一个 mock 存储库，需要实现 Product 属性以便返回 Product 对象的集合并通过 ClassData 特性传递给测试方法，如下所示：

```
...
mock.SetupGet(m => m.Products).Returns(products);
...
```

SetupGet 方法用于实现属性的读取器。该方法的参数是一个 Lambda 表达式，用于指定待实现的属性，在示例中是 Products。在 SetupGet 方法的返回结果上调用 Returns 方法，以指定当属性被读取时返回的结果。对第二个 mock 存储库使用同样的方法，但是指定固定值，如下所示：

```
...
mock.SetupGet(m => m.Products)
              .Returns(new[] { new Product { Name = "P1", Price = 100 } });
...
```

Mock 类定义了 Object 属性，用于返回实现了指定接口且带有所定义行为的对象。在这两个单元测试中，使用 Object 属性获取存储库以配置控制器，如下所示：

```
...
var controller = new HomeController { Repository = mock.Object };
...
```

使用的最后一项 Moq 功能是检查 Products 属性是否只被调用了一次，如下所示：

```
...
mock.VerifyGet(m => m.Products, Times.Once);
...
```

VerifyGet 是一个由 Mock 类定义的方法，用于在测试完成后检查 mock 对象的状态。在本例中，VerifyGet 方法允许检查 Products 属性被读取的次数。值 Times.Once 用于指定如果属性不止读取了一次，就抛出异常，这将导致测试失败（通常，测试中的 Assert 方法通过在测试失败时抛出异常来工作，这就是 VerifyGet 方法可以在使用 mock 对象时替换 Assert 方法的原因）。

7.4 小结

本章的大部分内容集中于单元测试，单元测试已经成为改进代码质量的强大工具。单元测试不适合每个开发者，但值得尝试。即使仅用于复杂功能或问题诊断也是有用的。本章介绍了 xUnit.net 框架的使用，说明了测试隔离组件的重要性，演示了简化单元测试代码的一些工具和技术。下一章将介绍如何开发更现实的 MVC 应用程序，以展示不同功能的组件如何协同工作。

第 8 章　SportsStore 应用程序

前几章构建了快速且简单的 MVC 应用程序，描述了 MVC 模式、重要的 C#功能以及优秀的 MVC 开发人员所需要的各种工具。现在是时候组合这些技术以建立简单而实用的电子商务应用程序了。

创建的应用程序名为 SportsStore，它遵循一般在线商店的经典做法。首先，创建在线产品目录（客户可以按类别和页面浏览）、购物车（用户可以添加和删除产品以及结账单（供用户输入邮寄详情）。然后，创建一个管理区域，用来管理目录，包括创建、读取、更新和删除（CRUD）功能，这个区域是受保护的，只有登录的管理员可以进行更改。

本章的目的是让你通过创建一个尽可能真实的例子来了解真正的 MVC 开发。本书只关注 ASP.NET Core MVC，所以简化了与数据库等外部系统的集成，并完全省略了其他部分，如支付处理等。

你可能会发现，建立所需的基础设施部分的过程有些慢，但对 MVC 应用程序的基础部分进行投入是很有好处的，可以实现可维护、可扩展、结构良好的代码，并对单元测试提供良好的支持。

单元测试

本书已经强调过 MVC 中单元测试的易用性，以及单元测试如何成为开发过程中非常重要和有用的部分。在本书的这一部分，你将看到这一点，因为这部分已经包含与关键 MVC 功能相关的单元测试和技术细节。

使用单元测试并不是普遍做法。如果不想进行单元测试，也是可以的。如果你对单元测试不感兴趣，可以跳过这部分，SportsStore 应用程序将正常工作。你不需要进行任何类型的单元测试来获得 ASP.NET Core MVC 的技术优势，当然，支持测试是采用 ASP.NET Core MVC 的关键原因。

SportsStore 应用程序中使用的大多数 MVC 功能在本书后面也会单独介绍。在这里并不复制所有的部分，这里讲的内容对示例应用程序已经足够了，但会给出对应的其他章节位置，以便你获取更详细的信息。

这里将演示构建应用程序所需的每个步骤，以便你可以看到 MVC 功能如何组合在一起。当创建视图时，你应该特别注意。如果不严格按照示例，也许会获得一些奇怪的结果。

8.1　准备开始

如果在阅读本书该部分时，你在自己的计算机上编写 SportsStore 应用程序，则需要安装 Visual Studio，并确保安装 LocalDB（这是持久存储数据所必需的）。如果遵循第 2 章介绍的操作，LocalDB 将自动安装在计算机上。

注　意

如果只想在不重新创建项目的情况下执行项目，可以访问 Apress 网站，上面提供了本书附带的免费源代码，可以从本书的 GitHub 仓库中下载 SportsStore 项目的源代码。当然，你不需要跟着做。作者已尽力使屏幕截图和代码尽可能容易理解。

8.1.1　创建 MVC 项目

这里将使用与前面章节相同的基本方法，从一个空项目开始，并添加所需的所有配置文件和组件。从 Visual Studio 的 File 菜单中选择 New→Project，然后选择 ASP.NET Core Web Application 项目模板，如图 8-1

所示。将项目的名称设置为 SportsStore，然后单击 OK 按钮。

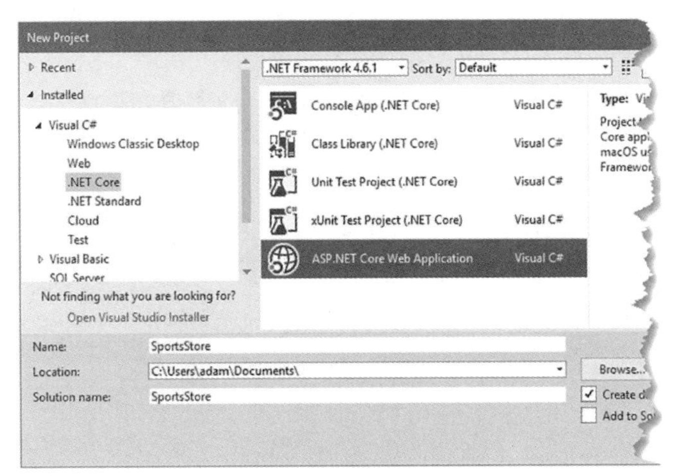

图 8-1　选择项目类型

选择 Empty 模板，如图 8-2 所示。在单击 OK 按钮创建 SportsStore 项目之前，确保在对话框顶部的下拉列表中选择了.NET Core 和 ASP.NET Core 2.0，并且未勾选 Enable Docker Support 复选框。

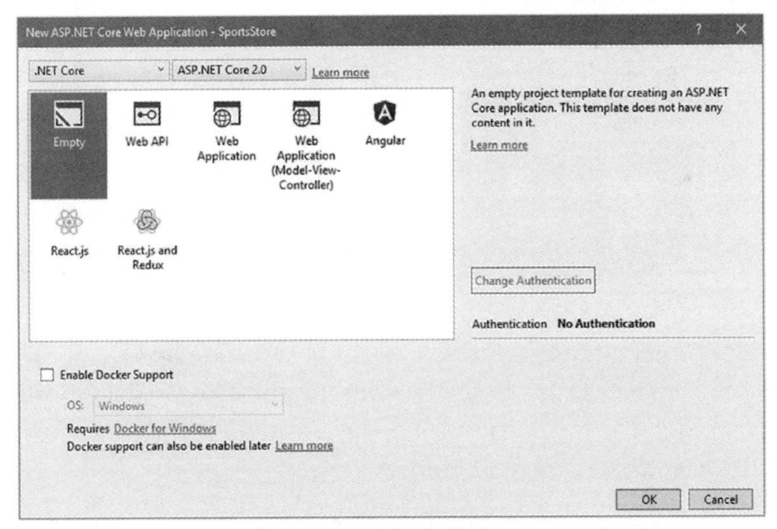

图 8-2　选择项目模板

1. 创建文件夹结构

下一步是添加一些文件夹，以包含 MVC 应用程序所需的应用程序组件——模型、控制器和视图。对于表 8-1 中描述的每个文件夹，右击 Solution Explorer 窗格中的 SportsStore 项目，从弹出的菜单中选择 Add→New Folder，然后设置文件夹的名称。稍后还需要其他文件夹，但这些文件夹反映了 MVC 应用程序的主要部分，并且足以开始使用。

表 8-1　SportsStore 项目所需的文件夹

名称	描述
Models	里面包含模型类
Controllers	里面包含控制器类
Views	里面包含与视图相关的所有内容，包括单个 Razor 文件、视图开始文件和视图导入文件

2. 配置应用

Startup 类负责配置 ASP.NET Core 应用程序。代码清单 8-1 显示了对 Startup 类所做的更改，以启用 MVC 框架和一些对开发有用的相关功能。

代码清单 8-1　在 SportsStore 文件夹下的 Startup.cs 文件中启用功能

```
using System;
using System.Collections.Generic;
using System.Linq;
using System.Threading.Tasks;
using Microsoft.AspNetCore.Builder;
using Microsoft.AspNetCore.Hosting;
using Microsoft.AspNetCore.Http;
using Microsoft.Extensions.DependencyInjection;

namespace SportsStore {

    public class Startup {

        public void ConfigureServices(IServiceCollection services) {
            services.AddMvc();
        }

        public void Configure(IApplicationBuilder app, IHostingEnvironment env) {
            app.UseDeveloperExceptionPage();
            app.UseStatusCodePages();
            app.UseStaticFiles();
            app.UseMvc(routes => {

            });
        }
    }
}
```

注　意

Startup 类是一项重要的 ASP.NET Core 功能，第 14 章会详细描述它。

ConfigureServices 方法用于设置共享对象，可通过依赖注入功能在整个应用程序中使用，第 18 章将详细介绍。在 ConfigureServices 方法中调用的 AddMvc 方法是一种扩展方法，用于设置 MVC 应用程序使用的共享对象。

Configure 方法用于设置接收和处理 HTTP 请求的功能。在 Configure 方法中调用的每种方法都是扩展方法，用于设置 HTTP 请求处理器，如表 8-2 所示。

表 8-2　　　　在 Start 类中调用的初始化方法

方法	描述
UseDeveloperExceptionPage()	显示发生在应用程序中的异常的详细信息，在开发过程中很有帮助。在发布的应用程序中不应该启用，在第 12 章中发布应用程序时会禁用该特性
UseStatusCodePages()	在 HTTP 响应中添加一条简单的信息，否则不会有响应体部分，例如，404 - Not Fund 响应
UseStaticFiles()	启用对 wwwroot 文件夹中静态内容的支持
UseMvc()	启用 ASP.NET Core MVC

接下来，需要为应用程序准备 Razer 视图。右击 Views 文件夹，从弹出菜单中选择 Add→New Item，从 ASP.NET 分类中选择 MVC View Imports Page 项目，如图 8-3 所示。

单击 Add 按钮以创建_ViewImports.cshtml 文件，如代码清单 8-2 所示，设置新文件的内容。

代码清单 8-2　Views 文件夹下的_ViewImports.cshtml 文件的内容

```
@using SportsStore.Models
@addTagHelper *, Microsoft.AspNetCore.Mvc.TagHelpers
```

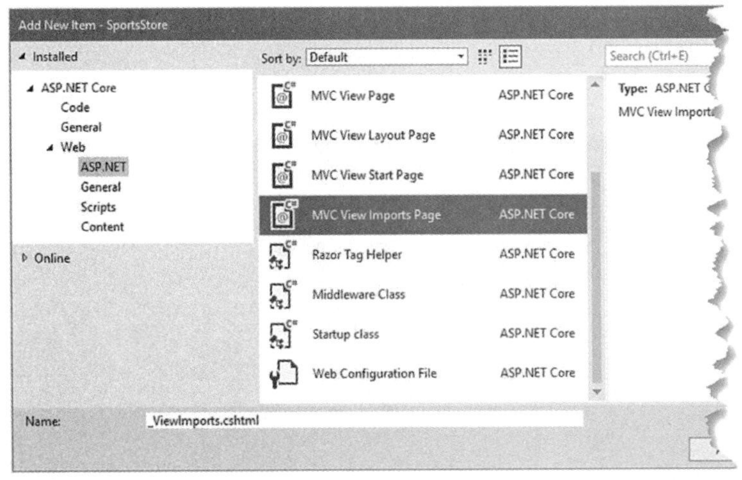

图 8-3 创建视图导入文件

语句@using 允许在视图中使用定义在命名空间 SportsStore.Models 中的类型，而不再需要引用命名空间。而语句@addTagHelper 启用了内置的标签助手特性，随后使用它来创建反映 SportsStore 应用程序的配置的 HTML 元素。

8.1.2 创建单元测试项目

创建单元测试项目的过程与第 7 章描述的相同。右击 Solution Explorer 窗格中的 SportsStore 项，然后从弹出的菜单中选择 Add→New Project。选择 xUnit Test Project（NET Core）项目模板，如图 8-4 所示，并将项目名称设置为 SportsStore.Tests。单击 OK 按钮即可创建单元测试项目。

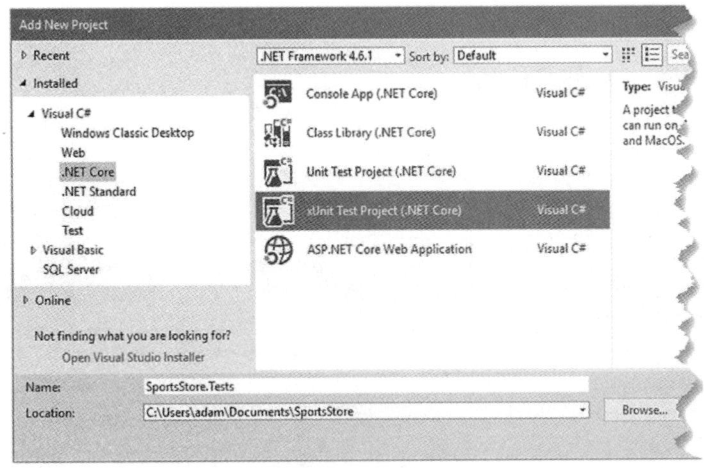

图 8-4 选择项目模板

一旦创建了单元测试项目，就右击解决方案资源管理器中的 SportsStore.Tests 项目并从弹出菜单中选择 Edit SportsStore.Tests.csproj。添加代码清单 8-3 中加粗显示的新元素，以将 Moq 包添加到测试项目中，并创建对 SportsStore 项目的引用。确保为 Moq 包指定了列表中显示的版本。

代码清单 8-3 在 SportsStore.Tests 项目的 SportsStore.Tests.csproj 文件中添加包

```
<Project Sdk="Microsoft.NET.Sdk">

  <PropertyGroup>
    <TargetFramework>netcoreapp2.0</TargetFramework>
```

```
    <IsPackable>false</IsPackable>
  </PropertyGroup>

  <ItemGroup>
    <ProjectReference Include="..\SportsStore\SportsStore.csproj" />
  </ItemGroup>

  <ItemGroup>
    <PackageReference Include="Microsoft.NET.Test.Sdk"
        Version="15.3.0-preview-20170628-02" />
    <PackageReference Include="xunit" Version="2.2.0" />
    <PackageReference Include="xunit.runner.visualstudio" Version="2.2.0" />
    <PackageReference Include="Moq" Version="4.7.99" />
  </ItemGroup>

</Project>
```

在将更改保存到.csproj 文件时，Visual Studio 将下载 Moq 包并将其安装到单元测试项目中，并创建对 SportsStore 项目的引用，以便其中包含的类可用于测试。

8.1.3　测试和启动应用程序

现在已经创建并配置了应用程序和单元测试项目，可以进行开发了。Solution Explorer 窗格应包含图 8-5 所示的项目。如果看到不同的项目或项目不在同一位置，你将会遇到问题，所以请花一点时间检查这些项目是否存在，并已位于正确的位置。

如果从 Debug 菜单中选择 Start Debugging（或者选择 Start Without Debugging），你将看到错误页面，如图 8-6 所示。显示错误消息是因为应用程序中没有控制器来处理当前的请求，作者将很快解决这个问题。

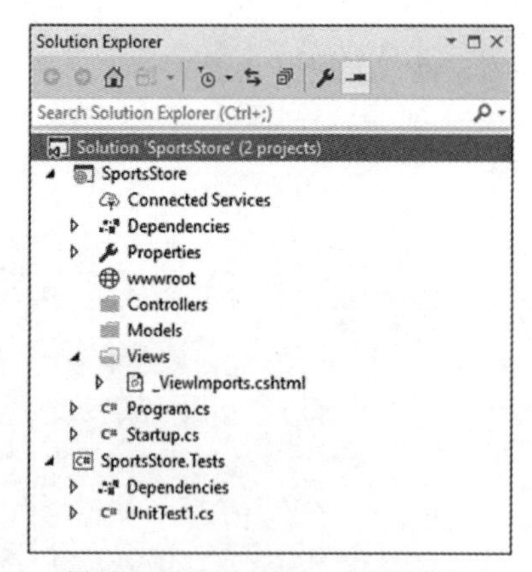

图 8-5　用于 SportsStore 应用程序和单元测试项目的 Solution Explorer 窗格

图 8-6　运行 SportsStore 应用程序后的错误页面

8.2　开始领域模型开发

所有项目都是从 MVC 应用程序的核心——领域模型开始的。这是一个电子商务应用程序，因此最需要的模型就是产品。添加名为 Product.cs 的类文件到 Models 文件夹中，并用它定义代码清单 8-4 所示的类。

代码清单 8-4　Models 文件夹下的 Product.cs 文件的内容

```
namespace SportsStore.Models {

    public class Product {
```

```
        public int ProductID { get; set; }
        public string Name { get; set; }
        public string Description { get; set; }
        public decimal Price { get; set; }
        public string Category { get; set; }
    }
}
```

8.2.1 创建存储库

现在需要一些从数据库获取 Product 对象的方法。正如第 3 章所解释的，模型包括用于从持久化数据库中存储和检索数据的逻辑。此时本书并不关心如何实现数据持久性，但本书将为之定义接口。在 Models 文件夹中添加名为 IProductRepository.cs 的 C#接口文件，并用它定义代码清单 8-5 所示的接口。

代码清单 8-5 Models 文件夹下的 IProductRepository.cs 文件的内容

```
using System.Linq;

namespace SportsStore.Models {

    public interface IProductRepository {

        IQueryable<Product> Products { get; }
    }
}
```

这个接口使用 IQueryable<T>来允许调用者获得一系列 Product 对象。IQueryable<T>接口来自你更熟悉的 IEnumerable<T>接口，表示可以查询的对象集合，例如由数据库管理的对象。

依赖于 IProductRepository 接口的类可以获取 Product 对象，而无须知道它们的存储方式或实现类将会如何传递它们的详细信息。

IEnumerable <T>和 IQueryable <T>接口

IQueryable<T>接口很有用，因为它允许有效地查询对象集合。在本章的后面，将添加对从数据库中检索产品对象子集的支持，并且 IQueryable<T>接口允许使用标准 LINQ 语句向数据库查询需要的对象，而无须知道数据库服务器存储数据或处理查询的方式。如果没有 IQueryable<T>接口，本书将不得不从数据库中检索所有 Product 对象，然后丢弃那些不想要的对象，随着应用程序使用的数据量的增加，代价将变得非常昂贵。正因为如此，数据库存储库接口和类通常使用 IQueryable<T>接口而不是 IEnumerable<T>接口。但是，必须小心使用 IQueryable<T>接口，因为每次枚举对象集合时，都会再次评估查询，这意味着将向数据库发送新的查询。这可能会破坏使用 IQueryable<T>的好处。在这种情况下，可以使用 ToList 或 ToArray 扩展方法将 IQueryable<T>转换为更可预测的形式。

8.2.2 创建虚拟存储库

既然已经定义了一个接口，就可以实现持久化机制并连接到一个数据库，但是这里希望先添加应用程序的其他部分。为此，创建 IProductRepository 接口的虚拟实现，等到再次处理数据存储的时候再真正实现。为了创建虚拟存储库，在 Models 文件夹中添加名为 FakeProductRepository.cs 的文件，并用它定义代码清单 8-6 所示的类。

代码清单 8-6 Models 文件夹下的 FakeProductRepository.cs 文件的内容

```
using System.Collections.Generic;
using System.Linq;

namespace SportsStore.Models {
```

```
    public class FakeProductRepository : IProductRepository {

        public IQueryable<Product> Products => new List<Product> {
            new Product { Name = "Football", Price = 25 },
            new Product { Name = "Surf board", Price = 179 },
            new Product { Name = "Running shoes", Price = 95 }
        }.AsQueryable<Product>();
    }
}
```

FakeProductRepository 类返回固定的 Product 对象集合，并作为 Products 属性的值，以此实现 IProductRepository 接口。实现 IProductRepository 接口时还需要 AsQueryable 方法，用于将固定的对象集合转换为 IQueryable<Product>，并允许创建兼容的虚拟存储库而不必处理真正的查询。

8.2.3　注册存储库服务

MVC 强调使用松耦合组件，这意味着在更改应用程序的某个部分后，无须在其他位置进行相应的更改。这会将应用程序的一部分变成服务，这些服务提供了应用程序其他部分需要使用的功能。这些类提供的服务可以更改或替换，而无须更改使用它们的类。第 18 章将深入解释这一点，但对于 SportsStore 应用程序，作者希望创建存储库服务，以允许控制器获取实现了 IProductRepository 接口的对象，而不需要知道正在使用哪个类。可以使用之前创建的 FakeProductRepository 类开发应用程序，然后替换为真正的存储库，而无须对需要访问存储库的所有类进行更改。服务已在 Startup 类的 ConfigureServices 方法中注册，在代码清单 8-7 中，为存储库定义了新的服务。

代码清单 8-7　在 SportsStore 文件夹下的 Startup.cs 文件中创建存储库服务

```
using System;
using System.Collections.Generic;
using System.Linq;
using System.Threading.Tasks;
using Microsoft.AspNetCore.Builder;
using Microsoft.AspNetCore.Hosting;
using Microsoft.AspNetCore.Http;
using Microsoft.Extensions.DependencyInjection;
using SportsStore.Models;

namespace SportsStore {

    public class Startup {

        public void ConfigureServices(IServiceCollection services) {
            services.AddTransient<IProductRepository, FakeProductRepository>();
            services.AddMvc();
        }

        public void Configure(IApplicationBuilder app, IHostingEnvironment env) {
            app.UseDeveloperExceptionPage();
            app.UseStatusCodePages();
            app.UseStaticFiles();
            app.UseMvc(routes => {

            });
        }
    }
}
```

添加到 ConfigureServices 方法的语句告诉 ASP.NET Core 当组件（如控制器）需要 IProductRepository 接口的实现时，应该会收到一个 FakeProductRepository 对象。AddTransient 方法指定了一个新的 FakeProductRepository 对象，每次需要 IProductRepository 接口时都应创建它。现在这段代码还没有任何意义，不要担心；你很快就会看到如何将之应用于应用程序，第 18 章将详细介绍这些情况。

8.3 显示产品清单

可以在本章的其余部分构建领域模型和存储库，而不触及应用程序的其余部分。不过，你可能会觉得无聊，所以想要切换一下，开始认真使用 MVC，并根据需要添加模型和存储库。

在本节中，将创建一个控制器和一个操作方法，可以在存储库中显示产品的详细信息。目前，这只是为了显示虚拟存储库中的数据，但稍后会对其进行排序。这里还将设置初始路由配置，以便 MVC 知道如何将应用程序的请求映射到创建的控制器。

使用 Visual Studio MVC 脚手架

在本书中，都是通过右击 Solution Explorer 窗格中的文件夹来创建 MVC 控制器和视图，方法是从弹出菜单中选择 Add→New Item，然后在 Add New Item 窗口中选择模板。还有一种替代方案，称为 scaffolding（脚手架），Visual Studio 在 Add 菜单中提供了专门用于创建控制器和视图的菜单项。当选择这些菜单项时，需要为创建的组件选择场景，例如具有读/写操作的控制器或包含用于创建特定模型对象的表单视图。

本书不使用脚手架。脚手架生成的代码和标记是非常通用的，但并不是很有用，因为这些代码和标记只能支持一组特定的场景，不能解决常见的开发问题。本书的目标不仅在于确保你知道如何创建 MVC 应用程序，还需要解释幕后的一切工作。如果将创建组件的责任交给脚手架，解释起来将会更加困难。

也就是说，你的开发风格可能与作者的不同，并且你可能更喜欢在自己的项目中使用脚手架，这将是另一种情况。这是非常合理的，但是建议你花时间了解脚手架的作用，以便在没有得到预期结果的情况下知道去哪里检查。

8.3.1 添加一个控制器

为了在应用程序中创建第一个控制器，将名为 ProductController.cs 的类文件添加到 Controllers 文件夹中，并定义代码清单 8-8 所示的类。

代码清单 8-8 Controller 文件夹下的 ProductController.cs 文件的内容

```
using Microsoft.AspNetCore.Mvc;
using SportsStore.Models;

namespace SportsStore.Controllers {

    public class ProductController : Controller {
        private IProductRepository repository;

        public ProductController(IProductRepository repo) {
            repository = repo;
        }
    }
}
```

当 MVC 需要创建一个新的 ProductController 实例来处理 HTTP 请求时，它将检查构造函数，并看到需要一个实现了 IProductRepository 接口的对象。要确定应该使用什么实现类，MVC 会参考 Startup 类中的配置，得知应该使用 FakeRepository，并且每次都应该创建一个新的实例。MVC 会创建一个新的 FakeRepository 对象，并使用它调用 ProductController 构造函数，以创建用于处理 HTTP 请求的控制器对象。

这称为依赖注入（dependency injection），这种方式允许 ProductController 通过 IProductRepository 接口访问应用程序的存储库，而无须了解配置了哪个实现类。之后，将使用真实存储库替换虚拟存储库，使用依赖注入意味着控制器将继续工作而无须做任何更改。

注　意

一些开发者不喜欢依赖注入，并认为它会使应用程序更加复杂。作者不这样认为，但如果你刚开始使用依赖注入，那么建议你看完第 18 章后，再来决定是否使用。

接下来，添加一个名为 List 的操作方法，它将呈现一个视图，以显示存储库中产品的完整列表，如代码清单 8-9 所示。

代码清单 8-9　在 Controller 文件夹下的 ProductController.cs 文件中添加一个操作方法

```
using Microsoft.AspNetCore.Mvc;
using SportsStore.Models;

namespace SportsStore.Controllers {

    public class ProductController : Controller {
        private IProductRepository repository;

        public ProductController(IProductRepository repo) {
            repository = repo;
        }

        public ViewResult List() => View(repository.Products);
    }
}
```

像这样调用 View 方法（不指定视图名称）会告诉 MVC 渲染操作方法的默认视图。将 List<Product>（Product 对象列表）传递给 View 方法，从而为框架提供在强类型视图中填充 Model 对象的数据。

8.3.2　添加并配置视图

需要创建视图以呈现用户的内容，还需要执行一些准备步骤来让视图更简单。首先，创建共享布局，定义将被包含在发送给客户端的所有 HTML 响应中的通用内容。共享布局是确保视图一致并包含重要的 JavaScript 文件和 CSS 样式表的有效方法，第 5 章解释了它们的工作原理。

创建 Views/Shared 文件夹，并添加名为_Layout.cshtml 的 MVC 视图布局页面，_Layout.cshtml 是 Visual Studio 分配给这类项目的默认名称。代码清单 8-10 显示了_Layout.cshtml 文件的内容，其中对默认内容进行了更改，将 title 元素的内容设置为 SportsStore。

代码清单 8-10　Views/Shared 文件夹下的_Layout.cshtml 文件的内容

```
<!DOCTYPE html>

<html>
<head>
    <meta name="viewport" content="width=device-width" />
    <title>SportsStore</title>
</head>
<body>
    <div>
        @RenderBody()
    </div>
</body>
</html>
```

接下来，需要配置应用程序，以便默认应用_Layout.cshtml 文件，这可通过将名为_ViewStart.cshtml 的 MVC 视图启动文件添加到 Views 文件夹来完成。Visual Studio 添加的默认内容如代码清单 8-11 所示，选择名为_Layout.cshtml 的布局，该布局对应代码清单 8-11 所示的文件。

代码清单 8-11 Views 文件夹下的_ViewStart.cshtml 文件的内容

```
@{
    Layout = "_Layout";
}
```

现在需要添加一个视图，当名为 List 的操作方法用于处理请求时将显示这个视图。创建 Views/Product 文件夹，并添加名为 List.cshtml 的 Razor 视图文件，然后添加代码清单 8-12 所示的标记。

代码清单 8-12 Views/Product 文件夹下的 List.cshtml 文件的内容

```
@model IEnumerable<Product>

@foreach (var p in Model) {
    <div>
        <h3>@p.Name</h3>
        @p.Description
        <h4>@p.Price.ToString("c")</h4>
    </div>
}
```

文件顶部的@model 表达式指定视图将从操作方法中接收一系列 Product 对象作为模型数据。使用@foreach 表达式来处理序列，并为接收的每个 Product 对象生成一组简单的 HTML 元素。

视图不知道 Product 对象来自哪里，如何获取，以及它们是否代表应用程序已知的所有产品。相反，视图仅涉及使用 HTML 元素显示每个产品的详细信息，这与第 3 章描述的关注点分离是一致的。

提 示

可使用.ToString（"c"）方法将 Price 属性转换为字符串，该方法会根据服务器上有效的 culture 设置将数值作为货币进行转换。例如，如果服务器设置为 en-US，（1002.3）.ToString（"c"）将返回$1,002.30；但如果服务器设置为 en-GB，那么相同的方法将返回£1,002.30。

8.3.3 设置默认路由

需要告诉 MVC 它应该将访问应用程序的根 URL（****://mysite/）的请求发送到 ProductController 类的名为 List 的操作方法。可通过编辑 Startup 类中的语句来设置处理 HTTP 请求的 MVC 类，如代码清单 8-13 所示。

代码清单 8-13 更改 SportStore 文件夹下的 Startup.cs 文件中的默认路由

```
using System;
using System.Collections.Generic;
using System.Linq;
using System.Threading.Tasks;
using Microsoft.AspNetCore.Builder;
using Microsoft.AspNetCore.Hosting;
using Microsoft.AspNetCore.Http;
using Microsoft.Extensions.DependencyInjection;
using SportsStore.Models;

namespace SportsStore {

    public class Startup {

        public void ConfigureServices(IServiceCollection services) {
            services.AddTransient<IProductRepository, FakeProductRepository>();
            services.AddMvc();
        }

        public void Configure(IApplicationBuilder app, IHostingEnvironment env) {
            app.UseDeveloperExceptionPage();
```

```
            app.UseStatusCodePages();
            app.UseStaticFiles();
            app.UseMvc(routes => {
                routes.MapRoute(
                    name: "default",
                    template: "{controller=Product}/{action=List}/{id?}");
            });
        }
    }
}
```

Startup 类的 Configure 方法用于设置请求管道，由检查 HTTP 请求并生成响应的类（称为中间件）组成。UseMvc 方法用于设置 MVC 中间件，其中一个配置选项用于将 URL 映射到控制器和操作方法。第 15 章和第 16 章将详细描述路由系统，但代码清单 8-13 告诉 MVC 将请求发送到 Product 控制器的名为 List 的操作方法，除非请求 URL 另有指定。

提　示

需要将代码清单 8-13 中的控制器名称设置为 Product 而不是控制器类的名称 ProductController，这是 MVC 命名约定的一部分，其中控制器类的名称通常以 Controller 结尾，但是在引用类时会忽略这部分名称。第 31 章将解释命名约定及其作用。

8.3.4　运行应用程序

现在所有基础工作已准备就绪。创建了一个控制器，其中包含一个操作方法，当请求应用程序的默认 URL 时，MVC 将使用这个操作方法。MVC 将创建一个 FakeRepository 实例，并使用它创建一个新的控制器对象来处理请求。虚拟存储库将为控制器提供一些简单的测试数据，将操作方法传递给 Razor 视图，以便在发送给浏览器的 HTML 响应中包含每个产品的详细信息。当生成 HTML 响应时，MVC 将来自操作方法所选视图的数据与来自共享布局的内容相结合，生成浏览器可以解析和显示的完整 HTML 文档。可以通过启动应用程序来查看结果，如图 8-7 所示。

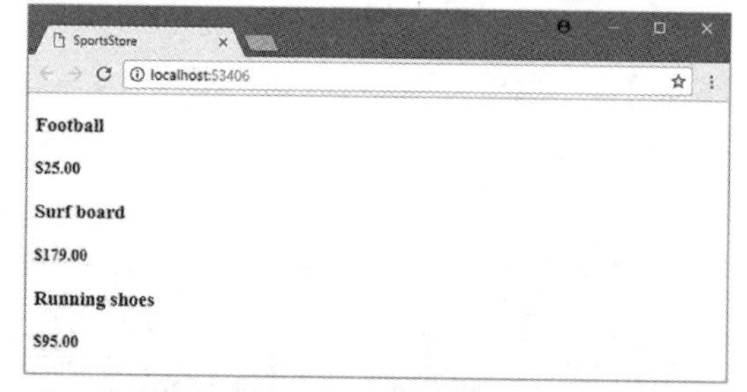

图 8-7　查看基本的应用程序功能

这是 ASP.NET Core MVC 的典型开发模式。虽然需要花费一点时间才能完成所有的初始设置，但这是必要的，这样应用程序的基本功能才能快速集成在一起。

8.4　准备数据库

可以显示包含产品详细信息的简单视图，但使用虚拟存储库提供的测试数据。在使用真实数据实现真正的存储库之前，需要建立数据库并用一些数据填充。

这里将使用 SQL Server 作为数据库，并且使用 Entity Framework Core（EF Core）（Microsoft.NET 对象关系映射（ORM）框架）访问数据库。ORM 框架通过常规 C#对象呈现关系数据库的表、列和行。

注　意

可以从各种工具和技术中进行选择。不仅可以使用不同的关系数据库，还可以使用对象存储库、文档存储和其他一些更复杂的备选方案。甚至可以使用其他的.NET ORM 框架，每个框架都采用稍微不同的实现方式；这些变化可能更适合你的项目。

这里选择 Entity Framework Core 有几个原因：使用起来简单，与 LINQ 的集成非常好，并且能够与 ASP.NET Core MVC 很好地协作。早期的版本多少有点问题，但是现在的版本非常优雅，而且功能丰富。

SQL Server 有一个很好的功能——LocalDB，它是专为开发人员设计的，具有基本的 SQL Server 功能，并且不需要管理功能。有了 LocalDB，就可以在构建项目时跳过设置数据库的步骤，稍后再部署到完整的 SQL Server 实例。大多数 MVC 应用程序被部署到由专业管理人员管理的托管环境中，因此 LocalDB 功能意味着数据库配置可以留给 DBA 来完成，开发人员可以继续进行编码。

提 示

如果在安装 Visual Studio 时没有选择 LocalDB，那么现在需要安装它，可以通过 Visual Studio 安装程序的 Individual Components 部分进行选择。如果按照第 2 章中的说明执行操作，那么 LocalDB 功能应该已经安装好并可以使用了。

8.4.1 安装 Entity Framework Core 工具包

当使用 Visual Studio 创建项目时，默认已将主要的 Entity Framework Core 功能添加到项目中。你还需要一个额外的 NuGet 包来提供命令行工具，这些工具用于创建一些类，从而准备数据库并存储应用程序的数据（称为迁移）。

要将包添加到项目中，请右击 Solution Explorer 窗格中的 SportsStore 项，从弹出菜单中选择 Edit SportsStore.csproj，然后根据代码清单 8-14 所示的文件进行更改。请注意，使用列表中指定的版本，并使用 DotNetCliToolReference 元素添加包，而不是使用 PackageReference 元素。

代码清单 8-14 在 SportsStore 文件夹下的 SportsStore.csproj 文件中添加包

```
<Project Sdk="Microsoft.NET.Sdk.Web">

  <PropertyGroup>
    <TargetFramework>netcoreapp2.0</TargetFramework>
  </PropertyGroup>

  <ItemGroup>
    <Folder Include="wwwroot\" />
  </ItemGroup>

  <ItemGroup>
    <PackageReference Include="Microsoft.AspNetCore.All" Version="2.0.0" />
    <DotNetCliToolReference Include="Microsoft.EntityFrameworkCore.Tools.DotNet"
        Version="2.0.0" />
  </ItemGroup>

</Project>
```

注 意

必须通过编辑文件来安装这个包。这种类型的包无法使用 NuGet Package Manager 或 dotnet 命令行工具进行添加。

当保存文件时，Visual Studio 将下载并安装 Entity Framework Core 命令行工具，然后添加到项目中。

8.4.2 创建数据库类

在 Models 文件夹中添加名为 ApplicationDbContext.cs 的类文件，并定义代码清单 8-15 所示的类。数据库上下文类（database context class）是应用程序和 Entity Framework Core 之间的桥梁，它使用模型对象提供对应用程序数据的访问。为了创建 SportsStore 应用程序的数据库上下文类，在 Models 文件夹中添加名为 ApplicationDbContext.cs 的类文件，并定义代码清单 8-15 所示的类。

代码清单 8-15 Models 文件夹下的 ApplicationDbContext.cs 文件的内容

```
using Microsoft.EntityFrameworkCore;
using Microsoft.EntityFrameworkCore.Design;
using Microsoft.Extensions.DependencyInjection;

namespace SportsStore.Models {

    public class ApplicationDbContext : DbContext {

        public ApplicationDbContext(DbContextOptions<ApplicationDbContext> options)
            : base(options) { }

        public DbSet<Product> Products { get; set; }
    }
}
```

DbContext 基类提供对 Entity Framework Core 的底层功能的访问，Products 属性将提供对数据库中 Product 对象的访问。ApplicationDbContext 类从 DbContext 派生，并添加了一些属性，用于读取和写入应用程序数据。目前只有一个属性，它将提供对 Product 对象的访问。

8.4.3 创建存储库类

目前建立数据库所需的大部分工作已完成。下一步是创建一个实现了 IProductRepository 接口并使用 Entity Framework Core 获取数据的类。在 Models 文件夹中添加名为 EFProductRepository.cs 的类文件，并定义代码清单 8-16 所示的存储库类。

代码清单 8-16 Models 文件夹下的 EFProductRepository.cs 文件的内容

```
using System.Collections.Generic;
using System.Linq;

namespace SportsStore.Models {

    public class EFProductRepository : IProductRepository {
        private ApplicationDbContext context;
        public EFProductRepository(ApplicationDbContext ctx) {
            context = ctx;
        }

        public IQueryable<Product> Products => context.Products;
    }
}
```

这里将向应用程序添加其他功能，但目前，存储库实现只是将由 IProductRepository 接口定义的 Products 属性映射到由 ApplicationDbContext 类定义的 Products 属性。上下文类中的 Products 属性会返回一个 DbSet <Product>对象，该对象实现了 IQueryable<T>接口，并且在使用 Entity Framework Core 时可以轻松实现 IProductRepository 接口。这可以确保对数据库的查询将只检索所需的对象，如本章前面所述。

8.4.4 定义连接字符串

连接字符串（connection string）用来指定数据库的位置和名称，并提供应用程序如何连接到数据库服务器的配置设置。连接字符串存储在名为 appsettings.json 的 JSON 文件中，在 SportsStore 项目中，可使用 Add New Item 窗口的 General 部分的 ASP.NET Configuration File 模板创建该文件。

Visual Studio 在创建文件时会向 appsettings.json 文件添加占位符连接字符串，可在代码清单 8-17 中进行替换。

代码清单 8-17 编辑 SportsStore 文件夹下的 appsettings.json 文件中的连接字符串

```
{
  "Data": {
    "SportStoreProducts": {
      "ConnectionString": "Server=(localdb)\\MSSQLLocalDB;Database=SportsStore;Trusted_Conne
      ction=True;MultipleActiveResultSets=true"
    }
  }
}
```

提 示

连接字符串必须表示为完整的行，这在 Visual Studio 编辑器中是没问题的，但不适合图书版面，因而代码清单 8-17 中的格式比较奇怪。当在自己的项目中定义连接字符串时，请确保 ConnectionString 的值位于同一行。

在配置文件的 Data 部分，已将连接字符串的名称设置为 SportsStoreProducts。ConnectionString 的值表示应将 LocalDB 功能用于名为 SportsStore 的数据库。

8.4.5 配置应用程序

接下来的步骤是读取连接字符串，并配置应用程序以连接到数据库。代码清单 8-18 显示了对 Startup 类所做的更改，以接收 appsettings.json 文件中包含的配置数据的详细信息，并用来配置 Entity Framework Core（读取 JSON 文件的工作由 Program 类处理，这将在第 14 章中介绍）。

代码清单 8-18 在 SportsStore 文件夹下的 Startup.cs 文件中配置应用程序

```
using System;
using System.Collections.Generic;
using System.Linq;
using System.Threading.Tasks;
using Microsoft.AspNetCore.Builder;
using Microsoft.AspNetCore.Hosting;
using Microsoft.AspNetCore.Http;
using Microsoft.Extensions.DependencyInjection;
using SportsStore.Models;
using Microsoft.Extensions.Configuration;
using Microsoft.EntityFrameworkCore;

namespace SportsStore {

    public class Startup {

        public Startup(IConfiguration configuration) =>
            Configuration = configuration;

        public IConfiguration Configuration { get; }

        public void ConfigureServices(IServiceCollection services) {
            services.AddDbContext<ApplicationDbContext>(options =>
                options.UseSqlServer(
                    Configuration["Data:SportStoreProducts:ConnectionString"]));
            services.AddTransient<IProductRepository, EFProductRepository>();
            services.AddMvc();
        }

        public void Configure(IApplicationBuilder app, IHostingEnvironment env) {
            app.UseDeveloperExceptionPage();
```

```
            app.UseStatusCodePages();
            app.UseStaticFiles();
            app.UseMvc(routes => {
                routes.MapRoute(
                    name: "default",
                    template: "{controller=Product}/{action=List}/{id?}");
            });
        }
    }
}
```

添加到 Startup 类的构造函数会接收从 appsettings.json 文件加载的配置数据，该文件可通过实现了 IConfiguration 接口的对象来呈现。Startup 构造函数将 IConfiguration 对象分配给名为 Configuration 的属性，以便 Startup 类的其余部分可以使用。

第 14 章将解释如何读取和访问配置数据。对于 SportsStore 应用程序，添加一系列方法调用，用于在 ConfigureServices 方法中设置 Entity Framework Core。

```
...
services.AddDbContext<ApplicationDbContext>(options =>
    options.UseSqlServer(Configuration["Data:SportStoreProducts:ConnectionString"]));
...
```

AddDbContext 扩展方法用于为代码清单 8-15 中创建的数据库上下文类设置由 Entity Framework Core 提供的服务。正如在第 14 章中解释的那样，在 Startup 类中使用的许多方法允许使用选项参数来配置服务和中间件功能。AddDbContext 方法的参数是一个 Lambda 表达式，它接收为上下文类配置数据库的可选对象。在这种情况下，使用 UseSqlServer 方法配置数据库，并指定从 Configuration 属性获取的连接字符串。

在 Startup 类中做出的下一个更改是用真实存储库替换虚拟存储库，如下所示：

```
...
services.AddTransient<IProductRepository, EFProductRepository>();
...
```

使用了 IProductRepository 接口的应用程序中的组件（目前只是 Product 控制器）将在创建时接收 EFProductRepository 对象，从而使它们可以访问数据库中的数据。第 18 章将详细解释这是如何工作的，结果是模拟数据将被数据库中的真实数据无缝替换，而无须更改 ProductController 类。

禁用范围验证

使用 Entity Framework Core 需要对依赖注入功能的配置进行更改，具体方法将在第 18 章中介绍。在将控制权交给 Startup 类之前，Program 类负责启动和配置 ASP.NET Core，代码清单 8-19 显示了需要执行的更改。如果不做更改，那么在尝试创建数据库架构时将引发异常。

代码清单 8-19 在 SportsStore 文件夹下的 Program.cs 文件中准备使用 Entity Framework Core

```
using System;
using System.Collections.Generic;
using System.IO;
using System.Linq;
using System.Threading.Tasks;
using Microsoft.AspNetCore;
using Microsoft.AspNetCore.Hosting;
using Microsoft.Extensions.Configuration;
using Microsoft.Extensions.Logging;
namespace SportsStore {
    public class Program {
        public static void Main(string[] args) {
            BuildWebHost(args).Run();
        }

        public static IWebHost BuildWebHost(string[] args) =>
```

```
            WebHost.CreateDefaultBuilder(args)
                .UseStartup<Startup>()
                .UseDefaultServiceProvider(options =>
                    options.ValidateScopes = false)
                .Build();
        }
    }
```

第 14 章将解释如何详细配置 ASP.NET Core，但这是 SportsStore 应用程序所需的对 Program 类要做的唯一更改。

8.4.6　创建数据库迁移

Entity Framework Core 能够通过称为迁移（migration）的功能，使用模型类生成数据库架构。准备迁移时，Entity Framework Core 会创建一个包含准备数据库所需的 SQL 命令的 C#类。如果需要修改模型类，则可以创建一个新的迁移，其中包含反映更改所需的 SQL 命令。通过这种方式，不必担心手动编写和测试 SQL 命令，而只需关注应用程序中的 C#模型类。

Entity Framework Core 命令从命令行执行。打开一个新的命令提示符或 PowerShell 窗口，导航到 SportsStore 项目文件夹（包含 Startup.cs 和 appsettings.json 文件的文件夹），然后运行以下命令来创建迁移类，以准备第一次使用数据库：

```
dotnet ef migrations add Initial
```

执行完以上命令后，你将在 Visual Studio 的 Solution Explorer 窗格中看到 Migrations 文件夹，这是 Entity Framework Core 存储其迁移类的地方。其中一个文件名将是一个时间戳，后跟_Initial.cs，这是用于为数据库创建初始架构的类。如果查看此文件的内容，可以看到如何使用 Product 模型类来创建数据库架构。

Add-Migration 和 Update-Database 命令

对于经验丰富的 Entity Framework 开发者，则可能习惯使用 Add-Migration 命令来创建数据库迁移，并使用 Update-Database 命令将其应用于数据库。

随着.NET Core 的推出，Entity Framework Core 添加了集成到 dotnet 命令行工具的命令，可通过添加到代码清单 8-14 中的 Microsoft.EntityFrameworkCore.Tools.DotNet 软件包来使用。本章将使用这些命令，因为它们与其他.NET 命令一致，可以在任何命令提示符或 PowerShell 窗口中使用，而不像 Add-Migration 和 Update-Database 命令，它们仅在特定 Visual Studio 窗口中使用。

8.4.7　创建种子数据

为了填充数据库并提供一些示例数据，在 Models 文件夹中添加名为 SeedData.cs 的类文件，并定义代码清单 8-20 所示的类。

代码清单 8-20　Models 文件夹下的 SeedData.cs 文件的内容

```
using System.Linq;
using Microsoft.AspNetCore.Builder;
using Microsoft.Extensions.DependencyInjection;
using Microsoft.EntityFrameworkCore;

namespace SportsStore.Models {

    public static class SeedData {

        public static void EnsurePopulated(IApplicationBuilder app) {
            ApplicationDbContext context = app.ApplicationServices
                .GetRequiredService<ApplicationDbContext>();
            context.Database.Migrate();
            if (!context.Products.Any()) {
```

```
                context.Products.AddRange(
                    new Product {
                        Name = "Kayak", Description = "A boat for one person",
                        Category = "Watersports", Price = 275 },
                    new Product {
                        Name = "Lifejacket",
                        Description = "Protective and fashionable",
                        Category = "Watersports", Price = 48.95m },
                    new Product {
                        Name = "Soccer Ball",
                        Description = "FIFA-approved size and weight",
                        Category = "Soccer", Price = 19.50m },
                    new Product {
                        Name = "Corner Flags",
                        Description = "Give your playing field a professional touch",
                        Category = "Soccer", Price = 34.95m },
                    new Product {
                        Name = "Stadium",
                        Description = "Flat-packed 35,000-seat stadium",
                        Category = "Soccer", Price = 79500 },
                    new Product {
                        Name = "Thinking Cap",
                        Description = "Improve brain efficiency by 75%",
                        Category = "Chess", Price = 16 },
                    new Product {
                        Name = "Unsteady Chair",
                        Description = "Secretly give your opponent a disadvantage",
                        Category = "Chess", Price = 29.95m },
                    new Product {
                        Name = "Human Chess Board",
                        Description = "A fun game for the family",
                        Category = "Chess", Price = 75 },
                    new Product {
                        Name = "Bling-Bling King",
                        Description = "Gold-plated, diamond-studded King",
                        Category = "Chess", Price = 1200
                    }
                );
                context.SaveChanges();
            }
        }
    }
}
```

静态的 EnsurePopulated 方法会接收 IApplicationBuilder 参数，该参数是 Startup 类的 Configure 方法中使用的接口，用来注册中间件以处理 HTTP 请求，这是确保数据库具有内容的地方。

EnsurePopulated 方法通过 IApplicationBuilder 接口获取 ApplicationDbContext 对象，并调用 Database.Migrate 方法以确保已应用迁移，这意味着将创建和准备数据库，以便可以存储 Product 对象。接下来，检查数据库中 Product 对象的数量。如果数据库中没有对象，则使用 AddRange 方法，用一组 Product 对象填充数据库，然后使用 SaveChanges 方法将它们写入数据库。

最后的更改是在应用程序启动时对数据库填充种子数据，可通过从 Startup 类中调用 EnsurePopulated 方法来完成，如代码清单 8-21 所示。

代码清单 8-21 在 SportsStore 文件夹下的 Startup.cs 文件中对数据库填充种子数据

```
using System;
using System.Collections.Generic;
using System.Linq;
using System.Threading.Tasks;
using Microsoft.AspNetCore.Builder;
```

```
using Microsoft.AspNetCore.Hosting;
using Microsoft.AspNetCore.Http;
using Microsoft.Extensions.DependencyInjection;
using SportsStore.Models;
using Microsoft.Extensions.Configuration;
using Microsoft.EntityFrameworkCore;

namespace SportsStore {

    public class Startup {

        public Startup(IConfiguration configuration) =>
            Configuration = configuration;

        public IConfiguration Configuration { get; }
        public void ConfigureServices(IServiceCollection services) {
            services.AddDbContext<ApplicationDbContext>(options =>
                options.UseSqlServer(
                    Configuration["Data:SportStoreProducts:ConnectionString"]));
            services.AddTransient<IProductRepository, EFProductRepository>();
            services.AddMvc();
        }

        public void Configure(IApplicationBuilder app, IHostingEnvironment env) {
            app.UseDeveloperExceptionPage();
            app.UseStatusCodePages();
            app.UseStaticFiles();
            app.UseMvc(routes => {
                routes.MapRoute(
                    name: "default",
                    template: "{controller=Product}/{action=List}/{id?}");
            });
            SeedData.EnsurePopulated(app);
        }
    }
}
```

启动应用程序，创建数据库并为其填充种子数据，用于向应用程序提供数据（请耐心等待，可能需要一些时间才能创建数据库）。

当浏览器请求应用程序的默认 URL 时，应用程序配置告知 MVC 需要创建一个 Product 控制器来处理请求。创建新的 Product 控制器意味着调用 ProductController 构造函数，该构造函数需要一个实现了 IProductRepository 接口的对象，新的配置告诉 MVC 应该为此创建并使用 EFProductRepository 对象。EFProductRepository 对象利用 Entity Framework Core 功能，从 SQL Server 加载数据并将其转换为 Product 对象。所有这些对 ProductController 类都是隐藏的，该类只接收一个实现了 IProductRepository 接口的对象，并使用其提供的数据。结果是在浏览器中显示数据库中的样本数据，如图 8-8 所示。

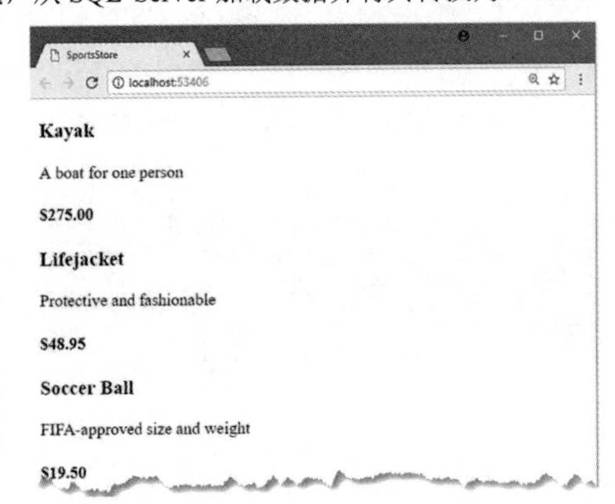

图 8-8　数据库中的样本数据

这种让 Entity Framework Core 将 SQL Server 数据库呈现为一系列模型对象的方法非常简单且易于使用，使你能够将注意力集中在 ASP.NET Core MVC 上，从而跳过 Entity Framework Core 运行的大量细节以及可用的大量配置选项。作者非常喜欢 Entity Framework Core，建议你花一些时间详细了解它。一个很好的起点是微软的 Entity Framework Core 网站。

8.5 添加分页

从图 8-8 可以看出，List.cshtml 视图文件在单个页面上显示了数据库中的产品。在本节中，将添加对分页的支持，以便视图在页面上显示较少数量的产品，用户可以翻页查看全部目录。为此，在 Product 控制器的 List 方法中添加一个参数，如代码清单 8-22 所示。

代码清单 8-22 在 Controller 文件夹下，为 ProductController.cs 文件中名为 List 的操作方法添加分页支持

```
using Microsoft.AspNetCore.Mvc;
using SportsStore.Models;
using System.Linq;
namespace SportsStore.Controllers {

    public class ProductController : Controller {
        private IProductRepository repository;
        public int PageSize = 4;

        public ProductController(IProductRepository repo) {
            repository = repo;
        }

        public ViewResult List(int productPage = 1)
            => View(repository.Products
                .OrderBy(p => p.ProductID)
                .Skip((productPage - 1) * PageSize)
                .Take(PageSize));
    }
}
```

PageSize 字段指定每页需要 4 个产品。这里向 List 方法添加了一个可选参数，从而意味着如果调用没有参数的方法（List()），调用将被视为已经提供了参数定义中指定的值（List（1））。结果是，当 MVC 调用不带参数的 List 方法时，将显示产品的第一页。在操作方法的主体中，得到了 Product 对象，可通过主键对它们进行排序，跳过当前页面之前出现的产品，并获取 PageSize 字段指定的产品数量。

单元测试——分页

可以通过创建模拟的存储库，再注入 ProductController 类的构造函数，并调用 List 方法来请求特定的页面来对分页功能进行单元测试。然后，可以对得到的 Product 对象与期望从模拟实现的测试数据中得到的结果进行比较。有关如何设置单元测试的详细信息，请参阅第 7 章。这里是为此创建的单元测试，已添加到 SportsStore 项目的名为 ProductControllerTests.cs 的类文件中。

```
using System.Collections.Generic;
using System.Linq;
using Moq;
using SportsStore.Controllers;
using SportsStore.Models;
using Xunit;

namespace SportsStore.Tests {

    public class ProductControllerTests {

        [Fact]
        public void Can_Paginate() {
```

```
            // Arrange
            Mock<IProductRepository> mock = new Mock<IProductRepository>();
            mock.Setup(m => m.Products).Returns((new Product[] {
                new Product {ProductID = 1, Name = "P1"},
                new Product {ProductID = 2, Name = "P2"},
                new Product {ProductID = 3, Name = "P3"},
                new Product {ProductID = 4, Name = "P4"},
                new Product {ProductID = 5, Name = "P5"}
            }).AsQueryable<Product>());

            ProductController controller = new ProductController(mock.Object);
            controller.PageSize = 3;

            // Act
            IEnumerable<Product> result =
                controller.List(2).ViewData.Model as IEnumerable<Product>;

            // Assert
            Product[] prodArray = result.ToArray();
            Assert.True(prodArray.Length == 2);
            Assert.Equal("P4", prodArray[0].Name);
            Assert.Equal("P5", prodArray[1].Name);
        }
    }
}
```

从操作方法返回的数据不太好处理。结果是一个 ViewResult 对象，必须将 ViewData.Model 属性的值转换为预期的数据类型。第 17 章将解释由操作方法返回的不同结果类型以及如何使用它们。

8.5.1　显示页面链接

如果运行应用程序，你将看到页面上显示了 4 个条目。如果要查看其他页面，可以将查询字符串参数附加到 URL 的末尾，如下所示：

```
http://localhost:5000/?productPage=2
```

你需要更改 URL 的端口部分，以匹配项目使用的端口。使用这些查询字符串，可以浏览产品目录。

客户无法确定这些查询字符串参数是否存在，即使它们存在，也不会以这种方式导航。相反，需要在每个产品列表的底部呈现一些页面链接，以便客户可以在页面之间导航。为了做到这一点，需要实现一个标签助手（tag helper），它可以为所需链接生成 HTML 标记。

1. 添加视图模型

为了使用标签助手，需要向视图传递可用页面的数量、当前页面以及存储库中产品总数的信息。最简单的方法是创建视图模型类，专门用于在控制器和视图之间传递数据。在 SportsStore 项目中创建 Models/ViewModels 文件夹，并添加名为 PagingInfo.cs 的类文件，如代码清单 8-23 所示。

代码清单 8-23　Models/ViewModels 文件夹下的 PagingInfo.cs 文件的内容

```
using System;

namespace SportsStore.Models.ViewModels {

    public class PagingInfo {
        public int TotalItems { get; set; }
        public int ItemsPerPage { get; set; }
        public int CurrentPage { get; set; }

        public int TotalPages =>
            (int)Math.Ceiling((decimal)TotalItems / ItemsPerPage);
```

```
    }
}
```

2. 添加标签助手类

现在已经有了视图模型，接下来可以创建标签助手类。在 SportsStore 项目中创建 Infrastructure 文件夹，并添加名为 PageLinkTagHelper.cs 的类文件，该文件用于定义代码清单 8-24 所示的类。标签助手是 ASP.NET Core MVC 的重要组成部分，第 23～25 章将解释它们的工作原理以及如何创建它们。

代码清单 8-24　Infrastructure 文件夹下的 PageLinkTagHelper.cs 文件的内容

```
using Microsoft.AspNetCore.Mvc;
using Microsoft.AspNetCore.Mvc.Rendering;
using Microsoft.AspNetCore.Mvc.Routing;
using Microsoft.AspNetCore.Mvc.ViewFeatures;
using Microsoft.AspNetCore.Razor.TagHelpers;
using SportsStore.Models.ViewModels;

namespace SportsStore.Infrastructure {

    [HtmlTargetElement("div", Attributes = "page-model")]
    public class PageLinkTagHelper : TagHelper {
        private IUrlHelperFactory urlHelperFactory;

        public PageLinkTagHelper(IUrlHelperFactory helperFactory) {
            urlHelperFactory = helperFactory;
        }

        [ViewContext]
        [HtmlAttributeNotBound]

        public ViewContext ViewContext { get; set; }

        public PagingInfo PageModel { get; set; }

        public string PageAction { get; set; }

        public override void Process(TagHelperContext context,
                TagHelperOutput output) {
            IUrlHelper urlHelper = urlHelperFactory.GetUrlHelper(ViewContext);
            TagBuilder result = new TagBuilder("div");
            for (int i = 1; i <= PageModel.TotalPages; i++) {
                TagBuilder tag = new TagBuilder("a");
                tag.Attributes["href"] = urlHelper.Action(PageAction,
                   new { productPage = i });
                tag.InnerHtml.Append(i.ToString());
                result.InnerHtml.AppendHtml(tag);
            }
            output.Content.AppendHtml(result.InnerHtml);
        }
    }
}
```

提　示

Infrastructure 文件夹是放置为应用程序提供基础设施的类的地方，这些类与应用程序的域无关。

标签助手使用与产品页面对应的元素填充 div 元素。现在不会详细介绍标签助手，知道它们是可以将 C#逻辑引入视图的最有用方法之一就足够了。标签助手的代码可能会比较难懂，因为 C#和 HTML 不容易混合。但是使用标签助手要比在视图中包含 C#代码块好一些，因为标签助手可以轻松进行单元测试。

大多数 MVC 组件（如控制器和视图）可以直接使用，但标签助手必须先注册。在代码清单 8-25 中，在 Views 文件夹下的_ViewImports.cshtml 文件中添加一条语句，告诉 MVC 在 SportsStore.Infrastructure 命名空间中查找标签助手类。这里还添加了一个@using 表达式，以便可以引用视图中的视图模型类，而无须使用命名空间限定名称。

代码清单 8-25 在 Views/Shared 文件夹下的 ViewImports.cshtml 文件中注册一个标签助手

```
@using SportsStore.Models
@using SportsStore.Models.ViewModels
@addTagHelper *, Microsoft.AspNetCore.Mvc.TagHelpers
@addTagHelper SportsStore.Infrastructure.*, SportsStore
```

单元测试——创建分页链接

为了测试 PageLinkTagHelper 标签助手类，可使用测试数据调用 Process 方法，并提供 TagHelperOutput 对象，以查看生成的 HTML，如下所示，在 SportsStore 项目的 PageLinkTagHelperTests.cs 文件中定义了这些内容：

```
using System.Collections.Generic;
using System.Threading.Tasks;
using Microsoft.AspNetCore.Mvc;
using Microsoft.AspNetCore.Mvc.Routing;
using Microsoft.AspNetCore.Razor.TagHelpers;
using Moq;
using SportsStore.Infrastructure;
using SportsStore.Models.ViewModels;
using Xunit;

namespace SportsStore.Tests {

    public class PageLinkTagHelperTests {

        [Fact]
        public void Can_Generate_Page_Links() {
            // Arrange
            var urlHelper = new Mock<IUrlHelper>();
            urlHelper.SetupSequence(x => x.Action(It.IsAny<UrlActionContext>()))
                .Returns("Test/Page1")
                .Returns("Test/Page2")
                .Returns("Test/Page3");

            var urlHelperFactory = new Mock<IUrlHelperFactory>();
            urlHelperFactory.Setup(f =>
                    f.GetUrlHelper(It.IsAny<ActionContext>()))
                        .Returns(urlHelper.Object);

            PageLinkTagHelper helper =
                    new PageLinkTagHelper(urlHelperFactory.Object) {
                PageModel = new PagingInfo {
                    CurrentPage = 2,
                    TotalItems = 28,
                    ItemsPerPage = 10
                },
                PageAction = "Test"
            };

            TagHelperContext ctx = new TagHelperContext(
                new TagHelperAttributeList(),
                new Dictionary<object, object>(), "");
```

```
            var content = new Mock<TagHelperContent>();
            TagHelperOutput output = new TagHelperOutput("div",
                new TagHelperAttributeList(),
                (cache, encoder) => Task.FromResult(content.Object));

            // Act
            helper.Process(ctx, output);

            // Assert
            Assert.Equal(@"<a href=""Test/Page1"">1</a>"
                + @"<a href=""Test/Page2"">2</a>"
                + @"<a href=""Test/Page3"">3</a>",
                 output.Content.GetContent());
        }
    }
}
```

以上测试的复杂性在于创建一些对象，用来创建和使用标签助手。标签助手使用 IUrlHelperFactory 对象来生成针对应用程序不同部分的 URL，并且已经使用 Moq 来创建对应接口的实现，以及提供测试数据的 IUrlHelper 接口。

以上测试的核心部分通过使用包含双引号的字符串来验证标签助手的输出。只要字符串的前缀为@，并使用两组双引号代替一组双引号，C#就可以很好地处理这些字符串。记住，不要将字符串分割成单独的行，除非希望折散正在比较的字符串。例如，在测试方法中使用的字符串已经被换行了，因为打印页面的宽度很窄。这里没有添加换行符；如果这样做，测试将会失败。

3. 添加视图模型数据

为了使用标签助手，还需要为视图提供 PagingInfo 视图模型类的实例。使用 view bag 功能可以做到这一点，但是作者常把控制器发送的所有数据包装到单个视图模型类的视图中。为此，将一个名为 ProductsListViewModel.cs 的类文件添加到 SportsStore 项目的 Models/ViewModels 文件夹中，代码清单 8-26 显示了这个文件的内容。

代码清单 8-26 Models/ViewModels 文件夹下的 ProductsListViewModel.cs 文件的内容

```
using System.Collections.Generic;
using SportsStore.Models;

namespace SportsStore.Models.ViewModels {

    public class ProductsListViewModel {
        public IEnumerable<Product> Products { get; set; }
        public PagingInfo PagingInfo { get; set; }
    }
}
```

可以更新 ProductController 类中的 List 操作方法来使用 ProductsListViewModel 类，以提供页面上显示的产品的详细信息以及分页的详细信息，如代码清单 8-27 所示。

代码清单 8-27 更新 Controllers 文件夹下的 ProductController.cs 文件中的 List 操作方法

```
using Microsoft.AspNetCore.Mvc;
using SportsStore.Models;
using System.Linq;
using SportsStore.Models.ViewModels;

namespace SportsStore.Controllers {

    public class ProductController : Controller {
        private IProductRepository repository;
```

```
        public int PageSize = 4;
        public ProductController(IProductRepository repo) {
            repository = repo;
        }

        public ViewResult List(int productPage = 1)
            => View(new ProductsListViewModel {
                Products = repository.Products
                    .OrderBy(p => p.ProductID)
                    .Skip((productPage - 1) * PageSize)
                    .Take(PageSize),
                PagingInfo = new PagingInfo {
                    CurrentPage = productPage,
                    ItemsPerPage = PageSize,
                    TotalItems = repository.Products.Count()
                }
            });
    }
}
```

这些更改会将 ProductsListViewModel 对象作为模型数据传递给视图。

单元测试——分页模型视图数据

以下是在测试项目中添加到 ProductControllerTests 类的单元测试，以确保控制器将正确的分页数据发送到视图：

```
...
[Fact]
public void Can_Send_Pagination_View_Model() {

    // Arrange
    Mock<IProductRepository> mock = new Mock<IProductRepository>();
    mock.Setup(m => m.Products).Returns((new Product[] {
        new Product {ProductID = 1, Name = "P1"},
        new Product {ProductID = 2, Name = "P2"},
        new Product {ProductID = 3, Name = "P3"},
        new Product {ProductID = 4, Name = "P4"},
        new Product {ProductID = 5, Name = "P5"}
    }).AsQueryable<Product>());

    // Arrange
    ProductController controller =
        new ProductController(mock.Object) { PageSize = 3 };

    // Act
    ProductsListViewModel result =
        controller.List(2).ViewData.Model as ProductsListViewModel;

    // Assert
    PagingInfo pageInfo = result.PagingInfo;
    Assert.Equal(2, pageInfo.CurrentPage);
    Assert.Equal(3, pageInfo.ItemsPerPage);
    Assert.Equal(5, pageInfo.TotalItems);
    Assert.Equal(2, pageInfo.TotalPages);
}
...
```

还需要修改早期的包含在 Can_Paginate 方法中的分页单元测试。它依赖 List 操作方法返回一个 ViewResult，其 Model 属性是 Product 对象的序列，并且需要将数据包含在另一视图模型类型中。下面是修改后的测试：

```
...
[Fact]
public void Can_Paginate() {
    // Arrange
    Mock<IProductRepository> mock = new Mock<IProductRepository>();
    mock.Setup(m => m.Products).Returns((new Product[] {
        new Product {ProductID = 1, Name = "P1"},
        new Product {ProductID = 2, Name = "P2"},
        new Product {ProductID = 3, Name = "P3"},
        new Product {ProductID = 4, Name = "P4"},
        new Product {ProductID = 5, Name = "P5"}
    }).AsQueryable<Product>());

    ProductController controller = new ProductController(mock.Object);
    controller.PageSize = 3;

    // Act
    ProductsListViewModel result =
        controller.List(2).ViewData.Model as ProductsListViewModel;

    // Assert
    Product[] prodArray = result.Products.ToArray();
    Assert.True(prodArray.Length == 2);
    Assert.Equal("P4", prodArray[0].Name);
    Assert.Equal("P5", prodArray[1].Name);
}
...
```

考虑到这两种测试方法之间存在重复，通常会创建一种通用的设置方法。

由于视图目前期望得到 Product 对象序列，因此需要更新 List.cshtml 文件，如代码清单 8-28 所示，以处理新的视图模型类型。

代码清单 8-28 更新 Views/Product 文件夹下的 List.cshtml 文件

```
@model ProductsListViewModel

@foreach (var p in Model.Products) {
    <div>
        <h3>@p.Name</h3>
        @p.Description
        <h4>@p.Price.ToString("c")</h4>
    </div>
}
```

这里已经更新了@model 指令，以告诉 Razor 现在正在使用不同的数据类型。此外，还更新了 foreach 循环，因此数据源要改为模型数据的 Products 属性。

4. 显示分页链接

现在，将页面链接添加到列表视图所需的代码已经写完，创建了包含分页信息的视图模型，更新了控制器以便将信息传递给视图，还更改了@model 指令以匹配新的模型视图类型。剩下的就是添加由标签助手生成的用于创建页面链接的 HTML 元素，如代码清单 8-29 所示。

代码清单 8-29 在 Views/Product 文件夹下的 List.cshtml 文件中添加分页链接

```
@model ProductsListViewModel

@foreach (var p in Model.Products) {
    <div>
        <h3>@p.Name</h3>
```

```
        @p.Description
        <h4>@p.Price.ToString("c")</h4>
    </div>
}

<div page-model="@Model.PagingInfo" page-action="List"></div>
```

如果运行应用程序，你将看到新的页面链接，如图 8-9 所示。使用的还是基本的样式，这将在本章的后面修正。现在重要的是，新的链接使用户能够翻页探索要销售的产品。当 Razor 在 div 元素上找到 page-model 属性时，要求 PageLinkTagHelper 类转换元素，进而产生图 8-9 所示的一组链接。

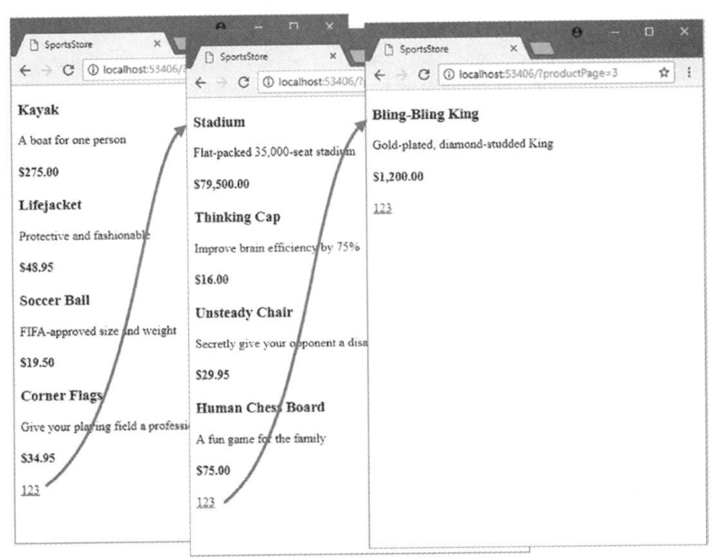

图 8-9 页面导航链接

为什么不使用 GridView？

如果以前曾经使用过 ASP.NET，那么可能会认为这是再普通不过的结果。如果使用的是 Web 窗体，那么可以通过 ASP.NET Web Forms 的 GridView 或 ListView 控件，直接连接到 Products 数据表，完成相同的操作。

本章完成的内容可能看起来不是很多，但与将控件拖到设计界面上相比有根本的区别。首先，这里正在创建一个具有可靠和可维护性架构的应用程序，涉及关注点分离问题。与简单地使用 ListView 控件不同，没有直接耦合 UI 和数据库，虽然这可以快速得到结果，但随着时间的推移会产生很多问题。其次，这里一直在创建单元测试，希望能够以一种自然的方式验证应用程序的行为，这对于复杂的 Web Forms 控件几乎是不可能的。最后，请记住，本章已经介绍了很多内容来创建正在构建的应用程序的基础架构。定义和实现存储库即可。例如，现在可以快速、轻松地构建和测试新功能。

当然，正如第 3 章解释的那样，即使在大型复杂项目中，即时展示数据可能昂贵且很痛苦，但所有这些都不会掩盖如下事实，即 Web 窗体可以快速地显示结果。

8.5.2 改进 URL

现在页面链接可以正常工作，但它们仍然使用查询字符串将页面信息传递到服务器，如下所示：

```
http://localhost/?productPage=2
```

可通过创建一种遵循可组合网址模式的方案，得到更具吸引力的 URL。可组合的 URL 是对用户有意义的 URL，就像下面这样：

```
http://localhost/Page2
```

MVC 可以轻松地更改应用程序中的 URL 结构，因为它使用了 ASP.NET 路由功能，该功能负责处理 URL，以确定要访问应用程序的哪一部分。你需要做的是，当你在 Startup 类的 Configure 方法中注册 MVC 中间件时，添加一个新的路由，如代码清单 8-30 所示。

代码清单 8-30 在 SportsStore 文件夹下的 Startup.cs 文件中添加一个新的路由

```
using System;
using System.Collections.Generic;
using System.Linq;
using System.Threading.Tasks;
using Microsoft.AspNetCore.Builder;
using Microsoft.AspNetCore.Hosting;
using Microsoft.AspNetCore.Http;
using Microsoft.Extensions.DependencyInjection;
using SportsStore.Models;
using Microsoft.Extensions.Configuration;
using Microsoft.EntityFrameworkCore;

namespace SportsStore {

    public class Startup {

        public Startup(IConfiguration configuration) =>
            Configuration = configuration;

        public IConfiguration Configuration { get; }

        public void ConfigureServices(IServiceCollection services) {
            services.AddDbContext<ApplicationDbContext>(options =>
                options.UseSqlServer(
                    Configuration["Data:SportStoreProducts:ConnectionString"]));
            services.AddTransient<IProductRepository, EFProductRepository>();
            services.AddMvc();
        }

        public void Configure(IApplicationBuilder app, IHostingEnvironment env) {
            app.UseDeveloperExceptionPage();
            app.UseStatusCodePages();
            app.UseStaticFiles();
            app.UseMvc(routes => {
                routes.MapRoute(
                    name: "pagination",
                    template: "Products/Page{productPage}",
                    defaults: new { Controller = "Product", action = "List" });

                routes.MapRoute(
                    name: "default",
                    template: "{controller=Product}/{action=List}/{id?}");
            });
            SeedData.EnsurePopulated(app);
        }
    }
}
```

重要的是，必须将这个路由添加到方法中已有的默认路由之前。如第 15 章所述，路由系统按照列出的顺序处理路由，新路由需要优先于现有路由。

这就是更改产品分页的 URL 方案所需的唯一更改。MVC 和路由功能是紧密集成的，因此应用程序会自动反映所使用 URL 中的更改，包括由标签助手生成的 URL，以及用于生成页面导航链接的 URL。如果现在理解不了路由，不要担心，第 15 章和第 16 章将详细解释路由。

如果运行应用程序并单击分页链接，你将看到新的 URL 结构，如图 8-10 所示。

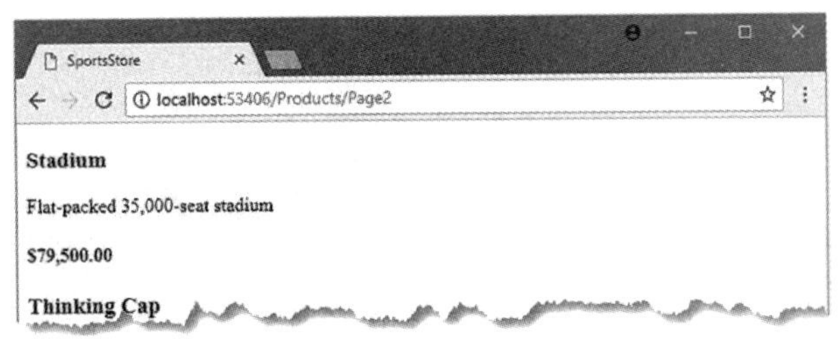

图 8-10 浏览器中显示的新的 URL 结构

8.6 更改内容样式

我们已经创建了大量的底层代码，并且开始集成应用程序的基本功能，但是我们没有在外观上花费太多精力。SportsStore 应用程序的设计如此糟糕，即使本书不是关于设计或 CSS 的，也会影响专业性。本节将修正这些问题，实现带有一个 header 的经典两列布局，如图 8-11 所示。

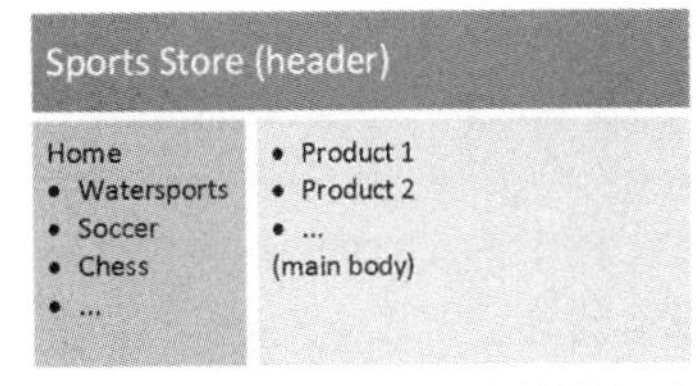

图 8-11 SportsStore 应用程序的设计目标

8.6.1 安装 Bootstrap 包

下面使用 Bootstrap 包来提供将要应用于应用程序的 CSS 样式。因为需要依赖 Visual Studio 对 Bower 的支持来安装 Bootstrap 包，所以在 Add New Item 对话框的 General 类别中选择 Bower Configuration File 模板，在 SportsStore 项目中创建名为 bower.json 的文件，如第 6 章所示。然后将 Bootstrap 包添加到创建的 bower.json 文件的 dependencies 部分，如代码清单 8-31 所示。就像之前所说的那样，在本书中为此例使用预发行版的 Bootstrap。

代码清单 8-31 将 Bootstrap 添加到 SportsStore 项目的 bower.json 文件中

```
{
  "name": "asp.net",
  "private": true,
  "dependencies": {
    "bootstrap": "4.0.0-alpha.6"
  }
}
```

当保存对 bower.json 文件所做的更改时，Visual Studio 使用 Bower 将 Bootstrap 软件包下载到 wwwroot/lib/bootstrap 文件夹中。Bootstrap 依赖于 jQuery 包，所以 jQuery 也将自动添加到项目中。

8.6.2 将 Bootstrap 样式应用于布局

第 5 章解释了 Razor 布局如何工作，如何使用它们以及它们如何组成布局。本章开头添加的视图启动文件指定使用名为_Layout.cshtml 的文件作为默认布局，这就是应用初始 Bootstrap 样式的地方，如代码清单 8-32 所示。

代码清单 8-32 将 Bootstrap CSS 应用于 Views/Shared 文件夹下的_Layout.cshtml 文件

```
<!DOCTYPE html>

<html>
<head>
```

```
    <meta name="viewport" content="width=device-width" />
    <link rel="stylesheet"
          asp-href-include="/lib/bootstrap/dist/**/*.min.css"
          asp-href-exclude="**/*-reboot*,**/*-grid*" />
    <title>SportsStore</title>
</head>
<body>
    <div class="navbar navbar-inverse bg-inverse" role="navigation">
        <a class="navbar-brand" href="#">SPORTS STORE</a>
    </div>
    <div class="row m-1 p-1">
        <div id="categories" class="col-3">
            Put something useful here later
        </div>
        <div class="col-9">
            @RenderBody()
        </div>
    </div>
</body>
</html>
```

上述代码中的 link 元素包含 asp-href-include 和 asp-href-exclude 属性，展示了一个内置标签助手类的例子。在这种情况下，标签助手将查看属性的值，并生成与指定路径匹配的所有文件的 link 元素，其中可包含通配符。这个功能很有用，能够确保可以在不破坏应用程序的情况下添加和删除 wwwroot 文件夹结构中的文件，但是正如第 25 章解释的那样，需要注意确保指定的路径仅选择期望的文件。

将 Bootstrap CSS 样式表添加到布局意味着可以在依赖于布局的任何视图中使用它定义样式。在代码清单 8-33 中，可以看到应用于 List.cshtml 文件的样式。

代码清单 8-33 Views/Product 文件夹下的 List.cshtml 文件中的样式内容

```
@model ProductsListViewModel

@foreach (var p in Model.Products) {
    <div class="card card-outline-primary m-1 p-1">
        <div class="bg-faded p-1">
            <h4>
                @p.Name
                <span class="badge badge-pill badge-primary" style="float:right">
                    <small>@p.Price.ToString("c")</small>
                </span>
            </h4>
        </div>
        <div class="card-text p-1">@p.Description</div>
    </div>
}

<div page-model="@Model.PagingInfo" page-action="List" page-classes-enabled="true"
    page-class="btn" page-class-normal="btn-secondary"
    page-class-selected="btn-primary" class="btn-group pull-right m-1">
</div>
```

需要为 PageLinkTagHelper 类生成的按钮设置样式，但是这里不会将 Bootstrap 样式硬编码到 C#代码中，因为当在应用程序的其他位置重用标签助手或更改按钮的外观时，这样会更加困难。相反，这里已经在 div 元素上定义了自定义属性，指定了需要的样式，它们对应添加到标签助手类中的属性，然后对生成的 a 元素设置样式，如代码清单 8-34 所示。

代码清单 8-34 将样式添加到 PageLinkTagHelper.cs 文件中生成的元素上

```
using Microsoft.AspNetCore.Mvc;
using Microsoft.AspNetCore.Mvc.Rendering;
using Microsoft.AspNetCore.Mvc.Routing;
```

```
using Microsoft.AspNetCore.Mvc.ViewFeatures;
using Microsoft.AspNetCore.Razor.TagHelpers;
using SportsStore.Models.ViewModels;

namespace SportsStore.Infrastructure {

    [HtmlTargetElement("div", Attributes = "page-model")]
    public class PageLinkTagHelper : TagHelper {
        private IUrlHelperFactory urlHelperFactory;

        public PageLinkTagHelper(IUrlHelperFactory helperFactory) {
            urlHelperFactory = helperFactory;
        }

        [ViewContext]
        [HtmlAttributeNotBound]
        public ViewContext ViewContext { get; set; }

        public PagingInfo PageModel { get; set; }

        public string PageAction { get; set; }

        public bool PageClassesEnabled { get; set; } = false;
        public string PageClass { get; set; }
        public string PageClassNormal { get; set; }
        public string PageClassSelected { get; set; }
        public override void Process(TagHelperContext context,
                TagHelperOutput output) {
            IUrlHelper urlHelper = urlHelperFactory.GetUrlHelper(ViewContext);
            TagBuilder result = new TagBuilder("div");
            for (int i = 1; i <= PageModel.TotalPages; i++) {
                TagBuilder tag = new TagBuilder("a");
                tag.Attributes["href"] = urlHelper.Action(PageAction,
                    new { productPage = i });
                if (PageClassesEnabled) {
                    tag.AddCssClass(PageClass);
                    tag.AddCssClass(i == PageModel.CurrentPage
                        ? PageClassSelected : PageClassNormal);
                }
                tag.InnerHtml.Append(i.ToString());
                result.InnerHtml.AppendHtml(tag);
            }
            output.Content.AppendHtml(result.InnerHtml);
        }
    }
}
```

这些属性的值将自动用于设置标签助手的属性值，并考虑 HTML 属性名称格式（page-class-normal）和 C#属性名称格式（PageClassNormal）之间的映射。这允许标签助手根据 HTML 元素的属性进行不同的响应，从而创建一种更灵活的方法在 MVC 应用程序中生成内容。

图 8-12 外观得到改进的 SportsStore 应用程序

如果运行应用程序，你将看到应用程序的外观已经得到一些改进，如图 8-12 所示。

8.6.3 创建分部视图

作为本章的结尾，下面重构应用程序以简化 List.cshtml 视图文件。创建一个分部视图（partial view），它是一个内容片段，可以嵌入另一个视图，有点类似于模板。第 21 章将详细描述分部视图，并且当你需要在应用程序的不同位置显示相同的内容时，它们有助于减少重复。可以在分部视图中定义一次，而不是将相同的 Razor 标记复制并粘贴到多个视图中。为了创建分部视图，在 Views/Shared 文件夹添加名为 ProductSummary.cshtml 的 Razor 视图文件，并添加代码清单 8-35 所示的代码。

代码清单 8-35 Views/Shared 文件夹下的 ProductSummary.cshtml 文件中的内容

```
@model Product

<div class="card card-outline-primary m-1 p-1">
    <div class="bg-faded p-1">
        <h4>
            @Model.Name
            <span class="badge badge-pill badge-primary" style="float:right">
                <small>@Model.Price.ToString("c")</small>
            </span>
        </h4>
    </div>
    <div class="card-text p-1">@Model.Description</div>
</div>
```

现在需要更新 Views/Products 文件夹中的 List.cshtml 文件，以便使用分部视图，如代码清单 8-36 所示。

代码清单 8-36 在 List.cshtml 文件中使用分部视图

```
@model ProductsListViewModel

@foreach (var p in Model.Products) {
    @Html.Partial("ProductSummary", p)
}

<div page-model="@Model.PagingInfo" page-action="List" page-classes-enabled="true"
     page-class="btn" page-class-normal="btn-secondary"
     page-class-selected="btn-primary" class="btn-group pull-right m-1">
</div>
```

这里已经将 List.cshtml 文件中的 foreach 循环中的标记移动到新的分部视图。使用 Html.Partial 辅助方法调用分部视图，其中包含视图名称和视图模型对象的参数。切换到这样的分部视图是比较好的做法，从而允许将相同的标记插入需要显示产品摘要的任何视图中。如图 8-13 所示，添加分部视图不会改变应用程序的外观；只是改变了 Razor 搜索生成内容的位置，从而生成发送到浏览器的响应。

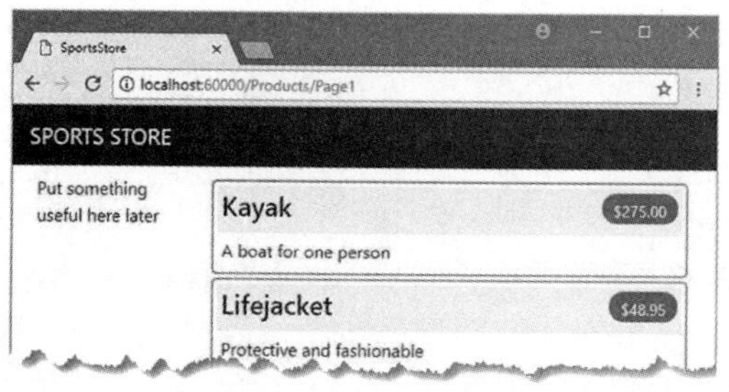

图 8-13 应用分部视图

8.7 小结

本章为 SportsStore 应用程序构建了核心基础架构。现在还没有很多功能可供你向客户展示，但我们已经开始了领域模型开发，其产品存储库由 SQL Server 和 Entity Framework Core 支持。另一个控制器 ProductController 可以生成分页的产品列表。我们还创建了一种简洁、友好的 URL 结构。

现在底层的基础设施代码已经到位，我们可以继续增加面向客户的所有功能——按类别导航、购物车以及结账。

第 9 章　SportsStore 的导航

本章将继续构建 SportsStore 应用程序，添加对应用程序导航的支持并开始构建购物车。

9.1　添加导航控件

如果客户可以按类别浏览产品，SportsStore 应用程序将更为方便，这可分 3 个阶段完成。

- 增强 ProductController 类中的 List 操作模型，以便能够过滤存储库中的 Product 对象。
- 重构并增强 URL 结构。
- 创建类别列表，展示在站点的边栏中，突出显示当前类别并链接到其他类别。

9.1.1　过滤产品列表

为了改进视图模型类（上一章添加到 SportsStore 项目中的 ProductsListViewModel），需要让当前类别与视图进行通信，以便渲染边栏，这是一个很好的起点。代码清单 9-1 显示了对 Models/ViewModels 文件夹中的 ProductsListViewModel.cs 文件所做的更改。

代码清单 9-1　在 Models/ViewModels 文件夹下的 ProductsListViewModel.cs 文件中添加一个属性

```
using System.Collections.Generic;
using SportsStore.Models;

namespace SportsStore.Models.ViewModels {

    public class ProductsListViewModel {
        public IEnumerable<Product> Products { get; set; }
        public PagingInfo PagingInfo { get; set; }
        public string CurrentCategory { get; set; }
    }
}
```

这里添加了一个名为 CurrentCategory 的属性。下一步是更新 Product 控制器，以便 List 操作方法按类别过滤 Product 对象，并使用添加到视图模型中的新属性来指示哪个类别被选中。代码清单 9-2 显示了所做的更改。

代码清单 9-2　在 Controllers 文件夹下的 ProductController.cs 文件的 List 操作方法中添加对类别的支持

```
using Microsoft.AspNetCore.Mvc;
using SportsStore.Models;
using System.Linq;
using SportsStore.Models.ViewModels;

namespace SportsStore.Controllers {

    public class ProductController : Controller {
        private IProductRepository repository;
        public int PageSize = 4;

        public ProductController(IProductRepository repo) {
            repository = repo;
        }

        public ViewResult List(string category, int productPage = 1)
```

```
            => View(new ProductsListViewModel {
                Products = repository.Products
                    .Where(p => category == null || p.Category == category)
                    .OrderBy(p => p.ProductID)
                    .Skip((productPage - 1) * PageSize)
                    .Take(PageSize),
                PagingInfo = new PagingInfo {
                    CurrentPage = productPage,
                    ItemsPerPage = PageSize,
                    TotalItems = repository.Products.Count()
                },
                CurrentCategory = category
            });
    }
}
```

这里对这个操作方法做了 3 处更改。首先，添加一个名为 category 的参数，category 参数由列表中的第二处更改使用，这是 LINQ 查询的增强功能：如果类别不为空，那么仅选择匹配 Category 属性的 Product 对象。最后一处更改是设置添加到 ProductsListViewModel 类的 CurrentCategory 属性的值。但是，这些更改意味着没有正确计算 PagingInfo.TotalItems 的值，因为没有考虑类别过滤器，稍后再解决这个问题。

单元测试——更新已存在的单元测试

这里更改了 List 操作方法的签名，这将影响编译一些现有的单元测试方法。为了解决这个问题，需要将 null 作为第一个参数传递给使用控制器的单元测试中的 List 方法。例如，在 ProductControllerTests.cs 文件的 Can_Paginate 测试中，单元测试的操作部分如下：

```
...
[Fact]
public void Can_Paginate() {
    // Arrange
    Mock<IProductRepository> mock = new Mock<IProductRepository>();
    mock.Setup(m => m.Products).Returns((new Product[] {
        new Product {ProductID = 1, Name = "P1"},
        new Product {ProductID = 2, Name = "P2"},
        new Product {ProductID = 3, Name = "P3"},
        new Product {ProductID = 4, Name = "P4"},
        new Product {ProductID = 5, Name = "P5"}
    }).AsQueryable<Product>());

    ProductController controller = new ProductController(mock.Object);
    controller.PageSize = 3;

    // Act
    ProductsListViewModel result =
        controller.List(null, 2).ViewData.Model as ProductsListViewModel;

    // Assert
    Product[] prodArray = result.Products.ToArray();
    Assert.True(prodArray.Length == 2);
    Assert.Equal("P4", prodArray[0].Name);
    Assert.Equal("P5", prodArray[1].Name);
}
...
```

通过对 category 参数使用 null，就能得到控制器从存储库中获取的所有 Product 对象，这与添加新参数之前的情况相同。此外，还需要对 Can_Send_Pagination_View_Model 测试做同样的修改。

```
...
[Fact]
public void Can_Send_Pagination_View_Model() {
```

```
        // Arrange
        Mock<IProductRepository> mock = new Mock<IProductRepository>();
        mock.Setup(m => m.Products).Returns((new Product[] {
            new Product {ProductID = 1, Name = "P1"},
            new Product {ProductID = 2, Name = "P2"},
            new Product {ProductID = 3, Name = "P3"},
            new Product {ProductID = 4, Name = "P4"},
            new Product {ProductID = 5, Name = "P5"}
        }).AsQueryable<Product>());

        // Arrange
        ProductController controller =
            new ProductController(mock.Object) { PageSize = 3 };

        // Act
        ProductsListViewModel result =
            controller.List(null, 2).ViewData.Model as ProductsListViewModel;

        // Assert
        PagingInfo pageInfo = result.PagingInfo;
        Assert.Equal(2, pageInfo.CurrentPage);
        Assert.Equal(3, pageInfo.ItemsPerPage);
        Assert.Equal(5, pageInfo.TotalItems);
        Assert.Equal(2, pageInfo.TotalPages);
    }
    ...
```

当进入测试模式时，使单元测试与代码更改快速同步成为第二条重要的原则。

要查看类别过滤的效果，请启动应用程序并使用以下查询字符串选择一个类别，将端口更改为 Visual Studio 为项目分配的端口（注意，Soccer 的首字母要大写）：

```
http://localhost:60000/?category=Soccer
```

你将只能看到足球类别中的商品，如图 9-1 所示。

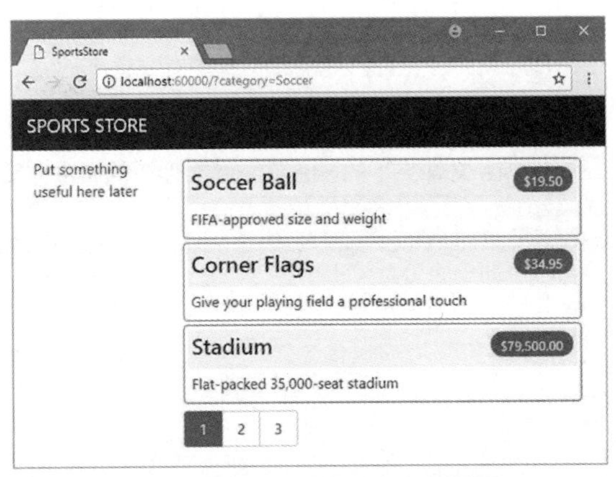

图 9-1　使用查询字符串按类别过滤商品

显然，用户不会使用 URL 导航到类别，但是一旦基础设施代码到位，就可以看到 MVC 应用程序中一个小的更改也可以产生很大的影响。

单元测试——类别过滤

你需要使用单元测试来测试类别过滤功能，以确保过滤器能够正确地生成指定类别的商品。下面是添加到 ProductControllerTests 类的测试方法：

```
...
[Fact]
public void Can_Filter_Products() {

    // Arrange
    // - create the mock repository
    Mock<IProductRepository> mock = new Mock<IProductRepository>();
    mock.Setup(m => m.Products).Returns((new Product[] {
        new Product {ProductID = 1, Name = "P1", Category = "Cat1"},
        new Product {ProductID = 2, Name = "P2", Category = "Cat2"},
        new Product {ProductID = 3, Name = "P3", Category = "Cat1"},
        new Product {ProductID = 4, Name = "P4", Category = "Cat2"},
        new Product {ProductID = 5, Name = "P5", Category = "Cat3"}
    }).AsQueryable<Product>());

    // Arrange - create a controller and make the page size 3 items
    ProductController controller = new ProductController(mock.Object);
    controller.PageSize = 3;

    // Action
    Product[] result =
        (controller.List("Cat2", 1).ViewData.Model as ProductsListViewModel)
            .Products.ToArray();

    // Assert
    Assert.Equal(2, result.Length);
    Assert.True(result[0].Name == "P2" && result[0].Category == "Cat2");
    Assert.True(result[1].Name == "P4" && result[1].Category == "Cat2");
}
...
```

以上测试将创建一个模拟的存储库，其中包含属于多个类别的 Product 对象。使用一个操作方法请求某个特定类别，并检查结果以确保结果是正确的对象并具有正确的顺序。

9.1.2 优化 URL 结构

没人想看到或使用丑陋的网址，如/?category = Soccer。为了解决这个问题，可在 Startup 类的 Configure 方法中更改路由配置，以创建一组更友好的 URL，如代码清单 9-3 所示。

代码清单 9-3 更改 SportsStore 文件夹下的 Startup.cs 文件中的路由方案

```
...
public void Configure(IApplicationBuilder app, IHostingEnvironment env) {
    app.UseDeveloperExceptionPage();
    app.UseStatusCodePages();
    app.UseStaticFiles();
    app.UseMvc(routes => {

        routes.MapRoute(
            name: null,
            template: "{category}/Page{productPage:int}",
            defaults: new { controller = "Product", action = "List" }
        );

        routes.MapRoute(
            name: null,
            template: "Page{productPage:int}",
            defaults: new { controller = "Product",
                action = "List", productPage = 1 }
        );
```

```
        routes.MapRoute(
            name: null,
            template: "{category}",
            defaults: new { controller = "Product",
                action = "List", productPage = 1 }
        );

        routes.MapRoute(
            name: null,
            template: "",
            defaults: new { controller = "Product", action = "List",
                productPage = 1 });

        routes.MapRoute(name: null, template: "{controller}/{action}/{id?}");
    });
    SeedData.EnsurePopulated(app);
}
...
```

警 告

按照代码清单 9-3 中的顺序添加新路由非常重要。路由按定义的顺序生效，如果更改顺序，你会得到一些奇怪的结果。

这些路由表示的 URL 结构如表 9-1 所示。第 15 章和第 16 章将详细解释路由系统。

表 9-1 路由摘要

URL	导航到
/	列出所有类别商品的第一页
/Page2	列出特定页面（在本例中为第 2 页），显示所有类别的商品
/Soccer	显示来自特定类别的商品的第一页（在本例中为 Soccer 类别）
/Soccer/Page2	显示指定类别（在这种情况下为 Soccer）的商品的指定页面（在本例中为第 2 页）

MVC 使用 ASP.NET Core 路由系统来处理来自客户端的传入请求，并且生成符合 URL 结构的可嵌入网页中的传出 URL。通过使用路由系统来处理传入请求和传出 URL，可以确保应用程序中的所有 URL 一致。

IUrlHelper 接口提供对 URL 生成功能的访问。你在上一章中创建的标签助手使用了这个接口，以及其中定义的操作方法。现在我们想开始生成更复杂的 URL，我们需要一种从视图中接收附加信息的方法，而不必为标签助手类添加额外的属性。幸运的是，标签助手有一个很好的功能，允许属性拥有公共前缀，从而可以在单个集合中接收，如代码清单 9-4 所示。

代码清单 9-4 在 Infrastructure 文件夹下的 PageLinkTagHelper.cs 文件中接收前缀属性值

```
using Microsoft.AspNetCore.Mvc;
using Microsoft.AspNetCore.Mvc.Rendering;
using Microsoft.AspNetCore.Mvc.Routing;
using Microsoft.AspNetCore.Mvc.ViewFeatures;
using Microsoft.AspNetCore.Razor.TagHelpers;
using SportsStore.Models.ViewModels;
using System.Collections.Generic;

namespace SportsStore.Infrastructure {

    [HtmlTargetElement("div", Attributes = "page-model")]
    public class PageLinkTagHelper : TagHelper {
```

```
        private IUrlHelperFactory urlHelperFactory;

        public PageLinkTagHelper(IUrlHelperFactory helperFactory) {
            urlHelperFactory = helperFactory;
        }

        [ViewContext]
        [HtmlAttributeNotBound]
        public ViewContext ViewContext { get; set; }

        public PagingInfo PageModel { get; set; }

        public string PageAction { get; set; }

        [HtmlAttributeName(DictionaryAttributePrefix = "page-url-")]
        public Dictionary<string, object> PageUrlValues { get; set; }
            = new Dictionary<string, object>();

        public bool PageClassesEnabled { get; set; } = false;
        public string PageClass { get; set; }
        public string PageClassNormal { get; set; }
        public string PageClassSelected { get; set; }

        public override void Process(TagHelperContext context,
                TagHelperOutput output) {
            IUrlHelper urlHelper = urlHelperFactory.GetUrlHelper(ViewContext);
            TagBuilder result = new TagBuilder("div");
            for (int i = 1; i <= PageModel.TotalPages; i++) {
                TagBuilder tag = new TagBuilder("a");
                PageUrlValues["productPage"] = i;
                tag.Attributes["href"] = urlHelper.Action(PageAction, PageUrlValues);
                if (PageClassesEnabled) {
                    tag.AddCssClass(PageClass);
                    tag.AddCssClass(i == PageModel.CurrentPage
                        ? PageClassSelected : PageClassNormal);
                }
                tag.InnerHtml.Append(i.ToString());
                result.InnerHtml.AppendHtml(tag);
            }
            output.Content.AppendHtml(result.InnerHtml);
        }
    }
}
```

通过使用 HtmlAttributeName 特性（attribute）装饰一个标签助手属性（property），可以为元素上的属性名称指定前缀，在这个例子中是 page-url-。名称以此前缀开头的任何属性的值都将被添加到分配给 PageUrlValues 属性的字典中，然后传递给 IUrlHelper.Action 方法，以生成由标签助手输出的 a 元素的 href 属性。

在代码清单 9-5 中，已经为标签助手处理的 div 元素添加了一个新的属性，指定了用于生成 URL 的类别。虽然只为视图添加了一个新的属性，但是具有相同前缀的任何属性都将被添加到字典中。

代码清单 9-5 在 Views/Home 文件夹下的 List.cshtml 文件中添加新属性

```
@model ProductsListViewModel

@foreach (var p in Model.Products) {
    @Html.Partial("ProductSummary", p)
}

<div page-model="@Model.PagingInfo" page-action="List" page-classes-enabled="true"
    page-class="btn" page-class-normal="btn-secondary"
```

```
    page-class-selected="btn-primary" page-url-category="@Model.CurrentCategory"
    class="btn-group pull-right m-1">
</div>
```

在进行上述更改之前，为分页链接生成的链接如下所示：

```
http://<myserver>:<port>/Page1
```

如果用户单击这样的页面链接，类别过滤器将会丢失，应用程序将显示一个包含所有类别产品的页面。通过添加从视图模型中获取的当前类别，就会生成下面这样的 URL：

```
http://<myserver>:<port>/Chess/Page1
```

当用户单击此类链接时，当前类别将被传递给 List 操作方法，并且会保留过滤功能。进行此更改后，可以访问/Chess 或/Soccer 等网址，你会看到页面底部的页面链接已正确包含该类别。

9.1.3 构建类别导航菜单

你需要为客户提供一种方式，无须用户在 URL 中输入类别。这意味着需要向他们呈现可用类别的列表，并指出当前选择了哪个类别（如果有的话）。当构建应用程序时，将在多个控制器中使用这个类别列表，所以需要一些自包含且可重用的东西。

ASP.NET Core MVC 具有视图组件的概念，这对于创建诸如可重复使用的导航控件之类的项目是非常合适的。视图组件是 C#类，可以提供少量可重用的应用程序逻辑，能够选择和显示 Razor 分部视图。第 22 章将详细描述视图组件。

在本节的例子中，将创建一个视图组件，通过从共享布局调用组件来呈现导航菜单，并将其集成到应用程序中。这种方法让我们能够使用一个常规的 C#类，可以包含我们所需要的任何应用程序逻辑，并且可以像任何其他类一样进行单元测试。这是一种在保留整体 MVC 方法的同时创建应用程序的较小片段的好方法。

1. 创建导航视图组件

创建名为 Components 的文件夹（一般用来存放视图组件），并在其中添加名为 NavigationMenuViewComponent.cs 的类，用于定义代码清单 9-6 所示的类。

代码清单 9-6 Components 文件夹下的 NavigationMenuViewComponent.cs 文件的内容

```
using Microsoft.AspNetCore.Mvc;

namespace SportsStore.Components {

    public class NavigationMenuViewComponent : ViewComponent {

        public string Invoke() {
            return "Hello from the Nav View Component";
        }
    }
}
```

在 Razor 视图中使用组件时，将调用视图组件的 Invoke 方法，并将 Invoke 方法的结果插入 HTML 中，发送到浏览器。我们已经开始使用一个返回字符串的简单视图组件，但我们很快就会用动态 HTML 内容替换它。

这里希望类别列表出现在所有页面上，因此我们将在共享布局中使用视图组件，而不是在特定视图中。在视图中，可通过@await Component.InvokeAsync 表达式使用视图组件，如代码清单 9-7 所示。

代码清单 9-7 在 Views/Shared 文件夹下的_Layout.cshtml 文件中使用视图组件

```
<!DOCTYPE html>

<html>
<head>
    <meta name="viewport" content="width=device-width" />
    <link rel="stylesheet"
```

```
            asp-href-include="/lib/bootstrap/dist/**/*.min.css"
            asp-href-exclude="**/*-reboot*,**/*-grid*" />
    <title>SportsStore</title>
</head>
<body>
    <div class="navbar navbar-inverse bg-inverse" role="navigation">
        <a class="navbar-brand" href="#">SPORTS STORE</a>
    </div>
    <div class="row m-1 p-1">
        <div id="categories" class="col-3">
            @await Component.InvokeAsync("NavigationMenu")
        </div>
        <div class="col-9">
            @RenderBody()
        </div>
    </div>
</body>
</html>
```

这里删除了占位符文本，并将其替换为对 Component.InvokeAsync 方法的调用。此方法的参数是组件类的名称，省略了类名的 ViewComponent 部分，因此 NavigationMenu 指定了 NavigationMenuViewComponent 类。如果运行该应用程序，你将看到 Invoke 方法的输出包含在发送给浏览器的 HTML 中，如图 9-2 所示。

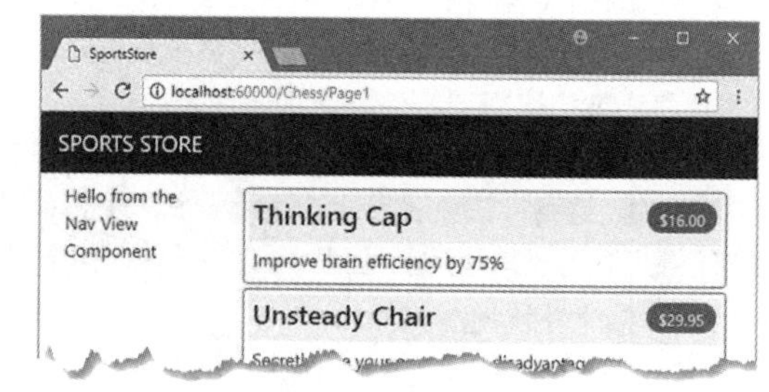

图 9-2 使用视图组件的输出

2. 生成类别列表

现在可以返回导航视图控制器并生成一组真实的类别。可以使用编程方式为类别构建 HTML，就像为页面标签助手所做的那样，但使用视图组件的一个好处是它们可以渲染 Razor 分部视图。这意味着可以使用视图组件生成组件列表，然后使用更具表现力的 Razor 语法来呈现用于显示它们的 HTML。第一步是更新视图组件，如代码清单 9-8 所示。

代码清单 9-8 在 Components 文件夹下的 NavigationMenuViewComponent.cs 文件中添加类别

```
using Microsoft.AspNetCore.Mvc;
using System.Linq;
using SportsStore.Models;

namespace SportsStore.Components {

    public class NavigationMenuViewComponent : ViewComponent {
        private IProductRepository repository;

        public NavigationMenuViewComponent(IProductRepository repo) {
            repository = repo;
        }

        public IViewComponentResult Invoke() {
            return View(repository.Products
                .Select(x => x.Category)
                .Distinct()
                .OrderBy(x => x));
        }
    }
}
```

代码清单 9-8 中的构造函数定义了一个 IProductRepository 参数。当 MVC 需要创建视图组件类的实例时，它将需要提供这个参数并检查 Startup 类中的配置以确定应使用哪个实现对象。这与第 8 章在控制器中使用的依赖注入功能相同，并且具有相同的效果，即允许视图组件在不知道将使用哪个存储库实现的情况下访问数据，如第 18 章所述。

在 Invoke 方法中，使用 LINQ 对存储库中的一组类别进行选择和排序，并将它们作为参数传递给 View 方法，View 方法会呈现默认的 Razor 分部视图，使用 IViewComponentResult 对象从方法返回其详细信息，第 22 章会详细描述此过程。

单元测试——生成类别列表

用于类别列表的单元测试相对简单。目标是创建一个按字母顺序排列的列表，并且不包含重复项。最简单的方法是提供一些具有重复类别且不按顺序排列的测试数据，将其传递给标签助手类，并判断数据是否已被正确清理。在 SportsStore.Tests 项目中，在名为 NavigationMenuViewComponentTests.cs 的类文件中定义的单元测试如下：

```
using System.Collections.Generic;
using System.Linq;
using Microsoft.AspNetCore.Mvc.ViewComponents;
using Moq;
using SportsStore.Components;
using SportsStore.Models;
using Xunit;
namespace SportsStore.Tests {

    public class NavigationMenuViewComponentTests {

        [Fact]
        public void Can_Select_Categories() {
            // Arrange
            Mock<IProductRepository> mock = new Mock<IProductRepository>();
            mock.Setup(m => m.Products).Returns((new Product[] {
                new Product {ProductID = 1, Name = "P1", Category = "Apples"},
                new Product {ProductID = 2, Name = "P2", Category = "Apples"},
                new Product {ProductID = 3, Name = "P3", Category = "Plums"},
                new Product {ProductID = 4, Name = "P4", Category = "Oranges"},
            }).AsQueryable<Product>());

            NavigationMenuViewComponent target =
                new NavigationMenuViewComponent(mock.Object);

            // Act = get the set of categories
            string[] results = ((IEnumerable<string>)(target.Invoke()
                as ViewViewComponentResult).ViewData.Model).ToArray();

            // Assert
            Assert.True(Enumerable.SequenceEqual(new string[] { "Apples",
                "Oranges", "Plums" }, results));
        }
    }
}
```

以上代码创建了一个模拟的存储库实现，其中包含重复的以及没有正确排序的类别。我们判断重复项将被删除，并强制按字母顺序排列。

3. 创建视图

Razor 使用不同的约定来处理视图组件选择的视图。视图的默认名称和搜索位置与用于控制器的不同。为此，创建 Views/Shared/Components/NavigationMenu 文件夹，并在其中添加名为 Default.cshtml 的视图文件，内容如代码清单 9-9 所示。

代码清单 9-9 Views/Shared/Components/NavigationMenu 文件夹下的 Default.cshtml 文件的内容

```
@model IEnumerable<string>

<a class="btn btn-block btn-secondary"
   asp-action="List"
   asp-controller="Product"
   asp-route-category="">
    Home
</a>
@foreach (string category in Model) {
    <a class="btn btn-block btn-secondary"
       asp-action="List"
       asp-controller="Product"
       asp-route-category="@category"
       asp-route-productPage="1">
        @category
    </a>
}
```

这个视图使用第 24 章和第 25 章描述的内置标签助手之一来创建 a 元素，其 href 属性包含选择不同产品类别的 URL。

如果运行应用程序，就可以看到类别链接，如图 9-3 所示。如果单击某个类别，就会更新条目列表，以仅显示所选类别中的条目。

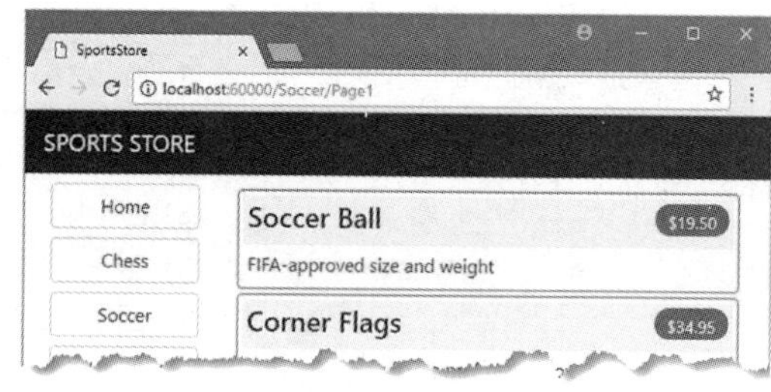

图 9-3 使用视图组件生成类别链接

4. 突出显示当前类别

现在还没有给用户发出反馈以指示选择了哪个类别。从列表中的条目推断出类别也是可能的，但是提供一些清晰的视觉反馈似乎更好。ASP.NET Core MVC 组件（如控制器和视图组件）可以通过请求上下文对象来接收有关当前请求的信息。大多数情况下，可以依赖用于创建组件的基类来获取上下文对象，例如在使用 Controller 基类创建控制器时。

ViewComponent 基类也不例外，它通过一组属性提供对上下文对象的访问。其中一个属性名为 RouteData，它提供有关路由系统如何处理请求 URL 的信息。

代码清单 9-10 使用 RouteData 属性来访问请求数据，以获取当前所选类别的值。可以通过创建另一个视图模型类将类别传递给视图（这就是在实际项目中要做的事情），但是为了展示更多方式，我们将使用第 2 章介绍的 view bag 功能。

代码清单 9-10 在 NavigationMenuViewComponent.cs 文件中传递所选类别

```
using Microsoft.AspNetCore.Mvc;
using System.Linq;
using SportsStore.Models;

namespace SportsStore.Components {

    public class NavigationMenuViewComponent : ViewComponent {
        private IProductRepository repository;

        public NavigationMenuViewComponent(IProductRepository repo) {
            repository = repo;
        }

        public IViewComponentResult Invoke() {
            ViewBag.SelectedCategory = RouteData?.Values["category"];
            return View(repository.Products
```

```
                .Select(x => x.Category)
                .Distinct()
                .OrderBy(x => x));
        }
    }
}
```

在 Invoke 方法中，已经为 ViewBag 对象动态分配了 SelectedCategory 属性，并将属性值设置为当前类别，当前类别是通过 RouteData 属性返回的上下文对象获得的。正如第 2 章解释的那样，ViewBag 是动态对象，可以简单地为它们分配值来定义新的属性。

单元测试——报告所选类别

可以通过在单元测试中读取 ViewBag 属性的值来测试视图组件是否正确添加了所选类别的详细信息，该值可通过第 22 章描述的 ViewViewComponentResult 类获得。下面是添加到 NavigationMenuViewComponentTests 类中的单元测试：

```
...
[Fact]
public void Indicates_Selected_Category() {

    // Arrange
    string categoryToSelect = "Apples";
    Mock<IProductRepository> mock = new Mock<IProductRepository>();
    mock.Setup(m => m.Products).Returns((new Product[] {
        new Product {ProductID = 1, Name = "P1", Category = "Apples"},
        new Product {ProductID = 4, Name = "P2", Category = "Oranges"},
    }).AsQueryable<Product>());
    NavigationMenuViewComponent target =
        new NavigationMenuViewComponent(mock.Object);
    target.ViewComponentContext = new ViewComponentContext {
        ViewContext = new ViewContext {
            RouteData = new RouteData()
        }
    };
    target.RouteData.Values["category"] = categoryToSelect;

    // Action
    string result = (string)(target.Invoke() as
        ViewViewComponentResult).ViewData["SelectedCategory"];

    // Assert
    Assert.Equal(categoryToSelect, result);
}
...
```

以上单元测试通过 ViewComponentContext 属性为视图组件提供路由数据，该属性指定了视图组件接收所有上下文数据的方式。ViewComponentContext 属性通过 ViewContext 子属性提供对特定于视图的上下文数据的访问，而后通过 RouteData 子属性提供对路由信息的访问。单元测试中的大多数代码用于创建上下文对象，这些对象将以与应用程序运行时呈现的相同方式提供所选类别，其上下文数据由 ASP.Net Core MVC 提供。

现在提供了有关选择哪个类别的信息，可以利用这一点更新视图组件选择的视图，并改变用于设置链接样式的 CSS 类，使代表当前类别的 CSS 类与其他类别不同。代码清单 9-11 显示了对 Default.cshtml 文件所做的更改。

代码清单 9-11　在 Views/Shared/Components/NavigationMenu 文件夹下的 Default.cshtml 文件中突出显示当前类别

```
@model IEnumerable<string>

<a class="btn btn-block btn-secondary"
```

```
        asp-action="List"
        asp-controller="Product"
        asp-route-category="">
         Home
    </a>

    @foreach (string category in Model) {
        <a class="btn btn-block
            @(category == ViewBag.SelectedCategory ? "btn-primary": "btn-secondary")"
            asp-action="List"
            asp-controller="Product"
            asp-route-category="@category"
            asp-route-productPage="1">
             @category
        </a>
    }
```

在 class 属性中使用了一个 Razor 表达式，从而将 btn-primary 类应用于表示所选类别的元素，其他类别的元素使用的是 btn-secondary 类。为这些类别应用不同的 Bootstrap 样式，并使活动按钮突出显示，如图 9-4 所示。

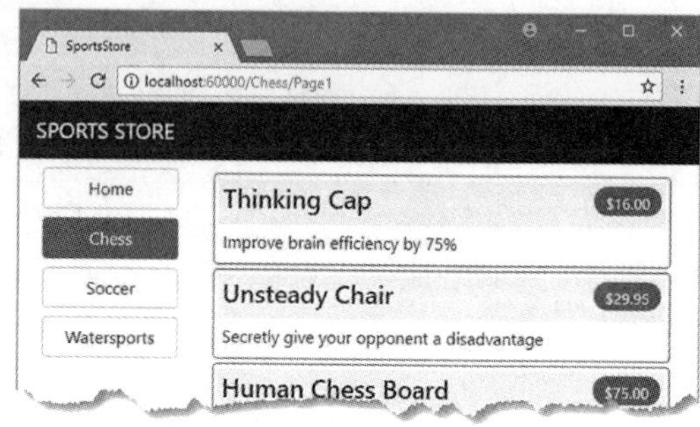

图 9-4　突出显示所选类别

9.1.4　更正页数

需要更正页面链接，以便在选择类别时它们能够正常工作。目前，页面链接的数量由存储库中的商品总数确定，而不是由所选类别中的商品数量确定。这意味着客户可以单击 Chess 类别第 2 页的链接，但会得到一个空白页面，因为没有足够的 Chess 商品来填充两个页面，如图 9-5 所示。

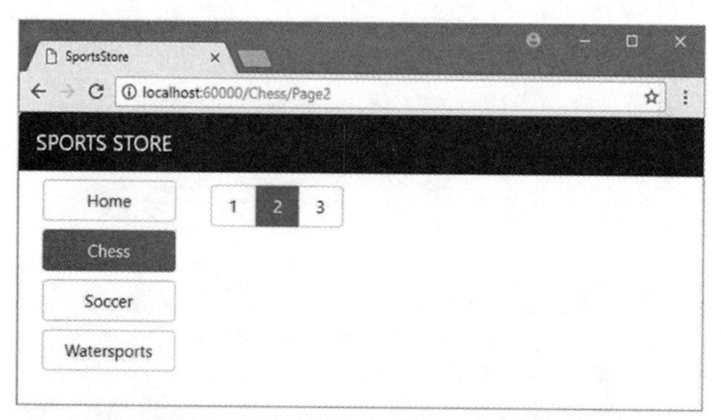

图 9-5　选择类别时显示错误的页面链接

可以通过更新 Product 控制器中的 List 操作方法来解决这个问题，让分页信息按类别计算，如代码清单 9-12 所示。

代码清单 9-12　在 Controllers 文件夹下的 ProductController.cs 文件中创建支持类别的分页数据

```
using Microsoft.AspNetCore.Mvc;
using SportsStore.Models;
using System.Linq;
using SportsStore.Models.ViewModels;

namespace SportsStore.Controllers {

    public class ProductController : Controller {
        private IProductRepository repository;
        public int PageSize = 4;

        public ProductController(IProductRepository repo) {
```

```
            repository = repo;
        }

        public ViewResult List(string category, int productPage = 1)
            => View(new ProductsListViewModel {
                Products = repository.Products
                    .Where(p => category == null || p.Category == category)
                    .OrderBy(p => p.ProductID)
                    .Skip((productPage - 1) * PageSize)
                    .Take(PageSize),
                    PagingInfo = new PagingInfo {
                        CurrentPage = productPage,
                        ItemsPerPage = PageSize,
                        TotalItems = category == null ?
                            repository.Products.Count() :
                            repository.Products.Where(e =>
                                e.Category == category).Count()
                    },
                    CurrentCategory = category
                });
    }
}
```

如果已选择某个类别，就返回该类别中的商品数量；如果没有，就返回商品总数。现在，当查看某个类别时，页面底部的链接正确反映了该类别中的商品数量，如图 9-6 所示。

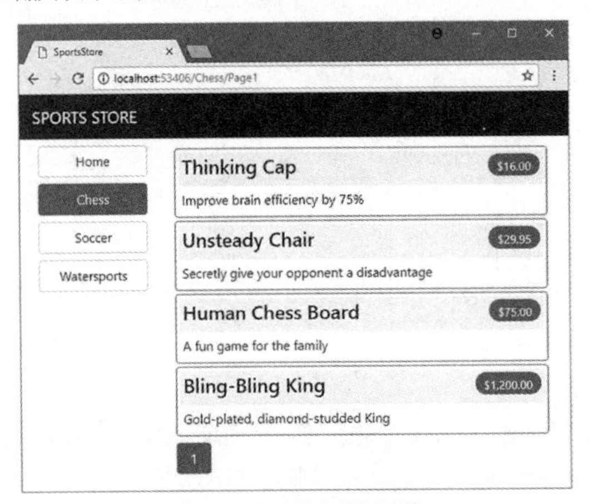

图 9-6　显示特定类别页数

测试特定类别的商品计数

为不同类别计算当前商品数量是很简单的。创建一个模拟的存储库，其中包含一系列类别中的已知数据，然后调用 List 操作方法依次请求每个类别。以下是添加到 ProductControllerTests 类中的单元测试方法：

```
...
[Fact]
public void Generate_Category_Specific_Product_Count() {
    // Arrange
    Mock<IProductRepository> mock = new Mock<IProductRepository>();
    mock.Setup(m => m.Products).Returns((new Product[] {
        new Product {ProductID = 1, Name = "P1", Category = "Cat1"},
        new Product {ProductID = 2, Name = "P2", Category = "Cat2"},
        new Product {ProductID = 3, Name = "P3", Category = "Cat1"},
        new Product {ProductID = 4, Name = "P4", Category = "Cat2"},
        new Product {ProductID = 5, Name = "P5", Category = "Cat3"}
    }).AsQueryable<Product>());
```

```
        ProductController target = new ProductController(mock.Object);
        target.PageSize = 3;

        Func<ViewResult, ProductsListViewModel> GetModel = result =>
            result?.ViewData?.Model as ProductsListViewModel;

        // Action
        int? res1 = GetModel(target.List("Cat1"))?.PagingInfo.TotalItems;
        int? res2 = GetModel(target.List("Cat2"))?.PagingInfo.TotalItems;
        int? res3 = GetModel(target.List("Cat3"))?.PagingInfo.TotalItems;
        int? resAll = GetModel(target.List(null))?.PagingInfo.TotalItems;

        // Assert
        Assert.Equal(2, res1);
        Assert.Equal(2, res2);
        Assert.Equal(1, res3);
        Assert.Equal(5, resAll);
    }
...
```

请注意，以上代码在没有指定类别的情况下调用了 List 操作方法，以确保得到正确的总数。

9.2 构建购物车

现在应用程序进展顺利，但在实现购物车之前，无法销售任何商品。图 9-7 展示了基本的购物流程。在网上购物的人都会熟悉这一流程。

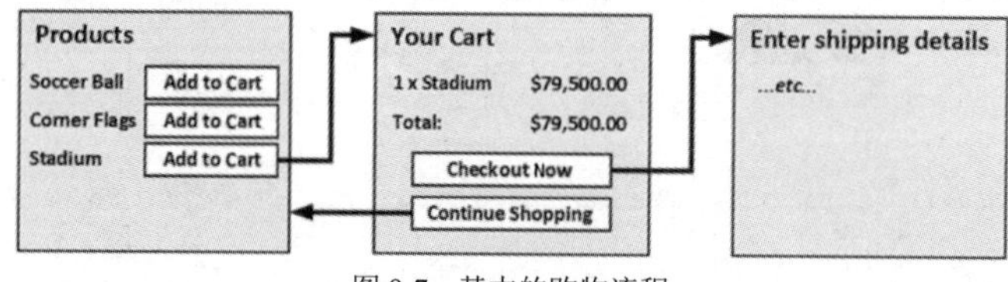

图 9-7　基本的购物流程

目录中每个产品的旁边都会显示 Add to Cart 按钮，单击此按钮将显示客户目前所选商品的摘要，包括总价。此时，用户可以单击 Continue Shopping 按钮返回商品目录，或单击 Checkout Now 按钮完成订单并结束购物流程。

9.2.1 定义购物车模型

将一个名为 Cart.cs 的类文件添加到 Models 文件夹中，并用它定义代码清单 9-13 所示的类。

代码清单 9-13　Models 文件夹下的 Cart.cs 文件的内容

```
using System.Collections.Generic;
using System.Linq;

namespace SportsStore.Models {

    public class Cart {
        private List<CartLine> lineCollection = new List<CartLine>();

        public virtual void AddItem(Product product, int quantity) {
            CartLine line = lineCollection
                .Where(p => p.Product.ProductID == product.ProductID)
                .FirstOrDefault();
```

```
            if (line == null) {
                lineCollection.Add(new CartLine {
                    Product = product,
                    Quantity = quantity
                });
            } else {
                line.Quantity += quantity;
            }
        }

        public virtual void RemoveLine(Product product) =>
            lineCollection.RemoveAll(l => l.Product.ProductID == product.ProductID);

        public virtual decimal ComputeTotalValue() =>
            lineCollection.Sum(e => e.Product.Price * e.Quantity);

        public virtual void Clear() => lineCollection.Clear();

        public virtual IEnumerable<CartLine> Lines => lineCollection;
    }

    public class CartLine {
        public int CartLineID { get; set; }
        public Product Product { get; set; }
        public int Quantity { get; set; }
    }
}
```

在以上代码中，Cart 类使用在同一文件中定义的 CartLine 类来表示客户选择的商品和用户想要购买的数量，还定义了以下功能：将商品添加到购物车，从购物车中删除以前添加的商品，计算购物车中商品的总价格，以及通过删除所有商品重置购物车。另外，还添加了一个属性，让你可以使用 IEnumerable<CartLine> 访问购物车的内容。这些内容比较简单，使用 C#的 LINQ 可以轻松实现。

单元测试——测试购物车

Cart 类相对简单，但它有一系列重要的行为，必须能够正常工作。功能糟糕的购物车会破坏整个 SportsStore 应用程序。我们已经分解了这些功能并单独测试了它们。在 SportsStore.Tests 项目中创建一个名为 CartTests.cs 的单元测试文件，其中包含这些测试。

第一个行为涉及何时将一个商品添加到购物车。如果这是第一次将给定的商品添加到购物车，那么需要添加一个新的 CartLine 对象。以下是包括单元测试类定义的测试：

```
using System.Linq;
using SportsStore.Models;
using Xunit;

namespace SportsStore.Tests {

    public class CartTests {
        [Fact]
        public void Can_Add_New_Lines() {

            // Arrange - create some test products
            Product p1 = new Product { ProductID = 1, Name = "P1" };
            Product p2 = new Product { ProductID = 2, Name = "P2" };

            // Arrange - create a new cart
            Cart target = new Cart();
```

```
            // Act
            target.AddItem(p1, 1);
            target.AddItem(p2, 1);
            CartLine[] results = target.Lines.ToArray();

            // Assert
            Assert.Equal(2, results.Length);
            Assert.Equal(p1, results[0].Product);
            Assert.Equal(p2, results[1].Product);
        }
    }
}
```

但是，如果客户已将商品添加到购物车，则需要增加相应的 CartLine 对象的数量，而不是创建一个新的。测试如下：

```
...
[Fact]
public void Can_Add_Quantity_For_Existing_Lines() {
    // Arrange - create some test products
    Product p1 = new Product { ProductID = 1, Name = "P1" };
    Product p2 = new Product { ProductID = 2, Name = "P2" };

    // Arrange - create a new cart
    Cart target = new Cart();

    // Act
    target.AddItem(p1, 1);
    target.AddItem(p2, 1);
    target.AddItem(p1, 10);
    CartLine[] results = target.Lines
        .OrderBy(c => c.Product.ProductID).ToArray();

    // Assert
    Assert.Equal(2, results.Length);
    Assert.Equal(11, results[0].Quantity);
    Assert.Equal(1, results[1].Quantity);
}
...
```

另外，还需要检查客户是否可以改变主意，并从购物车中删除商品。该功能由 RemoveLine 方法实现。测试如下：

```
...
[Fact]
public void Can_Remove_Line() {
    // Arrange - create some test products
    Product p1 = new Product { ProductID = 1, Name = "P1" };
    Product p2 = new Product { ProductID = 2, Name = "P2" };
    Product p3 = new Product { ProductID = 3, Name = "P3" };

    // Arrange - create a new cart
    Cart target = new Cart();
    // Arrange - add some products to the cart
    target.AddItem(p1, 1);
    target.AddItem(p2, 3);
    target.AddItem(p3, 5);
    target.AddItem(p2, 1);

    // Act
    target.RemoveLine(p2);
```

```
    // Assert
    Assert.Equal(0, target.Lines.Where(c => c.Product == p2).Count());
    Assert.Equal(2, target.Lines.Count());
}
...
```

这里想要测试的下一个行为是计算购物车中商品的总价。以下是对此行为的测试：

```
...
[Fact]
public void Calculate_Cart_Total() {
    // Arrange - create some test products
    Product p1 = new Product { ProductID = 1, Name = "P1", Price = 100M };
    Product p2 = new Product { ProductID = 2, Name = "P2", Price = 50M };

    // Arrange - create a new cart
    Cart target = new Cart();

    // Act
    target.AddItem(p1, 1);
    target.AddItem(p2, 1);
    target.AddItem(p1, 3);
    decimal result = target.ComputeTotalValue();

    // Assert
    Assert.Equal(450M, result);
}
...
```

最后的测试很简单。确保在重置购物车时正确删除了购物车的内容。测试如下：

```
...
[Fact]
public void Can_Clear_Contents() {
    // Arrange - create some test products
    Product p1 = new Product { ProductID = 1, Name = "P1", Price = 100M };
    Product p2 = new Product { ProductID = 2, Name = "P2", Price = 50M };

    // Arrange - create a new cart
    Cart target = new Cart();

    // Arrange - add some items
    target.AddItem(p1, 1);
    target.AddItem(p2, 1);

    // Act - reset the cart
    target.Clear();

    // Assert
    Assert.Equal(0, target.Lines.Count());
}
...
```

在本例中，有时测试类的功能所需的代码相比类本身更多、更复杂。不要因为复杂就放弃单元测试。简单类中的缺陷可能会产生巨大的影响，特别是在示例应用程序中扮演重要作用的一些类，比如 Cart 类。

9.2.2 添加 Add To Cart 按钮

这里需要编辑 Views/Shared/ProductSummary.cshtml 分部视图以将按钮添加到商品列表中。为准备此功能，这里将一个名为 UrlExtensions.cs 的类文件添加到 Infrastructure 文件夹中，并定义了扩展方法，如代码清单 9-14 所示。

代码清单 9-14　Infrastructure 文件夹下的 UrlExtensions.cs 文件的内容

```
using Microsoft.AspNetCore.Http;

namespace SportsStore.Infrastructure {

    public static class UrlExtensions {

        public static string PathAndQuery(this HttpRequest request) =>
            request.QueryString.HasValue
                ? $"{request.Path}{request.QueryString}"
                : request.Path.ToString();
    }
}
```

PathAndQuery 扩展方法在 HttpRequest 类上执行操作，ASP.NET 则使用 HttpRequest 类来描述 HTTP 请求。该扩展方法会生成一个 URL，浏览器将在购物车更新后返回，并将查询字符串考虑在内（如果有的话）。在代码清单 9-15 中，将包含扩展方法的命名空间添加到视图导入文件中，以便在分部视图中使用它。

代码清单 9-15　在 Views 文件夹下的_ViewImports.cshtml 文件中添加命名空间

```
@using SportsStore.Models
@using SportsStore.Models.ViewModels
@using SportsStore.Infrastructure
@addTagHelper *, Microsoft.AspNetCore.Mvc.TagHelpers
@addTagHelper SportsStore.Infrastructure.*, SportsStore
```

在代码清单 9-16 中，已经更新了描述每个商品的分部视图，以包含 Add To Cart 按钮。

代码清单 9-16　将按钮添加到 Views/Shared 文件夹下的 ProductSummary.cshtml 文件中

```
@model Product

<div class="card card-outline-primary m-1 p-1">
    <div class="bg-faded p-1">
        <h4>
            @Model.Name
            <span class="badge badge-pill badge-primary" style="float:right">
                <small>@Model.Price.ToString("c")</small>
            </span>
        </h4>
    </div>
    <form id="@Model.ProductID" asp-action="AddToCart"
            asp-controller="Cart" method="post">
        <input type="hidden" asp-for="ProductID" />
        <input type="hidden" name="returnUrl"
               value="@ViewContext.HttpContext.Request.PathAndQuery()" />
        <span class="card-text p-1">
            @Model.Description
            <button type="submit"
                class="btn btn-success btn-sm pull-right" style="float:right">
                Add To Cart
            </button>
        </span>
    </form>
</div>
```

这里添加了一个 form 元素，其中包含隐藏的 input 元素，用来指定视图模型中的 ProductID 值，以及更新购物车后浏览器应返回的 URL。form 元素和其中一个 input 元素是使用内置标签助手配置的，这是一种很有用的方式，可以生成包含模型值的表单，并在应用程序中定位控制器和操作，如第 24 章所述。另一个 input 元素使用这里创建的扩展方法来设置返回的 URL。这里还添加了一个 button 元素，用来将表单提交给应用程序。

注 意

这里已将 form 元素的 method 属性设置为 post，这指示浏览器使用 HTTP POST 请求提交表单数据。可以更改此设置，以便表单使用 GET 方法，但应该慎重考虑。HTTP 规范要求 GET 请求必须是幂等的，这意味着它们不能导致更改，但将商品添加到购物车绝对是一种更改。第 16 章将介绍有关这个主题的更多内容，包括解释如果忽略幂等 GET 请求的需求，可能会发生什么情况。

9.2.3 启用会话

这里将使用会话状态存储用户购物车的详细信息，会话状态是存储在服务器上并与用户发出的一系列请求相关联的数据。ASP.NET 提供了多种不同方式来存储会话状态，包括将其存储在内存中，这也是这里将要使用的方式。这种方式的优点是简单，但意味着当应用程序停止或重新启动时会话数据会丢失。启用会话需要在 Startup 类中添加服务和中间件，如代码清单 9-17 所示。

代码清单 9-17 在 SportsStore 文件夹下的 Startup.cs 文件中启用会话

```
using System;
using System.Collections.Generic;
using System.Linq;
using System.Threading.Tasks;
using Microsoft.AspNetCore.Builder;
using Microsoft.AspNetCore.Hosting;
using Microsoft.AspNetCore.Http;
using Microsoft.Extensions.DependencyInjection;
using SportsStore.Models;
using Microsoft.Extensions.Configuration;
using Microsoft.EntityFrameworkCore;

namespace SportsStore {

    public class Startup {

        public Startup(IConfiguration configuration) =>
            Configuration = configuration;

        public IConfiguration Configuration { get; }

        public void ConfigureServices(IServiceCollection services) {
            services.AddDbContext<ApplicationDbContext>(options =>
                options.UseSqlServer(
                    Configuration["Data:SportStoreProducts:ConnectionString"]));
            services.AddTransient<IProductRepository, EFProductRepository>();
            services.AddMvc();
            services.AddMemoryCache();
            services.AddSession();
        }
        public void Configure(IApplicationBuilder app, IHostingEnvironment env) {
            app.UseDeveloperExceptionPage();
            app.UseStatusCodePages();
            app.UseStaticFiles();
            app.UseSession();
            app.UseMvc(routes => {

                // ...routing configuration omitted for brevity...

            });
            SeedData.EnsurePopulated(app);
        }
    }
}
```

可调用 AddMemoryCache 方法来设置内存数据存储。AddSession 方法用来注册用于访问会话数据的服务，UseSession 方法允许会话系统在从客户端到达时自动将请求与会话相关联。

9.2.4 实现 Cart 控制器

这里需要一个控制器来处理 Add to Cart 按钮。在 Controllers 文件夹中添加一个名为 CartController.cs 的类文件，并用它定义代码清单 9-18 所示的类。

代码清单 9-18 Controllers 文件夹下的 CartController.cs 文件的内容

```
using System.Linq;
using Microsoft.AspNetCore.Http;
using Microsoft.AspNetCore.Mvc;
using SportsStore.Infrastructure;
using SportsStore.Models;

namespace SportsStore.Controllers {

    public class CartController : Controller {
        private IProductRepository repository;

        public CartController(IProductRepository repo) {
            repository = repo;
        }

        public RedirectToActionResult AddToCart(int productId, string returnUrl) {
            Product product = repository.Products
                .FirstOrDefault(p => p.ProductID == productId);

            if (product != null) {
                Cart cart = GetCart();
                cart.AddItem(product, 1);
                SaveCart(cart);
            }
            return RedirectToAction("Index", new { returnUrl });
        }

        public RedirectToActionResult RemoveFromCart(int productId,
        string returnUrl) {
            Product product = repository.Products
                .FirstOrDefault(p => p.ProductID == productId);

            if (product != null) {
                Cart cart = GetCart();
                cart.RemoveLine(product);
                SaveCart(cart);
            }
            return RedirectToAction("Index", new { returnUrl });
        }

        private Cart GetCart() {
            Cart cart = HttpContext.Session.GetJson<Cart>("Cart") ?? new Cart();
            return cart;
        }

        private void SaveCart(Cart cart) {
            HttpContext.Session.SetJson("Cart", cart);
        }
    }
}
```

关于这个控制器有几点需要注意。首先，使用 ASP.NET 会话状态功能来存储和检索 Cart 对象，这是 GetCart 方法的目的。前面在 9.2.3 节中注册的中间件使用 Cookie 或 URL 重写将来自一个用户的多个请求关联在一起以形成单个浏览会话。相关的功能是会话状态，用于将数据与会话相关联。这是 Cart 类的理想选择：希望每个用户都有自己的购物车，并且希望购物车在请求之间保持一致。会话过期时会删除与会话关联的数据（通常是因为用户在一段时间内没有发出任何请求），这意味着不需要管理 Cart 对象的存储或生命周期。

对于 AddToCart 和 RemoveFromCart 操作方法，可以使用与 ProductSummary.cshtml 视图文件中创建的 HTML 表单中的 input 元素匹配的参数名称。这允许 MVC 将表单传入的 POST 变量与这些参数相关联，进而意味着不需要自己处理表单。这称为模型绑定，模型绑定简化控制器类的强大工具，正如第 26 章解释的那样。

定义会话状态扩展方法

ASP.NET Core 中的会话状态功能仅存储 int、string 和 byte[]值。由于需要存储 Cart 对象，因此需要为 ISession 接口定义扩展方法，ISession 接口提供对会话状态数据的访问，以将 Cart 对象序列化为 JSON 并将其转换回来。在 Infrastructure 文件夹中添加名为 SessionExtensions.cs 的类文件，并定义代码清单 9-19 所示的扩展方法。

代码清单 9-19 Infrastructure 文件夹下的 SessionExtensions.cs 文件的内容

```
using Microsoft.AspNetCore.Http;
using Newtonsoft.Json;

namespace SportsStore.Infrastructure {

    public static class SessionExtensions {

        public static void SetJson(this ISession session, string key, object value) {
            session.SetString(key, JsonConvert.SerializeObject(value));
        }

        public static T GetJson<T>(this ISession session, string key) {
            var sessionData = session.GetString(key);
            return sessionData == null
            ? default(T) : JsonConvert.DeserializeObject<T>(sessionData);
        }
    }
}
```

这些方法依赖于 Json.Net 包并将对象序列化为 JavaScript Object Notation 格式，你将在第 20 章中再次遇到这种格式。Json.Net 包不必添加到项目中，因为 MVC 已经默认集成了它，用来提供 JSON 助手功能，如第 21 章所述（有关使用 Json.Net 包的信息，请参阅 newtonsoft 网站）。

扩展方法可以轻松存储和检索 Cart 对象。要在控制器中将 Cart 添加到会话状态，可以这样做：

```
...
HttpContext.Session.SetJson("Cart", cart);
...
```

HttpContext 属性由 Controller 基类提供，控制器通常从 Controller 基类派生，并返回一个 HttpContext 对象，该对象提供有关已接收的请求和正在准备的响应的上下文数据。HttpContext.Session 属性会返回一个实现了 ISession 接口的对象，ISession 接口定义了 SetJson 方法的类型，SetJson 方法接收一些参数，指定了一个键和要添加到会话状态的对象。扩展方法会序列化对象，并使用 ISession 接口提供的基础功能将其添加到会话状态中。

再次检索购物车，使用另一种扩展方法，但指定相同的键，如下所示：

```
...
Cart cart = HttpContext.Session.GetJson<Cart>("Cart");
...
```

类型参数用于指定要检索的类型，用于反序列化过程。

9.2.5 显示购物车的内容

关于 Cart 控制器的最后一点是，AddToCart 和 RemoveFromCart 方法都调用 RedirectToAction 方法。这具有向客户端浏览器发送 HTTP 重定向指令的效果，要求浏览器请求新的 URL。在这种情况下，我们已经要求浏览器请求一个 URL，该 URL 将调用 Cart 控制器的 Index 操作方法。

实现 Index 操作方法并用它来显示购物车的内容。如果回头参考图 9-7，你将看到这是用户单击 Add to

Cart 按钮时的工作流程。

需要将两条信息传递给显示购物车内容的视图，分别是 Cart 对象以及用户单击 Continue Shopping 按钮时显示的 URL。在 SportsStore 项目的 Models/ViewModels 文件夹中创建名为 CartIndexViewModel.cs 的类文件，并用它定义代码清单 9-20 所示的类。

代码清单 9-20 Models/ViewModels 文件夹下的 CartIndexViewModel.cs 文件的内容

```
using SportsStore.Models;

namespace SportsStore.Models.ViewModels {

    public class CartIndexViewModel {
        public Cart Cart { get; set; }
        public string ReturnUrl { get; set; }
    }
}
```

现在有了视图模型，可以在 Cart 控制器类中实现 Index 操作方法了，如代码清单 9-21 所示。

代码清单 9-21 在 Controllers 文件夹下的 CartController.cs 文件中实现 Index 操作方法

```
using System.Linq;
using Microsoft.AspNetCore.Http;
using Microsoft.AspNetCore.Mvc;
using SportsStore.Infrastructure;
using SportsStore.Models;
using SportsStore.Models.ViewModels;

namespace SportsStore.Controllers {

    public class CartController : Controller {
        private IProductRepository repository;

        public CartController(IProductRepository repo) {
            repository = repo;
        }

        public ViewResult Index(string returnUrl) {
            return View(new CartIndexViewModel {
                Cart = GetCart(),
                ReturnUrl = returnUrl
            });
        }

        // ...other methods omitted for brevity...
    }
}
```

Index 操作方法从会话状态检索 Cart 对象，并使用它创建 CartIndexViewModel 对象，然后将其作为视图模型传递给 View 方法。

显示购物车内容的最后一步是创建 Index 操作方法将要呈现的视图。创建 Views/Cart 文件夹，并在其中添加一个名为 Index.cshtml 的 Razor 视图文件，其中的内容如代码清单 9-22 所示。

代码清单 9-22 Views/Cart 文件夹下的 Index.cshtml 文件的内容

```
@model CartIndexViewModel

<h2>Your cart</h2>
<table class="table table-bordered table-striped">
    <thead>
        <tr>
            <th>Quantity</th>
            <th>Item</th>
            <th class="text-right">Price</th>
```

```
            <th class="text-right">Subtotal</th>
        </tr>
    </thead>
    <tbody>
        @foreach (var line in Model.Cart.Lines) {
            <tr>
                <td class="text-center">@line.Quantity</td>
                <td class="text-left">@line.Product.Name</td>
                <td class="text-right">@line.Product.Price.ToString("c")</td>
                <td class="text-right">
                    @((line.Quantity * line.Product.Price).ToString("c"))
                </td>
            </tr>
        }
    </tbody>
    <tfoot>
        <tr>
            <td colspan="3" class="text-right">Total:</td>
            <td class="text-right">
                @Model.Cart.ComputeTotalValue().ToString("c")
            </td>
        </tr>
    </tfoot>
</table>

<div class="text-center">
    <a class="btn btn-primary" href="@Model.ReturnUrl">Continue Shopping</a>
</div>
```

这个视图枚举了购物车中的行，并将其中的每行以及每行的总价和购物车的总价添加到 HTML 表格中。为了让表格和文本对齐，可以为这些元素分配与 Bootstrap 样式对应的 class 样式。

购物车的基本功能已完成。首先，列出的商品旁边有一个按钮，用来将它们添加到购物车中，如图 9-8 所示。其次，当用户单击 Add to Cart 按钮时，相应的商品将被添加到购物车中，并显示购物车的摘要，如图 9-9 所示。单击 Continue Shopping 按钮可让用户返回之前的商品页面。

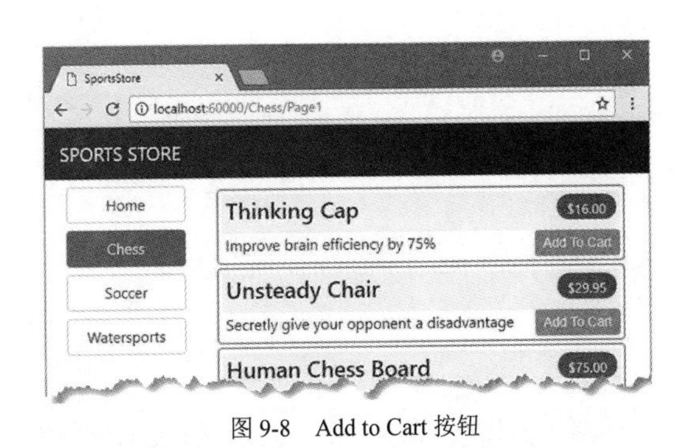

图 9-8 Add to Cart 按钮

图 9-9 显示购物车的内容

9.3 小结

本章介绍了如何实现 SportsStore 应用程序面向客户的部分。我们为用户提供了可以按类别导航的方法，并将基础构建模块准备就绪，以便将商品添加到购物车。此外，还有更多工作要做，我们将在下一章继续开发该应用程序。

第 10 章　完成购物车

本章将继续介绍如何构建 SportsStore 示例应用程序。在上一章中，我们添加了购物车的基本功能，现在将改进并完成这些功能。

10.1　使用服务优化购物车模型

上一章定义了 Cart 模型类，并展示了如何使用会话功能来存储它，从而允许用户保存一组已购买的商品。Cart 类的持久化是由 Cart 控制器管理的，Cart 控制器明确地定义了获取和存储 Cart 对象的方法。

这种方式的问题是，为了获取并存储 Cart 对象，必须在所有使用它们的组件中复制这些代码。在本节中，将使用 ASP.NET Core 的服务功能来简化 Cart 对象的管理方式，释放诸如 Cart 控制器的各个组件，从而无须直接处理细节。

服务的最常用方式是隐藏接口从依赖于它们的组件进行实例化的细节。你已经看到了一个例子，就是为 IProductRepository 接口创建一个服务，这使你能够无缝地将虚拟存储库类用 Entity Framework Core 存储库替换掉。但服务也可用于解决许多其他问题，并且可以用于组成和重构应用程序，即使正在使用具体的类，如 Cart 类。

10.1.1　创建支持存储感知的 Cart 类

整理 Cart 类的使用方式的第一步是创建一个子类，该子类能够知道如何使用会话状态来存储自己。添加一个名为 SessionCart.cs 的类文件到 Models 文件夹中，并用它定义代码清单 10-1 所示的类。

代码清单 10-1　Models 文件夹下的 SessionCart.cs 文件的内容

```
using System;
using Microsoft.AspNetCore.Http;
using Microsoft.Extensions.DependencyInjection;
using Newtonsoft.Json;
using SportsStore.Infrastructure;

namespace SportsStore.Models {

    public class SessionCart: Cart {
        public static Cart GetCart(IServiceProvider services) {
            ISession session = services.GetRequiredService<IHttpContextAccessor>()?
                .HttpContext.Session;
            SessionCart cart = session?.GetJson<SessionCart>("Cart")
                ?? new SessionCart();
            cart.Session = session;
            return cart;
        }

        [JsonIgnore]
        public ISession Session { get; set; }

        public override void AddItem(Product product, int quantity) {
            base.AddItem(product, quantity);
            Session.SetJson("Cart", this);
        }
```

```
        public override void RemoveLine(Product product) {
            base.RemoveLine(product);
            Session.SetJson("Cart", this);
        }

        public override void Clear() {
            base.Clear();
            Session.Remove("Cart");
        }
    }
}
```

SessionCart 类是 Cart 类的子类，并且重写了 AddItem、RemoveLine 和 Clear 方法，从而调用基本实现，然后使用第 9 章定义的 ISession 接口上的扩展方法将更新后的状态存储在会话中。静态方法 GetCart 是一个工厂，用于创建 SessionCart 对象并为它们提供一个 ISession 对象，以便它们可以存储自身。掌握 ISession 对象有点复杂。必须获得 IHttpContextAccessor 服务的一个实例，该服务提供对 HttpContext 对象的访问，而 HttpContext 对象又提供了 ISession。这种间接方法是必需的，因为会话不是作为常规服务提供的。

10.1.2 注册服务

下一步是为 Cart 类创建一个服务。目标是使用 SessionCart 对象来满足 Cart 对象的请求，这些对象将无缝地存储自身。代码清单 10-2 展示了如何创建服务。

代码清单 10-2 在 SportsStore 文件夹下的 Startup.cs 文件中创建购物车服务

```
...
public void ConfigureServices(IServiceCollection services) {
    services.AddDbContext<ApplicationDbContext>(options =>
        options.UseSqlServer(
            Configuration["Data:SportStoreProducts:ConnectionString"]));
    services.AddTransient<IProductRepository, EFProductRepository>();
    services.AddScoped<Cart>(sp => SessionCart.GetCart(sp));
    services.AddSingleton<IHttpContextAccessor, HttpContextAccessor>();
    services.AddMvc();
    services.AddMemoryCache();
    services.AddSession();
}
...
```

AddScoped 方法指定应使用相同的对象来满足 Cart 实例的相关请求。请求之间的关系是可以配置的，但默认情况下，这意味着处理相同 HTTP 请求的组件所需的任何 Cart 都将接收到相同的对象。

这里没有像为存储库那样为 AddScoped 方法提供类型映射的 AddScoped 方法，而是指定了一个 Lambda 表达式，这个 Lambda 表达式将被调用以满足 Cart 请求。这个 Lambda 表达式接收已注册的服务集合，并将集合传递给 SessionCart 类的 GetCart 方法。结果是，对 Cart 服务的请求将通过创建 SessionCart 对象来处理，当这些对象被修改时，会将自己序列化为会话数据。

这里还使用 AddSingleton 方法添加了一个服务，该方法指定应始终使用相同的对象，该服务告诉 MVC 在需要 IHttpContextAccessor 接口的实现时使用 HttpContextAccessor 类。该服务是必需的，因此在这里可以访问代码清单 10-1 中的 SessionCart 类中的当前会话。

10.1.3 简化 Cart 控制器

创建这种服务的好处是能够简化使用 Cart 对象的控制器。通过代码清单 10-3，重新设计 CartController 类以利用新服务。

代码清单 10-3 在 Controllers 文件夹下的 CartController.cs 文件中使用购物车服务

```
using System.Linq;
using Microsoft.AspNetCore.Mvc;
```

```
using SportsStore.Models;
using SportsStore.Models.ViewModels;

namespace SportsStore.Controllers {

    public class CartController : Controller {
        private IProductRepository repository;
        private Cart cart;

        public CartController(IProductRepository repo, Cart cartService) {
            repository = repo;
            cart = cartService;
        }

        public ViewResult Index(string returnUrl) {
            return View(new CartIndexViewModel {
                Cart = cart,
                ReturnUrl = returnUrl
            });
        }
        public RedirectToActionResult AddToCart(int productId, string returnUrl) {
            Product product = repository.Products
                .FirstOrDefault(p => p.ProductID == productId);
            if (product != null) {
                cart.AddItem(product, 1);
            }
            return RedirectToAction("Index", new { returnUrl });
        }

        public RedirectToActionResult RemoveFromCart(int productId,
                string returnUrl) {
            Product product = repository.Products
                .FirstOrDefault(p => p.ProductID == productId);

            if (product != null) {
                cart.RemoveLine(product);
            }
            return RedirectToAction("Index", new { returnUrl });
        }
    }
}
```

CartController 类通过声明构造函数的参数来指示这一个 Cart 对象，这样就可以删除从会话读取和写入数据的方法以及写入更新所需的步骤。其结果是一个更简单的控制器，并且仍然专注于它在应用程序中担负的角色，而不必担心如何创建或持久化 Cart 对象。此外，由于服务在整个应用程序中都可以使用，因此任何组件都可以使用相同的技术来处理用户的购物车。

10.2 完成购物车功能

既然上一节已经介绍了购物车服务，现在是时候通过添加两个新功能来完成购物车功能了。第一个功能将允许客户从购物车中删除商品。第二个功能将在页面顶部显示购物车的摘要。

10.2.1 从购物车中删除商品

由于已经在控制器中定义并测试了 RemoveFromCart 操作方法，因此只需要在视图中公开这个操作方法即可让客户从购物车中删除商品，可通过在购物车摘要的每一行中添加 Remove 按钮来完成此操作。代码清单 10-4 显示了对 Views/Cart/Index.cshtml 文件所做的更改。

代码清单 10-4 将 Remove 按钮引入 Views/Cart 文件夹下的 Index.cshtml 文件

```
@model CartIndexViewModel

<h2>Your cart</h2>
<table class="table table-bordered table-striped">
    <thead>
        <tr>
            <th>Quantity</th>
            <th>Item</th>
            <th class="text-right">Price</th>
            <th class="text-right">Subtotal</th>
        </tr>
    </thead>
    <tbody>
        @foreach (var line in Model.Cart.Lines) {
            <tr>
                <td class="text-center">@line.Quantity</td>
                <td class="text-left">@line.Product.Name</td>
                <td class="text-right">@line.Product.Price.ToString("c")</td>
                <td class="text-right">
                    @((line.Quantity * line.Product.Price).ToString("c"))
                </td>
                <td>
                    <form asp-action="RemoveFromCart" method="post">
                        <input type="hidden" name="ProductID"
                               value="@line.Product.ProductID" />
                        <input type="hidden" name="returnUrl"
                               value="@Model.ReturnUrl" />
                        <button type="submit" class="btn btn-sm btn-danger">
                            Remove
                        </button>
                    </form>
                </td>
            </tr>
        }
    </tbody>
    <tfoot>
        <tr>
            <td colspan="3" class="text-right">Total:</td>
            <td class="text-right">
                @Model.Cart.ComputeTotalValue().ToString("c")
            </td>
        </tr>
    </tfoot>
</table>

<div class="text-center">
    <a class="btn btn-primary" href="@Model.ReturnUrl">Continue shopping</a>
</div>
```

以上代码在表格的每一行中添加了一个新列，其中包含一个带有隐藏的 input 元素的 form 表单，用于指定要删除的商品和要返回的 URL，以及用于提交表单的按钮。

通过运行应用程序并将商品添加到购物车中，可以查看 Remove 按钮如何工作。请记住，购物车已经包含删除商品的功能，可以通过单击其中一个新的按钮来测试，如图 10-1 所示。

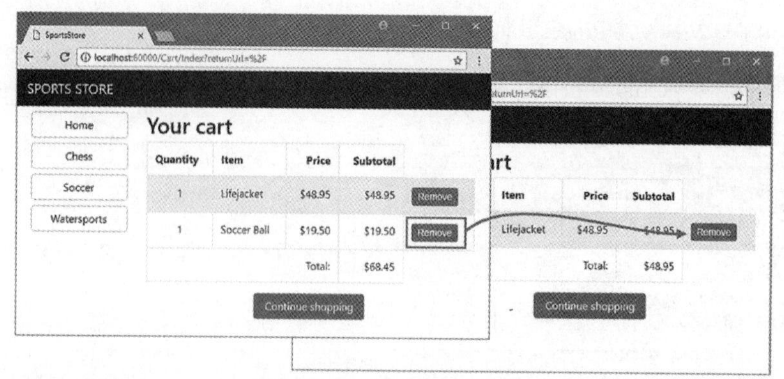

图 10-1 从购物车中删除商品

10.2.2 添加购物车摘要小部件

虽然已经有了功能正常的购物车，但也有一个问题：客户只能查看购物车摘要页面才能了解购物车中的商品，并且他们只能通过向购物车添加新商品来查看购物车摘要页面。

为了解决这个问题，可添加一个小部件，简要显示购物车的内容，并且单击该小部件就可以在整个应用程序中显示购物车的内容。我们以与添加导航小部件相同的方式执行此操作，作为视图组件，其输出可以包含在 Razor 共享布局中。

1. 添加 Font Awesome 包

作为购物车摘要的一部分，需要显示一个允许用户结账的按钮。这里想使用购物车图标，而不是在按钮上显示单词 Checkout。Font Awesome 包提供了一组优秀的开源图标，可以作为字体集成到应用程序中，字体中的每个字符都是不同的图像。可以在 GitHub 上了解有关 Font Awesome 包的更多信息，包括查看其中包含的图标。这里选择 SportsStore 项目，然后单击 Solution Explorer 窗格顶部的 Show All Items 按钮以显示 bower.json 文件。接下来，将 Font Awesome 包添加到 dependencies 部分，如代码清单 10-5 所示。

代码清单 10-5 在 SportsStore 文件夹下的 bower.json 文件中添加 Font Awesome 包

```
{
  "name": "asp.net",
  "private": true,
  "dependencies": {
    "bootstrap": "4.0.0-alpha.6",
    "fontawesome": "4.7.0"
  }
}
```

保存 bower.json 文件后，Visual Studio 使用 Bower 在 www/lib/fontawesome 文件夹中下载并安装 Font Awesome 包。

2. 创建视图组件类和视图

下面在 Components 文件夹中添加名为 CartSummaryViewComponent.cs 的类文件，并用它定义代码清单 10-6 所示的视图组件。

代码清单 10-6 Components 文件夹下的 CartSummaryViewComponent.cs 文件的内容

```
using Microsoft.AspNetCore.Mvc;
using SportsStore.Models;

namespace SportsStore.Components {

    public class CartSummaryViewComponent : ViewComponent {
        private Cart cart;
```

```
        public CartSummaryViewComponent(Cart cartService) {
            cart = cartService;
        }

        public IViewComponentResult Invoke() {
            return View(cart);
        }
    }
}
```

这个视图组件能够利用本章前面创建的服务，将 Cart 对象作为构造函数的参数接收。结果是一个简单的视图组件类，它将 Cart 对象传递给 View 方法，以便生成包含在布局中的 HTML 片段。为了创建布局，创建 Views/Shared/Components/CartSummary 文件夹，在其中添加名为 Default.cshtml 的 Razor 视图文件，并添加代码清单 10-7 所示的标记。

代码清单 10-7 Views/Shared/Components/CartSummary 文件夹下的 Default.cshtml 文件的内容

```
@model Cart

<div class="">
    @if (Model.Lines.Count() > 0) {
        <small class="navbar-text">
            <b>Your cart:</b>
            @Model.Lines.Sum(x => x.Quantity) item(s)
            @Model.ComputeTotalValue().ToString("c")
        </small>
    }
    <a class="btn btn-sm btn-secondary navbar-btn"
        asp-controller="Cart" asp-action="Index"
        asp-route-returnurl="@ViewContext.HttpContext.Request.PathAndQuery()">
         <i class="fa fa-shopping-cart"></i>
    </a>
</div>
```

视图将显示带有 Font Awesome 购物车图标的按钮，如果购物车中有商品，则会提供一个页面，详细说明商品的数量及总价。现在有了视图组件和视图，接下来可以修改共享布局，以将购物车摘要包含在应用程序控制器生成的响应中，如代码清单 10-8 所示。

代码清单 10-8 在 Views/Shared 文件夹下的_Layout.cshtml 文件中添加购物车摘要

```
<!DOCTYPE html>

<html>
<head>
    <meta name="viewport" content="width=device-width" />
    <link rel="stylesheet"
          asp-href-include="/lib/bootstrap/dist/**/*.min.css"
          asp-href-exclude="**/*-reboot*,**/*-grid*" />
    <link rel="stylesheet" asp-href-include="/lib/fontawesome/css/*.css" />
    <title>SportsStore</title>
</head>
<body>
    <div class="navbar navbar-inverse bg-inverse" role="navigation">
        <div class="row">
            <a class="col navbar-brand" href="#">SPORTS STORE</a>
            <div class="col-4 text-right">
                @await Component.InvokeAsync("CartSummary")
            </div>
        </div>
```

```
        </div>
        <div class="row m-1 p-1">
            <div id="categories" class="col-3">
                @await Component.InvokeAsync("NavigationMenu")
            </div>
            <div class="col-9">
                @RenderBody()
            </div>
        </div>
    </body>
    </html>
```

可以通过启动应用程序来查看购物车摘要。当购物车为空时，仅显示 Checkout 按钮。如果将商品添加到了购物车中，则会显示商品的数量及总价，如图 10-2 所示。通过以上修改，客户可以了解购物车中的内容，并可以方便地结账。

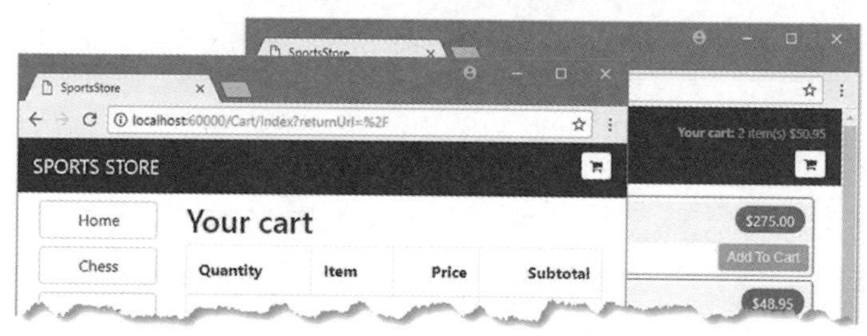

图 10-2 显示购物车摘要

10.3 提交订单

现在已经完成了 SportsStore 应用程序最终面向客户的功能——检查并完成订单。本节将介绍如何扩展域模型，支持获取用户的装运信息，并添加应用程序支持以处理它们。

10.3.1 创建模型类

在 Models 文件夹中添加一个名为 Order.cs 的类文件，内容如代码清单 10-9 所示。这是用于表示客户装运信息的类。

代码清单 10-9 Models 文件夹下的 Order.cs 文件的内容

```
using System.Collections.Generic;
using System.ComponentModel.DataAnnotations;
using Microsoft.AspNetCore.Mvc.ModelBinding;

namespace SportsStore.Models {

    public class Order {

        [BindNever]
        public int OrderID { get; set; }
        [BindNever]
        public ICollection<CartLine> Lines { get; set; }

        [Required(ErrorMessage = "Please enter a name")]
        public string Name { get; set; }

        [Required(ErrorMessage = "Please enter the first address line")]
```

```
        public string Line1 { get; set; }
        public string Line2 { get; set; }
        public string Line3 { get; set; }
        [Required(ErrorMessage = "Please enter a city name")]
        public string City { get; set; }

        [Required(ErrorMessage = "Please enter a state name")]
        public string State { get; set; }

        public string Zip { get; set; }

        [Required(ErrorMessage = "Please enter a country name")]
        public string Country { get; set; }

        public bool GiftWrap { get; set; }
    }
}
```

使用 System.ComponentModel.DataAnnotations 命名空间中的验证属性，就像在第 2 章中所做的那样。第 27 章将进一步描述验证。

这里还使用了 BindNever 特性，该特性会阻止用户在 HTTP 请求中为这些属性提供值。这是在第 26 章将要描述的模型绑定系统的特性之一——阻止 MVC 使用 HTTP 请求中的值来填充敏感或重要的模型属性。

10.3.2　添加结账流程

为了使用户能够输入装运信息并提交订单，首先需要在购物车的摘要视图中添加 Checkout 按钮。代码清单 10-10 显示了对 Views/Cart/Index.cshtml 文件所做的更改。

代码清单 10-10　将 Checkout 按钮添加到 Views/Cart 文件夹下的 Index.cshtml 文件中

```
...
<div class="text-center">
    <a class="btn btn-primary" href="@Model.ReturnUrl">Continue Shopping</a>
    <a class="btn btn-primary" asp-action="Checkout" asp-controller="Order">
        Checkout
    </a>
</div>
...
```

以上更改会生成一个链接，将其样式化为按钮，并在单击时调用 Order 控制器的 Checkout 操作方法。可以在图 10-3 中看到 Checkout 按钮的显示方式。

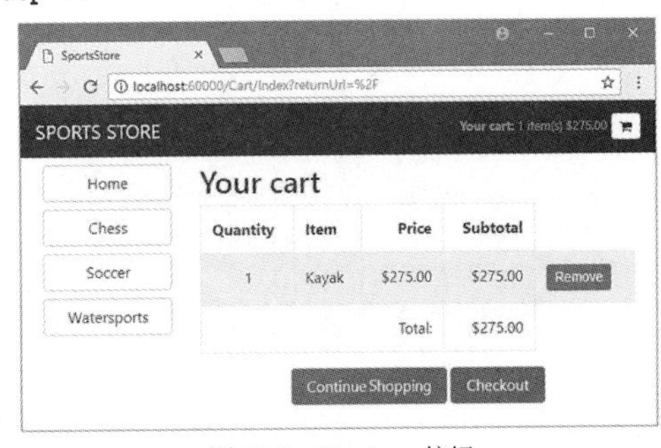

图 10-3　Checkout 按钮

现在需要定义 Order 控制器。在 Controllers 文件夹中添加一个名为 OrderController.cs 的类文件，并用它定义代码清单 10-11 所示的类。

代码清单 10-11　Controllers 文件夹下的 OrderController.cs 文件的内容

```
using Microsoft.AspNetCore.Mvc;
using SportsStore.Models;

namespace SportsStore.Controllers {

    public class OrderController : Controller {

        public ViewResult Checkout() => View(new Order());
    }
}
```

Checkout 方法返回默认视图，并传递一个新的 ShippingDetails 对象作为视图模型。为了创建视图，创建 Views/Order 文件夹并添加一个名为 Checkout.cshtml 的 Razor 视图文件，其中包含代码清单 10-12 所示的内容。

代码清单 10-12 Views/Order 文件夹下的 Checkout.cshtml 文件的内容

```
@model Order

<h2>Check out now</h2>
<p>Please enter your details, and we'll ship your goods right away!</p>

<form asp-action="Checkout" method="post">
    <h3>Ship to</h3>
    <div class="form-group">
        <label>Name:</label><input asp-for="Name" class="form-control" />
    </div>
    <h3>Address</h3>
    <div class="form-group">
        <label>Line 1:</label><input asp-for="Line1" class="form-control" />
    </div>
    <div class="form-group">
        <label>Line 2:</label><input asp-for="Line2" class="form-control" />
    </div>
    <div class="form-group">
        <label>Line 3:</label><input asp-for="Line3" class="form-control" />
    </div>
    <div class="form-group">
        <label>City:</label><input asp-for="City" class="form-control" />
    </div>
    <div class="form-group">
        <label>State:</label><input asp-for="State" class="form-control" />
    </div>
    <div class="form-group">
        <label>Zip:</label><input asp-for="Zip" class="form-control" />
    </div>
    <div class="form-group">
        <label>Country:</label><input asp-for="Country" class="form-control" />
    </div>
    <h3>Options</h3>
    <div class="checkbox">
        <label>
            <input asp-for="GiftWrap" /> Gift wrap these items
        </label>
    </div>
    <div class="text-center">
        <input class="btn btn-primary" type="submit" value="Complete Order" />
    </div>
</form>
```

对于模型中的每个属性，创建 Bootstrap 样式的 label 元素和 input 元素来捕获用户输入。input 元素的 asp-for 属性由内置的标签助手处理，标签助手将根据指定的模型属性生成 type、id、name 和 value 属性，如第 24 章所述。

你可以通过启动应用程序，单击页面顶部的购物车图标，然后单击 Checkout 按钮来查看操作方法和视图的效果，如图 10-4 所示。你也可以通过请求/Cart/Checkout 的 URL 来查看这个页面。

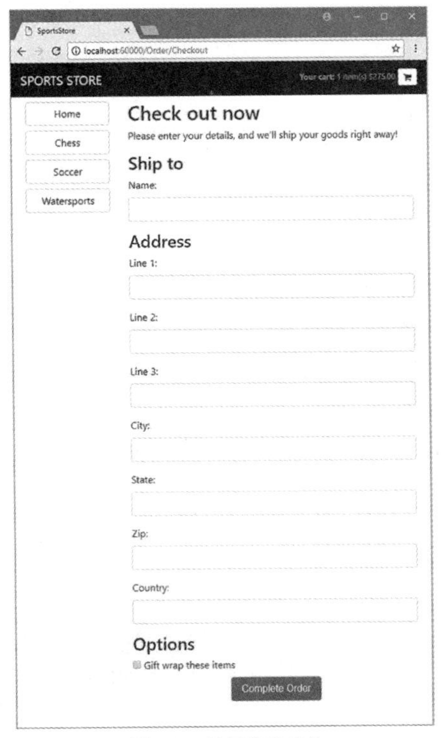

图 10-4 装运信息表单

10.3.3 实现订单处理

可通过将订单写入数据库来处理订单。当然，大多数电子商务网站不会简单地如此实现，而且这里没有提供处理信用卡或其他付款方式的功能。但是因为这里想把重点放在 MVC 上，所以一个简单的数据库条目就可以了。

1. 扩展数据库

如果在第 8 章中创建的基本管道程序就位了，就可以将新的模型添加到数据库中。首先，向数据库上下文类添加一个新的属性，如代码清单 10-13 所示。

代码清单 10-13 在 Models 文件夹下的 ApplicationDbContext.cs 文件中添加属性

```
using Microsoft.EntityFrameworkCore;
using Microsoft.EntityFrameworkCore.Design;
using Microsoft.EntityFrameworkCore.Infrastructure;
using Microsoft.Extensions.DependencyInjection;

namespace SportsStore.Models {

    public class ApplicationDbContext : DbContext {

        public ApplicationDbContext(DbContextOptions<ApplicationDbContext> options)
            : base(options) { }

        public DbSet<Product> Products { get; set; }
        public DbSet<Order> Orders { get; set; }
    }
}
```

上述更改能够让 Entity Framework Core 创建数据库迁移，以将 Order 对象存储在数据库中。要创建数据库迁移，请打开新的命令提示符或 PowerShell 窗口，导航到 SportsStore 项目文件夹（其中包含 Startup.cs

文件)，然后运行以下命令：

```
dotnet ef migrations add Orders
```

上述命令告诉 Entity Framework Core 获取应用程序数据模型的新快照，并对比与以前数据库版本的不同之处，生成名为 Orders 的数据库迁移。应用程序在启动时将自动应用新的数据库迁移，因为 SeedData 会调用 Entity Framework Core 提供的 Migrate 方法。

重置数据库

当频繁更改模型时，数据库迁移和数据库架构不同步时会出现问题。最简单的方法是删除数据库并重新开始。但是这仅适用于开发期间，因为你会丢失已存储的任何数据。

要删除数据库，请在 SportsStore 项目文件夹中运行以下命令：

```
dotnet ef database drop --force
```

删除数据库后，从 SportsStore 项目文件夹运行以下命令以重新创建数据库，并通过运行以下命令应用创建的数据库迁移：

```
dotnet ef database update
```

这将重置数据库，以准确反映模型的更改，这样你就可以继续开发应用程序了。

2. 创建订单存储库

下面沿用产品存储库的模式，以提供对 Order 对象的访问。在 Models 文件夹中添加一个名为 IOrderRepository.cs 的类文件，并用它定义代码清单 10-14 所示的接口。

代码清单 10-14 Models 文件夹下的 IOrderRepository.cs 文件的内容

```
using System.Linq;

namespace SportsStore.Models {

    public interface IOrderRepository {

        IQueryable<Order> Orders { get; }
        void SaveOrder(Order order);
    }
}
```

为了实现订单存储库接口，将一个名为 EFOrderRepository.cs 的类文件添加到 Models 文件夹中，并定义代码清单 10-15 所示的类。

代码清单 10-15 Models 文件夹下的 EFOrderRepository.cs 文件的内容

```
using Microsoft.EntityFrameworkCore;
using System.Linq;

namespace SportsStore.Models {

    public class EFOrderRepository : IOrderRepository {
        private ApplicationDbContext context;

        public EFOrderRepository(ApplicationDbContext ctx) {
            context = ctx;
        }
        public IQueryable<Order> Orders => context.Orders
                            .Include(o => o.Lines)
                            .ThenInclude(l => l.Product);

        public void SaveOrder(Order order) {
```

```
            context.AttachRange(order.Lines.Select(l => l.Product));
            if (order.OrderID == 0) {
                context.Orders.Add(order);
            }
            context.SaveChanges();
        }
    }
}
```

这个类使用 Entity Framework Core 实现了 IOrderRepository，从而能够检索已存储的 Order 对象集，并允许创建或更改订单。

了解订单存储库

为代码清单 10-15 中的订单实现存储库需要做一些额外的工作。Entity Framework Core 需要指令来加载相关数据（如果跨越多个表的话）。在代码中，使用 Include 和 ThenInclude 方法指定从数据库中读取 Order 对象时，还应该一起加载与 Lines 属性关联的集合以及与每个集合对象关联的 Product 对象。

```
...
public IQueryable<Order> Orders => context.Orders
    .Include(o => o.Lines)
    .ThenInclude(l => l.Product);
...
```

这将确保收到所需的所有数据对象，而无须执行查询和重新组装数据。

当在数据库中存储 Order 对象时，还需要执行一个额外的步骤。当从会话存储中反序列化用户的购物车数据时，JSON 包会创建 Entity Framework Core 不知道的新对象，然后尝试将所有对象写入数据库。对于 Product 对象，这意味着 Entity Framework Core 尝试写入已存储的对象，这会导致错误。为了避免这个问题，可通知 Entity Framework Core 对象已存在，除非它们被修改，否则不应该存储在数据库中，如下所示：

```
...
context.AttachRange(order.Lines.Select(l => l.Product));
...
```

这可确保 Entity Framework Core 不会尝试写入与 Order 对象关联的反序列化后的 Product 对象。

在代码清单 10-16 中，我们已在 Startup 类的 ConfigureServices 方法中将订单存储库注册为服务。

代码清单 10-16 在 SportsStore 文件夹下的 Startup.cs 文件中注册 Order 存储库服务

```
...
public void ConfigureServices(IServiceCollection services) {
    services.AddDbContext<ApplicationDbContext>(options =>
        options.UseSqlServer(
            Configuration["Data:SportStoreProducts:ConnectionString"]));
    services.AddTransient<IProductRepository, EFProductRepository>();
    services.AddScoped<Cart>(sp => SessionCart.GetCart(sp));
    services.AddSingleton<IHttpContextAccessor, HttpContextAccessor>();
    services.AddTransient<IOrderRepository, EFOrderRepository>();
    services.AddMvc();
    services.AddMemoryCache();
    services.AddSession();
}
...
```

10.3.4 完成 Order 控制器

为了完成 OrderController 类，还需要修改构造函数，以便接收处理订单所需的服务，并且需要添加一个新的操作方法。当用户单击 Complete Order 按钮时，这个操作方法将处理 HTTP 表单的 POST 请求。代码清单 10-17 显示了这两处更改。

代码清单 10-17 在 Controllers 文件夹下的 OrderController.cs 文件中完成 Order 控制器

```
using Microsoft.AspNetCore.Mvc;
using SportsStore.Models;
using System.Linq;

namespace SportsStore.Controllers {

    public class OrderController : Controller {
        private IOrderRepository repository;
        private Cart cart;

        public OrderController(IOrderRepository repoService, Cart cartService) {
            repository = repoService;
            cart = cartService;
        }

        public ViewResult Checkout() => View(new Order());

        [HttpPost]
        public IActionResult Checkout(Order order) {
            if (cart.Lines.Count() == 0) {
                ModelState.AddModelError("", "Sorry, your cart is empty!");
            }
            if (ModelState.IsValid) {
                order.Lines = cart.Lines.ToArray();
                repository.SaveOrder(order);
                return RedirectToAction(nameof(Completed));
            } else {
                return View(order);
            }
        }

        public ViewResult Completed() {
            cart.Clear();
            return View();
        }
    }
}
```

Checkout 操作方法使用 HttpPost 特性进行修饰，这意味着仅为 POST 请求调用这个操作方法。在本例中，也就是当用户提交表单时。再次使用模型绑定系统以接收 Order 对象，然后处理 Cart 中的数据并存储在存储库中。

MVC 使用数据注解属性检查应用于 Order 类的验证约束，并且任何验证相关的问题都会通过 ModelState 属性传递给操作方法。可以通过检查 ModelState.IsValid 属性来查看是否存在问题。如果购物车中没有商品，就调用 ModelState.AddModelError 方法来注册错误消息。稍后会解释如何显示此类错误，第 27 章和第 28 章将介绍更多关于模型绑定和验证的内容。

单元测试——订单处理

为了对 OrderController 类执行单元测试，需要测试 Checkout 方法的 POST 请求的行为。尽管 Checkout 方法看起来很简单，但使用 MVC 模型绑定意味着对测试来说还有很多事情要做。

我们想仅在购物车中有商品并且客户提供有效的装运信息时处理订单，而在其他情况下应向客户显示错误。在 SportsStore.Test 项目中，在名为 OrderControllerTests.cs 的类文件中定义第一个测试方法。

```
using Microsoft.AspNetCore.Mvc;
using Moq;
using SportsStore.Controllers;
using SportsStore.Models;
using Xunit;
```

```
namespace SportsStore.Tests {

    public class OrderControllerTests {

        [Fact]
        public void Cannot_Checkout_Empty_Cart() {
            // Arrange - create a mock repository
            Mock<IOrderRepository> mock = new Mock<IOrderRepository>();
            // Arrange - create an empty cart
            Cart cart = new Cart();
            // Arrange - create the order
            Order order = new Order();
            // Arrange - create an instance of the controller
            OrderController target = new OrderController(mock.Object, cart);

            // Act
            ViewResult result = target.Checkout(order) as ViewResult;

            // Assert - check that the order hasn't been stored
            mock.Verify(m => m.SaveOrder(It.IsAny<Order>()), Times.Never);
            // Assert - check that the method is returning the default view
            Assert.True(string.IsNullOrEmpty(result.ViewName));
            // Assert - check that I am passing an invalid model to the view
            Assert.False(result.ViewData.ModelState.IsValid);
        }
    }
}
```

以上测试可确保在购物车为空时不能结账。可通过确保不调用模拟IOrderRepository实现的SaveOrder方法来检查这一点，该方法返回的是默认视图（它将重新显示客户输入的数据并给他们提供纠正的机会），并且传递给视图的模型状态已被标记为无效。这可能看起来像是万无一失的断言，但需要确认3个部分以确保得到正确的结果。下一个测试方法的工作方式大致相同，但在视图模型中会注入错误，以模拟模型绑定器报告的问题（当客户输入无效的装运数据时会出现）：

```
...
[Fact]
public void Cannot_Checkout_Invalid_ShippingDetails() {

    // Arrange - create a mock order repository
    Mock<IOrderRepository> mock = new Mock<IOrderRepository>();
    // Arrange - create a cart with one item
    Cart cart = new Cart();
    cart.AddItem(new Product(), 1);
    // Arrange - create an instance of the controller
    OrderController target = new OrderController(mock.Object, cart);
    // Arrange - add an error to the model
    target.ModelState.AddModelError("error", "error");

    // Act - try to checkout
    ViewResult result = target.Checkout(new Order()) as ViewResult;

    // Assert - check that the order hasn't been passed stored
    mock.Verify(m => m.SaveOrder(It.IsAny<Order>()), Times.Never);
    // Assert - check that the method is returning the default view
    Assert.True(string.IsNullOrEmpty(result.ViewName));
    // Assert - check that I am passing an invalid model to the view
    Assert.False(result.ViewData.ModelState.IsValid);
}
...
```

除确保空的购物车或无效的装运信息能阻止继续处理订单之外，还需要确保在适当的时候处理订单。测试如下：

```
...
[Fact]
public void Can_Checkout_And_Submit_Order() {
    // Arrange - create a mock order repository
    Mock<IOrderRepository> mock = new Mock<IOrderRepository>();
    // Arrange - create a cart with one item
    Cart cart = new Cart();
    cart.AddItem(new Product(), 1);
    // Arrange - create an instance of the controller
    OrderController target = new OrderController(mock.Object, cart);

    // Act - try to checkout
    RedirectToActionResult result =
        target.Checkout(new Order()) as RedirectToActionResult;

    // Assert - check that the order has been stored
    mock.Verify(m => m.SaveOrder(It.IsAny<Order>()), Times.Once);
    // Assert - check that the method is redirecting to the Completed action
    Assert.Equal("Completed", result.ActionName);
}
...
```

不需要测试那些可以识别的有效的装运细节。这是由模型绑定器自动处理的，可通过在 Order 类的属性上应用属性注释来实现。

10.3.5　显示验证错误

MVC 将使用应用于 Order 类的验证属性来验证用户数据。但是，还需要进行简单的更改以显示所有问题。这依赖另一个内置的标签助手，它检查用户提供的数据的验证状态，并为已发现的每个问题添加警告消息。代码清单 10-18 显示了一个 HTML 元素的附加部分，该 HTML 元素将由标签助手在 Checkout.cshtml 文件中处理。

代码清单 10-18　将验证摘要添加到 Views/Order 文件夹下的 Checkout.cshtml 文件中

```
@model Order

<h2>Check out now</h2>
<p>Please enter your details, and we'll ship your goods right away!</p>

<div asp-validation-summary="All" class="text-danger"></div>

<form asp-action="Checkout" method="post">
    <h3>Ship to</h3>
...
```

通过进行上述简单更改，可以向用户报告验证错误。要查看效果，请转到/Order/Checkout URL，并在不选择任何商品或填写任何装运信息的情况下尝试结账，如图 10-5 所示。生成这些消息的标签助手是模型验证系统的一部分，将在第 27 章中详细介绍。

提　示

用户提交的数据在验证之前会被发送到服务器，这称为服务器端验证，MVC 对此提供良好的支持。服务器端验证的问题是，在将数据发送到服务器并进行处理以生成结果页面之前，不会向用户显示错误——在繁忙的服务器上有时可能需要几秒的时间。因此，服务器端验证通常需要补充使用客户端验证，在将表单数据发送到服务器之前，使用 JavaScript 检查用户输入的值。第 27 章将描述客户端验证。

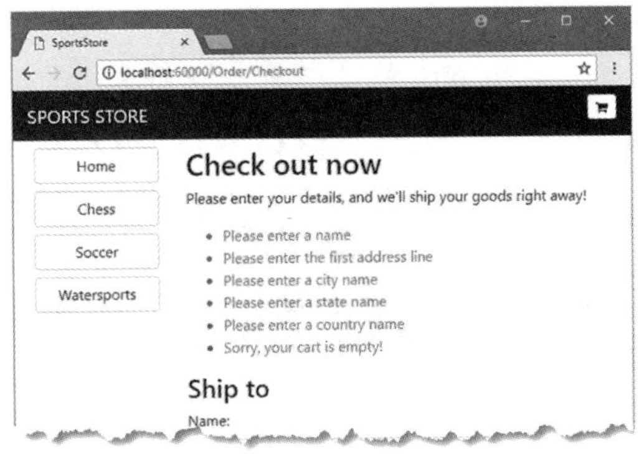

图 10-5　显示验证消息

10.3.6　显示摘要页面

为了完成结账流程，需要创建一个视图，在浏览器重定向到 Order 控制器上的 Completed action 时显示。在 Views/Order 文件夹中添加一个名为 Completed.cshtml 的 Razor 视图文件，并在其中添加代码清单 10-19 所示的代码。

代码清单 10-19　Views/Order 文件夹下的 Completed.cshtml 文件的内容

```
<h2>Thanks!</h2>
<p>Thanks for placing your order.</p>
<p>We'll ship your goods as soon as possible.</p>
```

在将这个视图集成到应用程序时，不需要更改任何代码，因为在定义 Completed 操作方法时已添加必需的语句。现在，客户可以完成从选择商品到结账的整个过程。如果提供有效的装运信息（并且购物车中有商品），在单击 Complete Order 按钮后就会看到摘要页面，如图 10-6 所示。

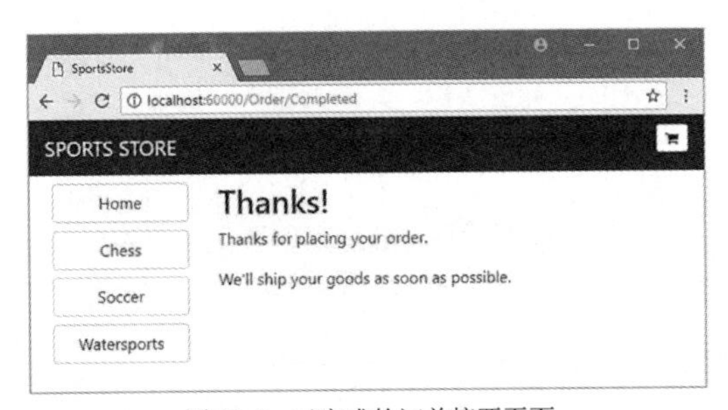

图 10-6　已完成的订单摘要页面

10.4　小结

本章介绍了如何实现 SportsStore 应用程序面向客户的所有主要功能。这里已经有了商品目录，可以按类别和页面浏览，还提供了简洁的购物车和方便的结账流程。

分层良好的架构意味着可以轻松地更改应用程序的任何部分，而不用担心在其他地方导致问题或不一致。例如，可以更改订单的存储方式，而不会对购物车、商品目录或应用程序的任何其他部分产生影响。在下一章中，将添加管理 SportsStore 应用程序所需的功能。

第 11 章　SportsStore 的管理

本章将继续介绍如何构建 SportsStore 应用程序，为站点管理员提供管理订单和商品的功能。

11.1　管理订单

在上一章中，添加了接收客户订单并将其存储在数据库中的功能。在本章中，将创建一个简单的管理工具，用于查看已收到的订单并将其标记为已发货。

11.1.1　增强模型

需要做的第一个更改是增强模型，用来记录已发货的订单。代码清单 11-1 在 Order 类中添加了一个新的属性，Order 类定义在 Models 文件夹下的 Order.cs 文件中。

代码清单 11-1　在 Models 文件夹下的 Order.cs 文件中添加一个新的属性

```
using System.Collections.Generic;
using System.ComponentModel.DataAnnotations;
using Microsoft.AspNetCore.Mvc.ModelBinding;

namespace SportsStore.Models {

    public class Order {

        [BindNever]
        public int OrderID { get; set; }
        [BindNever]
        public ICollection<CartLine> Lines { get; set; }

        [BindNever]
        public bool Shipped { get; set; }

        [Required(ErrorMessage = "Please enter a name")]
        public string Name { get; set; }

        [Required(ErrorMessage = "Please enter the first address line")]
        public string Line1 { get; set; }

        public string Line2 { get; set; }
        public string Line3 { get; set; }

        [Required(ErrorMessage = "Please enter a city name")]
        public string City { get; set; }

        [Required(ErrorMessage = "Please enter a state name")]
        public string State { get; set; }

        public string Zip { get; set; }

        [Required(ErrorMessage = "Please enter a country name")]
        public string Country { get; set; }
```

```
        public bool GiftWrap { get; set; }
    }
}
```

这种扩展和调整模型以支持不同功能的迭代方法是 MVC 开发的典型方式。理想情况下，可以在项目开始时完全定义模型类，并围绕它们构建应用程序，但这只发生在最简单的项目中。实际上，这种迭代开发是必需的，因为对需求的理解一直在发展和演变。

Entity Framework Core 迁移使这个过程更容易，因为不必通过编写 SQL 命令手动保持数据库架构与模型类同步。要更新数据库以将 Shipped 属性添加到 Order 类，请打开新的命令提示符或 PowerShell 窗口，导航到 SportsStore 项目文件夹（包含 Startup.cs 文件的那个文件夹）并运行以下命令：

```
dotnet ef migrations add ShippedOrders
```

当应用程序启动并且 SeedData 类调用 Entity Framework Core 提供的 Migrate 方法时，将自动应用迁移。

11.1.2 添加操作方法和视图

显示和更新数据库中的订单数据相对简单。代码清单 11-2 在 Order 控制器中添加了两个新的操作方法。

代码清单 11-2 在 Controllers 文件夹下的 OrderController.cs 文件中添加操作方法

```
using Microsoft.AspNetCore.Mvc;
using SportsStore.Models;
using System.Linq;

namespace SportsStore.Controllers {

    public class OrderController : Controller {
        private IOrderRepository repository;
        private Cart cart;
        public OrderController(IOrderRepository repoService, Cart cartService) {
            repository = repoService;
            cart = cartService;
        }

        public ViewResult List() =>
            View(repository.Orders.Where(o => !o.Shipped));

        [HttpPost]
        public IActionResult MarkShipped(int orderID) {
            Order order = repository.Orders
                .FirstOrDefault(o => o.OrderID == orderID);
            if (order != null) {
                order.Shipped = true;
                repository.SaveOrder(order);
            }
            return RedirectToAction(nameof(List));
        }

        public ViewResult Checkout() => View(new Order());

        [HttpPost]
        public IActionResult Checkout(Order order) {
            if (cart.Lines.Count() == 0) {
                ModelState.AddModelError("", "Sorry, your cart is empty!");
            }
            if (ModelState.IsValid) {
                order.Lines = cart.Lines.ToArray();
```

```
                repository.SaveOrder(order);
                return RedirectToAction(nameof(Completed));
            } else {
                return View(order);
            }
        }

        public ViewResult Completed() {
            cart.Clear();
            return View();
        }
    }
}
```

List操作方法会选择存储库中Shipped属性值为false的所有Order对象，并将它们传递给默认视图。这是用于向管理员显示未发货订单列表的操作方法。

MarkShipped操作方法将接收一个POST请求，该请求指定了订单的ID，用于从存储库中找到相应的Order对象，以便将其Shipped属性设置为true并保存。

为了显示未发货订单的列表，在Views/Order文件夹中添加一个名为List.cshtml的Razor视图文件，并添加代码清单11-3所示的代码。table元素用于显示一些细节，包括已购买商品的详细信息。

代码清单11-3　Views/Order文件夹下的List.cshtml文件的内容

```
@model IEnumerable<Order>

@{
    ViewBag.Title = "Orders";
    Layout = "_AdminLayout";
}

@if (Model.Count() > 0) {

    <table class="table table-bordered table-striped">
        <tr><th>Name</th><th>Zip</th><th colspan="2">Details</th><th></th></tr>
        @foreach (Order o in Model) {
            <tr>
                <td>@o.Name</td><td>@o.Zip</td><th>Product</th><th>Quantity</th>
                <td>
                    <form asp-action="MarkShipped" method="post">
                        <input type="hidden" name="orderId" value="@o.OrderID" />
                        <button type="submit" class="btn btn-sm btn-danger">
                            Ship
                        </button>
                    </form>
                </td>
            </tr>
            @foreach (CartLine line in o.Lines) {
                <tr>
                    <td colspan="2"></td>
                    <td>@line.Product.Name</td><td>@line.Quantity</td>
                    <td></td>
                </tr>
            }
        }
    </table>
} else {
    <div class="text-center">No Unshipped Orders</div>
}
```

每个订单都会显示一个Ship按钮，用于将表单提交给MarkShipped操作方法。使用Layout属性为List

视图指定不同的布局，以覆盖_ViewStart.cshtml 文件中指定的布局。

要添加布局，可使用 MVC View Layout Page 模板在 Views/Shared 文件夹中创建一个名为_AdminLayout.cshtml 的文件，并添加代码清单 11-4 所示的代码。

代码清单 11-4　Views/Shared 文件夹下的_AdminLayout.cshtml 文件的内容

```
<!DOCTYPE html>
<html>
<head>
    <meta name="viewport" content="width=device-width" />
    <link rel="stylesheet" asp-href-include="lib/bootstrap/dist/css/*.min.css" />
    <title>@ViewBag.Title</title>
</head>
<body class="m-1 p-1">
    <div class="bg-info p-2"><h4>@ViewBag.Title</h4></div>
    @RenderBody()
</body>
</html>
```

要查看和管理应用程序中的订单，请启动应用程序，选择一些商品，然后结账。导航到/Order/List URL，你将看到创建的订单摘要，如图 11-1 所示。单击 Ship 按钮，数据库将被更新，待处理的订单列表将为空。

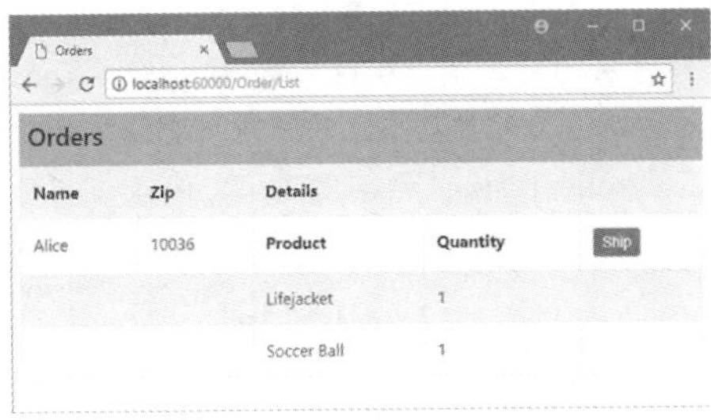

图 11-1　管理订单

注　意

目前没有办法阻止客户请求/Order/List URL 并管理他们自己的订单。本书第 12 章将解释如何限制对操作方法的访问。

11.2　添加目录管理

一般来说，在管理复杂的对象集合时，需要向用户呈现两种类型的页面——列表页面和编辑页面，如图 11-2 所示。

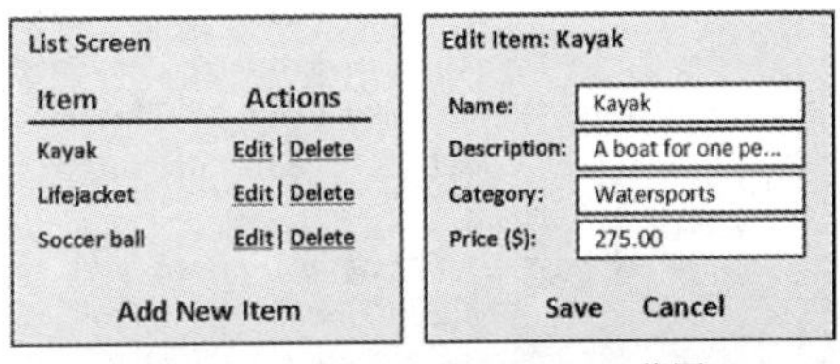

图 11-2　商品目录的 CRUD UI 草图

这些页面允许用户创建、读取、更新和删除集合中的对象。总的来说，这些动作称为 CRUD。开发人员经常需要实现 CRUD，Visual Studio 脚手架包括使用预定义操作方法创建 CRUD 控制器的场景（第 8 章解释了如何启用 Visual Studio 脚手架功能）。但是像所有 Visual Studio 模板一样，你最好学习如何直接使用 ASP.NET Core MVC 的功能。

11.2.1　创建 CRUD 控制器

这里将首先创建一个单独的控制器来管理商品目录。在 Controllers 文件夹中添加一个名为 AdminController.cs 的类文件，并添加代码清单 11-5 所示的代码。

代码清单 11-5 Controllers 文件夹下的 AdminController.cs 文件的内容

```
using Microsoft.AspNetCore.Mvc;
using SportsStore.Models;

namespace SportsStore.Controllers {

    public class AdminController : Controller {
        private IProductRepository repository;

        public AdminController(IProductRepository repo) {
            repository = repo;
        }

        public ViewResult Index() => View(repository.Products);
    }
}
```

这个控制器的构造函数声明了对 IProductRepository 接口的依赖关系，该接口将在创建实例时解析。这个控制器还定义了 Index 操作方法，这个操作方法会调用 View 方法以选择操作的默认视图，将数据库中的一组产品作为视图模型传递。

单元测试——Index 操作方法

Admin 控制器的 Index 操作方法的功能是正确返回存储库中的 Product 对象。你可以创建模拟的存储库实现，并将测试数据与操作方法返回的数据做比较以进行测试。以下是在 SportsStore.UnitTests 项目中添加的一个新的名为 AdminControllerTests.cs 的单元测试文件：

```
using System.Collections.Generic;
using System.Linq;
using Microsoft.AspNetCore.Mvc;
using Moq;
using SportsStore.Controllers;
using SportsStore.Models;
using Xunit;

namespace SportsStore.Tests {

    public class AdminControllerTests {

        [Fact]
        public void Index_Contains_All_Products() {
            // Arrange - create the mock repository
            Mock<IProductRepository> mock = new Mock<IProductRepository>();
            mock.Setup(m => m.Products).Returns(new Product[] {
                new Product {ProductID = 1, Name = "P1"},
                new Product {ProductID = 2, Name = "P2"},
                new Product {ProductID = 3, Name = "P3"},
            }.AsQueryable<Product>());

            // Arrange - create a controller
            AdminController target = new AdminController(mock.Object);

            // Action
            Product[] result
                = GetViewModel<IEnumerable<Product>>(target.Index())?.ToArray();

            // Assert
            Assert.Equal(3, result.Length);
            Assert.Equal("P1", result[0].Name);
```

```
                Assert.Equal("P2", result[1].Name);
                Assert.Equal("P3", result[2].Name);
            }

            private T GetViewModel<T>(IActionResult result) where T : class {
                return (result as ViewResult)?.ViewData.Model as T;
            }
        }
    }
```

在测试中添加 GetViewModel 方法，以解析操作方法的结果并获取视图模型数据。稍后将添加更多测试。

11.2.2 实现列表视图

下一步是为 Admin 控制器的 Index 操作方法添加视图。创建 Views/Admin 文件夹并在其中添加一个名为 Index.cshtml 的 Razor 文件，其内容如代码清单 11-6 所示。

代码清单 11-6 Views/Admin 文件夹下的 Index.cshtml 文件的内容

```
@model IEnumerable<Product>

@{
    ViewBag.Title = "All Products";
    Layout = "_AdminLayout";
}

<table class="table table-striped table-bordered table-sm">
    <tr>
        <th class="text-right">ID</th>
        <th>Name</th>
        <th class="text-right">Price</th>
        <th class="text-center">Actions</th>
    </tr>
    @foreach (var item in Model) {
        <tr>
            <td class="text-right">@item.ProductID</td>
            <td>@item.Name</td>
            <td class="text-right">@item.Price.ToString("c")</td>
            <td class="text-center">
                <form asp-action="Delete" method="post">
                    <a asp-action="Edit" class="btn btn-sm btn-warning"
                       asp-route-productId="@item.ProductID">
                        Edit
                    </a>
                    <input type="hidden" name="ProductID" value="@item.ProductID" />
                    <button type="submit" class="btn btn-danger btn-sm">
                        Delete
                    </button>
                </form>
            </td>
        </tr>
    }
</table>
<div class="text-center">
    <a asp-action="Create" class="btn btn-primary">Add Product</a>
</div>
```

列表视图包含一个表格，其中每个商品占据一行，包含商品名称、价格和两个按钮，这两个按钮将能够向 Edit 和 Delete 操作发送请求来编辑或删除商品。除表格之外，还有以 Create 操作为目标的 Add Product 按钮。后面还将添加 Edit、Delete 和 Create 操作，可以通过启动应用程序并请求/Admin/Index URL 来查看商品的显示方式，如图 11-3 所示。

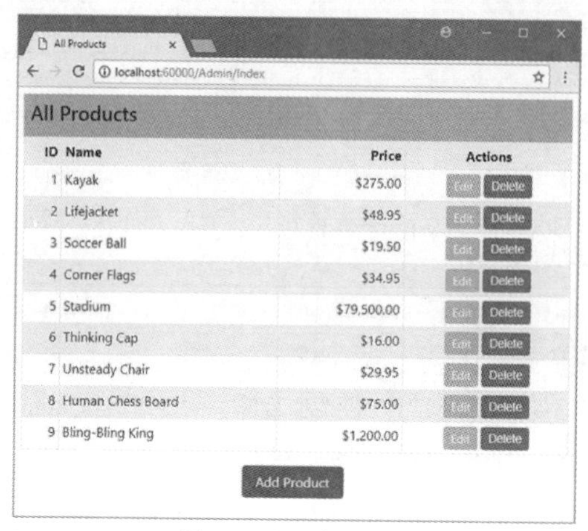

图 11-3 显示商品列表

提 示

Edit 按钮位于代码清单 11-6 中 form 元素的内部，因此 Edit 和 Delete 按钮彼此相邻，间距由 Bootstrap 确定。Edit 按钮将向服务器发送 HTTP GET 请求以获取商品的当前详细信息，这并不需要 form 元素。但是，由于 Delete 按钮将更改应用程序状态，需要使用 HTTP POST 请求，因此必须使用 form 元素。

11.2.3 编辑商品

为了提供创建和更新功能，添加图 11-2 所示的商品编辑页面。这项工作包括两部分：

- 显示一个允许管理员更改商品属性值的页面。
- 添加一个可以在提交时处理这些更改的操作方法。

1. 创建 Edit 操作方法

代码清单 11-7 显示了添加到 Admin 控制器的 Edit 操作方法，当用户单击 Edit 按钮时，它将接收浏览器发送的 HTTP 请求。

代码清单 11-7 在 Controllers 文件夹下的 AdminController.cs 文件中添加 Edit 操作方法

```
using Microsoft.AspNetCore.Mvc;
using SportsStore.Models;
using System.Linq;

namespace SportsStore.Controllers {

    public class AdminController : Controller {
        private IProductRepository repository;

        public AdminController(IProductRepository repo) {
            repository = repo;
        }

        public ViewResult Index() => View(repository.Products);

        public ViewResult Edit(int productId) =>
```

```
            View(repository.Products
                .FirstOrDefault(p => p.ProductID == productId));
        }
    }
```

这个简单的方法会查找具有与 productId 参数对应的 ID 的商品，并作为视图模型对象传递给 View 方法。

测试 Edit 操作方法

我们想在 Edit 操作方法中测试两个功能。首先，当提供有效的 id 时，能够得到正确的商品。其次，当请求存储库中不存在的 id 时，不会得到任何商品。以下是添加到 AdminControllerTests.cs 类文件的测试方法：

```
...
[Fact]
public void Can_Edit_Product() {
    // Arrange - create the mock repository
    Mock<IProductRepository> mock = new Mock<IProductRepository>();
    mock.Setup(m => m.Products).Returns(new Product[] {
        new Product {ProductID = 1, Name = "P1"},
        new Product {ProductID = 2, Name = "P2"},
        new Product {ProductID = 3, Name = "P3"},
    }.AsQueryable<Product>());
    // Arrange - create the controller
    AdminController target = new AdminController(mock.Object);

    // Act
    Product p1 = GetViewModel<Product>(target.Edit(1));
    Product p2 = GetViewModel<Product>(target.Edit(2));
    Product p3 = GetViewModel<Product>(target.Edit(3));

    // Assert
    Assert.Equal(1, p1.ProductID);
    Assert.Equal(2, p2.ProductID);
    Assert.Equal(3, p3.ProductID);
}

[Fact]
public void Cannot_Edit_Nonexistent_Product() {
    // Arrange - create the mock repository
    Mock<IProductRepository> mock = new Mock<IProductRepository>();
    mock.Setup(m => m.Products).Returns(new Product[] {
        new Product {ProductID = 1, Name = "P1"},
        new Product {ProductID = 2, Name = "P2"},
        new Product {ProductID = 3, Name = "P3"},
    }.AsQueryable<Product>());

    // Arrange - create the controller
    AdminController target = new AdminController(mock.Object);

    // Act
    Product result = GetViewModel<Product>(target.Edit(4));

    // Assert
    Assert.Null(result);
}
...
```

2. 创建 Edit 视图

现在有了一个操作方法，可以创建一个视图来显示它。在 Views/Admin 文件夹中添加一个名为 Edit.cshtml 的 Razor 视图文件，并添加代码清单 11-8 所示的标记。

代码清单 11-8 Views/Admin 文件夹下的 Edit.cshtml 文件的内容

```
@model Product
@{
    ViewBag.Title = "Edit Product";
    Layout = "_AdminLayout";
}

<form asp-action="Edit" method="post">
    <input type="hidden" asp-for="ProductID" />
    <div class="form-group">
        <label asp-for="Name"></label>
        <input asp-for="Name" class="form-control" />
    </div>
    <div class="form-group">
        <label asp-for="Description"></label>
        <textarea asp-for="Description" class="form-control"></textarea>
    </div>
    <div class="form-group">
        <label asp-for="Category"></label>
        <input asp-for="Category" class="form-control" />
    </div>
    <div class="form-group">
        <label asp-for="Price"></label>
        <input asp-for="Price" class="form-control" />
    </div>
    <div class="text-center">
        <button class="btn btn-primary" type="submit">Save</button>
        <a asp-action="Index" class="btn btn-secondary">Cancel</a>
    </div>
</form>
```

Edit 视图包含一个 HTML 表单，它使用标签助手生成大部分内容，包括设置 form 和 a 元素的目标，设置 label 元素的内容，以及为 input 和 textarea 元素生成 name、id 和 value 属性。

可以通过启动应用程序，导航到/Admin/Index URL，然后单击其中一个商品的 Edit 按钮来查看视图生成的 HTML，如图 11-4 所示。

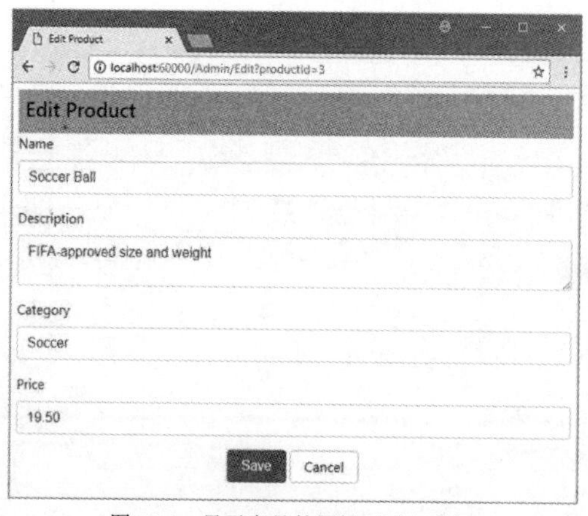

图 11-4 显示商品的属性以进行编辑

提 示

为了简单起见，为 ProductID 属性使用隐藏的 input 元素。当 Entity Framework Core 存储新对象时，由数据库生成 ProductID 的值并将其设置为主键，安全地更改它可能是一个复杂的过程。对于大多数应用程序，最简单的方法是防止用户更改值。

3. 更新 Product 存储库

在处理编辑之前，还需要改进 Product 存储库，以便能够保存更改。首先，向 IProductRepository 接口添加一个新的方法，如代码清单 11-9 所示。

代码清单 11-9 在 Models 文件夹下的 IProductRepository.cs 文件中添加一个新的方法

```
using System.Linq;

namespace SportsStore.Models {

    public interface IProductRepository {

        IQueryable<Product> Products { get; }

        void SaveProduct(Product product);
    }
}
```

然后，可以将这个新方法添加到由 Entity Framework Core 实现的存储库中，具体定义在 EFProductRepository.cs 文件中，如代码清单 11-10 所示。

代码清单 11-10 在 Models 文件夹下的 EFProductRepository.cs 文件中实现这个新方法

```
using System.Collections.Generic;
using System.Linq;

namespace SportsStore.Models {

    public class EFProductRepository : IProductRepository {
        private ApplicationDbContext context;

        public EFProductRepository(ApplicationDbContext ctx) {
            context = ctx;
        }

        public IQueryable<Product> Products => context.Products;

        public void SaveProduct(Product product) {
            if (product.ProductID == 0) {
                context.Products.Add(product);
            } else {
                Product dbEntry = context.Products
                    .FirstOrDefault(p => p.ProductID == product.ProductID);
                if (dbEntry != null) {
                    dbEntry.Name = product.Name;
                    dbEntry.Description = product.Description;
                    dbEntry.Price = product.Price;
                    dbEntry.Category = product.Category;
                }
            }
            context.SaveChanges();
        }
    }
}
```

如果 ProductID 为 0，SaveChanges 操作方法会将商品添加到存储库中；否则，就对数据库中的现有商品进行更新。

本章不会深入介绍 Entity Framework Core 的细节，因为如前所述，这本身就是一个较大的主题，而不是 ASP.NET Core MVC 的一部分。但是，SaveProduct 操作方法中有一些内容与 MVC 应用程序的设计有关。

当收到 ProductID 不为 0 的 Product 参数时，需要执行更新。可以通过以下步骤来实现：从存储库中获取具有相同 ProductID 的 Product 对象，并更新每个属性，使它们与参数对象匹配。

这样做是因为 Entity Framework Core 会跟踪从数据库创建的对象。传递给 SaveChanges 操作方法的对象是由 MVC 模型绑定功能创建的，这意味着 Entity Framework Core 对新的 Product 对象一无所知，在修改数据库时不会对数据库应用更新。有很多方法可以解决这个问题，这里采用最简单的方法，即找到 Entity Framework Core 跟踪的相应对象，并明确地更新它们。

在 IProductRepository 接口中添加新方法破坏了已在第 8 章中创建的虚拟存储库类 FakeProductRepository。可使用虚拟存储库启动开发过程，并演示如何使用服务来无缝替换接口实现，而无须修改依赖它们的组件。现在不再需要虚拟存储库，在代码清单 11-11 中，可以看到已从类声明中删除了接口，因此不必在添加存储库功能时继续修改该类。

代码清单 11-11 删除 Models 文件夹下的 FakeProductRepository.cs 文件中的接口

```
using System.Collections.Generic;
using System.Linq;

namespace SportsStore.Models {

    public class FakeProductRepository /* : IProductRepository */ {

        public IQueryable<Product> Products => new List<Product> {
            new Product { Name = "Football", Price = 25 },
            new Product { Name = "Surf board", Price = 179 },
            new Product { Name = "Running shoes", Price = 95 }
        }.AsQueryable<Product>();
    }
}
```

4. 处理编辑 POST 请求

现在可以在 Admin 控制器中实现 Edit 操作方法的重载，从而在管理员单击 Save 按钮时处理 POST 请求。代码清单 11-12 显示了新的 Edit 操作方法。

代码清单 11-12 在 Controllers 文件夹下的 AdminController.cs 文件中定义新的 Edit 操作方法

```
using Microsoft.AspNetCore.Mvc;
using SportsStore.Models;
using System.Linq;

namespace SportsStore.Controllers {

    public class AdminController : Controller {
        private IProductRepository repository;

        public AdminController(IProductRepository repo) {
            repository = repo;
        }

        public ViewResult Index() => View(repository.Products);

        public ViewResult Edit(int productId) =>
            View(repository.Products
                .FirstOrDefault(p => p.ProductID == productId));

        [HttpPost]
        public IActionResult Edit(Product product) {
            if (ModelState.IsValid) {
                repository.SaveProduct(product);
                TempData["message"] = $"{product.Name} has been saved";
                return RedirectToAction("Index");
            } else {
                // there is something wrong with the data values
                return View(product);
            }
```

```
        }
    }
}
```

可通过检查模型绑定过程是否能够读取 ModelState.IsValid 属性的值来验证用户提交的数据。如果一切正常，就将更改保存到存储库并将用户重定向到 Index 操作方法，以便他们看到修改后的商品列表。如果数据有问题，会再次渲染默认视图，以便用户进行更正。

保存存储库中的更改后，使用 temp data 功能存储消息，temp data 功能是 ASP.NET Core 会话状态功能的一部分。这是键/值字典，类似于之前使用的会话数据和 view bag 功能，与会话数据的主要区别在于临时数据在读取之前一直存在。

在这种情况下不能使用 ViewBag，因为 ViewBag 在控制器和视图之间传递数据，并且数据的保存时间不能超过当前的 HTTP 请求。编辑成功后，浏览器将重定向到新的 URL，因此 ViewBag 数据将丢失。可以使用会话数据功能，但是在明确删除之前，消息将一直存在，这里并不想这样做。

因此，temp data 功能是最合适的。数据仅限于单个用户的会话（因此用户看不到彼此的临时数据）并且会持续足够长的时间以供用户读取。这里将读取重定向用户的操作方法呈现的视图中的数据。

单元测试——编辑提交的数据

对于 Edit 操作方法处理 POST 请求的过程，需要确保将作为方法参数接收的 Product 对象的有效更新传递到 Product 存储库以进行保存。这里还需要检查无效更新（在模型验证错误的情况下）不会传递到存储库。以下是添加到 AdminControllerTests.cs 文件中的测试方法：

```
...
[Fact]
public void Can_Save_Valid_Changes() {
    // Arrange - create mock repository
    Mock<IProductRepository> mock = new Mock<IProductRepository>();
    // Arrange - create mock temp data
    Mock<ITempDataDictionary> tempData = new Mock<ITempDataDictionary>();
    // Arrange - create the controller
    AdminController target = new AdminController(mock.Object) {
        TempData = tempData.Object
    };
    // Arrange - create a product
    Product product = new Product { Name = "Test" };

    // Act - try to save the product
    IActionResult result = target.Edit(product);
    // Assert - check that the repository was called
    mock.Verify(m => m.SaveProduct(product));
    // Assert - check the result type is a redirection
    Assert.IsType<RedirectToActionResult>(result);
    Assert.Equal("Index", (result as RedirectToActionResult).ActionName);
}

[Fact]
public void Cannot_Save_Invalid_Changes() {
    // Arrange - create mock repository
    Mock<IProductRepository> mock = new Mock<IProductRepository>();
    // Arrange - create the controller
    AdminController target = new AdminController(mock.Object);
    // Arrange - create a product
    Product product = new Product { Name = "Test" };
    // Arrange - add an error to the model state
    target.ModelState.AddModelError("error", "error");

    // Act - try to save the product
    IActionResult result = target.Edit(product);
```

```
        // Assert - check that the repository was not called
        mock.Verify(m => m.SaveProduct(It.IsAny<Product>()), Times.Never());
        // Assert - check the method result type
        Assert.IsType<ViewResult>(result);
    }
...
```

5. 显示确认消息

在_AdminLayout.cshtml 布局文件中处理使用 TempData 存储的消息，如代码清单 11-13 所示。通过处理模板中的消息，可以在任何使用模板的视图中创建消息，而无须创建其他 Razor 表达式。

代码清单 11-13　处理_AdminLayout.cshtml 文件中的 ViewBag 消息

```
<!DOCTYPE html>
<html>
<head>
    <meta name="viewport" content="width=device-width" />
    <link rel="stylesheet" asp-href-include="lib/bootstrap/dist/css/*.min.css" />
    <title>@ViewBag.Title</title>
</head>
<body class="m-1 p-1">
    <div class="bg-info p-2"><h4>@ViewBag.Title</h4></div>
    @if (TempData["message"] != null) {
        <div class="alert alert-success">@TempData["message"]</div>
    }
    @RenderBody()
</body>
</html>
```

提　示

在模板中处理消息的好处是，在保存所做更改后的所有页面上，用户都能看到它们。目前，将它们返回商品列表中，但可以更改工作流以呈现其他视图，用户仍将看到消息（只要下一个视图也使用相同的布局）。

现在已经准备好用于编辑商品的所有代码。要查看具体是如何工作的，请启动应用程序，导航到/Admin/Index URL，单击 Edit 按钮，然后进行更改。单击 Save 按钮，你将被重定向到/Admin/Index URL，并显示 TempData 消息，如图 11-5 所示。

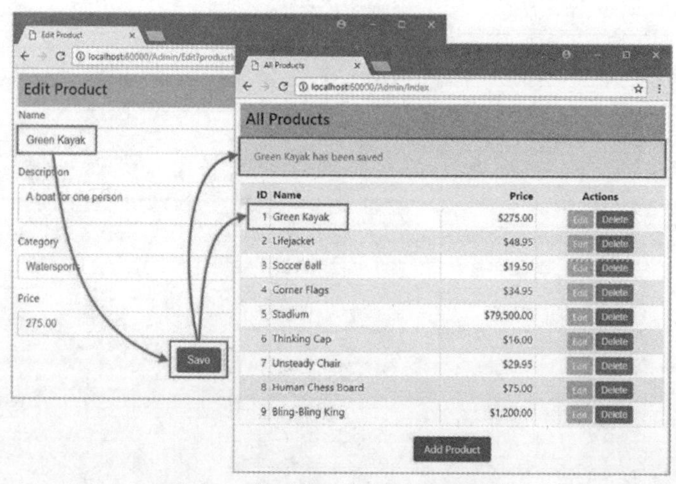

图 11-5　编辑商品并查看 TempData 消息

如果重新加载商品列表视图，消息将消失，因为 TempData 在读取时会被删除。这很方便，因为作者不希望一直显示旧的信息。

6. 添加模型验证

接下来需要向模型类添加验证规则。目前，管理员可以输入负的价格或空白描述，SportsStore 应用程序会很乐意将这些数据存储在数据库中。坏数据是否会成功保留将取决于它们是否符合 Entity Framework Core 创建的 SQL 表中的约束，这对于大多数应用程序来说保护得还不够。为了防止错误的数据值，使用特性修饰 Product 类的属性，如代码清单 11-14 所示，就像在第 10 章中对 Order 类所做的那样。

代码清单 11-14　在 Models 文件夹下的 Product.cs 文件中应用验证特性

```
using System.ComponentModel.DataAnnotations;
using Microsoft.AspNetCore.Mvc.ModelBinding;

namespace SportsStore.Models {

    public class Product {
        public int ProductID { get; set; }

        [Required(ErrorMessage = "Please enter a product name")]
        public string Name { get; set; }

        [Required(ErrorMessage = "Please enter a description")]
        public string Description { get; set; }

        [Required]
        [Range(0.01, double.MaxValue,
            ErrorMessage = "Please enter a positive price")]
        public decimal Price { get; set; }

        [Required(ErrorMessage = "Please specify a category")]
        public string Category { get; set; }
    }
}
```

第 10 章使用标签助手在表单顶部显示验证错误的摘要。在本例中，将使用类似的方法，但是将在 Edit 视图中的各个表单元素旁显示错误消息，如代码清单 11-15 所示。

代码清单 11-15　在 Views/Admin 文件夹下的 Edit.cshtml 文件中添加验证错误元素

```
@model Product
@{
    ViewBag.Title = "Edit Product";
    Layout = "_AdminLayout";
}

<form asp-action="Edit" method="post">
    <input type="hidden" asp-for="ProductID" />
    <div class="form-group">
        <label asp-for="Name"></label>
        <div><span asp-validation-for="Name" class="text-danger"></span></div>
        <input asp-for="Name" class="form-control" />
    </div>
    <div class="form-group">
        <label asp-for="Description"></label>
        <div><span asp-validation-for="Description" class="text-danger"></span></div>
        <textarea asp-for="Description" class="form-control"></textarea>
    </div>
    <div class="form-group">
        <label asp-for="Category"></label>
```

```
        <div><span asp-validation-for="Category" class="text-danger"></span></div>
        <input asp-for="Category" class="form-control" />
    </div>
    <div class="form-group">
        <label asp-for="Price"></label>
        <div><span asp-validation-for="Price" class="text-danger"></span></div>
        <input asp-for="Price" class="form-control" />
    </div>
    <div class="text-center">
        <button class="btn btn-primary" type="submit">Save</button>
        <a asp-action="Index" class="btn btn-secondary">Cancel</a>
    </div>
</form>
```

当应用于 span 元素时，为 asp-validation-for 属性应用标签助手，从而当存在任何验证问题时，为指定的属性添加验证错误消息。

标签助手将向 span 元素插入一条错误消息，并向 span 元素添加 input-validation-error 类，这样可以很容易地将 CSS 样式应用于错误消息元素，如代码清单 11-16 所示。

代码清单 11-16 将 CSS 添加到 Views/Shared 文件夹下的_AdminLayout.cshtml 文件中

```
<!DOCTYPE html>
<html>
<head>
    <meta name="viewport" content="width=device-width" />
    <link rel="stylesheet" asp-href-include="lib/bootstrap/dist/css/*.min.css" />
    <title>@ViewBag.Title</title>
    <style>
        .input-validation-error { border-color: red; background-color: #fee ; }
    </style>
</head>
<body class="m-1 p-1">
    <div class="bg-info p-2"><h4>@ViewBag.Title</h4></div>
    @if (TempData["message"] != null) {
        <div class="alert alert-success mt-1">@TempData["message"]</div>
    }
    @RenderBody()
</body>
</html>
```

定义的 CSS 样式将选择作为 input-validation-error 类成员的元素，并应用红色边框和背景颜色。

提 示

在使用 Bootstrap 这样的 CSS 库时，显式地设置样式会在应用内容时导致主题不一致。第 27 章将展示一种替代方法，该方法使用 JavaScript 代码将 Bootstrap 类应用于具有验证错误的元素，这虽然使得所有内容保持一致，但也更复杂。

可以在视图中的任何位置应用验证消息标签助手，但常规（并且合理的）做法是将标签助手放在问题元素附近的某个位置，以向用户提供一些上下文信息。图 11-6 显示了验证消息和提示，可以通过运行应用程序、编辑商品和提交无效数据来查看这些验证消息和提示。

7. 启用客户端验证

目前，仅当管理员用户向服务器提交编辑时才应用数据验证，但是大多数用户希望如果输入的数据存在问题，能够立即得到反馈。这就是开发人员经常想要执行客户端验证的原因，即使用 JavaScript 在浏览器中检查数据。MVC 应用程序可以基于应用于域模型类的数据注解来执行客户端验证。

第一步是向应用程序添加提供客户端功能的 JavaScript 库，这在 bower.json 文件中完成，如代码清单 11-17 所示。

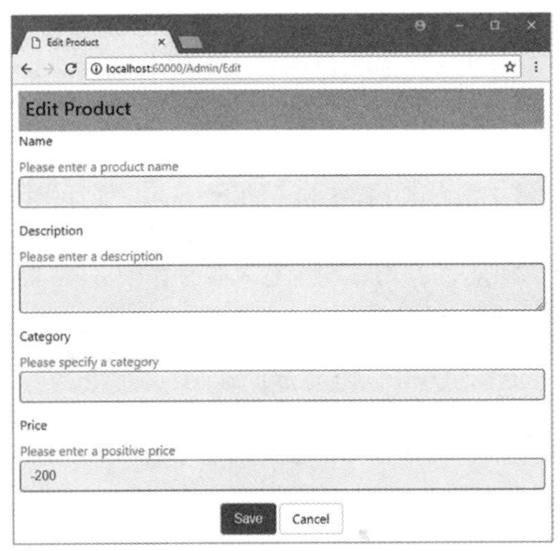

图 11-6 编辑商品时执行数据验证

代码清单 11-17 在 bower.json 文件中添加 JavaScript 包

```
{
  "name": "asp.net",
  "private": true,
  "dependencies": {
    "bootstrap": "4.0.0-alpha.6",
    "fontawesome": "4.7.0",
    "jquery": "3.2.1",
    "jquery-validation": "1.17.0",
    "jquery-validation-unobtrusive": "3.2.6"
  }
}
```

客户端验证建立在流行的 jQuery 库之上，简化了浏览器的 DOM API。下一步是将 JavaScript 文件添加到布局中，以便在使用 SportsStore 管理功能时加载它们，如代码清单 11-18 所示。

代码清单 11-18 将验证库添加到 Views/Shared 文件夹下的_AdminLayout.cshtml 文件中

```
<!DOCTYPE html>
<html>
<head>
    <meta name="viewport" content="width=device-width" />
    <link rel="stylesheet" asp-href-include="lib/bootstrap/dist/css/*.min.css" />
    <title>@ViewBag.Title</title>
    <style>
        .input-validation-error { border-color: red; background-color: #fee ; }
    </style>
    <script src="/lib/jquery/dist/jquery.min.js"></script>
    <script src="/lib/jquery-validation/dist/jquery.validate.min.js"></script>
    <script
      src="/lib/jquery-validation-unobtrusive/jquery.validate.unobtrusive.min.js">
    </script>
</head>
<body class="m-1 p-1">
    <div class="bg-info p-2"><h4>@ViewBag.Title</h4></div>
    @if (TempData["message"] != null) {
        <div class="alert alert-success mt-1">@TempData["message"]</div>
    }
    @RenderBody()
</body>
</html>
```

启用客户端验证不会导致任何可视化更改，但是由 C#模型类属性指定的约束将在浏览器中强制执行，从而阻止用户提交包含错误数据的表单，并在有问题时立即提供反馈。有关详细信息，请参见第 27 章。

11.2.4 创建新的商品

接下来，将实现 Create 操作方法，这个操作方法在主商品列表页面中由 Add Product 链接指定的。这将允许管理员将新的商品添加到商品目录中。添加创建新商品的功能需要对应用程序进行一点小的更改。这是一个很好的例子，展示了结构良好的 MVC 应用程序具有的强大功能和灵活性。首先，将 Create 操作方法（如代码清单 11-19 所示）添加到 Admin 控制器中。

代码清单 11-19 将 Create 操作方法添加到 Controllers 文件夹下的 AdminController.cs 文件中

```
using Microsoft.AspNetCore.Mvc;
using SportsStore.Models;
using System.Linq;

namespace SportsStore.Controllers {

    public class AdminController : Controller {
        private IProductRepository repository;

        public AdminController(IProductRepository repo) {
            repository = repo;
        }

        public ViewResult Index() => View(repository.Products);

        public ViewResult Edit(int productId) =>
            View(repository.Products
                .FirstOrDefault(p => p.ProductID == productId));

        [HttpPost]
        public IActionResult Edit(Product product) {
            if (ModelState.IsValid) {
                repository.SaveProduct(product);
                TempData["message"] = $"{product.Name} has been saved";
                return RedirectToAction("Index");
            } else {
                // there is something wrong with the data values
                return View(product);
            }
        }

        public ViewResult Create() => View("Edit", new Product());
    }
}
```

Create 操作方法不会渲染默认视图，而是指定应使用 Edit 视图。一个操作方法完全可以使用通常与另一个操作方法关联的视图。在这种情况下，可以提供一个新的 Product 对象作为视图模型，以便使用空字段填充 Edit 视图。

注 意

这里没有为 Create 操作方法添加单元测试。这样做只会测试 ASP.NET Core MVC 处理来自操作方法的结果的能力，这是理所当然的（除非怀疑存在缺陷，否则通常不会为框架功能编写测试）。

这是唯一需要执行的更改，因为 Edit 操作方法已设置为从模型绑定系统接收 Product 对象并将它们存储在数据库中。可以通过启动应用程序，导航到/Admin/Index，单击 Add Product 按钮，然后填充并提交表单来测试此功能。你在表单中指定的详细信息将用于在数据库中创建新的商品，然后新的商品将显示在列表中，如图 11-7 所示。

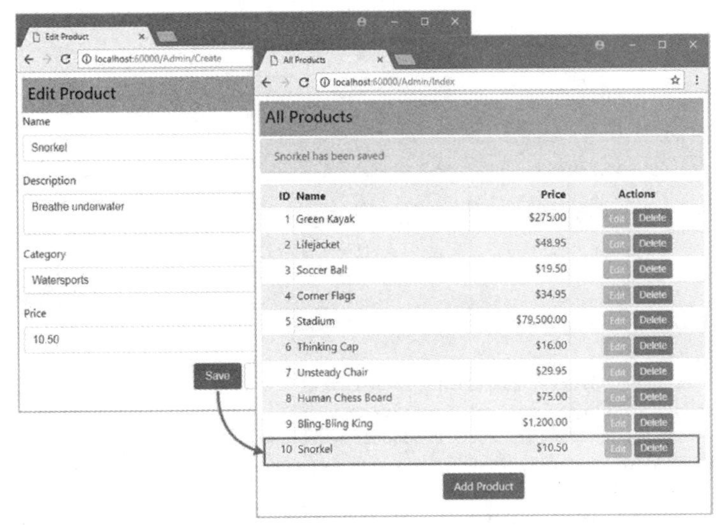

图 11-7　将新的商品添加到目录中

11.2.5　删除商品

添加删除商品的功能也很简单。第一步是向 IProductRepository 接口添加一个新的方法，如代码清单 11-20 所示。

代码清单 11-20　在 Models 文件夹下的 IProductRepository.cs 文件中添加商品删除方法

```
using System.Linq;

namespace SportsStore.Models {

    public interface IProductRepository {

        IQueryable<Product> Products { get; }

        void SaveProduct(Product product);

        Product DeleteProduct(int productID);
    }
}
```

接下来，在 Entity Framework Core 存储库类 EFProductRepository 中实现这个方法，如代码清单 11-21 所示。

代码清单 11-21　在 Models 文件夹下的 EFProductRepository.cs 文件中实现商品删除功能

```
using System.Collections.Generic;
using System.Linq;

namespace SportsStore.Models {

    public class EFProductRepository : IProductRepository {
        private ApplicationDbContext context;

        public EFProductRepository(ApplicationDbContext ctx) {
            context = ctx;
        }

        public IQueryable<Product> Products => context.Products;
```

```
        public void SaveProduct(Product product) {
            if (product.ProductID == 0) {
                context.Products.Add(product);
            } else {
                Product dbEntry = context.Products
                    .FirstOrDefault(p => p.ProductID == product.ProductID);
                if (dbEntry != null) {
                    dbEntry.Name = product.Name;
                    dbEntry.Description = product.Description;
                    dbEntry.Price = product.Price;
                    dbEntry.Category = product.Category;
                }
            }
            context.SaveChanges();
        }
        public Product DeleteProduct(int productID) {
            Product dbEntry = context.Products
                .FirstOrDefault(p => p.ProductID == productID);
            if (dbEntry != null) {
                context.Products.Remove(dbEntry);
                context.SaveChanges();
            }
            return dbEntry;
        }
    }
}
```

最后一步是在 Admin 控制器中实现 Delete 操作方法。这个操作方法应仅支持 POST 请求，因为删除对象不是幂等操作。正如在第 16 章中解释的那样，浏览器和缓存可以在未经用户明确同意的情况下自由发出 GET 请求，因此必须小心避免因 GET 请求而进行更改的情况。代码清单 11-22 显示了新的 Delete 操作方法。

代码清单 11-22 在 Controllers 文件夹下的 AdminController.cs 文件中添加 Delete 操作方法

```
using Microsoft.AspNetCore.Mvc;
using SportsStore.Models;
using System.Linq;

namespace SportsStore.Controllers {

    public class AdminController : Controller {
        private IProductRepository repository;

        public AdminController(IProductRepository repo) {
            repository = repo;
        }

        public ViewResult Index() => View(repository.Products);

        public ViewResult Edit(int productId) =>
            View(repository.Products
                .FirstOrDefault(p => p.ProductID == productId));

        [HttpPost]
        public IActionResult Edit(Product product) {
            if (ModelState.IsValid) {
                repository.SaveProduct(product);
                TempData["message"] = $"{product.Name} has been saved";
                return RedirectToAction("Index");
            } else {
                // there is something wrong with the data values
```

```
            return View(product);
        }
    }
    public IActionResult Create() => View("Edit", new Product());

    [HttpPost]
    public IActionResult Delete(int productId) {
        Product deletedProduct = repository.DeleteProduct(productId);
        if (deletedProduct != null) {
            TempData["message"] = $"{deletedProduct.Name} was deleted";
        }
        return RedirectToAction("Index");
    }
  }
}
```

单元测试——删除商品

下面测试 Delete 操作方法的基本功能：当把有效的 ProductID 作为参数传递时，调用存储库的 DeleteProduct 方法，并传递正确的 ProductID 值进行删除。下面是添加到 AdminControllerTests.cs 文件的测试：

```
...
[Fact]
public void Can_Delete_Valid_Products() {
    // Arrange - create a Product
    Product prod = new Product { ProductID = 2, Name = "Test" };

    // Arrange - create the mock repository
    Mock<IProductRepository> mock = new Mock<IProductRepository>();
    mock.Setup(m => m.Products).Returns(new Product[] {
        new Product {ProductID = 1, Name = "P1"},
        prod,
        new Product {ProductID = 3, Name = "P3"},
    }.AsQueryable<Product>());

    // Arrange - create the controller
    AdminController target = new AdminController(mock.Object);

    // Act - delete the product
    target.Delete(prod.ProductID);

    // Assert - ensure that the repository delete method was
    // called with the correct Product
    mock.Verify(m => m.DeleteProduct(prod.ProductID));
}
...
```

可以通过启动应用程序，导航到/Admin/Index，然后单击商品列表页面中的其中一个 Delete 按钮来查看删除功能，如图 11-8 所示。当从目录中删除商品时，这里使用 TempData 变量显示消息。

注　意

如果删除先前已创建订单的商品，则会发现错误。当 Order 对象存储在数据库中时，它将被转换为数据表中的一条记录，其中包含与之关联的 Product 对象的引用，称为外键关系。这意味着默认情况下，如果已为商品创建订单，数据库将不允许删除 Product 对象，因为这样做会在数据库中导致不一致。有许多方法可用来解决此问题，包括在删除商品时自动删除与之相关的订单对象或者更改商品和订单对象之间的关系。详细信息请参阅 Entity Framework Core 文档。

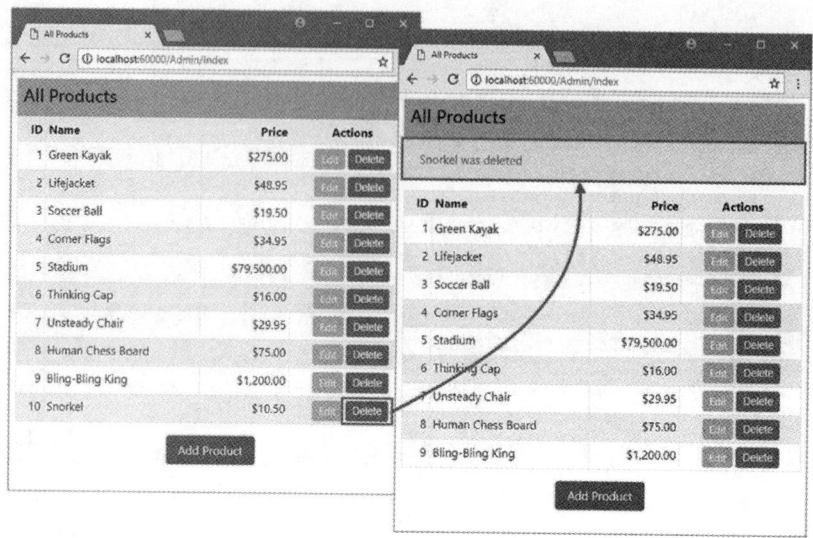

图 11-8 从目录中删除商品

11.3 小结

本章介绍了管理功能，并展示了如何实现 CRUD 操作，从而允许管理员在存储库中创建、读取、更新和删除商品，并将订单标记为已发货。下一章将展示如何保护管理功能，使其不对所有用户公开，并将 SportsStore 应用程序部署到生产环境中。

第 12 章　SportsStore 的安全和部署

在上一章中，添加了 SportsStore 应用程序的管理功能，如果按原样部署应用程序，可能没人注意到任何人都可以修改商品目录。他们需要知道的是，任何人使用 URL/Admin/Index 和/Order/List 都可以执行用管理功能。本章将展示如何通过密码保护来阻止任意用户使用管理功能。一旦具备了安全措施，就将展示如何准备和部署 SportsStore 应用程序到生产环境中。

12.1　保护管理功能

ASP.NET Core Identity 系统提供了身份验证和授权，该系统可以整合到 ASP.NET Core 平台和 MVC 应用程序中。在接下来的内容中，将创建基本的安全设置，允许名为 Admin 的用户对应用程序中的管理功能进行身份验证和访问。ASP.NET Core Identity 提供了许多用于验证用户身份和授权访问应用程序特性和数据的功能，可以在第 28～30 章中找到更详细的信息，其中将展示如何创建和管理用户账户、如何使用角色和策略，以及如何支持来自 Microsoft、Google、Facebook 和 Twitter 等第三方的身份验证。但是，在本章中，目标是获得足够的功能以防止客户访问 SportsStore 应用程序的敏感部分，并在此过程中，让你了解在 MVC 应用中如何使用身份验证和授权。

12.1.1　创建身份标识数据库

ASP.NET Core Identity 系统具有无限的可配置性和可扩展性，并支持多种用户数据存储方式。这里将使用最常见的使用 Entity Framework Core 访问的 Microsoft SQL Server 存储数据。

1. 创建上下文类

这里需要创建一个数据库上下文文件，以充当数据库与提供访问权限的 Identity 模型对象之间的桥梁。在 Model 文件夹中添加一个名为 AppIdentityDbContext.cs 的类文件，并用它定义代码清单 12-1 所示的类。

注　意

你可能习惯于将包添加到项目中以获得安全性等其他功能。但是，随着 ASP.NET Core 2 的发布，身份标识所需的 NuGet 包已经包含在项目中，也就是作为项目模板的一部分添加到 SportsStore.csproj 文件中的元数据包。

代码清单 12-1　Models 文件夹下的 AppIdentityDbContext.cs 文件的内容

```
using Microsoft.AspNetCore.Identity;
using Microsoft.AspNetCore.Identity.EntityFrameworkCore;
using Microsoft.EntityFrameworkCore;

namespace SportsStore.Models {

    public class AppIdentityDbContext : IdentityDbContext<IdentityUser> {

        public AppIdentityDbContext(DbContextOptions<AppIdentityDbContext> options)
            : base(options) { }
    }
}
```

AppIdentityDbContext 类派生自 IdentityDbContext，后者为 Entity Framework Core 提供特定于身份标识

的功能。对于 type 参数，使用 IdentityUser 类，它是用于表示用户的内置类。第 28 章将演示如何使用可扩展的自定义类来添加应用程序用户的额外信息。

2. 定义连接字符串

下一步是定义数据库的连接字符串。在代码清单 12-2 中，可以看到对 SportsStore 项目的 appsettings.json 文件所做的补充，格式与在第 8 章中为产品数据库定义的连接字符串相同。

代码清单 12-2　在 appsettings.json 文件中定义连接字符串

```
{
  "Data": {
    "SportStoreProducts": {
      "ConnectionString": " Server=(localdb)\\MSSQLLocalDB;Database=SportsStore;Trusted_
                          Connection=True;MultipleActiveResultSets=true"
    },
    "SportStoreIdentity": {
      "ConnectionString": " Server=(localdb)\\MSSQLLocalDB;Database=Identity;Trusted_
                          Connection=True;MultipleActiveResultSets=true"
    }
  }
}
```

请记住，连接字符串必须在 appsettings.json 文件中的单个连续行中定义（这里由于受版面宽度的限制换行显示）。代码清单 12-2 定义了一个名为 SportsStoreIdentity 的连接字符串，其中指定了名为 Identity 的 LocalDB 数据库。

3. 配置应用程序

与其他 ASP.NET Core 功能一样，ASP.NET Core Identity 在 Start 类中配置。代码清单 12-3 显示了使用上面定义的上下文类和连接字符串在 SportsStore 项目中为设置 ASP.NET Core Identity 所做的补充。

代码清单 12-3　在 Startup.cs 文件中配置身份标识

```
using System;
using System.Collections.Generic;
using System.Linq;
using System.Threading.Tasks;
using Microsoft.AspNetCore.Builder;
using Microsoft.AspNetCore.Hosting;
using Microsoft.AspNetCore.Http;
using Microsoft.Extensions.DependencyInjection;
using SportsStore.Models;
using Microsoft.Extensions.Configuration;
using Microsoft.EntityFrameworkCore;
using Microsoft.AspNetCore.Identity;

namespace SportsStore {

    public class Startup {

        public Startup(IConfiguration configuration) =>
            Configuration = configuration;

        public IConfiguration Configuration { get; }

        public void ConfigureServices(IServiceCollection services) {

            services.AddDbContext<ApplicationDbContext>(options =>
                options.UseSqlServer(
                    Configuration["Data:SportStoreProducts:ConnectionString"]));

            services.AddDbContext<AppIdentityDbContext>(options =>
```

```
                options.UseSqlServer(
                    Configuration["Data:SportStoreIdentity:ConnectionString"]));

            services.AddIdentity<IdentityUser, IdentityRole>()
                .AddEntityFrameworkStores<AppIdentityDbContext>()
                .AddDefaultTokenProviders();

            services.AddTransient<IProductRepository, EFProductRepository>();
            services.AddScoped<Cart>(sp => SessionCart.GetCart(sp));
            services.AddSingleton<IHttpContextAccessor, HttpContextAccessor>();
            services.AddTransient<IOrderRepository, EFOrderRepository>();
            services.AddMvc();
            services.AddMemoryCache();
            services.AddSession();
        }
        public void Configure(IApplicationBuilder app, IHostingEnvironment env) {
            app.UseDeveloperExceptionPage();
            app.UseStatusCodePages();
            app.UseStaticFiles();
            app.UseSession();
            app.UseAuthentication();
            app.UseMvc(routes => {

                // ...routes omitted for brevity...

            });
            SeedData.EnsurePopulated(app);
        }
    }
}
```

在 ConfigureServices 方法中，扩展 Entity Framework Core 配置以注册上下文类，并使用 AddIdentity 方法使用内置类来设置 ASP.NET Core Identity 服务以表示用户和角色。在 Configure 方法中，调用 UseAuthentication 方法来设置组件，用于拦截请求和响应以实现安全策略。

4. 创建和应用数据库迁移

基本配置已就位，现在可以使用 Entity Framework Core 迁移功能来定义数据库架构并将其应用于数据库。打开新的命令提示符或 PowerShell 窗口，并在 SportsStore 项目文件夹中运行以下命令，为 Identity 数据库创建新的迁移：

```
dotnet ef migrations add Initial --context AppIdentityDbContext
```

与以前的数据库命令的重要区别在于，使用-context 参数来指定与要使用的数据库关联的上下文类的名称，这里是 AppIdentityDbContext。当应用程序中有多个数据库时，确保使用正确的上下文类非常重要。

一旦 Entity Framework Core 生成了初始迁移，就运行以下命令来创建数据库并运行迁移命令：

```
dotnet ef database update --context AppIdentityDbContext
```

结果是生成名为 Identity 的 LocalDB 数据库，该数据库可以使用 Visual Studio SQL Server 对象资源管理器进行查看。

5. 定义种子数据

可通过在应用程序启动时对数据库进行填充来显式创建 Admin 用户。在 Models 文件夹中添加一个名为 IdentitySeedData.cs 的类文件，并定义代码清单 12-4 所示的静态类。

代码清单 12-4　Models 文件夹下的 IdentitySeedData.cs 文件的内容

```
using Microsoft.AspNetCore.Builder;
using Microsoft.AspNetCore.Identity;
using Microsoft.Extensions.DependencyInjection;
```

```
namespace SportsStore.Models {

    public static class IdentitySeedData {
        private const string adminUser = "Admin";
        private const string adminPassword = "Secret123$";

        public static async void EnsurePopulated(IApplicationBuilder app) {

            UserManager<IdentityUser> userManager = app.ApplicationServices
                .GetRequiredService<UserManager<IdentityUser>>();

            IdentityUser user = await userManager.FindByIdAsync(adminUser);
            if (user == null) {
                user = new IdentityUser("Admin");
                await userManager.CreateAsync(user, adminPassword);
            }
        }
    }
}
```

以上代码使用了 UserManager <T>类，它由 ASP.NET Core Identity 作为服务提供，用于管理用户，如第 28 章所述。数据库用来查找管理员账户，管理员账户是使用密码 Secret123 $创建的。请勿在此例中更改硬编码的密码，因为 ASP.NET Core Identity 具有验证策略，会要求密码包含数字和指定范围的字符。有关如何更改验证设置的详细信息，请参阅第 28 章。

警 告

在开发过程中经常需要对管理员账户的详细信息进行硬编码，以便在部署应用程序后立即登录并开始管理。执行此操作时，必须更改已创建账户的密码。有关如何使用身份标识更改密码的详细信息，请参阅第 28 章。

为了确保在应用程序启动时填充 Identity 数据库，将代码清单 12-5 所示的语句添加到 Startup 类的 Configure 方法中。

代码清单 12-5 在 SportsStore 文件夹下的 Startup.cs 文件中填充 Identity 数据库

```
...
public void Configure(IApplicationBuilder app, IHostingEnvironment env) {
    app.UseDeveloperExceptionPage();
    app.UseStatusCodePages();
    app.UseStaticFiles();
    app.UseSession();
    app.UseAuthentication();
    app.UseMvc(routes => {
        // ...routes omitted for brevity...

    });
    SeedData.EnsurePopulated(app);
    IdentitySeedData.EnsurePopulated(app);
}
...
```

12.1.2 应用基本授权策略

既然已经配置了 ASP.NET Core Identity，就可以将授权策略应用于想要保护的应用程序。可使用最基本的授权策略，允许任何经过身份验证的用户访问。虽然这在实际的应用程序中是一种有用的策略，但还可以选择创建更精细的授权控制（如第 28～30 章所述），但由于 SportsStore 应用程序只有一个用户，因此能够区分匿名请求和经过身份验证的请求就足够了。

Authorize 特性用于限制对操作方法的访问，在代码清单 12-6 中，可以看到已使用 Authorize 特性来保护对 Order 控制器中管理操作的访问。

代码清单 12-6 限制 OrderController.cs 文件中的访问权限

```
using Microsoft.AspNetCore.Mvc;
using SportsStore.Models;
using System.Linq;
using Microsoft.AspNetCore.Authorization;

namespace SportsStore.Controllers {

    public class OrderController : Controller {
        private IOrderRepository repository;
        private Cart cart;

        public OrderController(IOrderRepository repoService, Cart cartService) {
            repository = repoService;
            cart = cartService;
        }

        [Authorize]
        public ViewResult List() =>
            View(repository.Orders.Where(o => !o.Shipped));

        [HttpPost]
        [Authorize]
        public IActionResult MarkShipped(int orderID) {
            Order order = repository.Orders
                .FirstOrDefault(o => o.OrderID == orderID);
            if (order != null) {
                order.Shipped = true;
                repository.SaveOrder(order);
            }
            return RedirectToAction(nameof(List));
        }
        public ViewResult Checkout() => View(new Order());

        [HttpPost]
        public IActionResult Checkout(Order order) {
            if (cart.Lines.Count() == 0) {
                ModelState.AddModelError("", "Sorry, your cart is empty!");
            }
            if (ModelState.IsValid) {
                order.Lines = cart.Lines.ToArray();
                repository.SaveOrder(order);
                return RedirectToAction(nameof(Completed));
            } else {
                return View(order);
            }
        }

        public ViewResult Completed() {
            cart.Clear();
            return View();
        }
    }
}
```

由于不想阻止未经身份验证的用户访问 Order 控制器中的其他操作方法，因此仅将 Authorize 特性应用于

List 和 MarkShipped 操作方法。如果要保护 Admin 控制器定义的所有操作方法，那么可以通过将 Authorize 特性应用于控制器类来实现这一点，控制器类会将授权策略应用于它包含的所有操作方法，如代码清单 12-7 所示。

代码清单 12-7　限制 AdminController.cs 文件中的访问权限

```
using Microsoft.AspNetCore.Mvc;
using SportsStore.Models;
using System.Linq;
using Microsoft.AspNetCore.Authorization;

namespace SportsStore.Controllers {

    [Authorize]
    public class AdminController : Controller {
        private IProductRepository repository;

        public AdminController(IProductRepository repo) {
            repository = repo;
        }

        public ViewResult Index() => View(repository.Products);

        public ViewResult Edit(int productId) =>
            View(repository.Products
                .FirstOrDefault(p => p.ProductID == productId));
        [HttpPost]
        public IActionResult Edit(Product product) {
            if (ModelState.IsValid) {
                repository.SaveProduct(product);
                TempData["message"] = $"{product.Name} has been saved";
                return RedirectToAction("Index");
            } else {
                // there is something wrong with the data values
                return View(product);
            }
        }

        public ViewResult Create() => View("Edit", new Product());

        [HttpPost]
        public IActionResult Delete(int productId) {
            Product deletedProduct = repository.DeleteProduct(productId);
            if (deletedProduct != null) {
                TempData["message"] = $"{deletedProduct.Name} was deleted";
            }
            return RedirectToAction("Index");
        }
    }
}
```

12.1.3　创建账户控制器和视图

当未经身份验证的用户发送需要身份验证的请求时，他们会被重定向到/Account/Login URL，应用程序将提示用户需要提供凭据。在准备过程中，可添加视图模型来表示用户的凭据，方法是将名为 LoginModel.cs 的类文件添加到 Models/ViewModels 文件夹中，并用它定义代码清单 12-8 所示的类。

代码清单 12-8　Models/ViewModels 文件夹下的 LoginModel.cs 文件的内容

```
using System.ComponentModel.DataAnnotations;

namespace SportsStore.Models.ViewModels {
```

```
    public class LoginModel {

        [Required]
        public string Name { get; set; }

        [Required]
        [UIHint("password")]
        public string Password { get; set; }

        public string ReturnUrl { get; set; } = "/";
    }
}
```

Name 和 Password 属性已使用 Required 特性进行修饰，该特性使用模型验证来确保必须为其提供值。Password 属性已使用 UIHint 特性进行了修饰，因此，当在 Razor 视图 Login 中的 input 元素上使用 asp-for 属性时，标签助手会将 type 属性设置为 password；这样，用户输入的文本在屏幕上将是不可见的。第 24 章将描述 UIHint 特性的使用方法。

接下来，将一个名为 AccountController.cs 的类文件添加到 Controllers 文件夹中，并用它定义代码清单 12-9 所示的控制器，以响应对/Account/Login URL 的请求。

代码清单 12-9 Controllers 文件夹下的 AccountController.cs 文件的内容

```
using System.Threading.Tasks;
using Microsoft.AspNetCore.Authorization;
using Microsoft.AspNetCore.Identity;
using Microsoft.AspNetCore.Mvc;
using SportsStore.Models.ViewModels;

namespace SportsStore.Controllers {

    [Authorize]
    public class AccountController : Controller {
        private UserManager<IdentityUser> userManager;
        private SignInManager<IdentityUser> signInManager;

        public AccountController(UserManager<IdentityUser> userMgr,
                SignInManager<IdentityUser> signInMgr) {
            userManager = userMgr;
            signInManager = signInMgr;
        }

        [AllowAnonymous]
        public ViewResult Login(string returnUrl) {
            return View(new LoginModel {
                ReturnUrl = returnUrl
            });
        }

        [HttpPost]
        [AllowAnonymous]
        [ValidateAntiForgeryToken]
        public async Task<IActionResult> Login(LoginModel loginModel) {
            if (ModelState.IsValid) {
                IdentityUser user =
                    await userManager.FindByNameAsync(loginModel.Name);
                if (user != null) {
                    await signInManager.SignOutAsync();
                    if ((await signInManager.PasswordSignInAsync(user,
                            loginModel.Password, false, false)).Succeeded) {
```

```
                        return Redirect(loginModel?.ReturnUrl ?? "/Admin/Index");
                    }
                }
            }
            ModelState.AddModelError("", "Invalid name or password");
            return View(loginModel);
        }

        public async Task<RedirectResult> Logout(string returnUrl = "/") {
            await signInManager.SignOutAsync();
            return Redirect(returnUrl);
        }
    }
}
```

当用户被重定向到/Account/Login URL 时，Login 操作方法的 GET 版本将呈现页面的默认视图，并提供一个视图模型对象，该对象包括一个 URL。如果身份验证请求成功，浏览器将重定向到这个 URL。

身份验证凭据将被提交到 Login 操作方法的 POST 版本，该操作方法使用了 UserManager<IdentityUser>和 SignInManager<IdentityUser>服务，这两个服务是通过控制器的构造函数接收的，用于对用户进行身份验证并登录系统。第 28～30 章将解释这些类是如何工作的，但是现在知道以下内容就足够了：如果身份验证失败，那么将创建模型验证错误并呈现默认视图；但是如果身份验证成功，那么会在提示输入用户凭据之前将用户重定向到他们要访问的 URL。

警　告

一般情况下，使用客户端数据验证是一个好主意。这可以减轻服务器的工作量，并为用户提示有关他们所提供数据的即时反馈。但是，你不应该在客户端执行身份验证，因为这通常涉及将有效凭据发送到客户端，以便它们检查用户输入的用户名和密码，或者至少信任客户端关于他们是否已成功通过身份验证的报告。身份验证应始终在服务器端完成。

为了向 Login 操作方法提供渲染视图，创建 Views/Account 文件夹并在其中添加一个名为 Login.cshtml 的 Razor 视图文件，其内容如代码清单 12-10 所示。

代码清单 12-10　Views/Account 文件夹下的 Login.cshtml 文件的内容

```
@model LoginModel
@{
    ViewBag.Title = "Log In";
    Layout = "_AdminLayout";
}

<div class="text-danger" asp-validation-summary="All"></div>

<form asp-action="Login" asp-controller="Account" method="post">
    <input type="hidden" asp-for="ReturnUrl" />
    <div class="form-group">
        <label asp-for="Name"></label>
        <div><span asp-validation-for="Name" class="text-danger"></span></div>
        <input asp-for="Name" class="form-control" />
    </div>
    <div class="form-group">
        <label asp-for="Password"></label>
        <div><span asp-validation-for="Password" class="text-danger"></span></div>
        <input asp-for="Password" class="form-control" />
    </div>
    <button class="btn btn-primary" type="submit">Log In</button>
</form>
```

最后一步是更改共享的管理布局，可通过添加一个按钮，向 Logout 操作发送请求使当前用户注销登录，如代码清单 12-11 所示。这是一个非常有用的功能，可以让你更轻松地测试应用程序。如果没有这个功能，那么需要清除浏览器的 cookie 才能返回未经身份验证的状态。

代码清单 12-11 在_AdminLayout.cshtml 文件中添加 Log Out 按钮

```
<!DOCTYPE html>
<html>
<head>
    <meta name="viewport" content="width=device-width" />
    <link rel="stylesheet" asp-href-include="lib/bootstrap/dist/css/*.min.css" />
    <title>@ViewBag.Title</title>
    <style>
        .input-validation-error {
            border-color: red;
            background-color: #fee;
        }
    </style>
    <script src="/lib/jquery/dist/jquery.min.js"></script>
    <script src="/lib/jquery-validation/dist/jquery.validate.min.js"></script>
    <script
      src="/lib/jquery-validation-unobtrusive/jquery.validate.unobtrusive.min.js">
    </script>
</head>
<body class="m-1 p-1">
    <div class="bg-info p-2 row">
        <div class="col">
            <h4>@ViewBag.Title</h4>
        </div>
        <div class="col-2">
            <a class="btn btn-sm btn-primary"
               asp-action="Logout" asp-controller="Account">Log Out</a>
        </div>
    </div>
    @if (TempData["message"] != null) {
        <div class="alert alert-success mt-1">@TempData["message"]</div>
    }
    @RenderBody()
</body>
</html>
```

12.1.4 测试安全策略

一切就绪后，你就可以通过启动应用程序并请求/Admin/Index 来测试安全策略了。由于你目前尚未经过身份验证，并且尝试访问需要授权的操作，因此你的浏览器将被重定向到/Account/Login URL。输入 Admin 和 Secret123$（作为用户名和密码）并提交表单。Account 控制器将检查提供的凭据以及添加到 Identity 数据库中的种子数据，如果输入了正确的信息，就会对你进行身份验证并将浏览器重定向回/Account/Login URL，现在就可以进行访问了。图 12-1 说明了上述过程。

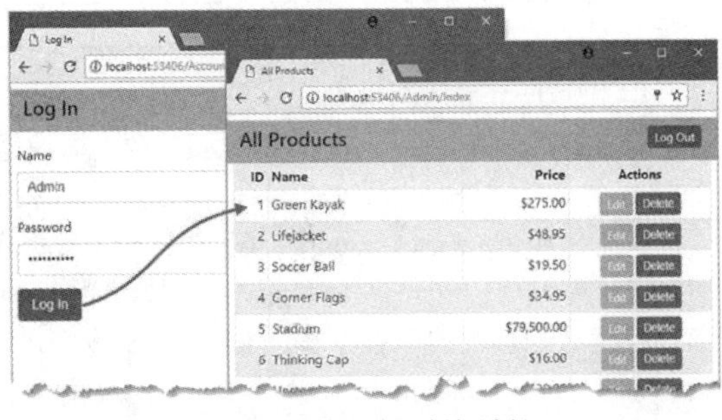

图 12-1 管理验证/授权过程

12.2 部署应用程序

SportsStore 应用程序的所有特性和功能都已到位，因此是时候准备将应用程序部署到生产环境中了。ASP.NET Core MVC 应用程序有很多托管选项，本章使用的是 Microsoft Azure 平台，选择它是因为它来自微软，而且提供免费账户，这意味着即使不想将 Microsoft Azure 用于自己的项目，也可以一直使用 SportsStore 示例。

注 意

你需要一个 Microsoft Azure 账户。如果还没有，可以在 Microsoft 官网上申请免费账户。

12.2.1 创建数据库

我们从创建 SportsStore 应用程序将在生产环境中使用的数据库开始。这个操作可以在 Visual Studio 部署过程中执行，但这有点像鸡与蛋的关系，由于需要在部署之前知道数据库的连接字符串，因此需要先完成创建数据库的过程。

警 告

微软会向 Microsoft Azure 添加新功能并修改现有功能，因此 Microsoft Azure 门户网站经常更新。在撰写这些内容时，本节中的说明是准确的，但是在你阅读本书时，所需的步骤可能会略有改变。基本方法应该仍然相同，但数据字段的名称和步骤的确切顺序可能需要通过一些实验才能获得正确的结果。

最简单的方法是使用 Microsoft Azure 账户登录 Azure 门户网站并手动创建数据库。登录后，选择 SQL Databases resource 类别，然后单击 Add 按钮以创建新的数据库。

对于第一个数据库，输入名称 products。单击 Configure Required Settings 链接，然后单击 Create a New Server 链接。输入新的服务器名称（在 Microsoft Azure 中必须是唯一的），并选择数据库管理员的用户名和密码。此处输入服务器名称 sportsstorecore2db，数据库管理员的用户名为 sportsstoreadmin，密码为 Secret123$。你需要使用不同的服务器名称，建议使用更复杂的密码。选择数据库的位置，单击 Select 按钮以关闭选项，然后单击 Create 按钮以创建数据库。Microsoft Azure 将花费几分钟时间来执行创建过程，之后它将出现在 SQL Databases resource 类别中。

创建另一个 SQL 服务器，这次输入名称 **identity**。可以使用刚刚创建的数据库服务器，而不需要创建新的数据库服务器。结果是 Microsoft Azure 托管的两个 SQL Server 数据库，其详细信息如表 12-1 所示。你将拥有不同的数据库服务器名称，理想情况下，你应该使用更健壮的密码。

表 12-1 SportsStore 应用程序的 Microsoft Azure 数据库

数据库名称	数据库服务器名称	数据库管理员	密码
products	sportsstorecore2db	sportsstoreadmin	Secret123$
identity	sportsstorecore2db	sportsstoreadmin	Secret123$

1. 配置防火墙访问权限

需要按照数据库架构填充数据，最简单的方法是打开 Microsoft Azure 防火墙，以便从开发机器运行 Entity Framework Core 命令。

选择 SQL Databases resource 类别中的任意数据库，单击 Tools 按钮，然后单击 Open in Visual Studio 链接。现在单击 Configure Your Firewall 链接，单击 Add Client IP 按钮，然后单击 Save 按钮，这将允许你从当前 IP 地址访问数据库服务器并执行配置命令（可以通过单击 Open In Visual Studio 按钮来检查数据库架构，这将打开 Visual Studio 并使用 SQL Server 对象资源管理器来检查数据库）。

2. 获取连接字符串

接下来需要新的数据库连接字符串。当通过 Show Database Connection Strings 链接单击 SQL Databases resource 类别中的数据库时，Microsoft Azure 会为不同的开发平台提供连接字符串，这是.NET 应用程序所需的 ADO.NET 字符串。以下是 Microsoft Azure 门户为 products 数据库提供的连接字符串：

```
Server=tcp:sportsstorecore2db.database.windows.net,1433;Initial Catalog=products;Persist
Security Info=False;User ID={your_username};Password={your_password};MultipleActiveResult
Sets=True;Encrypt=True;TrustServerCertificate=False;Connection Timeout=30;
```

你将看到不同的配置选项，具体取决于 Microsoft Azure 如何配置数据库。请注意，用户名和密码有占位符，这里已经用粗体标记了，当使用连接字符串配置应用程序时必须更改这些值。

12.2.2　准备应用程序

在部署应用程序之前，还有一些基本的准备工作要做，以便为生产环境做好准备。在接下来的内容中，将更改错误的显示方式并设置数据库的生产环境连接字符串。

1. 创建 Error 控制器和视图

目前，应用程序已配置为使用开发人员友好的错误页面，这些页面在出现问题时可以提供有用的信息。这不是最终用户应该看到的信息，因此将一个名为 ErrorController.cs 的类文件添加到 Controllers 文件夹中，并用它定义代码清单 12-12 所示的简单控制器。

代码清单 12-12　Controllers 文件夹下的 ErrorController.cs 文件的内容

```
using Microsoft.AspNetCore.Mvc;

namespace SportsStore.Controllers {

    public class ErrorController : Controller {

        public ViewResult Error() => View();
    }
}
```

Error 控制器定义了呈现默认视图的 Error 操作方法。为了给 Error 控制器提供视图，创建 Views/Error 文件夹，添加一个名为 Error.cshtml 的 Razor 视图文件，在其中编写代码清单 12-13 所示的代码。

代码清单 12-13　Views/Error 文件夹下的 Error.cshtml 文件的内容

```
@{
    Layout = null;
}
<!DOCTYPE html>
<html>
<head>
    <meta name="viewport" content="width=device-width" />
    <link rel="stylesheet" href="~/lib/bootstrap/dist/css/bootstrap.min.css" />
    <title>Error</title>
</head>
<body>
    <h2 class="text-danger">Error.</h2>
    <h3 class="text-danger">An error occurred while processing your request.</h3>
</body>
</html>
```

这种错误页面是万不得已的选择，最好尽可能简单，而不要依赖共享视图、视图组件或其他功能。在本例中，禁用共享布局并定义一个简单的 HTML 文档，该 HTML 文档说明出现了错误，但并未提供错误的任何信息。

2. 定义生产数据库设置

下一步是创建一个文件，该文件将为应用程序提供生产环境中的数据库连接字符串。向 SportsStore 项目添加一个名为 appsettings.production.json 的 ASP.NET 配置文件，并添加代码清单 12-14 所示的内容。

提　示

解决方案资源管理器会将此文件嵌套在文件列表中的 appsettings.json 中，如果稍后想再次编辑该文件，就必须先展开。

代码清单 12-14 appsettings.production.json 文件的内容

```
{
  "Data": {
    "SportStoreProducts": {
      "ConnectionString": "Server=tcp:sportsstorecore2db.database.windows.net,1433;Initial
      Catalog=products;Persist Security Info=False;User ID={your_username};Password={your_
      password};MultipleActiveResultSets=True;Encrypt=True;TrustServerCertificate=False;
      Connection Timeout=30;"

    },
    "SportStoreIdentity": {
      "ConnectionString": "Server=tcp:sportsstorecore2db.database.windows.net,1433;Initial
      Catalog=identity;Persist Security Info=False;User ID={your_username};Password={your_
      password};MultipleActiveResultSets=True;Encrypt=True;TrustServerCertificate=False;
      Connection Timeout=30;"
    }
  }
}
```

这个文件很难阅读，因为连接字符串不能跨多行。这个文件复制了 appsettings.json 文件的连接字符串部分，但使用了 Microsoft Azure 连接字符串（别忘了替换用户名和密码占位符）。这里还将 MultipleActiveResultSets 设置为 True，这允许多个并发查询，并避免在执行复杂 LINQ 查询时的常见错误情况。

注 意

在将用户名和密码插入连接字符串时需要删除括号字符，最终结果应该是 Password=MyPassword 而不是 Password={MyPassword}。

3. 配置应用程序

现在可以更改 Startup 类，这样应用程序在生产环境中就会执行不同的操作。代码清单 12-15 显示了所做的更改。

代码清单 12-15 在 Startup.cs 文件中配置应用程序

```
using System;
using System.Collections.Generic;
using System.Linq;
using System.Threading.Tasks;
using Microsoft.AspNetCore.Builder;
using Microsoft.AspNetCore.Hosting;
using Microsoft.AspNetCore.Http;
using Microsoft.Extensions.DependencyInjection;
using SportsStore.Models;
using Microsoft.Extensions.Configuration;
using Microsoft.EntityFrameworkCore;
using Microsoft.AspNetCore.Identity;

namespace SportsStore {

    public class Startup {

        public Startup(IConfiguration configuration) =>
            Configuration = configuration;

        public IConfiguration Configuration { get; }

        public void ConfigureServices(IServiceCollection services) {
            services.AddDbContext<ApplicationDbContext>(options =>
                options.UseSqlServer(
```

```
                Configuration["Data:SportStoreProducts:ConnectionString"]));

        services.AddDbContext<AppIdentityDbContext>(options =>
            options.UseSqlServer(
                Configuration["Data:SportStoreIdentity:ConnectionString"]));

        services.AddIdentity<IdentityUser, IdentityRole>()
            .AddEntityFrameworkStores<AppIdentityDbContext>()
            .AddDefaultTokenProviders();

        services.AddTransient<IProductRepository, EFProductRepository>();
        services.AddScoped<Cart>(sp => SessionCart.GetCart(sp));
        services.AddSingleton<IHttpContextAccessor, HttpContextAccessor>();
        services.AddTransient<IOrderRepository, EFOrderRepository>();
        services.AddMvc();
        services.AddMemoryCache();
        services.AddSession();
    }
    public void Configure(IApplicationBuilder app, IHostingEnvironment env) {

        if (env.IsDevelopment()) {
            app.UseDeveloperExceptionPage();
            app.UseStatusCodePages();
        } else {
            app.UseExceptionHandler("/Error");
        }

        app.UseStaticFiles();
        app.UseSession();
        app.UseAuthentication();
        app.UseMvc(routes => {
            routes.MapRoute(name: "Error", template: "Error",
                defaults: new { controller = "Error", action = "Error" });
            routes.MapRoute(name: null,
                template: "{category}/Page{productPage:int}",
                defaults: new { controller = "Product", action = "List" }
            );
            routes.MapRoute(name: null,template: "Page{productPage:int}",
                defaults: new { controller = "Product",
                action = "List", productPage = 1 }
            );
            routes.MapRoute(name: null, template: "{category}",
                defaults: new { controller = "Product",
                    action = "List", productPage = 1 }
            );
            routes.MapRoute(name: null,template: "",
                defaults: new { controller = "Product",
                    action = "List", productPage = 1 });
            routes.MapRoute(name: null, template: "{controller}/{action}/{id?}");
        });
        //SeedData.EnsurePopulated(app);
        //IdentitySeedData.EnsurePopulated(app);
    }
  }
}
```

IHostingEnvironment 接口用于提供有关运行应用程序的环境信息，例如，当前运行环境为开发环境还是生产环境。当托管环境设置为 Production 时，ASP.NET Core 将加载 appsettings.production.json 文件及其内容，用来覆盖 appsettings.json 文件中的设置，这意味着 Entity Framework Core 将连接到 Microsoft Azure 数据库而不是 LocalDB。可以使用很多选项在不同环境中定制应用程序的配置，这将在第 14 章中解释。

这里还注释了数据库的填充语句，12.2.4 节将解释这些语句。

12.2.3　应用数据库迁移

要使用应用程序所需的架构设置数据库，请打开新的命令提示符或 PowerShell 窗口，然后导航到 SportsStore 项目文件夹。首先需要设置环境变量，以便 dotnet 命令行工具使用 Microsoft Azure 连接字符串。如果使用的是 PowerShell，请使用以下命令设置环境变量：

```
$env:ASPNETCORE_ENVIRONMENT="Production"
```

如果使用的是命令提示符，请使用以下命令设置环境变量：

```
set ASPNETCORE_ENVIRONMENT=Production
```

在 SportsStore 项目文件夹中运行以下命令，以将项目中的迁移应用于 Microsoft Azure 数据库：

```
dotnet ef database update --context ApplicationDbContext
dotnet ef database update --context AppIdentityDbContext
```

环境变量指定了用于获取连接字符串以访问数据库的宿主环境。如果这些命令不起作用，请确保已将 Microsoft Azure 防火墙配置为允许开发计算机访问（如本章前面所述），并且已正确复制和修改了连接字符串。

12.2.4　管理数据库填充

代码清单 12-15 注释掉了 Startup 类中用于填充数据库的语句。这样做是因为用于将迁移应用于数据库的 Entity Framework Core 命令依赖于 Startup 类设置的服务，这意味着在启用这些语句的情况下，将在应用迁移之前调用为数据库设定填充数据的代码，这会导致错误并阻止迁移工作。设置数据库时，这不会导致问题。但对于生产数据库，因为 SeedData.EnsurePopulated 方法在填充数据之前应用迁移，而且还因为在迁移应用到数据库之前，还没有将 Identity 种子数据添加到应用程序中，所以会出错。

对于生产环境，建议采用不同的种子数据填充方法。对于用户账户，建议在登录时使用管理员账户填充数据库。这里将为管理工具添加一个功能，用于填充 products 数据库，以便生产系统可以填充测试数据，或者根据需要留空以获取实际数据。

注　意

在生产系统中填充验证数据时应谨慎处理，并且应用程序应使用第 28～30 章中描述的功能，在部署后立即更改密码。

1. 填充身份数据

更改用户数据填充方式的第一步是简化 IdentitySeedData 类中的代码，如代码清单 12-16 所示。

代码清单 12-16　简化 Models 文件夹下的 IdentitySeedData.cs 文件中的代码

```
using Microsoft.AspNetCore.Builder;
using Microsoft.AspNetCore.Identity;
using Microsoft.Extensions.DependencyInjection;
using System.Threading.Tasks;

namespace SportsStore.Models {

    public static class IdentitySeedData {
        private const string adminUser = "Admin";
        private const string adminPassword = "Secret123$";

        public static async Task EnsurePopulated(UserManager<IdentityUser>
                userManager) {

            IdentityUser user = await userManager.FindByIdAsync(adminUser);
            if (user == null) {
```

```
                user = new IdentityUser("Admin");
                await userManager.CreateAsync(user, adminPassword);
            }
        }
    }
}
```

EnsurePopulated 方法接收一个对象作为参数，而不是在内部获取 UserManager <IdentityUser>服务。这样就可以将数据库填充集成到 AccountController 类中，如代码清单 12-17 所示。

代码清单 12-17 在 Controllers 文件夹下的 AccountController.cs 文件中填充数据

```
using System.Threading.Tasks;
using Microsoft.AspNetCore.Authorization;
using Microsoft.AspNetCore.Identity;
using Microsoft.AspNetCore.Mvc;
using SportsStore.Models.ViewModels;
using SportsStore.Models;

namespace SportsStore.Controllers {

    [Authorize]
    public class AccountController : Controller {
        private UserManager<IdentityUser> userManager;
        private SignInManager<IdentityUser> signInManager;

        public AccountController(UserManager<IdentityUser> userMgr,
                SignInManager<IdentityUser> signInMgr) {
            userManager = userMgr;
            signInManager = signInMgr;
            IdentitySeedData.EnsurePopulated(userMgr).Wait();
        }

        // ...other methods omitted for brevity...
    }
}
```

这些更改将确保每次创建 AccountController 对象以处理 HTTP 请求时，都会为 Identity 数据库填充数据。当然，这并不是很理想，但是并没有很好的方法来为数据库填充数据，这种方法将确保应用程序可以在生产环境和开发环境中进行管理，尽管代价是产生一些额外的数据库查询。

2. 填充产品数据

对于产品数据，这里将向管理员提供一个按钮，该按钮将在数据库为空时为其提供初始数据。第一步是更改数据填充代码，以便使用一个接口，允许访问通过控制器而不是 Startup 类提供的服务，如代码清单 12-18 所示。这里还注释了自动应用挂起迁移的任何语句，这可能导致数据丢失，因此在生产环境中只能非常谨慎地使用。

代码清单 12-18 准备在 Models 文件夹下的 SeedData.cs 文件中手动填充数据

```
using System.Linq;
using Microsoft.AspNetCore.Builder;
using Microsoft.Extensions.DependencyInjection;
using Microsoft.EntityFrameworkCore;
using System;

namespace SportsStore.Models {

    public static class SeedData {

        public static void EnsurePopulated(IServiceProvider services) {
```

```
            ApplicationDbContext context =
                services.GetRequiredService<ApplicationDbContext>();
            //context.Database.Migrate();
            if (!context.Products.Any()) {
                context.Products.AddRange(

                    // ...statements omiited for brevity...

                );
                context.SaveChanges();
            }
        }
    }
}
```

下一步是更新 Admin 控制器以添加将触发数据填充操作的操作方法，如代码清单 12-19 所示。

代码清单 12-19 在 Controllers 文件夹下的 AdminController.cs 文件中填充数据库

```
using Microsoft.AspNetCore.Mvc;
using SportsStore.Models;
using System.Linq;
using Microsoft.AspNetCore.Authorization;

namespace SportsStore.Controllers {

    [Authorize]
    public class AdminController : Controller {
        private IProductRepository repository;

        public AdminController(IProductRepository repo) {
            repository = repo;
        }

        public ViewResult Index() => View(repository.Products);

        // ...other methods omitted for brevity...

        [HttpPost]
        public IActionResult SeedDatabase() {
            SeedData.EnsurePopulated(HttpContext.RequestServices);
            return RedirectToAction(nameof(Index));
        }
    }
}
```

对新的操作方法使用 HttpPost 特性进行修饰，以便可以使用 POST 请求进行定位，并且一旦数据库被填充，就将浏览器重定向到 Index 操作方法。剩下的就是创建一个按钮来填充数据库，如代码清单 12-20 所示。

代码清单 12-20 在 Views/Admin 文件夹下的 Index.cshtml 文件中添加一个按钮

```
@model IEnumerable<Product>

@{
    ViewBag.Title = "All Products";
    Layout = "_AdminLayout";
}

@if (Model.Count() == 0) {
    <div class="text-center m-2">
        <form asp-action="SeedDatabase" method="post">
            <button type="submit" class="btn btn-danger">Seed Database</button>
        </form>
```

```
        </div>
} else {
    <table class="table table-striped table-bordered table-sm">
        <tr>
            <th class="text-right">ID</th>
            <th>Name</th>
            <th class="text-right">Price</th>
            <th class="text-center">Actions</th>
        </tr>
        @foreach (var item in Model) {
            <tr>
                <td class="text-right">@item.ProductID</td>
                <td>@item.Name</td>
                <td class="text-right">@item.Price.ToString("c")</td>
                <td class="text-center">
                    <form asp-action="Delete" method="post">
                        <a asp-action="Edit" class="btn btn-sm btn-warning"
                           asp-route-productId="@item.ProductID">
                            Edit
                        </a>
                        <input type="hidden" name="ProductID"
                            value="@item.ProductID" />
                        <button type="submit" class="btn btn-danger btn-sm">
                            Delete
                        </button>
                    </form>
                </td>
            </tr>
        }
    </table>
}
<div class="text-center">
    <a asp-action="Create" class="btn btn-primary">Add Product</a>
</div>
```

12.2.5 部署应用程序

要部署应用程序，请在 Solution Explorer 窗格中右击 SportsStore 项目，然后从弹出的菜单中选择 Publish。Visual Studio 将为你提供一系列发布方法，如图 12-2 所示。

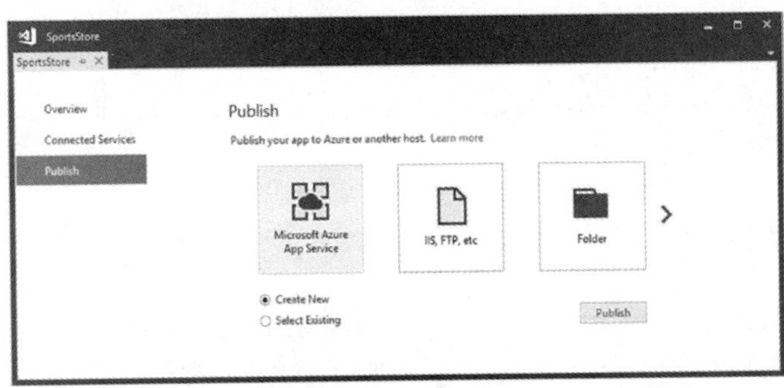

图 12-2 发布方法

如果部署失败，如何处理

部署失败的最可能原因是连接字符串错误，要么因为没有从 Microsoft Azure 中正确复制，要么因为编辑错误，没有正确插入用户名和密码。如果部署失败，首先应该检查连接字符串。如果在“应用数据库迁移”部分没有从 dotnet ef database update 命令获得预期结果，部署将失败。如果命令确实有效但部署失败，那么请确保已设置环境变量，因为你可能正在使用本地数据库而不是云中的数据库。

选择 Microsoft Azure App Service 选项，并确保选中 Create New（Select Existing 用于更新现有的已部署的应用程序）。系统将提示你提供部署的详细信息。首先单击 Add an Account，然后输入 Microsoft Azure 凭据。

输入凭据后，可以选择已部署应用程序的名称并输入服务的详细信息，具体取决于拥有的 Microsoft Azure 账户类型、要部署的区域以及所需的部署服务，如图 12-3 所示。

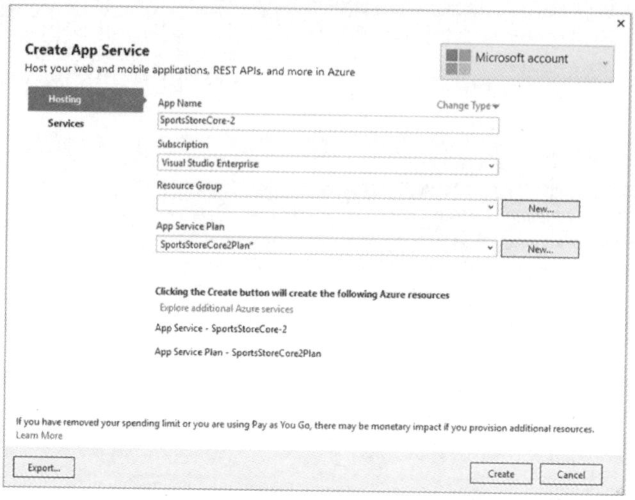

图 12-3　创建新的 Microsoft Azure App 服务

配置完服务后，单击 Create 按钮。设置服务后，系统将显示发布摘要，将应用程序发布到托管服务，如图 12-4 所示。

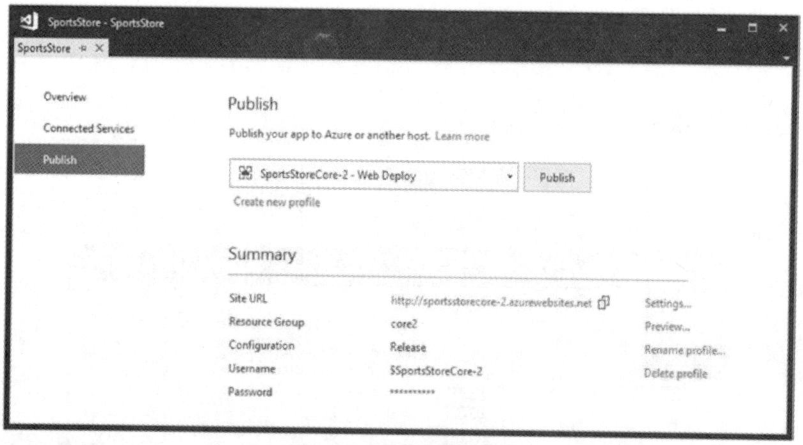

图 12-4　服务的发布摘要

单击 Publish 按钮开始部署过程。通过从 Visual Studio 的菜单栏中选择 View→Other Windows→Web Publish Activity，可以查看发布进度。在此过程中请耐心等待，因为将项目中的所有文件发送到 Azure 服务可能需要一段时间。后续更新将会更快，因为只会传输修改的文件。

部署完成后，Visual Studio 将为部署好的应用程序打开一个新的浏览器窗口。由于 products 数据库为空，你将看到图 12-5 所示的布局。

导航到/Admin/Index URL 并使用用户名 Admin 和密码 Secret123$进行身份验证。Identity 数据库将按需填充数据，允许你登录应用程序的管理后台，如图 12-6 所示。

单击 Seed Database 按钮以填充 products 数据库，这将产生图 12-7 所示的结果。然后，可以导航回应用程序的根 URL 并正常使用。

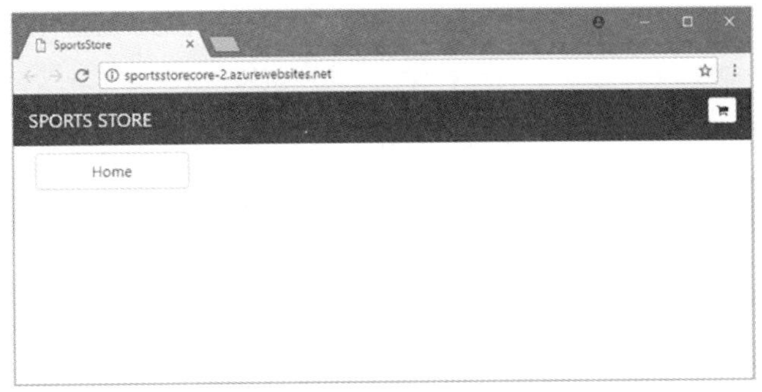

图 12-5　已部署应用程序的初始状态

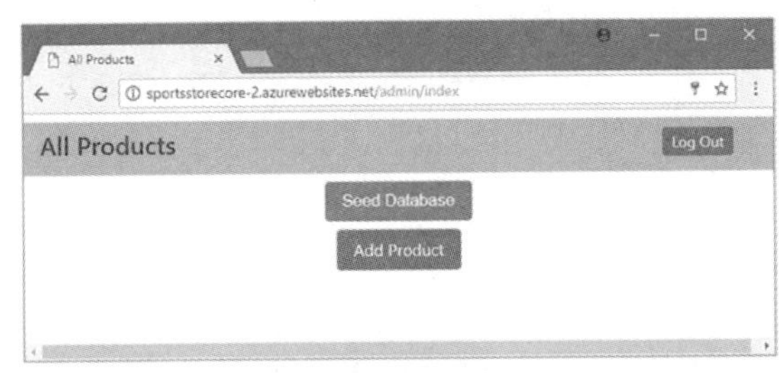

图 12-6　管理页面

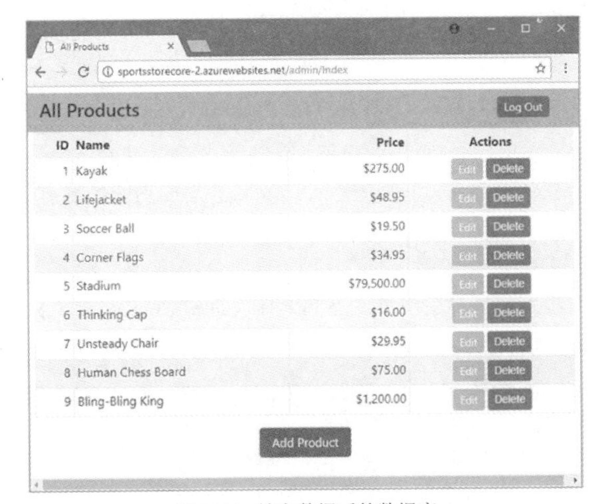

ID	Name	Price	Actions
1	Kayak	$275.00	Edit Delete
2	Lifejacket	$48.95	Edit Delete
3	Soccer Ball	$19.50	Edit Delete
4	Corner Flags	$34.95	Edit Delete
5	Stadium	$79,500.00	Edit Delete
6	Thinking Cap	$16.00	Edit Delete
7	Unsteady Chair	$29.95	Edit Delete
8	Human Chess Board	$75.00	Edit Delete
9	Bling-Bling King	$1,200.00	Edit Delete

图 12-7　填充数据后的数据库

12.3　小结

本章和前几章演示了如何通过 ASP.NET Core MVC 创建一个现实的电子商务应用程序。这个扩展的示例引入了许多关键的 MVC 特性——控制器、操作方法、路由、视图、元数据、验证、布局、身份验证等。本章还讲述了如何使用与 MVC 相关的一些关键技术，包括 Entity Framework Core、依赖注入和单元测试。这个最终的 SportsStore 应用程序拥有干净的、面向组件的体系结构，从而将各种关注点和易于扩展及维护的代码库分离开来。下一章将展示如何使用 Visual Studio 代码来创建 ASP.NET Core MVC 应用程序。

第13章　使用 Visual Studio Code

本章将展示如何使用 Visual Studio Code 创建 ASP.NET Core MVC 应用程序，Visual Studio Code 是一个由微软开发的开源跨平台编辑器。虽然名称如此，但 Visual Studio Code 与 Visual Studio 无关，Visual Studio Code 基于 Electron 框架，很多其他 Web 应用程序框架（如 Angular）的开发人员使用的 Atom 编辑器也基于这个框架。

Visual Studio Code 支持 Windows、macOS 和最流行的 Linux 发行版。Visual Studio Code 现已发展为强大且功能齐全的开发环境，尽管尚未提供 Visual Studio 的所有功能。作者自己也越来越多地使用 Visual Studio Code，因为它易于使用、速度很快，而且对其他语言（如 JavaScript 和 TypeScript）的支持也很好。

13.1　设置开发环境

设置 Visual Studio Code 需要执行一些额外步骤，因为有些功能在 Visual Studio 中是默认提供的，但对于 Visual Studio Code 来说，需要由外部工具处理。其中一些工具与 Visual Studio 使用的工具相同，但还有一些工具对.NET 开发世界来说是全新的，你可能还不熟悉。但好消息是，这些工具在其他 Web 应用程序框架的开发人员中已经得到广泛使用，质量和功能都很优秀。接下来将引导你安装 Visual Studio Code 以及 MVC 开发所需的基本工具和附件。

13.1.1　安装 Node.js

在客户端开发中，Node.js（也称为 Node）已经成为许多流行的开发工具所依赖的运行时。Node.js 是在 2009 年创建的，是用 JavaScript 为服务器端应用程序编写的简单高效的运行时。Node.js 基于 Chrome 浏览器中使用的 JavaScript 引擎，并提供了在浏览器环境之外执行 JavaScript 代码的 API。

Node.js 作为应用程序服务器已经取得了一些成功，对于本章来说，Node.js 为新一代跨平台的构建工具和包管理器提供了基础。Node.js 团队的一些明智的设计决策和 Chrome JavaScript 运行时提供的跨平台支持为那些热情的工具开发者，特别是那些想要支持 Web 应用程序开发的人提供了机会。

注　意

有两个可以使用的 Node.js 版本：长期支持（LTS）版本为生产环境中的部署提供了稳定的基础，在生产环境中要尽量减少变更，LTS 版本更新每 6 个月发布一次，并会维护 18 个月；当前版本是快速更新版本，注重新功能的实现，而不是稳定性。本章使用当前版本。

1. 在 Windows 系统上安装 Node.js

从 Node.js 网站下载并运行为 Windows 提供的 Node.js 安装程序。安装 Node.js 时，请确保将其添加到 PATH 系统变量中。图 13-1 显示了 Windows 安装程序，它提供了一个安装选项，可以将 Node.js 添加到 PATH 系统变量。

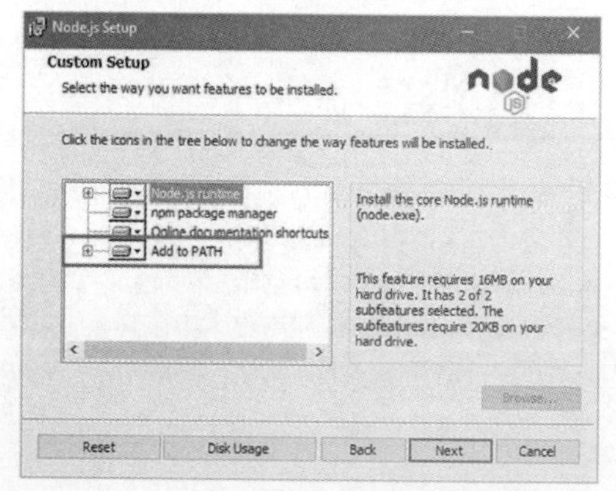

图 13-1　将 Node.js 添加到 PATH 系统变量中

2. 在 macOS 上安装 Node.js

可以从 Node.js 网站下载用于 macOS 的 Node.js 安装程序。运行安装程序并接受默认值。安装完成后，确保/usr/local/bin 位于$PATH 中。

3. 在 Linux 系统上安装 Node.js

在 Linux 系统上安装 Node.js 的最简单方法是使用包管

理器，Node.js 团队在 Node.js 网站上为主要发行版提供了说明。对于 Ubuntu 系统，可使用代码清单 13-1 中的命令下载并安装 Node.js。

代码清单 13-1 安装 Node.js

```
sudo curl -sL https***nodesource***/setup_6.x | sudo -E bash -
sudo apt-get install -y nodejs
```

13.1.2 检查 Node.js 安装状态

安装完成后，打开新的命令提示符并运行代码清单 13-2 所示的命令，检查 Node.js 是否正常工作，并显示已安装的版本。

代码清单 13-2 检查 Node.js 安装状态

```
node -v
```

如果安装成功并且 Node.js 已添加到 PATH 系统变量中，你将看到版本号。在撰写本书时，Node.js 的当前版本是 6.11.2。在根据本章中的示例执行操作时如果遇到问题，可尝试使用这个特定版本。

13.1.3 安装 Git

Visual Studio Code 已经集成了 Git 支持，但还需要单独安装 Git 以支持 Bower 工具，用于管理客户端软件包。

1. 在 Windows 系统或 macOS 上安装 Git

你可从 git-scm 网站下载并运行安装程序。

2. 在 Linux 系统上安装 Git

大多数 Linux 系统中已经安装了 Git。如果你仍想单独安装 Git，请参阅 git-scm 网站上的发行版安装说明。对于 Ubuntu 系统，使用代码清单 13-3 所示的命令安装 Git。

代码清单 13-3 在 Ubuntu 系统上安装 Git

```
sudo apt-get install git
```

13.1.4 检查 Git 安装状态

完成安装后，在新的命令提示符或 Terminal 中运行代码清单 13-4 所示的命令，检查 Git 是否已安装且可用。

代码清单 13-4 检查 Git 安装状态

```
git --version
```

以上命令会输出已安装的 Git 版本。在撰写本书时，用于 Windows 系统和 macOS 的最新版本的 Git 是 2.14.1，用于 Linux 系统的最新版本的 Git 是 2.7.4。

13.1.5 安装 Bower

Node.js 附带了 Node Package Manager（NPM），用于下载和安装使用 JavaScript 编写的软件包。本章所需的唯一软件包是 Bower，它用于管理客户端软件包，第 6 章对此进行了说明。运行代码清单 13-5 所示的命令，为 Windows 系统下载并安装 Bower。

代码清单 13-5 在 Windows 系统上安装 Bower

```
npm install -g bower@1.8.0
```

对于 Linux 系统和 macOS，可使用相同的命令，但需要 sudo，如代码清单 13-6 所示。

代码清单 13-6 在 Linux 系统或 macOS 上安装 Bower

```
sudo npm install -g bower@1.8.0
```

13.1.6 安装.NET Core

ASP.NET Core MVC 开发需要.NET Core 运行时。每个受支持的平台都有自己的安装过程，参见 Microsoft 官网。微软提供了用于 Windows 系统和 macOS 的安装程序，并使用 tar 存档提供 Linux 系统下的安装命令。

1. 在 Windows 系统和 macOS 上安装.NET Core

要在 Windows 系统或 macOS 上安装.NET Core，只需要下载并运行.NET Core SDK 安装程序即可。

2. 在 Linux 系统上安装.NET Core

微软在其官网上提供了在最流行的 Linux 发行版上安装.NET Core 的说明。本章使用的是 Ubuntu，首先使用代码清单 13-7 所示的命令为 apt-get 设置新的种子。

代码清单 13-7 准备在 Ubuntu Linux 系统上安装.NET Core

```
sudo sh -c 'echo "deb [arch=amd64] *****://apt-mo.trafficmanager.***/repos/dotnet-release/
xenial main" > /etc/apt/sources.list.d/dotnetdev.list'
sudo apt-key adv --keyserver ***://keyserver.ubuntu.***:80 --recv-keys 417A0893
sudo apt-get update
```

下一步是安装.NET Core，如代码清单 13-8 所示。

代码清单 13-8 在 Ubuntu Linux 系统上安装.NET Core

```
sudo apt-get install dotnet-sdk-2.0.0
```

13.1.7 检查.NET Core 安装状态

无论使用何种平台，都可以检查.NET Core 是否已安装并可以使用。打开新的命令提示符或 Terminal 并运行代码清单 13-9 所示的命令。

代码清单 13-9 检查.NET Core 版本

```
dotnet --version
```

dotnet 命令会启动.NET 运行时，并显示已安装的.NET Core 的版本号。在撰写本书时，当前版本是 2.0.0，但是到你阅读本书时，可能已经有更新的版本了。

13.1.8 安装 Visual Studio Code

最重要的一步是下载并安装 Visual Studio Code，安装包可从 Visual Studio 官网获得。安装包可用于 Windows 系统、macOS 和流行的 Linux 系统发行版。下载并安装所选平台的安装包。

注 意

微软每个月都会发布新版的 Visual Studio Code，这意味着你安装的版本可能会与本书的版本不同。虽然基本原理保持不变，但这意味着可能需要进行一些尝试才能完成本章中的一些示例。

1. 在 Windows 系统上安装 Visual Studio Code

要在 Windows 系统上安装 Visual Studio Code，只需要运行安装程序即可。完成安装过程后，Visual Studio Code 将启动，你将看到编辑器窗口，如图 13-2 所示。

2. 在 macOS 上安装 Visual Studio Code

Visual Studio Code 可作为 Mac 计算机的 zip 存档提供，可以从 Microsoft 官网下载。展开存档，然后双击其中包含的 Code.app 文件用于启动 Visual Studio Code，生成图 13-2 所示的编辑器窗口。

3. 在 Linux 系统上安装 Visual Studio Code

微软为 Debian 和 Ubuntu 系统提供了.deb 文件，为 Red Hat 系统、Fedora 系统和 CentOS 提供了.rpm 文件。下载并安装适用于对应 Linux 发行版的文件。由于本章使用的是 Ubuntu 系统，因此下载.deb 文件并使用 Ubuntu 软件工具进行安装。

安装完成后，运行代码清单 13-10 所示的命令以启动 Visual Studio Code，生成图 13-2 所示的编辑器窗口。

代码清单 13-10 在 Linux 系统上启动 Visual Studio Code

```
/usr/share/code/code
```

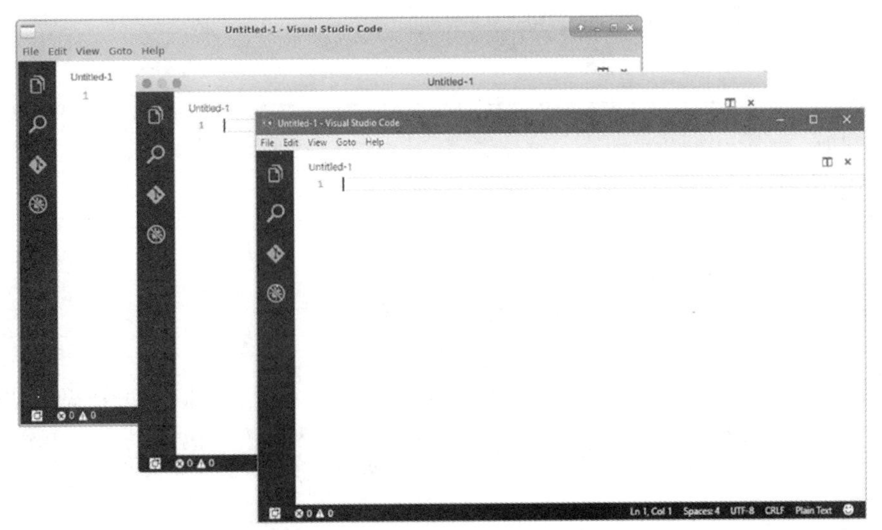

图 13-2 在 Windows 系统、macOS 和 Ubuntu Linux 系统上运行 Visual Studio Code

13.1.9 检查 Visual Studio Code 安装状态

成功安装 Visual Studio Code 后，就可以启动应用程序并查看编辑器了（这里更改了配色方案，因为默认的深色主题不适合图书出版）。

13.1.10 安装 Visual Studio Code 的 C#扩展

Visual Studio Code 通过扩展来支持特定于语言的功能，但这些扩展与 Visual Studio 支持的扩展不同。对 ASP.NET Core MVC 开发而言最重要的扩展是增加对 C#的支持，这看起来像是粗心的遗漏，但反映了如下事实：微软已将 Visual Studio Code 定位为支持尽可能广泛的语言和框架的通用跨平台编辑器。

要安装 C#扩展，请单击 Visual Studio Code 窗口左侧的 Extensions 图标。在搜索框中输入 csharp，在列表中找到 C# for Visual Studio Code 扩展，如图 13-3 所示。

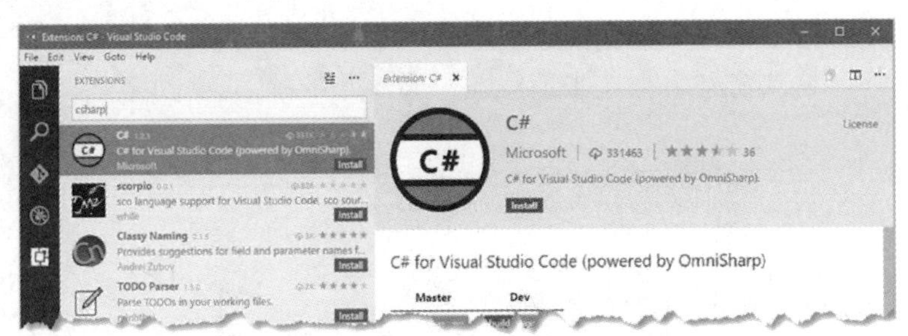

图 13-3 找到 C# for Visual Studio Code 扩展

单击 Install 按钮，Visual Studio Code 将下载并安装扩展。单击 Enable 按钮重新启动 Visual Studio Code 并启用 for Visual Studio Code 扩展，如图 13-4 所示。

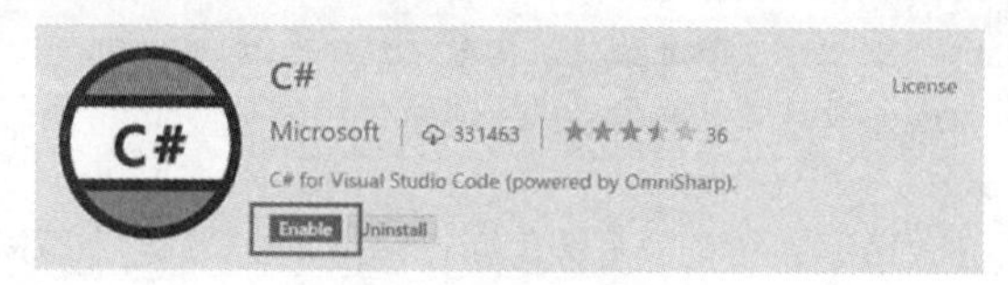

图 13-4 启用 C#扩展

13.2 创建 ASP.NET Code 项目

Visual Studio Code 没有集成创建 ASP.NET Core 项目的功能，但可以使用 dotnet 命令行创建新项目。

创建正确的文件夹结构以正确设置项目非常重要，尤其是在使用单元测试时。在合适的位置创建名为 InvitesProjects 的项目。

接下来，在 InvitesProjects 文件夹中创建一个名为 PartyInvites 的文件夹，使用命令提示符导航到该文件夹并运行代码清单 13-11 所示的命令。

代码清单 13-11 在 PartyInvites 文件夹中创建一个新的项目

```
dotnet new web --language C# --framework netcoreapp2.0
```

dotnet new 命令能够以命令行的方式访问项目模板，代码清单 13-11 指定的 Web 模板对应于前面章节中使用的 Visual Studio Empty 模板。表 13-1 描述了可用于 ASP.NET Core 开发的一组项目模板（其他模板也是可用的，但它们不适用于 ASP.NET Core。运行 dotnet new --help 命令可查看完整列表）。

表 13-1 用于 ASP.NET Core 开发的 dotnet new 模板

名称	描述
web	这是前面章节中使用的 Empty 模板，用于创建 ASP.NET Core 项目，但是没有启用 MVC 框架
mvc	这是第 2 章中使用的 Web Application(Model-View-Controller)模板，用于创建包含 MVC 框架、占位符控制器和视图的 ASP.NET Core 项目
xunit	这是 xUnit Test Project（.NET Core）模板，可使用 xUnit 包设置单元测试

13.3 使用 Visual Studio Code 准备项目

要在 Visual Studio Code 中打开项目，请从 File 菜单中选择 Open Folder，导航到 InvitesProjects 项目文件夹，然后单击 Select Folder 按钮。Visual Studio Code 将打开项目并自动安装编辑和调试 C#应用程序所需的一些软件包。第一次开始编辑文件几秒后，你将看到一条消息，提示将一些资源添加到项目中，如图 13-5 所示。

图 13-5 提示将资源添加到项目中

单击 Yes 按钮，Visual Studio Code 将创建.vscode 文件夹并添加一些用于配置构建过程的文件。默认情况下，Visual Studio Code 使用包含 3 个区域的布局。侧栏（在图 13-6 中已框选）提供对主要功能区域的访问。顶层的按钮用于打开资源管理器窗格，显示已打开文件夹的内容。其他按钮提供对搜索功能、集成源代码管理、调试器和已安装扩展集的访问。

单击资源管理器窗格中的文件，即可打开并进行编辑。Visual Studio Code 支持同时编辑多个文件，可以通过单击窗口右上角的 Split Editor 按钮来创建新的编辑器窗格。Visual Studio 代码编辑器相当不错，在输入 NuGet 和 Bower 软件包的名称和版本时，可以提供良好的智能感知支持。

除项目文件夹中的内容之外，资源管理器窗格还会显示当前正在编辑的文件，这使你可以轻松地将注意力集中在正在使用的文件集上，这在处理大型项目中的相关文件子集时非常有用。

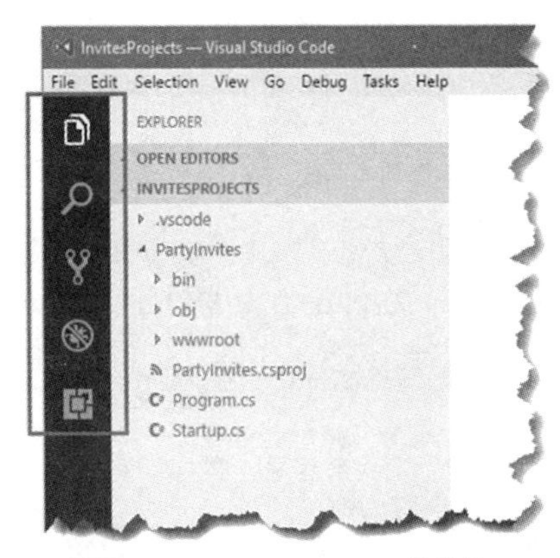

图 13-6　Visual Studio Code 的边栏

13.3.1　管理客户端软件包

Bower 用于管理 Visual Studio Code 项目中的客户端软件包，就像在 Visual Studio 中一样，但是还需要做一些额外的工作。

第一步是添加一个名为.bowerrc 的文件，该文件用于告诉 Bower 在哪里安装软件包。右击 PartyInvites 文件夹，从弹出菜单中选择 New File，创建新文件，如图 13-7 所示。

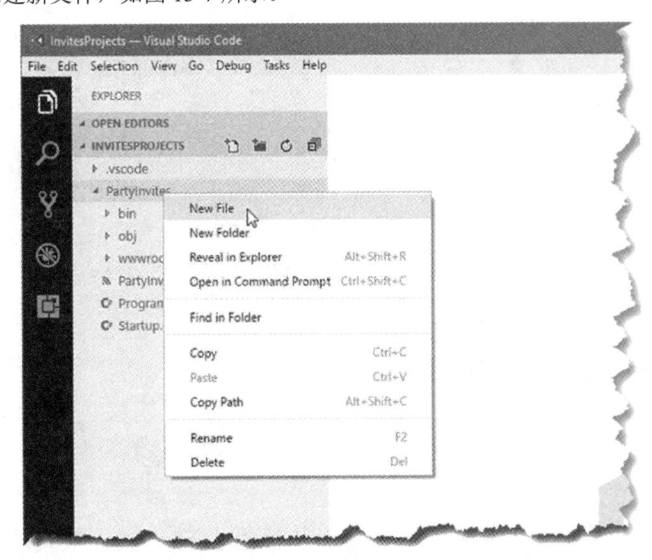

图 13-7　创建新文件

将文件名设置为.bowerrc（请注意，文件名中有两个 r）并添加代码清单 13-12 所示的内容。

代码清单 13-12　.bowerrc 文件的内容

```
{
  "directory": "wwwroot/lib"
}
```

接下来，创建一个名为 bower.json 的文件，并添加代码清单 13-13 所示的内容。

代码清单 13-13　bower.json 文件的内容

```
{
  "name": "PartyInvites",
  "private": true,
  "dependencies": {
    "bootstrap": "4.0.0-alpha.6"
  }
}
```

使用命令提示符或 Terminal 在 PartyInvites 文件夹中运行代码清单 13-14 所示的命令，从而使用 Bower 工具下载并安装 bower.json 文件中指定的客户端软件包。

代码清单 13-14 安装客户端软件包

```
bower install
```

13.3.2 配置应用程序

项目的初始化过程会创建一个不支持 MVC 的空项目。代码清单 13-15 显示了如何使用最基本配置在 Startup.cs 文件中添加对 MVC 的支持，其中的语句将在第 14 章中进行描述。

代码清单 13-15 在 Startup.cs 文件中添加对 MVC 的支持

```
using System;
using System.Collections.Generic;
using System.Linq;
using System.Threading.Tasks;
using Microsoft.AspNetCore.Builder;
using Microsoft.AspNetCore.Hosting;
using Microsoft.AspNetCore.Http;
using Microsoft.Extensions.DependencyInjection;

namespace PartyInvites {
    public class Startup {

        public void ConfigureServices(IServiceCollection services) {
            services.AddMvc();
        }

        public void Configure(IApplicationBuilder app, IHostingEnvironment env) {
            app.UseDeveloperExceptionPage();
            app.UseStatusCodePages();
            app.UseStaticFiles();
            app.UseMvcWithDefaultRoute();
        }
    }
}
```

13.3.3 构建和运行项目

要构建和运行项目，请在 PartyInvites 目录中运行代码清单 13-16 所示的命令。

代码清单 13-16 运行应用程序

```
dotnet run
```

Visual Studio Code 将编译项目中的代码，并使用第 14 章描述的 Kestrel 应用程序服务器来运行应用程序，等待端口 5000 上的 HTTP 请求。

Visual Studio Code 不支持检测 C#类文件发生的更改，因此在进行更改时必须停止应用程序并再次启动。

要测试应用程序，请使用浏览器导航到 http://localhost:5000。你将看到图 13-8 所示的响应，显示 404 错误是因为项目中没有控制器用来处理请求。

图 13-8 测试示例应用程序

13.4 重新创建 PartyInvites 应用程序

所有准备工作都已完成，这意味着可以将注意力转移到创建 MVC 应用程序。这里将从第 2 章开始重新创建简单的 PartyInvites 应用程序，但需要进行一些更改以重点展示如何使用 Visual Studio Code。

13.4.1　创建模型和存储库

请右击 PartyInvites 文件夹，然后从弹出菜单中选择 New Folder，创建新文件夹，如图 13-9 所示。将文件夹的名称设置为 Models。

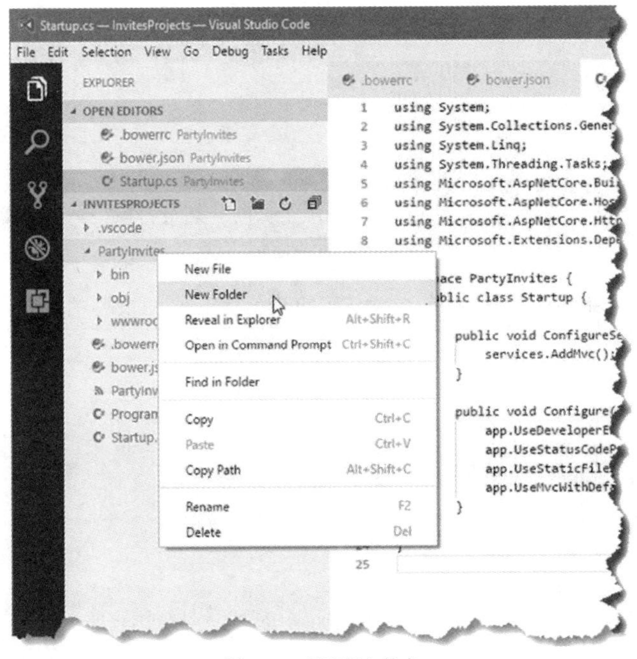

图 13-9　创建新文件夹

右击资源管理器中的 Models 文件夹，从弹出菜单中选择 New File，将文件名设置为 GuestResponse.cs，然后添加代码清单 13-17 所示的 C#代码。

使用 Visual Studio Code 编辑器

Visual Studio Code（以及本章前面安装的 C#扩展）为 C#文件以及 JavaScript、CSS 和普通 HTML 文件等常见 Web 格式提供了完整的编辑体验。在 Visual Studio Code 中编写 MVC 应用程序与使用 Visual Studio 编辑器有很多共同点，如智能感知支持、代码着色和高亮显示错误（以及提供修复建议）等。

Visual Studio Code 的主要缺点是缺乏自定义功能，特别是在格式化代码时。在写作本书时，还有其他语言可用的配置选项，但 C#扩展没有提供自定义功能。如果你喜欢的编码规范不是默认支持的，那么 C#扩展可能有点难以使用。但总的来说，编辑器响应迅速且易于使用，在 macOS 或 Linux 系统上编写 MVC 应用程序的体验并不差。

代码清单 13-17　PartyInvites/Models 文件夹下的 GuestResponse.cs 文件的内容

```
using System.ComponentModel.DataAnnotations;

namespace PartyInvites.Models {

    public class GuestResponse {

        public int id {get; set; }

        [Required(ErrorMessage = "Please enter your name")]
        public string Name { get; set; }
```

```
        [Required(ErrorMessage = "Please enter your email address")]
        [RegularExpression(".+\\@.+\\..+",
            ErrorMessage = "Please enter a valid email address")]
        public string Email { get; set; }

        [Required(ErrorMessage = "Please enter your phone number")]
        public string Phone { get; set; }

        [Required(ErrorMessage = "Please specify whether you'll attend")]
        public bool? WillAttend { get; set; }
    }
}
```

接下来，将名为 IRepository.cs 的文件添加到 Models 文件夹中，并用它定义代码清单 13-18 所示的接口。本章中的应用程序与第 2 章中的应用程序之间最重要的区别是将模型数据存储在持久化数据库中。IRepository 接口描述了应用程序如何访问模型数据而不指定实现。

代码清单 13-18 PartyInvites/Models 文件夹下的 IRepository.cs 文件的内容

```
using System.Collections.Generic;

namespace PartyInvites.Models {

    public interface IRepository {
        IEnumerable<GuestResponse> Responses {get; }

        void AddResponse(GuestResponse response);
    }
}
```

将一个名为 ApplicationDbContext.cs 的文件添加到 Models 文件夹中，并用它定义数据库上下文类，如代码清单 13-19 所示。

代码清单 13-19 PartyInvites/Models 文件夹下的 ApplicationDbContext.cs 文件的内容

```
using Microsoft.EntityFrameworkCore;

namespace PartyInvites.Models {
    public class ApplicationDbContext : DbContext {
        public ApplicationDbContext() {}

        protected override void OnConfiguring(DbContextOptionsBuilder builder) {
            builder.UseSqlite("Filename=./PartyInvites.db");
        }

        public DbSet<GuestResponse> Invites {get; set;}
    }
}
```

SQLite 将数据存储在由上下文类指定的文件中。对于示例应用程序，数据将存储在名为 PartyInvites.db 的文件中，该文件定义在 OnConfiguring 方法中。

要完成存储和访问模型数据所需的类，需要实现数据库上下文类的 IRepository 接口。将名为 EFRepository.cs 的新文件添加到 Models 文件夹中，并添加代码清单 13-20 所示的代码。

代码清单 13-20 PartyInvites/Models 文件夹下的 EFRepository.cs 文件的内容

```
using System.Collections.Generic;

namespace PartyInvites.Models {
    public class EFRepository : IRepository {
        private ApplicationDbContext context = new ApplicationDbContext();

        public IEnumerable<GuestResponse> Responses => context.Invites;
```

```
        public void AddResponse(GuestResponse response) {
            context.Invites.Add(response);
            context.SaveChanges();
        }
    }
}
```

EFRepository 类使用与第 8 章中对应的类似模式来设置 SportsStore 数据库。代码清单 13-21 在 Startup 类的 ConfigureServices 方法中添加了一条配置语句，该方法告诉 ASP.NET 在依赖注入功能要求实现 IRepository 接口时创建 EFRepository 类（参考第 18 章）。

代码清单 13-21 在 PartyInvites 文件夹下的 Startup.cs 文件中配置存储库

```
using System;
using System.Collections.Generic;
using System.Linq;
using System.Threading.Tasks;
using Microsoft.AspNetCore.Builder;
using Microsoft.AspNetCore.Hosting;
using Microsoft.AspNetCore.Http;
using Microsoft.Extensions.DependencyInjection;
using PartyInvites.Models;

namespace PartyInvites {
    public class Startup {
        public void ConfigureServices(IServiceCollection services) {
            services.AddTransient<IRepository, EFRepository>();
            services.AddMvc();
        }

        public void Configure(IApplicationBuilder app, IHostingEnvironment env) {
            app.UseDeveloperExceptionPage();
            app.UseStatusCodePages();
            app.UseStaticFiles();
            app.UseMvcWithDefaultRoute();
        }
    }
}
```

13.4.2 创建数据库

在本书的其他部分，每当需要演示数据持久化功能时，都会使用 LocalDB，LocalDB 是 Microsoft SQL Server 的简化版本。但是 LocalDB 仅在 Windows 平台上可用，这意味着在其他平台上创建 ASP.NET Core MVC 应用程序时需要替代方案。LocalDB 的最佳替代方案是 SQLite，SQLite 是跨平台的并且无须配置的数据库，可以嵌入应用程序中。在通常与 ASP.NET Core MVC 应用程序一起使用的数据访问层 Entity Framework Core 中，微软已默认包含对 SQLite 的支持。接下来介绍将 SQLite 添加到项目中并将其作为数据存储的过程。

使用 SQLite 进行开发

LocalDB 是一个非常有用的工具，原因之一是它允许使用 SQL Server 数据库引擎进行开发，这使得向 SQL Server 生产环境的过渡变得非常简单而且基本没有风险。SQLite 也是优秀的数据库，但它不适合大规模的 Web 应用程序，这意味着在部署 MVC 应用程序时需要转换到另一个数据库。使用第 14 章描述的项目配置功能可以简化配置更改，但是需要在暂存环境中彻底测试应用程序，以显示生产数据库引入的任何差异。

如果不确定是否在生产环境中使用 SQLite，请参阅 SQLite 官网，上面提供了有关 SQLite 的优势及缺陷的介绍。

> 需要注意的一个问题是，Entity Framework Core 可以为其他数据库生成完整的架构更改，但 SQLite 并不支持这个功能。在开发中使用 SQLite 时，这通常不是问题，因为可以删除数据库文件并重新生成具有干净架构的新文件。但是，如果考虑使用 SQLite 部署应用程序，则确实会使问题复杂化。
>
> 如果要在开发和生产中使用相同的数据库，请参阅 readthedocs 网站上受 Entity Framework Core 支持的数据库列表。在写作本书时这个列表还很短，但微软已经宣布支持比 SQLite 更适合部署的数据库，而且也可以在非 Windows 平台上运行。

1. 添加数据库包

必须手动将包含创建和应用数据库迁移命令行工具的 NuGet 包添加到项目中。打开 PartyInvites.csproj 文件并添加代码清单 13-22 所示的元素。

代码清单 13-22 将 NuGet 包添加到 PartyInvites 文件夹下的 PartyInvites.csproj 文件中

```
<Project Sdk="Microsoft.NET.Sdk.Web">

  <PropertyGroup>
    <TargetFramework>netcoreapp2.0</TargetFramework>
  </PropertyGroup>

  <ItemGroup>
    <Folder Include="wwwroot\" />
  </ItemGroup>

  <ItemGroup>
    <PackageReference Include="Microsoft.AspNetCore.All" Version="2.0.0" />
    <DotNetCliToolReference Include="Microsoft.EntityFrameworkCore.Tools.DotNet"
        Version="2.0.0" />
  </ItemGroup>

</Project>
```

保存更改，并在 PartyInvites 文件夹中运行代码清单 13-23 所示的命令，以确保下载并安装新的软件包。

代码清单 13-23 安装 NuGet 包

```
dotnet restore
```

2. 创建和应用数据库迁移

创建数据库与 Visual Studio 使用的过程相同。要创建数据库迁移，请在 PartyInvites 文件夹中运行代码清单 13-24 所示的命令。

代码清单 13-24 创建数据库迁移

```
dotnet ef migrations add Initial
```

Entity Framework Core 将创建一个名为 Migrations 的文件夹，其中包含将用于设置数据库架构的 C#类。要应用数据库迁移，请在 PartyInvites 文件夹中运行代码清单 13-25 所示的命令，从而在 PartyInvites 文件夹中创建数据库。

代码清单 13-25 应用数据库迁移

```
dotnet ef database update
```

Visual Studio Code 不支持查看 SQLite 数据库，但可以在 sqlitebrowser 网站上找到适用于 Windows 系统、macOS 和 Linux 系统的优秀开源工具。

13.4.3 创建控制器和视图

在本节中，将控制器和视图添加到应用程序中。首先创建 PartyInvites/Controllers 文件夹并添加一个名为 HomeController.cs 的文件，用来创建代码清单 13-26 所示的控制器。

代码清单 13-26 PartyInvites/Controllers 文件夹下的 HomeController.cs 文件的内容

```
using System;
using Microsoft.AspNetCore.Mvc;
using PartyInvites.Models;
using System.Linq;

namespace PartyInvites.Controllers {

    public class HomeController : Controller {
        private IRepository repository;

        public HomeController(IRepository repo) =>
            this.repository = repo;

        public ViewResult Index() {
            int hour = DateTime.Now.Hour;
            ViewBag.Greeting = hour < 12 ? "Good Morning" : "Good Afternoon";
            return View("MyView");
        }

        [HttpGet]
        public ViewResult RsvpForm() => View();

        [HttpPost]
        public ViewResult RsvpForm(GuestResponse guestResponse) {
            if (ModelState.IsValid) {
                repository.AddResponse(guestResponse);
                return View("Thanks", guestResponse);
            } else {
                // there is a validation error
                return View();
            }
        }
        public ViewResult ListResponses() =>
            View(repository.Responses.Where(r => r.WillAttend == true));
    }
}
```

为了设置内置的标签助手，创建 PartyInvites/Views 文件夹，并添加一个名为_ViewImports.cshtml 的文件，其中包含代码清单 13-27 所示的代码。

代码清单 13-27 PartyInvites/Views 文件夹下的_ViewImports.cshtml 文件的内容

```
@addTagHelper *, Microsoft.AspNetCore.Mvc.TagHelpers
```

接下来，创建 PartyInvites/Views/Home 文件夹，并添加一个名为 MyView.cshtml 的文件，作为代码清单 13-26 中 Index 操作方法选择的视图，其中的内容如代码清单 13-28 所示。

代码清单 13-28 PartyInvites/Views/Home 文件夹下的 MyView.cshtml 文件的内容

```
@{
    Layout = null;
}

<!DOCTYPE html>

<html>
<head>
    <meta name="viewport" content="width=device-width" />
    <title>Index</title>
    <link rel="stylesheet" href="/lib/bootstrap/dist/css/bootstrap.css" />
</head>
<body class="p-2">
    <div class="text-center">
```

```
        <h3>We're going to have an exciting party!</h3>
        <h4>And you are invited</h4>
        <a class="btn btn-primary" asp-action="RsvpForm">RSVP Now</a>
    </div>
</body>
</html>
```

在 PartyInvites/Views/Home 文件夹中添加一个名为 RsvpForm.cshtml 的文件，并添加代码清单 13-29 所示的内容。这个视图用于提供受邀者填写的 HTML 表单，以便他们接受或拒绝聚会邀请。

代码清单 13-29 PartyInvites/Views/Home 文件夹下的 RsvpForm.cshtml 文件的内容

```
@model PartyInvites.Models.GuestResponse

@{
    Layout = null;
}

<!DOCTYPE html>
<html>
<head>
    <meta name="viewport" content="width=device-width" />
    <title>RsvpForm</title>
    <link rel="stylesheet" href="/lib/bootstrap/dist/css/bootstrap.css" />
</head>
<body>
    <div class="m-2">
        <div class="text-center"><h4>RSVP</h4></div>
        <form class="p-1" asp-action="RsvpForm" method="post">
            <div asp-validation-summary="All"></div>
            <div class="form-group">
                <label asp-for="Name">Your name:</label>
                <input class="form-control" asp-for="Name" />
            </div>
            <div class="form-group">
                <label asp-for="Email">Your email:</label>
                <input class="form-control" asp-for="Email" />
            </div>
            <div class="form-group">
                <label asp-for="Phone">Your phone:</label>
                <input class="form-control" asp-for="Phone" />
            </div>
            <div class="form-group">
                <label>Will you attend?</label>
                <select class="form-control" asp-for="WillAttend">
                    <option value="">Choose an option</option>
                    <option value="true">Yes, I'll be there</option>
                    <option value="false">No, I can't come</option>
                </select>
            </div>
            <div class="text-center">
                <button class="btn btn-primary" type="submit">
                    Submit RSVP
                </button>
            </div>
        </form>
    </div>
</body>
</html>
```

下一个视图文件名为 Thanks.cshtml，也创建在 PartyInvites/Views/Home 文件夹中，内容如代码清单 13-30 所示，当来宾提交响应时将显示这个视图。

代码清单 13-30 PartyInvites/Views/Home 文件夹下的 Thanks.cshtml 文件的内容

```
@model PartyInvites.Models.GuestResponse

@{
    Layout = null;
}
<!DOCTYPE html>

<html>
<head>
    <meta name="viewport" content="width=device-width" />
    <title>Thanks</title>
    <link rel="stylesheet" href="/lib/bootstrap/dist/css/bootstrap.css" />
</head>
<body class="text-center">
    <p>
        <h1>Thank you, @Model.Name!</h1>
        @if (Model.WillAttend == true) {
            @:It's great that you're coming. The drinks are already in the fridge!
        } else {
            @:Sorry to hear that you can't make it, but thanks for letting us know.
        }
    </p>
    Click <a asp-action="ListResponses">here</a>
    to see who is coming.
</body>
</html>
```

最后一个视图名为 ListResponses.cshtml，与本例中的其他视图一样，也添加到 PartyInvites/Views/Home 文件夹中。这个视图使用代码清单 13-31 所示的标记显示来宾响应列表。

代码清单 13-31 PartyInvites/Views/Home 文件夹下的 ListResponses.cshtml 文件的内容

```
@model IEnumerable<PartyInvites.Models.GuestResponse>

@{
    Layout = null;
}

<!DOCTYPE html>
<html>
<head>
    <meta name="viewport" content="width=device-width" />
    <link rel="stylesheet" href="/lib/bootstrap/dist/css/bootstrap.css" />
    <title>Responses</title>
</head>
<body>
    <div class="m-1 p-1">
        <h2>Here is the list of people attending the party</h2>
        <table class="table table-sm table-striped table-bordered">
            <thead>
                <tr><th>Name</th><th>Email</th><th>Phone</th></tr>
            </thead>
            <tbody>
                @foreach (PartyInvites.Models.GuestResponse r in Model) {
                    <tr><td>@r.Name</td><td>@r.Email</td><td>@r.Phone</td></tr>
                }
            </tbody>
        </table>
    </div>
</body>
</html>
```

在 PartyInvites 项目中运行 dotnet run 命令以编译项目并启动 ASP.NET Core 运行时。应用程序启动后，可以通过导航到 http://localhost:5000 来查看已完成的应用程序，如图 13-10 所示。

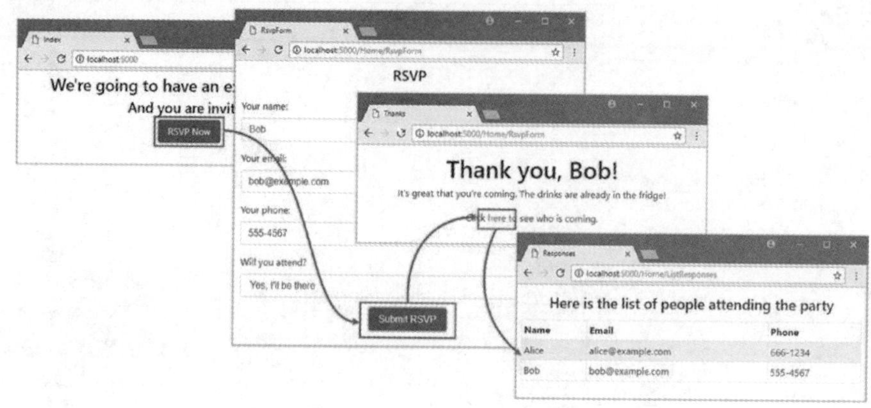

图 13-10 运行完成的应用程序

13.5 Visual Studio Code 中的单元测试

使用 Visual Studio Code 进行单元测试的过程与 Visual Studio 类似。第一步是为单元测试创建单独的项目。在 InvitesProject 文件夹中创建一个名为 Tests 的文件夹，并在这个文件夹中运行代码清单 13-32 所示的命令，以创建单元测试项目。

代码清单 13-32 创建单元测试项目

```
dotnet new xunit --language C# --framework netcoreapp2.0
```

在 Tests 文件夹中运行代码清单 13-33 所示的命令，添加对应用程序项目的引用，以引用其中包含的类用于测试。

代码清单 13-33 添加对应用程序项目的引用

```
dotnet add reference ../PartyInvites/PartyInvites.csproj
```

13.5.1 创建单元测试

创建单元测试的过程如第 7 章所述。将一个名为 HomeControllerTests.cs 的类文件添加到 Tests 文件夹中，其中的内容如代码清单 13-34 所示。

代码清单 13-34 Tests 文件夹下的 HomeControllerTests.cs 文件的内容

```
using System;
using System.Collections.Generic;
using PartyInvites.Controllers;
using PartyInvites.Models;
using Xunit;
using Microsoft.AspNetCore.Mvc;
using System.Linq;

namespace Tests {
    public class HomeControllerTests {

        [Fact]
        public void ListActionFiltersNonAttendees() {
            //Arrange
            HomeController controller = new HomeController(new FakeRepository());
            // Act
            ViewResult result = controller.ListResponses();
            // Assert
```

```
            Assert.Equal(2, (result.Model as IEnumerable<GuestResponse>).Count());
        }
    }

    class FakeRepository : IRepository {
        public IEnumerable<GuestResponse> Responses =>
            new List<GuestResponse> {
                new GuestResponse { Name = "Bob", WillAttend = true },
                new GuestResponse { Name = "Alice", WillAttend = true },
                new GuestResponse { Name = "Joe", WillAttend = false }
            };

        public void AddResponse(GuestResponse response) {
            throw new NotImplementedException();
        }
    }
}
```

这是一个标准的xUnit测试，用于检查Home控制器中的ListResponses操作，并正确过滤存储库中WillAttend属性为false的GuestResponse对象。

13.5.2 运行测试

要在项目中运行单元测试，请在Tests文件夹中运行代码清单13-35所示的命令。

代码清单13-35 运行单元测试

```
dotnet test
```

这将运行项目中的所有测试并显示结果，产生如下输出：

```
Starting test execution, please wait...
[xUnit.net 00:00:00.6731479]   Discovering: Tests
[xUnit.net 00:00:00.7900132]   Discovered:  Tests
[xUnit.net 00:00:00.8432715]   Starting:    Tests
[xUnit.net 00:00:00.9967614]   Finished:    Tests

Total tests: 2. Passed: 2. Failed: 0. Skipped: 0.
Test Run Successful.
Test execution time: 1.6974 Seconds
```

以上结果显示有两个测试，因为项目模板包含一个名为UnitTest1.cs的文件，而其中又包含一个空的单元测试。可以删除此文件，如第7章所述。

13.6 小结

本章简要介绍了如何使用Visual Studio Code，它是一个轻量级的开发工具，支持在Windows系统、macOS和Linux系统上进行ASP.NET Core MVC开发。Visual Studio Code还无法成为完整的Visual Studio产品的替代品，但它提供了创建MVC应用程序所需的核心功能，微软每月都会发布更新以增强其功能。

本书第一部分到此结束，第二部分将深入研究ASP.NET Core的细节，并详细展示如何使用这些功能来创建应用程序。

第二部分　ASP.NET Core MVC 详解

到目前为止，你已经了解了 ASP.NET Core MVC 存在的原因，并了解了其体系结构和底层设计目标。通过构建一个真实的电子商务应用程序，你已经掌握了测试驱动开发方式。现在是时候打开盖子并展示框架机制的全部细节了。

本书第二部分将深入讨论细节。该部分首先探讨 ASP.NET Core MVC 应用程序的结构以及处理请求的方式，然后重点关注各个功能，例如路由、控制器和 action、MVC 视图和标签助手系统，以及将 MVC 与域模型一起使用的方式。

第 14 章　配置应用程序

配置这个主题似乎并不是很有趣，但它揭示了很多 MVC 应用程序如何工作以及如何处理 HTTP 请求。你不应该跳过本章，而应该花时间了解配置系统构造 MVC Web 应用程序的方式，从而为理解后面的章节奠定坚实的基础。

本章将解释如何配置 MVC 应用程序，并展示 MVC 如何构建 ASP.NET Core 平台提供的功能。表 14-1 总结了在上下文中配置应用程序时的一些问题。

表 14-1　在上下文中进行配置时的一些问题

问题	答案
配置是什么？	Program 和 Startup 类以及 JSON 文件用于配置应用程序的工作方式以及依赖的包
有什么用？	配置系统允许应用程序根据运行环境进行定制，并管理依赖的
怎么用？	最重要的组件是 Startup 类，它用于创建服务（在整个应用程序中提供通用功能的对象）和中间件（用于处理 HTTP 请求）
有任何问题或限制吗？	在复杂的应用程序中，配置可能变得难以管理
还有其他选择吗？	没有。配置系统是 ASP.NET 和 MVC 应用程序设置方式的组成部分

表 14-2 列出了本章要介绍的操作。

表 14-2　本章要介绍的操作

操作	方法	代码清单
向应用程序添加功能	将 NuGet 包添加到.csproj 文件中	代码清单 14-5～代码清单 14-8
管理 ASP.NET 应用程序的初始化	使用 Program 类	代码清单 14-9～代码清单 14-11
配置应用程序	使用 Startup 类的 ConfigureServices 和 Configure 方法	代码清单 14-12 和代码清单 14-13
创建通用功能	使用 ConfigureServices 方法创建服务	代码清单 14-14～代码清单 14-16
生成内容响应	创建内容生成中间件	代码清单 14-17～代码清单 14-19
阻止请求遍历请求管道	创建短路中间件	代码清单 14-20 和代码清单 14-21
在其他中间件处理请求之前编辑请求	创建请求编辑中间件	代码清单 14-22～代码清单 14-24
编辑由其他中间件处理的响应	创建响应编辑中间件	代码清单 14-25 和代码清单 14-26
设置 MVC 功能	使用 UseMvc 或 UseMvcWithDefaultRoute 方法	代码清单 14-27
为不同环境改变应用程序配置	使用托管环境服务	代码清单 14-28
处理应用程序错误	使用开发环境或生产环境错误处理中间件	代码清单 14-29 和代码清单 14-30
在开发期间管理多个浏览器	使用浏览器链接	代码清单 14-31
启用图像、JavaScript 文件和 CSS 文件	启用静态内容中间件	代码清单 14-32
从 C#代码中分离配置数据	创建外部配置源，例如 JSON 文件	代码清单 14-33～代码清单 14-35
记录应用数据	使用日志记录中间件	代码清单 14-36～代码清单 14-38
准备依赖注入以与 Entity Framework Core 一起使用	禁用范围验证	代码清单 14-39
配置 MVC 服务	使用选项功能	代码清单 14-40
配置复杂的应用程序	使用多个外部文件或类	代码清单 14-41～代码清单 14-45

14.1　准备示例项目

本章中，我们将使用 Empty 模板创建一个名为 ConfigureApps 的新项目。稍后将配置应用程序，但是在做出更改之前，首先需要做一些基本的准备工作。

我们将在本章中使用 Bootstrap 来设置 HTML 内容的样式，因此使用 Bower Configuration File 模板创建 bower.json 文件，并添加代码清单 14-1 所示的包。

代码清单 14-1　在 ConfiguringApps 文件夹下的 bower.json 文件中添加 Bootstrap

```
{
  "name": "asp.net",
  "private": true,
  "dependencies": {
    "bootstrap": "4.0.0-alpha.6"
  }
}
```

接下来，创建 Controllers 文件夹并添加一个名为 HomeController.cs 的类文件，用来定义代码清单 14-2 所示的控制器。

代码清单 14-2　Controllers 文件夹下的 HomeController.cs 文件的内容

```
using System.Collections.Generic;
using Microsoft.AspNetCore.Mvc;

namespace ConfiguringApps.Controllers {

    public class HomeController : Controller {

        public ViewResult Index() => View(new Dictionary<string, string> {
                ["Message"] = "This is the Index action"
            });
    }
}
```

接下来，创建 Views/Home 文件夹并添加一个名为 Index.cshtml 的视图文件，其中的内容如代码清单 14-3 所示。

代码清单 14-3　Views/Home 文件夹下的 Index.cshtml 文件的内容

```
@model Dictionary<string, string>
@{ Layout = null; }

<!DOCTYPE html>
<html>
<head>
    <meta name="viewport" content="width=device-width" />
    <link asp-href-include="lib/bootstrap/dist/css/*.min.css" rel="stylesheet" />
    <title>Result</title>
</head>
<body class="p-1">
    <table class="table table-condensed table-bordered table-striped">
        @foreach (var kvp in Model) {
            <tr><th>@kvp.Key</th><td>@kvp.Value</td></tr>
        }
    </table>
</body>
</html>
```

视图中的 link 元素依赖于内置的标签助手来选择 Bootstrap CSS 文件。为了启用内置的标签助手，可使用 MVC View Imports Page 模板来创建 Views 文件夹中的_ViewImports.cshtml 文件，并添加代码清单 14-4 所示的代码。

代码清单 14-4　Views 文件夹下的_ViewImports.cshtml 文件的内容

```
@addTagHelper *, Microsoft.AspNetCore.Mvc.TagHelpers
```

启动应用程序，你将看到如图 14-1 所示的消息。

图 14-1 运行示例应用程序

14.2 配置项目

最重要的配置文件是<projectname>.csproj，以取代早期版本的 ASP.NET Core 中使用的 project.json 文件。这个文件在示例项目中名为 InstallingApps.csproj，Visual Studio 隐藏了该文件，必须通过右击 Solution Explorer 窗格中的项目并从弹出的菜单中选择 Edit ConfiguringApps.csproj 来访问。代码清单 14-5 显示了 ConfigureApps.csproj 文件的初始内容，该文件由 Visual Studio 作为 Empty 项目模板的一部分创建。

代码清单 14-5 ConfiguringApps 文件夹下的 ConfigureApps.csproj 文件的内容

```
<Project Sdk="Microsoft.NET.Sdk.Web">

  <PropertyGroup>
    <TargetFramework>netcoreapp2.0</TargetFramework>
  </PropertyGroup>

  <ItemGroup>
    <Folder Include="wwwroot\" />
  </ItemGroup>

  <ItemGroup>
    <PackageReference Include="Microsoft.AspNetCore.All" Version="2.0.0" />
  </ItemGroup>

</Project>
```

.csproj 文件用于配置 MSBuild 工具，MSBuild 工具用于构建.NET 项目。使用 XML 元素执行配置，表 14-3 描述了默认配置文件中的元素。尽管我们在后面的示例中使用了其他配置元素，但表 14-3 中的元素对于开发 ASP.NET Core MVC 项目已经足够了。

表 14-3 默认.csproj 文件中的 XML 配置元素

元素	描述
Project	此为根元素，表示这是一个 MSBuild 配置文件。Sdk 属性被设置为 Microsoft.NET.Sdk.Web，以导入构建项目所需的隐式包
PropertyGroup	此元素将相关配置属性分组，以将结构添加到文件中
TargetFramework	此元素指定构建过程所针对的.NET Framework 版本，必须在 PropertyGroup 元素中定义。默认值为 netcoreapp2.0，以.NET Core 2.0 为目标
ItemGroup	此元素将相关配置项分组，以将结构添加到文件中
Folder	此元素告诉 MSBuild 如何处理项目中的文件夹。列表中的元素告诉 MSBuild 在发布应用程序时应包含 wwwroot 文件夹
PackageReference	此元素用于指定 NuGet 包的依赖项，NuGet 包可通过 Include 和 Version 属性进行标识。Microsoft.AspNetCore.All 包用于提供对所有 ASP.NET Core 和 MVC Framework 功能的访问

14.2.1 将包添加到项目中

.csproj 文件最重要的作用是列出项目所依赖的包。当 Visual Studio 检测到对.csproj 文件进行的更改时，它会检查包的列表，下载新添加的内容，并删除所有不再需要的包。

随着 ASP.NET Core 2 的发布，ASP.NET Core MVC、MVC 框架和 Entity Framework Core 所需的所有基本功能都包含在 Microsoft.AspNetCore.All 元数据包中。这是一个很方便的功能，避免了开始新项目时需要向项目添加一长串 NuGet 包的问题。

即便如此，也需要为第三方或高级功能添加 NuGet 包。有 3 种方法可以将包添加到项目中。第一种是选择 Tools→NuGet Package Manager→Manage NuGet Packages for Solution，可以通过简单方便的界面管理 NuGet 包。如果你不熟悉.NET 开发，那么这是最好的方法，因为它可以减少选择包时出错的可能性。

例如，要添加 System.Net.Http 包，这个包提供了创建（不仅仅是接收）HTTP 请求的支持，可以转到包管理器的 Browse 部分，按名称搜索，并查看可用版本的完整列表，包括任何预发布版本，如图 14-2 所示。

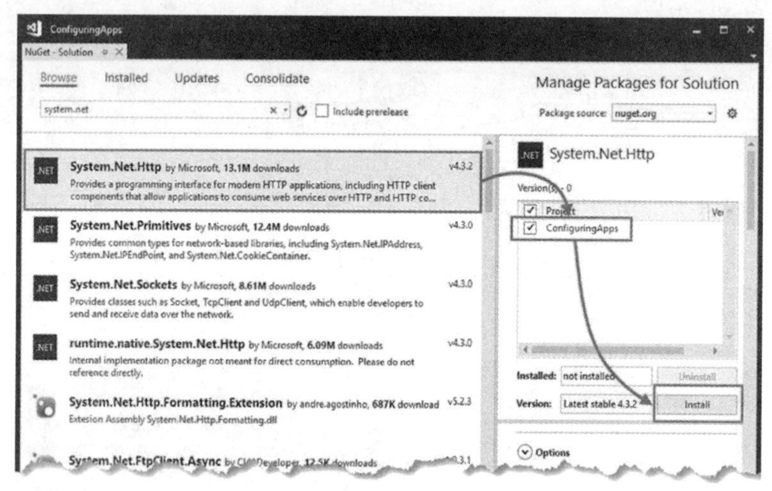

图 14-2　使用 NuGet 包管理器添加包

第二种方法是选择所需的软件包和版本，然后选择需要软件包的项目，单击 Install 按钮，Visual Studio 将下载软件包并更新.csproj 文件。

最后一种方法是使用命令行添加包，但需要知道包的名称（理想情况下还需要知道版本）。代码清单 14-6 显示了在 ConfiguringApps 文件夹中运行的命令，用于将 System.Net.Http 包添加到项目中。

代码清单 14-6　将包添加到项目中

```
dotnet add package System.Net.Http --version 4.3.2
```

NuGet 包管理器和 dotnet add package 命令都会将 PackageReference 元素添加到.csproj 文件中。如果愿意，可以编辑配置文件，通过手动添加 PackageReference 元素来添加包。这是最直接的方法，但需要注意避免错误输入包的名称或指定不存在的版本号。在代码清单 14-7 中，可以看到为了添加 System.Net.Http 包对.csproj 文件所做的修改。

代码清单 14-7　在 ConfiguringApps 文件夹下的 InstallingApps.csproj 文件中添加包

```
<Project Sdk="Microsoft.NET.Sdk.Web">

  <PropertyGroup>
    <TargetFramework>netcoreapp2.0</TargetFramework>
  </PropertyGroup>
  <ItemGroup>
    <Folder Include="wwwroot\" />
  </ItemGroup>

  <ItemGroup>
    <PackageReference Include="Microsoft.AspNetCore.All" Version="2.0.0" />
    <PackageReference Include="System.Net.Http" Version="4.3.2" />
  </ItemGroup>
```

PackageReference 元素包含用于指定包名称的 Include 属性和用于指定版本号的 Version 属性。

14.2.2　将工具包添加到项目中

虽然能够以不同的方式添加常规包，但是某些包扩展了 dotnet 命令行工具的功能，这些包需要.csproj 文件中的不同类型的元素，比如 DotNetCliToolReference 元素，而不是由应用程序直接使用的包所需的 PackageReference 元素。因此，只能通过直接编辑.csproj 文件将这些包添加到项目中。

代码清单 14-8 显示了添加的包，允许使用本书第一部分使用的 dotnet ef 命令创建和应用数据库迁移。

代码清单 14-8　将工具包添加到 ConfiguringApps 文件夹下的 InstallingApps.csproj 文件中

```
<Project Sdk="Microsoft.NET.Sdk.Web">

  <PropertyGroup>
    <TargetFramework>netcoreapp2.0</TargetFramework>
  </PropertyGroup>

  <ItemGroup>
    <Folder Include="wwwroot\" />
  </ItemGroup>

  <ItemGroup>
    <PackageReference Include="Microsoft.AspNetCore.All" Version="2.0.0" />
    <PackageReference Include="System.Net.Http" Version="4.3.2" />
    <DotNetCliToolReference Include="Microsoft.EntityFrameworkCore.Tools.DotNet"
        Version="2.0.0 " />
  </ItemGroup>

</Project>
```

在将工具包添加到项目中时，可以将 DotNetCliToolReference 元素包含在与常规 PackageReference 元素相同的 ItemGroup 中，如代码清单 14-8 所示，或创建单独的 ItemGroup 元素。将更改保存到.csproj 文件时，Visual Studio 将下载并安装软件包，然后使用它们来配置 dotnet 命令行工具。

14.3　理解 Program 类

Program 类定义在名为 Program.cs 的文件中，并提供运行应用程序的入口点。Program 类为.NET 提供了 Main 方法，可以执行 Main 方法来配置托管环境并选择完成 ASP.NET Core 应用程序的配置类。Program 类的默认内容足以支持大多数项目的启动和运行，代码清单 14-9 显示了 Visual Studio 添加到项目中的默认代码。

代码清单 14-9　ConfiguringApps 文件夹下的 Program.cs 文件的默认内容

```
using System;
using System.Collections.Generic;
using System.IO;
using System.Linq;
using System.Threading.Tasks;
using Microsoft.AspNetCore;
using Microsoft.AspNetCore.Hosting;
using Microsoft.Extensions.Configuration;
using Microsoft.Extensions.Logging;

namespace ConfiguringApps {
    public class Program {

        public static void Main(string[] args) {
            BuildWebHost(args).Run();
```

```
        }

        public static IWebHost BuildWebHost(string[] args) =>
            WebHost.CreateDefaultBuilder(args)
                .UseStartup<Startup>()
                .Build();
    }
}
```

Main 方法提供了所有.NET 应用程序必须提供的入口点，以便运行时可以执行它们。Program 类中的Main 方法会调用 BuildWebHost 方法，BuildWebHost 方法则负责配置 ASP.NET Core。

BuildWebHost 方法使用 WebHost 类定义的静态方法来配置 ASP.NET Core。随着 ASP.NET Core 2 的发布，可通过使用 CreateDefaultBuilder 方法简化配置，该方法使用适合大多数项目的设置来配置 ASP.NET Core。可调用 UseStartup 方法以标识特定于应用程序的配置类，但约定是使用名为 Startup 的类，详情将在本章后面介绍。Build 方法处理所有配置并创建一个实现了 IWebHost 接口的对象，然后返回给 Main 方法，Main 方法调用 Run 以开始处理 HTTP 请求。

深入配置细节

CreateDefaultBuilder 方法是一种快速启动 ASP.NET Core 配置的便捷方法，但它确实隐藏了许多重要细节，如果需要更改应用程序的配置方式，这可能会是一个问题。代码清单 14-10 将 CreateDefaultBuilder 方法替换成了调用创建默认配置的单条语句。

代码清单 14-10 ConfiguringApps 文件夹下的 Program.cs 文件中的详细配置语句

```
using System;
using System.Collections.Generic;
using System.IO;
using System.Linq;
using System.Threading.Tasks;
using Microsoft.AspNetCore;
using Microsoft.AspNetCore.Hosting;
using Microsoft.Extensions.Configuration;
using Microsoft.Extensions.Logging;
using System.Reflection;

namespace ConfiguringApps {
    public class Program {

        public static void Main(string[] args) {
            BuildWebHost(args).Run();
        }

        public static IWebHost BuildWebHost(string[] args) {
            return new WebHostBuilder()
                .UseKestrel()
                .UseContentRoot(Directory.GetCurrentDirectory())
                .ConfigureAppConfiguration((hostingContext, config) => {
                    var env = hostingContext.HostingEnvironment;
                    config.AddJsonFile("appsettings.json", optional: true,
                            reloadOnChange: true)
                        .AddJsonFile($"appsettings.{env.EnvironmentName}.json",
                            optional: true, reloadOnChange: true);

                    if (env.IsDevelopment()) {
                        var appAssembly =
                            Assembly.Load(new AssemblyName(env.ApplicationName));
                        if (appAssembly != null) {
                            config.AddUserSecrets(appAssembly, optional: true);
```

```
                }
            }

            config.AddEnvironmentVariables();

            if (args != null) {
                config.AddCommandLine(args);
            }
        })
        .ConfigureLogging((hostingContext, logging) => {
            logging.AddConfiguration(
                hostingContext.Configuration.GetSection("Logging"));
            logging.AddConsole();
            logging.AddDebug();
        })
            .UseIISIntegration()
            .UseDefaultServiceProvider((context, options) => {
                options.ValidateScopes =
                    context.HostingEnvironment.IsDevelopment();
            })
            .UseStartup<Startup>()
            .Build();
        }
    }
}
```

表 14-4 列出了添加到 BuildWebHost 方法的所有配置方法，并提供相应功能的简要说明。

表 14-4　添加到 BuildWebHost 方法的所有配置方法

方法	描述
UseKestrel	用于配置 Kestrel Web 服务器
UseContentRoot	用于配置应用程序的根目录，根目录用于加载配置文件并提供静态内容，如图像、JavaScript 和 CSS
ConfigureAppConfiguration	用于为应用程序准备配置数据
AddUserSecrets	用于在代码文件之外存储敏感数据。这是一个有点棘手的方法，本书没有使用
ConfigureLogging	用于配置应用程序的日志记录
UseIISIntegration	用于启用与 IIS 和 IIS Express 的集成
UseDefaultServiceProvider	用于配置依赖注入
UseStartup	指定将用于配置 ASP.NET 的类

代码清单 14-10 已经解释了一些更复杂的语句。现在，删除一些配置语句，只保留基本配置，如代码清单 14-11 所示。

代码清单 14-11　简化 ConfiguringApps 文件夹下的 Program.cs 文件中的配置

```
using System;
using System.Collections.Generic;
using System.IO;
using System.Linq;
using System.Threading.Tasks;
using Microsoft.AspNetCore;
using Microsoft.AspNetCore.Hosting;
using Microsoft.Extensions.Configuration;
using Microsoft.Extensions.Logging;
using System.Reflection;

namespace ConfiguringApps {
```

```
public class Program {

    public static void Main(string[] args) {
        BuildWebHost(args).Run();
    }

    public static IWebHost BuildWebHost(string[] args) {

        return new WebHostBuilder()
            .UseKestrel()
            .UseContentRoot(Directory.GetCurrentDirectory())
            .UseIISIntegration()
            .UseStartup<Startup>()
            .Build();
    }
}
}
```

这些语句提供了适用于大多数 ASP.NET Core MVC 应用程序的基本配置。当解释其他语句的功能时，再把它们添加回来。

直接使用 Kestrel

Kestrel 是跨平台的 Web 服务器，旨在运行 ASP.NET Core 应用程序。当使用 IIS Express（Visual Studio 提供的在开发期间使用的服务器）或完整版 IIS（.NET 应用程序的传统 Web 平台）运行 ASP.NET Core 应用程序时，会自动使用 Kestrel。

如有需要，也可以直接运行 Kestrel，这意味着可以在任何支持的平台上运行 ASP.NET Core MVC 应用程序，绕过 IIS 仅支持 Windows 平台的限制。有两种方法可以使用 Kestrel 运行应用程序。可以单击 Visual Studio 工具栏上 IIS Express 按钮右边缘的箭头，然后选择与项目名称匹配的选项。这将打开一个新的命令提示符并使用 Kestrel 运行应用程序。

也可以通过以下方法实现相同的效果：打开命令提示符，导航到包含应用程序配置文件的文件夹（包含.csproj 文件的那个文件夹）并运行以下命令。

```
dotnet run
```

默认情况下，Kestrel 服务器开始在端口 5000 上侦听 HTTP 请求。如果项目中有 Properties/launchSettings.json 文件，就从这个文件中读取应用程序的 HTTP 端口和环境。

14.4　了解 Startup 类

Program 类负责启动应用程序，但最重要的配置工作是通过 UseStartup 方法委派的，如下所示：

```
...
.UseStartup<Startup>()
...
```

UseStartup 方法依赖于 type 参数来标识用于配置 ASP.NET Core 的类。这个类的常规名称是 Startup，Startup 也是 ASP.NET Core MVC 项目模板（包括用于为本章创建示例项目的 Empty 模板）使用的名称。

研究 Startup 类的工作原理可以帮助你深入了解 HTTP 请求的处理方式，以及 MVC 如何集成到 ASP.NET Core 平台的其余部分。

在本节中，将从最简单的 Startup 类开始，并添加功能以演示不同配置选项的效果，最后使用适合大多数 MVC 项目的配置。作为起点，代码清单 14-12 显示了 Visual Studio 添加到 Empty 项目中的 Startup 类，这里为该类设置了足够的功能以使 ASP.NET Core 能够处理 HTTP 请求。

代码清单 14-12　Startup.cs 文件的初始内容

```
using System;
using System.Collections.Generic;
using System.Linq;
using System.Threading.Tasks;
using Microsoft.AspNetCore.Builder;
using Microsoft.AspNetCore.Hosting;
using Microsoft.AspNetCore.Http;
using Microsoft.Extensions.DependencyInjection;

namespace ConfiguringApps {
    public class Startup {

        public void ConfigureServices(IServiceCollection services) {
        }

        public void Configure(IApplicationBuilder app, IHostingEnvironment env) {

            if (env.IsDevelopment()) {
                app.UseDeveloperExceptionPage();
            }

            app.Run(async (context) => {
                await context.Response.WriteAsync("Hello World!");
            });
        }
    }
}
```

Startup 类定义了两个方法——ConfigureServices 和 Configure，它们用于设置应用程序所需的共享功能，并告诉 ASP.NET Core 应该如何使用它们。

当应用程序启动时，ASP.NET Core 会创建一个新的 Startup 实例并调用 ConfigureServices 方法，以便应用程序可以创建服务。正如 14.4.1 节所解释的，服务是为应用程序其他部分提供功能的对象。

以上描述过于模糊，因为服务可以用来提供任何功能。创建服务后，ASP.NET 将调用 Configure 方法。Configure 方法的目的是设置请求管道，请求管道是一组组件——称为中间件——用于处理传入的 HTTP 请求并为它们生成响应。其中解释请求管道的工作原理，并演示如何在 14.4.2 节中创建中间件。图 14-3 显示了 ASP.NET 使用 Startup 类的方式。

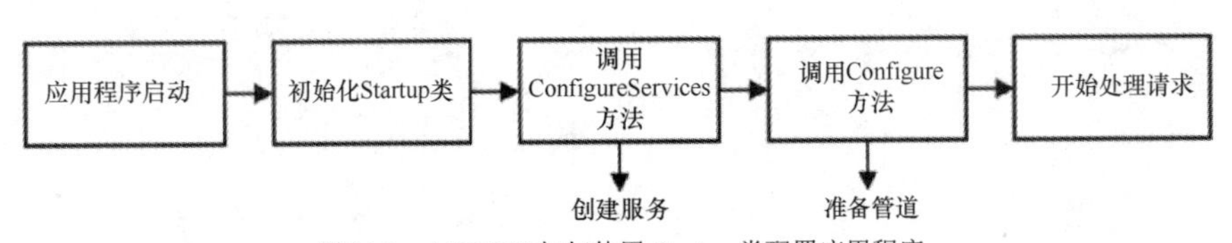

图 14-3　ASP.NET 如何使用 Startup 类配置应用程序

让 Startup 类为所有请求返回相同的“Hello，World”消息是没有意义的，所以在详细解释类中的方法之前，需要先启用 MVC，如代码清单 14-13 所示。

代码清单 14-13　在 ConfiguringApps 文件夹下的 Startup.cs 文件中启用 MVC

```
using System;
using System.Collections.Generic;
using System.Linq;
using System.Threading.Tasks;
using Microsoft.AspNetCore.Builder;
using Microsoft.AspNetCore.Hosting;
using Microsoft.AspNetCore.Http;
```

```
using Microsoft.Extensions.DependencyInjection;

namespace ConfiguringApps {
    public class Startup {

        public void ConfigureServices(IServiceCollection services) {
            services.AddMvc();
        }

        public void Configure(IApplicationBuilder app, IHostingEnvironment env) {
            app.UseMvcWithDefaultRoute();
        }
    }
}
```

后面的章节会解释这些修改，现在已经有足够的基础设施代码来处理 HTTP 请求并使用控制器和视图生成响应了。如果运行应用程序，你将看到图 14-4 所示的输出。

请注意，输出的内容尚未设置样式。代码清单 14-13 中的最小配置不支持提供静态内容，如 CSS 样式表和 JavaScript 文件，因此当 Index.cshtml 视图文件中的 link 元素请求 Bootstrap CSS 样式时，应用程序无法处理，导致浏览器无法获取所需的 CSS 样式。14.4.4 节将介绍如何解决这个问题。

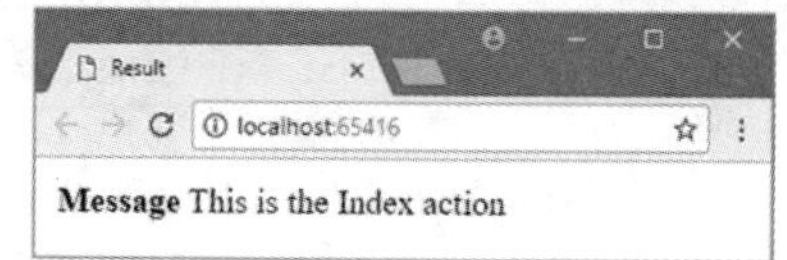

图 14-4 启用 MVC 的效果

14.4.1 了解 ASP.NET 服务

ASP.NET Core 调用 Startup.ConfigureServices 方法，以便应用程序可以设置所需的服务。服务是指为应用程序的其他部分提供功能的对象。如上所述，这样的描述过于模糊，因为服务可以执行应用程序所需的任何操作。举例来说，在项目中添加 Infrastructure 文件夹，并在其中添加一个名为 UptimeService.cs 的类文件，用来定义代码清单 14-14 所示的类。

代码清单 14-14 Infrastructure 文件夹下的 UptimeService.cs 文件的内容

```
using System.Diagnostics;

namespace ConfiguringApps.Infrastructure {

    public class UptimeService {
        private Stopwatch timer;

        public UptimeService() {
            timer = Stopwatch.StartNew();
        }

        public long Uptime => timer.ElapsedMilliseconds;
    }
}
```

创建 UptimeService 类时，UptimeService 构造函数会启动一个计时器，以跟踪应用程序的运行时间。这是一个很好的服务示例，因为它提供了可以在应用程序的其他部分使用的功能，并且可以在应用程序启动时创建。

ASP.NET 服务使用 Startup 类的 ConfigureServices 方法来注册，在代码清单 14-15 中，可以看到如何注册 UptimeService 类。

代码清单 14-15 在 ConfiguringApps 文件夹下的 Startup.cs 文件中注册自定义服务

```
using System;
using System.Collections.Generic;
using System.Linq;
using System.Threading.Tasks;
```

```
using Microsoft.AspNetCore.Builder;
using Microsoft.AspNetCore.Hosting;
using Microsoft.AspNetCore.Http;
using Microsoft.Extensions.DependencyInjection;
using ConfiguringApps.Infrastructure;

namespace ConfiguringApps {
    public class Startup {

        public void ConfigureServices(IServiceCollection services) {
            services.AddSingleton<UptimeService>();
            services.AddMvc();
        }

        public void Configure(IApplicationBuilder app, IHostingEnvironment env) {
            app.UseMvcWithDefaultRoute();
        }
    }
}
```

作为参数，ConfigureServices 方法会接收实现了 IServiceCollection 接口的对象。可使用在 IServiceCollection 接口上调用的扩展方法注册服务，这些扩展方法指定了不同的配置选项。第 18 章将描述可用于创建服务的选项，但目前使用了 AddSingleton 方法，这意味着整个应用程序将共享一个 UptimeService 对象。

服务与依赖注入功能密切相关，后者允许控制器等组件轻松获取服务，这将在第 18 章深入介绍。在 Startup.ConfigureServices 中注册的服务可以通过创建接收参数的构造函数来访问，参数需要为请求的服务类型。代码清单 14-16 显示了添加到 Home 控制器的构造函数，以接收代码清单 14-15 中创建的共享的 UptimeService 对象。这里还更新了控制器的 Index 操作方法，以便在生成的视图数据中包含服务的 Update 属性值。

代码清单 14-16　在 Controllers 文件夹下的 HomeController.cs 文件中访问服务

```
using System.Collections.Generic;
using Microsoft.AspNetCore.Mvc;
using ConfiguringApps.Infrastructure;

namespace ConfiguringApps.Controllers {

    public class HomeController : Controller {
        private UptimeService uptime;

        public HomeController(UptimeService up) => uptime = up;
        public ViewResult Index()
            => View(new Dictionary<string, string> {
                ["Message"] = "This is the Index action",
                ["Uptime"] = $"{uptime.Uptime}ms"
            });
    }
}
```

当 MVC 需要 Home 控制器类的实例来处理 HTTP 请求时，它会检查 HomeController 构造函数并发现需要 UptimeService 对象。然后，MVC 检查已在 Startup 类中配置的服务集，发现已经配置过 UptimeService，这样就可以将单例 UptimeService 对象用于所有请求，并在创建 HomeController 时传递该对象作为构造函数参数。

虽然可以使用更复杂的方式注册和使用服务，但此例演示了服务背后的核心思想，并说明了如何在 Startup 类中定义服务，以便定义整个应用程序中使用的功能或数据。

如果运行应用程序并请求默认 URL，你将看到一个响应，包括自应用程序启动以来的毫秒数，毫秒数是从 Startup 类中创建的 UptimeService 对象获得的，如图 14-5 所示（严格来说，这是自创建 UptimeService 服务对象以来的时间，但这与应用程序的启动时间非常接近，不会对本章的目的产生任何影响）。

每次收到对默认 URL 的请求时，MVC 都会创建一个新的 HomeController 对象，并为其提供共享的 UptimeService 对象作为构造函数参数。这允许 Home 控制器访问应用程序的正常运行时间，而不用关心如何提供或实现此信息。

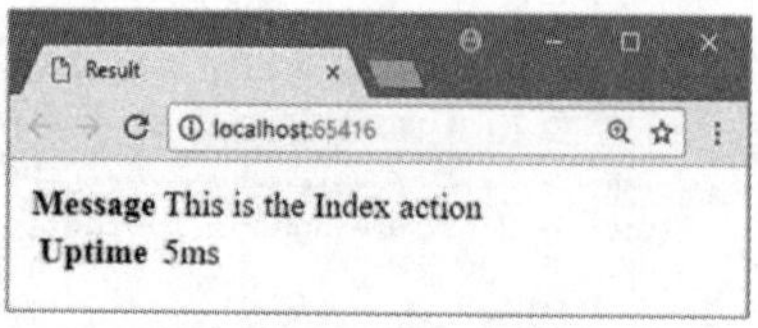

图 14-5 使用简单服务获得的响应

了解内置的 MVC 服务

像 MVC 这样的复杂软件包使用了许多服务。一些供内部使用，另一些为开发人员提供功能。包定义扩展方法，在单个方法调用中设置它们所需的所有服务。对于 MVC，这个方法名为 AddMvc，它是添加到 Startup 类以使 MVC 工作的两个方法之一。

```
...
public void ConfigureServices(IServiceCollection services) {
    services.AddSingleton<UptimeService>();
    services.AddMvc();
}
...
```

这个方法设置 MVC 所需的每个服务，无须在 ConfigureServices 方法中使用大量单个服务。

注 意

Visual Studio 的智能感知功能将显示可以在 ConfigureServices 方法中的 IServiceCollection 对象上调用的其他扩展方法的长列表。其中一些方法（如 AddSingleton 和 AddScoped）用于以不同方式注册服务，其他方法（如 AddRouting 或 AddCors）用于添加已由 AddMvc 方法使用的单个服务。结果是，对于大多数应用程序，ConfigureServices 方法仅包含少量自定义服务（对 AddMvc 方法的调用）以及一些可选的语句来配置内置服务，14.6 节将对此进行描述。

14.4.2 了解 ASP.NET 中间件

在 ASP.NET Core 中，中间件是指用来组合请求管道的组件。请求管道像链一样排列，当新请求到达时，新请求被传递给链中的第一个中间件。这个中间件检查请求并决定是否处理并生成响应，也可将请求传递给链中的下一个中间件。处理完请求后，将返回到客户端的响应沿链传回，这允许所有前面的中间件检查或修改响应。

中间件的工作方式似乎有点奇怪，但它允许应用程序组合在一起的方式具有很大的灵活性。了解如何使用中间件来构建应用程序非常重要，尤其是在没有得到预期的响应时。为了解释中间件系统的工作原理，下面创建一些自定义组件来演示你将遇到的 4 种类型的中间件。

1. 创建内容生成中间件

最重要的中间件类型用于为客户端生成内容，MVC 就属于这种类型。为了创建比 MVC 简单的内容生成中间件，在 Infrastructure 文件夹中添加一个名为 ContentMiddleware.cs 的类，用来定义代码清单 14-17 所示的类。

代码清单 14-17 Infrastructure 文件夹下的 ContentMiddleware.cs 文件的内容

```
using System.Text;
using System.Threading.Tasks;
using Microsoft.AspNetCore.Http;

namespace ConfiguringApps.Infrastructure {

    public class ContentMiddleware {
        private RequestDelegate nextDelegate;

        public ContentMiddleware(RequestDelegate next) => nextDelegate = next;
        public async Task Invoke(HttpContext httpContext) {
```

```
            if (httpContext.Request.Path.ToString().ToLower() == "/middleware") {
                await httpContext.Response.WriteAsync(
                "This is from the content middleware", Encoding.UTF8);
            } else {
                await nextDelegate.Invoke(httpContext);
            }
        }
    }
}
```

中间件既不实现接口，也不从公共基类派生。相反，它们定义了一个构造函数，这个构造函数接收 RequestDelegate 对象并定义了 Invoke 方法。RequestDelegate 对象表示链中的下一个中间件，当 ASP.NET 收到 HTTP 请求时调用 Invoke 方法。

HTTP 请求和返回给客户端的响应信息可通过 Invoke 方法的 HttpContext 参数提供。第 17 章将描述 HttpContext 类及其属性，但是在本章中，只需要知道代码清单 14-17 中的 Invoke 方法能够检查 HTTP 请求并验证请求是否已发送到/middleware URL 就可以了。如果有，就向客户端发送简单的文本响应；如果使用了不同的 URL，就将请求转发到链中的下一个中间件。

请求管道是在 Startup 类的 Configure 方法中设置的。在代码清单 14-18 中，从示例应用程序中删除了 MVC 方法，并使用 ContentMiddleware 类作为管道中的唯一组件。

代码清单 14-18　在 ConfiguringApps 文件夹下的 Startup.cs 文件中使用自定义中间件

```
using System;
using System.Collections.Generic;
using System.Linq;
using System.Threading.Tasks;
using Microsoft.AspNetCore.Builder;
using Microsoft.AspNetCore.Hosting;
using Microsoft.AspNetCore.Http;
using Microsoft.Extensions.DependencyInjection;
using ConfiguringApps.Infrastructure;

namespace ConfiguringApps {
    public class Startup {

        public void ConfigureServices(IServiceCollection services) {
            services.AddSingleton<UptimeService>();
            services.AddMvc();
        }

        public void Configure(IApplicationBuilder app, IHostingEnvironment env) {
            app.UseMiddleware<ContentMiddleware>();
        }
    }
}
```

可使用 Configure 方法中的 UseMiddleware 扩展方法注册自定义中间件。UseMiddleware 方法使用类型参数来指定中间件类。这样 ASP.NET Core 就可以构建一个将要使用的所有中间件的列表，然后将它们实例化以创建链。如果运行应用程序并请求/middleware URL，你将看到图 14-6 所示的结果。

图 14-7 展示了使用 ContentMiddleware 类创建的中间件管道。当 ASP.NET Core 收到 HTTP 请求时，它会将请求传递给 Startup 类中注册的唯一中间件。如果 URL 是/middleware，组件将生成结果，返回给 ASP.NET Core 并发送给客户端。

如果 URL 不是/middleware，ContentMiddleware 类就将请求传递给链中的下一个中间件。但由于没有其他组件，因此请求在创建管道时到达 ASP.NET Core 提供的后备处理程序，后备处理程序在另一个方向上沿着管道发回请求（如果能看到其他类型的中间件如何工作，这个过程就容易理解了）。

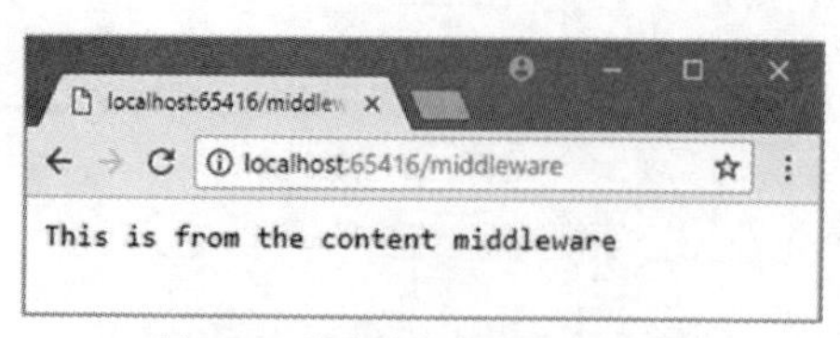

图 14-6　从自定义中间件生成内容

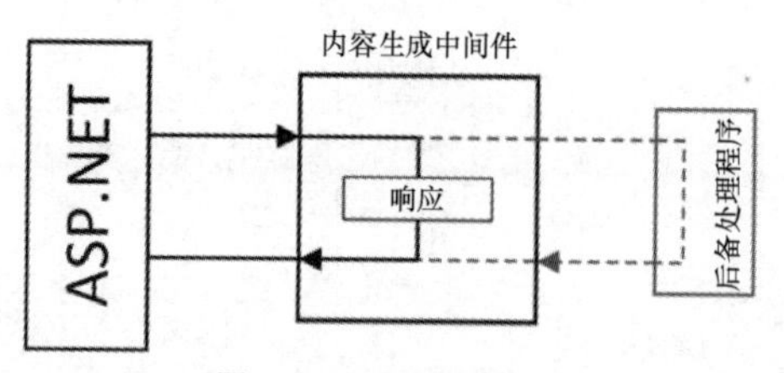

图 14-7　示例中间件管道

并非只有控制器能够使用 ConfigureServices 方法中设置的服务。ASP.NET Core 检查中间件类的构造函数，并使用服务为已定义的任何参数提供值。代码清单 14-19 向 ContentMiddleware 类的构造函数添加了一个参数，以告诉 ASP.NET Core 需要一个 UptimeService 对象。

代码清单 14-19　在 Infrastructure 文件夹下的 ContentMiddleware.cs 文件中使用服务

```
using System.Text;
using System.Threading.Tasks;
using Microsoft.AspNetCore.Http;

namespace ConfiguringApps.Infrastructure {

    public class ContentMiddleware {
        private RequestDelegate nextDelegate;
        private UptimeService uptime;

        public ContentMiddleware(RequestDelegate next, UptimeService up) {
            nextDelegate = next;
            uptime = up;
        }

        public async Task Invoke(HttpContext httpContext) {
            if (httpContext.Request.Path.ToString().ToLower() == "/middleware") {
                await httpContext.Response.WriteAsync(
                    "This is from the content middleware "+
                        $"(uptime: {uptime.Uptime}ms)", Encoding.UTF8);
            } else {
                await nextDelegate.Invoke(httpContext);
            }
        }
    }
}
```

能够使用服务意味着中间件可以共享通用功能并避免代码重复。运行应用程序并请求/middleware URL，你将看到图 14-8 所示的输出。

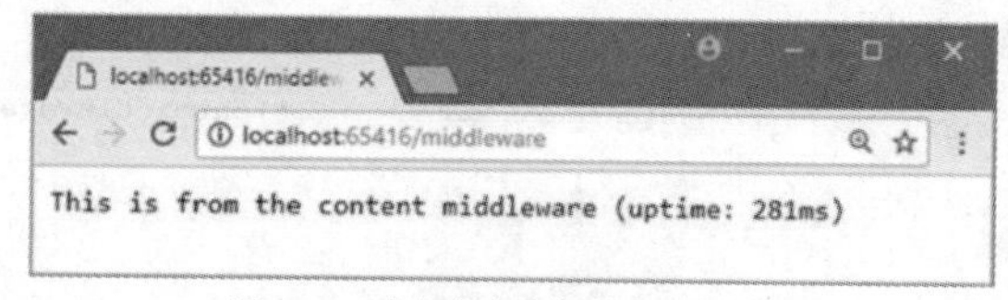

图 14-8　在自定义中间件中使用服务

2. 创建短路中间件

下一类中间件在请求到达内容生成中间件之前拦截它们，用来对管道处理过程进行短路，这通常是为了优化性能。代码清单 14-20 显示了添加到 Infrastructure 文件夹中的名为 ShortCircuitMiddleware.cs 的类文件的内容。

代码清单 14-20　Infrastructure 文件夹下的 ShortCircuitMiddleware.cs 文件的内容

```
using System.Linq;
using System.Threading.Tasks;
using Microsoft.AspNetCore.Http;

namespace ConfiguringApps.Infrastructure {
```

```
    public class ShortCircuitMiddleware {
        private RequestDelegate nextDelegate;

        public ShortCircuitMiddleware(RequestDelegate next) => nextDelegate = next;

        public async Task Invoke(HttpContext httpContext) {
            if (httpContext.Request.Headers["User-Agent"]
                    .Any(h => h.ToLower().Contains("edge"))) {
                httpContext.Response.StatusCode = 403;
            } else {
                await nextDelegate.Invoke(httpContext);
            }
        }
    }
}
```

这个中间件检查请求的 User-Agent 标头，浏览器使用该标头标识自身。使用 User-Agent 标头来标识特定的浏览器并不可靠，无法在实际应用程序中使用，但对此例来说已经足够了。

使用术语“短路”是因为这种类型的中间件并不总是将请求转发到链中的下一个中间件。在这种情况下，如果 User-Agent 标头包含术语 edge，那么组件会将状态码设置为 403-Forbidden，并且不将请求转发到下一个中间件。由于请求被拒绝，因此再由其他中间件处理请求是没有意义的，会不必要地消耗系统资源。相反，将请求处理提前终止，向客户端发送 403 响应。

中间件按照 Startup 类中设置的顺序接收请求，这意味着必须在内容生成中间件之前设置短路中间件，如代码清单 14-21 所示。

代码清单 14-21 在 ConfiguringApps 文件夹下的 Startup.cs 文件中注册短路中间件

```
using System;
using System.Collections.Generic;
using System.Linq;
using System.Threading.Tasks;
using Microsoft.AspNetCore.Builder;
using Microsoft.AspNetCore.Hosting;
using Microsoft.AspNetCore.Http;
using Microsoft.Extensions.DependencyInjection;
using ConfiguringApps.Infrastructure;

namespace ConfiguringApps {
    public class Startup {
        public void ConfigureServices(IServiceCollection services) {
            services.AddSingleton<UptimeService>();
            services.AddMvc();
        }

        public void Configure(IApplicationBuilder app, IHostingEnvironment env) {
            app.UseMiddleware<ShortCircuitMiddleware>();
            app.UseMiddleware<ContentMiddleware>();
        }
    }
}
```

如果使用 Microsoft Edge 浏览器运行应用程序并请求任何 URL，那么你将看到 403 错误。ShortCircuitMiddleware 组件会忽略来自其他浏览器的请求，并将请求传递给链中的下一个中间件，这意味着当请求的 URL 为/middleware 时将生成响应。将短路中间件添加到中间件管道中，如图 14-9 所示。

3. 创建请求编辑中间件

下一类中间件不会生成响应。相反，它们会在请求到达链中的其他中间件之前更改请求。这种中间件主要用于平台集成，还可用于准备请求，以便后续组件更容易处理它们。作为演示，将 BrowserTypeMiddleware.cs

文件添加到 Infrastructure 文件夹中，并用它定义代码清单 14-22 所示的中间件。

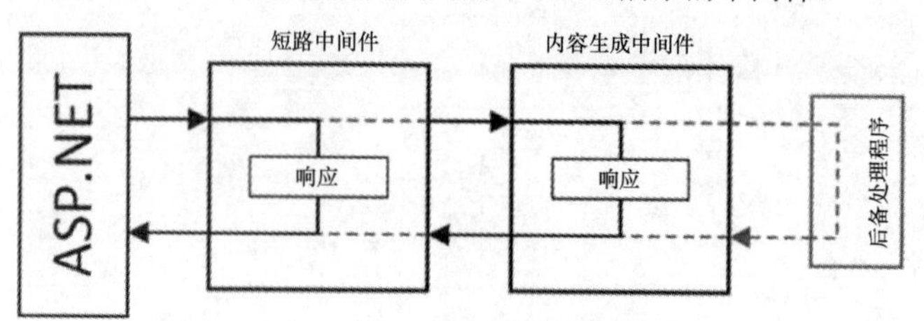

图 14-9　在中间件管道中添加短路中间件

代码清单 14-22　Infrastructure 文件夹下的 BrowserTypeMiddleware.cs 文件的内容

```
using System.Linq;
using System.Threading.Tasks;
using Microsoft.AspNetCore.Http;

namespace ConfiguringApps.Infrastructure {

    public class BrowserTypeMiddleware {
        private RequestDelegate nextDelegate;
        public BrowserTypeMiddleware(RequestDelegate next) => nextDelegate = next;

        public async Task Invoke(HttpContext httpContext) {
            httpContext.Items["EdgeBrowser"]
                = httpContext.Request.Headers["User-Agent"]
                    .Any(v => v.ToLower().Contains("edge"));
            await nextDelegate.Invoke(httpContext);
        }
    }
}
```

这个中间件检查请求的 User-Agent 标头并查找术语 edge，这表明请求可能是使用 Microsoft Edge 浏览器进行的。HttpContext 对象通过 Items 属性提供了一个字典，该字典用于在组件之间传递数据，请求将标头的搜索结果与键 EdgeBrowser 一起存储。

为了演示中间件如何协作，代码清单 14-23 显示了 ShortCircuitMiddleware 类，它拒绝来自 Microsoft Edge 浏览器的请求，并根据 BrowserTypeMiddleware 组件生成的数据做出决策。

代码清单 14-23　与 ShortCircuitMiddleware.cs 文件中的另一个中间件协作

```
using System.Linq;
using System.Threading.Tasks;
using Microsoft.AspNetCore.Http;

namespace ConfiguringApps.Infrastructure {

    public class ShortCircuitMiddleware {
        private RequestDelegate nextDelegate;

        public ShortCircuitMiddleware(RequestDelegate next) => nextDelegate = next;

        public async Task Invoke(HttpContext httpContext) {
            if (httpContext.Items["EdgeBrowser"] as bool? == true) {
                httpContext.Response.StatusCode = 403;
            } else {
                await nextDelegate.Invoke(httpContext);
            }
        }
    }
}
```

就本质而言，编辑请求中间件需要放置在与之合作的或者依赖于它们的中间件之前。在代码清单 14-24 中，将 BrowserTypeMiddleware 类注册为中间件管道中的第一个中间件。

代码清单 14-24　在 ConfiguringApps 文件夹下的 Startup.cs 文件中注册中间件

```
using System;
using System.Collections.Generic;
using System.Linq;
using System.Threading.Tasks;
using Microsoft.AspNetCore.Builder;
using Microsoft.AspNetCore.Hosting;
using Microsoft.AspNetCore.Http;
using Microsoft.Extensions.DependencyInjection;
using ConfiguringApps.Infrastructure;

namespace ConfiguringApps {
    public class Startup {

        public void ConfigureServices(IServiceCollection services) {
            services.AddSingleton<UptimeService>();
            services.AddMvc();
        }

        public void Configure(IApplicationBuilder app, IHostingEnvironment env) {
            app.UseMiddleware<BrowserTypeMiddleware>();
            app.UseMiddleware<ShortCircuitMiddleware>();
            app.UseMiddleware<ContentMiddleware>();
        }
    }
}
```

将中间件放置在中间件管道的开始处可确保在其他中间件收到请求之前已对请求进行了修改，如图 14-10 所示。

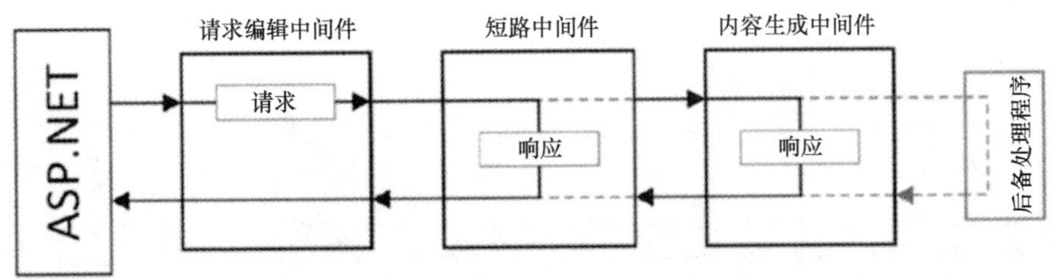

图 14-10　将请求编辑组件添加到中间件管道中

4. 创建响应编辑中间件

最后一类中间件用于对中间件管道中其他中间件生成的响应进行操作，这对于记录请求及响应的详细信息或处理错误非常有用。代码清单 14-25 显示了 ErrorMiddleware.cs 文件的内容，将这个文件添加到 Infrastructure 文件夹中以演示此类中间件。

代码清单 14-25　Infrastructure 文件夹下的 ErrorMiddleware.cs 文件的内容

```
using System.Text;
using System.Threading.Tasks;
using Microsoft.AspNetCore.Http;

namespace ConfiguringApps.Infrastructure {

    public class ErrorMiddleware {
```

```
        private RequestDelegate nextDelegate;

        public ErrorMiddleware(RequestDelegate next) {
            nextDelegate = next;
        }

        public async Task Invoke(HttpContext httpContext) {
            await nextDelegate.Invoke(httpContext);

            if (httpContext.Response.StatusCode == 403) {
                await httpContext.Response
                    .WriteAsync("Edge not supported", Encoding.UTF8);
            } else if (httpContext.Response.StatusCode == 404) {
                await httpContext.Response
                    .WriteAsync("No content middleware response", Encoding.UTF8);
            }
        }
    }
}
```

这个中间件在通过中间件管道并且生成响应之前对请求不感兴趣。如果状态码为 403 或 404，就向响应添加描述性消息。所有其他响应都被忽略。代码清单 14-26 显示了如何在 Startup 类中注册响应编辑中间件。

代码清单 14-26 在 Startup.cs 文件中注册响应编辑中间件

```
using System;
using System.Collections.Generic;
using System.Linq;
using System.Threading.Tasks;
using Microsoft.AspNetCore.Builder;
using Microsoft.AspNetCore.Hosting;
using Microsoft.AspNetCore.Http;
using Microsoft.Extensions.DependencyInjection;
using ConfiguringApps.Infrastructure;

namespace ConfiguringApps {
    public class Startup {
        public void ConfigureServices(IServiceCollection services) {
            services.AddSingleton<UptimeService>();
            services.AddMvc();
        }

        public void Configure(IApplicationBuilder app, IHostingEnvironment env) {
            app.UseMiddleware<ErrorMiddleware>();
            app.UseMiddleware<BrowserTypeMiddleware>();
            app.UseMiddleware<ShortCircuitMiddleware>();
            app.UseMiddleware<ContentMiddleware>();
        }
    }
}
```

提 示

你可能想知道产生 404 状态码的原因，因为它不是由这里创建的 3 个中间件中的任何一个设置的。答案是，这是当请求进入管道时由 ASP.NET 配置的响应，如果没有中间件更改响应，返回给客户端的状态码就是 404。

我们注册了 ErrorMiddleware 类，它占据中间件管道中的第一个位置。对于仅需要处理响应的中间件，这可能看起来很奇怪，但是在链的开始位置注册中间件可确保能够处理任何其他中间件生成的响应，如图 14-11 所示。如果放置在管道的后面位置，那么只能处理由某些其他中间件生成的响应。

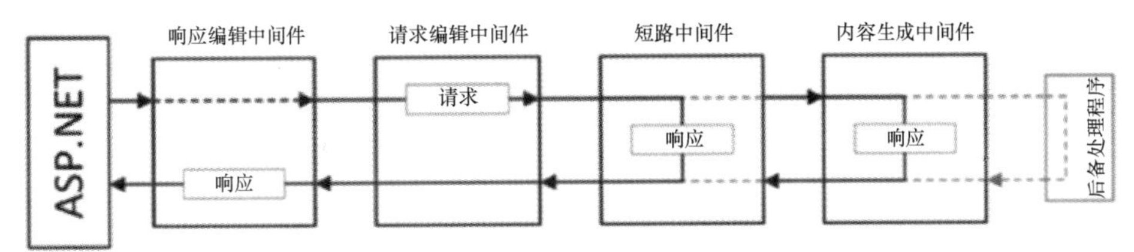

图 14-11 将响应编辑中间件添加到中间件管道中

可以通过启动应用程序并请求除//middleware URL 外的任何 URL 来查看新添加的中间件的效果。结果将显示图 14-12 所示的错误消息。

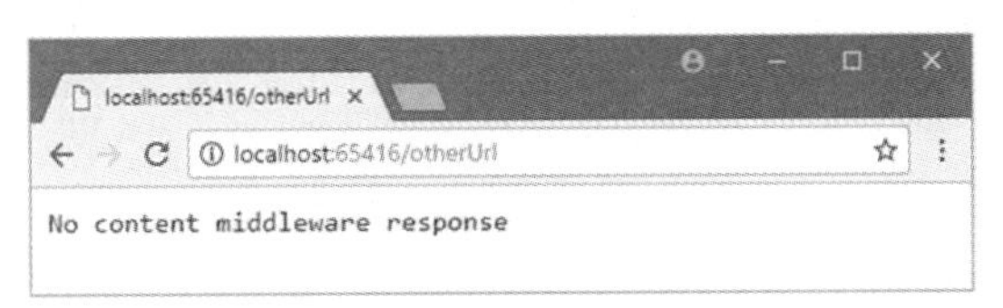

图 14-12 编辑其他中间件时的响应

14.4.3 了解如何调用 Configure 方法

ASP.NET Core 平台会在调用之前检查 Configure 方法，并获取其参数列表，该列表由 ConfigureServices 方法中设置的服务或表 14-5 中显示的特殊服务（可用作 Configure 方法的参数）提供。

表 14-5 特殊服务

特殊服务	描述
IApplicationBuilder	定义了设置应用程序中间件管道所需的功能
IHostingEnvironment	定义了区分不同类型环境（如开发环境和生产环境）所需的功能

1. 使用 Application Builder

虽然不必为 Configure 方法定义任何参数，但大多数 Startup 类至少使用 IApplicationBuilder，因为它允许创建中间件管道。对于自定义中间件来说，UseMiddleware 扩展方法用于注册类。复杂的内容生成中间件包提供了单一方法以在一个步骤中设置所有的中间件，就像它们提供单一方法来定义使用的服务一样。对于 MVC 来说，可以使用两种扩展方法，如表 14-6 所示。

表 14-6 两种扩展方法

扩展方法	描述
UseMvcWithDefaultRoute	使用默认路由设置 MVC 中间件
UseMvc	使用 Lambda 表达式指定的自定义路由配置设置 MVC 中间件

路由是将请求 URL 映射到控制器并由应用程序定义操作的过程，第 15 章和第 16 章将详细描述路由。UseMvcWithDefaultRoute 扩展方法对于开始使用 MVC 很有用，但是大多数应用程序会调用 UseMvc 扩展方法，即使结果是显式定义了由 UseMvcWithDefaultRoute 扩展方法创建的相同的路由配置，如代码清单 14-27 所示。这使得其他开发人员能够很容易理解应用程序使用的路由配置，并且以后可以轻松添加新的路由（几乎所有应用程序在某些时候都需要）。

代码清单 14-27 在 ConfiguringApps 文件夹下的 Startup.cs 文件中设置 MVC 中间件

```
using System;
using System.Collections.Generic;
using System.Linq;
using System.Threading.Tasks;
```

```
using Microsoft.AspNetCore.Builder;
using Microsoft.AspNetCore.Hosting;
using Microsoft.AspNetCore.Http;
using Microsoft.Extensions.DependencyInjection;
using ConfiguringApps.Infrastructure;

namespace ConfiguringApps {
    public class Startup {
        public void ConfigureServices(IServiceCollection services) {
            services.AddSingleton<UptimeService>();
            services.AddMvc();
        }

        public void Configure(IApplicationBuilder app, IHostingEnvironment env) {
            app.UseMiddleware<ErrorMiddleware>();
            app.UseMiddleware<BrowserTypeMiddleware>();
            app.UseMiddleware<ShortCircuitMiddleware>();
            app.UseMiddleware<ContentMiddleware>();

            app.UseMvc(routes => {
                routes.MapRoute(
                    name: "default",
                    template: "{controller=Home}/{action=Index}/{id?}");
            });
        }
    }
}
```

由于 MVC 设置了内容生成中间件，因此 UseMvc 扩展方法必须在注册了所有其他中间件之后调用。为了准备 MVC 所依赖的服务，必须在 ConfigureServices 方法中调用 AddMvc 扩展方法。

2. 使用托管环境

IHostingEnvironment 接口使用表 14-7 所示的属性提供有关运行应用程序的托管环境的一些基本但非常重要的信息。

表 14-7　IHostingEnvironment 属性

属性	描述
ApplicationName	返回应用程序的名称，该名称由托管平台设置
EnvironmentName	返回描述当前环境的字符串
ContentRootPath	返回包含应用程序的内容文件和配置文件的路径
WebRootPath	返回一个字符串，该字符串指定包含应用程序静态内容的文件夹，通常是 wwwroot 文件夹
ContentRootFileProvider	返回一个实现了 IFileProvider 接口的对象，该对象可用于从 ContentRootPath 属性指定的文件夹中读取文件
WebRootFileProvider	返回一个实现了 IFileProvider 接口的对象，该对象可用于从 WebRootPath 属性指定的文件夹中读取文件

ContentRootPath 和 WebRootPath 属性很有意思，但大多数应用程序中并不需要这两个属性，因为有一个内置的中间件可用于传递静态内容。

比较重要的属性是 EnvironmentName，它允许根据应用程序运行的环境修改应用程序的配置。有 3 种常规环境——开发环境、暂存环境和生产环境。

使用名为 ASPNETCORE_ENVIRONMENT 的环境变量设置当前托管环境。要设置环境变量，请从 Visual Studio 的 Project 菜单中选择 ConfiguringApps Properties，然后切换到 Debug 选项卡。双击环境变量的 Value 字段，默认情况下设置为 Development，将之更改为 Staging，如图 14-13 所示。保存更改以使新环境生效。

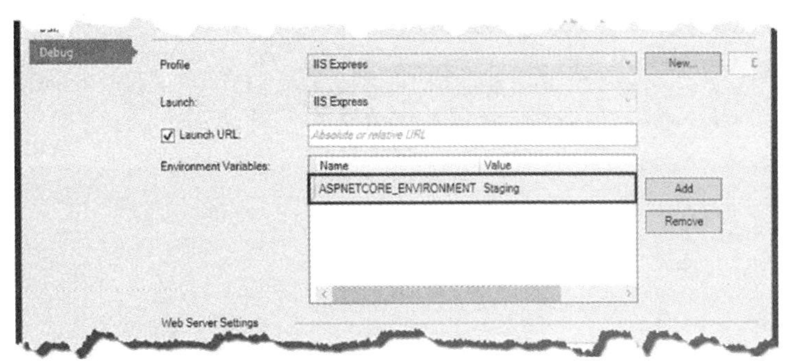

图 14-13　设置托管环境的名称

提　示

环境名称不区分大小写，因此 Staging 和 staging 可视为同一环境。虽然 Development、Staging 和 Production 是传统的环境名称，但你可以使用自己喜欢的任何名称。例如，如果项目中有多个开发人员，并且每个开发人员都需要不同的配置设置，那么这可能很有用。有关如何处理环境配置之间复杂差异的详细信息，请参阅 14.7 节。

在 Configure 方法中，可以通过读取 IHostingEnvironment.EnvironmentName 属性或使用对 IHostingEnvironment 对象进行操作的扩展方法之一来确定正在使用的托管环境，如表 14-8 所示。

表 14-8　　IhostingEnvironment 的扩展方法

扩展方法	描述
IsDevelopment()	如果托管环境的名称为 Development，返回 true
IsStaging()	如果托管环境的名称是 Staging，返回 true
IsProduction()	如果托管环境的名称为 Production，返回 true
IsEnvironment (env)	如果托管环境的名称与 env 参数匹配，返回 true

扩展方法用于更改中间件管道中的中间件集合，以定制应用程序到不同托管环境的行为。代码清单 14-28 使用一种扩展方法来确保本章前面创建的自定义中间件仅出现在 Development 托管环境的管道中。

代码清单 14-28　在 ConfiguringApps 文件夹下的 Startup.cs 文件中使用托管环境

```
using System;
using System.Collections.Generic;
using System.Linq;
using System.Threading.Tasks;
using Microsoft.AspNetCore.Builder;
using Microsoft.AspNetCore.Hosting;
using Microsoft.AspNetCore.Http;
using Microsoft.Extensions.DependencyInjection;
using ConfiguringApps.Infrastructure;

namespace ConfiguringApps {
    public class Startup {

        public void ConfigureServices(IServiceCollection services) {
            services.AddSingleton<UptimeService>();
            services.AddMvc();
        }

        public void Configure(IApplicationBuilder app, IHostingEnvironment env) {

            if (env.IsDevelopment()) {
```

```
            app.UseMiddleware<ErrorMiddleware>();
            app.UseMiddleware<BrowserTypeMiddleware>();
            app.UseMiddleware<ShortCircuitMiddleware>();
            app.UseMiddleware<ContentMiddleware>();
        }

        app.UseMvc(routes => {
            routes.MapRoute(
                name: "default",
                template: "{controller=Home}/{action=Index}/{id?}");
        });
    }
  }
}
```

这 3 个自定义中间件不会添加到使用当前配置的管道中，当前配置已将托管环境设置为 Staging。如果运行应用程序并请求/middleware URL，你将收到 404-Not Found 错误，因为唯一可用的中间件是使用 UseMvc 扩展方法设置的，它们没有可用于处理该 URL 的控制器。

注 意

一旦测试了更改托管环境的效果，就请务必更改回 Development；否则，本章其余部分的示例将无法正常工作。

14.4.4 添加其他中间件

有一组常用的中间件在大多数 MVC 项目中很有用，本书的示例中也使用了这些中间件。下面将这些中间件添加到请求管道中并解释它们的工作原理。

1. 启用异常处理

即使是精心编写的应用程序也会遇到异常，因此适当地处理它们非常重要。代码清单 14-29 添加了一些中间件，用于处理请求管道发生的异常。这里还删除了自定义中间件，以便可以专注于 MVC。

代码清单 14-29 在 ConfiguringApps 文件夹下的 Startup.cs 文件中添加异常处理中间件

```
using System;
using System.Collections.Generic;
using System.Linq;
using System.Threading.Tasks;
using Microsoft.AspNetCore.Builder;
using Microsoft.AspNetCore.Hosting;
using Microsoft.AspNetCore.Http;
using Microsoft.Extensions.DependencyInjection;
using ConfiguringApps.Infrastructure;

namespace ConfiguringApps {
    public class Startup {

        public void ConfigureServices(IServiceCollection services) {
            services.AddSingleton<UptimeService>();
            services.AddMvc();
        }

        public void Configure(IApplicationBuilder app, IHostingEnvironment env) {

            if (env.IsDevelopment()) {
                app.UseDeveloperExceptionPage();
                app.UseStatusCodePages();
```

```
            } else {
                app.UseExceptionHandler("/Home/Error");
            }

            app.UseMvc(routes => {
                routes.MapRoute(
                    name: "default",
                    template: "{controller=Home}/{action=Index}/{id?}");
            });
        }
    }
}
```

UseStatusCodePages 方法将描述性消息添加到不包含内容的响应中，例如 404-Not Found 响应，这可能很有用，因为并非所有浏览器都向用户显示自己的消息。

UseDeveloperExceptionPage 方法设置一个错误处理中间件，这个中间件在响应中显示异常的详细信息，包括异常跟踪。这些信息不应该向用户显示，因此只能在使用 IHostingEnvironment 对象检测到的开发托管环境中调用 UseDeveloperExceptionPage 方法。

对于 Staging 或 Production 环境，使用 UseExceptionHandler 方法。该方法设置错误处理，允许显示自定义错误消息，不会显示应用程序的内部错误细节。UseExceptionHandler 方法的参数是客户端应重定向到的 URL，以便接收错误消息。这可以是应用程序提供的任何 URL，但惯例是使用/Home/Error。

在代码清单 14-30 中，添加了根据需要为 Home 控制器的 Index 操作生成异常的功能，并添加了 Error 操作，以处理 UseExceptionHandler 生成的请求。

代码清单 14-30　在 Controllers 文件夹下的 HomeController.cs 文件中生成和处理异常

```
using System.Collections.Generic;
using Microsoft.AspNetCore.Mvc;
using ConfiguringApps.Infrastructure;

namespace ConfiguringApps.Controllers {

    public class HomeController : Controller {
        private UptimeService uptime;

        public HomeController(UptimeService up) => uptime = up;

        public ViewResult Index(bool throwException = false) {
            if (throwException) {
                throw new System.NullReferenceException();
            }
            return View(new Dictionary<string, string> {
                ["Message"] = "This is the Index action",
                ["Uptime"] = $"{uptime.Uptime}ms"
            });
        }

        public ViewResult Error() => View(nameof(Index),
            new Dictionary<string, string> {
                ["Message"] = "This is the Error action"});
            }
}
```

对 Index 操作的更改依赖于第 26 章描述的模型绑定功能，以从请求中获取 throwException 的值。如果 throwException 为 true，Index 操作抛出 NullReferenceException；如果为 false，就正常执行。

Error 操作使用 Index 视图显示简单消息。可以通过运行应用程序并请求/Home/Index?throwException=true URL 来查看不同异常处理中间件的效果。查询字符串提供了 Index 操作参数的值，你看到的响应将取决于托管环境的名称。图 14-14 显示了 UseDeveloperExceptionPage（用于开发托管环境）和 UseExceptionHandler 中间件

（用于所有其他托管环境）生成的输出结果。

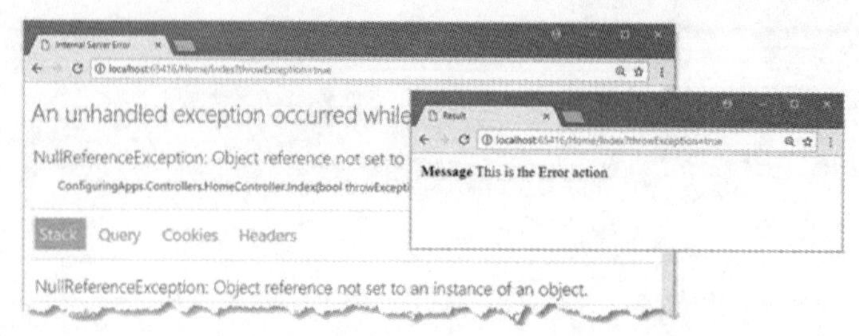

图 14-14 输出结果

开发人员异常页面提供了异常的详细信息，以及查看堆栈跟踪和导致异常的请求的选项。相比之下，用户异常页面应该仅用于表示发生了错误。

2. 启用浏览器链接

第 6 章介绍了浏览器链接功能，并演示了如何在开发过程中用来管理浏览器。浏览器链接的服务器端部分是作为中间件实现的，必须作为应用程序配置的一部分添加到 Startup 类中；否则，Visual Studio 集成将无法工作。浏览器链接仅在开发期间有用，不应在暂存环境或生产环境中使用，因为还会编辑其他中间件生成的响应，并插入 JavaScript 代码。打开到服务器端的 HTTP 连接，以接收重新加载的通知。在代码清单 14-31 中，可以看到如何仅为开发托管环境调用已注册中间件的 UseBrowserLink 方法。

代码清单 14-31 在 ConfiguringApps 文件夹下的 Startup.cs 文件中启用浏览器链接

```
using System;
using System.Collections.Generic;
using System.Linq;
using System.Threading.Tasks;
using Microsoft.AspNetCore.Builder;
using Microsoft.AspNetCore.Hosting;
using Microsoft.AspNetCore.Http;
using Microsoft.Extensions.DependencyInjection;
using ConfiguringApps.Infrastructure;

namespace ConfiguringApps {
    public class Startup {

        public void ConfigureServices(IServiceCollection services) {
            services.AddSingleton<UptimeService>();
            services.AddMvc();
        }
        public void Configure(IApplicationBuilder app, IHostingEnvironment env) {

            if (env.IsDevelopment()) {
                app.UseDeveloperExceptionPage();
                app.UseStatusCodePages();
                app.UseBrowserLink();
            } else {
                app.UseExceptionHandler("/Home/Error");
            }

            app.UseMvc(routes => {
                routes.MapRoute(
                    name: "default",
                    template: "{controller=Home}/{action=Index}/{id?}");
            });
        }
    }
}
```

3. 启用静态内容

最后一个中间件对大多数项目很有用，它提供了对 wwwroot 文件夹中文件的访问，使应用程序可以包含图像、JavaScript 文件和 CSS。使用 UseStaticFiles 方法添加一个组件，用于将静态文件的请求管道短路，如代码清单 14-32 所示。

代码清单 14-32　在 ConfiguringApps 文件夹下的 Startup.cs 文件中启用静态内容

```
using System;
using System.Collections.Generic;
using System.Linq;
using System.Threading.Tasks;
using Microsoft.AspNetCore.Builder;
using Microsoft.AspNetCore.Hosting;
using Microsoft.AspNetCore.Http;
using Microsoft.Extensions.DependencyInjection;
using ConfiguringApps.Infrastructure;

namespace ConfiguringApps {
    public class Startup {

        public void ConfigureServices(IServiceCollection services) {
            services.AddSingleton<UptimeService>();
            services.AddMvc();
        }

        public void Configure(IApplicationBuilder app, IHostingEnvironment env) {

            if (env.IsDevelopment()) {
                app.UseDeveloperExceptionPage();
                app.UseStatusCodePages();
                app.UseBrowserLink();
            } else {
                app.UseExceptionHandler("/Home/Error");
            }
            app.UseStaticFiles();
            app.UseMvc(routes => {
                routes.MapRoute(
                    name: "default",
                    template: "{controller=Home}/{action=Index}/{id?}");
            });
        }
    }
}
```

无论托管环境如何，通常都需要静态内容，这就是为所有环境调用 UseStaticFiles 方法的原因。这意味着 Index 视图中的 link 元素将正常工作，并允许浏览器加载 Bootstrap CSS 样式表。可以通过启动应用程序来查看效果，如图 14-15 所示。

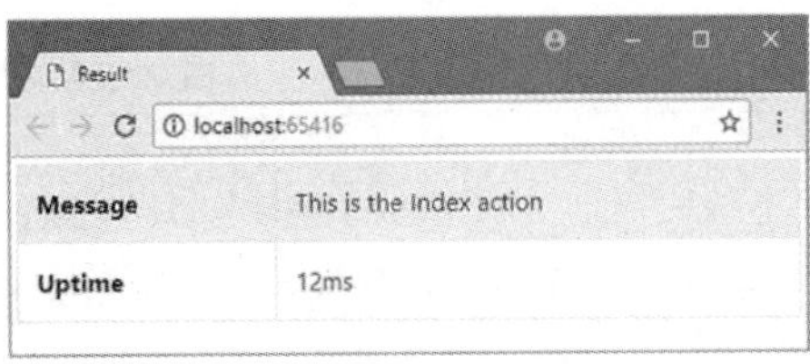

图 14-15　启用静态内容

14.5　配置应用程序

某些配置数据经常更改，例如，当应用程序从开发环境转移到生产环境时，数据库服务器需要不同的信息。ASP.NET Core 并不在 Startup 类中对此信息进行硬编码，而是从一系列更容易更改的源端提供配置数据，例如环境变量、命令行参数以及使用 JavaScript Object Notation（JSON）编写的文件。

配置数据通常是自动处理的，但由于已经替换了 Program 类中的默认设置，因此需要显式添加获取数据的代码，并使其可用于应用程序的其他部分，如代码清单 14-33 所示。

代码清单 14-33 在 ConfiguringApps 文件夹下的 Program.cs 文件中加载配置数据

```
using System;
using System.Collections.Generic;
using System.IO;
using System.Linq;
using System.Threading.Tasks;
using Microsoft.AspNetCore;
using Microsoft.AspNetCore.Hosting;
using Microsoft.Extensions.Configuration;
using Microsoft.Extensions.Logging;
using System.Reflection;

namespace ConfiguringApps {
    public class Program {

        public static void Main(string[] args) {
            BuildWebHost(args).Run();
        }

        public static IWebHost BuildWebHost(string[] args) {

            return new WebHostBuilder()
                .UseKestrel()
                .UseContentRoot(Directory.GetCurrentDirectory())
                .ConfigureAppConfiguration((hostingContext, config) => {
                    config.AddJsonFile("appsettings.json",
                        optional: true, reloadOnChange: true);
                    config.AddEnvironmentVariables();
                    if (args != null) {
                        config.AddCommandLine(args);
                    }
                })
                .UseIISIntegration()
                .UseStartup<Startup>()
                .Build();
        }
    }
}
```

ConfigureAppConfiguration 方法用于处理配置数据，其参数是 WebHostBuilderContext 对象和实现了 IConfigurationBuilder 接口的对象。WebBostBuilderContext 类定义了表 14-9 中描述的属性。

表 14-9 由 WebBostBuilderContext 类定义的属性

属性	描述
HostingEnvironment	返回一个实现了 IHostingEnvironment 接口的对象，并提供有关运行应用程序的托管环境的信息。有关详细信息
Configuration	返回一个实现了 IConfiguration 接口的对象，该接口提供对应用程序中配置数据的只读访问

IConfigurationBuilder 接口用于为应用程序的其余部分准备配置数据，这通常使用扩展方法完成。表 14-10 描述了代码清单 14-33 中用于添加配置数据的 3 个扩展方法。

表 14-10 用于添加配置数据的扩展方法

扩展方法	描述
AddJsonFile	用于从 JSON 文件（例如 appsettings.json）加载配置数据
AddEnvironmentVariables	用于从环境变量加载配置数据
AddCommandLine	用于从启动应用程序的命令行参数加载配置数据

在代码清单 14-33 所示的用于加载配置数据的 3 个方法中，最有趣的是 AddJsonFile 方法。该方法的参

数指定了文件名、文件是否可选以及在文件更改时是否应重新加载配置数据：

```
...
config.AddJsonFile("appsettings.json", optional: true, reloadOnChange: true);
...
```

以上代码指定了一个名为 appsettings.json 的文件，这是 JSON 配置文件的常规名称。appsettings.json 文件是可选的，也就是说，如果文件不存在，不会引发异常，并且将监视更改和自动刷新配置数据。

重新加载配置数据

ASP.NET Core 配置系统支持在配置文件更改时重新加载数据。某些内置中间件（如日志记录系统）支持此功能，这意味着可以在运行时更改日志记录级别，而无须重新启动应用程序。你也可以在自定义中间件中包含类似功能。

但仅仅因为支持这个功能而使用它并不意味着就是切合实际的。在生产系统中更改配置文件是导致停机的原因之一，很容易产生输入错误并导致创建错误的配置。即使成功进行了更改，也可能出现无法预料的后果，例如日志数据填满磁盘或导致性能下降。

建议避免实时编辑，并确保在部署到生产环境之前将所有更改推送到标准测试流程中。针对实时运行的系统来诊断问题可能很容易，但结果可能并不会尽如人意。如果你正在编辑生产环境中的配置文件，那么应该慎重考虑是否要将一个小问题变成一个更大的问题。

14.5.1 创建 JSON 配置文件

appsettings.json 文件的最常见用途是存储数据库连接字符串和日志记录设置，但也可以存储应用程序所需的任何数据。

要查看配置系统的工作方式，请将名为 appsettings.json 的新 JSON 文件添加到项目的根文件夹中，内容如代码清单 14-34 所示。

代码清单 14-34 ConfiguringApps 文件夹下的 appsettings.json 文件的内容

```
{
  "ShortCircuitMiddleware": {
    "EnableBrowserShortCircuit": true
  }
}
```

可以为 JSON 格式定义配置的结构。以上代码中的 JSON 内容定义了名为 ShortCircuitMiddleware 的配置类别，其中包含名为 EnableBrowserShortCircuit 的配置属性，它被设置为 true。

JSON 中的引号和逗号

如果不熟悉 JSON，那么值得花一些时间阅读 JSON 网站上的规范。JSON 格式易于使用，并且大多数平台对生成和解析 JSON 数据提供了良好支持，包括 MVC 应用程序和使用简单 Javascript API 的客户端。事实上，大多数 MVC 开发人员根本不会直接与 JSON 打交道，只有在配置文件中才需要手动编码 JSON。

关于 JSON 有两个很多开发人员容易遇到的陷阱，虽然仍然需要花时间阅读规范，但是当 Visual Studio 或 ASP.NET Core 无法解析 JSON 文件的时候，了解最常见的问题会让你更容易找到原因。下面是对 appsettings.json 文件的补充，显示了两个最常见的问题：

```
{
  "ShortCircuitMiddleware": {
    "EnableBrowserShortCircuit": true
  }
  mysetting : [ fast, slow ]
}
```

首先，JSON 的几乎所有内容都是用引号括起来的。如果正在编写 C#代码并期望在没有引号的情况下接收属性的名称和值，就会很容易忘记这一点。在 JSON 中，所有除布尔值和数字之外的其他内容必须用引号括起来，如下所示：

```
{
  "ShortCircuitMiddleware": {
    "EnableBrowserShortCircuit": true
  }
  "mysetting" : [ "fast", "slow"]
}
```

其次，在向对象的 JSON 描述添加一个新的属性时，必须记住在上一个大括号字符的后面添加一个逗号，如下所示：

```
{
  "ShortCircuitMiddleware": {
    "EnableBrowserShortCircuit": true
  },
  "mysetting" : [ "fast", "slow"]
}
```

即使这个错误已高亮显示也很难看出差异——这就是为什么这类错误如此常见——但这里已经在关闭 ShortCircuitMiddleware 部分的“}”后添加了一个逗号。另外要小心，引号后没有其他节点也是非法的。如果因为更改 JSON 导致问题发生，那么首先要检查这两个错误。

14.5.2 使用配置数据

Startup 类可以通过使用 IConfiguration 参数定义的构造函数来访问配置数据。在 Program 类中调用 UseStartup 方法时，会使用 ConfigureAppConfiguration 准备的配置数据创建 Startup 对象。代码清单 14-35 向 Startup 类添加了构造函数，并显示了如何访问配置数据。

代码清单 14-35 在 ConfiguringApps 文件夹下的 Startup.cs 文件中接收和使用配置数据

```
using System;
using System.Collections.Generic;
using System.Linq;
using System.Threading.Tasks;
using Microsoft.AspNetCore.Builder;
using Microsoft.AspNetCore.Hosting;
using Microsoft.AspNetCore.Http;
using Microsoft.Extensions.DependencyInjection;
using ConfiguringApps.Infrastructure;
using Microsoft.Extensions.Configuration;

namespace ConfiguringApps {
    public class Startup {

        public Startup(IConfiguration configuration) {
            Configuration = configuration;
        }

        public IConfiguration Configuration { get; }

        public void ConfigureServices(IServiceCollection services) {
            services.AddSingleton<UptimeService>();
            services.AddMvc();
        }
        public void Configure(IApplicationBuilder app, IHostingEnvironment env) {

            if ((Configuration.GetSection("ShortCircuitMiddleware")?
                    .GetValue<bool>("EnableBrowserShortCircuit")).Value) {
```

```
                app.UseMiddleware<BrowserTypeMiddleware>();
                app.UseMiddleware<ShortCircuitMiddleware>();
            }

            if (env.IsDevelopment()) {
                app.UseDeveloperExceptionPage();
                app.UseStatusCodePages();
                app.UseBrowserLink();
            } else {
                app.UseExceptionHandler("/Home/Error");
            }
            app.UseStaticFiles();
            app.UseMvc(routes => {
                routes.MapRoute(
                    name: "default",
                    template: "{controller=Home}/{action=Index}/{id?}");
            });
        }
    }
}
```

构造函数接收 IConfiguration 对象并将其分配给名为 Configuration 的属性，然后可以使用该属性访问配置数据，这些配置数据已从环境变量、命令行和 appsettings.json 文件加载。

要获取值，可以将数据结构导航到所需的配置部分，这部分由实现了 IConfiguration 接口的另一个对象表示，该接口提供了可用于 IConfigurationRoot 的成员，如表 14-11 所示。

表 14-11　可用于 IConfigurationRoot 的成员

成员	描述
[key]	该索引器用于获取特定键的字符串值
GetSection(name)	该方法返回一个 IConfiguration 对象，该对象表示配置数据的一个节点
GetChildren()	该方法返回 IConfiguration 对象的子节点条目，用来表示当前配置对象的子节点

还有一些扩展方法可用于对 IConfiguration 对象进行操作以获取值并将其从字符串转换为其他类型，如表 14-12 所示。

表 14-12　IConfiguration 接口的扩展方法

扩展方法	描述
GetValue<T>(keyName)	获取与指定键关联的值，并尝试将其转换为类型 T
GetValue<T>(keyName,defaultValue)	获取与指定键关联的值，并尝试将其转换为类型 T。如果配置数据中的键没有值，就使用默认值

重要的是，不要假设配置值一定会被指定。可使用 null 条件运算符来确保在尝试获取 EnableBrowser-ShortCircuit 值之前已获取到 ShortCircuitMiddleware 部分。结果是，只有在定义了 ShortCircuitMiddleware/EnableBrowserShortCircuit 值并将其设置为 true 时，才会将自定义中间件添加到请求管道中。

14.5.3 配置日志记录

ASP.NET Core 提供对捕获和处理日志记录数据的支持，并且已内置许多中间件以生成日志记录消息。大多数项目会自动设置日志记录，但由于在 Program 类中使用了单独的配置语句，因此需要添加代码清单 14-36 所示的语句来设置日志记录功能。

代码清单 14-36　在 ConfiguringApps 文件夹下配置 Program.cs 文件中的日志记录

```
using System;
using System.Collections.Generic;
using System.IO;
```

```
using System.Linq;
using System.Threading.Tasks;
using Microsoft.AspNetCore;
using Microsoft.AspNetCore.Hosting;
using Microsoft.Extensions.Configuration;
using Microsoft.Extensions.Logging;
using System.Reflection;

namespace ConfiguringApps {
    public class Program {

        public static void Main(string[] args) {
            BuildWebHost(args).Run();
        }

        public static IWebHost BuildWebHost(string[] args) {

            return new WebHostBuilder()
                .UseKestrel()
                .UseContentRoot(Directory.GetCurrentDirectory())
                .ConfigureAppConfiguration((hostingContext, config) => {
                    config.AddJsonFile("appsettings.json",
                        optional: true, reloadOnChange: true);
                    config.AddEnvironmentVariables();
                    if (args != null) {
                        config.AddCommandLine(args);
                    }
                })
                .ConfigureLogging((hostingContext, logging) => {
                    logging.AddConfiguration(
                        hostingContext.Configuration.GetSection("Logging"));
                    logging.AddConsole();
                    logging.AddDebug();
                })
                .UseIISIntegration()
                .UseStartup<Startup>()
                .Build();
        }
    }
}
```

ConfigureLogging方法使用Lambda函数设置日志记录系统，Lambda函数接收WebHostBuilderContext对象（可参考本章前面的内容）和实现了ILoggingBuilder接口的对象。ILoggingBuilder接口运行一组扩展方法来配置日志记录系统，如表14-13所示。

表14-13 ILoggingBuilder接口的扩展方法

扩展方法	描述
AddConfiguration	用于使用从appsettings.json文件、命令行或环境变量加载的配置数据来配置日志记录系统
AddConsole	将日志消息发送到控制台，这在使用dotnet run命令启动应用程序时非常有用
AddDebug	当Visual Studio调试器运行时，将日志消息发送到调试输出窗口
AddEventLog	将日志消息发送到Windows事件日志，如果已部署到Windows Server，并希望将来自ASP.NET Core MVC应用程序的日志消息与其他类型的应用程序中的日志消息合并，这将非常有用

1. 了解日志记录配置数据

AddConfiguration方法用于使用配置数据来配置日志记录系统，通常在appsettings.json文件中定义。代码清单14-37将名为Logging的配置部分添加到appsettings.json文件中，这部分配置对应于代码清单14-36

中用于 AddConfiguration 方法的名称。

代码清单 14-37 将配置部分添加到 ConfiguringApps 文件夹下的 appsettings.json 文件中

```
{
  "ShortCircuitMiddleware": {
    "EnableBrowserShortCircuit": true
  },
  "Logging": {
    "LogLevel": {
      "Default": "Debug",
      "System": "Information",
      "Microsoft": "Information"
    }
  }
}
```

Logging 配置部分指定应从不同的日志记录数据源显示的消息级别。日志记录系统支持 7 个级别的调试信息，表 14-14 按重要性对它们进行了排列。

表 14-14 调试级别

调试级别	描述
Trace	用于在开发期间有用但在生产环境中不需要的消息
Debug	用于开发人员因调试问题所需的详细消息
Information	用于描述应用程序常规操作的消息
Warning	用于描述意外但不中断应用程序的事件的消息
Error	用于描述中断应用程序的错误的消息
Critical	用于描述灾难性故障的消息
None	用于禁用日志记录消息

代码清单 14-37 中的 Default 条目设置了将日志消息显示到 Debug 级别的阈值，也就是说，只显示 Debug 级别或更高级别的消息。其余条目将覆盖从特定命名空间记录消息的默认值，以便仅在 Information 级别或更高级别时才显示源自 System 或 Microsoft 命名空间的日志记录消息。要查看启用日志记录的效果，请通过选择 Debug→Start Debugging，使用 Visual Studio 调试器启动应用程序。查看 Output 窗口，你将看到日志消息显示了如何处理每个 HTTP 请求，如下所示：

```
info: Microsoft.AspNetCore.Hosting.Internal.WebHost[1]
      Request starting HTTP/1.1 GET http://localhost:65417/
info: Microsoft.AspNetCore.Mvc.Internal.ControllerActionInvoker[1]
      Executing action method ConfiguringApps.Controllers.HomeController.Index
      (ConfiguringApps) with arguments (False) - ModelState is Valid
info: Microsoft.AspNetCore.Mvc.ViewFeatures.Internal.ViewResultExecutor[1]
      Executing ViewResult, running view at path /Views/Home/Index.cshtml.
info: Microsoft.AspNetCore.Mvc.Internal.ControllerActionInvoker[2]
      Executed action ConfiguringApps.Controllers.HomeController.Index
      (ConfiguringApps) in 1597.3535ms
info: Microsoft.AspNetCore.Hosting.Internal.WebHost[2]
      Request finished in 1695.6314ms 200 text/html; charset=utf-8
```

2. 创建自定义日志消息

日志消息是由处理 HTTP 请求并生成响应的 ASP.NET Core 和 MVC 组件生成的。这种消息可以提供有用的信息，但也可以生成针对应用程序自定义日志消息，如代码清单 14-38 所示。

代码清单 14-38 Controllers 文件夹下的 HomeController.cs 文件中的自定义日志消息

```
using System.Collections.Generic;
using Microsoft.AspNetCore.Mvc;
```

```
using ConfiguringApps.Infrastructure;
using Microsoft.Extensions.Logging;

namespace ConfiguringApps.Controllers {

    public class HomeController : Controller {
        private UptimeService uptime;
        private ILogger<HomeController> logger;

        public HomeController(UptimeService up, ILogger<HomeController> log) {
            uptime = up;
            logger = log;
        }

        public ViewResult Index(bool throwException = false) {
            logger.LogDebug($"Handled {Request.Path} at uptime {uptime.Uptime}");

            if (throwException) {
                throw new System.NullReferenceException();
            }
            return View(new Dictionary<string, string> {
                ["Message"] = "This is the Index action",
                ["Uptime"] = $"{uptime.Uptime}ms"
            });
        }

        public ViewResult Error() => View(nameof(Index),
            new Dictionary<string, string> {
                ["Message"] = "This is the Error action"});
        }
}
```

ILogger 接口定义了创建日志条目和获取实现该接口的对象所需的功能，HomeController 类具有类型为 ILogger 的构造函数参数<HomeController>。类型参数允许日志记录系统在日志消息中使用类的名称，构造函数的参数值是通过依赖注入功能自动提供的，可参考第 18 章中的描述。

拥有 ILogger 后，可以使用 Microsoft.Extensions.Logging 命名空间中定义的扩展方法创建日志消息。表 14-14 描述了每种日志记录级别的方法。HomeController 类使用 LogDebug 方法在 Debug 级别创建消息。要查看效果，请使用 Visual Studio 调试器运行应用程序，并检查日志消息的 Output 窗口，其中的内容如下所示：

```
dbug: ConfiguringApps.Controllers.HomeController[0]
  Handled / at uptime 12
```

启动应用程序时会显示很多消息，很难挑选有用的消息。如果单击 Output 窗口顶部的 Clear All 按钮，然后重新加载浏览器，就更容易看到日志消息——这将确保仅显示与单个请求相关的日志消息。

14.5.4 配置依赖注入

ASP.NET Core 应用程序的默认配置包括准备服务提供程序，第 18 章将详细介绍依赖项注入功能。代码清单 14-39 向 Program 类添加了配置语句。

代码清单 14-39 在 ConfiguringApps 文件夹下的 Program.cs 文件中配置服务

```
using System;
using System.Collections.Generic;
using System.IO;
using System.Linq;
using System.Threading.Tasks;
using Microsoft.AspNetCore;
using Microsoft.AspNetCore.Hosting;
using Microsoft.Extensions.Configuration;
```

```
using Microsoft.Extensions.Logging;
using System.Reflection;

namespace ConfiguringApps {
    public class Program {

        public static void Main(string[] args) {
            BuildWebHost(args).Run();
        }

        public static IWebHost BuildWebHost(string[] args) {

            return new WebHostBuilder()
                .UseKestrel()
                .UseContentRoot(Directory.GetCurrentDirectory())
                .ConfigureAppConfiguration((hostingContext, config) => {
                    config.AddJsonFile("appsettings.json",
                        optional: true, reloadOnChange: true);
                    config.AddEnvironmentVariables();
                    if (args != null) {
                        config.AddCommandLine(args);
                    }
                })
                .ConfigureLogging((hostingContext, logging) => {
                    logging.AddConfiguration(
                        hostingContext.Configuration.GetSection("Logging"));
                    logging.AddConsole();
                    logging.AddDebug();
                })
                .UseIISIntegration()
                .UseDefaultServiceProvider((context, options) => {
                    options.ValidateScopes =
                        context.HostingEnvironment.IsDevelopment();
                })
                .UseStartup<Startup>()
                .Build();
        }
    }
}
```

UseDefaultServiceProvider 方法使用内置的 ASP.NET Core 服务提供程序。还有一些其他可用的服务提供程序，但大多数项目可以使用内置功能，建议只有在遇到需要解决的特定问题时才使用第三方组件，并且需要对依赖注入有很好的理解，可参考第 18 章中的描述。

UseDefaultServiceProvider 接收一个 Lambda 函数，这个 Lambda 函数接收 WebHostBuilderContext 对象和 ServiceProviderOptions 对象，用来配置内置的服务提供程序。唯一的配置属性名为 ValidateScopes，在使用 Entity Framework Core 时需要禁用，如第 8 章所述。

14.6 配置 MVC 服务

在 Startup 类的 ConfigureServices 方法中调用 AddMvc 扩展方法时，会设置 MVC 应用程序所需的所有服务。这种方式非常方便，因为可以在一个步骤中注册所有 MVC 服务，但如果要改变默认的行为，那么确实需要做一些额外的工作来重新配置服务。

AddMvc 扩展方法返回一个实现了 IMvcBuilder 接口的对象，MVC 提供了一组可用于高级配置的扩展方法，其中一些有用的扩展方法参见表 14-15。另外，许多配置选项与后面章节中详细描述的功能有关。

表 14-15 有用的扩展方法

扩展方法	描述
AddMvcOptions	用于配置 MVC 使用的服务
AddFormatterMappings	用于配置允许客户端指定接收的数据格式的功能
AddJsonOptions	用于配置 JSON 数据的创建方式
AddRazorOptions	用于配置 Razor 视图引擎
AddViewOptions	用于配置 MVC 处理视图的方式，包括使用哪些视图引擎

AddMvcOptions 扩展方法用于配置最重要的 MVC 服务，它接收一个函数用以接收 MvcOptions 对象，该对象提供了一组配置属性，其中一些重要属性如表 14-16 所示。

表 14-16 MvcOptions 对象的一些重要属性

属性	描述
Conventions	返回模型约定的列表，这些约定用于自定义 MVC 如何创建控制器和操作
Filters	返回全局过滤器的列表
FormatterMappings	返回一个映射，以允许客户端指定接收的数据格式
InputFormatters	返回用于解析请求数据的对象列表
ModelValidatorProviders	返回用于验证数据的对象列表
OutputFormatters	返回类的列表，用于格式化从 API 控制器发送的数据
RespectBrowserAcceptHeader	指定在决定用于响应的数据格式时是否考虑 Accept 标头

这些配置属性用于微调 MVC 的运行方式，可以在指定的章节中找到与它们相关的功能的详细说明。但是，作为快速演示，代码清单 14-40 显示了如何使用 AddMvcOptions 方法更改配置选项。

代码清单 14-40 在 ConfiguringApps 文件夹下更改 Startup.cs 文件中的配置选项

```
...
public void ConfigureServices(IServiceCollection services) {
    services.AddSingleton<UptimeService>();
    services.AddMvc().AddMvcOptions(options => {
        options.RespectBrowserAcceptHeader = true;
    });
}
...
```

传递给 AddMvcOptions 方法的 Lambda 表达式接收一个 MvcOptions 对象，用它将 RespectBrowserAcceptHeader 属性设置为 true。这使客户端对内容协商过程中选择的数据格式有了更强的控制，如第 20 章所述。

14.7 处理复杂配置

如果需要支持大量的托管环境，或者托管环境之间存在很多差异，那么在 Startup 类中使用 if 语句进行分支配置可能会导致配置难以阅读且难以编辑，从而容易产生意外的变更。下面将介绍使用 Startup 类处理复杂配置的不同方法。

14.7.1 创建不同的外部配置文件

Program 类执行的应用程序的默认配置会查找特定于运行应用程序的托管环境的 JSON 配置文件，因此可以使用名为 appsettings.production.json 的文件来存储特定于生产环境的配置。代码清单 14-41 恢复了将 JSON 文件加载到 Program 类的语句。

代码清单 14-41 在 ConfiguringApps 文件夹下的 Program.cs 文件中加载托管环境文件

```
using System;
using System.Collections.Generic;
using System.IO;
using System.Linq;
using System.Threading.Tasks;
using Microsoft.AspNetCore;
using Microsoft.AspNetCore.Hosting;
using Microsoft.Extensions.Configuration;
using Microsoft.Extensions.Logging;
using System.Reflection;

namespace ConfiguringApps {
    public class Program {

        public static void Main(string[] args) {
            BuildWebHost(args).Run();
        }
        public static IWebHost BuildWebHost(string[] args) {

            return new WebHostBuilder()
                .UseKestrel()
                .UseContentRoot(Directory.GetCurrentDirectory())
                .ConfigureAppConfiguration((hostingContext, config) => {
                    var env = hostingContext.HostingEnvironment;
                    config.AddJsonFile("appsettings.json",
                            optional: true, reloadOnChange: true)
                        .AddJsonFile($"appsettings.{env.EnvironmentName}.json",
                            optional: true, reloadOnChange: true);
                    config.AddEnvironmentVariables();
                    if (args != null) {
                        config.AddCommandLine(args);
                    }
                })
                .ConfigureLogging((hostingContext, logging) => {
                    logging.AddConfiguration(
                        hostingContext.Configuration.GetSection("Logging"));
                    logging.AddConsole();
                    logging.AddDebug();
                })
                .UseIISIntegration()
                .UseDefaultServiceProvider((context, options) => {
                    options.ValidateScopes =
                        context.HostingEnvironment.IsDevelopment();
                })
                .UseStartup<Startup>()
                .Build();
        }
    }
}
```

从特定于平台的文件加载配置数据时，里面包含的配置设置将覆盖具有相同名称的任何现有数据。例如，使用 ASP.NET Configuration File 模板创建一个名为 appsettings.development.json 的文件，其中的配置数据如代码清单 14-42 所示。使用这个文件中的配置数据将 EnableBrowserShortCircuit 设置为 false。

提 示

appsettings.development.json 文件在创建后，好像会消失。在解决方案资源管理器中，如果在 appsettings.json 条目的左侧将箭头展开，你将看到 Visual Studio 对具有相似名称的项目进行了分组。

代码清单 14-42　ConfiguringApps 文件夹下的 appsettings.development.json 文件的内容

```
{
  "ShortCircuitMiddleware": {
    "EnableBrowserShortCircuit": false
  }
}
```

appsettings.json 文件将在应用程序启动时加载，如果应用程序在开发环境中运行，那么接着会加载 appsettings.development.json 文件。结果是，当应用程序在开发环境中运行时，EnableBrowserShortCircuit 将为 false，在暂存环境和生产环境中为 true。

14.7.2　创建不同的配置方法

选择不同的配置数据文件可能很有用，但无法为复杂配置提供完整的解决方案，因为数据文件不包含 C#语句。如果要更改用于创建服务或注册中间件的配置语句，则可以使用不同的方法，其中方法的名称包括托管环境，如代码清单 14-43 所示。

代码清单 14-43　在 Startup.cs 文件中使用不同的方法名称

```
using System;
using System.Collections.Generic;
using System.Linq;
using System.Threading.Tasks;
using Microsoft.AspNetCore.Builder;
using Microsoft.AspNetCore.Hosting;
using Microsoft.AspNetCore.Http;
using Microsoft.Extensions.DependencyInjection;
using ConfiguringApps.Infrastructure;
using Microsoft.Extensions.Configuration;

namespace ConfiguringApps {
    public class Startup {

        public Startup(IConfiguration configuration) {
            Configuration = configuration;
        }

        public IConfiguration Configuration { get; }

        public void ConfigureServices(IServiceCollection services) {
            services.AddSingleton<UptimeService>();
            services.AddMvc().AddMvcOptions(options => {
                options.RespectBrowserAcceptHeader = true;
            });
        }

        public void ConfigureDevelopmentServices(IServiceCollection services) {
            services.AddSingleton<UptimeService>();
            services.AddMvc();
        }

        public void Configure(IApplicationBuilder app, IHostingEnvironment env) {
            app.UseExceptionHandler("/Home/Error");
            app.UseStaticFiles();
            app.UseMvc(routes => {
                routes.MapRoute(
                    name: "default",
                    template: "{controller=Home}/{action=Index}/{id?}");
            });
        }
```

```
        public void ConfigureDevelopment(IApplicationBuilder app,
                IHostingEnvironment env) {
            app.UseDeveloperExceptionPage();
            app.UseStatusCodePages();
            app.UseBrowserLink();
            app.UseStaticFiles();
            app.UseMvcWithDefaultRoute();
        }
    }
}
```

当 ASP.NET Core 在 Startup 类中查找 ConfigureServices 和 Configure 方法时，首先检查是否存在包含托管环境名称的方法。以上代码添加了 ConfigureDevelopmentServices 方法，用于代替开发环境中的 ConfigureServices 方法；还添加了 ConfigureDevelopment 方法，用于代替开发环境中的 Configure 方法。可以为需要支持的每个环境定义单独的方法，如果没有可用的特定于环境的方法，则依赖要调用的默认方法。在该例中，ConfigureServices 和 Configure 方法将用于暂存环境与生产环境。

警　告

如果定义了特定于环境的方法，则不会调用默认方法。例如，在代码清单 14-43 中，ASP.NET Core 不会在开发环境中调用 Configure 方法，因为有 ConfigureDevelopment 方法。这意味着每个方法都负责环境所需的完整配置。

14.7.3 创建不同的配置类

使用不同的方法意味着不必使用 if 语句来检查托管环境的名称，但可能导致类变得很大，这本身就是问题。对于特别复杂的配置，最后的措施是为每个托管环境创建不同的配置类。当 ASP.NET 查找 Startup 类时，首先检查是否存在名称包含当前托管环境的类。为此，在项目中添加一个名为 StartupDevelopment.cs 的类文件，并用它定义代码清单 14-44 所示的类。

代码清单 14-44 ConfiguringApps 文件夹下的 StartupDevelopment.cs 文件的内容

```
using Microsoft.AspNetCore.Builder;
using Microsoft.AspNetCore.Hosting;
using Microsoft.Extensions.DependencyInjection;
using ConfiguringApps.Infrastructure;

namespace ConfiguringApps {
    public class StartupDevelopment {
        public void ConfigureServices(IServiceCollection services) {
            services.AddSingleton<UptimeService>();
            services.AddMvc();
        }

        public void Configure(IApplicationBuilder app, IHostingEnvironment env) {
            app.UseDeveloperExceptionPage();
            app.UseStatusCodePages();
            app.UseBrowserLink();
            app.UseStaticFiles();
            app.UseMvcWithDefaultRoute();
        }
    }
}
```

这个类包含特定于开发托管环境的 ConfigureServices 和 Configure 方法。要使 ASP.NET Core 能够找到特定于环境的 Startup 类，需要对 Program 类进行更改，如代码清单 14-45 所示。

代码清单 14-45　在 Program.cs 文件中启用特定于环境的 Startup

```
using System;
using System.Collections.Generic;
using System.IO;
using System.Linq;
using System.Threading.Tasks;
using Microsoft.AspNetCore;
using Microsoft.AspNetCore.Hosting;
using Microsoft.Extensions.Configuration;
using Microsoft.Extensions.Logging;
using System.Reflection;

namespace ConfiguringApps {
    public class Program {

        public static void Main(string[] args) {
            BuildWebHost(args).Run();
        }

        public static IWebHost BuildWebHost(string[] args) {

            return new WebHostBuilder()
                .UseKestrel()
                .UseContentRoot(Directory.GetCurrentDirectory())
                .ConfigureAppConfiguration((hostingContext, config) => {
                    var env = hostingContext.HostingEnvironment;
                    config.AddJsonFile("appsettings.json",
                        optional: true, reloadOnChange: true)
                    .AddJsonFile($"appsettings.{env.EnvironmentName}.json",
                            optional: true, reloadOnChange: true);
                    config.AddEnvironmentVariables();
                    if (args != null) {
                        config.AddCommandLine(args);
                    }
                })
                .ConfigureLogging((hostingContext, logging) => {
                    logging.AddConfiguration(
                        hostingContext.Configuration.GetSection("Logging"));
                    logging.AddConsole();
                    logging.AddDebug();
                })
                .UseIISIntegration()
                .UseDefaultServiceProvider((context, options) => {
                    options.ValidateScopes =
                        context.HostingEnvironment.IsDevelopment();
                })
                .UseStartup(nameof(ConfiguringApps))
                .Build();
        }
    }
}
```

UseStartup 方法不指定类，而是给出应该使用的程序集的名称。当应用程序启动时，ASP.NET 将查找名称中包含托管环境的类，例如 StartupDevelopment 或 StartupProduction。如果不存在，就返回使用常规的 Startup 类。

14.8　小结

本章解释了如何配置 MVC 应用程序，还描述了 Program 和 Startup 类的作用以及它们提供的默认配置选项，展示了如何使用管道处理请求以及如何使用不同类型的中间件来控制请求流和它们引发的响应。下一章将介绍路由系统，以便 MVC 将请求 URL 映射到控制器和操作方法。

第 15 章　URL 路由

早期版本的 ASP.NET 假设请求的 URL 与服务器硬盘上的文件之间存在直接关系。服务器的工作是从浏览器接收请求并从相应的文件传递输出。这种方法适用于 Web 窗体，其中每个 ASPX 页面都是文件和对请求的自包含响应。

这对 MVC 应用程序来说并无意义，请求由控制器类中的操作方法处理，而且与硬盘上的文件也没有一对一关联。

为了处理 MVC 的 URL，ASP.NET 平台使用了路由系统，路由系统已针对 ASP.NET Core 进行了大量修改。本章将展示如何使用路由系统为项目创建强大而灵活的 URL 处理方式。正如你将看到的，路由系统允许你创建所需的任何 URL 模式，并以清晰简洁的方式表达它们。路由系统有两个功能。

- 检查传入的 URL 并选择控制器和操作方法来处理请求。
- 生成传出的 URL。这些是在视图呈现的 HTML 中显示的 URL，以便在用户单击链接时调用特定的操作（此时将再次成为传入的 URL）。

本章将重点介绍路由并使用它们来处理传入的 URL，以便用户可以访问控制器和操作。在 MVC 应用程序中创建路由有两种方法——基于约定的路由和属性路由。本章将解释这两种方法。

然后，下一章将展示如何使用这些路由生成包含在视图中的传出 URL，以及如何自定义路由系统并使用称为 area（区域）的相关功能。表 15-1 总结了路由的相关问题。

表 15-1　路由的相关问题

问题	答案
路由是什么？	路由系统负责处理传入的请求并选择控制器和操作方法来处理它们。路由系统还用于在视图中生成路由，称为传出的 URL
路由有什么用？	路由系统能够灵活地处理请求，而不是将 URL 与 Visual Studio 项目中的类结构关联到一起
如何使用路由？	URL 与控制器和操作方法之间的映射定义在 Startup.cs 文件中，也可通过将 Route 属性应用于控制器来实现
有什么缺点或限制吗？	复杂应用程序的路由配置可能会变得难以管理
有其他代替方式吗？	没有。路由系统是 ASP.NET Core 的组成部分

表 15-2 列出了本章要介绍的操作。

表 15-2　本章要介绍的操作

操作	方法	代码清单
创建 URL 和操作方法之间的映射	创建路由	代码清单 15-9
省略 URL 片段	为路由片段定义默认值	代码清单 15-10～代码清单 15-12
匹配没有相应路由变量的 URL 片段	定义静态片段	代码清单 15-13～代码清单 15-16
将 URL 片段传递给操作方法	自定义片段变量	代码清单 15-17～代码清单 15-19
省略没有默认值的 URL 片段	定义可选片段	代码清单 15-20 和代码清单 15-21
定义路由来匹配任意数量的 URL 片段	使用 catchall 片段	代码清单 15-22 和代码清单 15-23
限制路由可以匹配的 URL	应用路由约束	代码清单 15-24～代码清单 15-33
在控制器中定义路由	使用属性路由	代码清单 15-34～代码清单 15-38

15.1　准备示例项目

本章将使用 ASP.NET Core Web Application（.NET Core）模板创建名为 UrlsAndRoutes 的 Empty 项目。为了

添加对 MVC 框架、开发人员错误页面和静态文件的支持，将代码清单 15-1 显示的语句添加到 Startup 类中。

代码清单 15-1 在 UrlsAndRoutes 文件夹下的 Startup.cs 文件中配置应用程序

```
using System;
using System.Collections.Generic;
using System.Linq;
using System.Threading.Tasks;
using Microsoft.AspNetCore.Builder;
using Microsoft.AspNetCore.Hosting;
using Microsoft.AspNetCore.Http;
using Microsoft.Extensions.DependencyInjection;

namespace UrlsAndRoutes {
    public class Startup {

        public void ConfigureServices(IServiceCollection services) {
            services.AddMvc();
        }

        public void Configure(IApplicationBuilder app, IHostingEnvironment env) {
            app.UseDeveloperExceptionPage();
            app.UseStatusCodePages();
            app.UseStaticFiles();
            app.UseMvc();
        }
    }
}
```

15.1.1 创建模型类

本章中的所有工作都是为了将请求 URL 与操作相匹配。需要的唯一模型类会传递有关处理请求的控制器和操作方法的详细信息。创建 Models 文件夹并添加一个名为 Result.cs 的类文件，用来定义代码清单 15-2 所示的类。

代码清单 15-2 Models 文件夹下的 Result.cs 文件的内容

```
using System.Collections.Generic;

namespace UrlsAndRoutes.Models {

    public class Result {
        public string Controller { get; set; }

        public string Action { get; set; }

        public IDictionary<string, object> Data { get; }
            = new Dictionary<string, object>();
    }
}
```

Controller 和 Action 属性将用于指示请求的处理方式，Data 字典将用于存储路由系统生成的请求的其他详细信息。

15.1.2 创建 Example 控制器

下面使用一些简单的控制器来演示路由的工作原理。创建 Controllers 文件夹并添加一个名为 HomeController.cs 的类文件，其中的内容如代码清单 15-3 所示。

代码清单 15-3 Controllers 文件夹下的 HomeController.cs 文件的内容

```
using Microsoft.AspNetCore.Mvc;
using UrlsAndRoutes.Models;
```

```
namespace UrlsAndRoutes.Controllers {

    public class HomeController : Controller {

        public ViewResult Index() => View("Result",
            new Result {
                Controller = nameof(HomeController),
                Action = nameof(Index)
            });
    }
}
```

使用 Home 控制器定义的 Index 操作方法调用 View 方法来渲染名为 Result 的视图（在 15.1.3 节中定义），并提供 Result 对象作为模型对象。模型对象的属性使用 nameof 函数设置，用来指示已使用哪个控制器和操作方法来处理请求。

按照相同的方式，将 CustomerController.cs 文件添加到 Controllers 文件夹中，并用它定义代码清单 15-4 所示的 Customer 控制器。

代码清单 15-4　Controllers 文件夹下的 CustomerController.cs 文件的内容

```
using Microsoft.AspNetCore.Mvc;
using UrlsAndRoutes.Models;

namespace UrlsAndRoutes.Controllers {

    public class CustomerController : Controller {

        public ViewResult Index() => View("Result",
            new Result {
                Controller = nameof(CustomerController),
                Action = nameof(Index)
            });

        public ViewResult List() => View("Result",
            new Result {
                Controller = nameof(CustomerController),
                Action = nameof(List)
            });
    }
}
```

最后一个控制器定义在名为 AdminController.cs 的文件中，将该文件添加到 Controllers 文件夹中，文件的内容如代码清单 15-5 所示。

代码清单 15-5　Controllers 文件夹下的 AdminController.cs 文件的内容

```
using Microsoft.AspNetCore.Mvc;
using UrlsAndRoutes.Models;

namespace UrlsAndRoutes.Controllers {

    public class AdminController : Controller {

        public ViewResult Index() => View("Result",
            new Result {
                Controller = nameof(AdminController),
                Action = nameof(Index)
            });
    }
}
```

15.1.3　创建视图

我们在 15.1.2 节定义的所有操作方法中指定了 Result 视图，从而允许创建一个将由所有控制器共享的视图。创建 Views/Shared 文件夹，并添加一个名为 Result.cshtml 的新视图，内容如代码清单 15-6 所示。

代码清单 15-6　Views/Shared 文件夹下的 Result.cshtml 文件的内容

```
@model Result
@{ Layout = null; }

<!DOCTYPE html>
<html>
<head>
    <meta name="viewport" content="width=device-width" />
    <title>Routing</title>
    <link rel="stylesheet" asp-href-include="lib/bootstrap/dist/css/*.min.css" />
</head>
<body class="m-1 p-1">
    <table class="table table-bordered table-striped table-sm">
        <tr><th>Controller:</th><td>@Model.Controller</td></tr>
        <tr><th>Action:</th><td>@Model.Action</td></tr>
        @foreach (string key in Model.Data.Keys) {
            <tr><th>@key :</th><td>@Model.Data[key]</td></tr>
        }
    </table>
</body>
</html>
```

Result 视图包含一个表格，其中显示了使用 Bootstrap 设置样式的模型对象的属性。为了向项目添加 Bootstrap，使用 Bower Configuration File 模板创建 bower.json 文件，并将 Bootstrap 包添加到 dependencies 部分，如代码清单 15-7 所示。

代码清单 15-7　在 UrlsAndRoutes 文件夹下的 bower.json 文件中添加 Bootstrap 包

```
{
  "name": "asp.net",
  "private": true,
  "dependencies": {
    "bootstrap": "4.0.0-alpha.6"
  }
}
```

最后的准备工作是在 Views 文件夹中创建_ViewImports.cshtml 文件，以设置用于 Razor 视图的内置标签助手，并导入 Modes 命名空间，如代码清单 15-8 所示。

代码清单 15-8　Views 文件夹下的_ViewImports.cshtml 文件的内容

```
@using UrlsAndRoutes.Models
@addTagHelper *, Microsoft.AspNetCore.Mvc.TagHelpers
```

Startup 类中的配置不包含任何有关 MVC 如何将 HTTP 请求映射到控制器和操作的说明。当运行应用程序时，请求的任何 URL 都将得到 404-Not Found 响应，如图 15-1 所示。

图 15-1　运行示例应用程序的响应

15.2 介绍 URL 模式

路由系统使用一组路由来工作。这些路由共同包含应用程序的 URL 架构或方案，这是应用程序将识别和响应的一组 URL。

不需要手动输入应用程序中支持的所有单个 URL。相反，每个路由都包含一种 URL 模式，用来与传入的 URL 进行比较。如果传入的 URL 与模式匹配，路由系统将使用匹配的模式来处理 URL。下面是一个简单的示例 URL：

```
http://mysite.com/Admin/Index
```

网址可以细分为片段。这些是 URL 的组成部分，不包括由/字符分隔的主机名和查询字符串。在上面的示例 URL 中，有两个片段，如图 15-2 所示。

```
http://mysite.com/Admin/Index
```

第一个片段　第二个片段

图 15-2　示例 URL 中的片段

第一个片段包含单词 Admin，第二个片段包含单词 Index。显然，第一个片段与控制器有关，而第二个片段涉及操作。但是，需要使用路由系统可以理解的 URL 模式来表达这种关系。以下是与示例 URL 匹配的网址格式：

```
{controller}/{action}
```

当处理传入的 HTTP 请求时，路由系统的功能是将请求的 URL 与模式匹配，并从 URL 中提取模式中定义的片段变量的值。

片段变量使用大括号括起来。示例模式具有两个名为 controller 和 action 的片段变量，因此 controller 片段变量的值将为 Admin，而 action 片段变量的值将为 Index。

MVC 应用程序通常会有多个路由，路由系统会将传入的 URL 与每个路由的 URL 模式进行比较，直到找到匹配为止。默认情况下，一个模式将匹配具有正确片段数量的任何 URL。例如，模式{controller}/{action}将匹配具有两个片段的任何 URL，如表 15-3 所示。

表 15-3　匹配的 URL

请求的 URL	片段变量
*****//mysite.***/Admin/Index	controller 片段变量的值为 Admin，action 片段变量的值为 Index
*****//mysite.***/Admin	没有匹配，片段太少
*****//mysite.***/Admin/Index/Soccer	没有匹配，片段太多

表 15-3 突出显示了 URL 模式的两个关键行为。

- URL 模式匹配的片段数是固定的。它们将仅匹配与模式具有相同数量片段的 URL。表 15-3 的第 2 个和第 3 个示例展示了这一点。
- URL 模式匹配的片段的内容是不固定的。如果 URL 具有正确的片段数，URL 模式将提取每个片段变量的值，无论它是什么。

这些默认行为是了解 URL 模式如何运作的关键。本章后面将介绍如何更改默认值。

15.3 创建和注册简单路由

一旦有了 URL 模式，就可以用它定义路由。路由定义在 Startup.cs 文件中，并作为参数传递给 UseMvc 方法，该方法用于在 Configure 方法中设置 MVC。代码清单 15-9 显示了一个基本路由，用于将请求映射到示例应用程序中的控制器。

代码清单 15-9　在 UrlsAndRoutes 文件夹下的 Startup.cs 文件中定义基本路由

```
using System;
using System.Collections.Generic;
using System.Linq;
using System.Threading.Tasks;
using Microsoft.AspNetCore.Builder;
using Microsoft.AspNetCore.Hosting;
using Microsoft.AspNetCore.Http;
using Microsoft.Extensions.DependencyInjection;

namespace UrlsAndRoutes {
    public class Startup {

        public void ConfigureServices(IServiceCollection services) {
            services.AddMvc();
        }
        public void Configure(IApplicationBuilder app, IHostingEnvironment env) {
            app.UseDeveloperExceptionPage();
            app.UseStatusCodePages();
            app.UseStaticFiles();
            app.UseMvc(routes => {
                routes.MapRoute(name: "default", template: "{controller}/{action}");
            });
        }
    }
}
```

路由是使用 Lambda 表达式创建的，Lambda 表达式则作为参数传递给 UseMvc 配置方法。Lambda 表达式接收一个对象，它实现了 Microsoft.AspNetCore.Routing 命名空间中的 IRouteBuilder 接口，并使用 MapRoute 扩展方法定义路由。为了使路由更容易理解，约定在调用 MapRoute 扩展方法时指定参数名称，这就是在代码中明确命名 name 和 template 参数的原因。name 参数指定了路由的名称，template 参数用于定义模式。

提　示

命名路由是可选的，但有观点认为，这样无法做到彻底的关注点分离。

可以通过启动示例应用程序来查看这里对路由所做更改的效果。应用程序首次启动时没有变化——你仍会看到 404 错误，但如果导航到与{controller}/{action}模式匹配的 URL，你将看到图 15-3 所示的结果，这表明已导航到/Admin/Index。

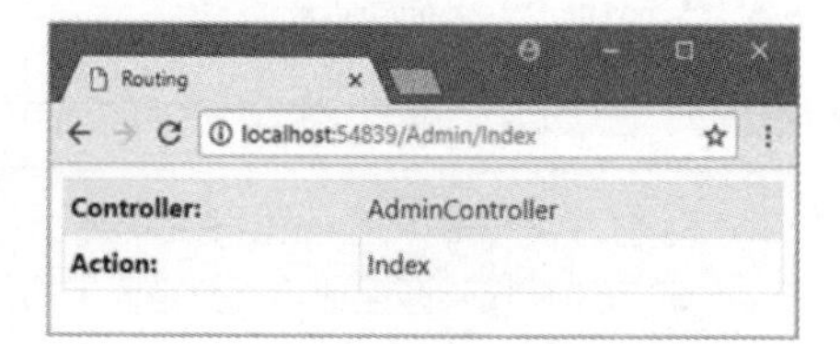

图 15-3　使用简单的路由进行导航

应用程序的根 URL 不起作用，原因是添加到 Startup.cs 文件的路由没有告诉 MVC，在请求的 URL 没有片段时如何选择控制器类和操作方法。15.4 节将介绍如何解决这个问题。

15.4　定义默认值

请求默认 URL 时，示例应用程序返回 404 错误，因为默认 URL 与 Startup 类中定义的路由模式不匹配。由于默认 URL 中没有可以与路由模式定义的 controller 和 action 变量匹配的片段，因此无法匹配路由系统。

之前解释过，URL 模式只匹配具有指定片段数的 URL。更改此行为的一种方法是使用默认值。当 URL 不包含可由路由模式匹配的片段时，将应用默认值。代码清单 15-10 定义了一个使用默认值的路由。

代码清单 15-10 在 UrlsAndRoutes 文件夹下的 Startup.cs 文件中提供默认值

```
using System;
using System.Collections.Generic;
using System.Linq;
using System.Threading.Tasks;
using Microsoft.AspNetCore.Builder;
using Microsoft.AspNetCore.Hosting;
using Microsoft.AspNetCore.Http;
using Microsoft.Extensions.DependencyInjection;

namespace UrlsAndRoutes {
    public class Startup {

        public void ConfigureServices(IServiceCollection services) {
            services.AddMvc();
        }

        public void Configure(IApplicationBuilder app, IHostingEnvironment env) {
            app.UseDeveloperExceptionPage();
            app.UseStatusCodePages();
            app.UseStaticFiles();
            app.UseMvc(routes => {
                routes.MapRoute(
                    name: "default",
                    template: "{controller}/{action}",
                    defaults: new { action = "Index" });
            });
        }
    }
}
```

默认值可作为匿名类型的属性提供，并作为 defaults 参数传递给 MapRoute 方法。在以上代码中，为 action 变量提供了默认值 Index。

这个路由将匹配所有两段式 URL，就像以前一样。例如，如果请求******//mydomain.***/Home/Index，路由将提取 Home 作为 controller 变量的值，并将提取 Index 作为 action 变量的值。

但是现在 action 变量有了默认值，路由也将匹配单片段 URL。当处理单片段 URL 时，路由系统将从 URL 中提取 controller 变量的值，并使用 action 变量的默认值。通过这种方式，用户在请求/Home 的时候，MVC 将在 Home 控制器上调用 Index 操作方法，如图 15-4 所示。

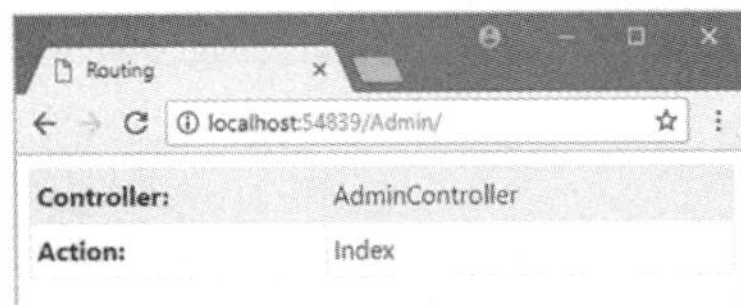

图 15-4 使用默认的操作方法

定义内联默认值

默认值也可以表示为 URL 模式的一部分，这是一种更简洁的表达路由的方式，如代码清单 15-11 所示。内联语法只能为 URL 模式的部分变量提供默认值，但是，正如你将了解到的那样，能够在这种模式之外提供默认值通常很有用。因此，了解表达默认值的两种方法是很有用的。

代码清单 15-11 在 UrlsAndRoutes 文件夹下的 Startup.cs 文件中定义内联默认值

```
using System;
using System.Collections.Generic;
using System.Linq;
using System.Threading.Tasks;
using Microsoft.AspNetCore.Builder;
using Microsoft.AspNetCore.Hosting;
using Microsoft.AspNetCore.Http;
using Microsoft.Extensions.DependencyInjection;

namespace UrlsAndRoutes {
```

```
    public class Startup {

        public void ConfigureServices(IServiceCollection services) {
            services.AddMvc();
        }

        public void Configure(IApplicationBuilder app, IHostingEnvironment env) {
            app.UseDeveloperExceptionPage();
            app.UseStatusCodePages();
            app.UseStaticFiles();
            app.UseMvc(routes => {
                routes.MapRoute(
                    name: "default",
                    template: "{controller}/{action=Index}");
            });
        }
    }
}
```

可以进一步匹配完全不包含任何片段变量的 URL，仅依靠默认值来标识 action 和 controller。作为示例，代码清单 15-12 显示了如何通过为两个片段提供默认值来映射应用程序的根 URL。

代码清单 15-12 在 UrlsAndRoutes 文件夹下的 Startup.cs 文件中提供默认值

```
using System;
using System.Collections.Generic;
using System.Linq;
using System.Threading.Tasks;
using Microsoft.AspNetCore.Builder;
using Microsoft.AspNetCore.Hosting;
using Microsoft.AspNetCore.Http;
using Microsoft.Extensions.DependencyInjection;

namespace UrlsAndRoutes {
    public class Startup {

        public void ConfigureServices(IServiceCollection services) {
            services.AddMvc();
        }

        public void Configure(IApplicationBuilder app, IHostingEnvironment env) {
            app.UseDeveloperExceptionPage();
            app.UseStatusCodePages();
            app.UseStaticFiles();
            app.UseMvc(routes => {
                routes.MapRoute(
                    name: "default",
                template: "{controller=Home}/{action=Index}");
            });
        }
    }
}
```

通过为 controller 和 action 变量提供默认值，路由将匹配具有零个、一个或两个片段的 URL，如表 15-4 所示。

表 15-4 匹配 URL

片段数	示例	匹配结果
0	/	controller = Home, action = Index
1	**/Customer**	controller = Customer, action = Index
2	**/Customer/List**	controller = Customer, action = List
3	**/Customer/List/All**	没有匹配，片段太多

传入的 URL 中接收的片段越少，路由越依赖于默认值，直到仅使用默认值匹配没有片段的 URL。

可以通过启动示例应用程序来查看默认值的效果。当浏览器请求应用程序的根 URL 时，将使用 controller 和 action 片段变量的默认值，这将导致 MVC 在 Home 控制器上调用 Index 操作方法，如图 15-5 所示。

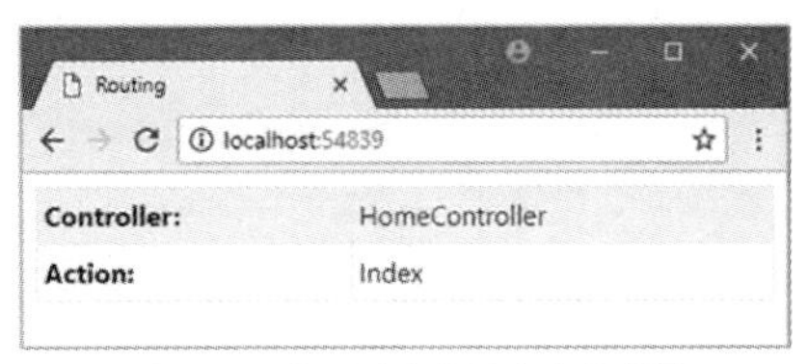

图 15-5 使用默认值扩大路由范围

15.5 使用静态 URL 片段

并非 URL 格式中的所有片段都必须是变量。还可以创建具有静态片段的模式。假设应用程序需要匹配以 Public 为前缀的 URL，如下所示：

```
****://mydomain.***/Public/Home/Index
```

这可以通过使用代码清单 15-13 所示的 URL 模式来完成。

代码清单 15-13 在 UrlsAndRoutes 文件夹下的 Startup.cs 文件中使用静态片段

```
using System;
using System.Collections.Generic;
using System.Linq;
using System.Threading.Tasks;
using Microsoft.AspNetCore.Builder;
using Microsoft.AspNetCore.Hosting;
using Microsoft.AspNetCore.Http;
using Microsoft.Extensions.DependencyInjection;

namespace UrlsAndRoutes {
    public class Startup {

        public void ConfigureServices(IServiceCollection services) {
            services.AddMvc();
        }

        public void Configure(IApplicationBuilder app, IHostingEnvironment env) {
            app.UseDeveloperExceptionPage();
            app.UseStatusCodePages();
            app.UseStaticFiles();
            app.UseMvc(routes => {
                routes.MapRoute(
                    name: "default",
                    template: "{controller=Home}/{action=Index}");

                routes.MapRoute(name: "",
                    template: "Public/{controller=Home}/{action=Index}");
            });
        }
    }
}
```

这种新模式将仅匹配包含 3 个片段的 URL，第一个片段必须为 Public。其他两个片段可以包含任何值，并将用于 controller 和 action 变量。如果省略最后两个片段，将使用默认值。

还可以创建包含静态和可变元素的片段的 URL 模式，如代码清单 15-14 所示。

代码清单 15-14 UrlsAndRoutes 文件夹下的 Startup.cs 文件中的混合片段

```
using System;
using System.Collections.Generic;
```

```
using System.Linq;
using System.Threading.Tasks;
using Microsoft.AspNetCore.Builder;
using Microsoft.AspNetCore.Hosting;
using Microsoft.AspNetCore.Http;
using Microsoft.Extensions.DependencyInjection;

namespace UrlsAndRoutes {
    public class Startup {

        public void ConfigureServices(IServiceCollection services) {
            services.AddMvc();
        }

        public void Configure(IApplicationBuilder app, IHostingEnvironment env) {
            app.UseDeveloperExceptionPage();
            app.UseStatusCodePages();
            app.UseStaticFiles();
            app.UseMvc(routes => {
                routes.MapRoute("", "X{controller}/{action}");

                routes.MapRoute(
                    name: "default",
                    template: "{controller=Home}/{action=Index}");

                routes.MapRoute(name: "",
                    template: "Public/{controller=Home}/{action=Index}");
            });
        }
    }
}
```

这个路由中的模式能匹配任何两片段的 URL，其中第一个片段以字母 X 开头。controller 的值取自第一个片段，不包括 X。action 的值取自第二个片段。如果启动应用程序并导航到/XHome/Index，路由效果如图 15-6 所示。

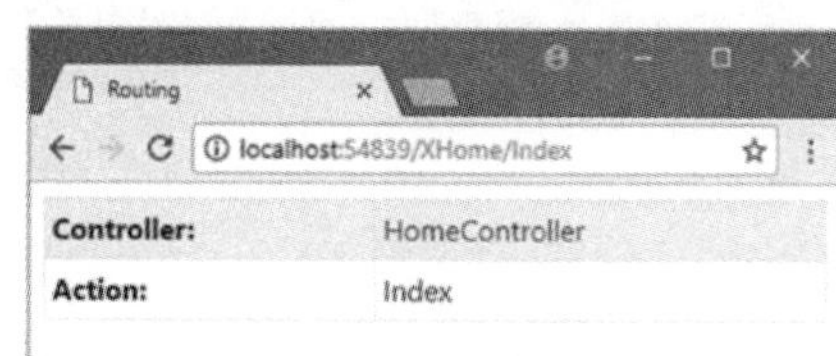

图 15-6 在单个片段中混合静态和可变元素的路由效果

路由排序

代码清单 15-14 定义了一个新的路由并将其放在所有其他路由之前。这样做是为了让路由按照定义的顺序应用。MapRoute 方法将路由添加到路由配置的末尾，这意味着路由通常会按定义的顺序应用。之所以说“通常”，是因为有一些方法还可以在特定位置插入路由。不建议使用这些方法，因为按照定义的顺序应用路由能够使应用程序的路由更容易理解。

路由系统尝试对传入的 URL 与首先定义的路由的 URL 模式进行匹配，只有在没有匹配时才进入下一个路由。这些路由按顺序尝试，直到找到匹配的路由或路由被用完。因此，必须首先定义最具体的路由。代码清单 15-14 添加的路由比后面的路由更具体。假设颠倒路由的顺序，如下所示：

```
...
routes.MapRoute("MyRoute", "{controller=Home}/{action=Index}");
routes.MapRoute("", "X{controller}/{action}");
...
```

然后，第一个路由能够匹配任何具有零个、一个或两个片段的 URL，因而将始终使用。更具体的路由，比如上述代码中的第二个路由，则永远不会使用。新路由排除了 URL 的前导 X，但旧路由不会这样做。因此，像下面这样的 URL：

```
http://mydomain.com/XHome/Index
```

> 将被定位到名为 XHome 的控制器，我们假设应用程序中有 XHomeController 类，并且有名为 Index 的操作方法。

你可以组合静态 URL 片段和默认值以创建特定 URL 的别名。使用的 URL 模式在部署应用程序时与用户形成了约定，如果随后重构应用程序，则需要保留以前的 URL 模式，以便用户创建的任何 URL 收藏夹、宏或脚本能够继续工作。

想象一下，以前有一个名为 Shop 的控制器，现在已被 Home 控制器取代了。代码清单 15-15 显示了如何创建保留旧 URL 模式的路由。

代码清单 15-15　UrlsAndRoutes 文件夹下的 Startup.cs 文件中的片段和默认值

```
using System;
using System.Collections.Generic;
using System.Linq;
using System.Threading.Tasks;
using Microsoft.AspNetCore.Builder;
using Microsoft.AspNetCore.Hosting;
using Microsoft.AspNetCore.Http;
using Microsoft.Extensions.DependencyInjection;

namespace UrlsAndRoutes {
    public class Startup {

        public void ConfigureServices(IServiceCollection services) {
            services.AddMvc();
        }

        public void Configure(IApplicationBuilder app, IHostingEnvironment env) {
            app.UseDeveloperExceptionPage();
            app.UseStatusCodePages();
            app.UseStaticFiles();
            app.UseMvc(routes => {
                routes.MapRoute(
                    name: "ShopSchema",
                    template: "Shop/{action}",
                    defaults: new { controller = "Home" });

                routes.MapRoute("", "X{controller}/{action}");

                routes.MapRoute(
                    name: "default",
                    template: "{controller=Home}/{action=Index}");

                routes.MapRoute(name: "",
                    template: "Public/{controller=Home}/{action=Index}");
            });
        }
    }
}
```

以上加粗显示的路由能匹配任何两个片段的 URL，其中第一个片段是 Shop。action 的值取自第二个 URL 片段。URL 模式不包含 controller 的变量片段，因此使用默认值。defaults 参数提供了 controller 的值，因为没有片段能够将值作为 URL 模式的一部分应用于该 URL 模式。

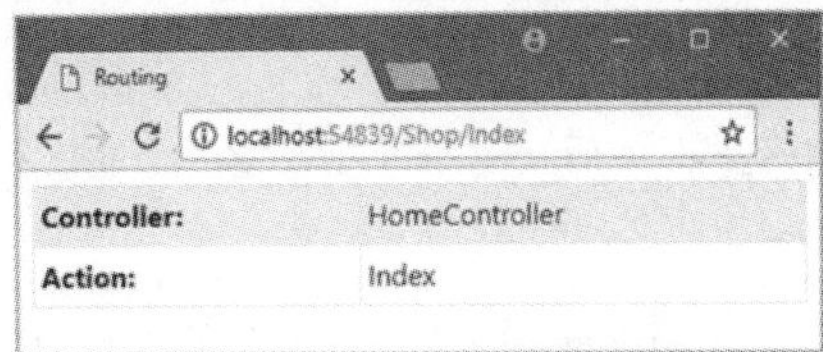

图 15-7　创建别名以保留 URL 模式

其结果是，Shop 控制器上的操作请求被转换为对 Home 控制器的请求。可以通过启动应用程序并导航到/Shop/Index 来查看路由效果。如图 15-7 所示，新路由使 MVC 在 Home 控制器中定位到 Index 操作方法。

可以更进一步，为已经重构并且控制器中已不存在的操作方法创建别名。为此，创建一个静态 URL，并提供 controller 和 action 的默认值，如代码清单 15-16 所示。

代码清单 15-16 在 UrlsAndRoutes 文件夹下的 Startup.cs 文件中为 controller 和 action 使用别名

```
using System;
using System.Collections.Generic;
using System.Linq;
using System.Threading.Tasks;
using Microsoft.AspNetCore.Builder;
using Microsoft.AspNetCore.Hosting;
using Microsoft.AspNetCore.Http;
using Microsoft.Extensions.DependencyInjection;

namespace UrlsAndRoutes {
    public class Startup {

        public void ConfigureServices(IServiceCollection services) {
            services.AddMvc();
        }

        public void Configure(IApplicationBuilder app, IHostingEnvironment env) {
            app.UseDeveloperExceptionPage();
            app.UseStatusCodePages();
            app.UseStaticFiles();
            app.UseMvc(routes => {

                routes.MapRoute(
                    name: "ShopSchema2",
                    template: "Shop/OldAction",
                    defaults: new { controller = "Home", action = "Index" });
                routes.MapRoute(
                    name: "ShopSchema",
                    template: "Shop/{action}",
                    defaults: new { controller = "Home" });

                routes.MapRoute("", "X{controller}/{action}");

                routes.MapRoute(
                    name: "default",
                    template: "{controller=Home}/{action=Index}");

                routes.MapRoute(name: "",
                    template: "Public/{controller=Home}/{action=Index}");
            });
        }
    }
}
```

请注意，首先要定义新路由，因为它比后面的路由更具体。例如，如果对 Shop/OldAction 的请求由下一个定义的路由处理，并且有一个带有 OldAction 操作方法的控制器，则可能无法得到想要的结果。

15.6 定义自定义片段变量

controller 和 action 片段变量在 MVC 应用程序中具有特殊含义，相当于为请求提供服务的控制器和操作方法。这些只是内置的片段变量，也可以定义自定义片段变量，如代码清单 15-17 所示（这里已经删除了已有路由，以便重新开始）。

代码清单 15-17 在 UrlsAndRoutes 文件夹下的 Startup.cs 文件中定义其他变量

```
using System;
using System.Collections.Generic;
using System.Linq;
using System.Threading.Tasks;
using Microsoft.AspNetCore.Builder;
using Microsoft.AspNetCore.Hosting;
using Microsoft.AspNetCore.Http;
using Microsoft.Extensions.DependencyInjection;

namespace UrlsAndRoutes {
    public class Startup {

        public void ConfigureServices(IServiceCollection services) {
            services.AddMvc();
        }

        public void Configure(IApplicationBuilder app, IHostingEnvironment env) {
            app.UseDeveloperExceptionPage();
            app.UseStatusCodePages();
            app.UseStaticFiles();
            app.UseMvc(routes => {
                routes.MapRoute(name: "MyRoute",
                    template: "{controller=Home}/{action=Index}/{id=DefaultId}");
            });
        }
    }
}
```

以上代码中的 URL 模式定义了标准的 controller 和 action 变量，以及名为 id 的自定义变量。路由将匹配 0～3 个片段的 URL。第 3 个片段的内容将分配给 id 变量，如果没有第 3 个片段，就使用默认值。

警 告

某些名称是保留的，不适合作为自定义片段变量的名称，它们是 controller、action、area 和 page，前两个名称的含义是显而易见的。area、page 由 Razor 页面的使用，本书将在下一章解释。

Controller 类是控制器的基础，它定义了 RouteData 属性，该属性返回一个 Microsoft.AspNetCore.Routing.RouteData 对象，该对象提供有关路由系统的详细信息以及当前请求的路由方式。在控制器中，可以使用 RouteData.Values 属性访问操作方法中的任何片段变量，该属性会返回一个包含片段变量的字典。为了演示，这里在 Home 控制器中添加一个名为 CustomVariable 的操作方法，如代码清单 15-18 所示。

代码清单 15-18 在 Controllers 文件夹下的 HomeController.cs 文件中访问片段变量

```
using Microsoft.AspNetCore.Mvc;
using UrlsAndRoutes.Models;

namespace UrlsAndRoutes.Controllers {

    public class HomeController : Controller {

        public ViewResult Index() => View("Result",
            new Result {
                Controller = nameof(HomeController),
                Action = nameof(Index)
            });
```

```
        public ViewResult CustomVariable() {
            Result r = new Result {
                Controller = nameof(HomeController),
                Action = nameof(CustomVariable),
            };
            r.Data["Id"] = RouteData.Values["id"];
            return View("Result", r);
        }
    }
}
```

操作方法 CustomVariable 使用 RouteData.Values 属性获取路由 URL 模式中的自定义片段变量 id 的值，RouteData.Values 属性则返回路由系统生成的变量字典。自定义片段变量将被添加到视图模型对象中，可以通过运行应用程序并请求以下 URL 来查看：

```
/Home/CustomVariable/Hello
```

路由模板对以上 URL 中的第 3 个片段与 id 变量的值进行匹配，产生图 15-8 所示的结果。

代码清单 15-17 中的 URL 模式定义了 id 变量的默认值，这意味着路由也可以匹配具有两个片段的 URL。可以通过请求如下 URL 来查看如何使用默认值：

```
/Home/CustomVariable
```

路由系统使用自定义片段变量的默认值，如图 15-9 所示。

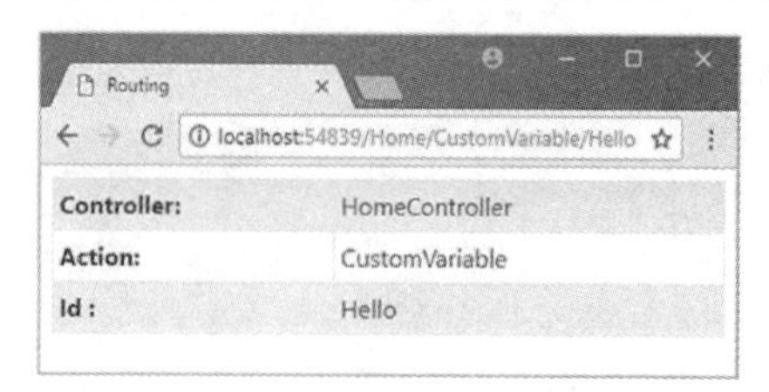

图 15-8 显示自定义片段变量的值

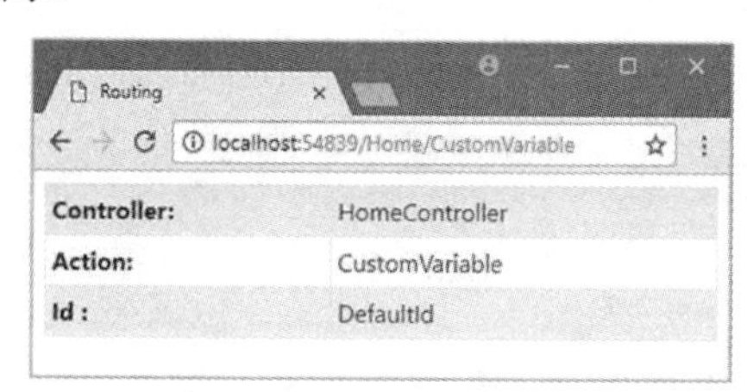

图 15-9 自定义片段变量的默认值

15.6.1 使用自定义片段变量作为操作方法的参数

使用 RouteData.Values 集合只是访问自定义片段变量的一种方式，另一种方式可能更优雅。如果操作方法定义了名称与 URL 模式变量匹配的参数，MVC 将把自动从 URL 获取的值作为参数传递给操作方法。

在代码清单 15-17 所示的路由中定义的自定义片段变量名为 id。这里可以修改 Home 控制器中的 CustomVariable 操作方法，使其具有相同名称的参数，如代码清单 15-19 所示。

代码清单 15-19 在 Controllers 文件夹下的 HomeController.cs 文件中添加 action 参数

```
using Microsoft.AspNetCore.Mvc;
using UrlsAndRoutes.Models;

namespace UrlsAndRoutes.Controllers {

    public class HomeController : Controller {

        public ViewResult Index() => View("Result",
            new Result {
                Controller = nameof(HomeController),
                Action = nameof(Index)
            });

        public ViewResult CustomVariable(string id) {
            Result r = new Result {
                Controller = nameof(HomeController),
                Action = nameof(CustomVariable),
```

```
            };
            r.Data["Id"] = id;
            return View("Result", r);
        }
    }
}
```

当路由系统对 URL 与代码清单 15-17 中定义的路由进行匹配时，URL 中第 3 个片段的值将被分配给自定义片段变量 id。MVC 对片段变量的列表与操作方法的参数列表进行比较，如果名称匹配，就将值从 URL 传递给操作方法。

id 参数的类型是字符串，但 MVC 会尝试将 URL 值转换为使用的任何参数类型。如果操作方法将 id 参数声明为 int 或 DateTime，那么 MVC 将从 URL 中接收值并转换为相应类型。这是一个优雅而实用的功能，无须自己处理转换。可以通过启动应用程序并请求/Home/CustomVariable/Hello 来查看操作方法参数的效果，从而生成图 15-10 所示的结果。如果省略第 3 个片段，将为操作方法提供默认片段值。

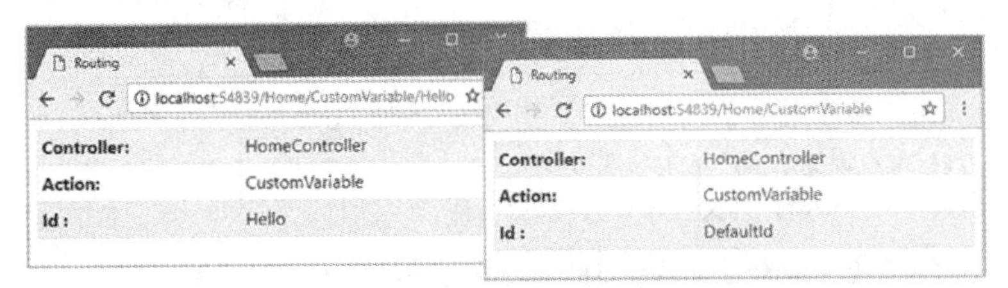

图 15-10 使用操作方法参数访问片段变量的结果

注 意

MVC 使用模型绑定功能将 URL 中包含的值转换为.NET 类型，模型绑定还可以处理比此例更复杂的情况。第 26 章将介绍模型绑定。

15.6.2 定义可选的 URL 片段

可选的 URL 片段是用户不需要指定的 URL 片段，而且没有指定默认值。可选片段由片段名称后面的问号（?字符）表示，如代码清单 15-20 所示。

代码清单 15-20 在 UrlsAndRoutes 文件夹下的 Startup.cs 文件中指定可选片段

```
using System;
using System.Collections.Generic;
using System.Linq;
using System.Threading.Tasks;
using Microsoft.AspNetCore.Builder;
using Microsoft.AspNetCore.Hosting;
using Microsoft.AspNetCore.Http;
using Microsoft.Extensions.DependencyInjection;

namespace UrlsAndRoutes {
    public class Startup {

        public void ConfigureServices(IServiceCollection services) {
            services.AddMvc();
        }

        public void Configure(IApplicationBuilder app, IHostingEnvironment env) {
            app.UseDeveloperExceptionPage();
            app.UseStatusCodePages();
            app.UseStaticFiles();
            app.UseMvc(routes => {
```

```
            routes.MapRoute(name: "MyRoute",
                template: "{controller=Home}/{action=Index}/{id?}");
            });
        }
    }
}
```

无论是否已提供 id 片段，以上路由都将匹配 URL。表 15-5 显示了可选片段如何适用于不同的 URL。

表 15-5　使用可选片段变量匹配 URL

片段编号	示例 URL	匹配结果
0	/	controller = Home, action = Index
1	/Customer	controller = Customer, action = Index
2	/Customer/List	controller = Customer, action = List
3	/Customer/List/All	controller = Customer, action = List id = All
4	/Customer/List/All/Delete	没有匹配，片段太多

从表 15-5 可以看出，只有当传入的 URL 中存在相应的片段时，才会将 id 变量添加到变量集合中。如果需要知道用户是否为片段变量提供了值，这个功能将非常有用。如果没有为可选片段变量提供值，相应参数的值将为 null。在代码清单 15-21 中，因为没有为 id 片段变量提供值，所以更新了 Home 控制器以进行响应。

代码清单 15-21　在 Controllers 文件夹下的 HomeController.cs 文件中检查片段

```
using Microsoft.AspNetCore.Mvc;
using UrlsAndRoutes.Models;

namespace UrlsAndRoutes.Controllers {

    public class HomeController : Controller {

        public ViewResult Index() => View("Result",
            new Result {
                Controller = nameof(HomeController),
                Action = nameof(Index)
            });

        public ViewResult CustomVariable(string id) {
            Result r = new Result {
                Controller = nameof(HomeController),
                Action = nameof(CustomVariable),
            };
            r.Data["Id"] = id ?? "<no value>";
            return View("Result", r);
        }
    }
}
```

图 15-11 显示了启动应用程序并导航到/Home/CustomVariable URL 的结果，这个 URL 不包含 id 片段变量的值。

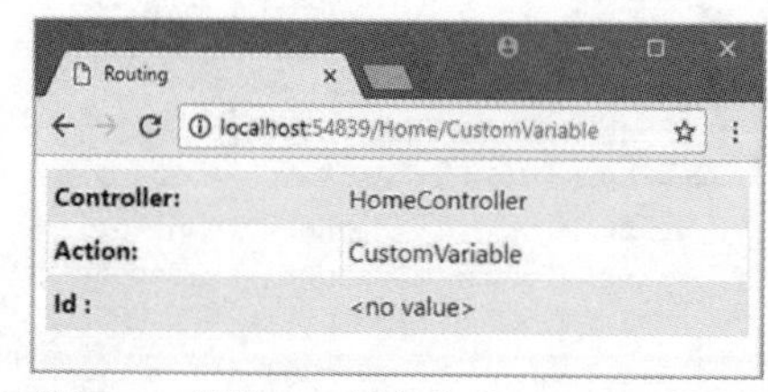

图 15-11　检测 URL 何时不包含可选片段变量的值

理解默认的路由配置

将添加 MVC 到 Startup 类时，可以使用 UseMvcWithDefaultRoute 方法执行此操作。这只是设置最常见路由配置的便捷方法，相当于以下代码：

```
...
app.UseMvc(routes => {
    routes.MapRoute(
        name: "default",
```

```
        template: "{controller=Home}/{action=Index}/{id?}");
});
...
```

默认配置匹配按名称定位控制器类和操作方法的 URL，并带有可选的 id 片段。如果控制器或操作片段丢失，则默认值分别用于定位 Home 控制器和 Index 操作方法。

15.6.3　定义可变长度路由

更改 URL 模式的默认保守性的另一种方法是接收可变数量的 URL 片段，这允许你在单个路由中匹配任意长度的 URL。可以通过将其中一个片段变量指定为 catchall 来定义对变量片段的支持，可通过在其前面添加星号（*字符）来完成，如代码清单 15-22 所示。

代码清单 15-22　在 UrlsAndRoutes 文件夹下的 Startup.cs 文件中指定 catchall 变量

```
using Microsoft.AspNetCore.Builder;
using Microsoft.Extensions.DependencyInjection;

namespace UrlsAndRoutes {

    public class Startup {
        public void ConfigureServices(IServiceCollection services) {
            services.AddMvc();
        }

        public void Configure(IApplicationBuilder app) {
            app.UseStatusCodePages();
            app.UseDeveloperExceptionPage();
            app.UseStaticFiles();
            app.UseMvc(routes => {
                routes.MapRoute(name: "MyRoute",
                template: "{controller=Home}/{action=Index}/{id?}/{*catchall}");
            });
        }
    }
}
```

这里通过扩展前一个例子中的路由来添加 catchall 片段变量，catchall 是一个十分富有想象力的名称。这个路由现在将匹配任何 URL，而无论其包含多少片段以及这些片段的值如何。前 3 个片段分别用于设置 controller、action 和 id 变量的值。如果 URL 包含其他片段，则把它们都分配给 catchall 变量，如表 15-6 所示。

表 15-6　使用 catchall 变量匹配 URL

片段编号	示例 URL	匹配结果
0	/	controller = Home, action = Index
1	/Customer	controller = Customer, action = Index
2	/Customer/List	controller = Customer, action = List
3	/Customer/List/All	controller = Customer, action = List id = All
4	/Customer/List/All/Delete	controller = Customer, action = List id = All catchall = Delete
5	/Customer/List/All/Delete/Perm	controller = Customer, action = List id = All catchall = Delete/Perm

代码清单 15-23 更新了 Customer 控制器，以便 List 操作通过模型对象将 catchall 变量的值传递给视图。

代码清单 15-23　更新 Controllers 文件夹下的 CustomerController.cs 文件中的操作方法

```
using Microsoft.AspNetCore.Mvc;
using UrlsAndRoutes.Models;
```

```
namespace UrlsAndRoutes.Controllers {
    public class CustomerController : Controller {

        public ViewResult Index() => View("Result",
            new Result {
                Controller = nameof(CustomerController),
                Action = nameof(Index)
            });
        public ViewResult List(string id) {
            Result r = new Result {
                Controller = nameof(HomeController),
                Action = nameof(List),
            };
            r.Data["Id"] = id ?? "<no value>";
            r.Data["catchall"] = RouteData.Values["catchall"];
            return View("Result", r);
        }
    }
}
```

要测试 catchall 片段，请运行应用程序并请求以下 URL：

```
/Customer/List/Hello/1/2/3
```

以上路由中的 URL 模式将匹配的片段数量没有上限。图 15-12 显示了 catchall 片段的效果。请注意，catchall 捕获的片段以 segment/segment/segment 的形式显示，并且需要处理字符串以分解各个片段。

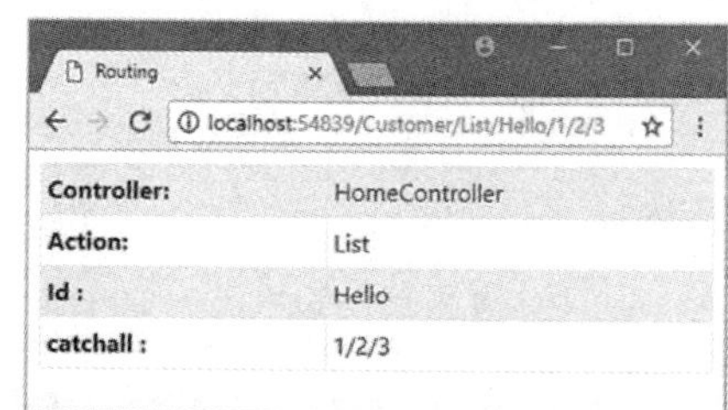

图 15-12 使用 catchall 片段

15.7 约束路由

URL 模式在匹配 URL 中的片段数时是保守的，当它们与片段的内容匹配时是自由的。前几节已经解释了控制保守程度的不同技术——使用默认值、可选变量等使路由匹配更多或更少的片段。

现在来看看如何控制匹配 URL 片段内容的自由度，也就是如何限制路由将匹配的 URL 集合。代码清单 15-24 演示了如何使用简单约束来限制路由匹配的 URL。

代码清单 15-24 在 UrlsAndRoutes 文件夹下的 Startup.cs 文件中约束路由

```
using System;
using System.Collections.Generic;
using System.Linq;
using System.Threading.Tasks;
using Microsoft.AspNetCore.Builder;
using Microsoft.AspNetCore.Hosting;
using Microsoft.AspNetCore.Http;
using Microsoft.Extensions.DependencyInjection;

namespace UrlsAndRoutes {
    public class Startup {

        public void ConfigureServices(IServiceCollection services) {
            services.AddMvc();
        }

        public void Configure(IApplicationBuilder app, IHostingEnvironment env) {
            app.UseDeveloperExceptionPage();
            app.UseStatusCodePages();
```

```
            app.UseStaticFiles();
            app.UseMvc(routes => {
                routes.MapRoute(name: "MyRoute",
                template: "{controller=Home}/{action=Index}/{id:int?}");
            });
        }
    }
}
```

使用冒号（：字符）将约束与片段变量名分开。以上代码中的约束是 int，已应用于 id 片段。这是一个内联约束，且被定义为 URL 模式的一部分，应用于单个片段：

```
...
template: "{controller}/{action}/{id:int?}",
...
```

int 约束仅允许 URL 模式匹配值可以解析为整数值的片段。id 片段是可选的，因此路由将匹配省略了 id 的片段，但如果片段存在，那么 id 必须是整数值，如表 15-7 所示。

表 15-7 使用约束匹配 URL

示例 URL	匹配结果
/	controller = Home, action = Index, id = null
/Home/CustomVariable/Hello	没有匹配，id 片段无法被解析为 int 值
/Home/CustomVariable/1	controller = Home, action = CustomVariable, id = 1
/Home/CustomVariable/1/2	没有匹配，片段太多

在定义路由时，也可以使用 MapRoute 方法的 constraints 参数在 URL 模式之外指定约束。如果希望将 URL 模式与其约束分开，或者更喜欢早期版本的 MVC（不支持内联约束）所使用的路由样式，那么这种方法很有用。代码清单 15-25 显示了相同的 id 片段变量的整数约束，可使用单独的约束来表示。使用这种格式时，默认值也在外部表示。

代码清单 15-25 在 UrlsAndRoutes 文件夹下的 Startup.cs 文件中表达约束

```
using System;
using System.Collections.Generic;
using System.Linq;
using System.Threading.Tasks;
using Microsoft.AspNetCore.Builder;
using Microsoft.AspNetCore.Hosting;
using Microsoft.AspNetCore.Http;
using Microsoft.Extensions.DependencyInjection;
using Microsoft.AspNetCore.Routing.Constraints;

namespace UrlsAndRoutes {
    public class Startup {

        public void ConfigureServices(IServiceCollection services) {
            services.AddMvc();
        }

        public void Configure(IApplicationBuilder app, IHostingEnvironment env) {
            app.UseDeveloperExceptionPage();
            app.UseStatusCodePages();
            app.UseStaticFiles();
            app.UseMvc(routes => {
                routes.MapRoute(name: "MyRoute",
                template: "{controller}/{action}/{id?}",
                defaults: new { controller = "Home", action = "Index" },
                constraints: new { id = new IntRouteConstraint() });
```

```
            });
        }
    }
}
```

MapRoute 方法的 constraints 参数是使用匿名类型定义的，匿名类型的属性名称对应于受约束的片段变量。Microsoft.AspNetCore.Routing.Constraints 命名空间包含一组可用于定义各个约束的类。在代码清单 15-25 中，constraints 参数被配置为对 id 片段使用 IntRouteConstraint 对象，以创建与代码清单 15-24 所示的内联约束相同的效果。

表 15-8 描述了 Microsoft.AspNetCore.Routing 命名空间中完整的约束类。Constraints 命名空间及等效的内联约束可以应用于 URL 模式中的单个片段，后面的章节将描述其中一些情况。

提 示

可以使用 MVC 提供的一组特性（例如 HttpGet 和 HttpPost 特性）对使用特定 HTTP 谓词（例如 GET 或 POST）请求的操作方法进行限制。有关使用这些特性处理控制器中表单的详细信息，请参阅第 7 章，有关此类可用特性的完整列表，请参阅第 20 章。

表 15-8 片段级路由约束

内联约束	描述	类名
alpha	匹配字母字符，无论大小写（A～Z、a～z）	AlphaRouteConstraint()
bool	匹配可以解析为 bool 类型的值	BoolRouteConstraint()
datetime	匹配可以解析为 DateTime 类型的值	DateTimeRouteConstraint()
decimal	匹配可以解析为 decimal 类型的值	DecimalRouteConstraint()
double	匹配可以解析为 double 类型的值	DoubleRouteConstraint()
float	匹配可以解析为 float 类型的值	FloatRouteConstraint()
guid	匹配 GUID 值	GuidRouteConstraint()
int	匹配可以解析为 int 类型的值	IntRouteConstraint()
length(len)/length(min,max)	匹配具有指定字符数量的值，或者匹配字符数量介于最小和最大字符数量之间的值	LengthRouteConstraint(len) LengthRouteConstraint(min, max)
long	匹配可以解析为 long 类型的值	LongRouteConstraint()
maxlength(len)	匹配字符数量不超过 len 的字符串	MaxLengthRouteConstraint(len)
max(val)	配置一个 int 值，该值小于 val	MaxRouteConstraint(val)
minlength(len)	匹配字符数量至少为 len 的字符串	MinLengthRouteConstraint(len)
min(val)	匹配一个 int 值，该值大于 val	MinRouteConstraint(val)
range(min,max)	匹配一个 int 值，该值介于 min 和 max 之间	RangeRouteConstraint(min, max)
regex(expr)	匹配正则表达式	RegexRouteConstraint(expr)

15.7.1 使用正则表达式约束路由

正则表达式提供了灵活性最大的约束，可使用正则表达式匹配片段。代码清单 15-26 对 controller 片段进行了约束以限制匹配的 URL 范围。

代码清单 15-26 在 UrlsAndRoutes 文件夹下的 Startup.cs 文件中使用正则表达式

```
using System;
using System.Collections.Generic;
using System.Linq;
using System.Threading.Tasks;
using Microsoft.AspNetCore.Builder;
using Microsoft.AspNetCore.Hosting;
using Microsoft.AspNetCore.Http;
```

```
using Microsoft.Extensions.DependencyInjection;
using Microsoft.AspNetCore.Routing.Constraints;

namespace UrlsAndRoutes {
    public class Startup {

        public void ConfigureServices(IServiceCollection services) {
            services.AddMvc();
        }

        public void Configure(IApplicationBuilder app, IHostingEnvironment env) {
            app.UseDeveloperExceptionPage();
            app.UseStatusCodePages();
            app.UseStaticFiles();
            app.UseMvc(routes => {
                routes.MapRoute(name: "MyRoute",
                template: "{controller:regex(^H.*)=Home}/{action=Index}/{id?}");
            });
        }
    }
}
```

以上代码使用约束来限制路由，使路由只匹配 controller 片段以字母 H 开头的 URL。

注 意

在检查约束之前会应用默认值。因此，举例来说，如果请求 URL /，就应用 controller 的默认值 Home。然后检查约束，并且由于 controller 的值以 H 开头，因此默认 URL 将匹配路由。

正则表达式可以约束路由，以便只有 URL 片段的特定值才能匹配。这是使用竖线（|）字符完成的，如代码清单 15-27 所示（这里将 URL 模式拆分成两行以适应纸面宽度，在实际项目中不用这么做）。

代码清单 15-27 在 UrlsAndRoutes 文件夹下的 Startup.cs 文件中约束路由

```
using System;
using System.Collections.Generic;
using System.Linq;
using System.Threading.Tasks;
using Microsoft.AspNetCore.Builder;
using Microsoft.AspNetCore.Hosting;
using Microsoft.AspNetCore.Http;
using Microsoft.Extensions.DependencyInjection;
using Microsoft.AspNetCore.Routing.Constraints;

namespace UrlsAndRoutes {
    public class Startup {

        public void ConfigureServices(IServiceCollection services) {
            services.AddMvc();
        }

        public void Configure(IApplicationBuilder app, IHostingEnvironment env) {
            app.UseDeveloperExceptionPage();
            app.UseStatusCodePages();
            app.UseStaticFiles();
            app.UseMvc(routes => {
                routes.MapRoute(name: "MyRoute",
                  template: "{controller:regex(^H.*)=Home}/"
                    + "{action:regex(^Index$|^About$)=Index}/{id?}");
            });
        }
```

```
    }
}
```

以上约束允许路由仅匹配 action 片段的值为 Index 或 About 的 URL。由于约束将被一起应用，因此需要组合 action 变量与 controller 变量的限制。这意味着仅当 controller 变量以字母 H 开头且 action 变量为 Index 或 About 时，代码清单 15-27 中的路由才能匹配 URL。

15.7.2　使用类型和值约束

大多数约束用于限制路由，因此它们仅匹配具有可转换为指定类型或具有特定格式的片段的 URL。本节开头使用的 int 约束是一个很好的例子：只有当受约束的片段的值可以解析为.NET 的 int 值时，才会匹配路由。代码清单 15-28 演示了 range 约束的用法，对路由进行限制，只有当一个片段值可以转换为 int 类型并且介于指定值之间时才匹配 URL。

代码清单 15-28　在 UrlsAndRoutes 文件夹下的 Startup.cs 文件中基于类型和值进行约束

```
using System;
using System.Collections.Generic;
using System.Linq;
using System.Threading.Tasks;
using Microsoft.AspNetCore.Builder;
using Microsoft.AspNetCore.Hosting;
using Microsoft.AspNetCore.Http;
using Microsoft.Extensions.DependencyInjection;
using Microsoft.AspNetCore.Routing.Constraints;

namespace UrlsAndRoutes {
    public class Startup {

        public void ConfigureServices(IServiceCollection services) {
            services.AddMvc();
        }

        public void Configure(IApplicationBuilder app, IHostingEnvironment env) {
            app.UseDeveloperExceptionPage();
            app.UseStatusCodePages();
            app.UseStaticFiles();
            app.UseMvc(routes => {
                routes.MapRoute(name: "MyRoute",
                template: "{controller=Home}/{action=Index}/{id:range(10,20)?}");
            });
        }
    }
}
```

此例中的约束已应用于可选的 id 片段。如果请求 URL 少于 3 个片段，就忽略约束。如果存在 id 片段，那么仅当片段值可以转换为 int 类型且介于 10～20 时，路径才会匹配 URL。range 约束包含边界值，也就是说，10 和 20 也被认为处在约束范围内。

15.7.3　组合约束

如果需要将多个约束应用于单个片段，可以将它们链接在一起，约束之间用冒号分隔，如代码清单 15-29 所示。

代码清单 15-29　在 UrlsAndRoutes 文件夹下的 Startup.cs 文件中组合内联约束

```
using System;
using System.Collections.Generic;
using System.Linq;
```

```
using System.Threading.Tasks;
using Microsoft.AspNetCore.Builder;
using Microsoft.AspNetCore.Hosting;
using Microsoft.AspNetCore.Http;
using Microsoft.Extensions.DependencyInjection;
using Microsoft.AspNetCore.Routing.Constraints;

namespace UrlsAndRoutes {
    public class Startup {

        public void ConfigureServices(IServiceCollection services) {
            services.AddMvc();
        }
        public void Configure(IApplicationBuilder app, IHostingEnvironment env) {
            app.UseDeveloperExceptionPage();
            app.UseStatusCodePages();
            app.UseStaticFiles();
            app.UseMvc(routes => {
                routes.MapRoute(name: "MyRoute",
                    template: "{controller=Home}/{action=Index}"
                        + "/{id:alpha:minlength(6)?}");
            });
        }
    }
}
```

在以上代码中，已将 alpha 和 minlength 约束应用于 id 片段。问号表示这是一个在所有约束之后应用的可选片段。这些约束的组合效果是：路由将仅匹配省略了 id 片段的 URL（因为 id 片段是可选的），或者当 id 片段存在且包含至少 6 个字母字符时才匹配。

如果不使用内联约束，则必须使用 Microsoft.AspNetCore.Routing.CompositeRouteConstraint 类，以允许多个约束与匿名类型对象中的单个属性相关联。代码清单 15-30 显示了在代码清单 15-29 中使用的组合约束。

代码清单 15-30 在 UrlsAndRoutes 文件夹下的 Startup.cs 文件中组合单独的约束

```
using System;
using System.Collections.Generic;
using System.Linq;
using System.Threading.Tasks;
using Microsoft.AspNetCore.Builder;
using Microsoft.AspNetCore.Hosting;
using Microsoft.AspNetCore.Http;
using Microsoft.Extensions.DependencyInjection;
using Microsoft.AspNetCore.Routing.Constraints;
using Microsoft.AspNetCore.Routing;

namespace UrlsAndRoutes {
    public class Startup {

        public void ConfigureServices(IServiceCollection services) {
            services.AddMvc();
        }

        public void Configure(IApplicationBuilder app, IHostingEnvironment env) {
            app.UseDeveloperExceptionPage();
            app.UseStatusCodePages();
            app.UseStaticFiles();
            app.UseMvc(routes => {
                routes.MapRoute(name: "MyRoute",
                    template: "{controller}/{action}/{id?}",
                    defaults: new { controller = "Home", action = "Index" },
```

```
                    constraints: new {
                        id = new CompositeRouteConstraint(
                            new IRouteConstraint[] {
                                new AlphaRouteConstraint(),
                                new MinLengthRouteConstraint(6)
                            })
                    });
            });
        }
    }
}
```

CompositeRouteConstraint 类的构造函数接收实现了 IRouteConstraint 对象的枚举，IRouteConstraint 对象是定义路由约束的接口。仅当满足所有约束时，路由系统才允许路由匹配 URL。

15.7.4 定义自定义约束

如果标准约束无法满足需求，那么可以通过实现 Microsoft.AspNetCore.Routing 命名空间中定义的 IRouteConstraint 接口来定义自定义约束。为了演示此功能，在示例项目中添加 Infrastructure 文件夹，并创建名为 WeekDayConstraint.cs 的类文件，内容如代码清单 15-31 所示。

代码清单 15-31 Infrastructure 文件夹下的 WeekDayConstraint.cs 文件的内容

```
using Microsoft.AspNetCore.Http;
using Microsoft.AspNetCore.Routing;
using System.Linq;

namespace UrlsAndRoutes.Infrastructure {
    public class WeekDayConstraint : IRouteConstraint {
        private static string[] Days = new[] { "mon", "tue", "wed", "thu",
                                                "fri", "sat", "sun" };

        public bool Match(HttpContext httpContext, IRouter route,
            string routeKey, RouteValueDictionary values,
            RouteDirection routeDirection) {

            return Days.Contains(values[routeKey]?.ToString().ToLowerInvariant());
        }
    }
}
```

IRouteConstraint 接口定义了 Match 方法，可调用该方法以允许约束来决定请求是否应该与路由匹配。Match 方法的参数提供对多种来源的请求的访问，包括客户端、路由、受约束片段的名称、从 URL 提取的片段变量以及请求是否检查传入或传出的 URL（第 16 章将解释传出的 URL）。

在该例中，使用 routeKey 参数从 values 参数获取已应用约束的片段变量的值，将它们转换为小写字符串，并查看是否匹配在静态字段 Days 中定义的一周中的那几天。代码清单 15-32 使用单独的方法将新约束应用于示例路由。

代码清单 15-32 在 UrlsAndRoutes 文件夹下的 Startup.cs 文件中应用自定义约束

```
using System;
using System.Collections.Generic;
using System.Linq;
using System.Threading.Tasks;
using Microsoft.AspNetCore.Builder;
using Microsoft.AspNetCore.Hosting;
using Microsoft.AspNetCore.Http;
using Microsoft.Extensions.DependencyInjection;
using Microsoft.AspNetCore.Routing.Constraints;
```

```
using Microsoft.AspNetCore.Routing;
using UrlsAndRoutes.Infrastructure;

namespace UrlsAndRoutes {
    public class Startup {

        public void ConfigureServices(IServiceCollection services) {
            services.AddMvc();
        }

        public void Configure(IApplicationBuilder app, IHostingEnvironment env) {
            app.UseDeveloperExceptionPage();
            app.UseStatusCodePages();
            app.UseStaticFiles();
            app.UseMvc(routes => {
                routes.MapRoute(name: "MyRoute",
                    template: "{controller}/{action}/{id?}",
                    defaults: new { controller = "Home", action = "Index" },
                    constraints: new { id = new WeekDayConstraint() });
            });
        }
    }
}
```

仅当 id 片段不存在（例如/Customer/List）或者与约束类中定义的星期几（例如/Customer/List/Fri）匹配时，路由才会匹配 URL。

定义内联自定义约束

为了设置自定义约束并使其可以内联使用，需要执行一个额外的配置步骤，如代码清单 15-33 所示。

代码清单 15-33 在 UrlsAndRoutes 文件夹下的 Startup.cs 文件中使用内联自定义约束

```
using System;
using System.Collections.Generic;
using System.Linq;
using System.Threading.Tasks;
using Microsoft.AspNetCore.Builder;
using Microsoft.AspNetCore.Hosting;
using Microsoft.AspNetCore.Http;
using Microsoft.Extensions.DependencyInjection;
using Microsoft.AspNetCore.Routing.Constraints;
using Microsoft.AspNetCore.Routing;
using UrlsAndRoutes.Infrastructure;

namespace UrlsAndRoutes {
    public class Startup {

        public void ConfigureServices(IServiceCollection services) {
            services.Configure<RouteOptions>(options =>
                options.ConstraintMap.Add("weekday", typeof(WeekDayConstraint)));
            services.AddMvc();
        }

        public void Configure(IApplicationBuilder app, IHostingEnvironment env) {
            app.UseDeveloperExceptionPage();
            app.UseStatusCodePages();
            app.UseStaticFiles();
            app.UseMvc(routes => {
                routes.MapRoute(name: "MyRoute",
                template: "{controller=Home}/{action=Index}/{id:weekday?}");
            });
```

```
        }
    }
}
```

以上代码在 ConfigureService 方法中配置了 RouteOptions 对象，以控制路由系统的一些行为。ConstraintMap 属性会返回一个字典，用于将内联约束的名称转换为提供约束逻辑的 IRouteConstraint 实现类。向字典添加一个新的映射，将 WeekDayConstraint 类内联引用为 weekday，如下所示：

```
...
template: "{controller=Home}/{action=Index}/{id:weekday?}",
...
```

约束的效果是相同的，但设置了映射以允许自定义类被内联使用。

15.8 使用特性路由

到目前为止，本章的所有示例都是使用一种称为“基于约定的路由”的技术定义的。MVC 还支持称为特性路由的技术，直接使用控制器类的 C#属性来定义路由。下面将展示如何使用特性创建和配置路由，这些特性可以与前面示例中的基于约定的路由自由混合使用。

15.8.1 准备特性路由

在 Startup.cs 文件中调用 UseMvc 方法时，将启用特性路由。MVC 检查应用程序中的控制器类，查找具有特性路由的任何控制器类，并为它们创建路由。

本节中，示例应用程序已还原到默认路由配置，如代码清单 15-34 所示。

代码清单 15-34 使用 UrlsAndRoutes 文件夹下的 Startup.cs 文件中的默认路由配置

```
using System;
using System.Collections.Generic;
using System.Linq;
using System.Threading.Tasks;
using Microsoft.AspNetCore.Builder;
using Microsoft.AspNetCore.Hosting;
using Microsoft.AspNetCore.Http;
using Microsoft.Extensions.DependencyInjection;
using Microsoft.AspNetCore.Routing.Constraints;
using Microsoft.AspNetCore.Routing;
using UrlsAndRoutes.Infrastructure;

namespace UrlsAndRoutes {
    public class Startup {

        public void ConfigureServices(IServiceCollection services) {
            services.Configure<RouteOptions>(options =>
                options.ConstraintMap.Add("weekday", typeof(WeekDayConstraint)));
            services.AddMvc();
        }

        public void Configure(IApplicationBuilder app, IHostingEnvironment env) {
            app.UseDeveloperExceptionPage();
            app.UseStatusCodePages();
            app.UseStaticFiles();
            app.UseMvcWithDefaultRoute();
        }
    }
}
```

默认路由将使用以下模式匹配 URL：

```
{controller}/{action}/{id?}
```

15.8.2 应用特性路由

Route 特性用于指定单个控制器和操作方法的路由。在代码清单 15-35 中，已将 Route 特性应用于 CustomerController 类。

代码清单 15-35 在 Controllers 文件夹下的 CustomerController.cs 文件中应用 Route 特性

```
using Microsoft.AspNetCore.Mvc;
using UrlsAndRoutes.Models;

namespace UrlsAndRoutes.Controllers {
    public class CustomerController : Controller {

        [Route("myroute")]
        public ViewResult Index() => View("Result",
            new Result {
                Controller = nameof(CustomerController),
                Action = nameof(Index)
            });

        public ViewResult List(string id) {
            Result r = new Result {
                Controller = nameof(HomeController),
                Action = nameof(List),
            };
            r.Data["id"] = id ?? "<no value>";
            r.Data["catchall"] = RouteData.Values["catchall"];
            return View("Result", r);
        }
    }
}
```

Route 特性可用于为操作方法或控制器定义路由。在以上代码中，已将 Route 特性应用于 Index 操作方法，并将 myroute 指定为应该使用的路由。效果是更改了用于访问 Customer 控制器定义的操作方法的路由集合，如表 15-9 所示。

表 15-9 路由集合

路由	描述
/Customer/List	以 List 操作方法为目标，依赖于 Startup.cs 中的默认路由
/myroute	以 Index 操作方法为目标

有两点需要注意。第一点是当使用 Route 特性时，你提供的用于配置特性的值将用于定义完整的路由，因此 myroute 成为访问 Index 操作方法的完整 URL。需要注意的第二点是，使用 Route 特性可以防止使用默认路由配置，因此无法再使用/Customer/Index URL 访问 Index 操作方法。

1. 更改操作方法的名称

为单个操作方法定义唯一路由在大多数应用程序中没什么用，但 Route 特性也可以更灵活地使用。代码清单 15-36 在路由中使用特殊的[controller]标记来引用控制器并设置路由的基本部分。

提 示

使用 ActionName 属性也可以更改操作的名称，具体将在第 31 章中介绍。

代码清单 15-36 重命名 Controllers 文件夹下的 CustomerController.cs 文件中的操作方法

```
using Microsoft.AspNetCore.Mvc;
using UrlsAndRoutes.Models;
```

```
namespace UrlsAndRoutes.Controllers {
    public class CustomerController : Controller {

        [Route("[controller]/MyAction")]
        public ViewResult Index() => View("Result",
            new Result {
                Controller = nameof(CustomerController),
                Action = nameof(Index)
            });

        public ViewResult List(string id) {
            Result r = new Result {
                Controller = nameof(HomeController),
                Action = nameof(List),
            };
            r.Data["id"] = id ?? "<no value>";
            r.Data["catchall"] = RouteData.Values["catchall"];
            return View("Result", r);
        }
    }
}
```

在 Route 特性的参数中使用[controller]标记就像使用 nameof 表达式一样，允许指定到控制器的路由，而无须对类名进行硬编码。表 15-10 描述了代码清单 15-36 中路由的效果。

表 15-10 Customer 控制器的路由

路由	描述
/Customer/List	以 List 操作方法为目标
/Customer/MyAction	以 Index 操作方法为目标

2. 创建更复杂的路由

Route 特性也可以应用于控制器类，允许定义路由的结构，如代码清单 15-37 所示。

代码清单 15-37 在 Controllers 文件夹下的 CustomerController.cs 文件中应用 Route 特性

```
using Microsoft.AspNetCore.Mvc;
using UrlsAndRoutes.Models;

namespace UrlsAndRoutes.Controllers {

    [Route("app/[controller]/actions/[action]/{id?}")]
    public class CustomerController : Controller {

        public ViewResult Index() => View("Result",
            new Result {
                Controller = nameof(CustomerController),
                Action = nameof(Index)
            });

        public ViewResult List(string id) {
            Result r = new Result {
                Controller = nameof(HomeController),
                Action = nameof(List),
            };
            r.Data["id"] = id ?? "<no value>";
            r.Data["catchall"] = RouteData.Values["catchall"];
            return View("Result", r);
        }
    }
}
```

以上路由混合使用了静态片段和变量片段，并使用[controller]和[action]标记来分别引用控制器类的名称和操作方法。表 15-11 显示了代码清单 15-37 中路由的效果。

表 15-11 代码清单 15-37 中路由的效果

路由	描述
app/customer/actions/index	以 Index 操作方法为目标
app/customer/actions/index/myid	以 Index 操作方法为目标，并将可选的 id 片段设置为 myid
app/customer/actions/list	以 List 操作方法为目标
app/customer/actions/list/myid	以 List 操作方法为目标，并将可选的 id 片段设置为 myid

15.8.3 应用路由约束

使用特性定义的路由可以像 Startup.cs 文件中定义的路由那样，使用与基于约定的路由相同的内联技术进行约束。在代码清单 15-38 中，已将本章前面创建的自定义约束应用于 Route 特性定义的可选 id 片段。

代码清单 15-38 在 Controllers 文件夹下的 CustomerController.cs 文件中约束路由

```
using Microsoft.AspNetCore.Mvc;
using UrlsAndRoutes.Models;

namespace UrlsAndRoutes.Controllers {

    [Route("app/[controller]/actions/[action]/{id:weekday?}")]
    public class CustomerController : Controller {

        public ViewResult Index() => View("Result",
            new Result {
                Controller = nameof(CustomerController),
                Action = nameof(Index)
            });

        public ViewResult List(string id) {
            Result r = new Result {
                Controller = nameof(HomeController),
                Action = nameof(List),
            };
            r.Data["id"] = id ?? "<no value>";
            r.Data["catchall"] = RouteData.Values["catchall"];
            return View("Result", r);
        }
    }
}
```

可以使用表 15-8 中描述的所有约束，也可如代码所示，使用已在 RouteOptions 服务中注册的自定义约束。可以通过将它们链接在一起并用冒号分隔它们来应用多个约束。

15.9 小结

本章深入介绍了路由系统。你已经了解了按约定或特性定义路由的方式。你还了解了如何匹配和处理传入的 URL，以及如何通过更改匹配 URL 片段的方式和使用默认值及可选片段来自定义路由。这里还展示了如何使用内置约束和自定义约束，用来约束路由以缩小它们匹配的请求范围。

下一章将展示如何通过路由在视图中生成传出的 URL，以及如何使用 area 特性，它依赖路由系统，可以用于管理大型的和复杂的 MVC 应用程序。

第 16 章　高级路由特性

上一章展示了如何使用路由系统来处理传入的 URL，我们还需要能够使用 URL 模式来生成可以签入视图中的输出 URL，以便用户可以单击链接，以正确的控制器和 action 为目标，将表单回递给应用程序。

本章将展示生成输出 URL 的不同技术，展示如何通过替换标准的 MVC 路由实现类来定制路由系统以及使用 MVC 区域（area）特性，area 特性让你能够将大型的复杂 MVC 应用程序分解为若干可管理的小型区域。最后，以一些有关 MVC 应用程序中 URL 模式最佳实践的建议结束本章，表 16-1 汇总了在上下文中使用高级路由特性时的一些问题。

表 16-1　在上下文中使用高级路由特性

问题	答案
路由是什么？	路由系统除了为 HTTP 请求匹配 URL 外，还提供了很多功能，比如支持在视图中生成 URL，使用自定义类替换内置的路由功能，在应用程序中构建相互隔离的模块，等等
为什么路由很有用？	路由的每个功能在它们各自的应用场景中都是很有用的，比如通过设置路由格式，我们不用更新视图，就可以很容易地更改 URL 的格式，使用自定义类可以让我们根据需要定制路由系统，更容易地利用应用程序构建复杂的系统
如何使用路由？	有关路由使用的详细信息，请参阅本章
路由有什么缺点或限制吗？	复杂应用程序的路由配置可能变得难以管理
还有其他的选择么？	没有，路由系统是 ASP.NET 的组成部分之一

表 16-2 列出了本章要介绍的操作。

表 16-2　本章要介绍的操作

操作	方法	代码清单
使用 URL 生成锚点元素	使用 asp-action 和 asp-controller 属性	代码清单 16-2～代码清单 16-5
提供路由片段的值	使用以 asp-route 为前缀的属性	代码清单 16-6 和代码清单 16-7
生成完全限定的 URL	使用 asp-procotol、asp-host 和 asp-fragment 属性	代码清单 16-8
选择生成 URL 的路由	使用 asp-route 属性	代码清单 16-9 和代码清单 16-10
生成没有 HTML 元素的 URL	在视图或操作方法中使用 Url.Action 方法	代码清单 16-11 和代码清单 16-12
自定义路由系统	在 Startup 类中使用 Configure 方法	代码清单 16-13
创建自定义路由类	实现 IRouter 接口	代码清单 16-14～代码清单 16-21
将应用程序分解为各个功能模块	利用 Area 属性创建区域	代码清单 16-22～代码清单 16-28

16.1　准备示例项目

本章继续使用前一章的 UrlsAndRoutes 项目，但是在 Startup 类中进行一些修改。在 Startup 类中，使用具有相同效果的显式路由替换 UseMvcWithDefaultRoute 方法，如代码清单 16-1 所示。

代码清单 16-1　修改 UrlsAndRoutes 文件夹下的 Startup 类中的路由配置

```
using System;
using System.Collections.Generic;
```

```
using System.Linq;
using System.Threading.Tasks;
using Microsoft.AspNetCore.Builder;
using Microsoft.AspNetCore.Hosting;
using Microsoft.AspNetCore.Http;
using Microsoft.Extensions.DependencyInjection;
using Microsoft.AspNetCore.Routing.Constraints;
using Microsoft.AspNetCore.Routing;
using UrlsAndRoutes.Infrastructure;

namespace UrlsAndRoutes {
    public class Startup {

        public void ConfigureServices(IServiceCollection services) {
            services.Configure<RouteOptions>(options =>
                options.ConstraintMap.Add("weekday", typeof(WeekDayConstraint)));
            services.AddMvc();
        }

        public void Configure(IApplicationBuilder app, IHostingEnvironment env) {
            app.UseDeveloperExceptionPage();
            app.UseStatusCodePages();
            app.UseStaticFiles();
            app.UseMvc(routes => {
                routes.MapRoute(
                    name: "default",
                    template: "{controller=Home}/{action=Index}/{id?}");
            });
        }
    }
}
```

当启动程序时，浏览器会请求默认的 URL，请求会被发送给 Home 控制的 Index 操作方法，如图 16-1 所示。

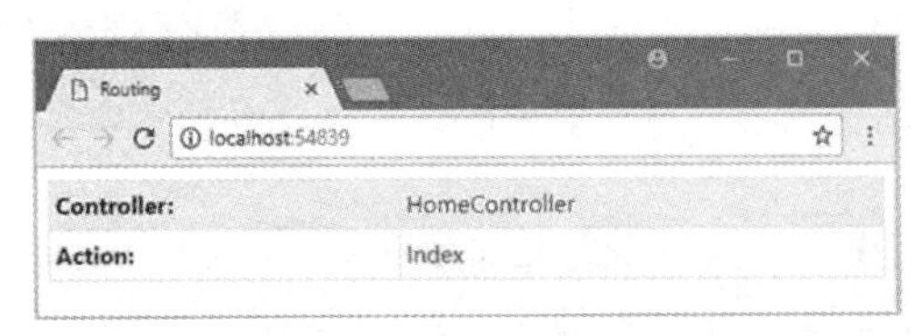

图 16-1　运行示例应用程序

16.2　在视图中生成传出的 URL

在几乎每个 MVC 应用程序中，都希望用户可以从一个视图跳转到另一个视图，这通常依赖于在第一个视图中包含一个针对生成第二个视图的操作方法的链接。方法是在页面中添加一个静态元素（称为锚点元素），用 href 属性定位到操作方法，如下所示：

```
<a href="/Home/CustomVariable">This is an outgoing URL</a>
```

如果应用程序使用默认的路由配置，这个 HTML 元素将创建一个链接，指向 Home 控制器中名为 CustomVariable 的操作方法。但是这样手动生成 URL 是非常危险的，当更改应用程序的 URL 格式时，必须查看所有视图来更新所有指向控制器和操作方法的链接，这将是一个非常冗长乏味且容易出错的过程，还难以测试。所以，更好的选择是使用路由系统来生成传出的 URL，这样可以确保使用程序的 URL 格式来动态生成 URL，并且还能在一定程度上反映程序的 URL 格式。

16.2.1 创建传出的链接

在视图中生成传出的 URL 的最简单方法是使用锚标签助手，它将为 HTML 元素生成 href 属性，如代码清单 16-2 所示，可在视图/Views/Shared/Result.cshtml 中使用锚标签助手来创建传出的链接。

提 示

第 23 章将解释标签助手的工作原理。

代码清单 16-2 在 Views/Shared 文件夹下的 Result.cshtml 文件中使用锚标签助手

```
@model Result
@{ Layout = null; }

<!DOCTYPE html>
<html>
<head>
    <meta name="viewport" content="width=device-width" />
    <title>Routing</title>
    <link rel="stylesheet" asp-href-include="lib/bootstrap/dist/css/*.min.css" />
</head>
<body class="m-1 p-1">
    <table class="table table-bordered table-striped table-sm">
        <tr><th>Controller:</th><td>@Model.Controller</td></tr>
        <tr><th>Action:</th><td>@Model.Action</td></tr>
        @foreach (string key in Model.Data.Keys) {
            <tr><th>@key :</th><td>@Model.Data[key]</td></tr>
        }
    </table>
    <a asp-action="CustomVariable">This is an outgoing URL</a>
</body>
</html>
```

asp-action 属性用来指定 href 属性中的 URL 指向的对应名称的操作方法。可以通过启动应用程序来查看结果，如图 16-2 所示。

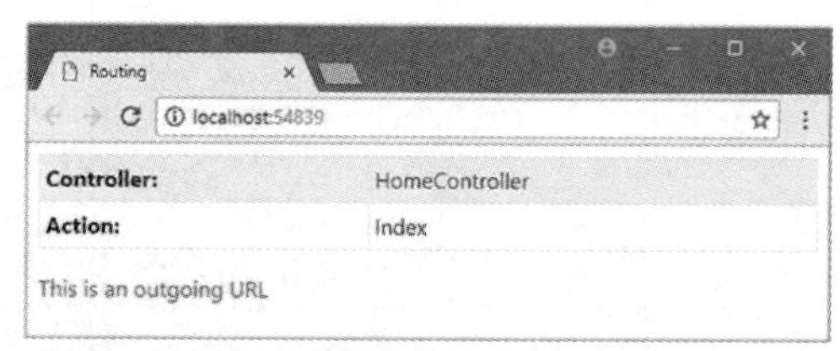

图 16-2 使用标签助手来生成链接

标签助手使用当前路由配置在 a 元素上设置 href 属性。如果查看浏览器中的 HTML 标记，你会看到其中包含以下元素：

```
<a href="/Home/CustomVariable">This is an outgoing URL</a>
```

这个链接看起来和之前手动编写的链接是一样的，而且为了生成这个链接，我们还做了很多其他的工作。但是这种方法的好处是链接会自动响应路由配置的更改。为了演示，在 Startup.cs 文件中添加一个新的路由配置，如代码清单 16-3 所示。

代码清单 16-3 在 UrlsAndRoutes 文件夹下的 Startup.cs 文件中添加路由配置

```
using System;
using System.Collections.Generic;
using System.Linq;
using System.Threading.Tasks;
using Microsoft.AspNetCore.Builder;
using Microsoft.AspNetCore.Hosting;
using Microsoft.AspNetCore.Http;
using Microsoft.Extensions.DependencyInjection;
using Microsoft.AspNetCore.Routing.Constraints;
using Microsoft.AspNetCore.Routing;
```

```
using UrlsAndRoutes.Infrastructure;

namespace UrlsAndRoutes {
    public class Startup {

        public void ConfigureServices(IServiceCollection services) {
            services.Configure<RouteOptions>(options =>
                options.ConstraintMap.Add("weekday", typeof(WeekDayConstraint)));
            services.AddMvc();
        }

        public void Configure(IApplicationBuilder app, IHostingEnvironment env) {
            app.UseDeveloperExceptionPage();
            app.UseStatusCodePages();
            app.UseStaticFiles();
            app.UseMvc(routes => {

                routes.MapRoute(
                    name: "NewRoute",
                    template: "App/Do{action}",
                    defaults: new { controller = "Home" });

                routes.MapRoute(
                    name: "default",
                    template: "{controller=Home}/{action=Index}/{id?}");
            });
        }
    }
}
```

新路由更改了针对 Home 控制器的请求的 URL 模式。如果启动应用，你将看到这些更改会反映在由 HTML 辅助方法 ActionLink 生成的 HTML 标记中，如下所示：

```
<a href="/App/DoCustomVariable">This is an outgoing URL</a>
```

使用标签助手生成链接可解决重要的维护问题。当更改路由格式时，视图中的传出链接会自动根据路由格式更改，无须在视图中手动编辑。

当单击链接时，传出的 URL 用于创建传入的 HTTP 请求，然后使用相同的路由来定位将用于处理请求的控制器和操作方法，如图 16-3 所示。

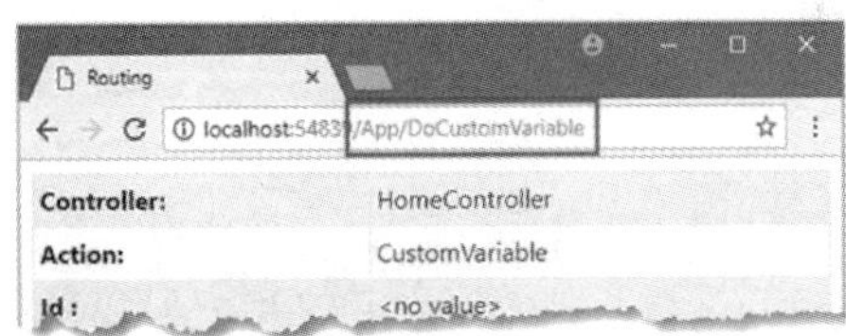

图 16-3　单击链接，将传出的 URL 转换为传入的请求

理解传出的 URL 的路由匹配

你已经看到了通过更改路由配置的 URL 模式，可以修改传出的 URL 的生成方式。应用程序通常会定义多个路由，所以了解如何为 URL 的生成选择路由规则很重要。路由系统按照定义的顺序处理路由，依次检查每个路由，查看是否匹配，这需要满足以下 3 个条件。

- URL 模式中定义的每个片段变量都必须有一个值。为了查找每个片段变量的值，路由系统首先查看你提供的值（使用匿名类型的属性），然后查看当前请求的变量值，最后查看路由中定义的默认值（本章稍后再回到这些值的第二个来源）。
- 为片段变量提供的任何值都不会与路由中定义的默认变量发生不一致。这些是已经提供了默认值但在 URL 模式中不会出现的变量。例如，在下面的路由定义中，myVar 是默认变量。

```
routes.MapRoute("MyRoute", "{controller}/{action}",
    new { myVar = "true" });
```

为了使以上路由匹配，必须注意不要为 myVar 提供值，或者确保提供的值与默认值匹配。

- 所有片段变量的值必须满足路由约束。有关不同种类约束的示例，请参阅 15.6 节。

需要明确的是，路由系统不会尝试选取最佳的路由规则来进行匹配，而只是寻找第一个可以匹配的规则，并根据该规则生成 URL，任何后续的路由规则都会被忽略。因此，必须首先定义那些特殊的路由规则，检查生成的传出 URL 也很重要。如果尝试生成找不到匹配路由的 URL，将创建如下包含空的 href 属性的链接：

```
<a href="">This is an outgoing URL</a>
```

以上链接将正确显示在视图中，但在用户单击时不会按预期方式运行。如果只生成 URL（本章稍后会展示如何实现），那么结果将为 null，在视图中作为空字符串呈现。可以使用命名路由对路由匹配进行一些控制。

1. 定位其他控制器

当在一个元素上指定 asp-action 属性时，标签助手会在渲染视图时假定你要定位到同一控制器中的 action。要创建以不同控制器为目标的传出 URL，可以使用 asp-controller 属性，如代码清单 16-4 所示。

代码清单 16-4　在 Views/Shared 文件夹下的 Result.cshtml 文件中定位不同的控制器

```
@model Result
@{ Layout = null; }

<!DOCTYPE html>
<html>
<head>
    <meta name="viewport" content="width=device-width" />
    <title>Routing</title>
    <link rel="stylesheet" asp-href-include="lib/bootstrap/dist/css/*.min.css" />
</head>
<body class="m-1 p-1">
    <table class="table table-bordered table-striped table-sm">
        <tr><th>Controller:</th><td>@Model.Controller</td></tr>
        <tr><th>Action:</th><td>@Model.Action</td></tr>
        @foreach (string key in Model.Data.Keys) {
            <tr><th>@key :</th><td>@Model.Data[key]</td></tr>
        }
    </table>
    <a asp-controller="Admin" asp-action="Index">
        This targets another controller
    </a>
</body>
</html>
```

在渲染视图时，你将看到生成了以下 HTML 标记：

```
<a href="/Admin">This targets another controller</a>
```

指向 Admin 控制器中 Index 操作方法的请求 URL 被标签助手表示为/Admin。路由系统知道应用程序中定义的路由默认使用 Index 操作方法，因而允许忽略不必要的片段变量。

路由系统包含那些使用 Route 特性给操作方法指定路由规则的路由。在代码清单 16-5 中，asp-controller 属性用于定位在第 15 章中使用了 Route 特性的 Customer 控制器中的 Index 操作方法。

代码清单 16-5　在 Views/Shared 文件夹下的 Result.cshtml 文件中定位操作方法

```
@model Result
@{ Layout = null; }

<!DOCTYPE html>
<html>
```

```
<head>
    <meta name="viewport" content="width=device-width" />
    <title>Routing</title>
    <link rel="stylesheet" asp-href-include="lib/bootstrap/dist/css/*.min.css" />
</head>
<body class="panel-body">
    <table class="table table-bordered table-striped table-sm">
        <tr><th>Controller:</th><td>@Model.Controller</td></tr>
        <tr><th>Action:</th><td>@Model.Action</td></tr>
        @foreach (string key in Model.Data.Keys) {
            <tr><th>@key :</th><td>@Model.Data[key]</td></tr>
        }
    </table>
    <a asp-controller="Customer" asp-action="Index">This is an outgoing URL</a>
</body>
</html>
```

生成的链接如下：

```
<a href="/app/Customer/actions/Index">This is an outgoing URL</a>
```

这对应于第 15 章中为 Customer 控制器设置的 Route 特性：

```
...
[Route("app/[controller]/actions/[action]/{id:weekday?}")]
public class CustomerController : Controller {
...
```

2. 传递额外的变量值

可以使用 asp-route-片段变量名格式将片段变量的值传递给路由系统，比如可以使用 asp-route-id 将值传给路由系统中的 id，如代码清单 16-6 所示。

代码清单 16-6　为 Views/Shared 文件夹下的 Result.cshtml 文件中的片段变量提供值

```
@model Result
@{ Layout = null; }

<!DOCTYPE html>
<html>
<head>
    <meta name="viewport" content="width=device-width" />
    <title>Routing</title>
    <link rel="stylesheet" asp-href-include="lib/bootstrap/dist/css/*.min.css" />
</head>
<body class="m-1 p-1">
    <table class="table table-bordered table-striped table-sm">
        <tr><th>Controller:</th><td>@Model.Controller</td></tr>
        <tr><th>Action:</th><td>@Model.Action</td></tr>
        @foreach (string key in Model.Data.Keys) {
            <tr><th>@key :</th><td>@Model.Data[key]</td></tr>
        }
    </table>
    <a asp-controller="Home" asp-action="Index" asp-route-id="Hello">
        This is an outgoing URL
    </a>
</body>
</html>
```

这里为名为 id 的片段变量设置了值，如果应用代码清单 16-6 所示的路由，在视图渲染的时候将会呈现以下 HTML 标记：

```
<a href="/App/DoIndex?id=Hello">This is an outgoing URL</a>
```

请注意，片段变量的值已经根据路由规则做了设置，成为 URL 中查询字符串的一部分。路由中没有设置对应于 id 的片段变量，为了解决这个问题，修改 Startup.cs 文件中的路由，只留下一个拥有 id 片段变量的路由，如代码清单 16-7 所示。

代码清单 16-7　在 UrlsAndRoutes 文件夹下的 Startup.cs 文件中修改路由

```
using System;
using System.Collections.Generic;
using System.Linq;
using System.Threading.Tasks;
using Microsoft.AspNetCore.Builder;
using Microsoft.AspNetCore.Hosting;
using Microsoft.AspNetCore.Http;
using Microsoft.Extensions.DependencyInjection;
using Microsoft.AspNetCore.Routing.Constraints;
using Microsoft.AspNetCore.Routing;
using UrlsAndRoutes.Infrastructure;

namespace UrlsAndRoutes {
    public class Startup {

        public void ConfigureServices(IServiceCollection services) {
            services.Configure<RouteOptions>(options =>
                options.ConstraintMap.Add("weekday", typeof(WeekDayConstraint)));
            services.AddMvc();
        }

        public void Configure(IApplicationBuilder app, IHostingEnvironment env) {
            app.UseDeveloperExceptionPage();
            app.UseStatusCodePages();
            app.UseStaticFiles();
            app.UseMvc(routes => {

                //routes.MapRoute(
                //    name: "NewRoute",
                //    template: "App/Do{action}",
                //    defaults: new { controller = "Home" });

                routes.MapRoute(
                    name: "default",
                    template: "{controller=Home}/{action=Index}/{id?}");
            });
        }
    }
}
```

再次运行应用，你将看到标签助手生成的 URL 中已经包含了 id 属性的值。

```
<a href="/Home/Index/Hello">This is an outgoing URL</a>
```

理解片段变量的重用

之前在描述匹配的路由传出 URL 的方式时，本书解释说，当试图为路由的 URL 模式中的每个片段变量查找值时，路由系统将查看当前请求的值。这种方式可能会困扰很多程序员，并且导致冗长的调试会话。

假设应用程序有如下路由：

```
...
app.UseMvc(routes => {
    routes.MapRoute(name: "MyRoute",
        template: "{controller}/{action}/{color}/{page}");
});
...
```

假设我们所在的页面 URL 为/Home/Index/Red/100，在页面中利用以下代码渲染链接：

```
...
<a asp-controller="Home" asp-action="Index" asp-route-page="789">
    This is an outgoing URL
</a>
...
```

你可能期望路由系统无法匹配路由，因为没有为 color 片段变量提供值，也没有定义默认值。期望是错误的，路由系统将与定义的路由匹配，还将生成以下 HTML 标记：

```
...
<a href="/Home/Index/Red/789">This is an outgoing URL</a>
...
```

路由系统热衷于对路由进行匹配，在生成传出的 URL 时，将在传入的 URL 中重用片段变量的值。在这种情况下，根据页面使用的 URL，路由系统将设置 color 变量的值为 Red。

这并不是最后的手段。路由系统将应用此技术作为对路由的常规评估的一部分，即使后面的路由不需要重用当前请求中的值，也会进行匹配。

强烈建议你不要依赖此行为，并为 URL 模式中的所有片段变量提供值。依赖这种行为不仅会使代码更难阅读，而且还会对用户发出请求的顺序做出假设，这将在应用程序进入维护时令人抓狂。

3. 生成完全限定的 URL

到目前为止，生成的所有链接都包含相对 URL，但是锚标签助手也可以生成完全限定的 URL，如代码清单 16-8 所示。

代码清单 16-8 在 Views/Shared 文件夹下的 Result.cshtml 文件中生成完全限定的 URL

```
@model Result
@{ Layout = null; }

<!DOCTYPE html>
<html>
<head>
    <meta name="viewport" content="width=device-width" />
    <title>Routing</title>
    <link rel="stylesheet" asp-href-include="lib/bootstrap/dist/css/*.min.css" />
</head>
<body class="m-1 p-1">
    <table class="table table-bordered table-striped table-sm">
        <tr><th>Controller:</th><td>@Model.Controller</td></tr>
        <tr><th>Action:</th><td>@Model.Action</td></tr>
        @foreach (string key in Model.Data.Keys) {
            <tr><th>@key :</th><td>@Model.Data[key]</td></tr>
        }
    </table>
    <a asp-controller="Home" asp-action="Index" asp-route-id="Hello"
       asp-protocol="https" asp-host="myserver.mydomain.com"
       asp-fragment="myFragment">
        This is an outgoing URL
    </a>
</body>
</html>
```

asp-protocol、asp-host 与 asp-fragment 属性分别用于指定协议（https）、服务器名称（myserver. mydomain.com）和 URL 片段（myFragment）。将这些值与路由系统的输出相结合，即可创建完全限定的 URL，可以运行应用以查看发送到浏览器的 HTML：

```
<a href="https://myserver.mydomain.com/Home/Index/Hello#myFragment">
    This is an outgoing URL
</a>
```

使用完全限定的 URL 时要小心，因为它们创建了与应用程序基础架构的依赖关系，并且基础架构发生变化时，必须记住对 MVC 视图进行相应的更改。

4. 从特定路由生成 URL

在前面的示例中，我们演示了路由系统如何选择路由来生成 URL。当需要生成特定格式的 URL 时，可以指定用于生成 URL 的路由。为了演示这是如何工作的，这里向 Startup.cs 文件添加一个新的路由，这时在示例应用程序中有两个路由，如代码清单 16-9 所示。

代码清单 16-9 在 UrlsAndRoutes 文件夹的 Startup.cs 文件中添加路由

```
using System;
using System.Collections.Generic;
using System.Linq;
using System.Threading.Tasks;
using Microsoft.AspNetCore.Builder;
using Microsoft.AspNetCore.Hosting;
using Microsoft.AspNetCore.Http;
using Microsoft.Extensions.DependencyInjection;
using Microsoft.AspNetCore.Routing.Constraints;
using Microsoft.AspNetCore.Routing;
using UrlsAndRoutes.Infrastructure;

namespace UrlsAndRoutes {
    public class Startup {

        public void ConfigureServices(IServiceCollection services) {
            services.Configure<RouteOptions>(options =>
                options.ConstraintMap.Add("weekday", typeof(WeekDayConstraint)));
            services.AddMvc();
        }

        public void Configure(IApplicationBuilder app, IHostingEnvironment env) {
            app.UseDeveloperExceptionPage();
            app.UseStatusCodePages();
            app.UseStaticFiles();
            app.UseMvc(routes => {

                //routes.MapRoute(
                //    name: "NewRoute",
                //    template: "App/Do{action}",
                //    defaults: new { controller = "Home" });

                routes.MapRoute(
                    name: "default",
                    template: "{controller=Home}/{action=Index}/{id?}");

                routes.MapRoute(
                    name: "out",
                    template: "outbound/{controller=Home}/{action=Index}");
            });
        }
    }
}
```

代码清单 16-10 所示的视图中包含两个锚元素，每个都指定了相同的控制器和操作。不同之处在于第二个元素使用 asp-route 属性来指定使用的路由。

代码清单 16-10 在 Views/Shared 文件夹下的 Result.Cshtml 文件中生成 URL

```
@model Result
@{ Layout = null; }

<!DOCTYPE html>
<html>
<head>
    <meta name="viewport" content="width=device-width" />
    <title>Routing</title>
    <link rel="stylesheet" asp-href-include="lib/bootstrap/dist/css/*.min.css" />
</head>
<body class="m-1 p-1">
    <table class="table table-bordered table-striped table-sm">
        <tr><th>Controller:</th><td>@Model.Controller</td></tr>
        <tr><th>Action:</th><td>@Model.Action</td></tr>
        @foreach (string key in Model.Data.Keys) {
            <tr><th>@key :</th><td>@Model.Data[key]</td></tr>
        }
    </table>
    <a asp-controller="Home" asp-action="CustomVariable">This is an outgoing URL</a>
    <a asp-route="out">This is an outgoing URL</a>
</body>
</html>
```

只有在 asp-controller 属性和 asp-action 属性不存在时，才可以使用 asp-route 属性，这意味着只能为渲染视图的控制器和操作选择特定的路由。如果运行示例程序并请求 URL /Home/CustomVariable，你将看到路由会生成两个不同的 URL：

```
<a href="/Home/CustomVariable">This is an outgoing URL</a>
<a href="/outbound">This is an outgoing URL</a>
```

针对命名路由的情况

依靠路由名称生成传出 URL 的问题在于，这样做会破坏对 MVC 设计模式至关重要的关注点分离原则。在视图或操作方法中生成链接或 URL 时，需要关注用户将被引导到的操作和控制器，而不是要使用的 URL 格式。通过将不同路由引入视图或控制器，可以避免创建依赖关系。在作者自己的项目中，作者倾向于避免命名路由（通过为 name 参数指定 null），并且更喜欢使用代码注释来提醒自己每个路由的目的。

16.2.2 创建非链接的 URL

标签助手的限制是：对于转换的 HTML 元素，当需要为应用程序生成 URL 而不想使用周围的 HTML 时，无法轻易地重用。

MVC 提供了一个辅助类，可以直接使用 Url.Action 方法创建 URL，如代码清单 16-11 所示。

代码清单 16-11 在 Views/Shared 文件夹下的 Result.cshtml 文件中生成 URL

```
@model Result
@{ Layout = null; }

<!DOCTYPE html>
<html>
<head>
    <meta name="viewport" content="width=device-width" />
    <title>Routing</title>
    <link rel="stylesheet" asp-href-include="lib/bootstrap/dist/css/*.min.css" />
</head>
<body class="m-1 p-1">
```

```
    <table class="table table-bordered table-striped table-sm">
        <tr><th>Controller:</th><td>@Model.Controller</td></tr>
        <tr><th>Action:</th><td>@Model.Action</td></tr>
        @foreach (string key in Model.Data.Keys) {
            <tr><th>@key :</th><td>@Model.Data[key]</td></tr>
        }
    </table>
    <p>URL: @Url.Action("CustomVariable", "Home", new { id = 100 })</p>
</body>
</html>
```

以上代码在 Url.Action 方法的参数中指定了操作方法、控制器和片段变量的值，生成的结果如下：

```
<p>URL: /Home/CustomVariable/100</p>
```

在操作方法中生成 URL

Url.Action 方法也可以用于在操作方法中使用 C#代码来创建 URL。代码清单 16-12 修改了 Home 控制器的一个操作方法，并使用 Url.Action 生成了一个 URL。

代码清单 16-12 在 Controllers 文件夹下的 HomeController.cs 文件里的操作方法中生成 URL

```
using Microsoft.AspNetCore.Mvc;
using UrlsAndRoutes.Models;

namespace UrlsAndRoutes.Controllers {

    public class HomeController : Controller {
        public ViewResult Index() => View("Result",
            new Result {
                Controller = nameof(HomeController),
                Action = nameof(Index)
            });

        public ViewResult CustomVariable(string id) {
            Result r = new Result {
                Controller = nameof(HomeController),
                Action = nameof(CustomVariable),
            };
            r.Data["Id"] = id ?? "<no value>";
            r.Data["Url"] = Url.Action("CustomVariable", "Home", new { id = 100 });
            return View("Result", r);
        }
    }
}
```

如果运行示例程序并且请求 URL /Home/CustomVariable/100，你将看到表格中有一行显示了这个 URL，如图 16-4 所示。

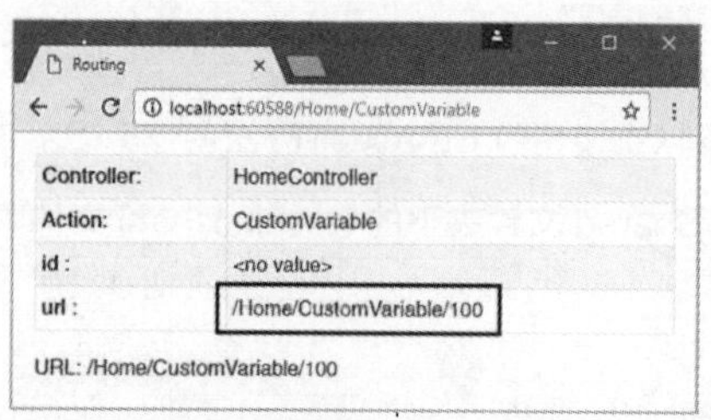

图 16-4 在操作方法中生成 URL

16.3 自定义路由系统

你已经看到了路由系统的灵活性和可配置性，如果这些还不符合要求，你也可以自定义行为。本节将展示集中不同的自定义路由的方法。

16.3.1　更改路由系统配置

第 15 章展示了如何在 Startup.cs 文件中配置 RouteOptions 对象以设置自定义路由约束。使用 RouteOptions 对象的属性（见表 16-3），配置一些路由功能。

表 16-3　RouteOptions 对象的属性

属性	描述
AppendTrailingSlash	当为 true 时，为路由系统生成的 URL 附加尾部斜线，默认值为 false
LowercaseUrls	当为 true 时，如果控制器、action 或片段变量的值包含大写字母，就将 URL 转换为小写，默认值为 false

代码清单 16-13 向 Startup.cs 文件添加了路由配置，以设置表 16-3 中描述的两个配置属性。

代码清单 16-13　在 UrlsAndRoutes 文件夹下的 Startup.cs 文件中配置路由系统

```
using System;
using System.Collections.Generic;
using System.Linq;
using System.Threading.Tasks;
using Microsoft.AspNetCore.Builder;
using Microsoft.AspNetCore.Hosting;
using Microsoft.AspNetCore.Http;
using Microsoft.Extensions.DependencyInjection;
using Microsoft.AspNetCore.Routing.Constraints;
using Microsoft.AspNetCore.Routing;
using UrlsAndRoutes.Infrastructure;

namespace UrlsAndRoutes {
    public class Startup {

        public void ConfigureServices(IServiceCollection services) {
            services.Configure<RouteOptions>(options => {
                options.ConstraintMap.Add("weekday", typeof(WeekDayConstraint));
                options.LowercaseUrls = true;
                options.AppendTrailingSlash = true;
            });
            services.AddMvc();
        }

        public void Configure(IApplicationBuilder app, IHostingEnvironment env) {
            app.UseDeveloperExceptionPage();
            app.UseStatusCodePages();
            app.UseStaticFiles();
            app.UseMvc(routes => {

                routes.MapRoute(
                    name: "default",
                    template: "{controller=Home}/{action=Index}/{id?}");

                routes.MapRoute(
                    name: "out",
                    template: "outbound/{controller=Home}/{action=Index}");
            });
        }
    }
}
```

如果运行应用并检查路由系统生成的 URL，你将看到 URL 全部为小写，并且附加了尾部斜杠，如图 16-5 所示。

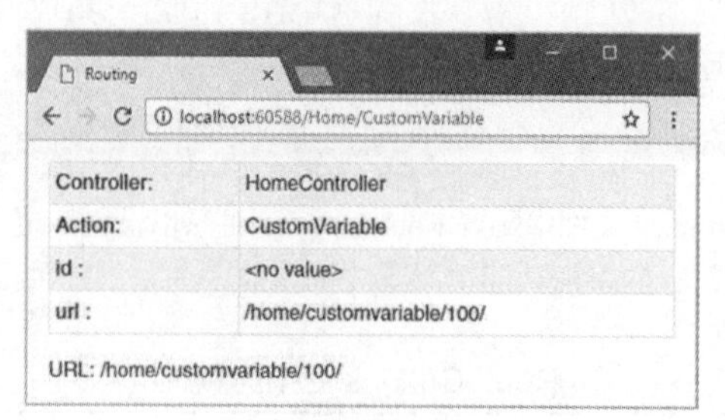

图 16-5 配置路由系统

16.3.2 创建自定义路由类

如果不喜欢路由系统匹配 URL 的方式，或者需要在应用中进行一些特殊的实现，可以创建自己的路由类并使用它们来处理 URL。ASP.NET 提供了 Microsoft.AspNetCore.Routing.IRouter 接口，可以通过实现此接口来创建自定义路由。以下是 IRouter 接口的定义：

```
using System.Threading.Tasks;

namespace Microsoft.AspNetCore.Routing {

    public interface IRouter {

        Task RouteAsync(RouteContext context);

        VirtualPathData GetVirtualPath(VirtualPathContext context);
    }
}
```

要创建自定义路由，可以实现 RouteAsync 方法来处理传入的请求，并通过实现 GetVirtualPath 方法来生成传出的 URL。

为了演示，我们将创建一个可以处理传统 URL 请求的自定义路由类。想象一下，这里已经将一个现有的应用程序迁移到 MVC，但是有些用户已经将之前的网址加入书签，或者硬编码到脚本中。为了仍然支持那些旧的 URL，可以使用常规路由系统，但是这个问题的解决方法值得参考。

1. 路由传入的 URL

要了解自定义路由如何工作，可以首先创建一个路由来处理请求，而不使用控制器和视图。在 Infrastructure 文件夹中创建一个名为 LegacyRoute.cs 的类文件，用于实现 IRouter 接口，如代码清单 16-14 所示。

代码清单 16-14 Infrastructure 文件夹下的 LegacyRoute.cs 文件的内容

```
using Microsoft.AspNetCore.Http;
using Microsoft.AspNetCore.Routing;
using System;
using System.Linq;
using System.Text;
using System.Threading.Tasks;

namespace UrlsAndRoutes.Infrastructure {
    public class LegacyRoute : IRouter {
        private string[] urls;

        public LegacyRoute(params string[] targetUrls) {
            this.urls = targetUrls;
        }

        public Task RouteAsync(RouteContext context) {
```

```
            string requestedUrl = context.HttpContext.Request.Path
                .Value.TrimEnd('/');

            if (urls.Contains(requestedUrl, StringComparer.OrdinalIgnoreCase)) {
                context.Handler = async ctx => {
                    HttpResponse response = ctx.Response;
                    byte[] bytes = Encoding.ASCII.GetBytes($"URL: {requestedUrl}");
                    await response.Body.WriteAsync(bytes, 0, bytes.Length);
                };
            }
            return Task.CompletedTask;
        }

        public VirtualPathData GetVirtualPath(VirtualPathContext context) {
            return null;
        }
    }
}
```

LecagyRoute 类实现了 IRouter 接口，但只定义了用于处理传入请求的 RouteAsync 方法；我们将在后面的内容中添加处理传出 URL 的方法。

RouteAsync 方法中只有几条语句，但它们的工作依赖于一些重要的 ASP.NET 类型。我们先从方法签名开始介绍：

```
...
public async Task RouteAsync(RouteContext context) {
...
```

RouteAsync 方法负责评估是否可以处理请求，如果可以，就可以通过生成响应并发送回客户端来管理整个过程。因为 RouteAsync 方法返回一个 Task 对象，所以这个过程是异步执行的。

RouteAsync 方法使用了 RouteContext 参数，从而提供对请求的所有已知信息的访问，并提供将响应发送回客户端所需的功能。RouteContext 类在命名空间 Microsoft.AspNetCore.Routing 中定义，有 3 个属性，如表 16-4 所示。

表 16-4 RouteContext 类中定义的属性

属性	描述
RouteData	返回一个 Microsoft.AspNetCore.Routing.RouteData 对象，当编写依赖于 MVC 功能的自定义路由时，该对象用于定义控制器、操作方法以及用于处理请求的参数
HttpContext	返回一个 Microsoft.AspNetCore.Http.HttpContext 对象，该对象提供了访问 HTTP 请求的详细信息以及生成 HTTP 响应的方法
Handler	用于为路由系统提供处理请求的 RequestDelegate。如果未设置该属性，路由系统将继续按照应用设置的路由集合工作

路由系统调用 RouteAsync 方法来处理应用中的每个路由，并在每次调用后检查 Handler 属性的值。如果 Handler 属性已设置为 RequestDelegate，路由就为路由系统提供可以处理请求的委托，并调用委托以生成响应。以下是 RequestDelegate 的签名，RequestDelegate 定义在 Microsoft.AspNetCore.Http 命名空间中。

```
using System.Threading.Tasks;

namespace Microsoft.AspNetCore.Http {
    public delegate Task RequestDelegate(HttpContext context);
}
```

RequestDelegate 接收 HttpContext 对象并返回一个任务来生成响应。如果没有为任何路由设置 Handler 属性，路由系统便得知应用程序无法处理请求，因而生成 404-Not Found 响应。

考虑到这一点，RouteAsync 方法的实现必须确定是否可以处理通常需要 HttpContext 的请求。在该例中，使用了 HttpContext.Request 属性，该属性返回描述请求的 Microsoft.AspNetCore.Http.HttpRequest 对象。HttpRequest

对象提供对有关请求的所有可用信息的访问，包括头文件、主体以及请求发起地址的详细信息。Path 属性最有趣，因为它提供了客户端发送的 URL 请求的详细信息。Path 属性会返回一个 PathString 对象，该对象提供了一些有用的方法来编写和比较 URL 路径。但这里使用了 Value 属性，因为它将 URL 的整个路径部分作为字符串提供了，从而可以与支持的 URL 集合进行比较。URL 可由 LegacyRoute 构造函数接收。

```
...
string requestedUrl = context.HttpContext.Request.Path.Value.TrimEnd('/');
if (urls.Contains(requestedUrl, StringComparer.OrdinalIgnoreCase)) {
...
```

可以由用户手动添加，也可以使用 AppendTrailingSlash 配置选项进行添加。因为在 URL 路径的最后会有斜杠，所以需要使用 TrimEnd 方法来去除 URL 尾部的斜杠。

如果请求的路径已由 LegacyRoute 配置为支持路径，那么将使用一个用于生成响应的 Lambda 函数来设置 Handler 属性，如下所示：

```
...
context.Handler = async ctx => {
    HttpResponse response = ctx.Response;
    byte[] bytes = Encoding.ASCII.GetBytes($"URL: {requestedUrl}");
    await response.Body.WriteAsync(bytes, 0, bytes.Length);
};
...
```

HttpContext.Response 属性会返回一个 HttpResponse 对象，该对象可用于创建对客户端的响应，以及提供对发送到客户端的头文件和内容的访问。这里使用 HttpResponse.Body.WriteAsync 方法异步写入一个简单的 ASCII 字符串作为响应，虽然在正式的项目中不应该这么做，但这允许生成响应，而不必选择和渲染视图（下一章会介绍如何在 MVC 中选择可渲染视图）。

当 Handler 属性被设置时，路由系统知道对路由的搜索已经完成，并且可以调用委托来生成对客户端的响应。

1）应用自定义路由类

到目前为止，一直用于创建路由的 MapRoute 扩展方法不支持使用自定义路由类。要使用 LegacyRoute 类，必须采取不同的方法，如代码清单 16-15 所示。

代码清单 16-15 在 UrlsAndRoutes 文件夹下的 Startup.cs 文件中应用自定义路由类

```
using System;
using System.Collections.Generic;
using System.Linq;
using System.Threading.Tasks;
using Microsoft.AspNetCore.Builder;
using Microsoft.AspNetCore.Hosting;
using Microsoft.AspNetCore.Http;
using Microsoft.Extensions.DependencyInjection;
using Microsoft.AspNetCore.Routing.Constraints;
using Microsoft.AspNetCore.Routing;
using UrlsAndRoutes.Infrastructure;

namespace UrlsAndRoutes {
    public class Startup {

        public void ConfigureServices(IServiceCollection services) {
            services.Configure<RouteOptions>(options => {
                options.ConstraintMap.Add("weekday", typeof(WeekDayConstraint));
                options.LowercaseUrls = true;
                options.AppendTrailingSlash = true;
            });
            services.AddMvc();
```

```
        }

        public void Configure(IApplicationBuilder app, IHostingEnvironment env) {
            app.UseDeveloperExceptionPage();
            app.UseStatusCodePages();
            app.UseStaticFiles();
            app.UseMvc(routes => {
                routes.Routes.Add(new LegacyRoute(
                    "/articles/Windows_3.1_Overview.html",
                    "/old/.NET_1.0_Class_Library"));

                routes.MapRoute(
                    name: "default",
                    template: "{controller=Home}/{action=Index}/{id?}");

                routes.MapRoute(
                    name: "out",
                    template: "outbound/{controller=Home}/{action=Index}");
            });
        }
    }
}
```

使用自定义路由类时，必须使用路由集合上的 Add 方法来注册 IRouter 实现类。在本例中，LegacyRoute 构造函数的参数是希望自定义路由支持的旧 URL。可以通过启动应用程序并请求/articles/Windows_3.1_Overview.html 来查看效果。自定义路由会显示请求的 URL，如图 16-6 所示。

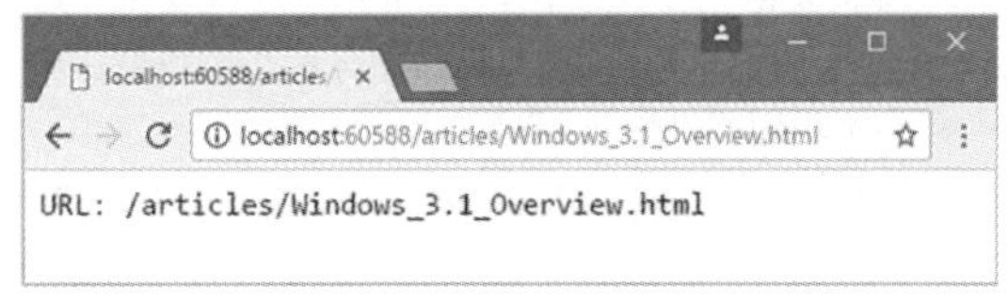

图 16-6 使用自定义路由

2）路由到 MVC 控制器

在简单的 URL 匹配字符串和使用 MVC 系统的控制器、操作和 Razor 视图之间还有很大的差距。但幸运的是，不必在创建自定义路由时实现这些功能，MVC 会在后台完成这些重要工作。为了使用 MVC 的基础架构，在 Controllers 文件夹中添加一个名为 LegacyController.cs 的类文件并用它定义控制器，如代码清单 16-16 所示。

代码清单 16-16 Controllers 文件夹下的 LegacyController.cs 文件的内容

```
using Microsoft.AspNetCore.Mvc;

namespace UrlsAndRoutes.Controllers {

    public class LegacyController : Controller {

        public ViewResult GetLegacyUrl(string legacyUrl)
            => View((object)legacyUrl);
    }
}
```

在这个控制器中，操作方法 GetLegacyUrl 用来接收客户端发来的包含旧 URL 的参数。在真实的项目中，在这个控制器的操作方法中应该取回请求的文件，但这里只是在视图中显示 URL。

提 示

在代码清单 16-16 中，为 View 方法传值的参数为 object 类型。View 方法的一个重载版本可以使用一个字符串来指定要呈现的视图的名称，并且没有发生转换，C#编译器认为这正是我们想要的重载版本。为了避免这种情况，需要转换为 object 类型，以便显式地调用传递视图模型的重载版本，并使用默认视图。也可以通过使用同时具有视图名称和视图模型的重载版本来解决这个问题，但是这里不希望在操作方法和视图之间明确地进行关联。详细信息请参见第 17 章。

创建 Views/Legacy 文件夹，并添加一个名为 GetLegacyUrl.cshtml 的视图，如代码清单 16-17 所示。视图将显示模型的值，在这里显示的是客户端请求的 URL。

代码清单 16-17 Views / Legacy 文件夹下的 GetLegacyUrl.cshtml 文件的内容

```
@model string
@{ Layout = null; }

<!DOCTYPE html>
<html>
<head>
    <meta name="viewport" content="width=device-width" />
    <title>Routing</title>
    <link rel="stylesheet" asp-href-include="lib/bootstrap/dist/css/*.min.css" />
</head>
<body class="m-1 p-1">
    <h2>GetLegacyURL</h2>
    The URL requested was: @Model
</body>
</html>
```

在代码清单 16-18 中，更新 LegacyRoute 类，从而将处理的 URL 路由到 Legacy 控制器中的 GetLegacyUrl 操作方法。

代码清单 16-18 在 LegacyRoute.cs 文件中将 URL 路由到控制器

```
using Microsoft.AspNetCore.Http;
using Microsoft.AspNetCore.Routing;
using System;
using System.Linq;
using System.Text;
using System.Threading.Tasks;
using Microsoft.AspNetCore.Mvc.Internal;
using Microsoft.Extensions.DependencyInjection;

namespace UrlsAndRoutes.Infrastructure {
    public class LegacyRoute : IRouter {
        private string[] urls;
        private IRouter mvcRoute;

        public LegacyRoute(IServiceProvider services, params string[] targetUrls) {
            this.urls = targetUrls;
            mvcRoute = services.GetRequiredService<MvcRouteHandler>();
        }

        public async Task RouteAsync(RouteContext context) {

            string requestedUrl = context.HttpContext.Request.Path
                .Value.TrimEnd('/');

            if (urls.Contains(requestedUrl, StringComparer.OrdinalIgnoreCase)) {
                context.RouteData.Values["controller"] = "Legacy";
                context.RouteData.Values["action"] = "GetLegacyUrl";
                context.RouteData.Values["legacyUrl"] = requestedUrl;
                await mvcRoute.RouteAsync(context);
            }
        }

        public VirtualPathData GetVirtualPath(VirtualPathContext context) {
            return null;
        }
    }
}
```

Microsoft.AspNetCore.Mvc.Internal.MvcRouteHandler 类提供了使用 controller 和 action 片段变量定位控制器类、执行操作方法并将结果返回客户端的机制。这个类可以由自定义的 IRouter 实现来调用，IRouter 实现提供了 controller 和 aciton 的值以及所需的任何其他值，比如操作方法的参数。

代码清单 16-18 创建了一个 MvcRouteHandler 实例，以把任务委派给控制器。为了实现这些，需要给路由提供数据，如下所示：

```
...
context.RouteData.Values["controller"] = "Legacy";
context.RouteData.Values["action"] = "GetLegacyUrl";
context.RouteData.Values["legacyUrl"] = requestedUrl;
...
```

RouteContext.RouteData.Values 属性会返回一个用于向 MvcRouteHandler 类提供数据值的字典。在默认路由系统中，可通过将 URL 模式应用到请求来创建数据值。但在自定义路由类中，已经对值进行了硬编码，以便始终将目标定位到旧版控制器上的 GetLegacyUrl 操作。不同请求之间唯一更改的是 legacyUrl 数据值，它被设置为请求 URL，并将被用作由操作方法接收的相同名称的参数。

在代码清单 16-18 的最后，查找和使用控制器类来处理请求：

```
...
await mvcRoute.RouteAsync(context);
...
```

包含 controller、action 和 legacyUrl 值的 RouteContext 对象将被传递给 MvcRouteHandler 对象的 RouteAsync 方法，该方法负责对请求做进一步处理，包括设置 Handler 属性。因此，LegacyRoute 类可以专注于决定处理哪些 URL，而不会陷入直接使用控制器的细节。

在本例中，MvcRouteHandler 对象必须作为服务进行请求，第 18 章会解释原因。为了向 LegacyRoute 构造函数提供用于创建 MvcRouteHandler 的 IServiceProvider 对象，可以更新路由的定义语句，以便在 Startup 类中提供对应用程序服务的访问，如代码清单 16-19 所示。

代码清单 16-19 在 UrlsAndRoutes 文件夹下的 Startup 类中提供对应用程序服务的访问

```
using System;
using System.Collections.Generic;
using System.Linq;
using System.Threading.Tasks;
using Microsoft.AspNetCore.Builder;
using Microsoft.AspNetCore.Hosting;
using Microsoft.AspNetCore.Http;
using Microsoft.Extensions.DependencyInjection;
using Microsoft.AspNetCore.Routing.Constraints;
using Microsoft.AspNetCore.Routing;
using UrlsAndRoutes.Infrastructure;

namespace UrlsAndRoutes {
    public class Startup {

        public void ConfigureServices(IServiceCollection services) {
            services.Configure<RouteOptions>(options => {
                options.ConstraintMap.Add("weekday", typeof(WeekDayConstraint));
                options.LowercaseUrls = true;
                options.AppendTrailingSlash = true;
            });
            services.AddMvc();
        }

        public void Configure(IApplicationBuilder app, IHostingEnvironment env) {
            app.UseDeveloperExceptionPage();
            app.UseStatusCodePages();
            app.UseStaticFiles();
```

```
            app.UseMvc(routes => {
                routes.Routes.Add(new LegacyRoute(
                    app.ApplicationServices,
                    "/articles/Windows_3.1_Overview.html",
                    "/old/.NET_1.0_Class_Library"));

                routes.MapRoute(
                    name: "default",
                    template: "{controller=Home}/{action=Index}/{id?}");

                routes.MapRoute(
                    name: "out",
                    template: "outbound/{controller=Home}/{action=Index}");
            });
        }
    }
}
```

如果再次启动应用程序并请求地址/articles/Windows_3.1_Overview.html，你将看到简单的文本响应现已被视图的输出替换，如图 16-7 所示。

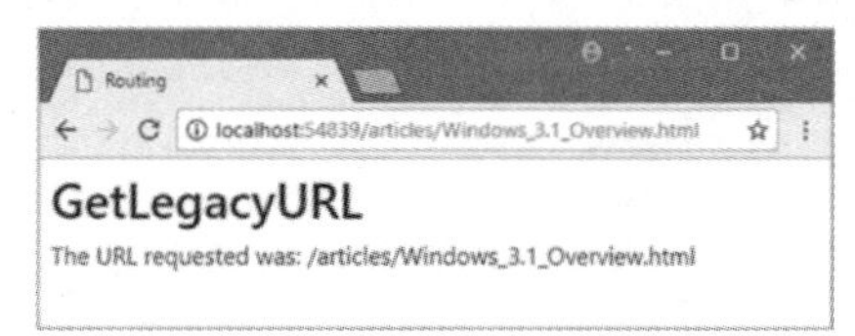

图 16-7 委托处理控制器和操作

2. 创建传出的 URL

为了支持生成传出的 URL，需要在 LegacyRoute 类中实现 GetVirtualPath 方法，如代码清单 16-20 所示。

代码清单 16-20 在 Infrastructure 文件夹下的 LegacyRoute.cs 文件中生成传出的 URL

```
using Microsoft.AspNetCore.Http;
using Microsoft.AspNetCore.Routing;
using System;
using System.Linq;
using System.Text;
using System.Threading.Tasks;
using Microsoft.AspNetCore.Mvc.Internal;
using Microsoft.Extensions.DependencyInjection;

namespace UrlsAndRoutes.Infrastructure {
    public class LegacyRoute : IRouter {
        private string[] urls;
        private IRouter mvcRoute;

        public LegacyRoute(IServiceProvider services, params string[] targetUrls) {
            this.urls = targetUrls;
            mvcRoute = services.GetRequiredService<MvcRouteHandler>();
        }

        public async Task RouteAsync(RouteContext context) {

            string requestedUrl = context.HttpContext.Request.Path
                .Value.TrimEnd('/');

            if (urls.Contains(requestedUrl, StringComparer.OrdinalIgnoreCase)) {
                context.RouteData.Values["controller"] = "Legacy";
                context.RouteData.Values["action"] = "GetLegacyUrl";
                context.RouteData.Values["legacyUrl"] = requestedUrl;
                await mvcRoute.RouteAsync(context);
            }
        }

        public VirtualPathData GetVirtualPath(VirtualPathContext context) {
            if (context.Values.ContainsKey("legacyUrl")) {
                string url = context.Values["legacyUrl"] as string;
```

```
                if (urls.Contains(url)) {
                    return new VirtualPathData(this, url);
                }
            }
            return null;
        }
    }
}
```

路由系统调用 Startup 类中定义的每个路由的 GetVirtualPath 方法，使每个路由都有机会生成应用程序需要的传出 URL。GetVirtualPath 方法的参数是一个 VirtualPathContext 对象，用于提供有关所需 URL 的信息。表 16-5 介绍了 VirtualPathContext 类的属性。

表 16-5 VirtualPathContext 类的属性

属性	描述
RouteName	返回路由的名称
Values	返回可用于片段变量的所有值的字典，并按名称进行索引
AmbientValues	返回有助于生成 URL 的值的字典，但不会将它们合并到结果中。当实现自己的路由类时，这个字典通常为空
HttpContext	返回一个 HttpContext 对象，用于提供有关请求及响应的信息

在本例中，使用 Values 属性获取一个名为 legacyUrl 的值，如果能匹配路由配置支持的 URL，就返回一个 VirtualPathData 对象，从而为路由系统提供 URL 的详细信息。VirtualPathData 构造函数的参数是生成 URL 和 URL 本身的 IRouter。

```
...
return new VirtualPathData(this, url);
...
```

代码清单 16-21 修改了 Result.cshtml 视图文件，以要求以自定义视图为目标的传出 URL。

代码清单 16-21 在 Views/Shared 文件夹下的 Result.cshtml 文件中使用自定义路由类生成传出的 URL

```
@model Result
@{ Layout = null; }

<!DOCTYPE html>
<html>
<head>
    <meta name="viewport" content="width=device-width" />
    <title>Routing</title>
    <link rel="stylesheet" asp-href-include="lib/bootstrap/dist/css/*.min.css" />
</head>
<body class="m-1 p-1">
    <table class="table table-bordered table-striped table-sm">
        <tr><th>Controller:</th><td>@Model.Controller</td></tr>
        <tr><th>Action:</th><td>@Model.Action</td></tr>
        @foreach (string key in Model.Data.Keys) {
            <tr><th>@key :</th><td>@Model.Data[key]</td></tr>
        }
    </table>
    <a asp-route-legacyurl="/articles/Windows_3.1_Overview.html"
       class="btn btn-primary">
        This is an outgoing URL
    </a>
    <p>
        URL: @Url.Action(null, null,
            new { legacyurl = "/articles/Windows_3.1_Overview.html" })
    </p>
</body>
</html>
```

在本例中，不需要为标签助手的传出路由指定控制器和操作，因为它们不在 URL 生成中使用。考虑到这一点，在 a 元素中省略了 asp-controller 和 asp-action 标签助手属性。当生成 URL 时，出于相同的原因，将 Url.Action 辅助方法的前两个参数设置为 null。

如果运行应用程序并在默认 URL 的响应中查看 HTML，你将看到自定义路由类已用于创建 URL，如下所示：

```
<a class="btn btn-primary" href="/articles/windows_3.1_overview.html/">
    This is an outgoing URL
</a>
<p>URL: /articles/windows_3.1_overview.html/</p>
```

附加到 URL 尾部的斜杠是将 Startend.cs 文件中的 AppendTrailingSlash 配置选项设置为 true 的结果，并且需要注意的是，匹配的传入路由能够匹配添加了斜杠的 URL。

提　示

如果在 HTML 响应中看到的 URL 是不同的格式，比如/?legacyurl=%2Farticles%2FWindows_3.1_Overview.html，那么说明自定义路由没有被用于生成 URL，应用程序已经调用了其他的路由规则。由于没有指定控制器或操作，因此 Home 控制器上的 Index 操作方法将成为目标，legacyUrl 值将被添加到 URL 查询字符串中。如果发生这种情况，要在 GetVirtualPath 方法中将 IsBound 属性设置为 true，检查 Startup.cs 文件中的配置是否为 LegacyRoute 构造函数指定了正确的 URL，并且确认自定义路由定义在其他任何路由之前。

16.4　使用区域

ASP.NET Core MVC 支持将 Web 应用程序组织到各个区域（area），每个区域代表应用程序的功能部分，例如管理、计费、客户支持等。这在大型项目中非常有用，因为所有控制器、视图和模型都有一组文件夹可能会变得难以管理。

每个 MVC 区域都有自己的文件夹结构，允许保持一切独立分开。这就可以更明确哪些项目元素与应用程序的每个功能区域相关，可帮助多个开发人员在项目上进行工作而不会相互冲突。区域主要通过路由系统进行支持，这就是将此功能与 URL 和路由一起描述的原因。本节将介绍如何在 MVC 项目中设置和使用区域。

16.4.1　创建区域

为了创建区域，需要将一些文件夹添加到项目中。最上层的文件夹名为 Areas，所有的区域都要包含在该文件夹中。每个区域都包含自己的 Controllers、Views 和 Models 文件夹。在本章中，我们将创建一个名为 Admin 的区域，这意味着需要创建一组文件夹。为了准备示例项目，请创建表 16-6 中的所有文件夹。

表 16-6　　准备区域所需的文件夹

文件夹	描述
Areas	该文件夹将包含 MVC 应用程序中的所有区域
Areas/Admin	该文件夹将包含 Admin 区域的类和视图
Areas/Admin/Controllers	该文件夹将包含 Admin 区域的控制器
Areas/Admin/Views	该文件夹将包含 Admin 区域的视图
Areas/Admin/Views/Home	该文件夹将包含 Admin 区域中 Home 控制器的视图
Areas/Admin/Models	该文件夹将包含 Admin 区域的模型

虽然每个区域都是相互独立的，但许多 MVC 功能依赖于标准的 C#或.NET 功能，如命名空间。为了让区域更容易使用，首先要在 Views 文件夹中引入文件，这样在视图中使用模型时就不需要再写命名空间名，并且可以使用标签助手。在 Areas/Admin/Views 文件夹中创建一个名为_ViewImports.cshtml 的视图并

引入文件，添加的语句如代码清单 16-22 所示。

代码清单 16-22 Areas/Admin/Views 文件夹下的_ViewImports.cshtml 文件的内容

```
@using UrlsAndRoutes.Areas.Admin.Models
@addTagHelper *, Microsoft.AspNetCore.Mvc.TagHelpers
```

16.4.2 创建区域路由

要使用区域，必须在 Startup.cs 文件中添加包含区域片段变量的路由，如代码清单 16-23 所示。

代码清单 16-23 在 UrlsAndRoutes 文件夹下的 Startup.cs 文件中添加区域片段路由

```
using System;
using System.Collections.Generic;
using System.Linq;
using System.Threading.Tasks;
using Microsoft.AspNetCore.Builder;
using Microsoft.AspNetCore.Hosting;
using Microsoft.AspNetCore.Http;
using Microsoft.Extensions.DependencyInjection;
using Microsoft.AspNetCore.Routing.Constraints;
using Microsoft.AspNetCore.Routing;
using UrlsAndRoutes.Infrastructure;

namespace UrlsAndRoutes {
    public class Startup {

        public void ConfigureServices(IServiceCollection services) {
            services.Configure<RouteOptions>(options => {
                options.ConstraintMap.Add("weekday", typeof(WeekDayConstraint));
                options.LowercaseUrls = true;
                options.AppendTrailingSlash = true;
            });
            services.AddMvc();
        }

        public void Configure(IApplicationBuilder app, IHostingEnvironment env) {
            app.UseDeveloperExceptionPage();
            app.UseStatusCodePages();
            app.UseStaticFiles();
            app.UseMvc(routes => {
                routes.MapRoute(
                    name: "areas",
                    template: "{area:exists}/{controller=Home}/{action=Index}");

                routes.Routes.Add(new LegacyRoute(
                    app.ApplicationServices,
                    "/articles/Windows_3.1_Overview.html",
                    "/old/.NET_1.0_Class_Library"));

                routes.MapRoute(
                    name: "default",
                    template: "{controller=Home}/{action=Index}/{id?}");

                routes.MapRoute(
                    name: "out",
                    template: "outbound/{controller=Home}/{action=Index}");
            });
        }
    }
}
```

区域片段变量用于匹配特定区域中控制器的 URL。在这里，按照标准的 URL 格式，可以设置成任何希望的格式。用于对区域添加支持的路由应该出现在不太具体的路由之前，以确保 URL 被正确匹配。exists 约束用于确保请求仅与应用程序中定义的区域匹配。

16.4.3　填充区域

可以像在 MVC 应用程序的主要部分一样，在区域中创建控制器、视图和模型。要创建模型，可右击 Areas/Admin/Models 文件夹，从弹出的菜单中选择 Add→Class 并创建一个名为 Person.cs 的类文件，其中的内容如代码清单 16-24 所示。

代码清单 16-24　Areas/Admin/Models 文件夹下的 Person.cs 文件的内容

```
namespace UrlsAndRoutes.Areas.Admin.Models {
    public class Person {
        public string Name { get; set; }
        public string City { get; set; }
    }
}
```

要创建控制器，可右击 Areas/Admin/Controllers 文件夹，从弹出的菜单中选择 Add→Class 并创建一个名为 HomeController.cs 的类文件，用于定义控制器的内容，如代码清单 16-25 所示。

代码清单 16-25　Areas/Admin/Controllers 文件夹下的 HomeController.cs 文件的内容

```
using Microsoft.AspNetCore.Mvc;
using UrlsAndRoutes.Areas.Admin.Models;

namespace UrlsAndRoutes.Areas.Admin.Controllers {

    [Area("Admin")]
    public class HomeController : Controller {
        private Person[] data = new Person[] {
            new Person { Name = "Alice", City = "London" },
            new Person { Name = "Bob", City = "Paris" },
            new Person { Name = "Joe", City = "New York" }
        };

        public ViewResult Index() => View(data);

    }
}
```

新的控制器是标准控制器，但和标准控制器不同的是，为了将控制器与区域相关联，必须在控制器上设置 Area 特性。

```
...
[Area("Admin")]
public class HomeController : Controller {
...
```

如果没有 Area 特性，控制器就不是区域的一部分，即使控制器是在应用程序的主要部分定义的也是如此。缺少 Area 特性可能导致奇怪的结果。在使用区域时，如果程序没有得到预期的结果，那么要做的第一件事就是检查是否正确添加了区域的 Area 特性。

提　示

如果使用特性设置路由（如第 15 章所述），则可以使用 Route 特性的参数中的[area]标记来引用 Area 特性指定的区域：

```
[Route("[area]/app/[controller]/actions/[action]/{id:weekday?}")]
```

在 Areas/Admin/Views/Home 文件夹中添加一个名为 Index.cshtml 的 Razor 视图，内容见代码清单 16-26。

代码清单 16-26　Areas/Admin/Views/Home 文件夹下的 Index.cshtml 文件的内容

```
@model Person[]
@{ Layout = null; }

<!DOCTYPE html>
```

```
<html>
<head>
    <meta name="viewport" content="width=device-width" />
    <title>Areas</title>
    <link rel="stylesheet" asp-href-include="lib/bootstrap/dist/css/*.min.css" />
</head>
<body class="m-1 p-1">
    <table class="table table-bordered table-striped table-sm">
        <tr><th>Name</th><th>City</th></tr>
        @foreach (Person p in Model) {
            <tr><td>@p.Name</td><td>@p.City</td></tr>
        }
    </table>
</body>
</html>
```

Index 视图的模型是一个 Person 对象数组。由于在代码清单 16-26 中已经为视图引入文件，因此在这里可以直接使用 Person 类型而不需要再写命名空间。运行应用程序并请求 URL /Admin 以测试创建的区域，结果如图 16-8 所示。

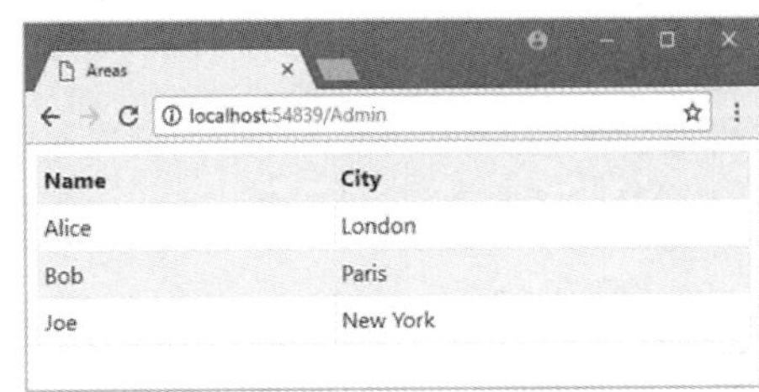

图 16-8 使用区域

> **了解区域对 MVC 应用程序的影响**
>
> 了解区域对其余应用程序的影响是非常重要的。这里创建了名为 Admin 的区域，但在应用程序中还有名为 Admin 的控制器。在创建区域之前，对/Admin URL 的请求将被指向程序主目录中 Admin 控制器的 Index 操作。在添加区域后，请求将被指向 Admin 区域中 Home 控制器的 Index 操作(区域为 controller 和 action 片段变量提供了默认值)。这种更改可能会带来很多意外的情况，使用区域的最佳实践是将其用于项目的初始控制器命名方案。如果要在已经建立的应用程序中添加区域，就必须仔细考虑区域可能对路由带来的影响。

16.4.4 生成区域中指向操作的链接

不需要采取任何特殊步骤就能创建指向当前请求所在的同一 MVC 区域中操作的链接。MVC 检测到请求与特定区域相关，并确保传出的 URL 只能在为该区域定义的路由中找到匹配项。例如，在 Areas/Admin/Views/Home 文件夹下的 Index.cshtml 文件中添加一个新的元素，如代码清单 16-27 所示。

代码清单 16-27 在 Areas/Admin/Views/Home 文件夹下的 Index.cshtml 文件中添加锚点

```
@model Person[]
@{ Layout = null; }

<!DOCTYPE html>
<html>
<head>
    <meta name="viewport" content="width=device-width" />
    <title>Areas</title>
    <link rel="stylesheet" asp-href-include="lib/bootstrap/dist/css/*.min.css" />
</head>
<body class="m-1 p-1">
    <table class="table table-bordered table-striped table-sm">
        <tr><th>Name</th><th>City</th></tr>
        @foreach (Person p in Model) {
            <tr><td>@p.Name</td><td>@p.City</td></tr>
        }
    </table>
```

```
    <a asp-action="Index" asp-controller="Home">Link</a>
</body>
</html>
```

如果运行应用程序并请求 URL /admin，你将看到响应包含以下元素：

```
<a href="/admin/">Link</a>
```

路由系统选择了用于生成传出 URL 的区域路由，并考虑了可用于 controller 和 action 片段变量的默认值。

必须为路由系统提供区域片段的值，以创建指向应用程序的不同区域或主要部分中的操作的链接，如代码清单 16-28 所示。

代码清单 16-28 在 Areas/Admin/Views/Home 文件夹下的 Index.cshtml 文件中生成指向不同区域的链接

```
@model Person[]
@{ Layout = null; }

<!DOCTYPE html>
<html>
<head>
    <meta name="viewport" content="width=device-width" />
    <title>Areas</title>
    <link rel="stylesheet" asp-href-include="lib/bootstrap/dist/css/*.min.css" />
</head>
<body class="m-1 p-1">
    <table class="table table-bordered table-striped table-sm">
        <tr><th>Name</th><th>City</th></tr>
        @foreach (Person p in Model) {
            <tr><td>@p.Name</td><td>@p.City</td></tr>
        }
    </table>
    <a asp-action="Index" asp-controller="Home">Link</a>
    <a asp-action="Index" asp-controller="Home" asp-route-area="">Link</a>
</body>
</html>
```

asp-route-area 属性用于设置区域片段变量的值。在这种情况下，将这个属性设置为空字符串，以指向应用程序的主要部分，并生成以下 HTML 元素：

```
<a href="/">Link</a>
```

如果控制器中有多个区域并且想要路由到它们，则需要使用区域名称代替上面的空字符串。

16.5 URL 模式最佳实践

在介绍了上面这些内容后，你可能还不知道从哪里开始设计自己的 URL 模式。近年来，应用程序的 URL 设计越来越受到重视，一些重要的设计原则也应运而生。如果遵循这些设计模式，将能够提高应用程序的可用性、兼容性和搜索引擎排名。

16.5.1 保持 URL 的整洁性

下面是提高 URL 整洁性的一些建议。

- 设计的 URL 应该用来描述内容而不是体现应用程序实现的细节，比如应该使用/Articles/AnnualReport 而不是 Website_v2/CachedContentServer/FromCache/AnnualReport。
- 尽量在 URL 中使用内容标题而不是数字 ID，比如应尽量使用/Articles/AnnualReport 而不是/Articles/2392。如果必须使用数字 ID（以区分具有相同标题的项，或避免使用标题查找项所需的额外数据库查询），则建议使用/Articles/2392/AnnualReport。虽然这个 URL 看起来需要更长的录入时间，但是它对用户来说更有意义，并且有助于提高搜索引擎排名。应用可以直接忽略标题，利用 ID 来查找与之匹配的页面。

- 不要使用 HTML 页面的文件扩展名（例如.aspx 或.mvc），但是要使用专门的文件类型（例如.jpg、.pdf 和.zip）。必须正确设置 MIME 类型，Web 浏览器不关心文件扩展名，但人们仍然希望 PDF 文件以.pdf 结尾。
- 创建的 URL 要具有层次感（例如/Products/Menswear/Shirts/Red），以便访问者可以猜测父类别的 URL。
- 不区分大小写。默认情况下，ASP.NET Core 路由系统不区分大小写。
- 避免使用符号、代码和字符序列。如果需要单词分隔符，可使用连字符（比如/my-great-article）。下画线并不友好。在 URL 编码中，空格的编码很奇怪（/my+great+article），也让人难受（/my%20great%20article）。
- 不要更改 URL。无效的链接会导致丢失业务。当更改 URL 时，可以通过重定向继续支持旧的 URL 模式。
- 保持一致。在整个应用程序中采用一种 URL 格式。

网址应该很简单，易于输入，并且人性化（可编辑）且具有持久性，还应该可视化网站结构。URL 可用性专家 Jakob Nielsen 扩展了这个主题。Web 专家 Tim Berners-Lee 也提供了类似的建议。

16.5.2 GET 方法和 POST 方法：选择最合适的方法

经验法则是，将 GET 请求应用于所有只读信息的检索，而将 POST 请求应用于任何更改应用程序状态的操作。在符合标准的条款中，GET 请求是为了安全交互（除信息检索以外没有副作用），而 POST 请求则用于不安全的交互（做出决定或更改某些内容）。这些约定是由万维网联盟（W3C）制定的。

GET 请求是可寻址的，所有的信息都包含在 URL 中，因此可以将这些网址收藏为书签或创建链接。

需要在更改状态时使用 GET 方法，当 Google Web Accelerator 在 2005 年向公众发布时，很多网站开发人员有过惨痛的教训。Google Web Accelerator 收录每个页面的所有链接，这在 HTTP 中是合法的，因为 GET 请求应该是安全的。但是，许多 Web 开发人员忽略了 HTTP 约定，并在应用程序中使用了“删除商品”或“添加到购物车”的链接，混乱也就随之而来。

一家公司认为自己的管理系统正在遭受攻击，因为所有的内容都在不断地被删除，后来该公司才发现原来是搜索引擎抓取工具抓取了管理页面的 URL，并且在抓取所有删除数据的链接。如果有身份验证，就可以保护你不受此影响，但无法让你免受搜索引擎爬虫的困扰。

16.6 小结

本章展示了路由系统的高级功能，如何生成出站链接和 URL，以及如何自定义路由系统。此外，本章还介绍了区域的概念，并讨论了如何创建有效且有意义的 URL 模式。下一章将介绍控制器和操作，这是 ASP.NET Core MVC 的核心内容，除详细解释这些内容之外，还将展示如何使用它们在应用中获得最佳效果。

第 17 章　控制器和操作

每个应用程序的请求都由控制器处理。在 ASP.NET Core MVC 中，控制器是包含处理请求所需逻辑的.NET 类。第 3 章介绍过控制器的作用是封装应用程序逻辑。这意味着控制器负责处理传入的请求，对模型执行操作，并选择要呈现给用户的视图。

只要不偏离属于模型和视图的责任区域，控制器就可以自由地处理自认为合适的任何方式的请求。这意味着控制器既不包含或存储数据，也不生成用户界面。

本章将展示如何实现控制器以及使用控制器接收和生成数据的不同方式。表 17-1 展示了在上下文中使用的控制器。

表 17-1　在上下文中使用的控制器

问题	答案
什么是控制器？	控制器包含用于接收请求、更新应用程序状态或模型以及选择将发送给客户端的响应的逻辑
控制器有什么用？	控制器是 MVC 项目的核心，并包含 Web 应用程序的逻辑
如何使用控制器？	控制器是调用公共方法来处理 HTTP 请求的 C#类。这些方法可以负责直接为客户生成响应，但更常见的方式是返回操作结果，告诉 MVC 应该如何去响应
使用控制器是否有一些隐患或局限性？	刚开始使用 MVC 时，可以轻松创建包含更适合于模型或视图功能的控制器。一个更为具体的问题是任何以 Controller 为名的公共类都会被假定为 MVC 的控制器，这意味着一些可能不是作为控制器的类被意外用于处理 HTTP 请求
是否有其他的替代选择？	没有，控制器是 MVC 应用程序的核心部分

表 17-2 总结了本章要介绍的操作。

表 17-2　本章要介绍的操作

操作	方法	代码清单
定义控制器	创建一个公共类，名称以 Controller 为结尾或派生自 Controller 类	代码清单 17-7～代码清单 17-9
获取 HTTP 请求的详细信息	使用 context 对象或定义操作方法参数	代码清单 17-10～代码清单 17-13
从操作方法生成结果	直接使用 context 结果对象或创建操作结果对象	代码清单 17-14～代码清单 17-16
生成 HTML 结果	创建视图结果	代码清单 17-17～代码清单 17-24
重定向客户端	创建重定向结果	代码清单 17-25～代码清单 17-30
向客户端返回内容	创建内容结果	代码清单 17-31～代码清单 17-35
返回 HTTP 状态码	创建 HTTP 结果	代码清单 17-36 和代码清单 17-37

17.1　准备示例项目

在本章中，将使用 ASP.NET Core Web Application（.NET Core）模板创建一个名为 ControllersAndActions 的新的 Empty 项目。代码清单 17-1 在 Startup 类中添加了一些语句，以启用 MVC 框架和其他中间件。

注　意

本章包括用于关键功能的单元测试。为简洁起见，没有将单元测试项目包含在创建示例项目的说明中。可以按照第 7 章中描述的过程创建测试项目，或从本书的 GitHub 存储库下载该项目。

代码清单 17-1 在 ControllersAndActions 文件夹下的 Startup.cs 文件中添加 MVC 和其他中间件

```
using System;
using System.Collections.Generic;
using System.Linq;
using System.Threading.Tasks;
using Microsoft.AspNetCore.Builder;
using Microsoft.AspNetCore.Hosting;
using Microsoft.AspNetCore.Http;
using Microsoft.Extensions.DependencyInjection;

namespace ControllersAndActions {
    public class Startup {
        public void ConfigureServices(IServiceCollection services) {
            services.AddMvc();
            services.AddMemoryCache();
            services.AddSession();
        }

        public void Configure(IApplicationBuilder app, IHostingEnvironment env) {
            app.UseStatusCodePages();
            app.UseDeveloperExceptionPage();
            app.UseStaticFiles();
            app.UseSession();
            app.UseMvcWithDefaultRoute();
        }
    }
}
```

AddMemoryCache 和 AddSession 方法将创建会话管理所需的服务。UseSession 方法用于将一个中间件添加到管道中，将会话数据与请求关联起来，并将 cookie 添加到响应中，以确保将来的请求能够被识别。必须在 UseMvc 方法之前调用 UseSession 方法，以便会话组件能够在到达 MVC 中间件之前截获请求，并且可以在生成后修改响应。其他几个方法用于设置第 14 章中描述的标准包。

准备视图

本章的重点是控制器及其操作方法，本章将定义控制器类。这里将定义一些视图，它们可以帮助展示控制器类的工作方式。这些视图创建在 Views/Shared 文件夹中，以便可以在本章后面创建的任何控制器中使用它们。创建 Views/Shared 文件夹，添加一个名为 Result.cshtml 的 Razor 视图文件，内容如代码清单 17-2 所示。

代码清单 17-2 Views/Shared 文件夹下的 Result.cshtml 文件中的内容

```
@model string
@{ Layout = null; }

<!DOCTYPE html>
<html>
<head>
    <meta name="viewport" content="width=device-width" />
    <title>Controllers and Actions</title>
    <link rel="stylesheet" asp-href-include="lib/bootstrap/dist/css/*.min.css" />
</head>
<body class="m-1 p-1">
    Model Data: @Model
</body>
</html>
```

用于 Result 视图的模型是一个字符串，这将允许显示简单的消息。接下来，在 Views/Shared 文件夹中创建一个名为 DictionaryResult.cshtml 的文件，内容如代码清单 17-3 所示，用于 DictionaryResult 视图的模型是一个字典，用于显示比前一个视图更复杂的数据。

代码清单 17-3 Views/Shared 文件夹下的 DictionaryResult.cshtml 文件的内容

```
@model IDictionary<string, string>
@{ Layout = null; }

<!DOCTYPE html>
<html>
<head>
    <meta name="viewport" content="width=device-width" />
    <title>Controllers and Actions</title>
    <link rel="stylesheet" asp-href-include="lib/bootstrap/dist/css/*.min.css" />
</head>
<body class="m-1 p-1">
    <table class="table table-bordered table-sm table-striped">
        <tr><th>Name</th><th>Value</th></tr>
        @foreach (string key in Model.Keys) {
            <tr><td>@key</td><td>@Model[key]</td></tr>
        }
    </table>
</body>
</html>
```

接下来，在 Views/Shared 文件夹中创建一个名为 SimpleForm.cshtml 的文件，内容如代码清单 17-4 所示。顾名思义，SimpleForm 视图包含一个简单的 HTML 表单，用于提交数据。

代码清单 17-4 Views/Shared 文件夹下的 SimpleForm.cshtml 文件的内容

```
@{ Layout = null; }
<!DOCTYPE html>
<html>
<head>
    <meta name="viewport" content="width=device-width" />
    <title>Controllers and Actions</title>
    <link rel="stylesheet" asp-href-include="lib/bootstrap/dist/css/*.min.css" />
</head>
<body class="m-1 p-1">
    <form method="post" asp-action="ReceiveForm">
        <div class="form-group">
            <label for="name">Name:</label>
            <input class="form-control" name="name" />
        </div>
        <div class="form-group">
            <label for="name">City:</label>
            <input class="form-control" name="city" />
        </div>
        <button class="btn btn-primary center-block" type="submit">Submit</button>
    </form>
</body>
</html>
```

这些视图使用内置的标签助手从路由系统生成 URL。要启用标签助手，可在 Views 文件夹中创建一个名为_ViewImports.cshtml 的视图导入文件，内容如代码清单 17-5 所示。

代码清单 17-5 Views 文件夹下的_ViewImports.cshtml 文件的内容

```
@addTagHelper *, Microsoft.AspNetCore.Mvc.TagHelpers
```

在 Views/Shared 文件夹中创建的视图都依赖于 Bootstrap CSS 包。要将 Bootstrap 添加到项目中，可使用 Bower Configuration File 模板创建 bower.json 文件，添加的内容如代码清单 17-6 所示。

代码清单 17-6 在 ControllersAndActions 文件夹下的 bower.json 文件中添加包

```
{
  "name": "asp.net",
  "private": true,
  "dependencies": {
    "bootstrap": "4.0.0-alpha.6"
  }
}
```

17.2 理解控制器

控制器是 C#类，其公共方法（称为 action 或操作方法）负责处理 HTTP 请求并准备将返回给客户端的响应。MVC 使用第 15 章和第 16 章描述的路由系统来确定处理请求所需的控制器类和操作方法。然后，MVC 创建一个控制器类的新实例，调用操作方法，并使用调用结果来向客户端输出响应。

MVC 提供了上下文数据的操作方法，以便它们能够理解如何处理请求。有大量的上下文数据，它们描述了当前请求的所有内容、正在准备的响应、由路由系统提取的数据以及用户身份的详细信息。

当 MVC 调用操作方法时，操作方法的响应描述了应该发送给客户端的响应。最常见的响应是创建并渲染 Razor 视图，所以操作方法使用响应来告诉 MVC 要使用哪个视图，以及应该提供哪些视图模型数据。此外还能提供其他类型的响应，并且操作方法可以从 MVC 中发送 HTTP 重定向到客户端以发送复杂的数据对象。

这意味着可从 3 个重要的方面来理解控制器。首先，了解如何定义控制器，以便 MVC 可以使用它们来处理请求。控制器只是 C#类，但和一般类的创建方法不同，了解它们之间的差异很重要。17.3 节将解释如何定义控制器。

其次，了解 MVC 如何提供上下文数据的操作方法也是很重要的。获取所需的上下文数据对于有效的 Web 应用开发非常重要，通过定义一组用于描述操作方法所需的所有内容的类，MVC 可以使 Web 应用开发变得容易。17.4 节将解释 MVC 如何描述请求和响应。

最后，了解操作方法如何产生响应也很重要。操作方法很少需要自己生成 HTTP 响应，你需要知道如何使用 MVC 生成所需的响应，17.5 节将对此进行解释。

17.3 创建控制器

到目前为止，几乎所有章节都使用了控制器。现在是时候回头来看看它们是如何定义的。本节将描述控制器的不同创建方式，并解释它们之间的差异。

17.3.1 创建 POCO 控制器

MVC 配置十分方便灵活，这意味着 MVC 应用程序中的控制器是自动发现的，而不用在配置文件中定义。基本的发现过程很简单：名称以 Controller 结尾的任何公共类都是控制器，其中定义的任何公共方法都是操作方法。为了演示这是如何工作的，在项目中添加 Controllers 文件夹，并在其中添加一个名为 PocoController.cs 的类文件，内容如代码清单 17-7 所示。

提 示

虽然约定是将控制器放在 Controllers 文件夹中，但是可以将它们放在项目的任何位置，MVC 仍然会自动找到它们。

代码清单 17-7 Controllers 文件夹下的 PocoController.cs 文件的内容

```
namespace ControllersAndActions.Controllers {

    public class PocoController {
```

```
        public string Index() => "This is a POCO controller";
    }
}
```

PocoController 类满足了 MVC 控制器的基础标准，它定义了一个名为 Index 的公共方法，可作为一个操作方法并返回一个字符串。

PocoController 类是 POCO 控制器的范例，其中 POCO 表示“普通的旧 CLR 对象”，并且指的是使用标准.NET 功能实现控制器，而不直接依赖于 ASP.NET Core MVC 提供的 API。

要测试 POCO 控制器，可启动应用程序并请求 URL /Poco/Index/。路由系统将使用默认 URL 模式匹配请求，并将请求指向 PocoController 类的 Index 方法，结果如图 17-1 所示。

图 17-1 使用 POCO 控制器

> **使用属性标识调整控制器**
>
> 对 POCO 控制器的支持并不总是按希望的方式工作。一个常见的问题是，MVC 将以单元测试中的假类作为控制器。避免这个问题的最简单方法是注重类名，避免名字类似于 FakeController。如果无法避免，那么可以使用命名空间 Microsoft.AspNetCore.Mvc 中定义的 NonController 属性，用于告诉 MVC 这个类不是控制器类。NonAction 属性可应用于方法，以阻止它们被作为操作方法使用。
>
> 在某些项目中，可能无法在作为 POCO 控制器的类上遵循命名约定。通过应用 Controller 属性（在命名空间 Microsoft.AspNetCore.Mvc 中定义），可以告诉 MVC 这是控制器类，即使不符合 POCO 选择标准。

使用 MVC API

PocoController 类是 MVC 识别控制器以及简单使用控制器的有用演示。但是，不依赖 Microsoft.AspnetCore 命名空间的纯 POCO 控制器并不是特别有用，因为它们无法访问 MVC 为处理请求提供的功能。

可以通过从 Microsoft. AspnetCore 命名空间创建新的实例来访问 MVC API 的某些功能。作为一个简单的例子，POCO 类可以通过从操作方法返回一个 ViewResult 对象来使 MVC 呈现 Razor 视图，如代码清单 17-8 所示（详见 17.5 节的 ViewResult 类）。

代码清单 17-8 在 Controllers 文件夹下的 PocoController.cs 文件中使用 ASP.NET API

```
using Microsoft.AspNetCore.Mvc;
using Microsoft.AspNetCore.Mvc.ModelBinding;
using Microsoft.AspNetCore.Mvc.ViewFeatures;
namespace ControllersAndActions.Controllers {

    public class PocoController {

        public ViewResult Index() => new ViewResult() {
                ViewName = "Result",
                ViewData = new ViewDataDictionary(
                    new EmptyModelMetadataProvider(),
                    new ModelStateDictionary()) {
                        Model = $"This is a POCO controller"
                    }
            };
    }
}
```

这不再是纯 POCO 控制器，因为它直接依赖于 MVC API。除纯粹之外，它比前一个例子有用得多，因为它要求 MVC 渲染一个 Razor 视图。

但是，代码很复杂。为了创建一个 ViewResult 对象，需要创建 ViewDataDictionary、EmptyModelMetadataProvider

和 ModelStateDictionary 对象，这需要访问 3 个不同的命名空间（后面的章节将描述这些类型所涉及的特性）。此例的重点是说明 MVC 提供的功能可以直接访问，即使结果有点混乱。

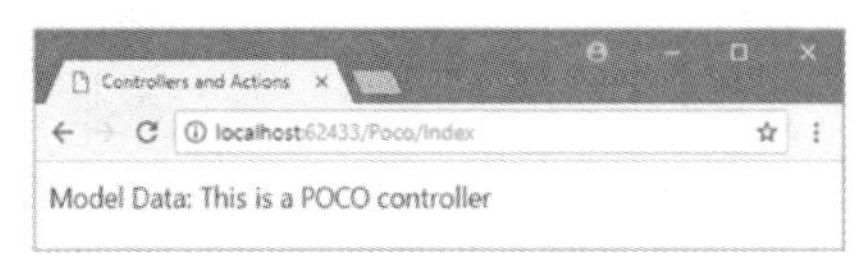

图 17-2 直接使用 MVC API 请求的结果

我们使用一个字符串作为视图模型来渲染 Result.cshtml，如果你运行应用程序并请求 URL /Poco/Index，你看到的结果将如图 17-2 所示。

17.3.2 使用控制器基类

之前的例子展示了如何从 POCO 控制器开始，并对其进行扩展来访问 MVC 功能。这种方法揭示了 MVC 的工作原理，这是很有用的知识，但是这样会使 POCO 控制器难以编写、阅读和维护。

创建控制器的一种更简单的方法是从 Microsoft.AspNetCore.Mvc.Controller 类派生类，基类定义了一些方法和属性，从而以更简洁有用的方式访问 MVC 功能。为了演示，将一个名为 DerivedController.cs 的类文件添加到 Controllers 文件夹中，内容如代码清单 17-9 所示。

代码清单 17-9 在 Controllers 文件夹下的由 Controller 类派生出的 DerivedController.cs 文件

```
using Microsoft.AspNetCore.Mvc;

namespace ControllersAndActions.Controllers {

    public class DerivedController : Controller {

        public ViewResult Index() =>
            View("Result", $"This is a derived controller");
    }
}
```

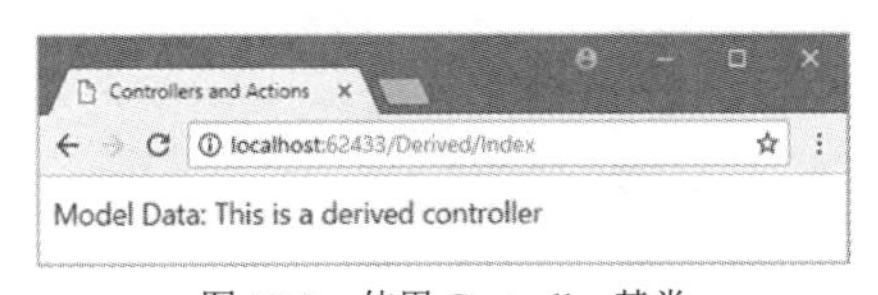

图 17-3 使用 Controller 基类

如果运行应用程序并访问/Derived/Index，你看到的结果将如图 17-3 所示。

代码清单 17-9 中的控制器执行与代码清单 17-8 中的控制器相同的操作（要求 MVC 渲染具有字符串视图模型的视图），但是使用 Controller 基类意味着可以更简单地实现结果。

关键的变化是，可以使用 View 方法创建渲染 Razor 视图所需的 ViewResult 对象，而不必直接在操作方法中进行实例化。View 方法继承自 Controller 基类，ViewResult 对象仍然以相同的方式创建，只是没有多余代码来扰乱操作方法。从 Controller 类进行派生不会改变控制器的工作方式，而只是简化了编写的代码以完成常见任务。

注 意

MVC 会为要求处理的每个请求创建一个控制器类的新实例，这意味着不需要同步对操作方法或实例属性和字段的访问。第 18 章将要描述的共享对象（包括数据库和单例服务）可以同时使用，并且必须根据规则来编写。

17.4 接收上下文数据

无论如何定义控制器，它们几乎都不会孤立存在，通常需要从传入的请求中访问数据，例如查询字符串值、表单值以及路由系统从 URL 中解析的参数，统称为上下文数据。访问上下文数据有 3 种主要方式：

- 从一组 Context 对象中提取。
- 接收数据作为操作方法的参数。

- 明确地调用框架的模型绑定功能。

这里将介绍如何在操作方法获取输入，重点是使用 Context 对象和操作方法参数。第 26 章将介绍模型绑定。

17.4.1 从 Context 对象中接收数据

使用 Controller 基类创建控制器的主要优点之一是可以方便地访问描述当前请求、正在准备的响应和应用程序状态的一组 Context 对象。表 17-3 描述了用于上下文数据的一些有用的控制器类属性。

表 17-3 用于上下文数据的一些有用的控制器类属性

属性	描述
Request	返回一个 HttpRequest 对象，用于描述从客户端收到的请求
Response	返回用于创建客户端响应的 HttpResponse 对象
HttpContext	返回一个 HttpContext 对象，它是由其他属性（如请求和响应）返回的许多对象的源，它还提供了有关可用的 HTTP 功能的信息以及对低级功能（如 Web 套接字）的访问
RouteData	返回由路由系统在匹配请求时生成的 RouteData 对象
ModelState	返回一个 ModelStateDictionary 对象，用于验证客户端发送的数据
User	返回一个 ClaimsPrincipal 对象，用于描述已发出请求的用户

许多控制器的编写不需要使用表 17-3 中的属性，因为上下文数据也可以通过后面几章描述的功能提供，这更符合 MVC 开发风格。例如，大多数控制器不需要使用 Request 属性来获取正在处理的 HTTP 请求的详细信息，因为通过第 26 章描述的模型绑定过程可以获得相同的数据。

但是，理解和使用 Context 对象仍然有用，而且它们对于调试很有用。代码清单 17-10 使用 Request 属性来访问 HTTP 请求中的标头。

代码清单 17-10 在 Controllers 文件夹下的 DerivedController.cs 文件中使用上下文数据

```
using Microsoft.AspNetCore.Mvc;
using System.Linq;

namespace ControllersAndActions.Controllers {

    public class DerivedController : Controller {

        public ViewResult Index() =>
            View("Result", $"This is a derived controller");

        public ViewResult Headers() => View("DictionaryResult",
                Request.Headers.ToDictionary(kvp => kvp.Key,
                    kvp => kvp.Value.First()));
    }
}
```

使用 Context 对象意味着浏览一系列不同的类型和命名空间。用来获取有关列表中 HTTP 请求的上下文数据的 Controller.Request 属性会返回一个 HttpRequest 对象。表 17-4 描述了在编写控制器时有用的 HttpRequest 属性。

表 17-4 有用的 HttpRequest 属性

属性	描述
Path	返回请求 URL 的路径部分
QueryString	返回请求 URL 的查询字符串部分
Headers	返回按名称索引的请求标头的字典
Body	返回可用于读取请求正文的流
Form	返回请求中表单数据的字典（按名称索引）
Cookies	返回按名称索引的请求 cookie 的字典

可使用 Request.Headers 来获取标头的字典，并使用 LINQ 对其进行处理。

```
...
View("DictionaryResult", Request.Headers.ToDictionary(kvp => kvp.Key,
      kvp => kvp.Value.First()));
...
```

由 Request.Headers 属性返回的字典使用 StringValues 结构存储每个标头的值，该结构在 ASP.NET 中用于表示字符串值序列。HTTP 客户端可以为 HTTP 标头发送多个值，但是这里只想显示第一个值。可使用 LINQ ToDictionary 方法为每个标题接收一个 KeyValuePair<String,StringValues>对象，并选择第一个值。结果是包含字符串值的字典，可以由 DictionaryResult 视图显示。如果运行应用程序并请求 URL /Derived/Headers，你将看到的输出类似于图 17-4（根据使用的浏览器，标头及值将有所不同）。

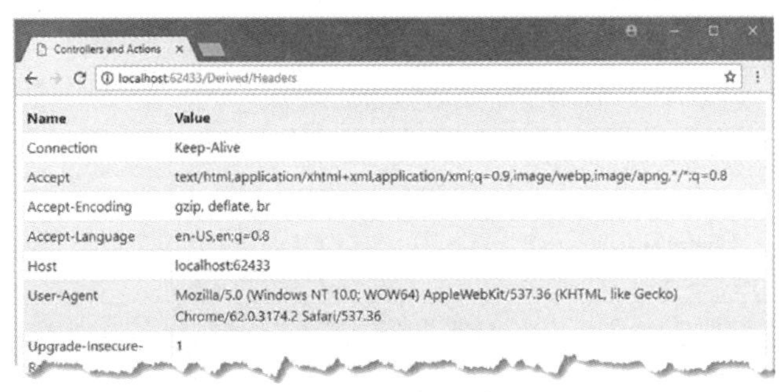

图 17-4　显示上下文数据

在 POCO 控制器中获取上下文数据

即使它们在常规项目中不是特别有用，POCO 控制器也能让我们在幕后查看 MVC 是如何操作的。在 POCO 控制器中获取上下文数据是一个问题，因为不能只实例化自己的 HttpRequest 或 HttpResponse 对象，还需要那些已由 ASP.NET 创建并由所有在处理请求时填充了数据字段的中间件更新的对象。

要获取上下文数据，POCO 控制器必须要求 MVC 提供在前。代码清单 17-11 更新了 PocoController 类以添加显示 HTTP 请求标头的操作方法。

代码清单 17-11　在 Controllers 文件夹下的 PocoController.cs 文件中显示上下文数据

```
using Microsoft.AspNetCore.Mvc;
using Microsoft.AspNetCore.Mvc.ModelBinding;
using Microsoft.AspNetCore.Mvc.ViewFeatures;
using System.Linq;

namespace ControllersAndActions.Controllers {

    public class PocoController {

        [ControllerContext]
        public ControllerContext ControllerContext { get; set; }

        public ViewResult Index() => new ViewResult() {
            ViewName = "Result",
            ViewData = new ViewDataDictionary(new EmptyModelMetadataProvider(),
                            new ModelStateDictionary()) {
                Model = $"This is a POCO controller"
            }
        };
        public ViewResult Headers() =>
            new ViewResult() {
                ViewName = "DictionaryResult",
                ViewData = new ViewDataDictionary(
```

```
                    new EmptyModelMetadataProvider(),
                    new ModelStateDictionary()) {
                        Model = ControllerContext.HttpContext.Request.Headers
                        .ToDictionary(kvp => kvp.Key, kvp => kvp.Value.First())
                }
            };
        }
    }
```

为了获取上下文数据，可定义一个名为ControllerContext的属性，类型为ControllerContext，并使用同名的ControllerContext属性进行装饰。

下面介绍3种不同的ControllerContext的用途。首先，Microsoft.AspNetCore.Mvc命名空间中定义的ControllerContext类用于使用ControllerContext的属性（见表17-5）汇集控制器的操作方法所需的所有Context对象。

表17-5 ControllerContext的属性

属性	描述
ActionDescriptor	返回一个ActionDescriptor对象，用于描述操作方法
HttpContext	返回一个HttpContext对象，用于提供HTTP请求的详细信息和发送回客户端的HTTP响应
ModelState	返回一个ModelStateDictionary对象，用于验证客户端发送的数据
RouteData	返回一个RouteData对象，用于描述路由系统处理请求的方式

通过ControllerContext.HttpContext属性可访问HTTP相关数据，该属性会返回一个Microsoft.AspNetCore.Http.HttpContext对象。HttpContext类合并了几个用于描述请求的不同方面的对象，可通过HttpContext的属性（见表17-6）进行访问。

表17-6 HttpContext的属性

属性	描述
Connection	返回一个ConnectionInfo对象，用于描述与客户端的低级别连接
Request	返回一个HttpRequest对象，用于描述从客户端接收的HTTP请求
Response	返回一个HttpResponse对象，用于创建将返回给客户端的响应
Session	返回一个用于描述与请求关联的会话的ISession对象
User	返回一个ClaimPrincipal对象，用于描述与请求关联的用户

其次，ControllerContext属性用于描述代码清单17-11中的属性，并告诉MVC使用描述当前请求的ControllerContext对象设置属性值。这使用了一种称为依赖注入的技术，这将在第18章中介绍，MVC将使用该属性，在使用操作方法处理请求之前为控制器提供上下文数据。

最后，ControllerContext是属性的名称。可以在自己的POCO控制器中使用任何合法的C#属性名称，在幕后，Controller类依赖于相同的ControllerContext类的上下文数据，该类使用相同的ControllerContext属性进行装饰。表17-3中描述的所有控制器属性都是直接使用ControllerContext属性的更方便、更简洁的替代方法，这正是控制器类提供的属性的使用情况。例如，下面是控制器类中的HttpContext属性的定义：

```
...
public HttpContext HttpContext {
    get {
        return ControllerContext.HttpContext;
    }
}
...
```

HttpContext属性只是获取ControllerContext. HttpContext属性值的一种手段。控制器基类没有魔法：它们只不过是更简单、更清晰的控制器，因为常见任务已被整合到方法和属性中。如有需要，可以在POCO控制器中重新创建自己所需的内容。ASP.NET Core MVC中的很多功能都是非常简单的，深入挖掘细节，你将发现并没有特别的内容，只有精心设计的功能，并且提供了一套详细的 NuGet 包。如果有时间，建议从 GitHub 下载 MVC 源代码并进行浏览，从而确认这一点。

17.4.2 使用操作方法参数

一些上下文数据也可以通过操作方法参数来接收，这可以让我们编写更自然且优雅的代码。一个常见的例子是当操作方法需要接收用户提交的表单数据时。为了比较，下面演示如何通过 Context 对象获取表单数据，然后通过操作方法获取参数。

表单数据可通过 Controller 类的 Request.Form 属性访问。为了演示，添加一个名为 HomeController.cs 的类文件，并用它定义派生控制器，如代码清单 17-12 所示。

代码清单 17-12 Controllers 文件夹下的 HomeController.cs 文件的内容

```
using Microsoft.AspNetCore.Mvc;

namespace ControllersAndActions.Controllers {

    public class HomeController : Controller {

        public ViewResult Index() => View("SimpleForm");

        public ViewResult ReceiveForm() {
            var name = Request.Form["name"];
            var city = Request.Form["city"];
            return View("Result", $"{name} lives in {city}");
        }
    }
}
```

以上控制器中的 Index 操作方法将呈现本章开头在 Views/Shared 中创建的 SimpleForm 视图。我们需要关注 ReceiveForm 方法，以使用 HttpRequest context 对象从请求中获取表单数据。

如表 17-4 所示，由 HttpRequest 类定义的 Form 属性将返回一个包含表单数据的集合，并由关联的 HTML 元素的名称进行索引。SimpleForm 视图中有两个输入元素（name 和 city），可从 Context 对象中提取它们的值，并使用它们创建一个字符串作为模型传递给 Result 视图。

如果运行应用程序并请求 URL——/Home，将会显示一个表单。如果填写字段并单击 Submit 按钮，浏览器将发送表单数据作为 HTTP POST 请求的一部分，由 ReceiveForm 方法处理，产生的结果如图 17-5 所示。

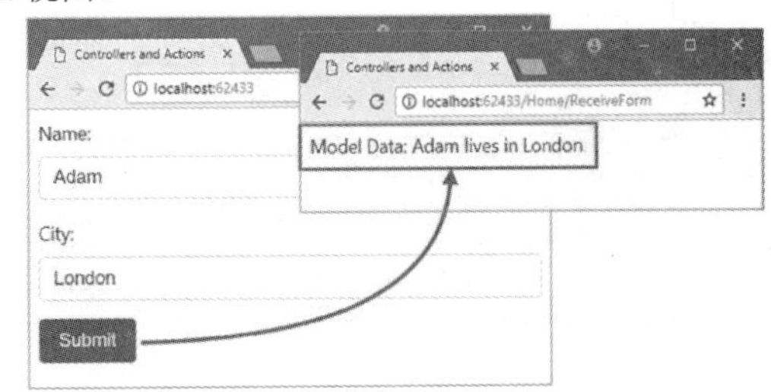

图 17-5 从 Context 对象获取表单数据

代码清单 17-12 所示的方法很有效，但是还有更优雅的选择。操作方法可以定义 MVC 使用的参数，将上下文数据传递给控制器，包括 HTTP 请求的详细信息。这比直接从 Context 对象提取更简单，并且能产生更容易阅读的操作方法。要接收表单数据，请在名称对应的表单数据的操作方法上声明参数，如代码清单 17-13 所示。

代码清单 17-13 在 Controllers 文件夹下的 HomeController.cs 文件中以参数方式接收上下文数据

```
using Microsoft.AspNetCore.Mvc;

namespace ControllersAndActions.Controllers {

    public class HomeController : Controller {

        public ViewResult Index() => View("SimpleForm");
```

```
        public ViewResult ReceiveForm(string name, string city)
            => View("Result", $"{name} lives in {city}");
    }
}
```

修改后的操作方法会产生相同的结果，但更易于阅读和理解。MVC将通过自动检查Context对象（包括请求字符串、表单和RouteData的值）来提供操作方法参数的值。参数的名称分情况处理，以便可以通过Request. Form["City"]中的值填充名为city的操作方法参数。以上方式还会生成更容易进行单元测试的操作方法，因为操作方法操作的值可作为常规的C#参数接收，并且不需要Context对象。

17.5 生成响应

操作方法在处理完请求之后，需要生成响应。有许多功能可用于从操作方法生成输出，下面进行详细介绍。

17.5.1 使用Context对象生成响应

生成输出的最简单方法是使用HttpResponse Context对象，该对象指定了ASP.NET Core如何访问发送到客户端的HTTP响应。表17-7描述了由HttpResponse类提供的基本属性，该类定义在Microsoft.AspNetCore.Http命名空间中。

表17-7 HttpResponse提供的基本属性

属性	描述
StatusCode	用于设置响应的HTTP状态码
ContentType	用于设置响应的Content-Type标头
Headers	返回将包含在响应中的HTTP标头的字典
Cookies	返回一个用于向响应添加Cookies的集合
Body	返回用于写入响应的主体数据的System.IO.Stream对象

代码清单17-14更新了Home控制器，从而在ReceivedForm操作方法中使用Controller.Request属性返回HttpResponse对象来生成响应。

代码清单17-14 在Controllers文件夹下的HomeController.cs文件中生成响应

```
using Microsoft.AspNetCore.Mvc;
using System.Text;

namespace ControllersAndActions.Controllers {

    public class HomeController : Controller {

        public ViewResult Index() => View("SimpleForm");

        public void ReceiveForm(string name, string city) {
            Response.StatusCode = 200;
            Response.ContentType = "text/html";
            byte[] content = Encoding.ASCII
                .GetBytes($"<html><body>{name} lives in {city}</body>");
            Response.Body.WriteAsync(content, 0, content.Length);
        }
    }
}
```

但这种生成响应的方式是很可怕的，因为它使用C#字符串对操作方法中的HTML进行硬编码，这容易出错且难以进行单元测试。但这种方式确实有助于了解如何在幕后创建响应。

还有比直接使用 HttpResponse 对象更好的替代方法。MVC 在底层响应的基础上提供了一个更有用的功能，并且也是控制器工作的核心——操作结果。

17.5.2 理解操作结果

MVC 使用操作结果来分离说明意图和执行意图。这个概念很简单，但是首先需要一些时间才能理解。

操作方法将返回一个从 Microsoft.AspNetCore.Mvc 命名空间实现了 IActionResult 接口的对象，而不是直接使用 HttpResponse 对象。IActionResult 对象又称为操作结果，它描述了控制器应该如何响应，例如呈现视图或将客户端重定向到另一个 URL。但是，它不会直接生成响应，而是通过 MVC 的操作结果来产生结果。

注 意

操作结果系统是命令模式的一个例子。命令模式描述了一种场景，以存储和传递这样的对象——这种对象描述了要执行的操作。

下面是来自 MVC 源代码的 IActionResult 接口的定义：

```
using System.Threading.Tasks;

namespace Microsoft.AspNetCore.Mvc {

    public interface IActionResult {
        Task ExecuteResultAsync(ActionContext context);
    }
}
```

这个接口可能看起来很简单，但这是因为 MVC 不会规定操作结果可以产生什么样的响应。当操作方法返回操作结果时，MVC 将调用 ExecuteResultAsync 方法，该方法负责生成响应以代表操作方法。ActionContext 参数提供了用于生成响应的上下文数据，包括 HttpResponse 对象（ActionContext 类是 ControllerContext 的超类，并定义了表 17-5 中描述的所有属性）。

为了演示操作结果如何工作，为项目添加 Infrastructure 文件夹，并添加一个名为 CustomHtmlResult.cs 的类文件，内容如代码清单 17-15 所示。

代码清单 17-15 Infrastructure 文件夹下的 CustomHtmlResult.cs 文件的内容

```
using Microsoft.AspNetCore.Mvc;
using System.Text;
using System.Threading.Tasks;

namespace ControllersAndActions.Infrastructure {

    public class CustomHtmlResult : IActionResult {

        public string Content { get; set; }
        public Task ExecuteResultAsync(ActionContext context) {
            context.HttpContext.Response.StatusCode = 200;
            context.HttpContext.Response.ContentType = "text/html";
            byte[] content = Encoding.ASCII.GetBytes(Content);
            return context.HttpContext.Response.Body.WriteAsync(content,
                0, content.Length);
        }
    }
}
```

CustomHtmlResult 类实现了 IActionResult 接口，其 ExecuteResultAsync 方法使用 HttpResponse 对象来编写 HTML 响应，其中包含 Content 属性的值。ExecuteResultAsync 方法必须返回一个 Task 对象，以便可

以异步生成响应；这适用于 CustomHtmlResult 类中的实现，该类依赖于表示响应体的 Stream 对象的 WriteAsync 方法，并返回可以用作操作结果的 Task 方法。

在代码清单 17-16 中，已通过在 Home 控制器中使用操作结果简化了 Home 控制器的 ReceiveForm 操作方法。

代码清单 17-16 在 Controllers 文件夹下的 HomeController.cs 文件中使用操作结果

```
using Microsoft.AspNetCore.Mvc;
using System.Text;
using ControllersAndActions.Infrastructure;

namespace ControllersAndActions.Controllers {

    public class HomeController : Controller {

        public ViewResult Index() => View("SimpleForm");

        public IActionResult ReceiveForm(string name, string city)
            => new CustomHtmlResult {
                Content = $"{name} lives in {city}"
            };
        }
}
```

发送响应的代码现在已与响应包含的数据分开定义，这简化了操作方法，并允许在其他操作方法中生成相同类型的响应，而不会产生重复的代码。

用于控制器和操作的单元测试

ASP.NET Core MVC 的许多部分是为了方便单元测试而设计的，对于操作和控制器来说尤其如此。ASP.NET Core MVC 支持单元测试，具体表现在以下几个方面。

- 可以在 Web 服务器之外测试操作和控制器。
- 不需要解析任何 HTML 来测试操作方法的结果。可以检查返回的 IActionResult 对象，以确保收到预期的结果。
- 不需要模拟客户端请求。MVC 模型绑定系统允许编写作为方法参数接收输入的操作方法。为了测试操作方法，只需要直接调用操作方法，并提供希望的参数值即可。

本章将介绍如何为不同类型的操作结果创建单元测试。有关设置单元测试项目的说明，请参阅第 7 章，或从 GitHub 下载示例项目。

17.5.3 生成 HTML 响应

刚才介绍了如何使用操作结果从控制器类中获取生成响应的代码。ASP.NET Core MVC 提供了更灵活的方法来生成响应——使用 ViewResult 类。

ViewResult 类是能够提供对 Razor 视图引擎的访问的操作结果，该引擎处理.cshtml 文件以合并模型数据，并通过 HttpResponse 上下文引擎将结果发送给客户端。第 21 章将解释视图引擎的工作原理，在本章中，重点是使用 ViewResult 类作为操作结果。

在代码清单 17-17 中，已经用 ViewResult 替换了自定义的操作结果类，操作结果类可通过 Controller 基类提供的 View 方法创建。

代码清单 17-17 使用 Controllers 文件夹下的 HomeController.cs 文件中的 ViewResult 类

```
using Microsoft.AspNetCore.Mvc;
using System.Text;
using ControllersAndActions.Infrastructure;
```

```
namespace ControllersAndActions.Controllers {

    public class HomeController : Controller {

        public ViewResult Index() => View("SimpleForm");

        public ViewResult ReceiveForm(string name, string city)
            => View("Result", $"{name} lives in {city}");
    }
}
```

可以直接创建 ViewResult 对象，正如本章开头的 POCO 控制器中演示的那样，但是使用 View 方法更简单、更简洁。Controller 类提供了几个不同版本的 View 方法，允许选择渲染的视图并提供模型数据，如表 17-8 所示。

表 17-8　Controller 类中的 View 方法

View 方法	描述
View()	为与操作方法关联的默认视图创建一个 ViewResult 对象，以便在名为 MyAction 的方法中调用 View()，呈现名为 MyAction.cshtml 的视图。不使用模型数据
View(view)	创建一个将呈现指定视图的 ViewResult 对象，调用 View（"MyView"）将呈现一个名为 MyView.cshtml 的视图。不使用模型数据
View(model)	为与操作方法关联的默认视图创建一个 ViewResult 对象，并使用指定对象作为模型数据
View(view,model)	为指定的视图创建一个 ViewResult 对象，并使用指定对象作为模型数据

如果运行应用程序并提交表单，你看到的结果将如图 17-6 所示。

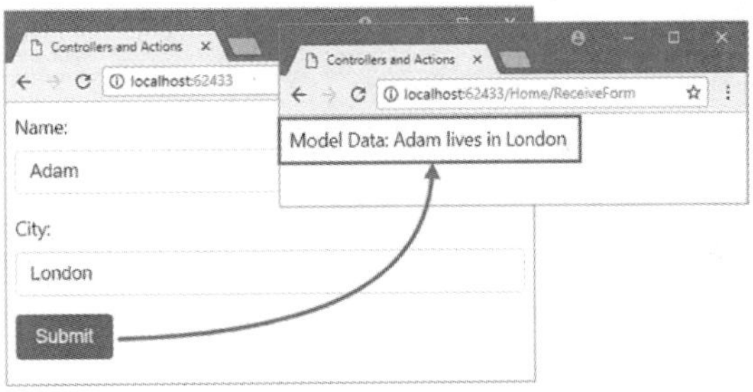

图 17-6　使用 ViewResult 对象生成 HTML 响应结果

1. 理解对视图文件的搜索

当 MVC 调用 ViewResult 对象的 ExecuteResult-Async 方法时，将从指定的视图开始搜索。MVC 搜索视图的目录顺序是与配置相关的约定示例。不需要使用框架注册视图文件。只需要将它们放到一组已知位置，框架即可找到它们。默认情况下，MVC 将在以下位置查找视图：

```
/Views/<ControllerName>/<ViewName>.cshtml
/Views/Shared/<ViewName>.cshtml
```

搜索从包含专用于当前控制器的视图的文件夹开始。该文件夹的名称中省略了类名的 Controller 部分，以便 HomeController 类的文件夹名为 Views/Home。

如果视图名未在 ViewResult 对象中指定，就使用路由数据中 action 变量的值。对于大多数控制器来说，这意味着将使用方法的名称，以使与 Index 方法关联的默认视图文件为 Index.cshtml。但是，如果已使用 Route 属性，那么与操作方法关联的视图名可能不同。

如果控制器是区域的一部分，如第 16 章所述，则搜索位置也会不同。

```
/Areas/<AreaName>/Views/<ControllerName>/<ViewName>.cshtml
/Areas/<AreaName>/Views/Shared/<ViewName>.cshtml
/Views/Shared/<ViewName>.cshtml
```

MVC 依次检查这些文件是否存在。一旦找到匹配，就会使用视图呈现操作方法的结果。这里因为没有在示例项目中使用区域，所以代码清单 17-18 中的操作方法会使 MVC 通过查找 Views/Home/Result.cshtml 文件来开始搜索。没有这样的文件，所以搜索继续，MVC 寻找 Views/Shared/Result.cshtml，它确实存在，因此它用于呈现 HTML 响应。

单元测试——渲染视图

要测试操作方法呈现的视图，可以检查返回的 ViewResult 对象。这和检查输出的视图并不完全相同（毕竟，你没有按照过程检查生成的最终 HTML），但只要 MVC 视图系统正常工作就足够了。为项目添加一个名为 ActionTests.cs 的单元测试文件，以保存本章执行的单元测试。想要测试的第一种情况是当操作方法选择特定的视图时，如下所示：

```
...
public ViewResult ReceiveForm(string name, string city)
    => View("Result", $"{name} lives in {city}");
...
```

可以通过读取 ViewResult 对象的 ViewName 属性来确定已选择哪个视图，如下所示：

```
using ControllersAndActions.Controllers;
using Microsoft.AspNetCore.Mvc;
using Xunit;

namespace ControllersAndActions.Tests {

    public class ActionTests {

        [Fact]
        public void ViewSelected() {
            // Arrange
            HomeController controller = new HomeController();

            // Act
            ViewResult result = controller.ReceiveForm("Adam", "London");

            // Assert
            Assert.Equal("Result", result.ViewName);
        }
    }
}
```

当测试选择默认视图的操作方法时，会发生变化，如下所示：

```
...
public ViewResult Result() => View();
...
```

在这种情况下，需要确保视图名为 null，如下所示：

```
...
Assert.Null(result.ViewName);
...
```

null 值是 ViewResult 对象向 MVC 发出与操作方法关联的默认视图已被选择的方式。

通过路径指定视图

视图的命名约定方法虽然方便简单，但它限制了可以呈现的视图。如果要呈现特定的视图，可以通过提供明确的路径并绕过搜索阶段来实现。以下是一个通过路径指定视图的例子。

```
using Microsoft.AspNetCore.Mvc;

namespace ControllersAndActions.Controllers {

    public class ExampleController : Controller {

        public ViewResult Index() {
```

```
            return View("/Views/Admin/Index");
        }
    }
}
```

当指定这样的视图时，路径必须以/或~/开头，并且可以包含文件扩展名（如果未指定，默认为.cshtml）。

如果发现自己使用了这个功能，建议花一点时间问问自己想要实现的目标。如果尝试呈现属于另一个控制器的视图，那么最好将用户重定向到该控制器中的操作方法（请参阅 17.5.4 节中的示例）。如果正在努力解决视图文件命名方案，那么请参阅第 21 章。

2. 将数据从操作方法传递给视图

当使用 ViewResult 选择视图时，在生成 HTML 内容时，你可以通过操作方法向其传递要使用的数据。MVC 为操作方法提供了将数据传递给视图的不同方式，这将在后面介绍。这些功能涉及第 21 章中的部分内容。本节仅讨论足够演示控制器功能的视图功能。

使用视图模型对象

通过将对象作为参数传递给 View 方法，你可以将创建的 ViewResult 对象传递给 ViewData.Model 属性。之前的章节直接设置了这个属性来解释 POCO 控制器是如何工作的，使用 View 方法会更简洁。代码清单 17-18 显示了新定义的 ExampleController 类，将它添加到 Controllers 文件夹中，并将视图模型对象传递给 View 方法。

代码清单 17-18 Controllers 文件夹下的 ExampleController.cs 文件的内容

```
using Microsoft.AspNetCore.Mvc;
using System;
namespace ControllersAndActions.Controllers {

    public class ExampleController : Controller {

        public ViewResult Index() => View(DateTime.Now);
    }
}
```

以上代码将一个 DateTime 对象传递给 View 方法以用作视图模型。要从视图内访问对象，可使用 Razor 的 Model 关键字。创建 Views/Example 文件夹，并添加一个名为 Index.cshtml 的视图文件，内容如代码清单 17-19 所示。

代码清单 17-19 Views/Example 文件夹下的 Index.cshtml 文件的内容

```
@{ Layout = null; }

<!DOCTYPE html>
<html>
<head>
    <meta name="viewport" content="width=device-width" />
    <title>Controllers and Actions</title>
    <link rel="stylesheet" asp-href-include="lib/bootstrap/dist/css/*.min.css" />
</head>
<body class="m-1 p-1">
    Model: @(((DateTime)Model).DayOfWeek)
</body>
</html>
```

这是一个非类型化或弱类型的视图。视图不了解视图模型对象的任何内容，并将其视为 Object 实例。为了获取 DayOfWeek 属性的值，需要将对象强制转换为 DateTime 实例，如下所示：

```
...
Model: @(((DateTime)Model).DayOfWeek)
...
```

虽然这样页面可以正常工作，但这可能产生混乱。可以通过创建强类型视图来整理这一切，其中的视图包括视图模型对象的类型细节，如代码清单 17-20 所示。

代码清单 17-20 在 Views/Example 文件夹下的 Index.cshtml 文件中添加强类型视图

```
@model DateTime
@{ Layout = null; }

<!DOCTYPE html>
<html>
<head>
    <meta name="viewport" content="width=device-width" />
    <title>Controllers and Actions</title>
    <link rel="stylesheet" asp-href-include="lib/bootstrap/dist/css/*.min.css" />
</head>
<body class="m-1 p-1">
    Model: @Model.DayOfWeek
</body>
</html>
```

这里使用 Razor 的 model 关键字指定了视图模型的类型。请注意，在指定类型时使用关键字 model，读取时使用关键字 Model。

强类型不仅有助于整理视图，而且 Visual Studio 支持强类型视图的智能感知功能，如图 17-7 所示。

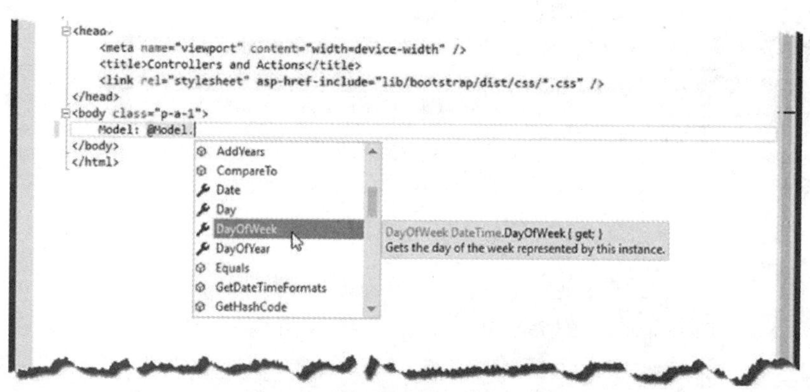

图 17-7 强类型视图的智能感知功能

单元测试——查看模型对象

视图模型对象被分配给 ViewResult.ViewData.Model 属性，这意味着可以在使用 View 方法时测试操作方法是否发送预期的数据。下面的测试方法用于检查代码清单 17-20 中的操作方法的模型类型。

```
...
[Fact]
public void ModelObjectType() {
    //Arrange
    ExampleController controller = new ExampleController();

    // Act
    ViewResult result = controller.Index();

    // Assert
    Assert.IsType<System.DateTime>(result.ViewData.Model);
}
...
```

Assert.IsType 方法用于检查视图模型对象是否是 DateTime 实例。

在使用 View 方法时，有一个需要注意的问题：当希望使用与某个操作关联的默认视图并向该视图提供字符串模型对象时，就会出现这种情况，如代码清单 17-21 所示。

代码清单 17-21　在 Controllers 文件夹下的 ExampleController.cs 文件中使用 View 方法

```
using Microsoft.AspNetCore.Mvc;
using System;

namespace ControllersAndActions.Controllers {

    public class ExampleController : Controller {

        public ViewResult Index() => View(DateTime.Now);

        public ViewResult Result() => View("Hello World");
    }
}
```

在新的 Result 操作方法中，作者希望使用呈现操作的默认视图的 View 方法，并指定模型数据。但是，如果运行应用程序并请求 URL /Example/Result，你将会看到类似下面的错误：

```
InvalidOperationException: The view 'Hello, World' was not found.
The following locations were searched:
/Views/Example/Hello, World.cshtml
/Views/Shared/Hello, World.cshtml
```

问题是，由于字符串参数被解释为要呈现的视图的名称，因此 MVC 尝试查找名为“Hello,World.cshtml”的视图文件，而不是查找 Result.cshtml。只需要通过将模型数据强制转换为对象，就可以很容易地解决这个问题，如代码清单 17-22 所示。

代码清单 17-22　在 Controllers 文件夹下的 ExampleController.cs 文件中选择正确的 View 方法

```
using Microsoft.AspNetCore.Mvc;
using System;

namespace ControllersAndActions.Controllers {

    public class ExampleController : Controller {

        public ViewResult Index() => View(DateTime.Now);

        public ViewResult Result() => View((object)"Hello World");
    }
}
```

将模型数据显式转换为对象可确保调用 View 方法的正确版本，并呈现 Result.cshtml 文件。

3. 通过 View Bag 传递数据

第 2 章介绍了 View Bag 功能。此功能允许定义动态对象的属性，并在视图中访问它们。通过 Controller 类提供的 ViewBag 属性可访问动态对象，如代码清单 17-23 所示。

代码清单 17-23　在 Controllers 文件夹下的 ExampleController.cs 文件中使用 View Bag 功能

```
using Microsoft.AspNetCore.Mvc;
using System;

namespace ControllersAndActions.Controllers {

    public class ExampleController : Controller {

        public ViewResult Index() {
            ViewBag.Message = "Hello";
            ViewBag.Date = DateTime.Now;
            return View();
```

```
        }

        public ViewResult Result() => View((object)"Hello World");
    }
}
```

上面通过对它们进行赋值定义了名为 Message 和 Date 的 ViewBag 属性。在此之前，没有这样的属性存在，我们也没有做任何创建它们的准备。为了在视图中读取数据，需要获得与操作方法中设置的相同的属性，如代码清单 17-24 所示。

代码清单 17-24 在 Views/Example 文件夹下的 Index.cshtml 文件中从 View Bag 中读取数据

```
@model DateTime
@{ Layout = null; }

<!DOCTYPE html>
<html>
<head>
    <meta name="viewport" content="width=device-width" />
    <title>Controllers and Actions</title>
    <link rel="stylesheet" asp-href-include="lib/bootstrap/dist/css/*.min.css" />
</head>
<body class="m-1 p-1">
    <p>The day is: @ViewBag.Date.DayOfWeek</p>
    <p>The message is: @ViewBag.Message</p>
</body>
</html>
```

View Bag 相比使用视图模型对象有一个优点，就是很容易将多个对象发送到视图。如果 MVC 仅支持视图模型，那么需要创建具有字符串和 DateTime 成员的新类型，才能获得相同的效果。

警 告

Visual Studio 无法为任何动态对象（包括 View Bag）提供智能感知支持，并且在呈现视图之前不会提示错误。

单元测试——View Bag

ViewResult.ViewData 属性会返回一个字典，其关键字是由操作方法定义的 ViewBag 属性的名称。以下是用于代码清单 17-24 中的操作方法的测试方法。

```
...
[Fact]
public void ModelObjectType() {
    //Arrange
    ExampleController controller = new ExampleController();

    // Act
    ViewResult result = controller.Index();

    // Assert
    Assert.IsType<string>(result.ViewData["Message"]);
    Assert.Equal("Hello", result.ViewData["Message"]);
    Assert.IsType<System.DateTime>(result.ViewData["Date"]);
}
...
```

这个测试方法使用 Assert.IsType 方法检查 Message 和 Date 属性的类型，并使用 Assert.Equal 方法检查 Message 属性的值。

17.5.4 执行重定向

操作方法的常见结果是不直接产生任何输出，而是将客户端重定向到另一个 URL。大多数情况下，这个 URL 是应用程序中的另一个操作方法，用于生成希望用户查看的输出。执行重定向时，可以向浏览器发送以下两个 HTTP 编码之一。

- HTTP 状态码 302，表示临时重定向。这是最常用的重定向类型，当使用 Post/Redirect/Get 模式时，HTTP 状态码 302 是将要发送的编码。
- HTTP 状态码 301，表示永久重定向。应该谨慎使用，因为它指示 HTTP 状态码的收件人不要再次请求原始 URL，并使用包含在重定向代码旁边的新 URL。如有疑问，请使用临时重定向——发送 HTTP 状态码 302。

有几个不同的操作结果可用于执行重定向，如表 17-9 所示。

表 17-9 用于执行重定向的操作结果

名称	控制器方法	描述
RedirectResult	Redirect RedirectPermanent	通过 HTTP 状态码 301 或 302 发送响应，将客户端重定向到新的 URL
LocalRedirectResult	LocalRedirect LocalRedirectPermanent	将客户端重定向到本地 URL
RedirectToActionResult	RedirectToAction RedirectionToActionPermanent	将客户端重定向到特定的操作和控制器
RedirectToRouteResult	RedirectToRoute RedirectToRoutePermanent	将客户端重定向到从特定路由生成的 URL

1. 重定向到文本 URL

重定向浏览器的最基本方法是调用由控制器类提供的重定向方法，返回 RedirectResult 类的一个实例，如代码清单 17-25 所示。

代码清单 17-25 在 Controllers 文件夹下的 ExampleController.cs 文件中重定向到文本 URL

```
using Microsoft.AspNetCore.Mvc;
using System;

namespace ControllersAndActions.Controllers {

    public class ExampleController : Controller {

        public ViewResult Index() {
            ViewBag.Message = "Hello";
            ViewBag.Date = DateTime.Now;
            return View();
        }

        public ViewResult Result() => View((object)"Hello World");

        public RedirectResult Redirect() => Redirect("/Example/Index");
    }
}
```

重定向 URL 表示为重定向方法的字符串参数，以产生临时重定向。可以使用 RedirectPermanent 方法执行永久重定向，如代码清单 17-26 所示。

提 示

LocalRedirectionResult 是可选的操作结果。如果控制器尝试重定向到任何非本地的 URL，则会引发异常。当重定向到用户提供的 URL 时，这是很有用的。在这种情况下，重定向攻击会将另一个用户重定向到不受信任的站点。此类操作结果可以通过继承 Controller 类的 LocalRedirect 方法来创建。

代码清单 17-26 在 Controllers 文件夹下的 ExampleController.cs 文件中永久重定到 URL

```
using Microsoft.AspNetCore.Mvc;
using System;

namespace ControllersAndActions.Controllers {

    public class ExampleController : Controller {

        public ViewResult Index() {
            ViewBag.Message = "Hello";
            ViewBag.Date = DateTime.Now;
            return View();
        }

        public ViewResult Result() => View((object)"Hello World");

        public RedirectResult Redirect() => RedirectPermanent("/Example/Index");
    }
}
```

单元测试——文本重定向

文本重定向很容易测试。可以读取 URL，并使用 RedirectResult 类的 Url 和 Permanent 属性来测试重定向是永久的还是临时的。下面是用于代码清单 17-26 所示的永久重定向的测试方法：

```
...
[Fact]
public void Redirection() {
    // Arrange
    ExampleController controller = new ExampleController();
    // Act
    RedirectResult result = controller.Redirect();
    // Assert
    Assert.Equal("/Example/Index", result.Url);
    Assert.True(result.Permanent);
}
...
```

请注意，当调用操作方法时，已经更新了接收 RedirectResult 的测试。

2. 重定向到路由系统 URL

如果要将用户重定向到应用程序的其他部分，那么需要确保发送的 URL 在 URL 模式中有效。使用文本 URL 进行重定向的问题是，路由模式中的任何更改都意味着需要查看代码并更新 URL。幸运的是，可以使用 RedirectToRoute 方法根据路由系统生成有效的 URL，该方法创建了 RedirectToRouteResult 实例，如代码清单 17-27 所示。

提 示

如果正在按照本章中的示例顺序执行，那么可能必须清除浏览器的历史记录，代码清单 17-27 中的代码才能正常工作。这是因为浏览器会记住代码清单 17-26 中的永久重定向，并将对/example/Redirect 的请求转换为对/Example/Index 的请求，而不必再联系服务器。

代码清单 17-27 在 Controllers 文件夹下的 ExampleController.cs 文件中重定向到路由系统 URL

```
using Microsoft.AspNetCore.Mvc;
using System;

namespace ControllersAndActions.Controllers {

    public class ExampleController : Controller {
```

```
        public ViewResult Index() {
            ViewBag.Message = "Hello";
            ViewBag.Date = DateTime.Now;
            return View();
        }

        public ViewResult Result() => View((object)"Hello World");

        public RedirectToRouteResult Redirect() =>
            RedirectToRoute(new { controller = "Example",
                                  action = "Index",
                                  ID = "MyID" });
    }
}
```

你可使用 RedirectToRoute 方法发出临时重定向，使用 RedirectToRoutePermanent 方法进行永久重定向。这两个方法都采用匿名类型，属性然后被传递给路由系统以生成 URL，如第 16 章所述。

单元测试——路由重定向

以下是用于代码清单 17-27 中的操作方法的单元测试：

```
...
[Fact]
public void Redirection() {
    // Arrange
    ExampleController controller = new ExampleController();
    // Act
    RedirectToRouteResult result = controller.Redirect();
    // Assert
    Assert.False(result.Permanent);
    Assert.Equal("Example", result.RouteValues["controller"]);
    Assert.Equal("Index", result.RouteValues["action"]);
    Assert.Equal("MyID", result.RouteValues["ID"]);
}
...
```

你可通过查看 RedirectToRouteResult 对象提供的路由信息来间接测试结果，这意味着不必解析 URL，这将需要单元测试来对应用程序使用的 URL 模式进行假设。

3. 重定向到操作方法

可以通过使用 RedirectTo 操作方法（用于临时重定向）或 RedirectToActionPermanent 方法（用于永久重定向），更加优雅地重定向到操作方法。这些都是对 RedirectToRoute 方法的包装，可以让你为操作方法和控制器指定值，而无须创建匿名类型，如代码清单 17-28 所示。

代码清单 17-28 在 Controllers 文件夹下的 ExampleController.cs 文件中使用 RedirectToAction 方法进行重定向

```
using Microsoft.AspNetCore.Mvc;
using System;

namespace ControllersAndActions.Controllers {

    public class ExampleController : Controller {

        public ViewResult Index() {
            ViewBag.Message = "Hello";
            ViewBag.Date = DateTime.Now;
            return View();
        }
```

```
        public RedirectToActionResult Redirect() => RedirectToAction(nameof(Index));
    }
}
```

如果只指定一个操作方法，那么默认是当前控制器中的操作方法。如果要重定向到另一个控制器，则需要提供控制器的名称作为参数，如下所示：

```
...
public RedirectToActionResult Redirect()
    => RedirectToAction(nameof(HomeController), nameof(HomeController.Index));
...
```

还有其他重载版本的方法，可以用来为 URL 提供其他的值。这些方法使用匿名类型，虽然这样会违背使用这些方法的目的，但它可以使代码更容易阅读。

注 意

为操作方法和控制器提供的值在传递到路由系统之前不会被验证。你有责任确保指定的目标实际存在。

单元测试——操作方法重定向

以下是用于代码清单 17-28 中的操作方法的单元测试：

```
...
[Fact]
public void Redirection() {
    // Arrange
    ExampleController controller = new ExampleController();
    // Act
    RedirectToActionResult result = controller.Redirect();
    // Assert
    Assert.False(result.Permanent);
    Assert.Equal("Index", result.ActionName);
}
...
```

RedirectToActionResult 类提供了 ControllerName 和 ActionName 属性，可以轻松地在控制器中创建重定向，而无须解析 URL。

4. 使用 Post/Redirect/Get 模式

重定向经常用于处理 HTTP POST 请求的操作方法。正如之前解释的那样，当需要更改应用程序的状态时，会使用 POST 请求。如果在处理 POST 请求后返回 HTML 响应，那么用户将有可能单击浏览器中的重新加载按钮并再次提交表单，这可能会产生不可预料的结果。

可以在示例应用程序的 Home 控制器中看到此问题。ReceiveForm 方法会从表单数据的参数中接收值，并使用 View 方法返回一个 ViewResult 对象：

```
...
public ViewResult ReceiveForm(string name, string city)
    => View("Result", $"{name} lives in {city}");
...
```

要查看问题，请运行应用程序并请求 URL——/Home。提交表单，然后单击浏览器中的重新加载按钮。可使用 F12 键来研究浏览器发出的 HTTP 请求，你将看到一个新的 POST 请求被发送到服务器。这在简单的应用程序中没有什么影响，但如果 POST 请求最终重复删除数据，提交订单或执行用户未打算的其他重要任务，那么这个问题可能会造成严重后果。

为了避免这个问题，可以按照 Post/Redirect/Get 模式进行操作。在这种模式中，你将收到 POST 请求，处理它，然后重定向浏览器，以便浏览器为另一个 URL 发出 GET 请求。GET 请求不应该修改应用程序的状态，因此在无意中重新提交请求不会导致任何问题。代码清单 17-29 添加了重定向，以便浏览器被重定向到具有 GET 请求的不同 URL。

代码清单 17-29 在 Controllers 文件夹下的 HomeController.cs 文件中使用 Post/Redirect/Get 模式

```
using Microsoft.AspNetCore.Mvc;
using System.Text;
using ControllersAndActions.Infrastructure;

namespace ControllersAndActions.Controllers {

    public class HomeController : Controller {

        public ViewResult Index() => View("SimpleForm");

        [HttpPost]
        public RedirectToActionResult ReceiveForm(string name, string city)
            => RedirectToAction(nameof(Data));

        public ViewResult Data() => View("Result");
    }
}
```

RedirectToActionResult 方法通过 POST 请求从用户接收数据，并将客户端重定向到 Data 操作方法。如果用户重新加载页面，无害的 GET 请求将被发送到 Data 操作方法。第 20 章将介绍使用 HttpPost 特性可确保只有 POST 请求可以发送到 ReceiveForm 操作方法。

5. 使用 temp 数据

重定向将导致浏览器发送全新的 HTTP 请求，这意味着无法从原始请求访问表单数据，还意味着 Data 方法不知道应该向用户显示的 name 和 city 值。

如果需要将数据从一个请求保存到另一个请求，则可以使用 temp 数据功能。temp 数据与第 9 章使用的会话数据类似，只是在处理请求时，temp 数据值在读取和删除数据存储区时被标记为删除。对于在 Post/Redirect/Get 模式中进行重定向工作所需的短期数据，这是理想的安排。temp 数据功能可通过名为 TempData 的控制器类属性获得，如代码清单 17-30 所示。

注 意

temp 数据依赖于会话中间件。有关 Startup 类中对应功能的配置语句，可参见本章的开始部分。

代码清单 17-30 在 Controllers 文件夹下的 HomeController.cs 文件中使用 temp 数据

```
using Microsoft.AspNetCore.Mvc;
using System.Text;
using ControllersAndActions.Infrastructure;

namespace ControllersAndActions.Controllers {

    public class HomeController : Controller {

        public ViewResult Index() {
            return View("SimpleForm");
        }

        [HttpPost]
        public RedirectToActionResult ReceiveForm(string name, string city) {
            TempData["name"] = name;
            TempData["city"] = city;
            return RedirectToAction(nameof(Data));
        }
```

```
        public ViewResult Data() {
            string name = TempData["name"] as string;
            string city = TempData["city"] as string;
            return View("Result", $"{name} lives in {city}");
        }
    }
}
```

ReceiveForm 方法使用 TempData 属性获取数据字典，以便在将客户端重定向到 Data 操作方法之前存储 name 和 city 值。Data 操作方法使用相同的 TempData 属性检索数据值，并使用它们创建将由视图显示的模型数据。

提 示

TempData 字典还提供了 Peek 方法，以允许获取数据值而不将它们标记为删除。TempData 字典还提供了 Keep 方法，可用于防止删除以前的读取值。Keep 方法不会永远保护值。如果再次读取该值，就会再次将其标记为删除。请使用会话数据存储值，这样在处理请求时这些值才不会被删除。

17.5.5 返回不同类型的内容

操作方法并不一定必须响应为 HTML，表 17-10 显示了可用于不同类型数据的内置操作结果。

表 17-10 操作结果

操作结果	控制器方法	描述
JsonResult	Json	将对象序列化为 JSON 并将其返回给客户端
ContentResult	Content	发送包含指定对象的响应
ObjectResult	Not Available	将使用内容协商将对象发送给客户端
OkObjectResult	Ok	如果内容协商成功，就使用内容协商将 HTTP 对象发送给客户端
NotFoundObjectResult	NotFound	如果内容协商成功，就使用内容协商将对象发送到具有 HTTP 404 状态码的客户端

1. 生成 JSON 响应

JSON（JavaScript 对象表示法）格式已成为在 Web 应用程序及其客户端之间传输数据的标准方法。作为一种数据交换格式，JSON 在很大程度上取代了 XML，因为 JSON 更易于使用，特别是在编写客户端 JavaScript 时，因为 JSON 与 JavaScript 用于定义文本数据值的语法密切相关。第 20 章将介绍 JSON 在 Web 应用程序中的作用，代码清单 17-31 显示了如何使用 Json 方法创建 JsonResult 对象。

代码清单 17-31 在 Controllers 文件夹下的 ExampleController.cs 文件中生成 JSON 响应

```
using Microsoft.AspNetCore.Mvc;
using System;

namespace ControllersAndActions.Controllers {

    public class ExampleController : Controller {

        public JsonResult Index() => Json(new[] { "Alice", "Bob", "Joe" });
    }
}
```

运行示例并请求 URL——/Example，你将看到操作方法的响应，响应为 JSON 格式的 C#字符串数组，如下所示：

```
["Alice","Bob","Joe"]
```

大多数浏览器可以直接显示 JSON 结果，但在其他浏览器（包括 Microsoft Internet Explorer）中需要将数据保存到文件中才能查看。

单元测试——非 HTML 操作结果

重要的是要记住，操作方法的单元测试应该集中于返回格式化的数据而不是格式本身，这是由 MVC 处理的，通常超出大多数测试项目的范围。例如，以下是用于代码清单 17-32 中的操作方法的单元测试：

```
...
[Fact]
public void JsonActionMethod() {
    // Arrange
    ExampleController controller = new ExampleController();
    // Act
    JsonResult result = controller.Index();
    // Assert
    Assert.Equal(new[] { "Alice", "Bob", "Joe" }, result.Value);
}
...
```

JsonResult 类提供了 Value 属性，以返回将被转换成 JSON 的数据，产生返回客户端的响应。在单元测试中，可对 Value 属性与预期的数据进行比较。

2. 使用对象生成响应

许多应用程序只需要来自控制器的 HTML 和 JSON 响应，其他类型的内容则依赖于静态文件来传递，比如图像、JavaScript 文件和 CSS 样式表。但是，当需要在响应中返回特定的内容类型时，就需要使用操作结果来帮助解决这些问题。最简单的是通过 Content 方法创建的 ContentResult 类，该类用于发送带有可选 MIME 内容类型的字符串值。在代码清单 17-32 中，使用 Content 方法重新创建了前面的 JSON 结果。

代码清单 17-32　在 Controllers 文件夹下的 ExampleController.cs 文件中手动创建 JSON 结果

```
using Microsoft.AspNetCore.Mvc;

namespace ControllersAndActions.Controllers {

    public class ExampleController : Controller {

        public ContentResult Index()
            => Content("[\"Alice\",\"Bob\",\"Joe\"]", "application/json");
    }
}
```

当内容方便使用字符串格式，并且知道客户端能够接收指定的 MIME 类型时，这种类型的操作结果非常有用。这种方法的问题在于不知道如何使用未知格式的数据向客户端发送响应。更健壮的方法依赖于内容协商，这是由 ObjectResult 执行的，如代码清单 17-33 所示。

代码清单 17-33　在 Controllers 文件夹下的 ExampleController.cs 文件中使用内容协商

```
using Microsoft.AspNetCore.Mvc;

namespace ControllersAndActions.Controllers {

    public class ExampleController : Controller {

        public ObjectResult Index() => Ok(new string[] { "Alice", "Bob", "Joe" });
    }
}
```

内容协商构建了一种复杂的系统，用于在浏览器和应用程序之间找出共同的格式。当浏览器发出 HTTP 请求时，已包括接收标头，用以指示可以处理的格式。下面是用来测试示例的 Google Chrome 版本的标头：

```
Accept: text/html,application/xhtml+xml,application/xml;q=0.9,image/webp,*/*;q=0.8
```

支持的格式已表示为 MIME 类型。MVC 有一组可用于数据值的格式，可与浏览器支持的格式做比较。

MVC 使用的首选格式是 JSON，大部分时间将使用这种格式，除非使用明文返回字符串值。有关内容协商的过程及实现方式的更多详细信息，请参见第 20 章。

17.5.6　响应文件的内容

大多数应用程序依赖于静态文件中间件传递文件的内容，但还有一组可用于将文件发送到客户端的操作结果，如表 17-11 所示。

警　告

使用这些操作结果时要小心，并确保不创建允许请求任意文件内容的应用程序。特别是，不要从请求的任何部分或用户可以通过请求修改的任何数据存储中获取要发送的文件的路径。

表 17-11　将文件发送到客户端的操作结果

操作结果	控制器方法	描述
FileContentResult	File	使用指定的 MIME 类型向客户端发送字节数组
FileStreamResult	File	读取流并将内容发送给客户端
VirtualFileResult	File	从虚拟路径读取流（相对于主机上的应用程序）
PhysicalFileResult	PhysicalFile	从指定的路径读取文件的内容，并将内容发送给客户端

代码清单 17-34 使用继承自 Controller 类的 File 方法返回 Bootstrap CSS 文件，并作为 Example 控制器上的 Index 操作方法的返回结果。

代码清单 17-34　在 Controllers 文件夹下的 ExampleController.cs 文件中使用文件作为响应

```
using Microsoft.AspNetCore.Mvc;

namespace ControllersAndActions.Controllers {

    public class ExampleController : Controller {

        public VirtualFileResult Index()
            => File("/lib/bootstrap/dist/css/bootstrap.css", "text/css");
    }
}
```

为了使用这个操作方法，这里已经使用 Url 辅助方法修改了 SimpleForm.cshtml 文件中的 link 元素，如代码清单 17-35 所示。

代码清单 17-35　在 SimplerForm.cshtml 文件中指向一个操作方法

```
@{ Layout = null; }

<!DOCTYPE html>
<html>
<head>
    <meta name="viewport" content="width=device-width" />
    <title>Controllers and Actions</title>
    <link rel="stylesheet" href="@Url.Action("Index", "Example")" />
</head>
<body class="m-1 p-1">
    <form method="post" asp-action="ReceiveForm">
        <div class="form-group">
            <label for="name">Name:</label>
            <input class="form-control" name="name" />
        </div>
        <div class="form-group">
            <label for="name">City:</label>
            <input class="form-control" name="city" />
        </div>
        <button class="btn btn-primary center-block" type="submit">Submit</button>
```

```
    </form>
</body>
</html>
```

如果运行示例并请求 URL——/Home，则发送到浏览器的 HTML 响应将包含以下元素：

```
<link rel="stylesheet" href="/Example" />
```

这将会使浏览器发送针对代码清单 17-35 中的操作方法的 HTTP 请求，进而发送在视图中为内容设置样式所需的 CSS 文件。

注 意

标签助手是提供 CSS 的非常有用的工具，将在第 25 章中介绍。

17.5.7 返回错误和 HTTP 状态码

内置的 ActionResult 类的最终集合（其成员见表 17-12）可用于向客户端发送特定的错误消息和 HTTP 状态码。大多数应用程序不需要这些功能，因为 ASP.NET Core 和 MVC 将自动生成这些类型的结果。但是，如果需要对发送给客户端的响应进行更直接的控制，它们将非常有用。

表 17-12 内置的 ActionResult 类的最终集合的成员

成员	控制器方法	描述
StatusCodeResult	StatusCode	向客户端发送指定的 HTTP 状态码
OkResult	Ok	向客户端发送 HTTP 200 状态码
CreatedResult	Created	向客户端发送 HTTP 201 状态码
CreatedAtActionResult	CreatedAtAction	将 HTTP 201 状态码发送到客户端，并在 Location 标头中指向 action 和控制器的 URL
CreatedAtRouteResult	CreatedAtRoute	将 HTTP 201 状态码发送到客户端，并将 URL 放在从特定路由生成的 Location 标头中
BadRequestResult	BadRequest	向客户端发送 HTTP 400 状态码
UnauthorizedResult	Unauthorized	向客户端发送 HTTP 401 状态码
NotFoundResult	NotFound	向客户端发送 HTTP 404 状态码
UnsupportedMediaTypeResult	None	向客户端发送 HTTP 415 状态码

1. 发送特定的 HTTP 状态码

若使用 StatusCode 方法向浏览器发送特定的 HTTP 状态码，将创建一个 StatusCodeResult 对象，如代码清单 17-36 所示。

代码清单 17-36 在 Controllers 文件夹下的 ExampleController.cs 文件中发送特定的状态码

```
using Microsoft.AspNetCore.Mvc;
using Microsoft.AspNetCore.Http;

namespace ControllersAndActions.Controllers {

    public class ExampleController : Controller {

        public StatusCodeResult Index()
            => StatusCode(StatusCodes.Status404NotFound);
    }
}
```

StatusCode 方法接收一个 int 值，可以使用它直接指定状态码。Microsoft.AspNetCore.Http 命名空间中的 StatusCodes 类定义了 HTTP 支持的所有状态码的字段。在以上代码中，已使用 Status404NotFound 字段返回 HTTP 状态码 404，这表示请求的资源不存在。

2. 发送 404 结果

表 17-12 中显示的其他操作结果扩展或依赖 StatusCodeResult 类，该类提供了一种更方便的发送特定状态码的方法。可以使用更方便的 NotFoundResult 类来实现与代码清单 17-36 相同的效果，该类是从 StatusCodeResult 类派生的，可以使用控制器的 NotFound 方法创建，如代码清单 17-37 所示。

代码清单 17-37 在 Controllers 文件夹下的 ExampleController.cs 文件中生成 404 结果

```
using Microsoft.AspNetCore.Mvc;
using Microsoft.AspNetCore.Http;

namespace ControllersAndActions.Controllers {

    public class ExampleController : Controller {

        public StatusCodeResult Index() => NotFound();
    }
}
```

测试 HTTP 状态码

StatusCodeResult 类遵循已为其他结果类型使用的模式，并通过一组属性使其状态可用。在这种情况下，StatusCode 属性返回 HTTP 状态码，StatusDescription 属性返回关联的描述性字符串。下面的测试方法可用于代码清单 17-37 中的操作方法。

```
...
[Fact]
public void NotFoundActionMethod() {
    // Arrange
    ExampleController controller = new ExampleController();
    // Act
    StatusCodeResult result = controller.Index();
    // Assert
    Assert.Equal(404, result.StatusCode);
}
...
```

17.5.8 理解其他操作结果类

一些其他的操作结果类与其他章节中描述的 MVC 功能紧密相连，表 17-13 列出了这些类。

表 17-13 其他操作结果类

名称	控制器方法	描述
PartialViewResult	PartialView	用于选择分部视图
ViewComponentResult	ViewComponent	用于选择视图组件
EmptyResult	None	不执行任何工作，并生成返回客户端的空响应
ChallengeResult	None	用于在请求中强制实施安全策略

17.6 小结

控制器是 MVC 设计模式中的关键构件之一，是 MVC 开发的核心。在本章，你已经了解了如何使用基本的 C#类创建 POCO 控制器，以及如何利用 Controller 基类进行快速开发。你看到了操作结果在 MVC 控制器中扮演的角色，以及它们如何简化单元测试。本章还展示了可以从操作方法接收输入和生成输出的不同方式，并演示了内置的操作结果。下一章将描述导致 ASP.NET 开发人员困惑的特性之一，即依赖注入，但它对于有效的 MVC 开发是必不可少的。

第 18 章　依赖注入

本章将介绍依赖注入（Dependency Injection，DI），这是一种有助于灵活创建应用和简化单元测试的技术。不论从依赖注入有什么作用看，还是从依赖注入如何执行看，依赖注入都是一个很难理解的主题。因此，本章从传统的构建应用程序组件的方式开始，逐步解释依赖注入的工作原理及其重要性。表 18-1 展示了在上下文中使用的依赖注入。

表 18-1　在上下文中使用的依赖注入

问题	答案
什么是依赖注入？	依赖注入可以轻松创建松散耦合的组件，这通常意味着组件使用接口定义的功能，而不需要关注功能具体由哪些类来实现
依赖注入有什么作用？	依赖注入可以通过更改实现了定义应用程序功能的接口的组件来更容易地更改应用程序的行为，还会导致更易于隔离单元测试的组件
如何使用依赖注入？	Startup 类用于指定哪些实现类用于传递由应用程序使用的接口指定的功能。当对象（如控制器）创建新的处理请求时，将自动提供它们所需的实现类的实例
有什么陷阱或限制吗？	主要的限制是，类将服务的使用声明为构造函数参数，这可能导致构造函数的唯一作用是接收依赖项并将它们分配给实例字段
有其他的选择吗？	不必在自己的代码中使用依赖注入，但是了解依赖注入的工作原理是很有帮助的，因为 MVC 使用依赖注入为开发人员提供功能

表 18-2 展示了本章要介绍的操作。

表 18-2　本章要介绍的操作

操作	方法	代码清单
创建松散耦合的组件	通过接口隔离类，并使用外部映射将它们连接在一起	代码清单 18-9～代码清单 18-16
在组件（比如控制器）中声明依赖项	定义组件需要的类型的构造函数参数	代码清单 18-17
配置服务映射	在 Startup 类中添加映射	代码清单 18-18，代码清单 18-20～代码清单 18-26
对具有依赖项的组件进行单元测试	创建服务接口的模拟实现，并在单元测试中创建组件时作为构造函数参数传递	代码清单 18-19
指定实现对象的创建方式	使用适合要管理的服务的生命周期方法创建服务映射	代码清单 18-27～代码清单 18-31
在控制器中接收单个操作方法的依赖项	使用操作注入	代码清单 18-32
在控制器中手动请求实现对象	使用 HttpContext.RequestServices 属性	代码清单 18-33

18.1　准备示例项目

在本章中，使用 ASP.NET Core Web Application（.NET Core）模板创建一个新的名为 DependencyInjection 的 Empty 项目。代码清单 18-1 显示了 Startup 类，用于配置项目的服务和中间件。

代码清单 18-1　DependencyInjection 文件夹下的 Startup.cs 文件的内容

```
using System;
using System.Collections.Generic;
using System.Linq;
using System.Threading.Tasks;
using Microsoft.AspNetCore.Builder;
```

```
using Microsoft.AspNetCore.Hosting;
using Microsoft.AspNetCore.Http;
using Microsoft.Extensions.DependencyInjection;

namespace DependencyInjection {
    public class Startup {

        public void ConfigureServices(IServiceCollection services) {
            services.AddMvc();
        }

        public void Configure(IApplicationBuilder app, IHostingEnvironment env) {
            app.UseStatusCodePages();
            app.UseDeveloperExceptionPage();
            app.UseStaticFiles();
            app.UseMvcWithDefaultRoute();
        }
    }
}
```

18.1.1 创建模型和存储库

创建 Models 文件夹并添加一个名为 Product.cs 的类文件，内容如代码清单 18-2 所示。

代码清单 18-2 Models 文件夹下的 Product.cs 文件的内容

```
namespace DependencyInjection.Models {

    public class Product {

        public string Name { get; set; }
        public decimal Price { get; set; }
    }
}
```

为了管理模型，在 Models 文件夹中添加一个名为 IRepository.cs 的类文件，其中定义的接口如代码清单 18-3 所示。

代码清单 18-3 Models 文件夹下的 IRepository.cs 文件的内容

```
using System.Collections.Generic;

namespace DependencyInjection.Models {

    public interface IRepository {

        IEnumerable<Product> Products { get; }

        Product this[string name] { get; }

        void AddProduct(Product product);
        void DeleteProduct(Product product);
    }
}
```

IRepository 接口定义了可对 Product 对象集合执行的操作。为了提供这个接口的实现，在 Models 文件夹中添加一个名为 MemoryRepository.cs 的类文件，内容如代码清单 18-4 所示。

代码清单 18-4 Models 文件夹下的 MemoryRepository.cs 文件的内容

```
using System.Collections.Generic;

namespace DependencyInjection.Models {
```

```
    public class MemoryRepository : IRepository {
        private Dictionary<string, Product> products;

        public MemoryRepository() {
            products = new Dictionary<string, Product>();
            new List<Product> {
                new Product { Name = "Kayak", Price = 275M },
                new Product { Name = "Lifejacket", Price = 48.95M },
                new Product { Name = "Soccer ball", Price = 19.50M }
            }.ForEach(p => AddProduct(p));
        }

        public IEnumerable<Product> Products => products.Values;

        public Product this[string name] => products[name];

        public void AddProduct(Product product) =>
            products[product.Name] = product;

        public void DeleteProduct(Product product) =>
            products.Remove(product.Name);
    }
}
```

MemoryRepository 类使用字典将模型对象存储在内存中。这意味着没有持久存储，在停止或重新启动应用程序时，会将模型重置为在构造函数中创建的示例数据对象。对于真正的项目来说，这不是合理的方法，但用在这里很合适，本章的重点在于介绍应用程序的各个不同方面是如何工作的。

18.1.2 创建控制器和视图

创建 Controllers 文件夹，添加一个名为 HomeController.cs 的类文件，内容如代码清单 18-5 所示。

代码清单 18-5 Controllers 文件夹下的 HomeController.cs 文件的内容

```
using Microsoft.AspNetCore.Mvc;

namespace DependencyInjection.Controllers {

    public class HomeController : Controller {

        public ViewResult Index() => View();
    }
}
```

Home 控制器只有一个操作方法，它使用 View 方法创建一个用于渲染默认视图的 ViewResult 对象。要创建与操作方法关联的视图，可创建 Views/Home 文件夹，并在其中添加一个名为 Index.cshtml 的 Razor 视图文件，内容如代码清单 18-6 所示。

代码清单 18-6 Views/Home 文件夹下的 Index.cshtml 文件的内容

```
@model IEnumerable<Product>
@{ Layout = null; }

<!DOCTYPE html>
<html>
<head>
    <meta name="viewport" content="width=device-width" />
```

```
        <title>Dependency Injection</title>
        <link rel="stylesheet" asp-href-include="lib/bootstrap/dist/css/*.min.css" />
    </head>
    <body class="m-1 p-1">
        @if (ViewData.Count > 0) {
            <table class="table table-bordered table-sm table-striped">
                @foreach (var kvp in ViewData) {
                    <tr><td>@kvp.Key</td><td>@kvp.Value</td></tr>
                }
            </table>
        }
        <table class="table table-bordered table-sm table-striped">
            <thead>
                <tr><th>Name</th><th>Price</th></tr>
            </thead>
            <tbody>
                @if (Model == null) {
                    <tr><td colspan="3" class="text-center">No Model Data</td></tr>
                } else {
                    @foreach (var p in Model) {
                        <tr>
                            <td>@p.Name</td>
                            <td>@string.Format("{0:C2}", p.Price)</td>
                        </tr>
                    }
                }
            </tbody>
        </table>
    </body>
    </html>
```

Index 视图使用 Product 对象的枚举进行强类型化，视图的主要内容是 HTML 表格。如果控制器没有提供任何类型的数据，则会显示一条消息作为表格的唯一内容。如果存在模型数据，则会为枚举中的每个 Product 对象向表中添加一行。还有一个表格，用于在视图中枚举 View Bag 的键和值（如果 View Bag 的键和值不存在，则隐藏这个表格）。

Index 视图依赖于 Bootstrap CSS 包来为 HTML 元素设置样式。要将 Bootstrap 添加到项目中，可使用 Bower Configuration File 模板创建 bower.json 文件，并将 Bootstrap 软件包添加到 bower.json 的 dependencies 部分，如代码清单 18-7 所示。

代码清单 18-7 在 DependencyInjection 文件夹下的 bower.json 文件中添加 Bootstrap

```
{
  "name": "asp.net",
  "private": true,
  "dependencies": {
    "bootstrap": "4.0.0-alpha.6"
  }
}
```

最后的准备工作是在 Views 文件夹中创建_ViewImports.cshtml 文件，该文件设置了用于 Razor 视图的内置标签助手，并导入模型命名空间，如代码清单 18-8 所示。

代码清单 18-8 Views 文件夹下的_ViewImports.cshtml 文件的内容

```
@using DependencyInjection.Models
@addTagHelper *, Microsoft.AspNetCore.Mvc.TagHelpers
```

18.1.3 创建单元测试项目

按照第7章中描述的步骤，使用 xUnit Test Project（.NET Core）模板创建一个名为 DependencyInjection.Tests 的项目。这里删除了 unittest1. cs 文件，因此项目中目前没有任何测试。如果运行应用程序，你将看到图 18-1 所示的结果。

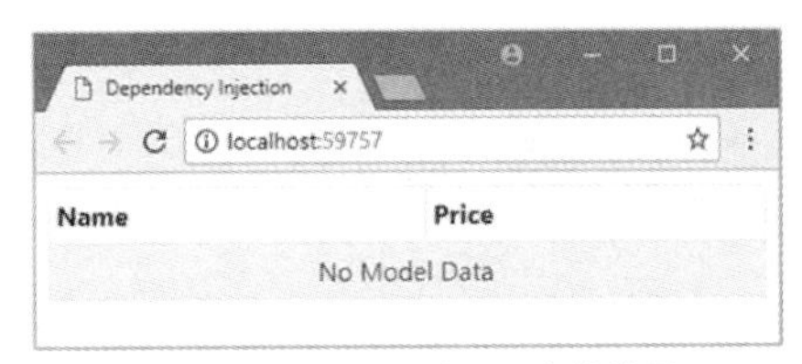

图 18-1 运行示例应用程序的结果

18.2 创建松散耦合的组件

图 18-1 没有显示模型数据的原因在于需要将模型数据传递给视图的 HomeController 类与包含模型数据的 MemoryRepository 类之间没有关系。在 MVC 应用程序中，将组件连接在一起的目标是希望能够使用相同功能的替代实现轻松替换组件。

通过更换组件可以进行有效的单元测试，轻松地在不同的托管环境（如开发和生产服务器）中更改应用程序的行为，并简化长期的应用程序维护。

接下来的几节将首先解释替代方法及其带来的问题。这似乎是解释依赖注入（DI）特性的一种间接方式，但是 DI 面临的挑战在于解决一些在编写代码时不明显，但在开发周期的末段才出现的问题。

了解依赖注入

依赖注入可能是一个难以理解的话题，DI 是一个有用的工具，但不是每个人都喜欢或需要。

如果没有进行单元测试，或者正在开展小型、独立且稳定的项目，DI 将只能提供有限的好处。了解 DI 的工作原理仍然有帮助，因为 DI 用于访问某些重要的 MVC 功能，但你并不总是需要在控制器和其他类中使用 DI。

推荐你在自己的项目中使用 DI，主要是因为作者发现项目经常出现一些意想不到的问题，如果能够轻松地用新的实现替换组件，就可以避免大量烦琐且容易出错的更改。

检查紧密耦合的组件

对于大多数开发人员来说，自然倾向是采取最直接的途径来解决问题。对于示例应用程序来说，这意味着使用 new 关键字创建控制器所需的存储库对象，以获取模型数据，如代码清单 18-9 所示。

代码清单 18-9 在 Controllers 文件夹下的 HomeController.cs 文件中实例化存储库

```
using Microsoft.AspNetCore.Mvc;
using DependencyInjection.Models;

namespace DependencyInjection.Controllers {

    public class HomeController : Controller {

        public ViewResult Index() => View(new MemoryRepository().Products);
    }
}
```

好消息是，这段代码是有效的。如果运行应用程序，你将看到浏览器中显示的模型对象的详细信息，如图 18-2 所示。

但坏消息是，Home 控制器和 MemoryRepository 类现在紧密耦合，这意味着无法在不改变 HomeController 类的前提下直接替换存储库。正如第 7 章所解释的，执行有效的单元测试意味着能够隔离单个组件，但是在测试代码清单 18-9 中的 Index 操作方法时会隐式地测试存储库类。如果单元测试失败，将无法确认问题出在控制器、存储库还是存储库依赖的其他组件上。所以根据当前实际情况，Home 控制器和 MemoryRepository 形成了单独的单元，紧耦合组件的影响如图 18-3 所示。

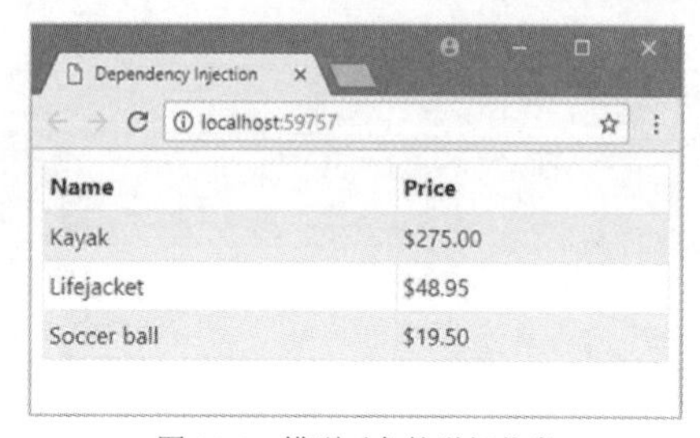

图 18-2 模型对象的详细信息

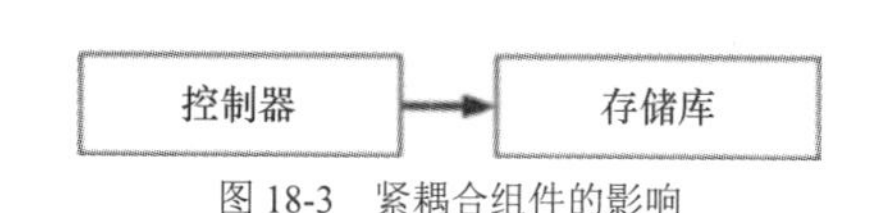

图 18-3 紧耦合组件的影响

1. 为单元测试解耦合

第 7 章使用一个属性，通过实现的接口来存储对存储库类的引用，从而允许为单元测试创建模拟库。代码清单 18-10 显示了在本章的示例应用程序中，如何在控制器中使用这种方法。

代码清单 18-10 在 Controllers 文件夹下的 HomeController.cs 文件中使用 Repository 属性

```
using Microsoft.AspNetCore.Mvc;
using DependencyInjection.Models;

namespace DependencyInjection.Controllers {

    public class HomeController : Controller {

        public IRepository Repository { get; set; } = new MemoryRepository();

        public ViewResult Index() => View(Repository.Products);
    }
}
```

如果要进行单元测试，那么这种方法非常适用，因为可以通过在单元测试中调用操作方法之前设置 Repository 属性来隔离控制器类。

将一个名为 DITests.cs 的类文件添加到 DependencyInjection.Tests 项目中，并用来定义代码清单 18-11 所示的单元测试，从而使用 Repository 属性在对控制器执行操作之前设置假的存储库。

代码清单 18-11 在单元测试项目中使用 DITests.cs 文件测试控制器

```
using DependencyInjection.Controllers;
using DependencyInjection.Models;
using Microsoft.AspNetCore.Mvc;
using Moq;
using Xunit;

namespace Tests {

    public class DITests {

        [Fact]
        public void ControllerTest() {
            // Arrange
            var data = new[] { new Product { Name = "Test", Price = 100 } };
```

```
            var mock = new Mock<IRepository>();
            mock.SetupGet(m => m.Products).Returns(data);
            HomeController controller = new HomeController {
                Repository = mock.Object
            };

            // Act
            ViewResult result = controller.Index();

            // Assert
            Assert.Equal(data, result.ViewData.Model);
        }
    }
}
```

Repository 属性允许隔离控制器并提供可以在操作方法创建的 ViewResult 中检查的测试数据。这仅提供紧密耦合的组件问题的部分解决方案，因为在应用程序运行时无法设置 Repository 属性。正如第 17 章所解释的那样，MVC 负责实例化控制器来处理请求，它并不知道 Repository 属性的特殊重要性。该技术产生的效果是在执行单元测试时控制器和存储库松散耦合，但在应用程序运行时控制器和存储库紧密耦合，如图 18-4 所示。

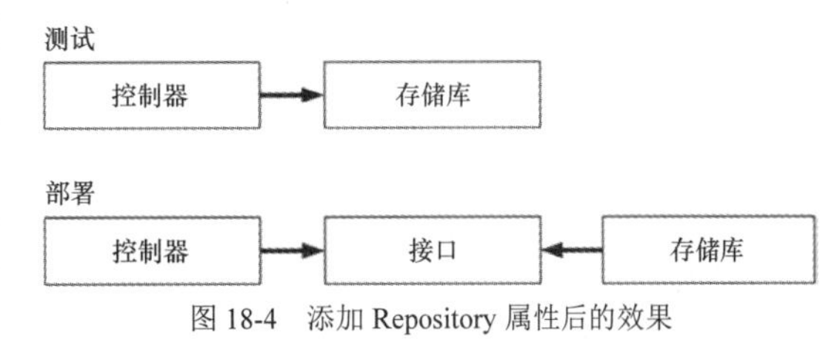

图 18-4　添加 Repository 属性后的效果

2. 使用类型代理

下一个合乎逻辑的步骤是决定将哪种存储库接口的实现用于控制器类，并放在应用程序的其他位置。为了演示这是如何工作的，在示例应用程序中添加 Infrastructure 文件夹，并在其中添加一个名为 TypeBroker.cs 的类文件，内容如代码清单 18-12 所示。

代码清单 18-12　Infrastructure 文件夹下的 TypeBroker.cs 文件的内容

```
using DependencyInjection.Models;
using System;

namespace DependencyInjection.Infrastructure {
    public static class TypeBroker {
        private static Type repoType = typeof(MemoryRepository);
        private static IRepository testRepo;

        public static IRepository Repository =>
            testRepo ?? Activator.CreateInstance(repoType) as IRepository;

        public static void SetRepositoryType<T>() where T : IRepository =>
            repoType = typeof(T);

        public static void SetTestObject(IRepository repo) {
            testRepo = repo;
        }
    }
}
```

TypeBroker 类定义了 Repository 属性，并返回实现了 IRepository 接口的新对象。Repository 属性使用的实现类由 repoType 字段的值确定，该字段默认为 MemoryRepository，但可以通过调用 SetRepositoryType 方法进行更改。

为了支持单元测试，SetTestObject 方法允许使用特定的对象。代码清单 18-13 更新了 Home 控制器，以便从代理获取存储库对象。

代码清单 18-13　在 Controllers 文件夹下的 HomeController.cs 文件中使用类型代理

```
using Microsoft.AspNetCore.Mvc;
using DependencyInjection.Models;
```

```
using DependencyInjection.Infrastructure;

namespace DependencyInjection.Controllers {

    public class HomeController : Controller {

        public IRepository Repository { get; } = TypeBroker.Repository;

        public ViewResult Index() => View(Repository.Products);
    }
}
```

示例应用程序中现在有一组更复杂的关系，如图 18-5 所示。需要注意的一点是，控制器类和存储库类之间没有直接的关系：一切都是通过接口和代理来调用的。这意味着可以更改存储库类，而不必对控制器进行任何更改。

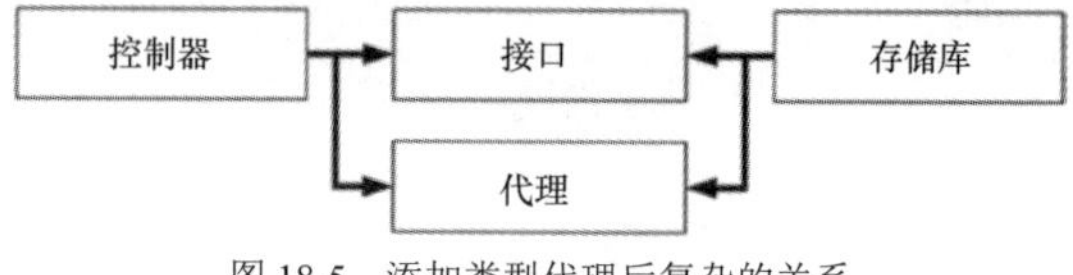

图 18-5 添加类型代理后复杂的关系

为了演示类型代理的使用，在 Models 文件夹中添加一个名为 AlternateRepository.cs 的类文件，并用来定义 IRepository 接口的另一个实现，如代码清单 18-14 所示。

代码清单 18-14 Models 文件夹下的 AlternateRepository.cs 文件的内容

```
using System.Collections.Generic;

namespace DependencyInjection.Models {
    public class AlternateRepository : IRepository {
        private Dictionary<string, Product> products;

        public AlternateRepository() {
            products = new Dictionary<string, Product>();
            new List<Product> {
                new Product { Name = "Corner Flags", Price = 34.95M },
                new Product { Name = "Stadium", Price = 79500M }
            }.ForEach(p => AddProduct(p));
        }
        public IEnumerable<Product> Products => products.Values;

        public Product this[string name] => products[name];

        public void AddProduct(Product product) =>
            products[product.Name] = product;

        public void DeleteProduct(Product product) =>
            products.Remove(product.Name);
    }
}
```

在实际应用中，备用存储库可能以不同的格式存储数据，或者使用不同类型的持久化数据。在本例中，AlternateRepository 类和 MemoryRepository 类之间的区别是在实例化类时创建的模型数据。要使用 AlternateRepository 类，可在 Startup 类的 ConfigureServices 方法中配置类型代理，如代码清单 18-15 所示。

代码清单 18-15 在 DependencyInjection 文件夹下的 Startup.cs 文件中配置类型代理

```
using System;
using System.Collections.Generic;
using System.Linq;
using System.Threading.Tasks;
using Microsoft.AspNetCore.Builder;
using Microsoft.AspNetCore.Hosting;
using Microsoft.AspNetCore.Http;
using Microsoft.Extensions.DependencyInjection;
using DependencyInjection.Infrastructure;
```

```
using DependencyInjection.Models;

namespace DependencyInjection {
    public class Startup {

        public void ConfigureServices(IServiceCollection services) {
            TypeBroker.SetRepositoryType<AlternateRepository>();
            services.AddMvc();
        }

        public void Configure(IApplicationBuilder app, IHostingEnvironment env) {
            app.UseStatusCodePages();
            app.UseDeveloperExceptionPage();
            app.UseStaticFiles();
            app.UseMvcWithDefaultRoute();
        }
    }
}
```

可以通过启动应用程序来查看更改的效果，这会显示新的存储库类提供的数据，如图 18-6 所示。

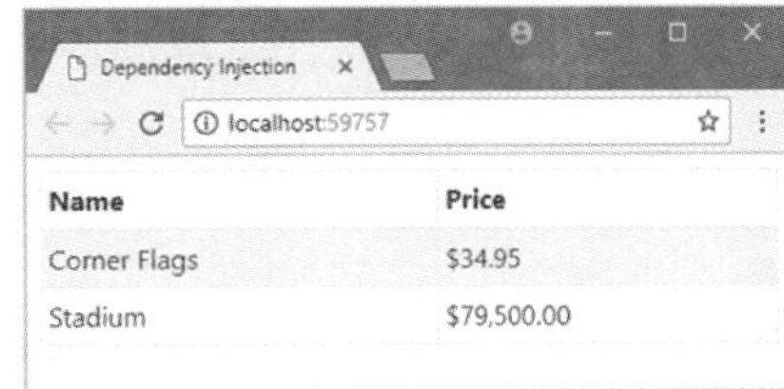

图 18-6 更换存储库类

类型代理允许将特定对象用作存储库，这样就可以像代码清单 18-16 那样编写单元测试。

代码清单 18-16 使用测试项目的 DITests.cs 文件中的代理进行测试

```
using DependencyInjection.Controllers;
using DependencyInjection.Infrastructure;
using DependencyInjection.Models;
using Microsoft.AspNetCore.Mvc;
using Moq;
using Xunit;

namespace Tests {

    public class DITests {

        [Fact]
        public void ControllerTest() {
            // Arrange
            var data = new[] { new Product { Name = "Test", Price = 100 } };
            var mock = new Mock<IRepository>();
            mock.SetupGet(m => m.Products).Returns(data);
            TypeBroker.SetTestObject(mock.Object);
            HomeController controller = new HomeController();

            // Act
            ViewResult result = controller.Index();

            // Assert
            Assert.Equal(data, result.ViewData.Model);
        }
    }
}
```

18.3 ASP.NET 的依赖注入

18.2 节介绍了分离控制器类和提供模型数据的存储库的过程。HomeController 类现在可以获取 IRepository 接口的实现，而不知道正在使用哪个类或如何实例化。有关 Irepository 接口的信息包含在 TypeBroker 类中，可由任何其他需要访问存储库的控制器使用，也可用于其他测试对象。

虽然依赖注入可以构建更加灵活的应用，但也有一些不方便的地方。最大的缺点是开发人员必须为每个用来管理代理的新类型添加新的方法和属性。可以重写 TypeBroker 类来使其更通用，但是没有任何需要，因为 ASP.NET Core 提供相同功能的版本，并且更易于使用，不需要任何其他的类。

18.3.1 准备依赖注入

依赖注入描述了创建松散耦合组件的另一种方法，已被集成到 ASP.NET Core 平台中，由 MVC 自动使用，这意味着控制器和其他组件不需要知道如何创建它们所需的类型。代码清单 18-17 展示了应如何在 Home 控制器中使用 DI。

代码清单 18-17 准备在 Controllers 文件夹下的 HomeController.cs 文件中使用依赖注入

```
using Microsoft.AspNetCore.Mvc;
using DependencyInjection.Models;
using DependencyInjection.Infrastructure;

namespace DependencyInjection.Controllers {

    public class HomeController : Controller {
        private IRepository repository;

        public HomeController(IRepository repo) => repository = repo;

        public ViewResult Index() => View(repository.Products);
    }
}
```

控制器将其依赖项声明为构造函数参数。这是依赖注入的第一部分，依赖注入中的依赖项是创建类的新实例所需的对象。在这种情况下，控制器类已声明对 IRepository 接口的依赖项。

在 ASP.NET Core 中，名为服务提供者的组件负责将接口映射到用于满足依赖关系的实现类型。

当需要新的控制器时，MVC 会要求服务提供者创建一个新的 HomeController 实例。服务提供者检查 HomeController 构造函数以确定其依赖性，创建所需的服务对象，并将它们注入 HomeController 构造函数以创建可用于处理请求的新控制器。这是依赖注入的核心过程，所以再详细介绍一遍。

（1）MVC 接收针对 Home 控制器上的操作方法的传入请求。

（2）MVC 向 ASP.NET 的服务提供者组件请求 HomeController 类的新实例。

（3）服务提供者检查 HomeController 构造函数，并发现它与 IRepository 接口有依赖关系。

（4）服务提供者查询其映射，以查找已被告知要用于依赖于 IRepository 接口的实现类。

（5）服务提供者创建实现类的新实例。

（6）服务提供者创建一个新的 HomeController 对象，使用实现对象作为构造函数参数。

（7）服务提供者将新创建的 HomeController 对象返回给 MVC，MVC 使用它处理传入的 HTTP 请求。

整体效果与前面的自定义类 TypeBroker 相同，但重要的优点是将依赖注入过程集成到了 MVC 中，这意味着在创建控制器类时将使用服务提供者组件。这允许控制器类声明依赖关系，而不需要知道如何解决它们。你只需要编写将其依赖性声明为构造函数参数的控制器类，并让 MVC 和服务提供者组件去完成其他的工作。

注 意

本章的所有示例都使用 ASP.NET 内置的依赖注入，它是 ASP.NET Core 的一部分。你可使用第三方软件包替换内置的依赖注入，这些包可以提供一些更强或更多的功能。热门的包括 Autofac 和 StructureMap 等，但在使用时需要将它们集成到 ASP. NET Core 中，你可以在 GitHub 网站上找到更多详细信息。

18.3.2 配置服务提供者

如果运行示例项目，你会发现通过 HomeController 构造函数声明依赖项已经破坏了应用程序。如果 MVC 尝试创建 HomeController 类的实例来为请求提供服务，就会遇到图 18-7 所示的错误。

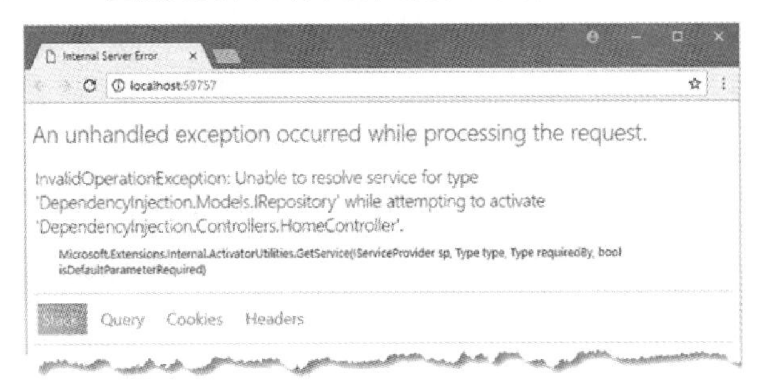

图 18-7 运行示例项目出现的错误

要解决依赖项，必须配置服务提供者，以便它知道如何解决服务依赖关系。当前的服务提供者没有这些信息，当要求创建 HomeController 对象时，会抛出异常，因为不知道如何解决 IRepository 接口上的依赖关系。

服务提供者的配置是在 Startup 类中定义的，这样才能保证在应用程序开始接收请求之前服务已经准备就绪。代码清单 18-18 配置了服务提供者，从而使它知道如何处理 IRepository 接口上的依赖关系。

代码清单 18-18 在 DependencyInjection 文件夹下的 Startup.cs 文件中配置服务提供者

```
using System;
using System.Collections.Generic;
using System.Linq;
using System.Threading.Tasks;
using Microsoft.AspNetCore.Builder;
using Microsoft.AspNetCore.Hosting;
using Microsoft.AspNetCore.Http;
using Microsoft.Extensions.DependencyInjection;
using DependencyInjection.Infrastructure;
using DependencyInjection.Models;

namespace DependencyInjection {
    public class Startup {

        public void ConfigureServices(IServiceCollection services) {
            services.AddTransient<IRepository, MemoryRepository>();
            services.AddMvc();
        }

        public void Configure(IApplicationBuilder app, IHostingEnvironment env) {
            app.UseStatusCodePages();
            app.UseDeveloperExceptionPage();
            app.UseStaticFiles();
            app.UseMvcWithDefaultRoute();
        }
    }
}
```

依赖注入是使用你在 ConfigureServices 方法接收的 IServiceCollection 对象上调用的扩展方法进行配置的。以上代码中使用的 AddTransient 扩展方法会告诉服务提供者如何处理依赖关系（本章后面会详细描述）。映射使用类型参数表示，第一个类型是接口，第二个类型是实现类。

```
...
services.AddTransient<IRepository, MemoryRepository>();
...
```

上面的语句告诉服务提供者通过创建一个 MemoryRepository 对象来解析 IRepository 接口上的依赖关系。如果运行示例应用程序，你将看到由 HomeController 构造函数声明的依赖项被解析，并且控制器提供了对模型数据的访问，依赖注入的配置如图 18-8 所示。

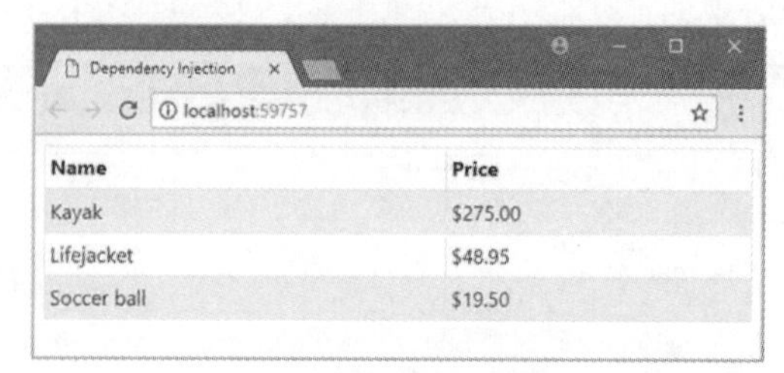

图 18-8 依赖注入的配置

18.3.3 对具有依赖项的控制器进行单元测试

使用构造函数接收依赖关系可以使控制器的单元测试更容易。代码清单 18-19 显示了代码清单 18-18 中控制器的单元测试。

代码清单 18-19 在单元测试项目中使用 DITests.cs 文件测试控制器

```
using DependencyInjection.Controllers;
using DependencyInjection.Models;
using Microsoft.AspNetCore.Mvc;
using Moq;
using Xunit;

namespace Tests {

    public class DITests {

        [Fact]
        public void ControllerTest() {
            // Arrange
            var data = new[] { new Product { Name = "Test", Price = 100 } };
            var mock = new Mock<IRepository>();
            mock.SetupGet(m => m.Products).Returns(data);
            HomeController controller = new HomeController(mock.Object);

            // Act
            ViewResult result = controller.Index();

            // Assert
            Assert.Equal(data, result.ViewData.Model);
        }
    }
}
```

只要实现的接口正确，控制器就不需要知道或关心要将什么样的对象传递给构造函数。这允许使用测试用的存储库，而不必依赖任何可能影响测试结果的外部类（例如 TypeBroker 类）。

18.3.4 使用依赖关系链

当服务提供者需要解析依赖关系时，将检查已配置用于查看是否具有解析依赖关系的类型。结果是可以创建依赖关系链，所有这些都是在运行时解析的，并且可以通过 Startup 类中的配置进行管理。为了演示依赖关系链，在 Models 文件夹中创建一个名为 IModelStorage.cs 的类文件，内容如代码清单 18-20 所示。

代码清单 18-20 Models 文件夹下的 IModelStorage.cs 文件的内容

```
using System.Collections.Generic;

namespace DependencyInjection.Models {

    public interface IModelStorage {
        IEnumerable<Product> Items { get; }
        Product this[string key] { get; set; }
        bool ContainsKey(string key);
        void RemoveItem(string key);
    }
}
```

IModelStorage 接口定义了简单的对 Product 对象进行存储的行为。为了实现这个接口，在 Models 文件夹中添加一个名为 DictionaryStorage.cs 的类文件，内容如代码清单 18-21 所示。

代码清单 18-21 Models 文件夹下的 DictionaryStorage.cs 文件的内容

```
using System.Collections.Generic;

namespace DependencyInjection.Models {
    public class DictionaryStorage : IModelStorage {
        private Dictionary<string, Product> items
            = new Dictionary<string, Product>();

        public Product this[string key] {
            get { return items[key]; }
            set { items[key] = value; }
        }

        public IEnumerable<Product> Items => items.Values;
        public bool ContainsKey(string key) => items.ContainsKey(key);
        public void RemoveItem(string key) => items.Remove(key);
    }
}
```

DictionaryStorage 类通过使用强类型字典来存储模型对象，从而实现了 IModelStorage 接口。这是当前包含在 MemoryRepository 类中的功能，在实际的项目中使用接口来分隔几乎没有什么价值，但它可以提供一个有用的示例，以说明如何使用依赖注入而不会使示例应用程序的复杂性大大增加。

代码清单 18-22 更新了 MemoryRepository 类，以便声明该类依赖 IModelStorage 接口，但它并不了解在实际运行时使用的实现类。

代码清单 18-22 在 Models 文件夹下的 MemoryRepository.cs 文件中声明依赖关系

```
using System.Collections.Generic;

namespace DependencyInjection.Models {
    public class MemoryRepository : IRepository {
        private IModelStorage storage;

        public MemoryRepository(IModelStorage modelStore) {
            storage = modelStore;
            new List<Product> {
                new Product { Name = "Kayak", Price = 275M },
                new Product { Name = "Lifejacket", Price = 48.95M },
                new Product { Name = "Soccer ball", Price = 19.50M }
            }.ForEach(p => AddProduct(p));
        }

        public IEnumerable<Product> Products => storage.Items;
```

```
        public Product this[string name] => storage[name];

        public void AddProduct(Product product) =>
            storage[product.Name] = product;

        public void DeleteProduct(Product product) =>
            storage.RemoveItem(product.Name);
    }
}
```

如果运行示例应用程序，你将看到服务提供者会提示以下异常信息：

```
InvalidOperationException: Unable to resolve service for type
'DependencyInjection.Models.IModelStorage' while attempting to activate
'DependencyInjection.Models.MemoryRepository'.
```

这表明服务提供者正在使用依赖关系链。当要求创建一个新的控制器时，服务提供者检查 HomeController 构造函数，并发现了具有依赖关系的 IRepository 接口，该接口知道应该使用一个 MemoryRepository 对象来解决问题。服务提供者随后检查 MemoryRepository 构造函数，该构造函数依赖于 IModelStorage 接口。但配置中没有指定如何解析 IModelStorage 依赖关系，这意味着无法创建 MemoryRepository 对象，反过来这意味着无法创建 HomeController 对象。所以服务提供者无法向 MVC 提供处理请求所需的对象，抛出异常。

因此，需要添加类型映射，告诉服务提供者如何解决依赖于 IModelStorage 的依赖关系，这里已经将类型映射添加到代码清单 18-23 所示的应用程序配置中。

代码清单 18-23 在 DependencyInjection 文件夹下的 Startup.cs 文件中配置类型映射

```
using System;
using System.Collections.Generic;
using System.Linq;
using System.Threading.Tasks;
using Microsoft.AspNetCore.Builder;
using Microsoft.AspNetCore.Hosting;
using Microsoft.AspNetCore.Http;
using Microsoft.Extensions.DependencyInjection;
using DependencyInjection.Infrastructure;
using DependencyInjection.Models;

namespace DependencyInjection {
    public class Startup {

        public void ConfigureServices(IServiceCollection services) {
            services.AddTransient<IRepository, MemoryRepository>();
            services.AddTransient<IModelStorage, DictionaryStorage>();
            services.AddMvc();
        }

        public void Configure(IApplicationBuilder app, IHostingEnvironment env) {
            app.UseStatusCodePages();
            app.UseDeveloperExceptionPage();
            app.UseStaticFiles();
            app.UseMvcWithDefaultRoute();
        }
    }
}
```

通过上面的代码，服务提供者可以满足依赖关系链中的两个依赖关系，并且能够创建服务请求所需的一组对象——将注入 MemoryRepository 构造函数的 DictionaryStorage 对象注入 HomeController 构造函数。依赖关系链不仅很智能，它们还允许通过组合轻松隔离测试的组件来组合复杂的功能，以及轻松地更改以适应项目逐渐成熟的不断变化的需求。

18.3.5 对具体类型使用依赖注入

依赖注入也可以用于不能通过接口访问的具体类型。虽然这并没有提供使用接口的松散耦合优势，但它是十分有用的技术，因为允许在应用程序中的任何地方访问对象，并将具体的类型放在生命周期管理中，这将在后面具体介绍。

为了演示，在 Models 文件夹添加一个名为 ProductTotalizer.cs 的类文件，内容如代码清单 18-24 所示。

代码清单 18-24 Models 文件夹下的 ProductTotalizer.cs 文件的内容

```
using System.Linq;

namespace DependencyInjection.Models {
    public class ProductTotalizer {

        public ProductTotalizer(IRepository repo) => Repository = repo;

        public IRepository Repository { get; set; }

        public decimal Total => Repository.Products.Sum(p => p.Price);
    }
}
```

ProductTotalizer 类没有什么特别的用途，但它对 IRepository 接口具有依赖性，这意味着通过使用依赖注入，可使用适用于示例应用程序的其余部分的配置来解析依赖关系。代码清单 18-25 已将 ProductTotalizer 类声明为 HomeController 类的依赖项。

代码清单 18-25 在 Controllers 文件夹下的 HomeController.cs 文件中添加依赖关系

```
using Microsoft.AspNetCore.Mvc;
using DependencyInjection.Models;
using DependencyInjection.Infrastructure;

namespace DependencyInjection.Controllers {

    public class HomeController : Controller {
        private IRepository repository;
        private ProductTotalizer totalizer;

        public HomeController(IRepository repo, ProductTotalizer total) {
            repository = repo;
            totalizer = total;
        }

        public ViewResult Index() {
            ViewBag.Total = totalizer.Total;
            return View(repository.Products);
        }
    }
}
```

Index 操作方法会添加 ViewBag 属性，该属性包含 ProductTotalizer 类生成的产品总数，该类将显示在表格中，以查看本章开头部分添加到 Index.cshtml 文件中的 ViewBag 值。最后一步是告诉服务提供者如何处理 ProductTotalizer 请求，如代码清单 18-26 所示。

代码清单 18-26 在 DependencyInjection 文件夹下的 Startup.cs 文件中配置服务提供者

```
using System;
using System.Collections.Generic;
```

```
using System.Linq;
using System.Threading.Tasks;
using Microsoft.AspNetCore.Builder;
using Microsoft.AspNetCore.Hosting;
using Microsoft.AspNetCore.Http;
using Microsoft.Extensions.DependencyInjection;
using DependencyInjection.Infrastructure;
using DependencyInjection.Models;

namespace DependencyInjection {
    public class Startup {

        public void ConfigureServices(IServiceCollection services) {
            services.AddTransient<IRepository, MemoryRepository>();
            services.AddTransient<IModelStorage, DictionaryStorage>();
            services.AddTransient<ProductTotalizer>();
            services.AddMvc();
        }

        public void Configure(IApplicationBuilder app, IHostingEnvironment env) {
            app.UseStatusCodePages();
            app.UseDeveloperExceptionPage();
            app.UseStaticFiles();
            app.UseMvcWithDefaultRoute();
        }
    }
}
```

在这种情况下，服务类型和实现类型之间不存在映射，所以需要有 AddTransient 扩展方法。该方法接收单个类型参数，该参数的值通知服务提供者应将 ProductTotalizer 类实例化以解析此类型的依赖项。

这种方法的优点是服务提供者将解析任何具体类声明的依赖项，而不是简单地实例化控制器中的具体类，因此更改配置以便使用专门的子类来解析具体类的依赖项。具体类由服务提供者管理，并且受生命周期功能的控制。如果运行示例应用程序，你将看到模型中的 Product 对象的总价格将显示出来，如图 18-9 所示。

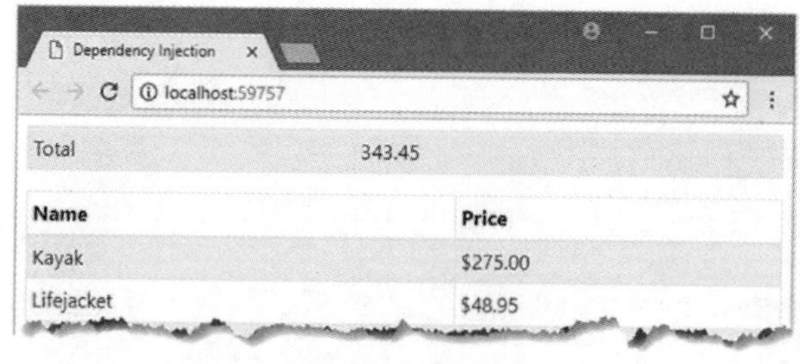

图 18-9 对类使用依赖注入后的总价格

18.4 服务的生命周期

前面使用 AddTransient 扩展方法告诉服务提供者如何处理 IRepository 和 IModelStorage 接口上的依赖关系。AddTransient 扩展方法是可以定义类型映射的 4 种不同方式之一。表 18-3 描述了用于告诉服务提供者如何解决依赖关系的扩展方法。表 18-3 中的扩展方法都使用类型参数，但也有可用于将类型对象作为参数接收的扩展方法，如果需要在运行时生成映射，就可以使用这些扩展方法。

表 18-3 服务提供者的依赖注入扩展方法

扩展方法	描述
AddTransient<service,implType>()	告诉服务提供者为服务类型上的每个依赖关系创建实现类型的新实例
AddTransient<service>()	用于注册单个类型（将为每个依赖项进行实例化）
AddTransient<service>(factoryFunc)	用于注册将被调用的工厂模式，以便为服务类型的每个依赖项创建实现对象
AddScoped<service,implType>() AddScoped<service>() AddScoped<service>(factoryFunc)	这些扩展方法告诉服务提供者重用实现类型的实例，以便与公共作用域关联的组件所做的所有服务请求（通常是单个 HTTP 请求）共享同一对象。这些扩展方法遵循与相应的 AddTransient 扩展方法相同的模式
AddSingleton<service,implType>() AddSingleton<service>() AddSingleton<service(factoryFunc)	这些扩展方法告诉服务提供者为第一个服务请求创建实现类型的新实例，然后为每个后续服务请求重用
AddSingleton<service>(instance)	为服务提供者提供可应用于服务所有服务请求的对象，服务提供者将不会创建任何新对象

18.4.1 使用瞬态生命周期

开始使用依赖注入的最简单方法是使用 AddTransient 扩展方法，该方法告诉服务提供者在需要解析依赖关系时创建实现类型的新实例。这是 Startup 类中已经存在的配置，如下所示：

```
...
public void ConfigureServices(IServiceCollection services) {
    services.AddTransient<IRepository, MemoryRepository>();
    services.AddTransient<IModelStorage, DictionaryStorage>();
    services.AddTransient<ProductTotalizer>();
    services.AddMvc();
}
...
```

表 18-3 对生命周期做了介绍。瞬态生命周期会产生在每次解决依赖关系时创建实现类的新实例的成本，但优点在于不必担心管理并发访问或确保对象可以安全地重用于多个请求。

为了演示瞬态生命周期，在 MemoryRepository 类中重写 ToString 方法，以便生成全局唯一标识符（GUID），如代码清单 18-27 所示。

代码清单 18-27 重写 Models 文件夹下的 MemoryRepository.cs 文件中的 ToString 方法

```
using System.Collections.Generic;

namespace DependencyInjection.Models {
    public class MemoryRepository : IRepository {
        private IModelStorage storage;
        private string guid = System.Guid.NewGuid().ToString();

        public MemoryRepository(IModelStorage modelStore) {
            storage = modelStore;
            new List<Product> {
                new Product { Name = "Kayak", Price = 275M },
                new Product { Name = "Lifejacket", Price = 48.95M },
                new Product { Name = "Soccer ball", Price = 19.50M }
            }.ForEach(p => AddProduct(p));
        }
        public IEnumerable<Product> Products => storage.Items;

        public Product this[string name] => storage[name];
```

```
        public void AddProduct(Product product) =>
            storage[product.Name] = product;

        public void DeleteProduct(Product product) =>
            storage.RemoveItem(product.Name);

        public override string ToString() {
            return guid;
        }
    }
}
```

GUID 使你可以轻松地识别 MemoryRepository 类的特定实例，并了解不同的生命周期方法如何改变服务提供者的行为方式。代码清单 18-28 更新了 Home 控制器上的 Index 操作方法，以便为通过存储库设置为 GUID 的 View Bag 创建 Controller 属性。

代码清单 18-28 在 Controllers 文件夹下的 HomeController.cs 文件中使用 View Bag

```
using Microsoft.AspNetCore.Mvc;
using DependencyInjection.Models;
using DependencyInjection.Infrastructure;

namespace DependencyInjection.Controllers {

    public class HomeController : Controller {
        private IRepository repository;
        private ProductTotalizer totalizer;

        public HomeController(IRepository repo, ProductTotalizer total) {
            repository = repo;
            totalizer = total;
        }

        public ViewResult Index() {
            ViewBag.HomeController = repository.ToString();
            ViewBag.Totalizer = totalizer.Repository.ToString();
            return View(repository.Products);
        }
    }
}
```

Index 操作方法会将值添加到 View Bag 中，其中包含通过 ProductTotalizer 类的构造函数直接接收的 repository 对象的 GUID，可以在运行应用程序时看到它们。两个 GUID 不同，因为服务提供者已使用 AddTransient 扩展方法进行了配置，这意味着将创建一个新的 MemoryRepository 对象来解决 HomeController 的依赖项和 ProductTotalizer 的另一个属性。瞬态生命周期的作用如图 18-10 所示。

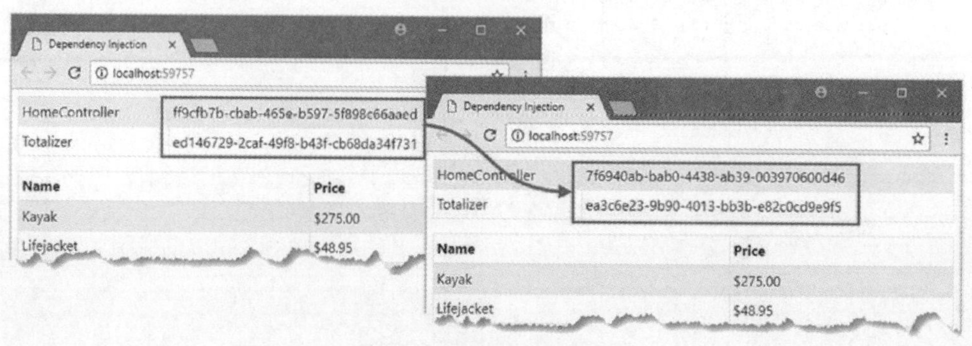

图 18-10 瞬态生命周期的作用

每次重新加载 Web 页面时，新的 HTTP 请求都会导致 MVC 创建一个新的 HomeController，这将导致创建两个新的 MemoryRepository 对象，每一个都有自己的 GUID。

提 示

GUID 是唯一的，或是接近唯一的。因此，当在自己的计算机上运行应用程序时，你将看到不同的值。

使用工厂模式

AddTransient 扩展方法的其中一个版本接收每次对服务类型有依赖性时调用的工厂模式。这允许创建的对象发生变化，以便不同的依赖项接收不同类型或实例的配置。代码清单 18-29 使用工厂模式根据应用程序运行的宿主环境来选择 IRepository 接口的不同实现。

代码清单 18-29 在 DependencyInjection 文件夹下的 Startup.cs 文件中使用工厂模式

```
using System;
using System.Collections.Generic;
using System.Linq;
using System.Threading.Tasks;
using Microsoft.AspNetCore.Builder;
using Microsoft.AspNetCore.Hosting;
using Microsoft.AspNetCore.Http;
using Microsoft.Extensions.DependencyInjection;
using DependencyInjection.Infrastructure;
using DependencyInjection.Models;

namespace DependencyInjection {
    public class Startup {
        private IHostingEnvironment env;
        public Startup(IHostingEnvironment hostEnv) => env = hostEnv;

        public void ConfigureServices(IServiceCollection services) {
            services.AddTransient<IRepository>(provider => {
                if (env.IsDevelopment()) {
                    var x = provider.GetService<MemoryRepository>();
                    return x;
                } else {
                    return new AlternateRepository();
                }
            });
            services.AddTransient<MemoryRepository>();
            services.AddTransient<IModelStorage, DictionaryStorage>();
            services.AddTransient<ProductTotalizer>();
            services.AddMvc();
        }

        public void Configure(IApplicationBuilder app, IHostingEnvironment env) {
            app.UseStatusCodePages();
            app.UseDeveloperExceptionPage();
            app.UseStaticFiles();
            app.UseMvcWithDefaultRoute();
        }
    }
}
```

第 14 章描述了 ASP.NET 如何为 Startup 类提供用来帮助设置应用程序的服务，包括用于确定宿主环境的 IHostingEnvironment 接口的实现。可以将这些服务作为参数传给 Configure 方法而不是 ConfigureServices 方法。为此，对 Startup 类添加一个构造函数，以提供对 IHostingEnvironment 对象的访问，并将其分配给一个名为 env 的字段。

在 ConfigureServices 方法中，使用 AddTransient 扩展方法定义一个使用了 Lambda 表达式的工厂模式。Lambda 表达式接收一个 System. IServiceProvider 对象，该对象可用于创建使用表 18-4 所示方法向服务提供者注册的其他类型的实例。

表 18-4 IserviceProvider 的方法

方法	描述
GetService<service>()	使用服务提供者创建服务类型的新实例。如果请求的类型没有映射，就返回 null
GetRequiredService<service>()	使用服务提供者创建服务类型的新实例。如果请求的类型没有映射，就抛出异常

在工厂模式中，使用 IHostingEnvironment 来确定应用程序是否在开发环境中运行。如果在开发环境中运行，就使用 GetService 方法创建 MemoryRepository 类的实例并从工厂返回作为用于 IRepository 依赖项的对象的函数。这里使用 GetService 方法创建对象，因为 MemoryRepository 在 IModelStorage 接口上有自己的依赖项，并使用服务提供者创建对象，这意味着检测和解决依赖关系将自动进行管理，但也意味着必须指定应用于 MemoryRepository 对象的生命周期，如下所示：

```
...
services.AddTransient<MemoryRepository>();
...
```

如果没有以上声明，服务提供者将不具备创建和管理 MemoryRepository 对象所需的信息。

如果应用程序未在开发环境中运行，那么 factory 函数将返回一个新的 AlternateRepository 实例。可以使用 new 关键字直接创建，因为不会在构造函数中声明任何依赖关系。

18.4.2 使用作用域的生命周期

作用域的生命周期会从实现类创建对象，用于解决与单个作用域关联的所有依赖项，通常表示单个 HTTP 请求（可以创建自己的作用域，但这在大多数应用程序中无效）。

由于默认作用域是 HTTP 请求，因此作用域的生命周期允许处理请求的所有组件共享单个对象，并且在编写自定义类（如路由）时，共享公共上下文数据通常非常有用。作用域的生命周期是通过使用 AddScoped 扩展方法以配置服务提供者来创建的，如代码清单 18-30 所示。

提 示

如表 18-4 所示，还有其他接收工厂模式的 AddScoped 版本，可用于注册具体类型。这些方法的工作方式与前面演示的 AddTransient 方法相同，主要的区别在于它们创建的对象的生命周期不同。

代码清单 18-30 在 DependencyInjection 文件夹下的 Startup.cs 文件中使用作用域的生命周期

```
using System;
using System.Collections.Generic;
using System.Linq;
using System.Threading.Tasks;
using Microsoft.AspNetCore.Builder;
using Microsoft.AspNetCore.Hosting;
using Microsoft.AspNetCore.Http;
using Microsoft.Extensions.DependencyInjection;
using DependencyInjection.Infrastructure;
using DependencyInjection.Models;
namespace DependencyInjection {
    public class Startup {
        private IHostingEnvironment env;
```

```
        public Startup(IHostingEnvironment hostEnv) => env = hostEnv;

        public void ConfigureServices(IServiceCollection services) {
            services.AddScoped<IRepository, MemoryRepository>();
            services.AddTransient<IModelStorage, DictionaryStorage>();
            services.AddTransient<ProductTotalizer>();
            services.AddMvc();
        }

        public void Configure(IApplicationBuilder app, IHostingEnvironment env) {
            app.UseStatusCodePages();
            app.UseDeveloperExceptionPage();
            app.UseStaticFiles();
            app.UseMvcWithDefaultRoute();
        }
    }
}
```

在示例应用程序中，HomeController 和 ProductTotalizer 将一起实例化以处理请求，并且都要求存储库服务解决 IRepository 接口上的依赖关系。使用 AddScoped 方法可确保两个对象的依赖项都通过单个 MemoryRepository 对象来解决。可以通过运行示例应用程序来查看效果。浏览器显示的两个 GUID 是相同的，如图 18-11 所示。重新加载页面后，将创建一个新的 HTTP 请求，这意味着创建了一个新的 MemoryRepository 对象。

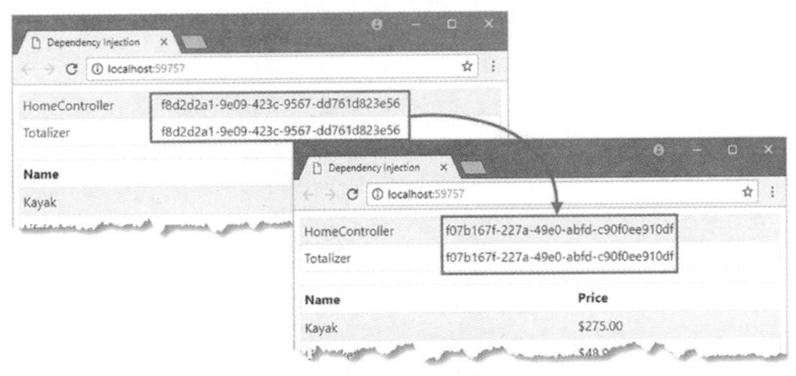

图 18-11　作用域的生命周期产生的影响

18.4.3　使用单例生命周期

单例生命周期可确保单个对象用于解决给定服务类型的所有依赖项。使用单例生命周期时，必须确保用于解析依赖项的实现类对于并发访问是安全的。代码清单 18-31 更改了 IRepository 配置的作用域。

代码清单 18-31　在 DependencyInjection 文件夹下的 Startup.cs 文件中使用作用域的生命周期

```
using System;
using System.Collections.Generic;
using System.Linq;
using System.Threading.Tasks;
using Microsoft.AspNetCore.Builder;
using Microsoft.AspNetCore.Hosting;
using Microsoft.AspNetCore.Http;
using Microsoft.Extensions.DependencyInjection;
using DependencyInjection.Infrastructure;
using DependencyInjection.Models;

namespace DependencyInjection {
```

```
    public class Startup {
        private IHostingEnvironment env;

        public Startup(IHostingEnvironment hostEnv) => env = hostEnv;

        public void ConfigureServices(IServiceCollection services) {
            services.AddSingleton<IRepository, MemoryRepository>();
            services.AddTransient<IModelStorage, DictionaryStorage>();
            services.AddTransient<ProductTotalizer>();
            services.AddMvc();
        }

        public void Configure(IApplicationBuilder app, IHostingEnvironment env) {
            app.UseStatusCodePages();
            app.UseDeveloperExceptionPage();
            app.UseStaticFiles();
            app.UseMvcWithDefaultRoute();
        }
    }
}
```

AddSingleton 方法在第一次创建 MemoryRepository 类的新实例时，必须在 IRepository 接口上解析依赖项，然后重用到任何后续依赖项，即使它们是不同的 HTTP 请求，如图 18-12 所示。

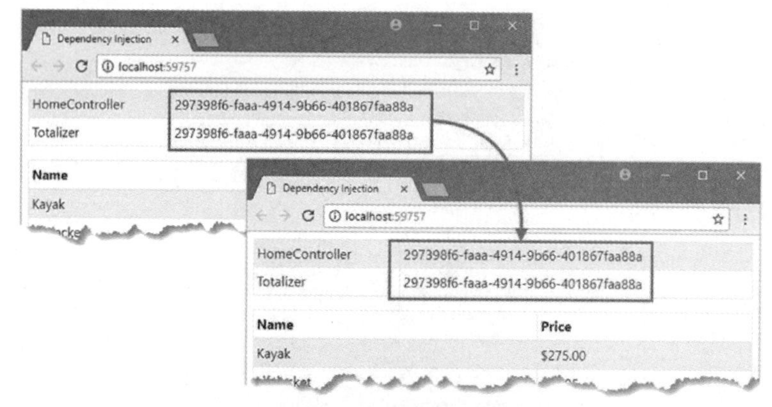

图 18-12 单例生命周期的影响

18.5 使用操作注入

声明依赖关系的标准方法是使用构造函数，这是一种可以在任何类中使用的技术，依赖于依赖注入功能，依赖注入功能是 ASP.NET Core 平台的一部分。

MVC 使用一种名为操作注入的替代方法来补充标准功能，允许通过参数将依赖关系声明为操作方法。严格说来，操作注入是由第 26 章介绍的模型绑定系统提供的，允许以不同的方式使用服务。可使用 FromServices 属性执行操作注入，如代码清单 18-32 所示。

代码清单 18-32 在 Controllers 文件夹下的 HomeController.cs 文件中使用操作注入

```
using Microsoft.AspNetCore.Mvc;
using DependencyInjection.Models;
using DependencyInjection.Infrastructure;

namespace DependencyInjection.Controllers {
```

```
    public class HomeController : Controller {
        private IRepository repository;

        public HomeController(IRepository repo) {
            repository = repo;
        }

        public ViewResult Index([FromServices]ProductTotalizer totalizer) {
            ViewBag.HomeController = repository.ToString();
            ViewBag.Totalizer = totalizer.Repository.ToString();
            return View(repository.Products);
        }
    }
}
```

MVC 使用服务提供者获取 ProductTotalizer 类的实例，并在调用 Index 操作方法时将其作为参数提供。操作注入相比构造函数注入更不常见，但是，当需要创建的对象具有依赖关系，并且仅在控制器定义的操作方法中需要时，操作注入会很有用。使用构造函数注入可以解决所有操作方法的依赖关系，即使用于处理请求的对象也不使用实现目标。使用 FromServices 属性修饰操作方法可缩小依赖项的焦点，并确保只有在需要时才实例化实现类型。

18.6 使用属性注入特性

第 17 章解释了如何通过声明属性并使用 ControllerContext 来修饰 POCO 控制器中的上下文数据。在阅读本章后，你就会明白这是依赖注入的一种特殊形式，称为属性注入。

MVC 提供了一组专用属性，这些属性可用于通过控制器中的属性注入和视图组件（参见第 22 章）接收特定类型。如果从控制器基类派生控制器，就不需要使用这些属性，因为上下文信息是通过 convenience 属性公开的，表 18-5 列出了可在 POCO 控制器中使用的属性。

表 18-5 专用属性

名称	描述
ControllerContext	提供 ActionContext 类的功能超集
ActionContext	为操作方法提供上下文信息。控制器类会通过 ActionContext 属性公开上下文信息，第 31 章将介绍一组 convenience 属性
ViewContext	为视图操作提供上下文数据，包括标签助手
ViewComponentContext	为 View 组件设置 ViewComponentContext 属性
ViewDataDictionary	设置 ViewDataDictionary 属性以提供对模型绑定数据的访问

18.7 手动请求实现对象

ASP.NET 的依赖注入功能以及 MVC 为属性注入和 action 注入提供的附加属性，提供了为大多数应用程序创建松散耦合组件所需的所有支持。但是有些时候，在不使用依赖注入的情况下获取接口的实现也是有用的。在这些情况下，可以直接与服务提供者一起工作，如代码清单 18-33 所示。

代码清单 18-33 在 Controllers 文件夹下的 HomeController.cs 文件中直接使用服务提供者

```
using Microsoft.AspNetCore.Mvc;
using DependencyInjection.Models;
```

```
using DependencyInjection.Infrastructure;
using Microsoft.Extensions.DependencyInjection;

namespace DependencyInjection.Controllers {

    public class HomeController : Controller {

        public ViewResult Index([FromServices]ProductTotalizer totalizer) {

            IRepository repository =
                HttpContext.RequestServices.GetService<IRepository>();

            ViewBag.HomeController = repository.ToString();
            ViewBag.Totalizer = totalizer.Repository.ToString();
            return View(repository.Products);
        }
    }
}
```

由同名属性返回的 HttpContext 对象定义了 RequestServices 方法，该方法返回一个 IServiceProvider 对象，可以调用表 18-4 中描述的扩展方法。以上代码删除了使用属性注入设置的 Repository 属性，并使用 HttpContext.RequestServices 属性获取 IRepository 接口的实现。这就是服务定位器模式，一些开发者认为应该避免使用这种模式。Mark Seemann 在其博客上详细描述了这种模式可能导致的问题。当通过构造函数接收依赖关系的常规技术由于某种原因不能使用时，以这种方式获取服务是完全合理的。

18.8　小结

本章介绍了依赖注入在 MVC 应用程序中扮演的角色，依赖注入有助于创建松散耦合的组件，以便能够轻松地进行替换和隔离测试。此外，本章还演示了 ASP.NET 依赖注入功能和 MVC 为将依赖项注入属性和操作方法而提供的属性，介绍了配置服务提供者时可用的不同生命周期选项，并解释了它们如何影响对象的创建方式。下一章将介绍过滤器，它会在请求处理过程中添加额外的逻辑。

第 19 章　过滤器

过滤器用于向 MVC 请求处理注入额外的逻辑。它们提供了一种简单而优雅的方式来实现交叉关注。所谓交叉关注（cross-cutting concern），是指可用于整个应用程序，但又不适合放置在某个局部位置的功能，否则会打破关注点分离原则。典型的交叉关注例子是日志记录、授权和缓存。本章将展示 MVC 支持的不同类别的过滤器、如何创建和使用自定义过滤器以及如何控制它们的执行。

表 19-1 展示了上下文中使用的过滤器。

表 19-1　　上下文中使用的过滤器

问题	答案
什么是过滤器？	过滤器可以将逻辑应用于操作方法，而不必在控制器类中添加代码
过滤器有什么作用？	过滤器允许应用不属于操作的经典 MVC 模式定义的代码。可以实现更简单的控制器类和可重用的功能，并且可在整个应用程序中应用
如何使用过滤器？	MVC 可通过不同的方式使用不同类型的过滤器。创建过滤器的最常见方法是创建一个类，该类将 MVC 为所需的筛选类型提供的属性作为子类
过滤器有什么陷阱或限制吗？	不同类型的过滤器提供的功能有重叠，有时候很难选择需要哪种类型的过滤器
除了过滤器还有其他的选择吗？	没有，过滤器是 MVC 的核心功能，用于实现日常所需的功能，比如用户授权

表 19-2 展示了本章要介绍的操作。

表 19-2　　本章要介绍的操作

操作	方法	代码清单
在请求处理中注入额外的逻辑	在控制器或操作方法上应用过滤器	代码清单 19-6～代码清单 19-9
限制对操作的访问	使用授权过滤器	代码清单 19-10 和代码清单 19-11
将通用逻辑注入请求处理过程	使用操作过滤器	代码清单 19-12～代码清单 19-14
检查或改变操作方法产生的结果	使用结果过滤器	代码清单 19-15～代码清单 19-19
处理错误	使用异常过滤器	代码清单 19-20 和代码清单 19-21
在过滤器中使用服务	声明过滤器构造函数中的依赖关系，在 Startup 类中注册服务，并使用 TypeFilter 属性应用过滤器	代码清单 19-22～代码清单 19-26
将过滤器置于生命周期管理之下	使用依赖注入的生命周期在 Startup 类中注册过滤器，并使用 ServiceFilter 属性应用过滤器	代码清单 19-27～代码清单 19-29
对应用程序中的每个操作方法应用过滤器	使用全局过滤器	代码清单 19-30～代码清单 19-32
更改执行过滤器的顺序	使用 Order 参数	代码清单 19-33～代码清单 19-36

19.1　准备示例项目

在本章中，将遵循前几章使用的相同方法来创建示例应用程序。使用 ASP.NET Core Web Application（.NET Core）模板创建一个名为 Filters 的 Empty 项目。代码清单 19-1 显示了对 Startup 类所做的更改，用于启用 MVC 框架和开发所需的其他中间件。

代码清单 19-1　Filters 文件夹下的 Startup.cs 文件的内容

```
using System;
using System.Collections.Generic;
```

```
using System.Linq;
using System.Threading.Tasks;
using Microsoft.AspNetCore.Builder;
using Microsoft.AspNetCore.Hosting;
using Microsoft.AspNetCore.Http;
using Microsoft.Extensions.DependencyInjection;

namespace Filters {
    public class Startup {
        public void ConfigureServices(IServiceCollection services) {
            services.AddMvc();
        }
        public void Configure(IApplicationBuilder app, IHostingEnvironment env) {
            app.UseStatusCodePages();
            app.UseDeveloperExceptionPage();
            app.UseStaticFiles();
            app.UseMvcWithDefaultRoute();
        }
    }
}
```

19.1.1 启用 SSL

本章中的一些示例需要使用 SSL，默认情况下 SSL 是禁用的。要启用 SSL，请从 Visual Studio 菜单中选择 Filter Properties，然后在 Debug 选项卡中勾选“Enable SSL https://localhost:44318/”复选框，如图 19-1 所示。请注意记录分配的端口，这对于每个项目将是不同的。

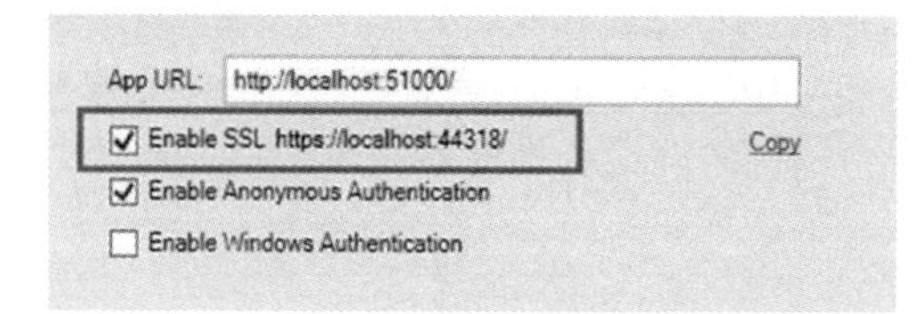

图 19-1 启用 SSL

19.1.2 创建控制器和视图

本章中的控制器很简单，因为本章的重点是应用程序的其他位置。这里创建 Controllers 文件夹，添加一个名为 HomeController.cs 的类文件，内容如代码清单 19-2 所示。

代码清单 19-2 Controllers 文件夹下的 HomeController.cs 文件的内容

```
using Microsoft.AspNetCore.Mvc;

namespace Filters.Controllers {

    public class HomeController : Controller {

        public ViewResult Index() => View("Message",
            "This is the Index action on the Home controller");
    }
}
```

Index 操作方法将呈现一个名为 Message 的视图，并将一个字符串传递给该视图。这里创建 Views/Shared 文件夹，并添加一个名为 Message.cshtml 的 Razor 视图文件，内容如代码清单 19-3 所示。

代码清单 19-3 Views/Shared 文件夹下的 Message.cshtml 文件的内容

```
@{ Layout = null; }

<!DOCTYPE html>
<html>
<head>
    <meta name="viewport" content="width=device-width" />
    <title>Filters</title>
```

```
    <link asp-href-include="lib/bootstrap/dist/css/*.min.css" rel="stylesheet" />
</head>
<body class="m-1 p-1">
    @if (Model is string) {
        @Model
    } else if (Model is IDictionary<string, string>) {
        var dict = Model as IDictionary<string, string>;
        <table class="table table-sm table-striped table-bordered">
            <thead><tr><th>Name</th><th>Value</th></tr></thead>
            <tbody>
                @foreach (var kvp in dict) {
                    <tr><td>@kvp.Key</td><td>@kvp.Value</td></tr>
                }
            </tbody>
        </table>
    }
</body>
</html>
```

Message 视图是弱类型视图，将显示字符串或字典 Dictionary<string,string>，在当前情况下，会显示一个表格。

Message 视图依赖于 Bootstrap CSS 包来为 HTML 元素设置样式。要将 Bootstrap 包添加到项目中，可以使用 Bower Configuration File 模板，在项目的根文件夹中创建 bower.json 文件，并将 Bootstrap 包添加到 dependencies 部分，如代码清单 19-4 所示。

代码清单 19-4　在 Filters 文件夹下的 bower.json 文件中添加 Bootstrap 包中

```
{
  "name": "asp.net",
  "private": true,
  "dependencies": {
    "bootstrap": "4.0.0-alpha.6"
  }
}
```

最后的准备工作是在 Views 文件夹中创建_ViewImports.cshtml 文件，用于设置内置的标签助手，以便在 Razor 视图中使用，如代码清单 19-5 所示。

代码清单 19-5　Views 文件夹下的_ViewImports.cshtml 文件的内容

```
@addTagHelper *, Microsoft.AspNetCore.Mvc.TagHelpers
```

如果运行示例应用程序，看到的输出将如图 19-2 所示。

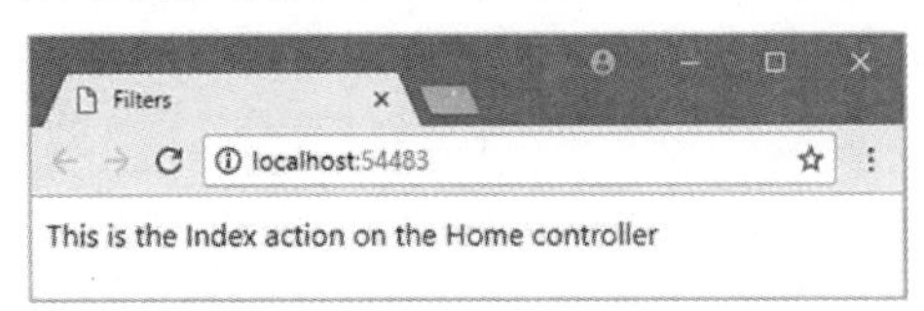

图 19-2　运行示例应用程序的输出

提　示

系统可能会提示需要信任 Visual Studio 生成的证书。请接受建议，这与本章中的示例依赖于 SSL 有关。

19.2　使用过滤器

过滤器允许从控制器中删除在操作方法中应用的一些逻辑，并在可重用类中重新定义。例如，假设要确保只能使用 HTTPS 访问操作方法，而不能使用常规的未加密 HTTP。HttpRequest 上下文对象提供了是否使用 HTTPS 的信息，如代码清单 19-6 所示。

代码清单 19-6 在 Controllers 文件夹下的 HomeController.cs 文件中验证是否使用 HTTPS

```
using Microsoft.AspNetCore.Mvc;
using Microsoft.AspNetCore.Http;

namespace Filters.Controllers {

    public class HomeController : Controller {

        public IActionResult Index() {
            if (!Request.IsHttps) {
                return new StatusCodeResult(StatusCodes.Status403Forbidden);
            } else {
                return View("Message",
                    "This is the Index action on the Home controller");
            }
        }
    }
}
```

这演示了如何在没有使用过滤器的情况下处理 HTTPS 问题。如果运行示例应用程序，浏览器将请求 Index 操作方法并处理不带 HTTPS 的默认 URL，还会返回一个 StatusCodeResult，用以在响应中发送 HTTP 403 状态编码（如第 17 章所述）。如果要求带 HTTPS 的默认 URL（比如 https://localhost:44318），Index 操作方法将通过渲染 Message 视图进行响应（可能需要在浏览器显示结果之前确认安全警告）。图 19-3 显示了这两种结果。

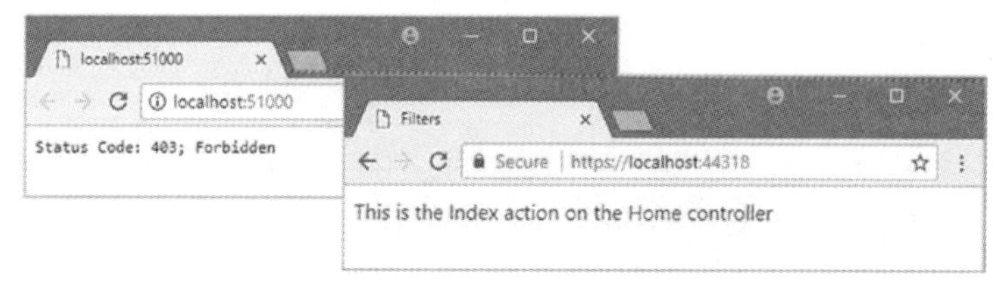

图 19-3 限制对 HTTPS 请求的访问的结果

提 示

如果没有从示例中获得期望的结果，请清除浏览器的历史记录。浏览器通常会拒绝向已经生成 SSL 错误的服务器发送请求，这是很好的安全设置，但在开发过程中可能不是很方便。

代码清单 19-6 中的代码可以正常工作，但也存在一些问题。其中一个问题是，操作方法包含的代码更多的是实现安全策略，而不是处理请求、更新模型和选择响应。另一个更严重的问题是，在操作方法中包括的 HTTP 检测代码不能很好地进行扩展，并且必须在控制器中的每个操作方法中重复，如代码清单 19-7 所示。

代码清单 19-7 在 Controllers 文件夹下的 HomeController.cs 文件中添加一个操作方法

```
using Microsoft.AspNetCore.Http;
using Microsoft.AspNetCore.Mvc;

namespace Filters.Controllers {

    public class HomeController : Controller {

        public IActionResult Index() {
```

```
            if (!Request.IsHttps) {
                return new StatusCodeResult(StatusCodes.Status403Forbidden);
            } else {
                return View("Message",
                    "This is the Index action on the Home controller");
            }
        }

        public IActionResult SecondAction() {
            if (!Request.IsHttps) {
                return new StatusCodeResult(StatusCodes.Status403Forbidden);
            } else {
                return View("Message",
                    "This is the SecondAction action on the Home controller");
            }
        }
    }
}
```

必须记住，在每个需要 HTTPS 的控制器的每个操作方法中都要执行相同的检查。实现安全策略的代码是控制器的重要组成部分，虽然看起来实现很简单，但会使控制器更难理解，忘记将它们添加到新的操作方法中只是时间问题，这是安全漏洞。这个问题可以通过使用过滤器来解决，如代码清单 19-8 所示。

代码清单 19-8　在 Controllers 文件夹下的 HomeController.cs 文件中使用过滤器

```
using Microsoft.AspNetCore.Http;
using Microsoft.AspNetCore.Mvc;

namespace Filters.Controllers {

    public class HomeController : Controller {

        [RequireHttps]
        public ViewResult Index() => View("Message",
            "This is the Index action on the Home controller");

        [RequireHttps]
        public ViewResult SecondAction() => View("Message",
            "This is the SecondAction action on the Home controller");
    }
}
```

RequireHttps 特性会将一个内置的过滤器应用于 HomeController 类。这将限制对操作方法的访问，仅支持 HTTPS 请求，并允许从每个方法中删除安全代码，使方法可以专注于处理成功接收到的请求。

注　意

RequireHttps 过滤器的工作方式与代码清单 19-7 中的自定义代码完全相同。对于 GET 请求，RequireHttps 特性将客户端重定向到原始请求的 URL，但是要求通过使用 HTTPS 方案来执行操作，因此对 http://localhost/Home/Index 的请求会被重定向到 https://localhost/Home/Index。这对于部署的大多数应用程序来说是有意义的，但在开发过程中不是这样，因为 HTTP 和 HTTPS 使用的是不同的端口。RequireHttpsAttribute 类定义了一个名为 HandleNonHttpsRequest 的受保护方法，可以重写该方法以更改其行为。也可参考 19.4 节，从头开始创建原始的功能。

当然，还需要将 RequireHttps 特性应用到每个操作方法，有时候你可能会忘记这件事。但是过滤器有如下很有用的技巧：将特性应用于控制器类与将特性应用于每个单独的操作方法具有相同的效果，如代码清单 19-9 所示。

代码清单19-9　将过滤器应用于HomeController.cs文件中的所有操作方法

```
using Microsoft.AspNetCore.Http;
using Microsoft.AspNetCore.Mvc;

namespace Filters.Controllers {

    [RequireHttps]
    public class HomeController : Controller {

        public ViewResult Index() => View("Message",
            "This is the Index action on the Home controller");

        public ViewResult SecondAction() => View("Message",
            "This is the SecondAction action on the Home controller");
    }
}
```

过滤器有不同的级别。如果要限制访问某些操作而不影响其他操作，可以将RequireHttps特性应用于这些方法。如果要保护所有操作方法，包括未来添加到控制器中的任何操作，那么可以将RequireHttps特性应用于控制器类。如果要对应用程序中的每个操作应用过滤器，可以使用全局过滤器，这将在本章后面进行介绍。

19.3　实现过滤器

你已经看到如何使用过滤器，现在介绍过滤器在幕后是如何实现的。过滤器实现了IFilterMetadata接口，该接口定义在Microsoft.AspNetCore.Mvc.Filters命名空间中：

```
namespace Microsoft.AspNetCore.Mvc.Filters {
    public interface IFilterMetadata { }
}
```

该接口是空的，不需要过滤器类来实现任何特定的行为。这是因为有几种不同类型的过滤器，它们的工作方式各不相同，并用于不同的目的。

表19-3列出了各种类型的过滤器、定义它们的接口以及它们所做的事情（MVC还支持一些其他类型的过滤器，但它们不是直接使用的，而是集成到功能中，并通过特定属性进行应用，包括第20章将要介绍的Produces和Consumes属性。）

表19-3　不同类型的过滤器

过滤器	接口	描述
授权过滤器	IAuthorizationFilter IAsyncAuthorizationFilter	使用应用程序的安全策略，包括用户授权
操作过滤器	IActionFilter IAsyncActionFilter	在执行操作方法之前或之后立即执行工作
结果过滤器	IResultFilter IAsyncResultFilter	在处理操作方法的结果之前或之后立即执行工作
异常过滤器	IExceptionFilter IAsyncExceptionFilter	处理异常

表19-3中的描述有些模糊，因为可以使用过滤器进行很多工作，仅受想象力和需要解决的问题的限制。如果你了解过滤器是如何工作的，这将变得更清楚，现在有两个要点需要理解。

首先，表19-3中的每种类型的过滤器都有两种不同的接口。过滤器可以同步或异步地执行它们的工作，比如，同步实现的过滤器实现了IResultFilter接口，而异步实现的过滤器则实现了IAsyncResultFilter接口。

其次，过滤器按特定的顺序执行。首先执行授权过滤器，然后执行文件操作，接下来执行结果过滤器。只有抛出异常才会执行异常过滤器，这会打乱正常的执行序列。

获取上下文数据

过滤器以 FilterContext 对象的形式提供上下文数据。FilterContext 类派生自 ActionContext，ActionContext 也是第 17 章描述的 ControllerContext 类的基类。为方便起见，表 19-4 列出了从 ActionContext 类继承的属性以及 FilterContext 定义的附加属性。

表 19-4 FilterContext 类的属性

属性	描述
ActionDescriptor	返回一个 ActionDescriptor 对象，该对象描述了操作方法
HttpContext	返回一个 HttpContext 对象，该对象提供 HTTP 请求的详细信息以及将要作为回复发送的 HTTP 响应
ModelState	返回一个 ModelStateDictionary 对象，用于验证客户端发送的数据
RouteData	返回一个 RouteData 对象，描述路由系统处理请求的方式
Filters	返回已应用于操作方法的过滤器列表，表示为 IList <IFilterMetadata>

19.4 使用授权过滤器

授权过滤器用于实现应用程序的安全策略。授权过滤器在其他类型的过滤器之前和执行操作方法之前执行。以下是 IAuthorizationFilter 接口的定义：

```
namespace Microsoft.AspNetCore.Mvc.Filters {

    public interface IAuthorizationFilter : IFilterMetadata {

        void OnAuthorization(AuthorizationFilterContext context);
    }
}
```

可调用 OnAuthorization 方法来为过滤器提供验证授权请求。对于异步授权过滤器，以下是 IAsyncAuthorizationFilter 接口的定义：

```
using System.Threading.Tasks;

namespace Microsoft.AspNetCore.Mvc.Filters {

    public interface IAsyncAuthorizationFilter : IFilterMetadata {

        Task OnAuthorizationAsync(AuthorizationFilterContext context);
    }
}
```

调用 OnAuthorizationAsync 方法，以便过滤器可以授权请求。无论使用哪个接口，过滤器都会通过 AuthorizationFilterContext 对象（从 FilterContext 类派生）接收描述请求的上下文数据，并添加一个重要属性——Result，如表 19-5 所示。

表 19-5 AuthorizationFilterContext 类的属性

属性	描述
Result	当请求不符合应用程序的授权策略时，IActionResult 属性由授权过滤器设置，如果设置了这个属性，MVC 将会呈现 IActionResult 而不是调用操作方法

创建授权过滤器

为了演示授权过滤器的工作原理，在示例项目中创建 Infrastructure 文件夹，并添加一个名为 HttpsOnlyAttribute.cs 的类文件（见代码清单 19-10），用它定义过滤器。

代码清单 19-10 Infrastructure 文件夹下的 HttpsOnlyAttribute.cs 文件的内容

```
using System;
using Microsoft.AspNetCore.Http;
using Microsoft.AspNetCore.Mvc;
using Microsoft.AspNetCore.Mvc.Filters;
namespace Filters.Infrastructure {
    public class HttpsOnlyAttribute : Attribute, IAuthorizationFilter {

        public void OnAuthorization(AuthorizationFilterContext context) {
            if (!context.HttpContext.Request.IsHttps) {
                context.Result =
                    new StatusCodeResult(StatusCodes.Status403Forbidden);
            }
        }
    }
}
```

如果请求符合授权策略，则授权过滤器不执行任何操作，MVC 会进入下一个过滤器，最后执行操作方法。

注 意

在之前的 MVC 中，可以用于限制特定用户和用户组访问的 Authorize 属性是使用过滤器实现的，但 ASP.NET Core MVC 中不再是这样。Authorize 属性仍然使用，但工作方式不同。在幕后，全局过滤器（本章后面将介绍全局过滤器）用于检测 Authorize 属性并执行 ASP.NET Core Identity 系统定义的策略，但 Authorize 属性不再是过滤器，不能实现 IAuthorizationFilter 接口。第 29 章将介绍如何使用 ASP.NET Core Identity 系统和 Authorize 属性。

如果存在问题，则过滤器会设置传递给 OnAuthorization 方法的 AuthorizationFilterContext 对象的 Result 属性。这可以阻止事件的进一步执行，并为 MVC 提供返回到客户端的结果。在以上代码中，HttpsOnlyAttribute 类检查 HttpRequest Context 对象的 IsHttps 属性，并将 Result 属性设置为在没有 HTTPS 的情况下发出请求时中断执行。代码清单 19-11 展示了如何在 Home 控制器中使用新的过滤器。

代码清单 19-11 在 Controllers 文件夹下的 HomeController.cs 文件中使用自定义过滤器

```
using Microsoft.AspNetCore.Mvc;
using Filters.Infrastructure;

namespace Filters.Controllers {

    [HttpsOnly]
    public class HomeController : Controller {

        public ViewResult Index() => View("Message",
            "This is the Index action on the Home controller");

        public ViewResult SecondAction() => View("Message",
            "This is the SecondAction action on the Home controller");
    }
}
```

这个自定义过滤器重新实现了代码清单 19-7 中的操作方法的功能。在实际项目中，与执行重定向（如内置的 RequireHttps 过滤器）相比，这并没有提供更多的功能，因为用户不理解 403 状态码的含义，但这个示例对于演示授权过滤器如何工作十分有用。

过滤器的单元测试

过滤器的单元测试的大部分工作是设置传递给过滤器方法的 Context 对象。所需的模仿数据取决于过滤器使用的上下文信息。例如，下面是用于代码清单 19-10 中的 HttpsOnly 过滤器的单元测试。

```
using System.Linq;
using Filters.Infrastructure;
using Microsoft.AspNetCore.Http;
using Microsoft.AspNetCore.Mvc;
using Microsoft.AspNetCore.Mvc.Abstractions;
using Microsoft.AspNetCore.Mvc.Filters;
using Moq;
using Xunit;

namespace Tests {

    public class FilterTests {

        [Fact]
        public void TestHttpsFilter() {

            // Arrange
            var httpRequest = new Mock<HttpRequest>();
            httpRequest.SetupSequence(m => m.IsHttps).Returns(true)
                                                     .Returns(false);
            var httpContext = new Mock<HttpContext>();
            httpContext.SetupGet(m => m.Request).Returns(httpRequest.Object);

            var actionContext = new ActionContext(httpContext.Object,
                new Microsoft.AspNetCore.Routing.RouteData(),
                new ActionDescriptor());
            var authContext = new AuthorizationFilterContext(actionContext,
                Enumerable.Empty<IFilterMetadata>().ToList());

            HttpsOnlyAttribute filter = new HttpsOnlyAttribute();

            // Act and Assert
            filter.OnAuthorization(authContext);
            Assert.Null(authContext.Result);

            filter.OnAuthorization(authContext);
            Assert.IsType(typeof(StatusCodeResult), authContext.Result);
            Assert.Equal(StatusCodes.Status403Forbidden,
                (authContext.Result as StatusCodeResult).StatusCode);
        }
    }
}
```

首先模仿 HttpRequest 和 HttpContext Context 对象，这样就可以使用或不使用 HTTPS 呈现请求。如果想测试这两个条件，可以这样做：

```
...
httpRequest.SetupSequence(m => m.IsHttps).Returns(true).Returns(false);
...
```

以上语句设置了 HttpRequest.IsHttps 属性，使它返回一系列测试值：该属性在第一次读取时返回 true，并在第二次读取时返回 false。一旦有了一个 HttpContext 对象，就可以用它创建一个 ActionContext 对象，从而创建用于执行单元测试的 AuthorizationContext 对象。通过检查 AuthorizationFilterContext 对象的 Result 属性，可以测试过滤器如何响应非 HTTPS 请求，然后测试 HTTP 请求发生的情况。设置 AuthorizationFilterContext 对象时需要很多类型，并且它们依赖于许多 ASP.NET Core 和 MVC 命名空间，但是一旦有了 Context 对象，那么编写其余的测试就会比较简单。

19.5 使用操作过滤器

了解操作过滤器的最佳方式是查看它们的定义。以下是 IActionFilter 接口：

```
namespace Microsoft.AspNetCore.Mvc.Filters {

    public interface IActionFilter : IFilterMetadata {

        void OnActionExecuting(ActionExecutingContext context);

        void OnActionExecuted(ActionExecutedContext context);
    }
}
```

当操作过滤器已应用于操作方法时，可在调用操作方法之前调用 OnActionExecuting 方法，之后再调用 OnActionExecuted 方法。操作过滤器通过两个不同的上下文类提供了上下文数据，它们分别是 OnActionExecuting 方法的 ActionExecutingContext 类和 OnActionExecuted 方法的 ActionExecutedContext 类。这两个上下文类都扩展了 FilterContext 类，详见表 19-4。

ActionExecutingContext 类用于描述将要调用的操作，表 19-6 描述了该类的属性。

表 19-6 ActionExecutingContext 类的属性

属性	描述
Controller	返回要调用操作方法的控制器（操作方法的详细信息可通过从基类继承的 ActionDescriptor 属性获得）
ActionArguments	返回将传递给操作方法的参数字典（按名称索引），过滤器可以插入、移除或更改参数
Result	如果过滤器为 Result 属性分配 IActionResult，请求进程将被跳过，并且操作结果将用于在不调用操作方法的情况下生成返回给客户端的响应

ActionExecutedContext 类用于表示已执行的操作，其中定义了表 19-7 描述的属性。

表 19-7 ActionExecutedContext 类的属性

属性	描述
Controller	返回将调用操作方法的控制器
Canceled	如果另一个操作过滤器通过将操作结果分配给 ActionExecutingContext 对象的 Result 属性来将请求处理过程短路，就将这个属性设置为 true
Exception	包含由操作方法抛出的所有异常
ExceptionDispatchInfo	返回一个 ExceptionDispatchInfo 对象，其中包含由操作方法抛出的任何异常的堆栈跟踪详细信息
ExceptionHandled	若设置为 true，表示过滤器已处理异常，异常不会再进一步传播
Result	返回由操作方法返回的 IActionResult。如果需要，过滤器可以更改或替换操作结果

19.5.1 创建操作过滤器

操作过滤器是一种通用工具，可用于实现应用程序中的任何横切问题。操作过滤器可用于在调用操作之前中断请求进程，并可在执行操作后更改结果。创建操作过滤器的最简单方法是从 ActionFilterAttribute 类派生一个类，并让它实现 IActionFilter 接口。为了演示，在 Infrastructure 文件夹中添加一个名为 ProfileAttribute.cs 的类文件，内容如代码清单 19-12 所示。

代码清单 19-12 Infrastructure 文件夹下的 ProfileAttribute.cs 文件的内容

```
using System.Diagnostics;
using System.Text;
```

```
using Microsoft.AspNetCore.Mvc.Filters;
namespace Filters.Infrastructure {

    public class ProfileAttribute : ActionFilterAttribute {
        private Stopwatch timer;

        public override void OnActionExecuting(ActionExecutingContext context) {
            timer = Stopwatch.StartNew();
        }

        public override void OnActionExecuted(ActionExecutedContext context) {
            timer.Stop();
            string result = "<div>Elapsed time: "
                + $"{timer.Elapsed.TotalMilliseconds} ms</div>";
            byte[] bytes = Encoding.ASCII.GetBytes(result);
            context.HttpContext.Response.Body.Write(bytes, 0, bytes.Length);
        }
    }
}
```

以上代码使用 Stopwatch 对象来测量通过在 OnActionExecuting 方法中启动计时器来执行操作方法所需的毫秒数，并在 OnActionExecuted 方法中停止计数。为了观察结果，可使用 Context 对象获取 HttpResponse 对象，并在响应中包含简单的 HTML 片段。

代码清单 19-13 显示了如何在 Home 控制器上应用 Profile 特性（这里删除了以前的过滤器，以便可以接收 HTTP 请求）。

提 示

有一种很特别的情况，控制器也是操作过滤器。控制器基类实现了 IActionFilter 和 IAsyncActionFilter 接口，这意味着可以通过重写这些接口定义的方法来创建操作过滤器功能。对于 POCO 控制器，MVC 检查这些类并检查它们是否实现了其中一个操作过滤器接口，并自动将它们作为操作过滤器使用。

代码清单 19-13 在 Controllers 文件夹下的 HomeController.cs 文件中应用过滤器

```
using Microsoft.AspNetCore.Mvc;
using Filters.Infrastructure;

namespace Filters.Controllers {

    [Profile]
    public class HomeController : Controller {

        public ViewResult Index() => View("Message",
            "This is the Index action on the Home controller");

        public ViewResult SecondAction() => View("Message",
            "This is the SecondAction action on the Home controller");
    }
}
```

如果运行示例应用程序，你将看到图 19-4 所示的结果。你看到的毫秒数与开发时使用的计算机的性能有关。

图 19-4 运行结果

注 意

将 HTML 片段直接写入响应依赖于浏览器能否支持显示格式不正确的 HTML 文档。你在过滤器中生成的 div 元素出现在响应正文的开头，在 Razor 视图生成表示 HTML 文档开头的 DOCTYPE 和 html 元素之前。这种技术可以用于生成诊断信息，但是不应该在生产环境使用。

19.5.2 创建异步操作过滤器

IAsyncActionFilter 接口用于定义异步操作的操作过滤器，定义如下：

```
using System.Threading.Tasks;

namespace Microsoft.AspNetCore.Mvc.Filters {

    public interface IAsyncActionFilter : IFilterMetadata {

        Task OnActionExecutionAsync(ActionExecutingContext context,
            ActionExecutionDelegate next);
    }
}
```

在执行操作方法前后，存在依赖于任务延续的单一方法，该方法允许过滤器运行。代码清单 19-14 显示了如何在 Profile 过滤器中使用 OnActionExecutionAsync 方法。

代码清单 19-14 在 Infrastructure 文件夹下的 ProfileAttribute.cs 文件中创建异步操作过滤器

```
using System.Diagnostics;
using System.Text;
using System.Threading.Tasks;
using Microsoft.AspNetCore.Mvc.Filters;
namespace Filters.Infrastructure {

    public class ProfileAttribute : ActionFilterAttribute {

        public override async Task OnActionExecutionAsync(
                ActionExecutingContext context,
                ActionExecutionDelegate next) {

            Stopwatch timer = Stopwatch.StartNew();

            await next();

            timer.Stop();
            string result = "<div>Elapsed time: "
                + $"{timer.Elapsed.TotalMilliseconds} ms</div>";
            byte[] bytes = Encoding.ASCII.GetBytes(result);
            await context.HttpContext.Response.Body.WriteAsync(bytes,
                0, bytes.Length);
        }
    }
}
```

ActionExecutingContext 对象为过滤器提供了上下文数据，ActionExectionDelegate 对象表示要执行的操作方法（或下一个过滤器）。过滤器在调用委托之前做准备工作，然后在委托完成后完成这些工作。代理会返回一个 Task 对象，因而上面的代码中使用了 await 关键字。

19.6 使用结果过滤器

结果过滤器在 MVC 处理操作方法返回的操作结果的前后应用。结果过滤器能够更改或替换操作结果或完全取消请求（即使已调用操作方法）。下面是定义结果过滤器的 IResultFilter 接口：

```
namespace Microsoft.AspNetCore.Mvc.Filters {

    public interface IResultFilter : IFilterMetadata {

        void OnResultExecuting(ResultExecutingContext context);

        void OnResultExecuted(ResultExecutedContext context);
    }
}
```

结果过滤器遵循与操作过滤器相同的模式。在处理操作方法产生的操作结果之前调用 OnResultExecuting 方法，并通过 ResultExecutingContext 对象提供上下文信息。ResultExecutingContext 类是从 FilterContext 派生的，并定义了表 19-8 中的属性。

表 19-8 ResultExecutingContext 类定义的属性

属性	描述
Controller	返回执行操作方法的控制器
Cancel	将这个属性设置为 true，就会停止处理操作结果以生成响应
Result	返回由操作方法返回的 IActionResult 对象

调用 OnResultExecuted 方法后，MVC 已经处理了操作结果，并通过 ResultExecutedContext 类的实例提供了上下文数据，ResultExecutedContext 类继承自 FilterContext，并定义了表 19-9 中的属性。

表 19-9 ResultExecutedContext 类定义的属性

属性	描述
Controller	返回执行操作方法的控制器
Canceled	这个属性指示请求是否被取消
Exception	包含由操作方法抛出的所有异常
ExceptionDispatchInfo	返回一个 ExceptionDispatchInfo 对象，其中包含由操作方法抛出的任何异常的堆栈跟踪详细信息
ExceptionHandled	将这个属性设置为 true 表示过滤器已处理异常，异常不会再进一步传播
Result	返回用于生成对客户端响应的 IActionResult 对象

19.6.1 创建结果过滤器

ResultFilterAttribute 类实现了结果过滤器接口，并提供了创建结果过滤器的最简单方法。为了演示结果过滤器的工作原理，将一个名为 ViewResultDetailsAttribute.cs 的类文件添加到 Infrastructure 文件夹中，内容如代码清单 19-15 所示。

代码清单 19-15 Infrastructure 文件夹下的 ViewResultDetailsAttribute.cs 文件的内容

```
using System.Collections.Generic;
using Microsoft.AspNetCore.Mvc;
using Microsoft.AspNetCore.Mvc.Filters;
using Microsoft.AspNetCore.Mvc.ModelBinding;
using Microsoft.AspNetCore.Mvc.ViewFeatures;

namespace Filters.Infrastructure {

    public class ViewResultDetailsAttribute : ResultFilterAttribute {

        public override void OnResultExecuting(ResultExecutingContext context) {
            Dictionary<string, string> dict = new Dictionary<string, string> {
```

```
                ["Result Type"] = context.Result.GetType().Name,
            };

            ViewResult vr;
            if ((vr = context.Result as ViewResult) != null) {
                dict["View Name"] = vr.ViewName;
                dict["Model Type"] = vr.ViewData.Model.GetType().Name;
                dict["Model Data"] = vr.ViewData.Model.ToString();
            }

            context.Result = new ViewResult {
                ViewName = "Message",
                ViewData = new ViewDataDictionary(
                        new EmptyModelMetadataProvider(),
                        new ModelStateDictionary()) { Model = dict }
                };
        }
    }
}
```

这个类只重写了 OnResultExecuting 方法，并使用 Context 对象更改了用于发送响应到客户端的操作结果。过滤器创建了一个 ViewResult 对象，该对象使用包含简单诊断信息的字典作为视图模型传递给 Message 视图。

OnResultExecuting 方法在操作方法生成了操作结果之后，但在处理生成的结果之前调用，并且更改了 Context 对象的 Result 对象的值，从而允许为应用过滤器的操作方法提供不同类型的结果。代码清单 19-16 显示了应用于 Home 控制器的结果过滤器。

代码清单 19-16 在 Controllers 文件夹下的 HomeController.cs 文件中使用结果过滤器

```
using Microsoft.AspNetCore.Mvc;
using Filters.Infrastructure;

namespace Filters.Controllers {

    [ViewResultDetails]
    public class HomeController : Controller {

        public ViewResult Index() => View("Message",
            "This is the Index action on the Home controller");

        public ViewResult SecondAction() => View("Message",
            "This is the SecondAction action on the Home controller");
    }
}
```

如果运行示例应用程序，你将看到结果过滤器的效果，如图 19-5 所示。

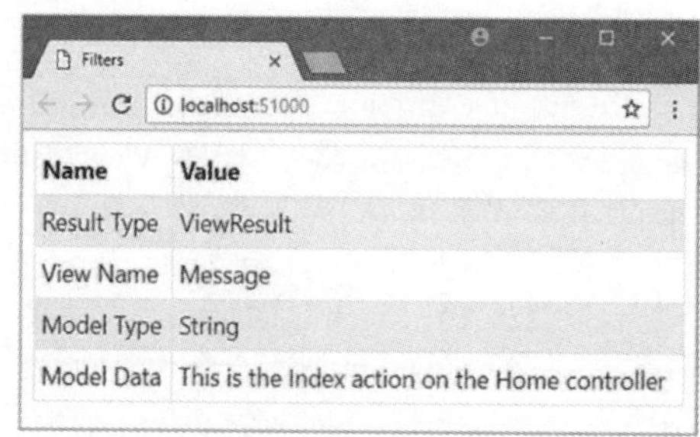

图 19-5 结果过滤器的效果

19.6.2 创建异步结果过滤器

IAsyncResultFilter 接口可用于创建异步结果过滤器，定义如下：

```
using System.Threading.Tasks;

namespace Microsoft.AspNetCore.Mvc.Filters {

    public interface IAsyncResultFilter : IFilterMetadata {

        Task OnResultExecutionAsync(ResultExecutingContext context,
            ResultExecutionDelegate next);
    }
}
```

这个接口类似于异步操作过滤器。代码清单 19-17 重写了 ViewResultDetailsAttribute 类以实现 IAsyncResultFilter 接口。

代码清单 19-17 在 Infrastructure 文件夹下的 ViewResultDetailsAttribute.cs 文件中创建异步结果过滤器

```
using System.Collections.Generic;
using System.Threading.Tasks;
using Microsoft.AspNetCore.Mvc;
using Microsoft.AspNetCore.Mvc.Filters;
using Microsoft.AspNetCore.Mvc.ModelBinding;
using Microsoft.AspNetCore.Mvc.ViewFeatures;

namespace Filters.Infrastructure {
    public class ViewResultDetailsAttribute : ResultFilterAttribute {

        public override async Task OnResultExecutionAsync(
                ResultExecutingContext context,
                ResultExecutionDelegate next) {

            Dictionary<string, string> dict = new Dictionary<string, string> {
                ["Result Type"] = context.Result.GetType().Name,
            };

            ViewResult vr;
            if ((vr = context.Result as ViewResult) != null) {
                dict["View Name"] = vr.ViewName;
                dict["Model Type"] = vr.ViewData.Model.GetType().Name;
                dict["Model Data"] = vr.ViewData.Model.ToString();
            }

            context.Result = new ViewResult {
                ViewName = "Message",
                ViewData = new ViewDataDictionary(
                        new EmptyModelMetadataProvider(),
                        new ModelStateDictionary()) {
                            Model = dict
                        }
                };

            await next();
        }
    }
}
```

请注意，需要负责调用作为 OnResultExecutionAsync 方法的参数接收的委托。如果不调用委托，请求处理管道将不会完成，并且不会呈现操作结果。

19.6.3 创建混合操作/结果过滤器

区分请求处理的操作阶段和结果阶段并不总是有帮助。这可能是因为你希望将两个阶段视为一个步骤，或是因为你希望过滤器影响响应执行操作的方式，而不是通过干预结果来实现。因此，能够创建既可以是操作过滤器又可以是结果过滤器，并且能够在每个阶段执行工作的过滤器是非常有用的。

以下要求很常见：要求 ActionFilterAttribute 类实现两种类型的过滤器接口，这意味着可以在单个属性中混合和匹配过滤器类型。为了演示这是如何工作的，修改代码清单 19-18 中的 ProfileAttribute 类的代码，以便将操作过滤器与结果过滤器相结合。

代码清单 19-18 在 Infrastructure 文件夹下的 ProfileAttribute.cs 文件中创建混合过滤器

```
using System.Diagnostics;
using System.Text;
```

```
using System.Threading.Tasks;
using Microsoft.AspNetCore.Mvc.Filters;
namespace Filters.Infrastructure {

    public class ProfileAttribute : ActionFilterAttribute {
        private Stopwatch timer;
        private double actionTime;

        public override async Task OnActionExecutionAsync(
                ActionExecutingContext context,
                ActionExecutionDelegate next) {

            timer = Stopwatch.StartNew();

            await next();

            actionTime = timer.Elapsed.TotalMilliseconds;
        }

        public override async Task OnResultExecutionAsync(
                ResultExecutingContext context,
                ResultExecutionDelegate next) {

            await next();

            timer.Stop();
            string result = "<div>Action time: "
                + $"{actionTime} ms</div><div>Total time: "
                + $"{timer.Elapsed.TotalMilliseconds} ms</div>";
            byte[] bytes = Encoding.ASCII.GetBytes(result);

            await context.HttpContext.Response.Body.WriteAsync(bytes,
                0, bytes.Length);
        }
    }
}
```

这里对两种类型的过滤器使用了异步方法，但也可以混合使用它们，从而获取所需的功能，因为这些方法的默认实现会调用它们的同步对应版本。在过滤器中，可使用 Stopwatch 来测量要处理的操作所需的时间，以及总的经历时间，并将结果写入响应。代码清单 19-19 已将混合过滤器应用于 Home 控制器。

代码清单 19-19 在 Controllers 文件夹下的 HomeController.cs 文件中使用混合过滤器

```
using Microsoft.AspNetCore.Mvc;
using Filters.Infrastructure;

namespace Filters.Controllers {

    [Profile]
    [ViewResultDetails]
    public class HomeController : Controller {
        public ViewResult Index() => View("Message",
            "This is the Index action on the Home controller");

        public ViewResult SecondAction() => View("Message",
            "This is the SecondAction action on the Home controller");
    }
}
```

如果运行示例应用程序，你将看到类似于图 19-6 的输出。输出显示在 ViewResultDetails 提供的内容之后，因为是在结果过滤器的最后处理阶段写入的，而不是来自先前版本中使用的操作过滤器。

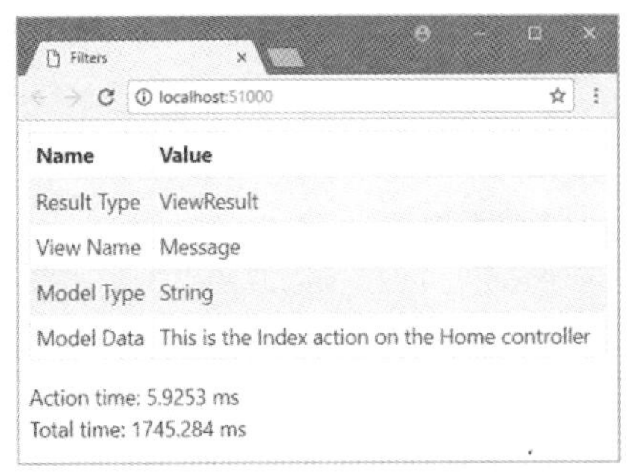

图 19-6 混合操作/结果过滤器的输出

19.7 使用异常过滤器

异常过滤器允许响应异常，而无须在每个操作方法中写入 try … catch 代码块。异常过滤器可以应用于控制器类或操作方法。当操作方法或已应用于操作方法的操作过滤器或结果过滤器未处理异常时，将会调用它们（操作过滤器和结果过滤器可以通过将 Context 对象的 ExceptionHandled 属性设置为 true 来处理未处理的异常）。异常过滤器实现了 IExceptionFilter 接口，定义如下：

```
namespace Microsoft.AspNetCore.Mvc.Filters {

    public interface IExceptionFilter : IFilterMetadata {

        void OnException(ExceptionContext context);
    }
}
```

如果遇到未处理的异常，则调用 OnException 方法。IAsyncExceptionFilter 接口可用于创建异步异常过滤器，如果需要使用异步 API 响应异常，这将非常有用。以下是异步接口的定义：

```
using System.Threading.Tasks;

namespace Microsoft.AspNetCore.Mvc.Filters {

    public interface IAsyncExceptionFilter : IFilterMetadata {

        Task OnExceptionAsync(ExceptionContext context);
    }
}
```

OnExceptionAsync 方法是来自 IExceptionFilter 接口的 OnException 方法的异步对象，当存在未处理的异常时调用。

IAsyncExceptionFilter 接口和 IExceptionFilter 接口都通过 ExceptionContext 类提供上下文数据，该类派生自 FilterContext，并定义了表 19-10 中描述的属性。

表 19-10 ExceptionContext 类定义的属性

属性	描述
Exception	包含抛出的任意异常
ExceptionDispatchInfo	返回一个 ExceptionDispatchInfo 对象，该对象包含异常的栈跟踪详细信息
ExceptionHandled	这个属性用于指示异常是否已被处理
Result	设置将用于生成响应的 IActionResult

创建异常过滤器

ExceptionFilterAttribute 类实现了两个异常过滤器接口，也是创建过滤器的最简单方法，可以应用为属性。

异常过滤器的常见用途是为特定异常类型提供自定义错误页面，以便为用户提供相比标准错误处理功能所能提供的更有用的信息。例如，将一个名为 RangeExceptionAttribute.cs 的类文件添加到 Infrastructure 文件夹中，文件内容如代码清单 19-20 所示。

代码清单 19-20 Infrastructure 文件夹下的 RangeExceptionAttribute.cs 文件的内容

```
using System;
using Microsoft.AspNetCore.Mvc;
using Microsoft.AspNetCore.Mvc.Filters;
using Microsoft.AspNetCore.Mvc.ModelBinding;
using Microsoft.AspNetCore.Mvc.ViewFeatures;

namespace Filters.Infrastructure {
    public class RangeExceptionAttribute : ExceptionFilterAttribute {

        public override void OnException(ExceptionContext context) {
            if (context.Exception is ArgumentOutOfRangeException) {
                context.Result = new ViewResult() {
                    ViewName = "Message",
                    ViewData = new ViewDataDictionary(
                        new EmptyModelMetadataProvider(),
                        new ModelStateDictionary()) {
                            Model = @"The data received by the
                                application cannot be processed"
                    }
                };
            }
        }
    }
}
```

这个过滤器使用 ExceptionContext 对象获取未处理异常的类型，如果类型为 ArgumentOutOfRangeException，则会创建用于向用户显示消息的操作结果。代码清单 19-21 向 Home 控制器添加一个操作方法，并将异常过滤器应用到这个操作方法。

代码清单 19-21 在 Controllers 文件夹下的 HomeController.cs 文件中应用异常过滤器

```
using Filters.Infrastructure;
using Microsoft.AspNetCore.Mvc;
using System;

namespace Filters.Controllers {

    [Profile]
    [ViewResultDetails]
    [RangeException]
    public class HomeController : Controller {

        public ViewResult Index() => View("Message",
            "This is the Index action on the Home controller");

        public ViewResult SecondAction() => View("Message",
            "This is the SecondAction action on the Home controller");

        public ViewResult GenerateException(int? id) {
            if (id == null) {
                throw new ArgumentNullException(nameof(id));
            } else if (id > 10) {
                throw new ArgumentOutOfRangeException(nameof(id));
            } else {
                return View("Message", $"The value is {id}");
            }
        }
```

```
        }
    }
```

GenerateException 操作方法依赖默认路由模式从请求 URL 中接收可空的 int 值。如果值为 null，操作方法将抛出 ArgumentNullException 异常；如果值大于 50，则抛出 ArgumentOutOfRangeException 异常。如果有值且在指定范围内，那么操作方法返回一个 ViewResult 对象。

可以通过运行示例应用程序并请求 URL——/Home/GenerateException/100 来测试异常过滤器。最后一段已超出操作方法的预期范围，这将抛出由过滤器处理的异常类型，产生图 19-7 所示的结果。如果请求 URL——/Home/GenerateException，那么由操作方法抛出的异常将不会被过滤器处理，并且将使用默认错误处理。

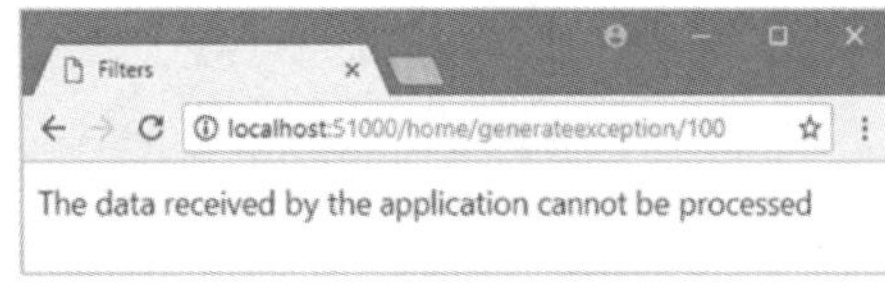

图 19-7 使用异常过滤器

19.8 为过滤器使用依赖注入

当从 convenience 属性类（如 ExceptionFilterAttribute）派生过滤器时，MVC 将创建过滤器类的新实例来处理每个请求。这是一种合理的方法，因为可以避免任何可能的重用或并发问题，并且满足了开发人员需要的大多数过滤器类的需求。

另一种方法是使用依赖注入系统为过滤器选择不同的生命周期。在过滤器中使用依赖注入有两种不同的方法，下面进行介绍。

19.8.1 解决过滤器依赖项

第一种方法是使用依赖注入来管理过滤器的上下文数据，以允许不同类型的过滤器共享数据，或者为单个过滤器共享数据，而实例本身用于处理其他请求。为了演示这是如何工作的，在 Infrastructure 文件夹中添加一个名为 FilterDiagnostics.cs 的类文件，内容如代码清单 19-22 所示。

代码清单 19-22 Infrastructure 文件夹下的 FilterDiagnostics.cs 文件的内容

```
using System.Collections.Generic;

namespace Filters.Infrastructure {

    public interface IFilterDiagnostics {
        IEnumerable<string> Messages { get; }
        void AddMessage(string message);
    }
    public class DefaultFilterDiagnostics : IFilterDiagnostics {
        private List<string> messages = new List<string>();

        public IEnumerable<string> Messages => messages;

        public void AddMessage(string message) =>
            messages.Add(message);
    }
}
```

IFilterDiagnostics 接口定义了一个简单的模型，用于在筛选过程中收集诊断消息。这里将使用 DefaultFilterDiagnostics 类来实现。代码清单 19-23 更新了 Startup 类，以使用新的接口及其实现来配置服务提供者。

代码清单 19-23 在 Filters 文件夹下的 Startup.cs 文件中配置服务提供者

```
using System;
using System.Collections.Generic;
using System.Linq;
```

```
using System.Threading.Tasks;
using Microsoft.AspNetCore.Builder;
using Microsoft.AspNetCore.Hosting;
using Microsoft.AspNetCore.Http;
using Microsoft.Extensions.DependencyInjection;
using Filters.Infrastructure;

namespace Filters {
    public class Startup {
        public void ConfigureServices(IServiceCollection services) {
            services.AddScoped<IFilterDiagnostics, DefaultFilterDiagnostics>();
            services.AddMvc();
        }

        public void Configure(IApplicationBuilder app, IHostingEnvironment env) {
            app.UseStatusCodePages();
            app.UseDeveloperExceptionPage();
            app.UseStaticFiles();
            app.UseMvcWithDefaultRoute();
        }
    }
}
```

以上代码使用 AddScoped 扩展方法来配置服务提供者，这意味着所有实例化的处理单个请求的过滤器都将收到相同的 DefaultFilterDiagnostics 对象。这是在过滤器之间共享自定义上下文数据的基础。

1. 创建具有依赖关系的过滤器

下一步是创建在 IFilterDiagnostics 接口上声明依赖关系的过滤器。在 Infrastructure 文件夹中创建一个名为 TimeFilter.cs 的类文件，内容如代码清单 19-24 所示。

代码清单 19-24 Infrastructure 文件夹下的 TimeFilter.cs 文件的内容

```
using System.Diagnostics;
using System.Threading.Tasks;
using Microsoft.AspNetCore.Mvc.Filters;

namespace Filters.Infrastructure {

    public class TimeFilter : IAsyncActionFilter, IAsyncResultFilter {
        private Stopwatch timer;
        private IFilterDiagnostics diagnostics;

        public TimeFilter(IFilterDiagnostics diags) {
            diagnostics = diags;
        }

        public async Task OnActionExecutionAsync(
                ActionExecutingContext context,
                ActionExecutionDelegate next) {

            timer = Stopwatch.StartNew();
            await next();
            diagnostics.AddMessage($@"Action time:
                {timer.Elapsed.TotalMilliseconds}");
        }

        public async Task OnResultExecutionAsync(
                ResultExecutingContext context,
                ResultExecutionDelegate next) {

            await next();
```

```
            timer.Stop();
            diagnostics.AddMessage($@"Result time:
                {timer.Elapsed.TotalMilliseconds}");
        }
    }
}
```

TimeFilter 类是混合操作/结果过滤器，可从先前的示例重新创建计时器功能，但使用 IFilterDiagnostics 接口存储定时信息，该接口被声明为构造函数参数，并在创建过滤器时由依赖注入系统提供。

请注意，TimeFilter 类直接实现了过滤器接口，而不是从 convenience 属性类派生。正如你将看到的，依赖于依赖注入的过滤器是通过不同的属性应用的，并且不用于直接修饰控制器或 action。

为了演示过滤器如何使用依赖项注入来共享上下文数据，在 Infrastructure 文件夹中添加一个名为 DiagnosticsFilter.cs 的类文件，内容如代码清单 19-25 所示。

代码清单 19-25　Infrastructure 文件夹下的 DiagnosticsFilter.cs 文件的内容

```
using System.Text;
using System.Threading.Tasks;
using Microsoft.AspNetCore.Mvc.Filters;

namespace Filters.Infrastructure {

    public class DiagnosticsFilter : IAsyncResultFilter {
        private IFilterDiagnostics diagnostics;

        public DiagnosticsFilter(IFilterDiagnostics diags) {
            diagnostics = diags;
        }

        public async Task OnResultExecutionAsync(
                ResultExecutingContext context,
                ResultExecutionDelegate next) {

            await next();

            foreach (string message in diagnostics?.Messages) {
                byte[] bytes = Encoding.ASCII
                    .GetBytes($"<div>{message}</div>");
                await context.HttpContext.Response.Body
                    .WriteAsync(bytes, 0, bytes.Length);
            }
        }
    }
}
```

DiagnosticsFilter 类是结果过滤器，可将 IFilterDiagnostics 接口的实现作为构造函数参数接收，并将其中包含的消息写入响应。

2. 应用过滤器

最后一步是将过滤器应用于控制器类。标准的 C#属性不具有解析构造函数依赖项的整体能力，这就是前面的过滤器不是属性的原因。相反，而是应用 TypeFilter 特性，并使用所需的过滤器类型进行配置，如代码清单 19-26 所示。

代码清单 19-26　在 Controllers 文件夹下的 HomeController.cs 文件中应用使用依赖注入的过滤器

```
using Microsoft.AspNetCore.Mvc;
using Filters.Infrastructure;
using System;

namespace Filters.Controllers {
```

```
    [TypeFilter(typeof(DiagnosticsFilter))]
    [TypeFilter(typeof(TimeFilter))]
    public class HomeController : Controller {

        public ViewResult Index() => View("Message",
            "This is the Index action on the Home controller");

        public ViewResult SecondAction() => View("Message",
            "This is the SecondAction action on the Home controller");

        public ViewResult GenerateException(int? id) {
            if (id == null) {
                throw new ArgumentNullException(nameof(id));
            } else if (id > 10) {
                throw new ArgumentOutOfRangeException(nameof(id));
            } else {
                return View("Message", $"The value is {id}");
            }
        }
    }
}
```

提 示

代码清单 19-26 中过滤器的应用顺序很重要，正如 19.10 节中解释的那样。

TypeFilter 特性为每个请求创建过滤器类的新实例，但使用依赖注入功能，这样可以创建松散耦合的组件，并将用于解决依赖关系的对象置于生命周期管理之下。

在该例中，这意味着代码清单 19-26 中应用的两个过滤器都将接收到相同的 IFilterDiagnostics 实现对象，因此 TimeFilter 类写入的消息将被 DiagnosticsFilter 类写入响应。图 19-8 显示了启动应用程序并请求应用程序的默认 URL 后可以看到的效果。

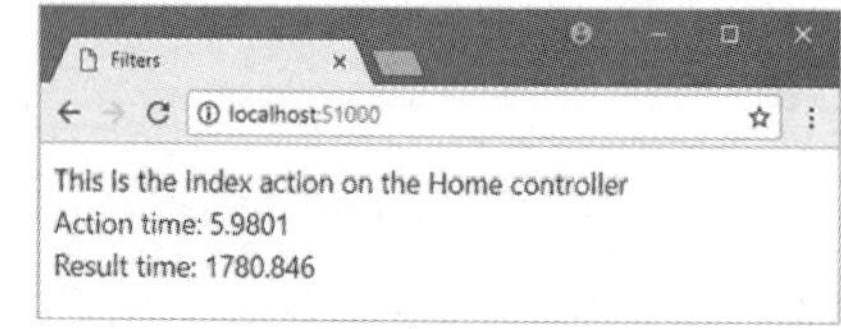

图 19-8 启动应用程序并请求应用程序的默认 URL 的效果

19.8.2 管理过滤器的生命周期

当使用 TypeFilter 特性时，会为每个请求创建过滤器类的新实例。这与将过滤器作为属性直接应用的行为相同，只是 TypeFilter 特性允许过滤器类声明通过服务提供者解析的依赖项。

相比使用 ServiceFilter 特性更进一步，可使用服务提供者创建过滤器对象。这使得过滤器对象也可以置于生命周期管理之下。作为演示，代码清单 19-27 修改了 TimeFilter 类，以便保持简单的平均时间值。

代码清单 19-27 在 Infrastructure 文件夹下 TimeFilter.cs 文件中保持简单的平均时间值

```
using System.Collections.Concurrent;
using System.Diagnostics;
using System.Linq;
using System.Threading.Tasks;
using Microsoft.AspNetCore.Mvc.Filters;

namespace Filters.Infrastructure {

    public class TimeFilter : IAsyncActionFilter, IAsyncResultFilter {
        private ConcurrentQueue<double> actionTimes = new ConcurrentQueue<double>();
        private ConcurrentQueue<double> resultTimes = new ConcurrentQueue<double>();
        private IFilterDiagnostics diagnostics;

        public TimeFilter(IFilterDiagnostics diags) {
            diagnostics = diags;
```

```
        }

        public async Task OnActionExecutionAsync(
                ActionExecutingContext context, ActionExecutionDelegate next) {

            Stopwatch timer = Stopwatch.StartNew();
            await next();
            timer.Stop();
            actionTimes.Enqueue(timer.Elapsed.TotalMilliseconds);
            diagnostics.AddMessage($@"Action time:
                {timer.Elapsed.TotalMilliseconds}
                Average: {actionTimes.Average():F2}");
        }

        public async Task OnResultExecutionAsync(
                ResultExecutingContext context, ResultExecutionDelegate next) {

            Stopwatch timer = Stopwatch.StartNew();
            await next();
            timer.Stop();
            resultTimes.Enqueue(timer.Elapsed.TotalMilliseconds);
            diagnostics.AddMessage($@"Result time:
                {timer.Elapsed.TotalMilliseconds}
                Average: {resultTimes.Average():F2}");
        }
    }
}
```

过滤器现在使用线程安全的集合来存储为请求处理的操作阶段和结果阶段记录的时间，并在每次处理请求时使用单独的 Stopwatch。代码清单 19-28 已经在 Startup 类中将 TimeFilter 类注册为服务提供者的单例。

代码清单 19-28 在 Filters 文件夹下的 Startup.cs 文件中配置服务提供者

```
using System;
using System.Collections.Generic;
using System.Linq;
using System.Threading.Tasks;
using Microsoft.AspNetCore.Builder;
using Microsoft.AspNetCore.Hosting;
using Microsoft.AspNetCore.Http;
using Microsoft.Extensions.DependencyInjection;
using Filters.Infrastructure;

namespace Filters {
    public class Startup {
        public void ConfigureServices(IServiceCollection services) {
            services.AddSingleton<IFilterDiagnostics, DefaultFilterDiagnostics>();
            services.AddSingleton<TimeFilter>();
            services.AddMvc();
        }

        public void Configure(IApplicationBuilder app, IHostingEnvironment env) {
            app.UseStatusCodePages();
            app.UseDeveloperExceptionPage();
            app.UseStaticFiles();
            app.UseMvcWithDefaultRoute();
        }
    }
}
```

请注意，这里还改变了 IFilterDiagnostics 的生命周期，它是一个单例。如果继续为每个请求创建一个新的实例，那么单例 TimeFilter 将从 DiagnosticsFilter 接收一个不同的 IFilterDiagnostics 对象，该对象继续通过 TypeFilter 特性进行实例化，并为每个请求创建一个。

应用过滤器

使用 ServiceType 特性将过滤器应用于控制器，如代码清单 19-29 所示。

代码清单 19-29 在 Controllers 文件夹下的 HomeController.cs 文件中使用过滤器

```
using Microsoft.AspNetCore.Mvc;
using Filters.Infrastructure;
using System;

namespace Filters.Controllers {
    [TypeFilter(typeof(DiagnosticsFilter))]
    [ServiceFilter(typeof(TimeFilter))]
    public class HomeController : Controller {

        public ViewResult Index() => View("Message",
            "This is the Index action on the Home controller");

        public ViewResult SecondAction() => View("Message",
            "This is the SecondAction action on the Home controller");

        public ViewResult GenerateException(int? id) {
            if (id == null) {
                throw new ArgumentNullException(nameof(id));
            } else if (id > 10) {
                throw new ArgumentOutOfRangeException(nameof(id));
            } else {
                return View("Message", $"The value is {id}");
            }
        }
    }
}
```

可以通过运行示例应用程序并请求默认 URL 来查看效果。因为IFilterDiagnostics接口的单个实现对象用于解析所有依赖关系，所以显示的消息集合会随着每个请求而建立，如图 19-9 所示。

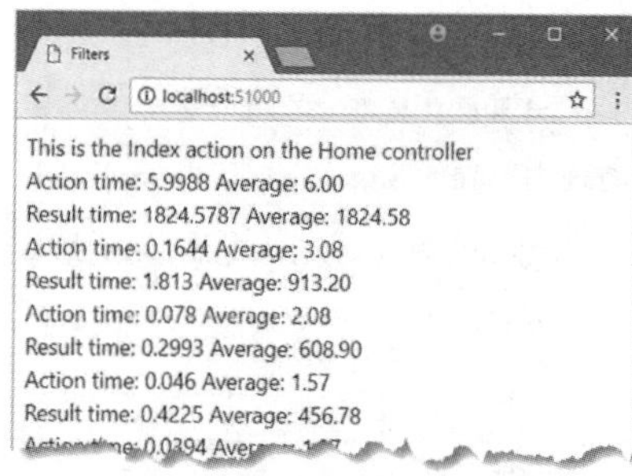

图 19-9 使用 service provider 管理过滤器的生命周期

19.9 创建全局过滤器

你可以将过滤器应用于控制器类，而不必将它们应用于各个操作方法。全局过滤器更进一步，只在 Startup 类中应用一次，就可自动应用于应用程序的每个控制器中的每个操作方法。任何过滤器都可以用作全局过滤器，为了演示，创建一个名为 ViewResultDiagnostics.cs 的类文件并保存到 Infrastructure 文件夹中，文件的内容如代码清单 19-30 所示。

代码清单 19-30 Infrastructure 文件夹下的 ViewResultDiagnostics.cs 文件的内容

```
using Microsoft.AspNetCore.Mvc;
using Microsoft.AspNetCore.Mvc.Filters;

namespace Filters.Infrastructure {
    public class ViewResultDiagnostics : IActionFilter {
        private IFilterDiagnostics diagnostics;

        public ViewResultDiagnostics(IFilterDiagnostics diags) {
            diagnostics = diags;
        }

        public void OnActionExecuting(ActionExecutingContext context) {
            // do nothing - not used in this filter
        }
```

```
        public void OnActionExecuted(ActionExecutedContext context) {
            ViewResult vr;
            if ((vr = context.Result as ViewResult) != null) {
                diagnostics.AddMessage($"View name: {vr.ViewName}");
                diagnostics.AddMessage($@"Model type:
                    {vr.ViewData.Model.GetType().Name}");
            }
        }
    }
}
```

这个过滤器使用 IFilterDiagnostics 对象来存储有关 ViewResult action 结果的视图名称和模型类型方面的消息。代码清单 19-31 将这个过滤器与 DiagnosticsFilter 类一起应用于全局，过滤器依赖于 DiagnosticsFilter 类写入诊断消息。

代码清单 19-31 在 Filters 文件夹下的 Startup.cs 文件中注册全局过滤器

```
using System;
using System.Collections.Generic;
using System.Linq;
using System.Threading.Tasks;
using Microsoft.AspNetCore.Builder;
using Microsoft.AspNetCore.Hosting;
using Microsoft.AspNetCore.Http;
using Microsoft.Extensions.DependencyInjection;
using Filters.Infrastructure;
namespace Filters {
    public class Startup {
        public void ConfigureServices(IServiceCollection services) {
            services.AddScoped<IFilterDiagnostics, DefaultFilterDiagnostics>();
            services.AddScoped<TimeFilter>();
            services.AddScoped<ViewResultDiagnostics>();
            services.AddScoped<DiagnosticsFilter>();
            services.AddMvc().AddMvcOptions(options => {
                options.Filters.AddService(typeof(ViewResultDiagnostics));
                options.Filters.AddService(typeof(DiagnosticsFilter));
            });
        }

        public void Configure(IApplicationBuilder app, IHostingEnvironment env) {
            app.UseStatusCodePages();
            app.UseDeveloperExceptionPage();
            app.UseStaticFiles();
            app.UseMvcWithDefaultRoute();
        }
    }
}
```

全局过滤器是通过配置 MVC 服务包来设置的。本例使用了全局过滤器的 MvcOptions.Filters.AddService 方法。AddService 方法接收将使用 ConfigureServices 方法中其他位置指定的生命周期规则实例化的.NET 类型。可将其他过滤器类型的生命周期更改为范围受限，以便为每个请求创建新的实例。因此，将会为每个控制器的每个请求创建并应用 ViewResultDiagnostics 和 DiagnosticsFilter 过滤器的新实例。

提 示

还可以使用 Add 方法而不是 AddService 方法添加全局过滤器，这样就可以将过滤器对象注册为全局过滤器，而不依赖于依赖注入和服务提供者。

将一个名为 GlobalController.cs 的类文件添加到 Controllers 文件夹中，内容如代码清单 19-32 所示。

代码清单 19-32 Controllers 文件夹下的 GlobalController.cs 文件的内容

```
using Microsoft.AspNetCore.Mvc;

namespace Filters.Controllers {

    public class GlobalController : Controller {

        public ViewResult Index() => View("Message",
            "This is the global controller");
    }
}
```

没有将任何过滤器应用于 Global 控制器，但如果启动应用程序并请求/global，你将看到两个全局过滤器的输出，如图 19-10 所示。

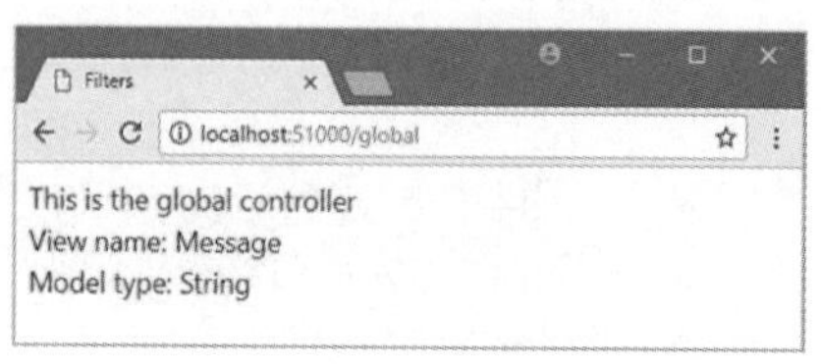

图 19-10 两个全局过滤器的输出

19.10 理解和更改过滤器的执行顺序

过滤器以特定顺序运行，即授权过滤器→操作过滤器→结果过滤器。但是，如果有多个给定类型的过滤器，则它们的应用顺序由已应用过滤器的范围驱动。为了演示这如何工作，将一个名为 MessageAttribute.cs 的类文件添加到 Infrastructure 文件夹中，文件内容如代码清单 19-33 所示。

代码清单 19-33 Infrastructure 文件夹下的 MessageAttribute.cs 文件的内容

```
using System.Text;
using Microsoft.AspNetCore.Mvc.Filters;

namespace Filters.Infrastructure {

    public class MessageAttribute : ResultFilterAttribute {
        private string message;

        public MessageAttribute(string msg) {
            message = msg;
        }

        public override void OnResultExecuting(ResultExecutingContext context) {
            WriteMessage(context, $"<div>Before Result:{message}</div>");
        }

        public override void OnResultExecuted(ResultExecutedContext context) {
            WriteMessage(context, $"<div>After Result:{message}</div>");
        }

        private void WriteMessage(FilterContext context, string msg) {
            byte[] bytes = Encoding.ASCII
                .GetBytes($"<div>{msg}</div>");
            context.HttpContext.Response
                .Body.Write(bytes, 0, bytes.Length);
        }
    }
}
```

这是一个结果过滤器，但并非在处理操作结果前后将 HTML 片段写入响应。过滤器写入的消息是通过构造函数参数来配置的，该参数在应用为属性时可以使用。代码清单 19-34 简化了 Home 控制器，并将以前示例中的过滤器替换为 Message 过滤器的多个实例。

代码清单 19-34 在 Controllers 文件夹下的 HomeController.cs 文件中使用过滤器

```
using Microsoft.AspNetCore.Mvc;
using Filters.Infrastructure;

namespace Filters.Controllers {

    [Message("This is the Controller-Scoped Filter")]
    public class HomeController : Controller {

        [Message("This is the First Action-Scoped Filter")]
        [Message("This is the Second Action-Scoped Filter")]
        public ViewResult Index() => View("Message",
            "This is the Index action on the Home controller");
    }
}
```

上面已经修改了一组全局过滤器，以便在这里使用 Message 过滤器，如代码清单 19-35 所示。

代码清单 19-35 在 Filters 文件夹下的 Startup.cs 文件中创建全局过滤器

```
using System;
using System.Collections.Generic;
using System.Linq;
using System.Threading.Tasks;
using Microsoft.AspNetCore.Builder;
using Microsoft.AspNetCore.Hosting;
using Microsoft.AspNetCore.Http;
using Microsoft.Extensions.DependencyInjection;
using Filters.Infrastructure;

namespace Filters {
    public class Startup {
        public void ConfigureServices(IServiceCollection services) {
            services.AddScoped<IFilterDiagnostics, DefaultFilterDiagnostics>();
            services.AddScoped<TimeFilter>();
            services.AddScoped<ViewResultDiagnostics>();
            services.AddScoped<DiagnosticsFilter>();
            services.AddMvc().AddMvcOptions(options => {
                options.Filters.Add(new
                    MessageAttribute("This is the Globally-Scoped Filter"));
            });
        }

        public void Configure(IApplicationBuilder app, IHostingEnvironment env) {
            app.UseStatusCodePages();
            app.UseDeveloperExceptionPage();
            app.UseStaticFiles();
            app.UseMvcWithDefaultRoute();
        }
    }
}
```

当 Index 操作方法响应请求时，将使用 4 个过滤器实例。如果运行应用程序并请求默认 URL，你将在浏览器中看到以下输出：

```
Before Result:This is the Globally-Scoped Filter
Before Result:This is the Controller-Scoped Filter
Before Result:This is the First Action-Scoped Filter
Before Result:This is the Second Action-Scoped Filter
After Result:This is the Second Action-Scoped Filter
After Result:This is the First Action-Scoped Filter
After Result:This is the Controller-Scoped Filter
After Result:This is the Globally-Scoped Filter
```

默认情况下，MVC 运行全局过滤器，然后将过滤器应用于控制器，最后将过滤器应用于操作方法。

一旦调用操作方法或处理操作结果，过滤器的栈就被释放了，因而输出中的 After Result 消息以相反的顺序显示。

修改过滤器的执行顺序

可以通过实现 IOrderedFilter 接口来更改默认顺序，MVC 在指定如何按顺序堆叠过滤器的过程中会查找这个接口。这个接口的定义如下：

```
namespace Microsoft.AspNetCore.Mvc.Filters {

    public interface IOrderedFilter : IFilterMetadata {
        int Order { get; }
    }
}
```

Order 属性会返回一个 int 值，较低的值将通知 MVC 在执行具有较高顺序值的筛选之前应用过滤器。convenience 属性已经实现了 IOrder 值，代码清单 19-36 为应用于 Home 控制器的过滤器设置了 Order 属性。

提 示

TypeFilter 和 ServiceFilter 特性还实现了 IOrderedFilter 接口，这意味着可以在使用依赖注入时更改过滤器的执行顺序。

代码清单 19-36 在 Controllers 文件夹下的 HomeController.cs 文件中设置过滤器的执行顺序

```
using Filters.Infrastructure;
using Microsoft.AspNetCore.Mvc;

namespace Filters.Controllers {

    [Message("This is the Controller-Scoped Filter", Order = 10)]
    public class HomeController : Controller {

        [Message("This is the First Action-Scoped Filter", Order = 1)]
        [Message("This is the Second Action-Scoped Filter", Order = -1)]
        public ViewResult Index() => View("Message",
            "This is the Index action on the Home controller");
    }
}
```

过滤器的 Order 值也可以为负数，这是确保在具有默认执行顺序的任何全局过滤器之前应用过滤器的有用方法（尽管也可以在创建全局过滤器时设置执行顺序）。如果运行示例应用程序，你将看到消息的输出顺序已更改，以反映新设置的优先级。

```
Before Result:This is the Second Action-Scoped Filter
Before Result:This is the Globally-Scoped Filter
Before Result:This is the First Action-Scoped Filter
Before Result:This is the Controller-Scoped Filter
After Result:This is the Controller-Scoped Filter
After Result:This is the First Action-Scoped Filter
After Result:This is the Globally-Scoped Filter
After Result:This is the Second Action-Scoped Filter
```

19.11 小结

本章讲述了如何将交叉关注问题的逻辑封装为过滤器，展示了各种不同的过滤器以及它们的实现方法。此外，本章还讨论了如何将过滤器作为属性应用于控制器和操作方法，以及如何将它们作为全局过滤器来应用。下一章将展示如何使用控制器来创建 Web 服务。

第 20 章　API 控制器

并非所有控制器都用于向客户端发送 HTML 文档，还有 API 控制器用于提供对应用程序数据的访问。这是以前通过单独的 Web API 框架提供的功能，但现在已经集成到 ASP.NET Core MVC 中。本章将解释 API 控制器在 Web 应用程序中的作用，描述它们解决的问题，并演示如何创建、测试和使用它们。表 20-1 描述了 API 控制器的背景。

表 20-1　API 控制器的背景

问题	答案
什么是 API 控制器？	API 控制器与常规控制器相似，除了它们的操作方法产生的响应是发送到客户端的数据对象而不是 HTML 标记之外
API 控制器有什么作用？	API 控制器允许客户端访问应用程序中的数据，而不会收到将内容呈现给用户所需的 HTML 标记。并不是所有客户端都是浏览器，也并不是所有客户端都向用户显示数据。API 控制器使应用程序变成开放的，可以支持新类型的客户端或由第三方开发的客户端
如何使用 API 控制器？	API 控制器可像常规 HTML 控制器一样使用
使用 API 控制器时有哪些陷阱或局限性？	最常见的问题涉及数据对象被序列化以发送到客户端的方式。有关详细信息，请参阅 20.4 节
是否有其他的替代方案？	不必在项目中使用 API 控制器，但这样做可以增加平台对客户端的价值

表 20-2 列出了本章要介绍的操作。

表 20-2　本章要介绍的操作

操作	方法	代码清单
提供对应用程序数据的访问	创建 API 控制器	代码清单 20-10
从 API 控制器请求数据	使用 Ajax 查询，直接使用浏览器 API 或通过库，比如 jQuery	代码清单 20-11～代码清单 20-13
覆盖内容协商过程	使用 Produces 属性	代码清单 20-14～代码清单 20-16
允许客户端通过 URL 中指定的数据格式覆盖 Accept 标头	在 Startup 类中添加格式化映射，添加捕获数据格式的片段变量，并可选择应用 FormatFilter 特性	代码清单 20-17 和代码清单 20-18
为内容协商过程提供全面支持	启用 HttpNotAcceptableOutputFormatter 格式化程序并设置 RespectBrowserAcceptHeader 配置属性	代码清单 20-19 和代码清单 20-20
使用不同的操作方法接收不同格式的数据	使用 Consumes 属性	代码清单 20-21

20.1　准备示例项目

对于本章，使用 ASP.NET Core Web Application（.NET Core）模板创建一个名为 ApiControllers 的 Empty 项目。

20.1.1　创建模型和存储库

首先创建 Models 文件夹，添加一个名为 Reservation.cs 的类文件，并用它定义如代码清单 20-1 所示的模型类。

代码清单 20-1　Models 文件夹下的 Reservation.cs 文件的内容

```
namespace ApiControllers.Models {

    public class Reservation {
```

```
        public int ReservationId { get; set; }
        public string ClientName { get; set; }
        public string Location { get; set; }
    }
}
```

这里还将一个名为 IRepository.cs 的文件添加到 Models 文件夹中，并用它定义模型存储库的接口，如代码清单 20-2 所示。

代码清单 20-2 Models 文件夹下的 IRepository.cs 文件的内容

```
using System.Collections.Generic;

namespace ApiControllers.Models {

    public interface IRepository {

        IEnumerable<Reservation> Reservations { get; }
        Reservation this[int id] { get; }

        Reservation AddReservation(Reservation reservation);
        Reservation UpdateReservation(Reservation reservation);
        void DeleteReservation(int id);
    }
}
```

添加一个名为 MemoryRepository.cs 的类文件到 Models 文件夹中，并用它定义 IRepository 接口的非持久性实现，如代码清单 20-3 所示。

代码清单 20-3 Models 文件夹下的 MemoryRepository.cs 文件的内容

```
using System.Collections.Generic;

namespace ApiControllers.Models {

    public class MemoryRepository : IRepository {
        private Dictionary<int, Reservation> items;

        public MemoryRepository() {
            items = new Dictionary<int, Reservation>();
            new List<Reservation> {
                new Reservation { ClientName = "Alice", Location = "Board Room" },
                new Reservation { ClientName = "Bob", Location = "Lecture Hall" },
                new Reservation { ClientName = "Joe", Location = "Meeting Room 1" }
            }.ForEach(r => AddReservation(r));
        }

        public Reservation this[int id] => items.ContainsKey(id) ? items[id] : null;

        public IEnumerable<Reservation> Reservations => items.Values;

        public Reservation AddReservation(Reservation reservation) {
            if (reservation.ReservationId == 0) {
                int key = items.Count;
                while (items.ContainsKey(key)) { key++; };
                reservation.ReservationId = key;
            }
            items[reservation.ReservationId] = reservation;
            return reservation;
        }

        public void DeleteReservation(int id) => items.Remove(id);
```

```
        public Reservation UpdateReservation(Reservation reservation)
            => AddReservation(reservation);

    }
}
```

存储库在实例化时会创建一组简单的模型对象，并且由于没有持久存储，因此当应用程序停止或重新启动时，任何更改都将丢失。有关如何在 SportsStore 示例应用程序中创建持久存储库的示例，请参阅第 8 章。

20.1.2 创建控制器和视图

在本章的后面部分，将创建 REST 控制器，但在准备时，需要创建常规控制器，从而为以后的例子提供基础。创建 Controllers 文件夹，并添加一个名为 HomeController.cs 的文件，Home 控制器的定义如代码清单 20-4 所示。

代码清单 20-4 Controllers 文件夹下的 HomeController.cs 文件的内容

```
using Microsoft.AspNetCore.Mvc;
using ApiControllers.Models;

namespace ApiControllers.Controllers {

    public class HomeController : Controller {
        private IRepository repository { get; set; }

        public HomeController(IRepository repo) => repository = repo;

        public ViewResult Index() => View(repository.Reservations);

        [HttpPost]
        public IActionResult AddReservation(Reservation reservation) {
            repository.AddReservation(reservation);
            return RedirectToAction("Index");
        }
    }
}
```

Home 控制器定义了 Index 操作，这是应用程序的默认值，用于渲染数据模型。另外，还定义了 AddReservation 操作，但只能用于接收 HTTP POST 请求，并且将用于从用户接收表单数据。这些操作遵循第 17 章中描述的 Post/Redirect/Get 模式，这样重新加载网页时就不会创建重复提交的表单。

创建布局，以便可以将 HTML 内容与 HTML 文档标头分离出来，这将简化本章稍后部分的一些更改。创建 Views/Shared 文件夹，添加一个名为_Layout.cshtml 的布局文件，并添加代码清单 20-5 所示的标记。

代码清单 20-5 Views/Shared 文件夹下的_Layout.cshtml 文件的内容

```
<!DOCTYPE html>
<html>
<head>
    <meta name="viewport" content="width=device-width" />
    <title>RESTful Controllers</title>
    <link asp-href-include="lib/bootstrap/dist/css/*.min.css" rel="stylesheet" />
</head>
<body class="m-1 p-1">
    @RenderBody()
</body>
</html>
```

接下来，创建 Views/Home 文件夹，添加一个名为 Index.cshtml 的视图文件，并添加代码清单 20-6 所示的内容。

代码清单 20-6 Views/Home 文件夹下的 Views.cshtml 文件的内容

```
@model IEnumerable<Reservation>
@{  Layout = "_Layout"; }

<form id="addform" asp-action="AddReservation" method="post">
    <div class="form-group">
        <label for="ClientName">Name:</label>
        <input class="form-control" name="ClientName" />
    </div>
    <div class="form-group">
        <label for="Location">Location:</label>
        <input class="form-control" name="Location" />
    </div>
    <div class="text-center panel-body">
        <button type="submit" class="btn btn-sm btn-primary">Add</button>
    </div>
</form>

<table class="table table-sm table-striped table-bordered m-2">
    <thead><tr><th>ID</th><th>Client</th><th>Location</th></tr></thead>
    <tbody>
        @foreach (var r in Model) {
            <tr>
                <td>@r.ReservationId</td>
                <td>@r.ClientName</td>
                <td>@r.Location</td>
            </tr>
        }
    </tbody>
</table>
```

这种强类型视图接收一系列 Reservation 对象作为模型，并使用 Razor foreach 循环来填充表格。另外，还有一个表单，已配置为将 POST 请求发送到 AddReservation 操作。

本章中的示例依赖于 Bootstrap CSS 包。要将 Bootstrap 包添加到项目中，可使用 Bower Configuration File 模板在项目的根目录中创建 bower.json 文件，并将 Bootstrap 包添加到 dependencies 部分，如代码清单 20-7 所示。

代码清单 20-7 在 bower.json 文件中添加 Bootstrap 包

```
{
  "name": "asp.net",
  "private": true,
  "dependencies": {
    "bootstrap": "4.0.0-alpha.6"
  }
}
```

接下来，在 Views 文件夹中创建_ViewImports.cshtml 文件，并在其中设置用于 Razor 视图的内置标签助手，然后导入模型命名空间，如代码清单 20-8 所示。

代码清单 20-8 Views 文件夹下的_ViewImports.cshtml 文件的内容

```
@using ApiControllers.Models
@addTagHelper *, Microsoft.AspNetCore.Mvc.TagHelpers
```

为了启用开发所需的 MVC 框架和中间件，对 Startup 类进行修改，如代码清单 20-9 所示。另外，还使用 AddSingleton 方法为模型存储库设置了服务映射。

代码清单 20-9 在 ApiControllers 文件夹下的 Startup.cs 文件中启用中间件

```
using System;
using System.Collections.Generic;
```

```
using System.Linq;
using System.Threading.Tasks;
using Microsoft.AspNetCore.Builder;
using Microsoft.AspNetCore.Hosting;
using Microsoft.AspNetCore.Http;
using Microsoft.Extensions.DependencyInjection;
using ApiControllers.Models;

namespace ApiControllers {
    public class Startup {

        public void ConfigureServices(IServiceCollection services) {
            services.AddSingleton<IRepository, MemoryRepository>();
            services.AddMvc();
        }

        public void Configure(IApplicationBuilder app, IHostingEnvironment env) {
            app.UseStatusCodePages();
            app.UseDeveloperExceptionPage();
            app.UseStaticFiles();
            app.UseMvcWithDefaultRoute();
        }
    }
}
```

设置 HTTP 端口

本章中的一些示例需要通过手动输入 URL 进行测试。为了描述更容易，下面设置用于接收 HTTP 请求的端口。从 Visual Studio 的 Project 菜单中选择 ApiControllers Properties，显示 Debug 选项卡，将 App URL 的值更改为 http://localhost:7000/，如图 20-1 所示。确保在设置端口号后保存更改。

启动应用程序，填写表单，然后单击 Add 按钮；应用程序将为模型添加新的数据，如图 20-2 所示。但你对存储库所做的更改不会持久保留，在应用程序停止或重新启动时将会丢失。

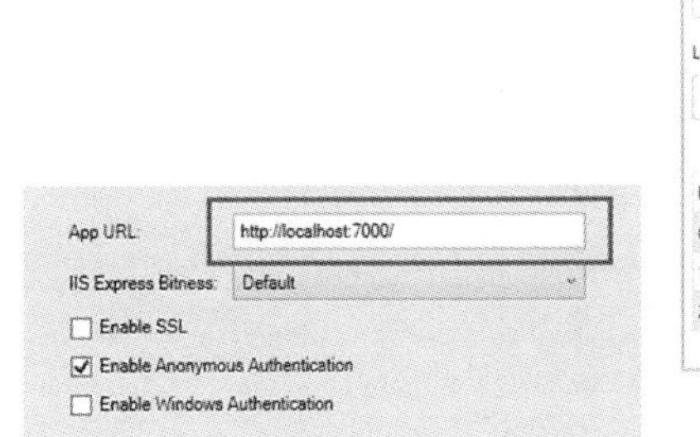

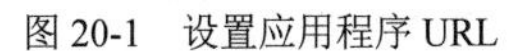

图 20-1　设置应用程序 URL

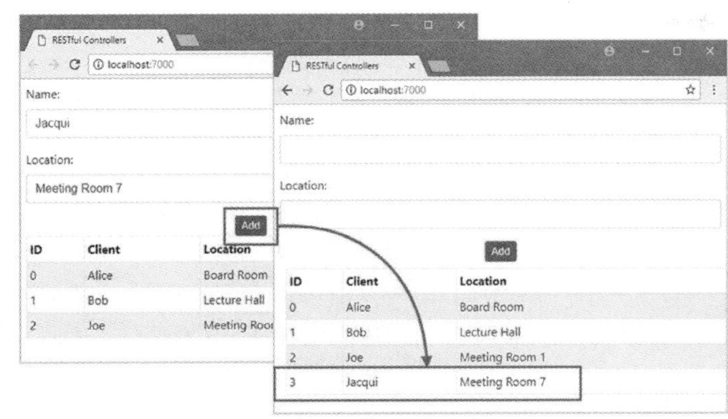

图 20-2　启动示例应用程序

20.2　REST 控制器的作用

这个示例应用程序是经典的 Web 应用程序。示例应用程序中的所有逻辑都存在于服务器上，包含在 C#类中，这使得它们易于管理、测试和维护。但是，以这种方式设计的应用程序在速度、效率和开放性方面可能会有严重的缺陷。

20.2.1　速度问题

此时，示例应用程序是同步的 Web 应用程序。当用户单击 Add 按钮时，浏览器将 POST 请求发送到服务器，等待响应，然后呈现接收到的HTML。在此期间，用户无法做任何事情，只能等待。当浏览器和服务器位于同一台计算机上时，等待时间在开发过程中是不可察觉的；然而，部署的应用程序受实际容量和延迟的限制，并且同步应用程序要求用户等待响应的时间可能很长。

同步应用程序不会总有速度问题。例如，如果你正在编写一个线路业务应用程序，用于单个位置，其中所有客户端都通过快速可靠的LAN连接，那么不会有需要解决的问题。如果你正在基础设施薄弱的地区为移动客户端编写应用程序，则速度问题可能很严重。

提 示

某些浏览器可让你模拟不同类型的网络，这可以作为有用的工具，用于查看用户是否可能接受将同步应用程序用于各种场景。例如，Google Chrome提供了名为网络限制的功能，可按F12键，在开发人员工具的Network部分找到。有一系列预定义的网络可用，也可以通过指定上传/下载速率和延迟来创建自己的网络。

20.2.2 效率问题

效率问题源于同步Web应用程序将浏览器视为仅用于显示服务器发送的HTML文档的HTML呈现引擎的方式。

当用户首先请求示例应用程序的默认URL时，发回的HTML文档将包含浏览器需要显示应用程序内容的所有信息，包括以下信息：

- 内容依赖于Bootstrap CSS文件，如果缓存副本不可用，就应该下载该文件。
- 内容包含配置为向AddReservation操作发送POST请求的表单。
- 内容包含一个表格，这个表格包含3行。

示例应用程序很简单，初始请求导致服务器向客户端发送约1.3KB数据。但是，当用户提交表单时，客户端将再次重定向到Index操作，这将导致另外1.3KB数据反映单个表行的添加。浏览器已经呈现了表单和表格，但是这些表单和表格都被丢弃，并被替换为在很大程度上内容相同的全新表示形式。

你可能认为1.3KB数据并不多，但是，如果考虑到有用内容与重复内容的比例，你将看到发送到浏览器的绝大多数数据被浪费了。示例应用程序是故意简化的；很少有应用程序需要这么少的HTML标记，并且随着应用程序复杂性的增加，重复内容的数量将显著增加。

20.2.3 开放性问题

传统Web应用程序提出的最终问题是设计是关闭的，这意味着模型中的数据只能通过Home控制器提供的操作进行访问。当需要在另一个应用程序中使用底层数据时，封闭的应用程序成为问题，特别是当应用程序由不同的团队甚至不同的组织开发时。开发人员经常认为，应用程序的价值在于为用户提供互动，主要原因在于这些是我们花时间思考和写作的部分。一旦建立了应用程序并具有活跃的用户群，应用程序包含的数据往往将变得重要。

20.3 REST和API控制器

API控制器是MVC控制器，负责提供对应用程序数据的访问，而不将它们封装在HTML中。这允许检索或修改模型中的数据，而不必使用常规控制器提供的操作，例如示例中的Home控制器。

从应用程序传递数据的常见方法是使用称为REST的表示状态传输模式。没有关于REST的详细规范，这导致许多不同的方法包含在REST风格之下。但是，这在客户端Web应用程序开发中是有用的。

Web服务的核心前提是采用HTTP的特征，以便请求方法（也称为谓词）指定服务器执行的操作，请求URL指定一个或多个数据对象，以应用操作。

作为示例，下面是一个可能在示例应用程序中引用的特别预留的 URL：

```
/api/reservations/1
```

URL 的第一部分——api 用于将应用程序的数据部分与生成 HTML 的标准控制器分离。下一部分——reservations 指示将要操作的对象集合。最后一部分——1 指定对象集合中的单个对象。在示例应用程序中，这就是唯一标识对象并将在 URL 中使用的 ReservationId 属性值。

请识别对象的 URL 与 HTTP 方法组合以指定操作。表 20-3 列出了常用的 HTTP 方法以及与示例 URL 结合使用的方法，还列出了每个方法和 URL 组合的请求及响应中包含哪些数据（有效负载）的详细信息。处理这些请求的 API 控制器使用响应状态代码来报告请求的结果。

表 20-3　常用的 HTTP 方法以及与示例 URL 结合使用的方法

请求方法	URL 地址	描述	载荷
GET	/api/reservations	检索所有对象	包含预留对象的完整集合
GET	/api/reservations/1	检索 ReservationId 为 1 的 Reservation 对象	包含指定的 Reservation 对象
POST	/api/reservation	创建新的 Reservation 对象	包含创建 Reservation 对象所需的其他属性的值。响应中包含存储的对象，确保客户端接收保存的数据
PUT	/api/reservation	更新现有的 Reservation 对象	包含更改指定 Reservation 对象的属性所需的值。响应中包含存储的对象，确保客户端接收保存的数据
PATCH	/api/reservation/1	修改现有的 Reservation 对象，该对象的 ReservationId 为 1	包含一组应该应用于指定的 Reservation 对象的修改，是对更改已应用的确认
DELETE	/api/reservation/1	删除 ReservationId 为 1 的 Reservation 对象	请求或响应中没有有效负载

遵循 REST 约定不是必需的，但这么做有助于使应用程序更容易使用，因为许多已建立的 Web 应用程序广泛采用了相同的方法。

20.3.1 创建 API 控制器

创建 API 控制器的过程基于创建标准控制器的方法，还有一些其他功能可以帮助指定提供给客户端的 API。为了演示，添加一个名为 ReservationController.cs 的类文件到 Controllers 文件夹中，并用它定义代码清单 20-10 所示的类。稍后将介绍这个控制器提供的功能。

代码清单 20-10　Controllers 文件夹下的 ReservationController.cs 文件的内容

```
using System.Collections.Generic;
using Microsoft.AspNetCore.Mvc;
using ApiControllers.Models;
using Microsoft.AspNetCore.JsonPatch;

namespace ApiControllers.Controllers {

    [Route("api/[controller]")]
    public class ReservationController : Controller {
        private IRepository repository;

        public ReservationController(IRepository repo) => repository = repo;

        [HttpGet]
        public IEnumerable<Reservation> Get() => repository.Reservations;

        [HttpGet("{id}")]
        public Reservation Get(int id) => repository[id];
```

```
        [HttpPost]
        public Reservation Post([FromBody] Reservation res) =>
            repository.AddReservation(new Reservation {
                ClientName = res.ClientName,
                Location = res.Location
            });

        [HttpPut]
        public Reservation Put([FromBody] Reservation res) =>
            repository.UpdateReservation(res);

        [HttpPatch("{id}")]
        public StatusCodeResult Patch(int id,
                [FromBody]JsonPatchDocument<Reservation> patch) {
            Reservation res = Get(id);
            if (res != null) {
                patch.ApplyTo(res);
                return Ok();
            }
            return NotFound();
        }

        [HttpDelete("{id}")]
        public void Delete(int id) => repository.DeleteReservation(id);
    }
}
```

提 示

请记住，控制器类可以在项目中的任何位置进行定义，而不仅仅是在 Controllers 文件夹中。对于大型和复杂项目，与常规 HTML 控制器分开，将定义的 API 控制器完全放置在子文件夹甚至单独的文件夹中是有帮助的。

API 控制器的工作方式与常规控制器相同，这意味着可以创建 POCO 控制器，或者从控制器基类派生类，这样可以更方便地访问请求上下文数据。

适应 REST 模式

REST 模式在如何将 Web 应用程序 API 呈现给客户端方面，在一定程度上鼓励教条主义。REST 不是标准的甚至不是定义明确的模式，但 REST 提供了一些有用的方法，可以让 ASP.NET Core MVC 应用程序更容易采用，但是有一种趋势会刺激那些对作为 REST 计数有固定观点的程序员。

在表 20-3 中，为 POST 和 PUT 操作列出的 URL 不唯一标识资源，有些人认为这是一个基本的 REST 特性。在 POST 操作的情况下，预留对象的唯一标识符由模型分配，这意味着客户端无法将其作为 URL 的一部分提供。在 PUT 操作的情况下，第 26 章描述的 MVC 模型绑定功能是应用代码清单 20-10 中的 FromBody 特性的原因，使得从请求体接收到要修改的 Reservation 对象的详细信息更加容易。所以，这就是 Reservation 控制器期望找到 ReservationId 值的地方，该值用于标识要修改的模型对象。

与所有模式一样，REST 模式是起点，这不是必须不惜一切代价遵守的严格标准，唯一重要的是编写可以被理解、测试和维护的代码。通过适应 MVC 应用程序的性质和存储库的设计可造就更简单的应用程序，同时仍然为客户提供有用的 API。建议将模式视为按自己需求所选的指导原则，这与 REST 模式一样，对于 MVC 本身也是如此。

1. 定义路由

到达 API 控制器的路由只能使用 Route 特性进行定义，无法在 Startup 类的应用程序配置中定义。API 控制器的惯例是使用前缀为 api 的路由，后跟控制器的名称，以便通过 URL——/api/reservation 访问如代码清单 20-10 所示的 ReservationController 控制器，如下所示：

```
...
[Route("api/[controller]")]
public class ReservationController : Controller {
...
```

2. 声明依赖关系

API 控制器以与常规控制器相同的方式实例化，这意味着它们可以声明将使用 service provider 解析的依赖关系。ReservationController 类在 IRepository 接口上声明了构造函数依赖关系，它将被解析为提供对模型中数据的访问：

```
...
public ReservationController(IRepository repo) => repository = repo;
...
```

3. 定义操作方法

每个操作方法都使用一个特性进行装饰，该特性指定了能够接受的 HTTP 方法，如下所示：

```
...
[HttpGet]
public IEnumerable<Reservation> Get() => repository.Reservations;
...
```

HttpGet 特性是用于将对操作方法的访问限制为具有特定 HTTP 方法或谓词的请求集合之一。HTTP 特性参见表 20-4。

表 20-4 HTTP 特性

HTTP 特性	描述
HttpGet	指定只能由使用 GET 方法的 HTTP 请求调用
HttpPost	指定只能由使用 POST 方法的 HTTP 请求调用
HttpDelete	指定只能由使用 DELETE 方法的 HTTP 请求调用
HttpPut	指定只能由使用 PUT 方法的 HTTP 请求调用
HttpPatch	指定只能由使用 PATCH 方法的 HTTP 请求调用
HttpHead	指定只能由使用 HEAD 方法的 HTTP 请求调用

通过将路由片段作为 HTTP 方法特性的参数包括在内，可以进一步细化路由，如下所示：

```
...
[HttpGet("{id}")]
public Reservation Get(int id) => repository[id];
...
```

路由片段{id}可与应用于控制器的 Route 特性定义的路由以及基于 HTTP 方法的约束相结合。在这种情况下，这意味着可以通过发送一个 URL 匹配/api/reservations/{id}路由模式的 GET 请求来执行此操作，然后使用 id 片段来标识应该检索的 Reservation 对象。

请注意，为 API 控制器生成的路由不包含{action}片段变量，这意味着操作方法的名称不是目标特定方法所需 URL 的一部分。API 控制器中的所有操作都通过相同的基本 URL（例如/api/reservation）达成，HTTP 方法和可选参数片段可用于区分它们。

4. 定义操作结果

API 控制器的操作方法不依赖于 ViewResult 对象来呈现结果，因此在传递数据时不需要视图。API 控制器的操作方法直接返回数据对象，如下所示：

```
...
[HttpGet]
```

```
public IEnumerable<Reservation> Get() => repository.Reservations;
...
```

以上操作返回一系列 Reservation 对象，并使 MVC 负责将其序列化为可由客户端处理的格式。20.4 节将更详细地解释这个过程。

自定义 API 结果

API 控制器最有吸引力的是可以从操作方法返回 C#对象，并让 MVC 弄清楚如何处理它们。MVC 很擅长处理，例如，如果从 API 控制器的操作方法返回 null，客户端将被发送 204-No Content 响应。

但 API 控制器也可以使用常规控制器的功能，这意味着可以通过从指定要发送何种结果的操作方法返回 IActionResult 来覆盖默认行为。作为示例，这里是来自示例控制器的操作方法的实现，示例控制器向与模型中的对象匹配失败的查询发送 404-Not Found 响应：

```
...
[HttpGet("{id}")]
public IActionResult Get(int id) {
    Reservation result = repository[id];
    if (result == null) {
        return NotFound();
    } else {
        return Ok(result);
    }
}
...
```

如果指定 ID 的存储库中没有对象，那么将调用 NotFound 方法，该方法将创建一个 NotFoundResult 对象，这反过来又导致发送到客户端的 404-Not Found 响应。如果存储库中有对象，就调用 Ok 方法来创建 ObjectResult 对象。Ok 方法允许在返回 IActionResult 的操作中向客户端发送对象，如第 17 章所述。通常不需要覆盖默认的 API 控制器响应，但是如果需要，则可以使用全部 action 结果。

20.3.2　测试 API 控制器

这里有很多可用于帮助测试 Web 应用程序 API 的工具。好的例子包括 Fiddler，它是一个独立的 HTTP 调试工具；还有 Swashbuckle，它是一个 NuGet 软件包，可将摘要页面添加到描述其 API 操作并允许对它们在应用程序中进行测试。

最简单的测试 API 控制器的方法是使用 PowerShell，这样可以轻松地从 Windows 命令行创建 HTTP 请求，并且可以让你专注于 API 操作的结果，而无须挖掘细节。PowerShell 起源于 Windows，但现在也可用于 Linux 系统和 macOS。

下面的内容将告诉你如何使用 PowerShell 来测试由 Reservation 控制器提供的每个操作。可以打开一个新的 PowerShell 窗口来运行测试命令或使用 PowerShell 的 Visual Studio 包管理器控制台窗口。

1. 测试 GET 操作

要测试 Reservation API 控制器提供的 GET 操作，请从 Visual Studio 的 Debug 菜单中选择 Start Without Debugging 以启动应用程序，然后等待，直到看到 Home 控制器提供的同步响应。应用程序运行后，打开 PowerShell 窗口并键入以下命令：

```
Invoke-RestMethod http://localhost:7000/api/reservation -Method GET
```

上述命令使用 Invoke-RestMethod PowerShell cmdlet 将 GET 请求发送到 URL——/api/reservation。结果被解析和格式化，以使数据容易阅读，如下所示：

```
reservationId clientName location
------------- ---------- --------
            0 Alice      Board Room
            1 Bob        Lecture Hall
            2 Joe        Meeting Room 1
```

服务器使用模型中包含的 Reservation 对象的 JSON 表示形式来响应 GET 请求，Invoke-RestMethod cmdlet 以表格形式显示。

理解 JSON

JSON（JavaScript 对象表示法）已成为 Web 应用程序的标准数据格式。JSON 之所以变得流行，是因为它简单且易于使用。由于 JSON 格式与 JavaScript 代码中的文字对象类似，因此 JavaScript 代码中的 JSON 数据处理尤其容易。现代浏览器提供生成和解析 JSON 数据的内置支持，流行的 JavaScript 库（如 jQuery）会自动与 JSON 互相转换。虽然 JSON 从 JavaScript 演变而来，但它的结构很容易让 C#开发人员阅读和理解。举例来说，以下是示例应用程序中 API 控制器的响应：

```
...
[{"reservationId":0,"clientName":"Alice","location":"Board Room"},
 {"reservationId":1,"clientName":"Bob","location":"Lecture Hall"},
 {"reservationId":2,"clientName":"Joe","location":"Meeting Room 1"}]
...
```

上述 JSON 字符串描述了一个对象数组。该数组由[和]字符表示，每个对象使用{和}字符表示。对象是键/值对的集合，用冒号分隔键和值，并且用逗号分隔键值对。这在广义上与 MemoryRepository 类中用于定义代码清单 20-3 所示数据的 C#文字语法类似。

```
...
new List<Reservation> {
    new Reservation { ClientName = "Alice", Location = "Board Room" },
    new Reservation { ClientName = "Bob", Location = "Lecture Hall" },
    new Reservation { ClientName = "Joe", Location = "Meeting Room 1" }
...
```

但请注意，MVC 会将 C#约定（ClientName，具有初始大写字母）的属性名中的初始大写字母更改为 JavaScript 约定（clientName，带有初始小写字母）。

即使这些格式不完全相同，也仍有很多相似之处，C#开发人员读取和理解 JSON 数据并不费力。你不需要了解大多数 Web 应用程序的 JSON 详细信息，因为 MVC 做了很大提升，可以从 www.json.org 了解更多有关 JSON 的信息。

Reservation 控制器提供两种 GET 操作。当把 GET 请求发送到/api/reservation 时，会返回包含所有对象的响应。为了检索单个对象，可将这个对象的 ReservationId 值指定为 URL 中的最后一个片段，如下所示：

```
Invoke-RestMethod http://localhost:7000/api/reservation/1 -Method GET
```

以上命令请求 ReservationId 值为 1 的 Reservation 对象，并产生以下结果：

```
reservationId clientName location
------------- ---------- --------
            1 Bob        Lecture Hall
```

2. 测试 POST 操作

API 控制器提供的所有操作都可以使用 PowerShell 进行测试，尽管命令的格式可能有点怪异。以下命令向 API 控制器发送 POST 请求，以在存储库中创建一个新的 Reservation 对象，并在响应中接收发回的数据：

```
Invoke-RestMethod http://localhost:7000/api/reservation -Method POST -Body
(@{clientName="Anne"; location="Meeting Room 4"} | ConvertTo-Json) -ContentType
"application/json"
```

以上命令使用-Body 参数来指定请求的主体，请求的主体被编码为 JSON。-ContentType 参数用于设置请求的 Content-Type 标头。以上命令将产生以下结果：

```
reservationId clientName location
------------- ---------- --------
            3 Anne       Meeting Room 4
```

POST 操作使用 clientName 和 location 的值创建了一个 Reservation 对象，并将这个新对象的 JSON 表示返回给客户端，其中包括已分配给新对象的 ReservationId 值。这可能看起来像客户端只是接收到请求中发送给服务器的数据值，但这种方法可以确保客户端正在使用与服务器正在使用的相同的数据，并且可以满足服务器对从客户端接收的数据执行任何格式化或翻译操作的需求。要查看 POST 请求的效果，请将另一个 GET 请求发送到 API /api/reservation，如下所示：

```
Invoke-RestMethod http://localhost:7000/api/reservation -Method GET
```

客户端返回的数据反映了新的 Reservation 对象已添加。

```
reservationId clientName location
------------- ---------- --------
            0 Alice      Board Room
            1 Bob        Lecture Hall
            2 Joe        Meeting Room 1
            3 Anne       Meeting Room 4
```

3. 测试 PUT 操作

PUT 操作用于更改模型中的现有对象。对象的 ReservationId 值被指定为请求 URL 的一部分，并且 clientName 和 location 的值在请求的主体中提供。以下 PowerShell 命令将发送一个 PUT 请求来修改一个 Reservation 对象：

```
Invoke-RestMethod http://localhost:7000/api/reservation -Method PUT -Body
(@{reservationId="1"; clientName="Bob"; location="Media Room"} | ConvertTo-Json)
-ContentType "application/json"
```

上述 PUT 请求会更改 ReservationId 值为 1 的 Reservation 对象，并为 location 属性指定新值。如果运行上述命令，你将看到以下响应，这表示更改已完成。

```
reservationId clientName location
------------- ---------- --------
            1 Bob        Media Room
```

为了查看 PUT 请求的效果，需要向 API /api/reservation 发送另一个 GET 请求，如下所示：

```
Invoke-RestMethod http://localhost:7000/api/reservation -Method GET
```

客户端返回的数据反映了新的 Reservation 对象已添加。

```
reservationId clientName location
------------- ---------- --------
            0 Alice      Board Room
            1 Bob        Media Room
            2 Joe        Meeting Room 1
            3 Anne       Meeting Room 4
```

4. 测试 PATCH 操作

PATCH 操作用于修改模型中的现有对象。许多应用程序使用 PUT 请求并完全忽略 PATCH，如果客户端可以访问模型中对象定义的所有属性，那么这将是一种合理的方法。但是在复杂的应用程序中，出于安全原因，客户端可能会收到一组特定的属性值，这会阻止它们作为 PUT 请求的一部分发送完整的对象。PATCH 请求更具选择性，允许客户端为对象指定一组精细更改。

ASP.NET Core MVC 支持使用 JSON Patch 标准，从而允许以统一的方式指定更改。这里不打算详细介绍 JSON Patch 标准，可以从 IETF 网站查看。对于示例应用程序，客户端将发送 HTTP PATCH 请求中的数据，如下所示：

```
[
{ "op": "replace", "path": "clientName", "value": "Bob"},
{ "op": "replace", "path": "location", "value": "Lecture Hall"}
]
```

JSON Patch 文档将被表示为一系列操作。每个操作都有 op 属性，用于指定操作的类型。另外，每个操作还有 path 属性，用于指定操作的应用位置。

对于示例应用程序（事实上，对于大多数应用程序），仅需要替换操作，替换操作用于更改属性的值。为 clientName 和 location 属性设置新值，而要修改的对象由请求 URL 标识。ASP.NET Core MVC 将自动处理 JSON 数据并将其作为 JsonPatchDocument<T>对象呈现给操作方法，其中 T 是要修改的模型对象的类型。然后，可以使用 JsonPatchDocument<T>对象的 ApplyTo 方法从存储库中修改对象。以下是发送 PATCH 请求的 PowerShell 命令：

```
Invoke-RestMethod http://localhost:7000/api/reservation/2 -Method PATCH -Body (@
{ op="replace"; path="clientName"; value="Bob"},@{ op="replace"; path="location";
value="Lecture Hall"} | ConvertTo-Json) -ContentType "application/json"
```

上述请求要求服务器修改 ReservationId 为 2 的 Reservation 对象的 clientName 和 location 属性。要查看 PUT 请求的效果，请向 API /api/ reservation 发送 GET 请求，如下所示：

```
Invoke-RestMethod http://localhost:7000/api/reservation -Method GET
```

客户端返回的数据反映了新的 Reservation 对象已添加。

```
reservationId clientName location
------------- ---------- --------
            0 Alice      Board Room
            1 Bob        Media Room
            2 Bob        Lecture Hall
            3 Anne       Meeting Room 4
```

5. 测试 DELETE 操作

发送一个 DELETE 请求，用于从存储库中删除一个 Reservation 对象，如下所示：

```
Invoke-RestMethod http://localhost:7000/api/reservation/2 -Method DELETE
```

Reservation 控制器中接收 DELETE 请求的操作不会返回结果，因此在命令完成后不会显示任何数据。要查看删除效果，请使用以下命令请求存储库中的内容：

```
Invoke-RestMethod http://localhost:7000/api/reservation -Method GET
```

ReservationId 为 2 的 Reservation 对象已从存储库中删除。

```
reservationId clientName location
------------- ---------- --------
            0 Alice      Board Room
            1 Bob        Media Room
            3 Anne       Meeting Room 4
```

20.3.3 在浏览器中使用 API 控制器

通过定义 API 控制器虽然解决了应用程序的开放性问题，但是没有解决速度或效率问题。为此，这里需要更新应用程序的 HTML 部分，以便靠 JavaScript 来向 API 控制器发出 HTTP 请求以执行数据操作。

在浏览器中，异步 HTTP 请求通常称为 Ajax 请求，其中 Ajax 是 Asynchronous JavaScript and XML 的缩写。XML 数据格式近年来已不那么受欢迎，但 Ajax 仍然用于引用异步 HTTP 请求，即使它们返回 JSON 数据。更广泛地说，本节描述的技术是单页应用程序的基础，其中 HTML 单页应用程序中的 JavaScript 用于为应用程序的多个部分提取数据，以生成动态显示的内容。

注 意

客户端开发是本书范围内的讨论主题。这里只创建了基本的异步 HTTP 请求，而不进行详细解释。要想了解更多的信息，请参阅由 Apress 出版的 *Pro ASP.NET Core MVC Client Development* 一书，其中详细介绍了如何使用 JavaScript 和 jQuery 来创建从 API 控制器提供服务的单页应用程序。

浏览器提供了用于创建 Ajax 请求的 JavaScript API，但处理方式有点怪异，不同的浏览器在实现一些可选功能的方式上有一些差异。创建 Ajax 请求的最简单方法是使用 jQuery 库，jQuery 是客户端开发中非常有用的工具。代码清单 20-11 将 jQuery 包添加到了 bower.json 文件中。

代码清单 20-11 在 ApiControllers 文件夹下的 bower.json 文件中添加 jQuery 包

```
{
  "name": "asp.net",
  "private": true,
  "dependencies": {
    "bootstrap": "4.0.0-alpha.6",
    "jquery": "3.2.1"
  }
}
```

实际上，由于某些 Bootstrap 功能取决于 jQuery，Bower 将会在 wwwroot/lib 文件夹中安装 jQuery 包。为了使用 jQuery 提供的功能，这里创建 wwwroot/js 文件夹，并添加一个名为 client.js 的 JavaScript 文件，内容如代码清单 20-12 所示。

代码清单 20-12 wwwroot/js 文件夹下的 client.js 文件的内容

```
$(document).ready(function () {

    $("form").submit(function (e) {
        e.preventDefault();
        $.ajax({
            url: "api/reservation",
            contentType: "application/json",
            method: "POST",
            data: JSON.stringify({
                clientName: this.elements["ClientName"].value,
                location: this.elements["Location"].value
            }),
            success: function(data) {
                addTableRow(data);
            }
        })
    });
});

var addTableRow = function (reservation) {
    $("table tbody").append("<tr><td>" + reservation.reservationId + "</td><td>"
        + reservation.clientName + "</td><td>"
        + reservation.location + "</td></tr>");
}
```

当用户在浏览器中提交表单，将表单数据编码为 JSON，并使用 HTTP POST 请求将它们发送到服务器时，client.js 文件中的 JavaScript 文件将会响应。服务器返回的 JSON 数据将自动由 jQuery 解析，然后用于向 HTML 表格中添加一行。代码清单 20-13 更新了布局以包含 jQuery 库和 client.js 文件。

代码清单 20-13 在_Layout.cshtml 文件中添加 JavaScript 引用

```
<!DOCTYPE html>
<html>
<head>
    <meta name="viewport" content="width=device-width" />
    <title>RESTful Controllers</title>
    <link asp-href-include="lib/bootstrap/dist/css/*.min.css" rel="stylesheet" />
    <script src="lib/jquery/dist/jquery.js"></script>
    <script src="js/client.js"></script>
</head>
```

```
<body class="m-1 p-1">
    @RenderBody()
</body>
</html>
```

第一个 script 元素告诉浏览器加载 jQuery 库，第二个 script 元素指定包含自定义代码的文件。如果运行应用程序，你会发现这和使用 HTML 表单在应用程序的存储库中创建 Reservation 没有什么区别。但是，如果检查浏览器发送的 HTTP 请求，你会看到需要的数据少于同步版本的应用程序。在这里的测试中，异步请求需要 480 字节的数据，大约是同步请求所需的 40%。数据的大小倾向于比用于显示数据的 HTML 文档小得多，在这样的实际应用中，改进更为显著。

20.4 内容格式

当操作方法返回 C#对象作为结果时，MVC 必须弄清楚应该使用哪种数据格式对对象进行编码并发送给客户端。本节将解释什么是默认进程，以及默认进程如何受到客户端发送的请求和应用程序的配置的影响。为了帮助理解工作原理，这里将一个名为 ContentController.cs 的类文件添加到 Controllers 文件夹中，并用来定义代码清单 20-14 所示的 API 控制器。

代码清单 20-14 Controllers 文件夹下的 ContentController.cs 文件的内容

```
using Microsoft.AspNetCore.Mvc;
using ApiControllers.Models;

namespace ApiControllers.Controllers {

    [Route("api/[controller]")]
    public class ContentController : Controller {

        [HttpGet("string")]
        public string GetString() => "This is a string response";

        [HttpGet("object")]
        public Reservation GetObject() => new Reservation {
            ReservationId = 100,
            ClientName = "Joe",
            Location = "Board Room"
        };
    }
}
```

这里指定静态的片段变量作为 HttpGet 特性的参数，用于控制器中的两个操作，这意味着它们可以通过请求 URL /api/controller/string 和/api/controller/object 来访问。Content 控制器不会松散地遵循 REST 模式，但可以使你能够轻松了解内容协商的工作原理。

MVC 选择的内容格式取决于 4 个因素——客户端接收的格式，MVC 可以生成的格式，操作指定的内容策略以及操作方法返回的类型。了解清楚这一切如何融合在一起是很复杂的，但好消息是，默认策略对大多数应用程序适用，你只需要了解当需要进行更改或未获得更改时幕后发生的情况，即可得到期望的格式。

20.4.1 默认内容策略

首先采用标准应用程序配置，当客户端和操作方法都不对可以使用的数据格式应用任何限制时，返回的结果是简单和可预测的。

- 如果操作方法返回一个字符串，那么不对这个字符串做修改并发送给客户端，将响应的 Content-Type 标头设置为 text/plain。

- 对于所有其他数据类型，包括其他简单类型（如 int），数据格式为 JSON，并且将响应的 Content-Type 标头设置为 application/json。

字符串得到特殊处理的原因是它们在编码为 JSON 时会导致问题。当编码其他简单类型（例如 C#中的 int 类型，值为 2）时，结果是引用的字符串，比如"2"。当编码字符串时，会得到两组引号，所以"Hello"变成""Hello""。并不是所有客户端都能应对这种双重编码，所以使用 text/plain 格式更可靠，可以全面避免这个问题。这基本不是问题，因为极少有应用程序直接发送字符串值；以 JSON 格式发送对象更常见。可以通过使用 PowerShell 查看这两个结果。以下命令将调用 GetString 方法并返回一个字符串：

```
Invoke-WebRequest http://localhost:7000/api/content/string | select
@{n='Content-Type';e={ $_.Headers."Content-Type" }}, Content
```

以上命令将向 URL /api/content/string 发送 GET 请求，并处理响应以显示 Content-Type 标头和响应的内容。

提 示

如果尚未执行 Internet Explorer 的初始设置，那么在使用 Invoke-WebRequest cmdlet 时可能会收到错误提示。这极可能发生在 Edge 已经替换的 Windows 10 计算机上。可以通过运行 IE 并选择所需的初始配置来修复此问题。

以上命令产生的输出结果如下：

```
Content-Type              Content
------------              -------
text/plain; charset=utf-8 This is a string response
```

同样的命令也可以通过改变请求的 URL 来显示 JSON 格式，如下所示：

```
Invoke-WebRequest http://localhost:7000/api/content/object | select
@{n='Content-Type';e={ $_.Headers."Content-Type" }}, Content
```

以上命令产生的输出格式清晰，表明响应已编码为 JSON：

```
Content-Type                    Content
------------                    -------
application/json; charset=utf-8 {"reservationId":100,
                                 "clientName":"Joe",
                                 "location":"Board Room"}
```

20.4.2 内容协商

大多数客户端将在请求中包含 Accept 标头，用于指定它们希望在响应中接收的数据格式，以一组 MIME 类型表示。以下是 Google Chrome 在请求中发送的 Accept 标头：

```
Accept: text/html,application/xhtml+xml,application/xml;q=0.9,image/webp,*/*;q=0.8
```

以上 Accept 标头表示 Google Chrome 可以处理 HTML、XHTML 格式（XHTML 是 XML 兼容的 HTML 格式）、XML 和 WEBP 图像格式。标头中的 q 值指定了有关偏好，默认值为 1.0。为 application/xml 指定 0.9 的 q 值会告诉服务器 Google Chrome 可以接收 XML 数据，但 Google Chrome 通常会处理 HTML 或 XHTML 数据。最后一项*/*会告诉服务器 Google Chrome 可以接收任何数据格式，但 q 值表明这是所有指定类型的最低优先级。

(1) Google Chrome 通常会接收 HTML/XHTML 数据或 WEBP 图像。

(2) 如果这些格式不可用，下一种优选的格式是 XML。

(3) 如果以上格式都不可用，Google Chrome 将接收任何格式的数据。

由此假设可以通过设置 Accept 标头来更改请求从 MVC 应用程序接收数据的格式，但不能以这种方式工作，或者更确切地说，现在还不能这样工作，因为需要做一些准备工作。首先需要一个 PowerShell 命令，使用 Accept 标头向 GetObject 方法发送 GET 请求，Accept 标头指定客户端将仅接收 XML 数据。

```
Invoke-WebRequest http://localhost:7000/api/content/object -Headers @{Accept="application/
xml"} | select @{n='Content-Type';e={ $_.Headers."Content-Type" }}, Content
```

以下是请求返回的结果，服务器返回了 application/json 响应：

```
Content-Type                     Content
------------                     -------
application/json; charset=utf-8 {"reservationId":100,
                                  "clientName":"Joe",
                                  "location":"Board Room"}
```

包括 Accept 标头的请求对返回数据的格式没有影响，服务器已经向客户端发送了未指定格式的数据。因为在默认情况下 MVC 仅支持 JSON，所以没有其他可以使用的格式。MVC 发送 JSON 数据是希望客户端可以处理数据而不是返回错误信息，即使 JSON 格式不是请求的 Accept 标头指定的格式之一。

配置 JSON Serializer

ASP.NET Core MVC 使用流行的第三方 JSON 包 Json.Net 将对象序列化为 JSON。默认配置适用于大多数项目，但如果需要以特定方式创建 JSON，则可以对它们进行更改。可以在 Startup 类中使用 AddMvc().AddJsonOptions 扩展方法，以提供对 MvcJsonOptions 对象的访问，进而配置 Json.Net 包。有关可用配置选项的详细信息，请参见 newtonsoft 网站。

启用 XML 格式

为了在工作中看到内容协商，必须给 MVC 一些用于对响应数据进行编码的格式选择。虽然 JSON 已经成为 Web 应用程序的默认格式，但 MVC 也可以支持 XML 作为编码数据，如代码清单 20-15 所示。

提 示

从 Microsoft.AspNetCore.Mvc.Formatters.OutputFormatter 类可以派生和创建自己的内容格式，但很少有人这么做，因为创建自定义数据格式不是在应用程序中展示数据的有效方式，而且最常见的格式 JSON 和 XML 已经实现了。

代码清单 20-15 在 ApiControllers 文件夹下的 Startup.cs 文件中启用 XML 格式

```
using System;
using System.Collections.Generic;
using System.Linq;
using System.Threading.Tasks;
using Microsoft.AspNetCore.Builder;
using Microsoft.AspNetCore.Hosting;
using Microsoft.AspNetCore.Http;
using Microsoft.Extensions.DependencyInjection;
using ApiControllers.Models;

namespace ApiControllers {
    public class Startup {

        public void ConfigureServices(IServiceCollection services) {
            services.AddSingleton<IRepository, MemoryRepository>();
            services.AddMvc().AddXmlDataContractSerializerFormatters();
        }

        public void Configure(IApplicationBuilder app, IHostingEnvironment env) {
            app.UseStatusCodePages();
            app.UseDeveloperExceptionPage();
            app.UseStaticFiles();
            app.UseMvcWithDefaultRoute();
```

```
        }
    }
}
```

当 MVC 只有 JSON 格式可用时，别无选择，只能将响应编码为 JSON。现在有了另一选择，可以看到内容协商过程更加完善。

提　示

这里使用了代码清单 20-15 中的 AddXmlDataContractSerializerFormatter 扩展方法，但是你也可以使用 AddXmlSerializerFormatters 扩展方法，以提供对旧的序列化类的访问。如果需要为较早版本的.NET 客户端生成 XML 内容，这可能会很有帮助。

以下是再次请求 XML 数据的 PowerShell 命令：

```
Invoke-WebRequest http://localhost:7000/api/content/object -Headers @{Accept="application/
xml"} | select @{n='Content-Type';e={ $_.Headers."Content-Type" }}, Content
```

运行上述命令，你将看到服务器返回 XML 数据而不是 JSON，如下所示（为了简洁起见，这里省略了 XML 命名空间特性）：

```
Content-Type                  Content
------------                  -------
application/xml; charset=utf-8 <Reservation>
                                  <ClientName>Joe</ClientName>
                                  <Location>Board Room</Location>
                                  <ReservationId>100</ReservationId>
                              </Reservation>
```

20.4.3　指定 action 数据格式

你可以覆盖内容协商系统，并通过使用 Produces 特性直接在操作方法上指定数据格式，如代码清单 20-16 所示。

代码清单 20-16　在 Controllers 文件夹下的 ContentController.cs 文件中指定数据格式

```
using Microsoft.AspNetCore.Mvc;
using ApiControllers.Models;

namespace ApiControllers.Controllers {

    [Route("api/[controller]")]
    public class ContentController : Controller {

        [HttpGet("string")]
        public string GetString() => "This is a string response";

        [HttpGet("object")]
        [Produces("application/json")]
        public Reservation GetObject() => new Reservation {
            ReservationId = 100,
            ClientName = "Joe",
            Location = "Board Room"
        };
    }
}
```

Produces 特性是过滤器，可以更改 ObjectResult 对象的内容类型，MVC 使用这些对象来表示 API 控制器中的 action 结果。Produces 特性的参数不仅可以指定用于 action 结果的格式，还可以指定其他的类型。Produces 特性将强制响应使用的格式，你可以通过运行以下 PowerShell 命令看到。

```
(Invoke-WebRequest http://localhost:7000/api/content/object -Headers
@{Accept="application/xml"}).Headers."Content-Type"
```

以上命令将显示 GET 请求 URL /api/content/object 返回的响应中的 Content-Type 标头的值。运行该命令后你会发现，请求的 Accept 标头指定应使用 XML，返回的数据使用 Produces 特性指定的 JSON 格式。

20.4.4 从路由或查询字符串获取数据格式

Accept 标头并不总是在编写客户端的程序员的控制之下，特别是在使用旧的浏览器或工具包进行开发的情况下。对于这种情况，允许通过用于定位操作方法或请求 URL 的查询字符串部分的路由来请求响应的数据格式是有帮助的。第一步是在 Startup 类中定义可用于引用路由或查询字符串格式的速记值。默认情况下有一个映射，其中 json 用作 application/json 的缩写。代码清单 20-17 添加了一个额外的 XML 映射。

代码清单 20-17 在 ApiControllers 文件夹下的 Startup.cs 文件中添加格式

```
using System;
using System.Collections.Generic;
using System.Linq;
using System.Threading.Tasks;
using Microsoft.AspNetCore.Builder;
using Microsoft.AspNetCore.Hosting;
using Microsoft.AspNetCore.Http;
using Microsoft.Extensions.DependencyInjection;
using ApiControllers.Models;
using Microsoft.Net.Http.Headers;

namespace ApiControllers {
    public class Startup {

        public void ConfigureServices(IServiceCollection services) {
            services.AddSingleton<IRepository, MemoryRepository>();
            services.AddMvc()
                .AddXmlDataContractSerializerFormatters()
                .AddMvcOptions(opts => {
                    opts.FormatterMappings.SetMediaTypeMappingForFormat("xml",
                        new MediaTypeHeaderValue("application/xml"));
                });
        }

        public void Configure(IApplicationBuilder app, IHostingEnvironment env) {
            app.UseStatusCodePages();
            app.UseDeveloperExceptionPage();
            app.UseStaticFiles();
            app.UseMvcWithDefaultRoute();
        }
    }
}
```

MvcOptions.FormatterMappings 属性用于设置和管理映射。在以上代码中，使用 SetMediaTypeMapping-ForFormat 方法创建了一个新的映射，以便简写的 xml 代表 application/xml 格式。下一步是将 FormatFilter 属性应用于操作方法，并可选地调整操作的路由，使它包含 format 变量，如代码清单 20-18 所示。

代码清单 20-18 在 Controllers 文件夹下的 ContentController.cs 文件中应用 FormatFilter 特性

```
using Microsoft.AspNetCore.Mvc;
using ApiControllers.Models;

namespace ApiControllers.Controllers {

    [Route("api/[controller]")]
```

```
    public class ContentController : Controller {

        [HttpGet("string")]
        public string GetString() => "This is a string response";

        [HttpGet("object/{format?}")]
        [FormatFilter]
        [Produces("application/json", "application/xml")]
        public Reservation GetObject() => new Reservation {
            ReservationId = 100,
            ClientName = "Joe",
            Location = "Board Room"
        };
    }
}
```

这里已经将 FormatFilter 特性应用于 GetObject 方法，并修改了操作的路由，使之包含可选的 format 字段。不必将 Produces 特性与 FormatFilter 特性一起使用，但如果这样做，那么只有指定 Produces 特性配置格式的请求才能正常工作。指定 Produces 特性尚未配置的格式的请求将收到 404-Not Found 响应。如果不应用 Produces 特性，那么请求可以使用 MVC 被配置的任何格式。

这里还将 application/xml 格式添加到 Produces 特性中，以使操作方法支持对 JSON 和 XML 的请求。以下 PowerShell 命令将 xml 格式指定为请求 URL 的一部分：

```
(Invoke-WebRequest http://localhost:7000/api/content/object/xml).Headers."Content-Type"
```

运行以上命令显示响应的内容类型，如下所示：

```
application/xml; charset=utf-8
```

使用 FormatFilter 特性查找名为 format 的路由片段变量，获取其中包含的速记值，并从应用程序配置中检索关联的数据格式，然后用于响应。如果没有可用的路由数据，就检查查询字符串。以下是使用查询字符串请求 XML 的 PowerShell 命令：

```
(Invoke-WebRequest http://localhost:7000/api/content/object?format=xml).Headers."Content-Type"
```

FormatFilter 特性找到的格式将覆盖 Accept 标头指定的任何格式，即使在使用不允许 Accept 标头设置的工具包和浏览器的情况下，也可以将格式选择权掌握在客户端开发人员手中。

20.4.5 启用完成内容协商

对于大多数应用程序来说，当没有其他格式可用时发送 JSON 数据是明智的，因为 Web 应用程序的客户端很可能错误设置了接收标头，而不是不能处理 JSON。也就是说，如果不管 Accept 标头是什么 JSON 都被返回，一些应用程序将不得不处理那些导致问题的客户端。获取内容协商需要在 Startup 类中更改两个配置，如代码清单 20-19 所示。

代码清单 20-19 在 ApiControllers 文件夹下的 Startup.cs 文件中启用完成内容协商

```
using System;
using System.Collections.Generic;
using System.Linq;
using System.Threading.Tasks;
using Microsoft.AspNetCore.Builder;
using Microsoft.AspNetCore.Hosting;
using Microsoft.AspNetCore.Http;
using Microsoft.Extensions.DependencyInjection;
using ApiControllers.Models;
using Microsoft.Net.Http.Headers;

namespace ApiControllers {
    public class Startup {
```

```
        public void ConfigureServices(IServiceCollection services) {
            services.AddSingleton<IRepository, MemoryRepository>();
            services.AddMvc()
                .AddXmlDataContractSerializerFormatters()
                .AddMvcOptions(opts => {
                    opts.FormatterMappings.SetMediaTypeMappingForFormat("xml",
                        new MediaTypeHeaderValue("application/xml"));
                    opts.RespectBrowserAcceptHeader = true;
                    opts.ReturnHttpNotAcceptable = true;
                });
        }

        public void Configure(IApplicationBuilder app, IHostingEnvironment env) {
            app.UseStatusCodePages();
            app.UseDeveloperExceptionPage();
            app.UseStaticFiles();
            app.UseMvcWithDefaultRoute();
        }
    }
}
```

RespectBrowserAcceptHeader 选项用于控制是否完全遵守 Accept 标头。如果没有合适的格式可用，ReturnHttpNotAcceptable 选项用于控制是否将 406-Not Acceptable 响应发送到客户端。

这里还必须从操作方法中删除 Produces 特性，以使内容协商过程不被覆盖，如代码清单 20-20 所示。

代码清单 20-20 在 Controllers 文件夹下的 ContentController.cs 文件中删除 Produces 特性

```
using Microsoft.AspNetCore.Mvc;
using ApiControllers.Models;

namespace ApiControllers.Controllers {

    [Route("api/[controller]")]
    public class ContentController : Controller {

        [HttpGet("string")]
        public string GetString() => "This is a string response";

        [HttpGet("object/{format?}")]
        [FormatFilter]
        //[Produces("application/json", "application/xml")]
        public Reservation GetObject() => new Reservation {
            ReservationId = 100,
            ClientName = "Joe",
            Location = "Board Room"
        };
    }
}
```

这里有一个 PowerShell 命令，用于向/api/content/object 发送 GET 请求，并使用 Accept 标头指定应用程序无法提供的内容类型：

```
Invoke-WebRequest http://localhost:7000/api/content/object -Headers
@{Accept="application/custom"}
```

如果运行以上命令，你将看到 406 错误消息，指示服务器无法提供请求的格式。

20.4.6 接收不同的数据格式

当客户端向控制器发送数据（例如 POST 请求）时，可以使用 Consumes 特性指定不同的操作方法来处理特定的数据格式，如代码清单 20-21 所示。

代码清单 20-21　在 Controllers 文件夹下的 ContentController.cs 文件中处理不同的数据格式

```
using Microsoft.AspNetCore.Mvc;
using ApiControllers.Models;

namespace ApiControllers.Controllers {
    [Route("api/[controller]")]
    public class ContentController : Controller {

        [HttpGet("string")]
        public string GetString() => "This is a string response";

        [HttpGet("object/{format?}")]
        [FormatFilter]
        //[Produces("application/json", "application/xml")]
        public Reservation GetObject() => new Reservation {
            ReservationId = 100,
            ClientName = "Joe",
            Location = "Board Room"
        };

        [HttpPost]
        [Consumes("application/json")]
        public Reservation ReceiveJson([FromBody] Reservation reservation) {
            reservation.ClientName = "Json";
            return reservation;
        }

        [HttpPost]
        [Consumes("application/xml")]
        public Reservation ReceiveXml([FromBody] Reservation reservation) {
            reservation.ClientName = "Xml";
            return reservation;
        }
    }
}
```

ReceiveJson 和 ReceiveXml 操作都接收 POST 请求，它们之间的区别在于使用 Consumes 特性指定的数据格式，检查 Content-Type 标头，以确定操作方法是否可以处理请求。当有请求的 Content-Type 设置为 application/json 时，使用 ReceiveJson 方法；当 ContentType 设置为 application/xml 时，使用 ReceiveXml 方法。

20.5　小结

本章介绍了 API 控制器在 MVC 应用程序中的作用，演示了如何创建和测试 API 控制器，还简要演示了如何使用 jQuery 进行异步 HTTP 请求，并解释了内容的格式化过程。下一章将更详细地解释视图和视图引擎的工作原理。

第 21 章　视图

第 17 章讲述了操作方法如何返回 ViewResult 对象，该对象用于告诉 MVC 渲染视图并向客户端返回 HTML 响应。

在本书中，你已经看到许多示例中使用的视图，大致了解了它们的作用，本章将对视图进行更深入的介绍。

本章首先展示 MVC 如何使用视图引擎处理 ViewResult 对象，包括演示如何创建自定义视图引擎，然后介绍如何有效地使用内置的 Razor 视图引擎，包括使用分部视图和布局部分，这是涉及有效进行 MVC 开发的重要内容。表 21-1 介绍了视图的背景。

表 21-1　视图的背景

问题	答案
什么是视图？	视图是用于向用户显示内容的 MVC 模式的一部分。在 ASP.NET Core MVC 应用程序中，视图是包含 HTML 元素和 C#代码的文件，它们将被处理以生成响应
视图有什么作用？	视图允许将数据的呈现与处理请求的逻辑分开。视图还允许在整个应用程序中应用相同的演示文稿，因为许多控制器可以使用相同的视图
如何使用视图？	大多数 MVC 应用程序使用 Razor 视图引擎，这可以轻松地混合 HTML 和 C#内容。如第 17 章所述，视图可通过返回 ViewResult 对象作为操作方法的结果得以选择
在使用过程中是否有什么陷阱或局限性？	你可能需要一段时间才能习惯使用 Razor。本章将解释 Razor 如何工作，这有助于揭示一些常用操作
是否有其他的替代方案？	有一些第三方视图引擎可用于 MVC，但是它们的使用很有限

表 21-2 列出了本章要介绍的操作。

表 21-2　本章要介绍的操作

操作	方法	代码清单
创建自定义视图引擎	实现 IViewEngine 和 IView 接口	代码清单 21-3～代码清单 21-6
轻松创建混合了 HTML 和 C#代码的响应	使用 Razor 视图引擎	代码清单 21-7～代码清单 21-11
定义用于布局的内容区域	使用 Razor 部分	代码清单 21-12～代码清单 21-18
创建可重用的标记片段	使用分部视图	代码清单 21-19～代码清单 21-22
在视图中添加 JSON 内容	使用@ Json.Serialze 表达式	代码清单 21-23～代码清单 21-25
更改 Razor 搜索视图的位置	创建视图位置扩展	代码清单 21-26～代码清单 21-30

21.1　准备示例项目

在本章中，使用 ASP.NET Core Web Application（.NET Core）模板创建一个名为 Views 的 Empty 项目。为了启用 MVC 框架和其他对开发有用的中间件，对 Startup 类进行代码清单 21-1 所示的更改。

代码清单 21-1　Startup.cs 文件的内容

```
using System;
using System.Collections.Generic;
using System.Linq;
using System.Threading.Tasks;
using Microsoft.AspNetCore.Builder;
using Microsoft.AspNetCore.Hosting;
```

```
using Microsoft.AspNetCore.Http;
using Microsoft.Extensions.DependencyInjection;

namespace Views {
    public class Startup {
        public void ConfigureServices(IServiceCollection services) {
            services.AddMvc();
        }

        public void Configure(IApplicationBuilder app, IHostingEnvironment env) {
            app.UseStatusCodePages();
            app.UseDeveloperExceptionPage();
            app.UseStaticFiles();
            app.UseMvcWithDefaultRoute();
        }
    }
}
```

创建 Controllers 文件夹，添加一个名为 HomeController.cs 的类文件，并用它定义控制器，如代码清单 21-2 所示。

代码清单 21-2 Controllers 文件夹下的 HomeController.cs 文件的内容

```
using System;
using Microsoft.AspNetCore.Mvc;

namespace Views.Controllers {

    public class HomeController : Controller {

        public ViewResult Index() {
            ViewBag.Message = "Hello, World";
            ViewBag.Time = DateTime.Now.ToString("HH:mm:ss");
            return View("DebugData");
        }

        public ViewResult List() => View();
    }
}
```

21.2 创建自定义视图引擎

本节将深入剖析如何创建自定义视图引擎。在绝大多数项目中并不需要执行此操作，因为 MVC 提供了 Razor 视图引擎，第 5 章介绍了 Razor 的用法，Razor 将出现在本书的所有示例中。

创建自定义视图引擎的价值在于查看后台发生的情况，并扩展你对 MVC 运行方式的了解，包括了解视图引擎在将 ViewResult 转换为对客户端的响应时有多少自由度。

视图引擎是实现了 IViewEngine 接口的类，定义在 Microsoft.AspNetCore.Mvc.ViewEngines 命名空间中。以下是 IViewEngine 接口的定义：

```
namespace Microsoft.AspNetCore.Mvc.ViewEngines {
    public interface IViewEngine {

        ViewEngineResult GetView(string executingFilePath, string viewPath,
            bool isMainPage);

        ViewEngineResult FindView(ActionContext context, string viewName,
            bool isMainPage);
    }
}
```

视图引擎的作用是将对视图的请求转换为 ViewEngineResult 对象。当 MVC 需要视图时，将调用 GetView 方法，使视图引擎根据视图名称来提供视图。

视图引擎所做的工作是为 MVC 提供可用于生成响应的 ViewEngineResult 对象。ViewEngineResult 类不能直接实例化，而是提供用于创建实例的静态方法，如表 21-3 所示。

表 21-3 ViewEngineResult 类的静态方法

静态方法	描述
Found(name,view)	为 MVC 提供请求的视图，该视图是使用 view 参数设置的。视图实现了 IView 接口
NotFound(name,locations)	创建一个 ViewEngineResult 对象，以告诉 MVC 找不到请求的视图。locations 参数是描述视图引擎查看视图位置的字符串值的枚举

在编写视图引擎时，可以选择表 21-3 中描述的方法之一来指示视图请求的结果。Found 方法会创建一个指示请求成功的 ViewEngineResult 对象，并为 MVC 提供要处理的视图。NotFound 方法会创建一个 ViewEngineResult 对象，指示不成功的请求，并向 MVC 提供视图引擎在查找视图时搜索过的位置清单（还将作为错误消息的一部分显示给开发人员）。

视图引擎系统的另一个构建模块是 IView 接口，用于描述视图提供的功能，而不考虑创建它们的视图引擎。以下为 IView 接口的定义：

```
using Microsoft.AspNetCore.Mvc.Rendering;
using System.Threading.Tasks;

namespace Microsoft.AspNetCore.Mvc.ViewEngines {

    public interface IView {

        string Path { get; }
        Task RenderAsync(ViewContext context);
    }
}
```

Path 属性返回视图的路径，它假定视图为定义在磁盘上的文件。RenderAsync 方法由 MVC 调用以生成对客户端的响应。通过从 ActionContext 派生的 ViewContext 类的实例将 Context 数据提供给视图。除从父类继承的 Context 属性（用于提供对请求的访问、路由数据、控制器等）之外，ViewContext 类还提供了渲染响应中有用的属性，见表 21-4。

表 21-4 ViewContext 有用的属性

属性	描述
ViewData	返回一个 ViewDataDictionary 对象，该对象包含控制器提供的视图数据
TempData	返回包含临时数据的字典
Writer	返回一个 TextWriter 对象，用于写入视图的输出

这些属性中最有用的是 ViewData，它返回一个 ViewDataDictionary 对象。ViewDataDictionary 类定义了一些有用的属性，可以访问视图模型、视图包和视图模型元数据。表 21-5 列出了这些有用的属性。

表 21-5 ViewDataDictionary 有用的属性

属性	描述
Model	返回由控制器提供的模型数据
ModelMetadata	返回一个 ModelMetadata 对象，该对象可用于反映模型数据的类型
ModelState	返回模型的状态
Keys	返回可用于访问 ViewBag 数据的键值枚举

查看 IviewEngine、ViewEngineResult 和 IView 如何结合在一起工作的最简单方法是创建视图引擎。这里将创建一个简单的视图引擎，返回一种视图，以呈现包含有关请求的信息以及由操作方法生成的视图数据的结果。这种方法有助于演示视图引擎的运行方式，而不会影响解析视图模板和重新创建 Razor 提供的其他功能。

21.2.1 创建自定义 IView

本节将从创建 IView 接口的实现开始。在示例项目中创建 Infrastructure 文件夹，并添加一个名为 DebugDataView.cs 的类文件，如代码清单 21-3 所示。

代码清单 21-3 Infrastructure 文件夹下的 DebugDataView.cs 文件的内容

```
using System;
using System.Text;
using System.Threading.Tasks;
using Microsoft.AspNetCore.Mvc.Rendering;
using Microsoft.AspNetCore.Mvc.ViewEngines;

namespace Views.Infrastructure {

    public class DebugDataView : IView {
        public string Path => String.Empty;

        public async Task RenderAsync(ViewContext context) {
            context.HttpContext.Response.ContentType = "text/plain";

            StringBuilder sb = new StringBuilder();

            sb.AppendLine("---Routing Data---");
            foreach (var kvp in context.RouteData.Values) {
                sb.AppendLine($"Key: {kvp.Key}, Value: {kvp.Value}");
            }
            sb.AppendLine("---View Data---");
            foreach (var kvp in context.ViewData) {
                sb.AppendLine($"Key: {kvp.Key}, Value: {kvp.Value}");
            }

            await context.Writer.WriteAsync(sb.ToString());
        }
    }
}
```

当渲染这个视图时，使用 ViewContext 参数向 RenderAsync 方法写入路由数据和视图数据的详细信息。响应是简单的文本，以上代码已经使用 context 对象设置响应的 Content-Type 标头为 text/plain。不这么做的话，ASP.NET 默认会使用 text/html，这将导致浏览器将数据显示为单个不间断的字符行。

21.2.2 创建 IViewEngine 实现

视图引擎的目的是生成一个 ViewEngineResult 对象，该对象包含一个 IView 或者一个用于搜索合适视图的位置清单。既然有了 IView 实现，就可以创建视图引擎。在 Infrastructure 文件夹中添加一个名为 DebugDataViewEngine.cs 的类文件，内容如代码清单 21-4 所示。

代码清单 21-4 Infrastructure 文件夹下的 DebugDataViewEngine.cs 文件的内容

```
using Microsoft.AspNetCore.Mvc;
using Microsoft.AspNetCore.Mvc.ViewEngines;

namespace Views.Infrastructure {
```

```
    public class DebugDataViewEngine : IViewEngine {

        public ViewEngineResult GetView(string executingFilePath, string viewPath,
                bool isMainPage) {
            return ViewEngineResult.NotFound(viewPath,
                new string[] { "(Debug View Engine - GetView)" });
        }

        public ViewEngineResult FindView(ActionContext context, string viewName,
                bool isMainPage) {
            if (viewName == "DebugData") {
                return ViewEngineResult.Found(viewName, new DebugDataView());
            } else {
                return ViewEngineResult.NotFound(viewName,
                    new string[] { "(Debug View Engine - FindView)" });
            }
        }
    }
}
```

这个视图引擎中的 GetView 方法始终返回 NotFound 响应。FindView 方法仅支持单个视图，名为 DebugData。当接收到具有这一名称的视图请求时，就返回一个新的 DebugDataView 实例，如下所示：

```
...
if (viewName == "DebugData") {
    return ViewEngineResult.Found(viewName, new DebugDataView());
}
...
```

如果正在实现一个更完整的视图引擎，那么可以利用这个机会来搜索模板。但在这个示例中，只需要一个新的 DebugDataView 实例。如果收到除 DebugData 之外的视图请求，将创建 NotFound 响应，如下所示：

```
...
return ViewEngineResult.NotFound(viewName,
    new string[] { "(Debug View Engine - FindView)" });
...
```

ViewEngineResult.NotFound 方法假定视图引擎具有查找视图所需的位置。这是一个合理的假设，因为视图通常是作为项目中的文件存储的模板文件。在这种情况下，没有任何地方可以查找视图，所以只返回一个虚拟位置，以指示调用哪个方法来定位视图。

21.2.3 注册自定义视图引擎

视图引擎可通过配置 MvcViewOptions 对象在 Startup 类中注册，如代码清单 21-5 所示。

代码清单 21-5 在 Startup.cs 文件中注册自定义视图引擎

```
using System;
using System.Collections.Generic;
using System.Linq;
using System.Threading.Tasks;
using Microsoft.AspNetCore.Builder;
using Microsoft.AspNetCore.Hosting;
using Microsoft.AspNetCore.Http;
using Microsoft.Extensions.DependencyInjection;
using Microsoft.AspNetCore.Mvc;
using Views.Infrastructure;

namespace Views {
    public class Startup {
```

```
        public void ConfigureServices(IServiceCollection services) {
            services.AddMvc();
            services.Configure<MvcViewOptions>(options => {
                options.ViewEngines.Insert(0, new DebugDataViewEngine());
            });
        }
        public void Configure(IApplicationBuilder app, IHostingEnvironment env) {
            app.UseStatusCodePages();
            app.UseDeveloperExceptionPage();
            app.UseStaticFiles();
            app.UseMvcWithDefaultRoute();
        }
    }
}
```

MvcViewOptions 类定义了 ViewEngines 属性，其值是一组 IViewEngine 对象。Razor 可通过 AddMvc 方法添加到 ViewEngine 集合中，这里使用自定义类补充了默认视图引擎。

当 MVC 从一个操作方法接收到一个 ViewResult 对象时，它会调用 MvcViewOptions.ViewEngines 集合中包含的每个视图引擎的 FindView 方法，直到收到使用 Found 方法创建的 ViewEngineResult 对象。

如果两个或多个视图引擎能够为视图名称相同的请求提供服务，那么将视图引擎添加到 ViewEngines.Engines 集合的顺序非常重要。要使你的视图优先，应将其插入视图引擎集合的开头，如代码清单 21-5 所示。

21.2.4 测试视图引擎

当应用程序启动时，浏览器将自动导航到项目的根 URL，根 URL 将映射到 Home 控制器中的 Index 操作方法。这个操作方法使用 View 方法返回用于指定 DebugData 视图的 ViewResult 对象。

MVC 将转向视图引擎的集合，并开始调用它们的 FindView 方法。由于请求的视图是自定义视图引擎设置为将要处理的视图，因此为 MVC 提供一个视图，该视图将产生图 21-1 所示的结果。

要查看当没有视图引擎可以提供视图时会发生什么，可请求 URL——/Home/List，这将创建一个 ViewResult 对象，指向一个名为 List 的视图，List 视图既不是 Razor 提供的，也不是自定义视图引擎提供的。你将看到图 21-2 所示的错误。

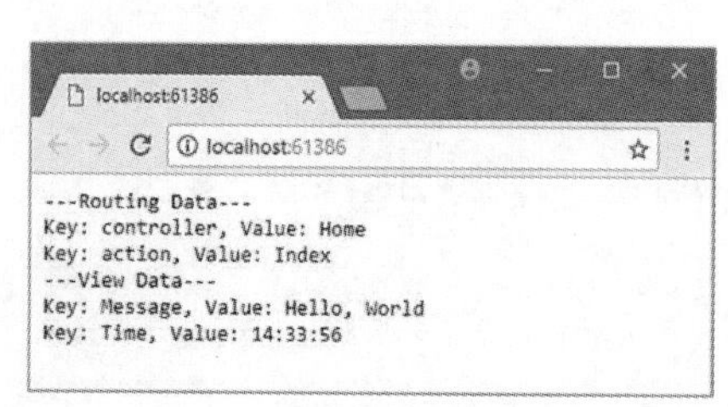

图 21-1 使用自定义视图引擎

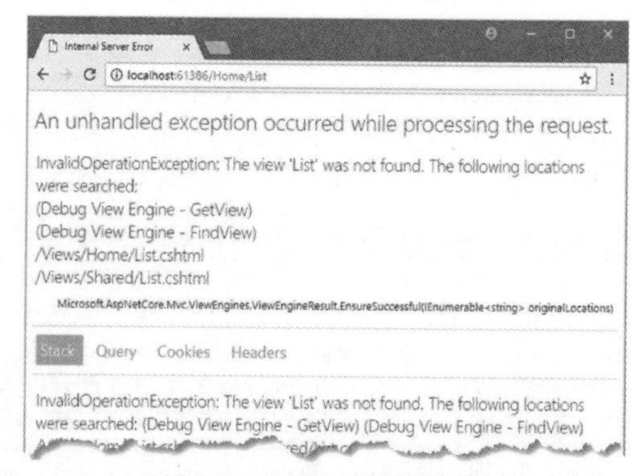

图 21-2 请求无法提供的视图

可以看到，自定义视图引擎报告了寻找 List 视图的位置列表，同时还报告了 Razor 已检查的位置。

如果想确保只有你的视图引擎被使用，那就必须对视图引擎的集合调用 Clear 方法来删除 Razor，如代码清单 21-6 所示。

代码清单 21-6 在 Views 项目的 Startup.cs 文件中删除其他视图引擎

```
using System;
using System.Collections.Generic;
using System.Linq;
```

```
using System.Threading.Tasks;
using Microsoft.AspNetCore.Builder;
using Microsoft.AspNetCore.Hosting;
using Microsoft.AspNetCore.Http;
using Microsoft.Extensions.DependencyInjection;
using Microsoft.AspNetCore.Mvc;
using Views.Infrastructure;

namespace Views {
    public class Startup {
        public void ConfigureServices(IServiceCollection services) {
            services.AddMvc();
            services.Configure<MvcViewOptions>(options => {
                options.ViewEngines.Clear();
                options.ViewEngines.Insert(0, new DebugDataViewEngine());
            });
        }
        public void Configure(IApplicationBuilder app, IHostingEnvironment env) {
            app.UseStatusCodePages();
            app.UseDeveloperExceptionPage();
            app.UseStaticFiles();
            app.UseMvcWithDefaultRoute();
        }
    }
}
```

如果启动应用程序并再次导航到/Home/List，将只会使用自定义视图引擎，如图 21-3 所示。

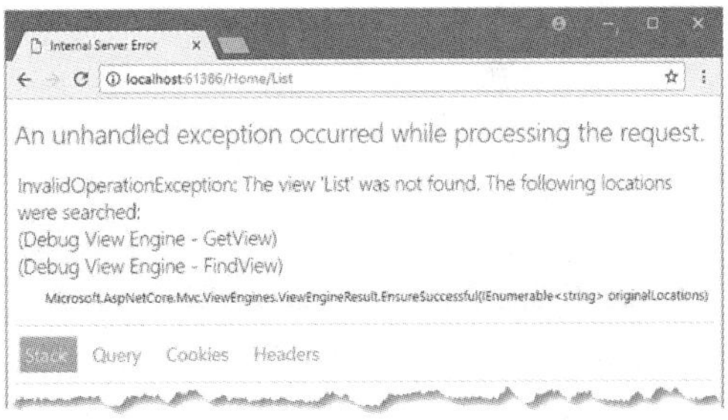

图 21-3 在示例应用程序中仅使用自定义视图引擎

21.3 使用 Razor 引擎

可以通过仅实现两个接口来创建自定义视图引擎，MVC 让添加或替换核心功能变得十分轻松。

视图引擎的复杂性来自视图模板系统，包括代码片段、支持布局和性能优化。这里在简单的自定义视图引擎中没有做以上这些事情，也没有太多需要，因为内置的 Razor 引擎提供了所有这些功能。实际上，几乎所有 MVC 应用程序所需的功能都可以在 Razor 中使用。只有极少量的项目需要自定义视图引擎。

第 5 章给出了关于 Razor 语法的介绍，本节将展示如何使用其他功能来创建和渲染 Razor 视图，还将讲述如何自定义 Razor 引擎。

21.3.1 准备示例项目

为了使用 Razor，需要对示例应用做一些修改。首先，更改 Home 控制器的 Index 操作方法，以便选择默认视图并提供一些模型数据，如代码清单 21-7 所示。

代码清单 21-7 修改 Controllers 文件夹下的 HomeController.cs 文件中的 Index 操作方法

```
using System;
using Microsoft.AspNetCore.Mvc;

namespace Views.Controllers {

    public class HomeController : Controller {

        public ViewResult Index() =>
            View(new string[] { "Apple", "Orange", "Pear" });
```

```
        public ViewResult List() => View();
    }
}
```

为了向 Index 操作方法提供视图，创建 Views/Home 文件夹，并添加一个名为 index.cshtml 的视图文件，内容如代码清单 21-8 所示。

代码清单 21-8 Views/Home 文件夹下的 index.cshtml 文件的内容

```
@model string[]
@{ Layout = null; }

<!DOCTYPE html>
<html>
<head>
    <meta name="viewport" content="width=device-width" />
    <title>Razor</title>
    <link asp-href-include="lib/bootstrap/dist/css/*.min.css" rel="stylesheet" />
</head>
<body class="m-1 p-1">
    This is a list of fruit names:
    @foreach (string name in Model) {
        <span><b>@name</b></span>
    }
</body>
</html>
```

Index 视图依赖于 Bootstrap CSS 库，要将 Bootstrap 添加到示例项目中，可使用 Bower Configuration File 模板在项目的根文件夹中创建 bower.json 文件，内容如代码清单 21-9 所示。

代码清单 21-9 bower.json 文件的内容

```
{
  "name": "asp.net",
  "private": true,
  "dependencies": {
    "bootstrap": "4.0.0-alpha.6"
  }
}
```

在 Views 文件夹中添加一个名为_ViewImports.cshtml 的视图文件，以启用内置的标签助手，如代码清单 21-10 所示。

代码清单 21-10 Views 文件夹下的_ViewImports.cshtml 文件的内容

```
@addTagHelper *, Microsoft.AspNetCore.Mvc.TagHelpers
```

最后的准备步骤是重新启动 Startup 类中的视图引擎以删除自定义引擎，并删除用来禁用 Razor 的 Clear 方法调用，如代码清单 21-11 所示。

代码清单 21-11 重启 Startup.cs 文件中的视图引擎

```
using System;
using System.Collections.Generic;
using System.Linq;
using System.Threading.Tasks;
using Microsoft.AspNetCore.Builder;
using Microsoft.AspNetCore.Hosting;
using Microsoft.AspNetCore.Http;
using Microsoft.Extensions.DependencyInjection;
using Microsoft.AspNetCore.Mvc;
using Views.Infrastructure;
```

```
namespace Views {
    public class Startup {
        public void ConfigureServices(IServiceCollection services) {
            services.AddMvc();
            //services.Configure<MvcViewOptions>(options => {
            //    options.ViewEngines.Clear();
            //    options.ViewEngines.Insert(0, new DebugDataViewEngine());
            //});
        }

        public void Configure(IApplicationBuilder app, IHostingEnvironment env) {
            app.UseStatusCodePages();
            app.UseDeveloperExceptionPage();
            app.UseStaticFiles();
            app.UseMvcWithDefaultRoute();
        }
    }
}
```

如果运行示例项目，你将看到图 21-4 所示的结果。

图 21-4 运行示例项目

21.3.2 Razor 视图

了解 Razor 的工作原理可以帮助将大量功能引入上下文中，并揭开 CSHTML 文件处理方式的谜团。

那么，Razor 如何将 HTML 元素和 C#语句相结合，并产生 HTTP 响应的内容？答案简单明了，它们建立在 MVC 功能的基础之上，你已经在前面的章节中了解了这些功能。Razor 将 CSHTML 文件转换为 C#类并编译，然后在每次需要视图生成结果时创建新的实例。下面是 Razor 为 Index.cshtml 创建的 C#类：

```
using System.Threading.Tasks;
using Microsoft.AspNetCore.Mvc;
using Microsoft.AspNetCore.Mvc.Razor;
using Microsoft.AspNetCore.Mvc.Razor.Internal;
using Microsoft.AspNetCore.Mvc.Rendering;

namespace Asp {

    public class ASPV_Views_Home_Index_cshtml : RazorPage<string[]> {

        public IUrlHelper Url { get; private set; }

        public IViewComponentHelper Component { get; private set; }

        public IJsonHelper Json { get; private set; }

        public IHtmlHelper<string[]> Html { get; private set; }

        public override async Task ExecuteAsync() {
            Layout = null;

            WriteLiteral(@"<!DOCTYPE html><html><head>
                <meta name=""viewport"" content=""width=device-width"" />
                <title>Razor</title>
                <link asp-href-include=""lib/bootstrap/dist/css/*.min.css""
                    rel=""stylesheet"" />
                </head><body class=""m-1 p-1"">This is a list of fruit names:");
            foreach (string name in Model) {
                WriteLiteral("<span><b>");
                Write(name);
                WriteLiteral("</b></span>");
            }
```

```
            WriteLiteral("</body></html>");
        }
    }
}
```

这里对类中的代码进行了整理，以便于阅读，并删除了Razor在生成类时为检测而添加的一些C#语句。后面将深入剖析类，并解释编译视图的工作原理。

注 意

以前很容易查看早期版本的Razor创建的类，因为每个视图都会在磁盘上生成一个C#文件，然后进行编译以便在应用程序中使用。检查类的内容只是为了找到正确的文件。当前版本的Razor依赖于C#编译器的进步，允许在内存中生成和编译代码，从而提高性能，但很难看到正在执行的事情。要获得以前显示的类，将不得不重新使用ASP.NET Core MVC源代码中包含的一些单元测试，其中提供了Razor依赖于查找和处理视图文件的类的模拟实现。这不是日常开发中需要做的事情，但揭示了视图如何工作。

1. 理解类的名称

我们先从Razor创建的类的名称开始。

```
...
public class ASPV_Views_Home_Index_cshtml : RazorPage<string[]> {
...
```

Razor需要使用一些方法来将CSHTML文件的名称和路径转换为分析文件时创建的类，并通过对类名称中的信息进行编码来实现。Razor使用ASPV作为类名前缀，后跟项目名称和控制器名称，最后是视图文件名；这种组合使得当MVC通过本章前面描述的IViewEngine请求视图时，可以轻松检查类是否可用。

2. 基类

Razor的很多核心功能，比如能够以@Model的形式引用视图模型，缘于它们派生自基类。

```
...
public class ASPV_Views_Home_Index_cshtml : RazorPage<string[]> {
...
```

由于@model指令已用于指定模型类型，因此View类继承自RazorPage类或RazorPage<T>类。RazorPage类提供了让CSHTML文件访问MVC功能的方法和属性，其中有用的RazorPage属性和方法分别如表21-6和表21-7所示。

表21-6 视图开发中有用的RazorPage属性

属性	描述
Model	返回由操作方法提供的模型数据
ViewData	返回一个ViewDataDictionary对象，该对象提供对其他视图数据功能的访问
ViewContext	返回一个ViewContext对象
Layout	用于指定布局
ViewBag	用于访问ViewBag对象
TempData	提供对Temp数据的访问
Context	返回一个描述当前请求和正在准备的响应的HttpContext对象
User	返回与请求相关联的用户的配置文件

表21-7 视图开发中有用的RazorPage方法

方法	描述
RenderSection()	用于将视图中的一部分内容插入布局
RenderBody()	用于将不包含在部分（section）中的视图的所有内容插入布局
IsSectionDefined()	用于确定视图是否定义部分的内容

Razor Pages

随着 ASP.NET Core 2 的发布，微软增加了对 Razor Pages 的支持，Razor Pages 打破了 MVC 模型，并将支持视图所需的代码与 Razor 视图相关联。这类似于 ASP.NET Web Forms，是微软尝试的一种设计方法，用于重新获得 Web Pages 平台的简单性，而不会出现第 1 章所述的缺点。

不要将本节中描述的 RazorPage 基类与 Razor Pages 的功能弄混淆。虽然它们使用相似的名称，但 RazorPage 基类为 MVC 框架使用的 Razor 视图引擎提供了基础。本书没有介绍 Razor Pages，因为它不符合 MVC 模型，也不是 MVC 平台的一部分。

Razor 还提供了一些可以在视图中用于生成内容的辅助属性，如表 21-8 所示。

表 21-8　　Razor 辅助属性

属性	描述
HtmlEncoder	返回一个 HtmlEncoder 对象，可用于在视图中安全地对 HTML 内容进行编码
Component	返回一个视图组件助手
Json	返回一个 JSON 助手
Url	返回一个 URL 辅助器，可用于使用路由配置生成 URL
Html	返回一个 HTML 助手，可用于生成动态内容。此功能已被标签助手取代，但仍然用于分部视图

表 21-6 和表 21-8 中描述的属性在日常 MVC 开发中访问模型数据、配置视图和执行其他重要任务时会用到。这些属性揭开了 Razor 的神秘面纱，并将 Razor 牢牢地置于为人熟知的 C#世界中。例如，当使用@Model 指令访问视图模型对象或使用@TempData 检索临时数据值时，就会用到由 RazorPage 类定义的属性。

3. 理解视图渲染

除了向开发人员提供属性和方法外，RazorPage 类还负责通过 ExecuteAsyc 方法生成响应内容。该方法显示了 Razor 如何将 Index.cshtml 文件处理成一组 C#语句：

```
...
public override async Task ExecuteAsync() {
    Layout = null;
    WriteLiteral(@"<!DOCTYPE html><html><head>
            <meta name=""viewport"" content=""width=device-width"" />
            <title>Razor</title>
            <link asp-href-include=""lib/bootstrap/dist/css/*.min.css""
                rel=""stylesheet"" />
            </head><body class=""m-1 p-1"">This is a list of fruit names:");
    foreach (string name in Model) {
        WriteLiteral("<span><b>");
        Write(name);
        WriteLiteral("</b></span>");
    }
    WriteLiteral("</body></html>");
}
...
```

数据值（如 Model 属性的值）将使用 Write 方法发送给客户端，该方法会转义字符串，以便浏览器不将其解释为 HTML 元素。这很重要，因为可以防止恶意数据值向应用程序的输出添加内容。WriteLiteral 方法不会转义字符串，并且被用于 Index.cshtml 文件中的静态内容。当然，浏览器应该将其解释为 HTML 元素。结果是 CSHTML 文件的静态和动态内容包含在常规 C#类中，并通过简单的方法发出。

21.4　将动态内容添加到 Razor 视图中

视图的整个目的是让你能够将域模型的一部分呈现给用户。为此，你需要能够向视图添加动态内容。

动态内容在运行时生成，每个请求都可以是不同的。这与你在编写应用程序时创建的对于每个请求都相同的静态内容（例如 HTML）相反。可以按照表 21-9 中描述的不同方式向视图中添加动态内容。

表 21-9 将动态内容添加到视图中的方式

方式	何时使用
内联代码	用于小型、独立的视图逻辑，如 if 和 foreach 语句。这是在视图中创建动态内容的基础工具，其他一些方法便以此为基础。第 5 章介绍了这种技术，你在以后的章节中也将看到很多这样的例子
标签助手	用于生成 HTML 元素的属性
部分（section）	用于创建将在特定位置插入布局的内容部分，如本节后面所述
分部视图	用于在视图之间共享视图标记。分部视图可以包含内联代码、HTML 辅助方法和其他分部视图的引用。分部视图不会调用操作方法，因此不能用于执行业务逻辑
视图组件	用于创建可重用的 UI 控件或需要包含业务逻辑的窗口小部件

21.4.1 使用布局部分

Razor 视图引擎支持部分（section）的概念，以允许在布局中提供内容区域。部分可以更好地控制将视图的哪些内容插入布局以及放置在哪里。为了演示部分的功能，编辑/Views/Home/Index.cshtml 文件，如代码清单 21-12 所示。

代码清单 21-12 在 Views/Home 文件夹下的 Index.cshtml 文件中定义部分

```
@model string[]
@{ Layout = "_Layout"; }

@section Header {
    <div class="bg-success">
        @foreach (string str in new [] {"Home", "List", "Edit"}) {
            <a class="btn btn-sm btn-primary" asp-action="str">@str</a>
        }
    </div>
}
This is a list of fruit names:
@foreach (string name in Model) {
    <span><b>@name</b></span>
}

@section Footer {
    <div class="bg-success">
        This is the footer
    </div>
}
```

这里从视图中删除了一些 HTML 元素，并设置了 Layout 属性，以指定使用名为_Layout.cshtml 的布局文件来呈现内容。

这里还在视图中添加了一些部分。部分使用 Razor @section 表达式定义，后跟部分的名称。此外创建了标题和页脚部分，内容是 HTML 标记和 Razor 表达式的组合，其他示例中已经展示过这些组合。

部分在视图中定义，但应用于带有@RenderSection 表达式的布局。为了演示原理，创建 Views/Shared 文件夹，并添加一个名为_Layout.cshtml 的布局文件，内容如代码清单 21-13 所示。

代码清单 21-13 Views/Shared 文件夹下的_Layout.cshtml 文件的内容

```
<!DOCTYPE html>
<html>
<head>
    <meta name="viewport" content="width=device-width" />
    <title>@ViewBag.Title</title>
    <link asp-href-include="lib/bootstrap/dist/css/*.min.css" rel="stylesheet" />
```

```
    </head>
    <body class="m-1 p-1">
        @RenderSection("Header")

        <div class="bg-info">
            This is part of the layout
        </div>

        @RenderBody()

        <div class="bg-info">
            This is part of the layout
        </div>

        @RenderSection("Footer")

        <div class="bg-info">
            This is part of the layout
        </div>
    </body>
    </html>
```

当 Razor 解析布局时，RenderSection 辅助程序将被视图中具有指定名称的部分的内容替换。不包含部分的视图内容已使用 RenderBody 助手插入布局。

可以通过启动应用程序来看到这些部分的效果，如图 21-5 所示。这里使用一些 Bootstrap 样式来帮助看清楚哪些部分（section）来自视图，而哪些来自输出。结果不是很漂亮，但它演示了如何将视图中的内容区域放在布局中的特定位置。

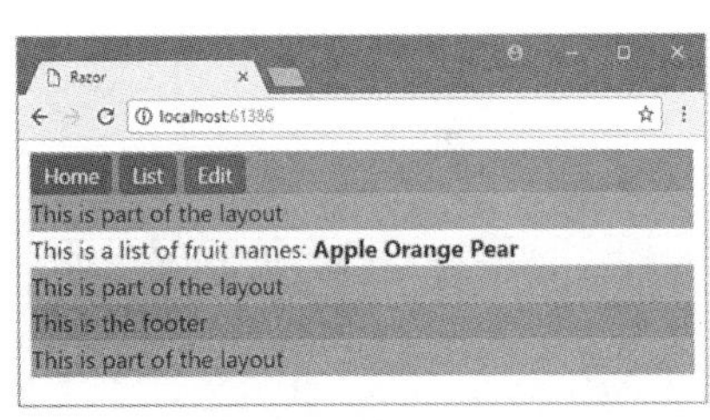

图 21-5　使用视图中的部分来定位布局中的内容

注　意

视图只能定义布局中引用的部分。如果尝试在视图中定义布局中没有对应的@RenderSection 表达式的部分，MVC 将显示异常。

将部分与其余视图混合是不正常的。约定是在视图的开始或结尾定义部分，以便更容易看到哪些内容区域将被视为部分，哪些内容将被 RenderBody 助手捕获。另一种方式是仅使用部分来定义视图，包括用于正文的部分，如代码清单 21-14 所示。

代码清单 21-14　在 Views/Home 文件夹下的 Index.cshtml 文件中根据 Razor 部分定义视图

```
@model string[]
@{ Layout = "_Layout"; }

@section Header {
    <div class="bg-success">
        @foreach (string str in new [] {"Home", "List", "Edit"}) {
            <a class="btn btn-sm btn-primary" asp-action="str">@str</a>
        }
    </div>
}
@section Body {
    This is a list of fruit names:
    @foreach (string name in Model) {
        <span><b>@name</b></span>
    }
}

@section Footer {
    <div class="bg-success">
        This is the footer
```

```
    </div>
}
```

这样做可以提供更清晰的视图，并减少 RenderBody 捕获无关内容的机会。要使用这种方法，就必须将 RenderBody 调用替换为 RenderBody("body")，如代码清单 21-15 所示。

代码清单 21-15 在 Views/Shared 文件夹下的_Layout.cshtml 文件中将 Body 渲染为部分

```
<!DOCTYPE html>
<html>
<head>
    <meta name="viewport" content="width=device-width" />
    <title>@ViewBag.Title</title>
    <link asp-href-include="lib/bootstrap/dist/css/*.min.css" rel="stylesheet" />
</head>
<body class="m-1 p-1">
    @RenderSection("Header")

    <div class="bg-info">
        This is part of the layout
    </div>

    @RenderSection("Body")

    <div class="bg-info">
        This is part of the layout
    </div>

    @RenderSection("Footer")

    <div class="bg-info">
        This is part of the layout
    </div>
</body>
</html>
```

1. 测试部分

可以检查视图是否从布局中定义了特定的部分。如果视图不需要或想要提供特定的内容，可以为部分提供默认内容。修改_Layout.cshtml 文件，以检查是否定义了页脚部分，如代码清单 21-16 所示。

代码清单 21-16 在 Views/Shared 文件夹下的_Layout.cshtml 文件中检查是否定义了某个部分

```
<!DOCTYPE html>
<html>
<head>
    <meta name="viewport" content="width=device-width" />
    <title>@ViewBag.Title</title>
    <link asp-href-include="lib/bootstrap/dist/css/*.min.css" rel="stylesheet" />
</head>
<body class="m-1 p-1">
    @RenderSection("Header")

    <div class="bg-info">
        This is part of the layout
    </div>

    @RenderSection("Body")

    <div class="bg-info">
        This is part of the layout
    </div>

    @if (IsSectionDefined("Footer")) {
        @RenderSection("Footer")
```

```
    } else {
        <h4>This is the default footer</h4>
    }

    <div class="bg-info">
        This is part of the layout
    </div>
</body>
</html>
```

IsSectionDefined 助手用于获取要检查的部分的名称，如果在渲染的视图中定义该部分，则返回 true。在该例中，可使用该助手来确定在视图未定义页脚部分时是否应该呈现一些默认内容。

2. 渲染可选部分

默认情况下，视图必须包含布局中有 RenderSection 调用的所有部分。如果缺少部分，那么 MVC 会向用户报告异常。为了演示，为_Layout.cshtml 文件添加一个新的 RenderSection 调用，用于名为 scripts 的部分，如代码清单 21-17 所示。

代码清单 21-17 在 Views/Shared 文件夹下的_Layout.cshtml 文件中渲染某个不存在的部分

```
<!DOCTYPE html>
<html>
<head>
    <meta name="viewport" content="width=device-width" />
    <title>@ViewBag.Title</title>
    <link asp-href-include="lib/bootstrap/dist/css/*.min.css" rel="stylesheet" />
</head>
<body class="m-1 p-1">
    @RenderSection("Header")

    <div class="bg-info">
        This is part of the layout
    </div>

    @RenderSection("Body")

    <div class="bg-info">
        This is part of the layout
    </div>

    @if (IsSectionDefined("Footer")) {
        @RenderSection("Footer")
    } else {
        <h4>This is the default footer</h4>
    }

    @RenderSection("scripts")

    <div class="bg-info">
        This is part of the layout
    </div>
</body>
</html>
```

当启动应用程序并且 Razor 引擎尝试渲染布局和视图时，会显示图 21-6 所示的错误。

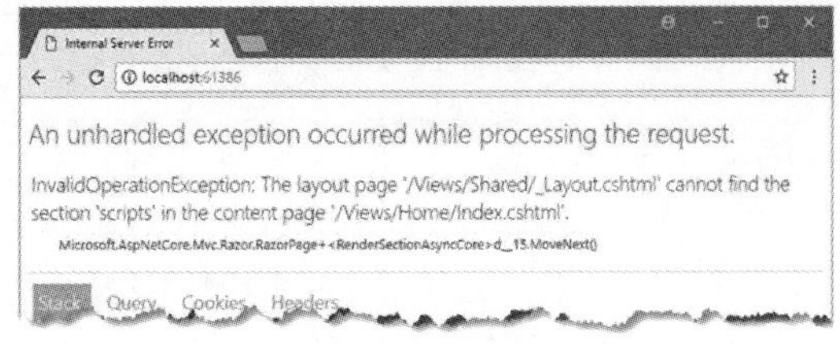

图 21-6 当缺少部分时显示错误

可以使用 IsSectionDefined 方法避免对视图未定义的部分进行 RenderSection 调用，但更好的方法是使用可选部分，也就是向 RenderSection 方法传递一个额外的虚假参数，如代码清单 21-18 所示。

代码清单 21-18 制作可选部分

```
...
@RenderSection("scripts", false)
...
```

这将创建一个可选部分，如果视图定义了该部分，就把内容插入结果中；否则，不会引发异常。

21.4.2 使用分部视图

通常需要在应用程序的多个不同位置使用相同的 Razor 标签和 HTML 标记。可以使用分部视图，而并非总是将内容复制一份，分部视图是单独的视图文件，其中包含可包含在其他视图中的标记。本节将展示如何创建和使用分部视图，解释它的工作原理，并演示将视图数据传递到分部视图的方法。

1. 创建分部视图

分部视图只是常规的 CSHTML 文件，你需要将它们与常规 Razor 视图区分开来。Visual Studio 为创建预制的分部视图提供了一些支持工具，但创建分部视图的最简单方法是使用 MVC View Page 模板创建常规视图。为了演示，在 Views/Home 文件夹中添加一个名为 MyPartial.cshtml 的文件，并添加代码清单 21-19 所示的内容。

代码清单 21-19 Views/Home 文件夹下的 MyPartial.cshtml 文件的内容

```
<div class="bg-info">
    <div>This is the message from the partial view.</div>
    <a asp-action="Index">This is a link to the Index action</a>
</div>
```

为了演示如何在分部视图中混合使用静态和动态内容，这里定义了一条简单的消息，并添加了一个使用标签助手生成的锚点元素。

2. 应用分部视图

可通过从另一个视图中调用@ Html.Partial 表达式来使用分部视图。为了演示，在 Views/Home 文件夹中添加一个名为 List.cshtml 的新文件，并添加代码清单 21-20 所示的内容。

代码清单 21-20 Views/Home 文件夹下的 List.cshtml 文件的内容

```
@{ Layout = null; }

<!DOCTYPE html>
<html>
<head>
    <meta name="viewport" content="width=device-width" />
    <title>Razor</title>
    <link asp-href-include="lib/bootstrap/dist/css/*.min.css" rel="stylesheet" />
</head>
<body class="m-1 p-1">
    This is the List View
    @Html.Partial("MyPartial")
</body>
</html>
```

Partial 方法是一种扩展方法，被应用于添加到 Razor 从视图文件生成的类的 Html 属性。这是一个 HTML 助手，在 MVC 早期版本中曾是在视图中生成动态内容的方式，但大部分已被标签助手替代。传递给 Partial 方法的参数是分部视图的名称，分部视图的内容被插入发送到客户端的输出。

提 示

Razor 查找分部视图与查看常规视图的方式相同（搜索 Views/<controller>和 Views/Shared 文件夹），这意味着可以创建特定于特定控制器的专用分部视图，并覆盖 Shared 文件夹中同名的分部视图。

可以通过启动应用程序并导航到 URL /Home/List 来查看分部视图的使用效果，如图 21-7 所示。

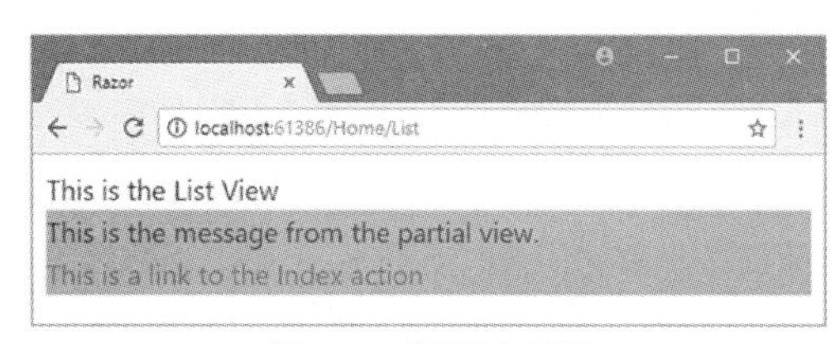

图 21-7　使用分部视图

3. 使用强类型的分部视图

可以创建强类型的分部视图，并提供在呈现分部视图时使用的视图模型对象。为了演示此功能，可在 Views/Home 文件夹中创建一个新的名为 MyStronglyTypedPartial.cshtml 的视图文件并添加代码清单 21-21 所示的内容。

代码清单 21-21　Views/Home 文件夹下的 MyStronglyTypedPartial.cshtml 文件的内容

```
@model IEnumerable<string>

<div class="bg-info">
    This is the message from the partial view.
    <ul>
        @foreach (string str in Model) {
            <li>@str</li>
        }
    </ul>
</div>
```

视图模型类型使用标准@model 表达式来定义，这里使用@foreach 循环将视图模型对象的内容显示为 HTML 代码清单中的条目。为了演示如何使用这个分部视图，更新/Views/Common/List.cshtml 文件，如代码清单 21-22 所示。

代码清单 21-22　Views/Common 文件夹下的 List.cshtml 文件的内容

```
@{ Layout = null; }

<!DOCTYPE html>
<html>
<head>
    <meta name="viewport" content="width=device-width" />
    <title>Razor</title>
    <link asp-href-include="lib/bootstrap/dist/css/*.min.css" rel="stylesheet" />
</head>
<body class="m-1 p-1">
    This is the List View
    @Html.Partial("MyStronglyTypedPartial",
        new string[] { "Apple", "Orange", "Pear" })
</body>
</html>
```

与上一个示例的区别在于，这里将一个额外的参数传递给提供视图模型的 Partial 辅助方法。可以通过启动应用程序和导航到 URL——/Home/List 来查看使用的强类型的分部视图，如图 21-8 所示。

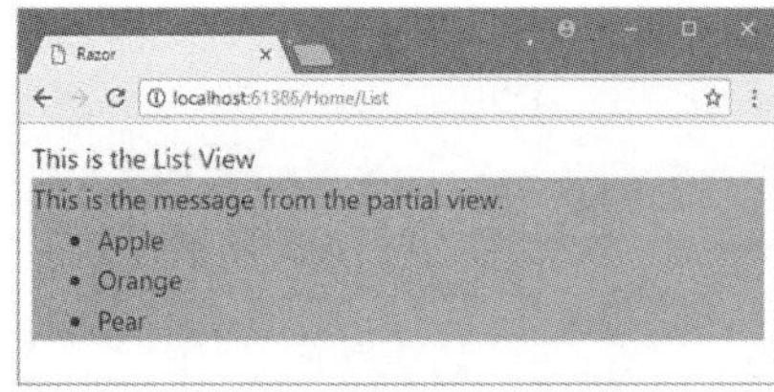

图 21-8　使用强类型的分部视图

21.4.3　将 JSON 内容添加到视图中

视图中经常包含 JSON，以便为客户端 JavaScript 代码提供可用于动态生成内容的数据。为了准备这个例子，可通过编辑 bower.json 文件将 jQuery 包添加到应用程序中，如代码清单 21-23 所示，这将使浏览器更容易处理作为 HTML 文档的一部分接收的 JSON 数据。

代码清单 21-23　在 bower.json 文件中添加 jQuery

```
{
  "name": "asp.net",
  "private": true,
```

```
  "dependencies": {
    "bootstrap": "4.0.0-alpha.6",
    "jquery": "3.2.1"
  }
}
```

代码清单 21-24 显示了 List.cshtml 视图文件的新增功能，该视图使用 Razor 在发送到浏览器的响应中包含 JSON 数据。

代码清单 21-24 在 Views/Common 文件夹下的 List.cshtml 文件中使用 JSON 数据

```
@{ Layout = null; }

<!DOCTYPE html>
<html>
<head>
    <meta name="viewport" content="width=device-width" />
    <title>Razor</title>
    <link asp-href-include="lib/bootstrap/dist/css/*.min.css" rel="stylesheet" />
    <script id="jsonData" type="application/json">
        @Json.Serialize(new string[] { "Apple", "Orange", "Pear" })
    </script>
</head>
<body class="m-1 p-1">
    This is the List View
    <ul id="list"></ul>
</body>
</html>
```

@Json.Serialize 表达式接收一个对象并将其序列化为 JSON 格式。以上代码已经在包含 JSON 数据的视图中添加了一个 script 元素。当视图被呈现并发送到浏览器时，其中包含如下元素：

```
...
<script id="jsonData" type="application/json">["Apple","Orange","Pear"]</script>
...
```

为了利用 JSON 数据，代码清单 21-25 添加了 jQuery 库和一些 JavaScript 代码，这些代码使用 jQuery 来解析 JSON 数据并动态创建一些 HTML 元素。

代码清单 21-25 在 Views/Common 文件夹下的 List.cshtml 文件中使用 JSON 数据

```
@{ Layout = null; }
<!DOCTYPE html>
<html>
<head>
    <meta name="viewport" content="width=device-width" />
    <title>Razor</title>
    <link asp-href-include="lib/bootstrap/dist/css/*.min.css" rel="stylesheet" />
    <script id="jsonData" type="application/json">
        @Json.Serialize(new string[] { "Apple", "Orange", "Pear" })
    </script>
    <script asp-src-include="lib/jquery/dist/*.min.js"></script>
    <script type="text/javascript">
        $(document).ready(function () {
            var list = $("#list")
            JSON.parse($("#jsonData").text()).forEach(function (val) {
                console.log("Val: " + val);
                list.append($("<li>").text(val));
            });
        });
    </script>
</head>
<body class="m-1 p-1">
```

```
    This is the List View
    <ul id="list"></ul>
</body>
</html>
```

如果运行示例应用程序并请求 URL /Home/List，就会显示图 21-9 所示的内容。这不是 JSON 数据最重要的使用方法，但它显示了如何将 JSON 数据包含在视图中。

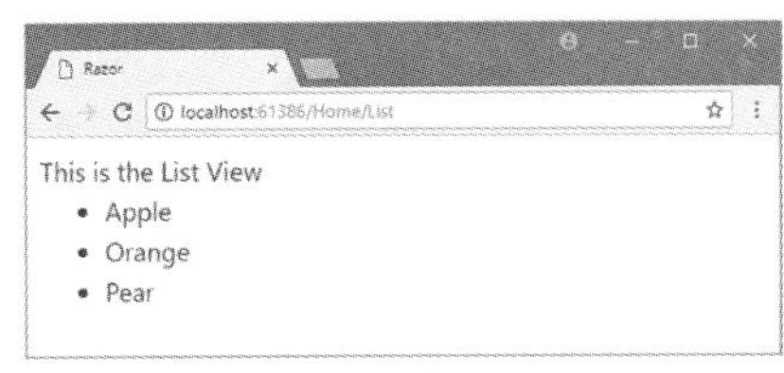

图 21-9 在视图中使用 JSON 数据

21.5 配置 Razor

可以使用 Microsoft.AspNetCore.Mvc.Razor 命名空间中定义的 RazorViewEngineOptions 类来配置 Razor。该类定义了两个配置属性，如表 21-10 所示。

表 21-10 RazorViewEngineOptions 类定义的配置属性

配置属性	描述
FileProvider	用于设置为文件和目录提供 Razor 的对象，由 Microsoft.AspNetCore.FileProviders.IFileProvider 接口定义，默认由从磁盘读取文件的 PhysicalFileProvider 实现
ViewLocationExpanders	用于配置视图扩展，用于更改 Razor 查找视图的方式

提 示

如果想深入学习，那么可以通过创建实现了 Microsoft.AspNetCore.Mvc.Razor 命名空间中接口的类来替换内部 Razor 组件，并注册到 Startup 类中的 service provider。但大多数开发人员并不需要这么做，但如果希望完全控制应用程序中内容的处理方式，那么这是一个很有用的方式。可以从 GitHub 下载 Razor 的源代码作为开始。

许多应用程序不需要更改 FileProvider 属性，因为从磁盘读取视图文件正是大多数项目所需要的，而 Razor 仅使用 service provider 来加载视图，以便在应用程序首次运行时进行编译。ViewLocationExpanders 属性更有用，因为它允许应用程序将自定义逻辑应用于 Razor 查看视图的方式。

视图位置扩展

Razor 使用视图位置扩展来构建应该搜索视图位置的清单。视图位置扩展实现了 IViewLocationExpander 接口，该接口的定义如下：

```
using System.Collections.Generic;

namespace Microsoft.AspNetCore.Mvc.Razor {

    public interface IViewLocationExpander {

        void PopulateValues(ViewLocationExpanderContext context);

        IEnumerable<string> ExpandViewLocations(ViewLocationExpanderContext context,
            IEnumerable<string> viewLocations);
    }
}
```

接下来解释视图位置扩展的工作原理，并创建 IViewLocationExpander 接口的自定义实现。为了准备创建视图位置扩展，代码清单 21-26 已经更改了 Home 控制器的 Index 操作方法，以便请求不存在的视图。视图不存在的错误消息将显示 Razor 搜索视图的位置以及视图位置扩展带来的影响。

代码清单 21-26 在 Controllers 文件夹下的 HomeController.cs 文件中请求不存在的视图

```
using System;
using Microsoft.AspNetCore.Mvc;
```

```
namespace Views.Controllers {

    public class HomeController : Controller {

        public ViewResult Index() =>
            View("MyView", new string[] { "Apple", "Orange", "Pear" });

        public ViewResult List() => View();
    }
}
```

如果启动应用程序并请求默认 URL，你将看到错误消息中显示的默认视图搜索位置，如下所示：

```
/Views/Home/MyView.cshtml
/Views/Shared/MyView.cshtml
```

1. 创建简单的视图位置扩展

最简单的视图位置扩展只需要更改 Razor 查找视图的位置集合。这可以通过实现 ExpandViewLocations 方法并返回要支持的位置清单来完成。为了演示，在 Infrastructure 文件夹中添加一个名为 SimpleExpander.cs 的类文件，内容如代码清单 21-27 所示。

代码清单 21-27 Infrastructure 文件夹下的 SimpleExpander.cs 文件的内容

```
using System.Collections.Generic;
using Microsoft.AspNetCore.Mvc.Razor;

namespace Views.Infrastructure {

    public class SimpleExpander : IViewLocationExpander {

        public void PopulateValues(ViewLocationExpanderContext context) {
            // do nothing - not required
        }

        public IEnumerable<string> ExpandViewLocations(
                ViewLocationExpanderContext context,
                IEnumerable<string> viewLocations) {

            foreach (string location in viewLocations) {
                yield return location.Replace("Shared", "Common");
            }
            yield return "/Views/Legacy/{1}/{0}/View.cshtml";
        }
    }
}
```

当需要搜索位置清单时，Razor 会调用 ExpandViewLocations 方法，并且使用 viewLocations 参数中的一系列字符串提供默认位置。位置表示为带有占位符的模板，占位符用于指代操作方法和控制器的名称。以下是在不使用路由区域的应用程序中默认使用的位置模板：

```
"/Views/{1}/{0}.cshtml"
"/Views/Shared/{0}.cshtml"
```

占位符{0}用于指代操作方法的名称，占位符用于指代{1}控制器的名称。视图位置扩展的作用是返回应被搜索的位置集合，可使用 string.Replace 方法在默认位置中更改 Shared 为 Common，并添加自己的位置，位置将遵循不同的文件和文件夹结构。

代码清单 21-28 通过在 Startup 类中配置 Razor 来设置视图位置扩展。ViewLocationExpanders 属性会返回一个在其中调用 Add 方法的 List <IViewLocationExpander>对象。

代码清单 21-28　在 Startup.cs 文件中配置 Razor

```
using System;
using System.Collections.Generic;
using System.Linq;
using System.Threading.Tasks;
using Microsoft.AspNetCore.Builder;
using Microsoft.AspNetCore.Hosting;
using Microsoft.AspNetCore.Http;
using Microsoft.Extensions.DependencyInjection;
using Microsoft.AspNetCore.Mvc;
using Views.Infrastructure;
using Microsoft.AspNetCore.Mvc.Razor;

namespace Views {
    public class Startup {
        public void ConfigureServices(IServiceCollection services) {
            services.AddMvc();
            services.Configure<RazorViewEngineOptions>(options => {
                options.ViewLocationExpanders.Add(new SimpleExpander());
            });
        }

        public void Configure(IApplicationBuilder app, IHostingEnvironment env) {
            app.UseStatusCodePages();
            app.UseDeveloperExceptionPage();
            app.UseStaticFiles();
            app.UseMvcWithDefaultRoute();
        }
    }
}
```

如果运行示例，错误消息将显示自定义的视图位置扩展提供给 Razor 的位置集合。

```
/Views/Home/MyView.cshtml
/Views/Common/MyView.cshtml
/Views/Legacy/Home/MyView/View.cshtml
```

2. 为请求选择特定视图

视图位置扩展可以轻松更改所有请求的检索位置，也可以更改单个请求的搜索位置。前面的例子仅实现了 ExpandViewLocations 方法，但实际作用来自 PopulateValues 方法，这是 IViewLocationExpander 接口中的另一个方法。

Razor 每次需要一个视图时，就会调用视图位置扩展的 PopulateValues 方法，为上下文数据提供一个 ViewLocationExpanderContext 对象。表 21-11 显示了 ViewLocationExpanderContext 类定义的属性。

表 21-11　ViewLocationExpanderContext 类定义的属性

属性	描述
ActionContext	返回一个 ActionContext 对象，用于描述已请求视图的操作方法，并包含有关请求和响应的详细信息
ViewName	返回操作方法请求的视图的名称
ControllerName	返回包含操作方法的控制器的名称
AreaName	如果已定义区域，就返回包含控制器的区域的名称
IsMainPage	如果 Razor 正在寻找分部视图，就返回 false；否则，返回 true
Values	返回一个 IDictionary<string,string>对象，视图位置扩展会添加唯一标识请求类别的键/值对，具体将在下面的内容中介绍

PopulateValues 方法的目的是通过为 Context 对象的 Values 属性返回的字典添加键/值对来对请求分类。Razor 不管请求如何分类，用于填充字典的方法都完全留给视图位置扩展，这很容易通过一个例子来解释。将一个名为 ColorExpander.cs 的类文件添加到 Infrastructure 文件夹中，并用它定义代码清单 21-29 所示的类。

代码清单 21-29　Infrastructure 文件夹下的 ColorExpander.cs 文件的内容

```
using System.Collections.Generic;
using Microsoft.AspNetCore.Mvc.Razor;

namespace Views.Infrastructure {

    public class ColorExpander : IViewLocationExpander {
        private static Dictionary<string, string> Colors
            = new Dictionary<string, string> {
                ["red"] = "Red", ["green"] = "Green", ["blue"] = "Blue"
            };

        public void PopulateValues(ViewLocationExpanderContext context) {

            var routeValues = context.ActionContext.RouteData.Values;
            string color;

            if (routeValues.ContainsKey("id")
                    && Colors.TryGetValue(routeValues["id"] as string, out color)
                    && !string.IsNullOrEmpty(color)) {
                context.Values["color"] = color;
            }
        }

        public IEnumerable<string> ExpandViewLocations(
                ViewLocationExpanderContext context,
                IEnumerable<string> viewLocations) {
            string color;
            context.Values.TryGetValue("color", out color);
            foreach (string location in viewLocations) {
                if (!string.IsNullOrEmpty(color)) {
                    yield return location.Replace("{0}", color);
                } else {
                    yield return location;
                }
            }
        }
    }
}
```

PopulateValues 方法使用 ActionContext 获取路由数据，并查找 URL 参数片段 id 的值。如果有一个 id 片段的值为红色、绿色或蓝色，视图位置扩展就会向值字典添加 Color 属性。在分类的过程中，id 片段匹配颜色的请求可使用颜色键进行分类，颜色键的值由参数片段的值定义。

接下来，Razor 调用 ExpandViewLocations 方法并提供与 PopulateValues 方法相同的 Context 对象。该操作允许视图位置扩展查看先前执行的分类，并生成 Razor 应该查看的视图的位置集合。以上示例使用 string.Replace 方法来替换具有颜色名称的{0}占位符。

提　示

Razor 为每个视图请求调用 PopulateValues 方法，但缓存由 ExpandViewLocations 方法返回的搜索位置集合。这意味着 PopulateValues 方法生成的同一组分类键和值的后续请求不需要调用 ExpandViewLocations 方法。

代码清单 21-30 已将 Razor 配置为使用 ColorExpander 类。

代码清单 21-30　在 Views 文件夹下的 Startup.cs 文件中添加视图位置扩展

```
using System;
using System.Collections.Generic;
using System.Linq;
using System.Threading.Tasks;
```

```
using Microsoft.AspNetCore.Builder;
using Microsoft.AspNetCore.Hosting;
using Microsoft.AspNetCore.Http;
using Microsoft.Extensions.DependencyInjection;
using Microsoft.AspNetCore.Mvc;
using Views.Infrastructure;
using Microsoft.AspNetCore.Mvc.Razor;

namespace Views {
    public class Startup {
        public void ConfigureServices(IServiceCollection services) {
            services.AddMvc();
            services.Configure<RazorViewEngineOptions>(options => {
                options.ViewLocationExpanders.Add(new SimpleExpander());
                options.ViewLocationExpanders.Add(new ColorExpander());
            });
        }

        public void Configure(IApplicationBuilder app, IHostingEnvironment env) {
            app.UseStatusCodePages();
            app.UseDeveloperExceptionPage();
            app.UseStaticFiles();
            app.UseMvcWithDefaultRoute();
        }
    }
}
```

通过启动应用程序并请求 URL /Home/Index/red 可以查看新的视图位置扩展的效果，并使 Razor 同时在以下位置进行搜索：

```
/Views/Home/Red.cshtml
/Views/Common/Red.cshtml
/Views/Legacy/Home/Red/View.cshtml
```

类似地，对 URL /Home/Index/green 的请求会使 Razor 在以下位置进行搜索：

```
/Views/Home/Green.cshtml
/Views/Common/Green.cshtml
/Views/Legacy/Home/Green/View.cshtml
```

视图位置扩展的注册顺序很重要，因为一个视图位置扩展的 ExpandViewLocations 方法生成的位置集合，会被用作清单中下一个视图位置扩展的 viewLocations 参数。从之前显示的位置可以看到这一点，其中，Views/Common 和 Views/Legacy 位置由 SimpleExpander 类生成，该类出现在 Startup 类中的 ColorExpander 之前。

21.6 小结

本章演示了如何创建自定义视图引擎，并通过将 CSHTML 文件转换为 C#类来解释 Razor 的工作原理。本章还展示了如何使用布局部分（section）和分部视图，并讨论了如何更改 Razor 用于查找视图文件的位置。下一章将介绍用于提供支持分部视图的逻辑的视图组件。

第 22 章　视图组件

本章描述视图组件，这是 ASP.NET Core MVC 新增的功能，以替代以前版本的子 action 功能。视图组件是提供操作样式逻辑以支持分部视图的类，这意味着可以将复杂内容嵌入视图，同时可以方便维护 C# 代码和支持单元测试。表 22-1 介绍了视图组件的背景。

表 22-1　视图组件的背景

问题	答案
什么是视图组件?	视图组件是提供应用程序逻辑以支持分部视图或将小部分的 HTML 或 JSON 数据注入父视图的类
视图组件有什么作用?	没有视图组件，将难以创建易于维护和支持单元测试的功能，如购物车或登录面板
如何使用视图组件?	视图组件通常派生自 ViewComponent 类，并使用@await Component.InvokeAsync 表达式应用于父视图
在使用中有什么陷阱或局限性?	没有，视图组件能提供简单可靠的功能，如果不使用视图组件，视图中的应用程序逻辑将会难以维护和测试
是否有其他替代方法?	可以将数据访问和处理逻辑直接放在分部视图中，但是结果很难使用，难以有效测试

表 22-2 列出了本章要介绍的操作。

表 22-2　本章要介绍的操作

操作	方法	代码清单
使用自己的逻辑和数据提供分部视图	使用视图组件	代码清单 22-12
调用视图组件	在视图中使用@await Component.InvokeAsync 表达式	代码清单 22-13
简化对上下文数据和结果的访问	从 ViewComponent 类派生	代码清单 22-14～代码清单 22-16
选择分部视图	使用 View 方法创建并返回 ViewViewComponentResult 对象	代码清单 22-17～代码清单 22-19
创建 HTML 片段	如果不想要对 HTML 片段进行编码，那么返回 Content 方法创建的 ContentViewComponentResult 对象或者明确地创建 HtmlContentViewComponentResult 对象	代码清单 22-20 和代码清单 22-21
使用请求的详细信息生成结果	使用视图组件上下文数据	代码清单 22-22
在调用视图组件时提供上下文数据	提供 InvokeAsync 方法的参数	代码清单 22-23～代码清单 22-25
创建异步视图组件	实现 InvokeAsync 方法并返回 Tast 对象以得到想要的结果	代码清单 22-26～代码清单 22-29
创建混合的控制器/视图组件	将 ViewComponent 属性应用于控制器类	代码清单 22-30～代码清单 22-33

22.1　准备示例项目

在本章中，将使用 ASP.NET Core Web Application（.NET Core）模板创建一个名为 UsingViewComponents 的 Empty 项目。

22.1.1　创建模型和存储库

这里需要两个不同的数据源来演示视图组件的工作原理。为了使部分应用程序对一组 Product 对象进行操作，创建 Models 文件夹，并添加一个名为 Product.cs 的文件，用来定义代码清单 22-1 所示的类。

代码清单 22-1　Models 文件夹下的 Product.cs 文件的内容

```
namespace UsingViewComponents.Models {

    public class Product {
        public string Name { get; set; }
        public decimal Price { get; set; }
    }
}
```

要为 Product 对象创建存储库，可将一个名为 ProductRepository.cs 的文件添加到 Models 文件夹中，并定义代码清单 22-2 所示的接口和实现类。

代码清单 22-2　Models 文件夹下的 ProductRepository.cs 文件的内容

```
using System.Collections.Generic;

namespace UsingViewComponents.Models {

    public interface IProductRepository {
        IEnumerable<Product> Products { get; }
        void AddProduct(Product newProduct);
    }

    public class MemoryProductRepository : IProductRepository {
        private List<Product> products = new List<Product> {
                new Product { Name = "Kayak", Price = 275M },
                new Product { Name = "Lifejacket", Price = 48.95M },
                new Product { Name = "Soccer ball", Price = 19.50M }
        };

        public IEnumerable<Product> Products => products;

        public void AddProduct(Product newProduct) {
            products.Add(newProduct);
        }
    }
}
```

IProductRepository 接口定义了一组有限的存储库功能，MemoryProductRepository 类使用 List 实现了该接口。为了使应用程序的其他部分将对 City 进行操作，在 Models 文件夹中添加一个名为 City.cs 的类文件，并用它定义代码清单 22-3 所示的类。

代码清单 22-3　Models 文件夹下的 City.cs 文件的内容

```
namespace UsingViewComponents.Models {

    public class City {
        public string Name { get; set; }
        public string Country { get; set; }
        public int Population { get; set; }
    }
}
```

对于 City 对象的存储库，创建一个名为 CityRepository.cs 的类文件，并用它定义代码清单 22-4 所示的接口和实现类。

代码清单 22-4　Models 文件夹下的 CityRepository.cs 文件的内容

```
using System.Collections.Generic;

namespace UsingViewComponents.Models {
```

```
    public interface ICityRepository {
        IEnumerable<City> Cities { get; }

        void AddCity(City newCity);
    }

    public class MemoryCityRepository : ICityRepository {

        private List<City> cities = new List<City> {
            new City { Name = "London", Country = "UK", Population = 8539000},
            new City { Name = "New York", Country = "USA", Population = 8406000 },
            new City { Name = "San Jose", Country = "USA", Population = 998537 },
            new City { Name = "Paris", Country = "France", Population = 2244000 }
        };

        public IEnumerable<City> Cities => cities;

        public void AddCity(City newCity) {
            cities.Add(newCity);
        }
    }
}
```

ICityRepository 接口提供了一组有限的存储库功能，MemoryCityRepository 类使用 List 来实现接口。

22.1.2 创建控制器和视图

由于这里只需要一个控制器，因此创建 Controllers 文件夹，将一个名为 HomeController.cs 的文件添加到 Controllers 文件夹中，并用它定义代码清单 22-5 所示的类。

代码清单 22-5 Controllers 文件夹下的 HomeController.cs 文件的内容

```
using Microsoft.AspNetCore.Mvc;
using UsingViewComponents.Models;

namespace UsingViewComponents.Controllers {

    public class HomeController : Controller {
        private IProductRepository repository;

        public HomeController(IProductRepository repo) {
            repository = repo;
        }
        public ViewResult Index() => View(repository.Products);

        public ViewResult Create() => View();

        [HttpPost]
        public IActionResult Create(Product newProduct) {
            repository.AddProduct(newProduct);
            return RedirectToAction("Index");
        }
    }
}
```

Home 控制器使用构造函数来声明对 IProductRepository 接口的依赖关系，当控制器用于处理请求时，service provider 将解析该接口。Index 操作方法从存储库中检索所有 Product 对象，并使用默认视图进行渲染。两个 Create 方法使用的是 Post/Redirect/Get 方式，用客户端提供的表单数据生成新的对象并添加到存储库中。

该例中的视图将共享一种通用布局。创建 Views/Shared 文件夹，并添加一个名为_Layout.cshtml 的文件，内容如代码清单 22-6 所示。

代码清单 22-6 Views/Shared 文件夹下的_Layout.cshtml 文件的内容

```
<!DOCTYPE html>
<html>
<head>
    <meta name="viewport" content="width=device-width" />
    <title>@ViewBag.Title</title>
    <link asp-href-include="lib/bootstrap/dist/css/*.min.css" rel="stylesheet" />
</head>
<body class="m-1 p-1">
    <div class="bg-primary m-1 p-1">
        <div class="row text-white">
            <div class="col-7"><h1>Products</h1></div>
            <div class="col-5">
                <div class="bg-info text-center m-1 p-1">City Placeholder</div>
            </div>
        </div>
    </div>
    <div class="m-1 p-1">@RenderBody()</div>
</body>
</html>
```

以上布局定义了标题，其中包含一个占位符，用于显示本章后面使用 City 存储库创建的内容。创建 Views/Home 文件夹，并添加一个名为 Index.cshtml 的视图文件，内容如代码清单 22-7 所示，其中列出了表格中 Product 对象的详细信息。

代码清单 22-7 Views/Home 文件夹下的 Index.cshtml 文件的内容

```
@model IEnumerable<Product>
@{
    ViewData["Title"] = "Products";
    Layout = "_Layout";
}
<table class="table table-sm table-striped table-bordered">
    <thead>
        <tr><th>Name</th><th>Price</th></tr>
    </thead>
    <tbody>
        @foreach (var product in Model) {
            <tr>
                <td>@product.Name</td>
                <td>@product.Price</td>
            </tr>
        }
    </tbody>
</table>
<a asp-action="Create" class="btn btn-primary">Create</a>
```

Index 视图中的最后一个元素被定义为按钮，且链接指向 Create 操作方法，用来在存储库中新建一个 Product 对象。要提供用户填写的表单，可在 Views/Home 文件夹新建 Create.cshtml 文件，并添加代码清单 22-8 所示的内容。

代码清单 22-8 Views/Home 文件夹下的 Create.cshtml 文件的内容

```
@model Product
@{
    ViewData["Title"] = "Create Product";
    Layout = "_Layout";
}
```

```
<form method="post" asp-action="Create">
    <div class="form-group">
        <label asp-for="Name">Name:</label>
        <input class="form-control" asp-for="Name" />
    </div>
    <div class="form-group">
        <label asp-for="Price">Price:</label>
        <input class="form-control" asp-for="Price" />
    </div>
    <button type="submit" class="btn btn-primary">Create</button>
    <a class="btn btn-secondary" asp-action="Index">Cancel</a>
</form>
```

这些视图需要使用内置的标签助手，可通过在 Views 文件夹中创建_ViewImports.cshtml 文件并添如代码清单 22-9 所示的表达式来启用标签助手，这可以使 Models 文件夹中的类在没有命名空间的情况下也可用。

代码清单 22-9 Views 文件夹下的_ViewImports.cshtml 文件的内容

```
@using UsingViewComponents.Models
@addTagHelper *, Microsoft.AspNetCore.Mvc.TagHelpers
```

这些视图还依赖于 Bootstrap CSS 包来提供样式。这里使用 Bower Configuration File 模板在项目的根目录中创建 bower.json 文件，并将 Bootstrap 添加到 dependencies 部分，如代码清单 22-10 所示。

代码清单 22-10 在 UsingViewComponents 文件夹下的 bower.json 文件中添加 Bootstrap

```
{
  "name": "asp.net",
  "private": true,
  "dependencies": {
    "bootstrap": "4.0.0-alpha.6"
  }
}
```

22.1.3 配置应用程序

最后的准备步骤是配置应用程序，如代码清单 22-11 所示。除了配置 MVC 服务和中间件之外，还为两个数据存储库创建了单例服务。

代码清单 22-11 UsingViewComponents 文件夹下的 Startup.cs 文件的内容

```
using System;
using System.Collections.Generic;
using System.Linq;
using System.Threading.Tasks;
using Microsoft.AspNetCore.Builder;
using Microsoft.AspNetCore.Hosting;
using Microsoft.AspNetCore.Http;
using Microsoft.Extensions.DependencyInjection;
using UsingViewComponents.Models;

namespace UsingViewComponents {

    public class Startup {

        public void ConfigureServices(IServiceCollection services) {
            services.AddSingleton<IProductRepository, MemoryProductRepository>();
            services.AddSingleton<ICityRepository, MemoryCityRepository>();
            services.AddMvc();
        }
```

```
        public void Configure(IApplicationBuilder app, IHostingEnvironment env) {
            app.UseStatusCodePages();
            app.UseDeveloperExceptionPage();
            app.UseStaticFiles();
            app.UseMvcWithDefaultRoute();
        }
    }
}
```

如果运行应用程序，你将看到 Product 存储库中的 Product 对象列表。可以通过单击 Create 按钮、填写表单并提交到服务器来添加新的商品，添加完成后链接会重定向到商品列表，如图 22-1 所示。由于应用程序中的视图使用共享布局，因此整个过程中会显示城市数据的占位符。

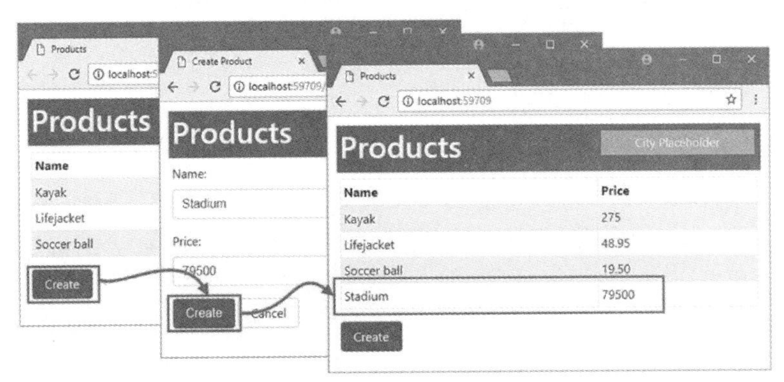

图 22-1 运行示例应用程序

22.2 视图组件

应用程序通常需要在视图中嵌入一些与视图主要功能无关的内容。常见的示例包括网站导航工具、以及让用户不用访问单独的页面就能登录的身份验证面板等。

所有这些示例的共同线索是显示嵌入式内容所需的数据不是从操作传递到视图的模型数据的一部分。因此，这里在示例应用程序中创建了两个存储库。使用 City 存储库生成的一些内容，将很难从 Product 存储库的用于接收数据的操作中获取。

第 21 章描述了如何使用分部视图来创建视图中需要的可重用标记，避免在应用程序的多个位置复制相同的内容。分部视图虽然有用，但它们只包含 HTML 和 Razor 指令的片段，并且操作的数据是从父视图接收的。如果需要显示不同的数据，就会遇到问题。可以直接从分部视图访问需要的数据，但这会破坏支持 MVC 模式的关注点分离原则，并导致将数据检索和处理逻辑放置在不能进行单元测试的视图文件中。也可以扩展应用程序使用的视图模型，以使其包含所需的数据，但这意味着必须更改每个操作方法，并使对操作方法的测试无法被隔离。

这正是需要使用视图组件的地方，视图组件是 C#类，提供了包含所需数据的分部视图，独立于父视图和操作。在这方面，视图组件负责专门的操作，但是仅用于提供数据的分部视图，不能接收 HTTP 请求，并且提供的内容将始终包含在父视图中。

22.3 创建视图组件

视图组件可以通过 3 种不同的方式创建，分别是通过定义 POCO 视图组件、通过从 ViewComponent 基类派生以及使用 ViewComponent 特性。本节介绍前两种方式，22.4 节将介绍第三种方式。

22.3.1 创建 POCO 视图组件

POCO 视图组件是一种不依赖于任何 MVC API 来提供视图组件功能的类。与 POCO 控制器一样，这种视图组件非常难以使用，但有助于了解它们的工作原理。POCO 视图组件是任何一种名称以 ViewComponent 结尾并定义了 Invoke 方法的类。视图组件类可以在应用程序中的任何位置进行定义，但是约定将它们组合在名为 Components 的文件夹中，该文件夹位于项目的根目录下。创建这个文件夹，并添加一个名为 PocoViewComponent.cs 的类文件，定义代码清单 22-12 所示的类。

代码清单 22-12 Components 文件夹下的 PocoViewComponent.cs 文件的内容

```
using System.Linq;
using UsingViewComponents.Models;

namespace UsingViewComponents.ViewComponents {

    public class PocoViewComponent {
        private ICityRepository repository;

        public PocoViewComponent(ICityRepository repo) {
            repository = repo;
        }

        public string Invoke() {
            return $"{repository.Cities.Count()} cities, "
                + $"{repository.Cities.Sum(c => c.Population)} people";
        }
    }
}
```

视图组件可以利用依赖注入来接收所需的服务。在此例中，POCO 视图组件声明了对 ICityRepository 接口的从属关系，然后在 Invoke 方法中使用该接口来创建描述城市数量和人口总数的字符串。

为了使用视图组件，需要使用 Razor @await Component.Invoke 表达式。可通过提供类的名称（不包含末尾的 ViewComponent）作为参数来选择视图组件。代码清单 22-13 已经删除了共享布局中的占位符，而使用 POCO 视图组件取代。

代码清单 22-13 在 Views/Shared 文件夹下的_Layout.cshtml 文件中使用视图组件

```
<!DOCTYPE html>
<html>
<head>
    <meta name="viewport" content="width=device-width" />
    <title>@ViewBag.Title</title>
    <link asp-href-include="lib/bootstrap/dist/css/*.min.css" rel="stylesheet" />
</head>
<body class="m-1 p-1">
    <div class="bg-primary m-1 p-1">
        <div class="row text-white">
            <div class="col-7"><h1>Products</h1></div>
            <div class="col-5">
                @await Component.InvokeAsync("Poco")
            </div>
        </div>
    </div>
    <div class="m-1 p-1">@RenderBody()</div>
</body>
</html>
```

为了应用视图组件，指定 Poco 作为 Invoke 方法的参数。当视图使用布局时，将定位 PocoViewComponent 类，调用 Invoke 方法，并将结果插入父视图的输出，如图 22-2 所示。

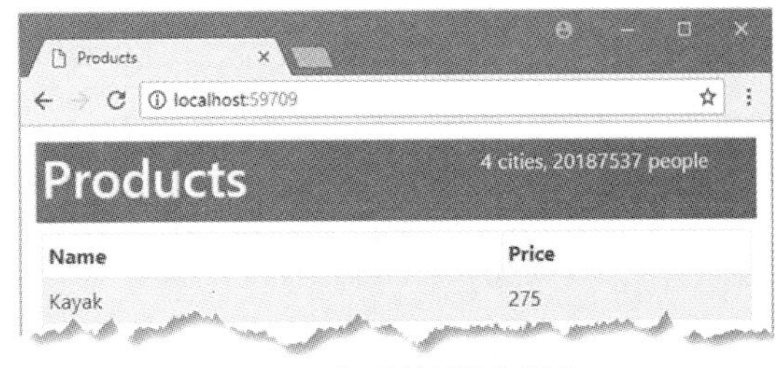

图 22-2　使用简单的视图组件

这只是一个简单的例子，但它说明了视图组件的一些重要特性。首先，PocoViewComponent 类能够访问所需的数据，而不依赖于处理 HTTP 请求或父视图的操作。其次，在 C#类中定义获取和处理城市信息所需的逻辑意味着可以轻松进行单元测试（参见 22.3.4 节）。最后，应用程序没有分裂，视图在关注 Product 对象的视图模型中包含 City 对象。

警　告

在视图中应用视图组件时，必须包含 await 关键字。如果只调用@Component.Invoke，虽然不会报错，但只会显示任务的字符串表示形式，类似于 System.Threading.Tasks.Task`1[Microsoft.AspNetCore.Html.IHtmlContent]。

22.3.2　从 ViewComponent 基类派生

POCO 视图组件的功能有限，除非它们可以利用 MVC API，这就需要执行一些更多的操作，需要从 ViewComponent 基类派生。Microsoft.AspNetCore.Mvc 命名空间中定义的 ViewComponent 类可以方便地访问上下文数据，从而更容易生成结果。代码清单 22-14 显示了添加到 Components 文件夹下的 CitySummary.cs 文件的内容。

代码清单 22-14　Components 文件夹下的 CitySummary.cs 文件的内容

```
using System.Linq;
using Microsoft.AspNetCore.Mvc;
using UsingViewComponents.Models;

namespace UsingViewComponents.Components {

    public class CitySummary : ViewComponent {
        private ICityRepository repository;

        public CitySummary(ICityRepository repo) {
            repository = repo;
        }

        public string Invoke() {
            return $"{repository.Cities.Count()} cities, "
            + $"{repository.Cities.Sum(c => c.Population)} people";
        }
    }
}
```

在从基类派生时，不需要在类名中包含 ViewComponent。除了使用基类之外，POCO 视图组件与 POCO 功能相同。下面将展示如何使用基类提供的便利功能来实现不同的视图组件功能。

提　示

注意，代码清单 22-14 没有重写 Invoke 方法。ViewComponent 类不提供 Invoke 方法的默认实现，因而必须明确定义。

为了演示视图组件的特性，更改共享布局中使用的组件，如代码清单 22-15 所示，使用 nameof 而不是直接用字符串指定视图组件的名称，第 4 章介绍过，这样可以减少类名错误的可能性。

代码清单 22-15 在 Views/Shared 文件夹下的_Layout.cshtml 文件中修改视图组件

```
<!DOCTYPE html>
<html>
<head>
    <meta name="viewport" content="width=device-width" />
    <title>@ViewBag.Title</title>
    <link asp-href-include="lib/bootstrap/dist/css/*.min.css" rel="stylesheet" />
</head>
<body class="m-1 p-1">
    <div class="bg-primary m-1 p-1">
        <div class="row text-white">
            <div class="col-7"><h1>Products</h1></div>
            <div class="col-5">
                @await Component.InvokeAsync(nameof(CitySummary))
            </div>
        </div>
    </div>
    <div class="m-1 p-1">@RenderBody()</div>
</body>
</html>
```

修改视图导入文件，从而可以在没有命名空间的 nameof 表达式中引用 CitySummary 类，如代码清单 22-16 所示。

代码清单 22-16 在 Views 文件夹下的_ViewImports.cshtml 文件中添加命名空间

```
@using UsingViewComponents.Models
@using UsingViewComponents.Components
@addTagHelper *, Microsoft.AspNetCore.Mvc.TagHelpers
```

22.3.3 视图组件结果

将简单的字符串值插入父视图的功能并不是特别有用，但幸运的是，视图组件能够做更多。通过使 Invoke 方法返回实现了 IViewComponentResult 接口的对象，可以实现更复杂的效果。3 个内置的类实现了 IViewComponentResult 接口，如表 22-3 所示，ViewComponent 基类提供了创建它们的便利方法。接下来介绍每种结果类型的用法。

注 意

如果使用 POCO 视图组件，就可以直接创建这些类的实例，尽管它们可能很难处理，因为它们都具有很复杂的构造函数。ViewComponent 类则提供了创建它们的简便方法。

表 22-3 内置的 IViewComponentResult 实现类

名称	描述
ViewViewComponentResult	用于指定 Razor 视图，并且具有可选的视图模型数据。该类的实例可使用 View 方法创建
ContentViewComponentResult	用于指定包含在 HTML 文档中且被安全编码的文本结果。该类的实例可使用 Content 方法创建
HtmlContentViewComponentResult	用于指定包含在 HTML 文档中的 HTML 片段，无须进一步编码。没有可用于创建这种类型结果的 ViewComponent 方法

有两种结果类型需要做特殊处理。如果视图组件返回一个字符串，那么它用于创建 ContentViewComponentResult 对象。如果视图组件返回 IHtmlContent 对象，那么它用于创建 HtmlContentViewComponentResult 对象。

1. 返回分部视图

最常使用的响应为 ViewViewComponentResult 对象，它告诉 Razor 呈现局部视图，并将结果包含在父视图中。ViewComponent 基类提供了用于创建 ViewViewComponentResult 对象的 View 方法，并且有 4 个版本的 View 方法可用，如表 22-4 所示。

表 22-4 ViewComponent.View 方法

名称	描述
View()	为视图组件选择默认视图，并且不提供视图模型
View(model)	选择默认视图并使用指定的对象作为视图模型
View(viewName)	可以选择指定的视图，而不提供视图模型
View(viewName,model)	选择指定的视图并使用指定的对象作为视图模型

这些 View 方法对应于由 Controller 基类提供的方法，并且以相同的方式使用。在 Models 文件夹中添加一个名为 CityViewModel.cs 的类文件，用来定义视图模型，如代码清单 22-17 所示。

代码清单 22-17 Models 文件夹下的 CityViewModel.cs 文件的内容

```
namespace UsingViewComponents.Models {

    public class CityViewModel {
        public int Cities { get; set; }
        public int Population { get; set; }
    }
}
```

代码清单 22-18 修改了 CitySummary 视图组件的 Invoke 方法，以便能够使用 View 方法来选择分部视图，并使用 CityViewModel 对象提供视图数据。

代码清单 22-18 在 Controller 文件夹下的 CitySummary.cs 文件中选择分部视图

```
using System.Linq;
using Microsoft.AspNetCore.Mvc;
using UsingViewComponents.Models;
namespace UsingViewComponents.Components {

    public class CitySummary : ViewComponent {
        private ICityRepository repository;

        public CitySummary(ICityRepository repo) {
            repository = repo;
        }

        public IViewComponentResult Invoke() {
            return View(new CityViewModel{
                Cities = repository.Cities.Count(),
                Population = repository.Cities.Sum(c => c.Population)
            });
        }
    }
}
```

在视图组件中选择分部视图与在控制器中选择视图类似，但有两个重要区别：Razor 在不同位置查找视图，并且如果未指定视图，就使用不同的默认视图名称查找视图。

因为没有为视图组件创建分部视图，所以在运行显示 Razor 所要查找的文件的应用程序时，你将会看到一条错误消息。

如果没有指定名称，那么 Razor 会尝试寻找名为 Default.cshtml 的文件。Razor 会在两个位置（/Views /Home/Components/CitySummary/Default.cshtml 与/Views/Shared/Components/CitySummary/Default.cshtml）

查找分部视图。第一个位置考虑了处理HTTP请求的控制器的名称，以允许每个控制器拥有自己的自定义视图；第二个位置可在所有控制器之间共享。

提 示

请注意，共享的分部视图仍然由视图组件区分，这意味着视图组件不能共享分部视图。可以通过在调用 View 方法时，在视图名称中包含路径来覆盖这种行为，比如调用 View ("Views/Shared/Components/Common/Default. html") 将覆盖分部视图的默认搜索位置。

为完成此例，创建Views/Home/Components/CitySummary文件夹，并添加一个名为Default.cshtml的视图文件，内容如代码清单22-19所示。

代码清单22-19 Views/Home/Components/CitySummary文件夹下的Default.cshtml文件的内容

```
@model CityViewModel

<table class="table table-sm table-bordered">
    <tr>
        <td>Cities:</td>
        <td class="text-right">
            @Model.Cities
        </td>
    </tr>
    <tr>
        <td>Population:</td>
        <td class="text-right">
            @Model.Population.ToString("#,###")
        </td>
    </tr>
</table>
```

视图组件的分部视图与控制器的工作方式相同。在这种情况下，创建一个强类型的视图，它可以包含一个CityViewModel对象，并在表格中显示Cities和Population值，如图22-3所示。

图22-3 使用视图组件渲染视图

2. 返回HTML片段

ContentViewComponentResult类可以在不使用视图的情况下，在父视图中包含HTML片段。ContentViewComponentResult类的实例可使用继承自ViewComponent基类的Content方法创建，该方法接收字符串值。代码清单22-20演示了如何使用Content方法。除了Content方法之外，Invoke方法可以返回一个字符串，MVC将自动把它转换为ContentViewComponentResult对象。

代码清单22-20 在Components文件夹下的CitySummary.cs文件中使用Content方法

```
using System.Linq;
using Microsoft.AspNetCore.Mvc;
using UsingViewComponents.Models;

namespace UsingViewComponents.Components {

    public class CitySummary : ViewComponent {
        private ICityRepository repository;
        public CitySummary(ICityRepository repo) {
            repository = repo;
        }

        public IViewComponentResult Invoke() {
```

```
            return Content("This is a <h3><i>string</i></h3>");
        }
    }
}
```

对 Content 方法接收到的字符串进行编码，使其安全地包含在 HTML 文档中。当处理由用户或外部系统提供的内容时，这特别重要，因为可以阻止 JavaScript 内容被嵌入应用程序生成的 HTML 中。在此例中，传递给 Content 方法的字符串包含一些基本的 HTML 标签，如果运行应用程序，你将看到它们已被安全地编码，如图 22-4 所示。

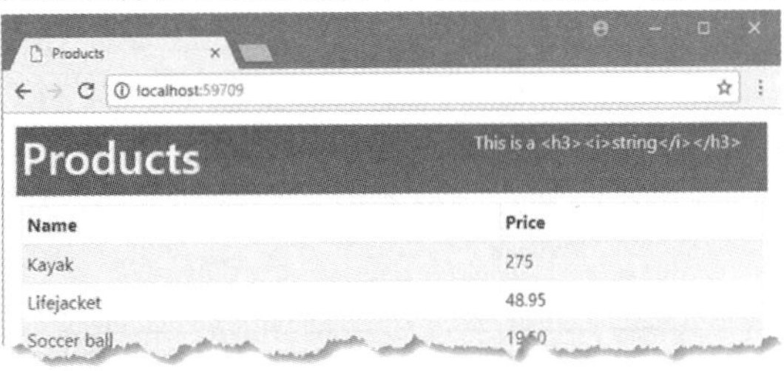

图 22-4 使用视图组件返回编码的 HTML 片段

如果查看视图组件生成的 HTML，你会发现尖括号已被替换，以使浏览器不将内容解释为 HTML 元素，如下所示：

```
...
<div class="col-5">This is a &lt;h3&gt;&lt;i&gt;string&lt;/i&gt;&lt;/h3&gt;</div>
...
```

如果信任它们的来源并希望解释为 HTML，就不需要对内容进行编码。Content 方法总是对参数进行编码，因此必须直接创建 HtmlContentViewComponentResult 对象，并为其构造函数提供一个 HtmlString 对象，该对象会显示你认为可以安全显示的字符串，因为来源值得信任，或者因为你确信它们已经编码，如代码清单 22-21 所示。

代码清单 22-21 在 Components 文件夹下的 CitySummary.cs 文件中返回信任的 HTML 片段

```
using System.Linq;
using Microsoft.AspNetCore.Mvc;
using UsingViewComponents.Models;
using Microsoft.AspNetCore.Mvc.ViewComponents;
using Microsoft.AspNetCore.Html;

namespace UsingViewComponents.Components {

    public class CitySummary : ViewComponent {
        private ICityRepository repository;

        public CitySummary(ICityRepository repo) {
            repository = repo;
        }

        public IViewComponentResult Invoke() {
            return new HtmlContentViewComponentResult(
                new HtmlString("This is a <h3><i>string</i></h3>"));
        }
    }
}
```

以上方法应慎用，只有当使用不能篡改的内容来源，并执行自己的编码时才可使用。如果运行应用程序，你将看到尖括号已包含在父视图中而未修改，这允许浏览器将视图组件的输出解释为 HTML 元素，如图 22-5 所示。

图 22-5 使用视图组件返回未编码的 HTML 片段

22.3.4 获取上下文数据

有关当前请求和父视图的详细信息会通过 ViewComponentContext 类的属性提供给视图组件，表 22-5 描述了该类提供的一些有用的属性。

表 22-5 ViewComponentContext 类定义的属性

属性	描述
Arguments	返回视图提供的参数字典，也可以通过 Invoke 方法接收
HtmlEncoder	返回一个可用于安全地编码 HTML 片段的 HtmlEncoder 对象
ViewComponentDescriptor	返回一个 ViewComponentDescriptor 对象，以提供视图组件的描述
ViewContext	从父视图返回 ViewContext 对象
ViewData	返回一个 ViewDataDictionary 对象，以提供对视图组件的视图数据的访问

ViewComponent 基类提供了一组便于访问特定上下文信息的属性，如表 22-6 所示。

表 22-6 ViewComponent 基类定义的属性

属性	描述
ViewComponentContext	返回一个 ViewComponentContext 对象
HttpContext	返回一个描述当前请求和正在准备的响应的 HttpContext 对象
Request	返回一个描述当前 HTTP 请求的 HttpRequest 对象
User	返回一个描述当前用户的 IPrincipal 对象
RouteData	返回一个描述当前请求的路由数据的 RouteData 对象
ViewBag	返回动态的 ViewBag 对象，可用于在视图组件和视图之间传递数据
ModelState	返回一个 ModelStateDictionary 对象，以提供模型绑定过程的详细信息
ViewContext	返回一个提供给父视图的 ViewContext 对象
ViewData	返回一个 ViewDataDictionary 对象，以提供对视图组件的视图数据的访问
Url	返回一个可用于生成 URL 的 IUrlHelper 对象

上下文数据可以各种方式用于帮助视图组件执行工作，包括改变数据的选择方式以及呈现不同的内容或视图。代码清单 22-22 使用路由数据来缩小 City 对象的选择范围。

代码清单 22-22 在 Components 文件夹下的 CitySummary.cs 文件中使用上下文数据

```
using System.Linq;
using Microsoft.AspNetCore.Mvc;
using UsingViewComponents.Models;
using Microsoft.AspNetCore.Mvc.ViewComponents;
using Microsoft.AspNetCore.Mvc.Rendering;

namespace UsingViewComponents.Components {

    public class CitySummary : ViewComponent {
        private ICityRepository repository;

        public CitySummary(ICityRepository repo) {
            repository = repo;
        }

        public IViewComponentResult Invoke() {
            string target = RouteData.Values["id"] as string;
            var cities = repository.Cities
                .Where(city => target == null ||
                    string.Compare(city.Country, target, true) == 0);
            return View(new CityViewModel{
                Cities = cities.Count(),
                Population = cities.Sum(c => c.Population)
            });
        }
    }
}
```

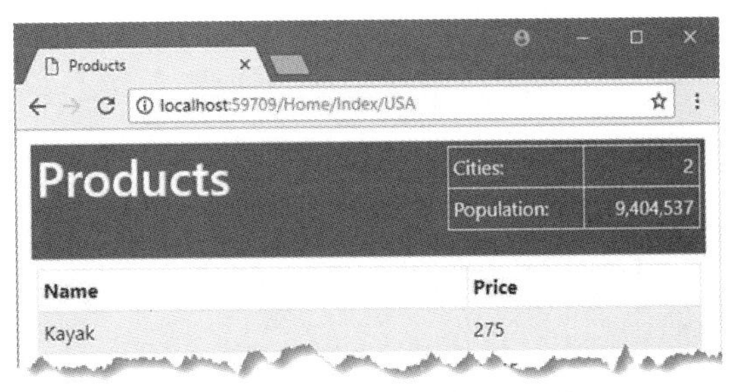

图 22-6　在视图组件中使用上下文数据

浏览器使用路由中的 id 片段来指定国家数据，使用 LINQ 来过滤存储库中的对象。如果启动应用程序并请求默认 URL，就会显示所有城市。可以通过请求 URL（比如/Home/Index/USA）来缩小选择范围，这样可以将选择范围缩小到美国的城市，如图 22-6 所示。

使用参数从父视图提供上下文

父视图可以提供附加的上下文数据作为@await Component.Invoke 表达式的参数。此功能可用于从父视图模型提供数据，或指导视图组件生成有关类型的内容。为了演示此功能，在 Views/Home/Component/CitySummary 文件夹中创建一个名为 CityList.cshtml 的视图文件，内容如代码清单 22-23 所示。

代码清单 22-23　Views/Home/Component/CitySummary 文件夹下的 CityList.cshtml 文件的内容

```
@model IEnumerable<City>

<table class="table table-sm table-bordered">
    @foreach (var city in Model) {
        <tr>
            <td>@city.Name</td>
            <td class="text-right">
                @city.Population.ToString("#,###")
            </td>
        </tr>
    }
    <tr>
        <th>Total:</th>
        <td class="text-right">
            @Model.Sum(p => p.Population).ToString("#,###")
        </td>
    </tr>
</table>
```

添加第二个视图以允许在视图之间选择视图组件，根据添加到 Invoke 方法的参数完成该操作，如代码清单 22-24 所示。

代码清单 22-24　在 Components 文件夹下的 CitySummary.cs 文件中选择视图

```
using System.Linq;
using Microsoft.AspNetCore.Mvc;
using UsingViewComponents.Models;
using Microsoft.AspNetCore.Mvc.ViewComponents;
using Microsoft.AspNetCore.Mvc.Rendering;

namespace UsingViewComponents.Components {

    public class CitySummary : ViewComponent {
        private ICityRepository repository;

        public CitySummary(ICityRepository repo) {
            repository = repo;
        }
        public IViewComponentResult Invoke(bool showList) {
            if (showList) {
                return View("CityList", repository.Cities);
            } else {
                return View(new CityViewModel {
                    Cities = repository.Cities.Count(),
                    Population = repository.Cities.Sum(c => c.Population)
                });
```

```
            }
        }
    }
}
```

如果 Invoke 方法的 showList 参数为 true，视图组件就选择 CityList，并将存储库中的所有 City 对象作为视图模型传递。如果 showList 参数为 false，就选择默认视图，并为视图模型提供一个 CitySummary 对象。

最后一步是在父视图中应用视图组件时提供上下文数据，方法是将匿名对象传递给 Invoke 方法，如代码清单 22-25 所示。

代码清单 22-25 在 Views/Shared 文件夹下的_Layout.cshtml 文件中提供上下文数据

```
<!DOCTYPE html>
<html>
<head>
    <meta name="viewport" content="width=device-width" />
    <title>@ViewBag.Title</title>
    <link asp-href-include="lib/bootstrap/dist/css/*.min.css" rel="stylesheet" />
</head>
<body class="m-1 p-1">
    <div class="bg-primary m-1 p-1">
        <div class="row text-white">
            <div class="col-7"><h1>Products</h1></div>
            <div class="col-5">
                @await Component.InvokeAsync("CitySummary", new { showList = true })
            </div>
        </div>
    </div>
    <div class="m-1 p-1">@RenderBody()</div>
</body>
</html>
```

如果运行应用程序，视图组件将接收父视图指定的值，并相应地进行响应，结果如图 22-7 所示。

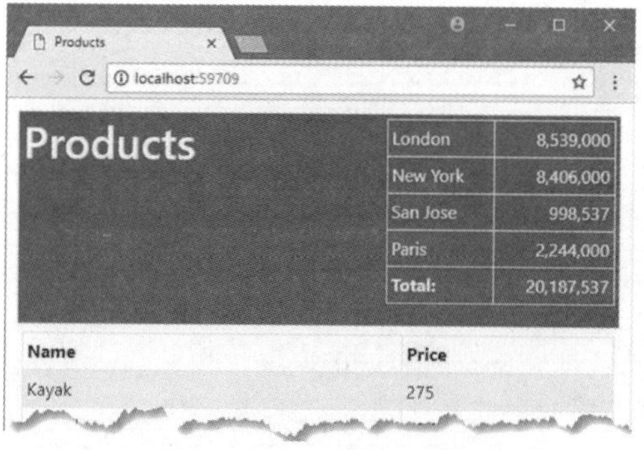

图 22-7 向视图组件提供上下文数据

测试视图组件

视图组件遵循一般的 MVC 方法，从格式化和呈现模型数据的视图标记中分离出选择模型数据并进行处理的逻辑，从而可以轻松地执行单元测试。以下是示例应用程序中用于 CitySummary 的单元测试：

```
using System.Collections.Generic;
using Microsoft.AspNetCore.Mvc.ViewComponents;
using Moq;
using UsingViewComponents.Models;
using UsingViewComponents.Components;
using Xunit;

namespace UsingViewComponents.Tests {
```

```
    public class SummaryViewComponentTests {

        [Fact]
        public void TestSummary() {

            // Arrange
            var mockRepository = new Mock<ICityRepository>();
            mockRepository.SetupGet(m => m.Cities).Returns(new List<City> {
                new City { Population = 100 },
                new City { Population = 20000 },
                new City { Population = 1000000 },
                new City { Population = 500000 }
            });
            var viewComponent
                = new CitySummary(mockRepository.Object);

            // Act
            ViewViewComponentResult result
                = viewComponent.Invoke(false) as ViewViewComponentResult;

            // Assert
            Assert.IsType(typeof(CityViewModel), result.ViewData.Model);
            Assert.Equal(4, ((CityViewModel)result.ViewData.Model).Cities);
            Assert.Equal(1520100,
                ((CityViewModel)result.ViewData.Model).Population);
        }
    }
}
```

为了安排测试，创建假的存储库并传递给 CitySummary 类的构造函数，以创建视图组件的新实例。对于测试的 act 部分，调用 Invoke 方法以提供结果对象。视图组件选择了 Razor 视图，因此将结果转换为 ViewViewComponentResult，并通过提供的 ViewData.Model 属性访问视图模型对象。对于测试的 Assert 部分，检查视图模型数据的类型及其包含的值。

22.3.5 创建异步视图组件

本章迄今为止的所有示例使用的都是同步视图组件，可以识别它们，因为它们定义了 Invoke 方法。如果视图组件依赖于异步 API，则可以通过定义返回任务的 InvokeAsync 方法来创建异步视图组件。当 Razor 从 InvokeAsync 方法接收到任务时，将等待任务完成，然后将结果插入主视图。为了准备此例，在 Solution Explorer 窗格中右击 UsingViewComponents 项目，在弹出的菜单中选择 Edit UsingViewComponents.csproj，并进行代码清单 22-26 所示的更改，以向项目添加包。

代码清单 22-26 在 UsingViewComponents.csproj 文件中添加包

```
<Project Sdk="Microsoft.NET.Sdk.Web">

  <PropertyGroup>
    <TargetFramework>netcoreapp2.0</TargetFramework>
  </PropertyGroup>

  <ItemGroup>
    <Folder Include="wwwroot\" />
  </ItemGroup>

  <ItemGroup>
    <PackageReference Include="Microsoft.AspNetCore.All" Version="2.0.0 " />
    <PackageReference Include="System.Net.Http" Version="4.3.2" />
  </ItemGroup>

</Project>
```

System.Net.Http 包提供了一个用于进行异步 HTTP 请求的 API，这里将用它查询 Apress 网站。代码清单 22-27 显示了名为 PageSize.cs 的类文件的内容，将这个类文件添加到 Components 文件夹中，用于创建异步视图组件。

代码清单 22-27 Components 文件夹下的 PageSize.cs 文件的内容

```
using System.Net.Http;
using System.Threading.Tasks;
using Microsoft.AspNetCore.Mvc;

namespace UsingViewComponents.Components {

    public class PageSize : ViewComponent {

        public async Task<IViewComponentResult> InvokeAsync() {
            HttpClient client = new HttpClient();
            HttpResponseMessage response
                = await client.GetAsync("http://apress.com");
            return View(response.Content.Headers.ContentLength);
        }
    }
}
```

InvokeAsync 方法通过第 4 章描述的 async 和 await 关键字来使用 HttpClient 类提供的异步 API，并通过向 Apress.com 发送 GET 请求来获取所返回内容的长度。将长度传递给 View 方法，选择与视图组件关联的默认分部视图。

为了创建视图，在项目中添加 Views/Shared/Components/PageSize 文件夹，并在其中添加一个名为 Default.cshtml 的视图文件，内容如代码清单 22-28 所示。

代码清单 22-28 Views/Shared/Components/PageSize 文件夹下的 Default.cshtml 文件的内容

```
@model long
<div class="m-1 p-1 bg-info text-white">Page size: @Model</div>
```

最后一步是在_Layout.cshtml 文件中使用异步视图组件，如代码清单 22-29 所示。

代码清单 22-29 在 Views/Shared 文件夹下的_Layout.cshtml 文件中使用异步视图组件

```
<!DOCTYPE html>
<html>
<head>
    <meta name="viewport" content="width=device-width" />
    <title>@ViewBag.Title</title>
    <link asp-href-include="lib/bootstrap/dist/css/*.min.css" rel="stylesheet" />
</head>
<body class="m-1 p-1">
    <div class="bg-primary m-1 p-1">
        <div class="row text-white">
            <div class="col-7"><h1>Products</h1></div>
            <div class="col-5">
                @await Component.InvokeAsync("CitySummary",
                    new { showList = true })
            </div>
        </div>
    </div>
    <div class="m-1 p-1">@RenderBody()</div>
    @await Component.InvokeAsync("PageSize")
</body>
</html>
```

如果启动应用程序，你将在浏览器中看到新增的内容，如图 22-8 所示。当运行示例时，显示的数字可能会更改，因为 Apress 经常更新其网站。

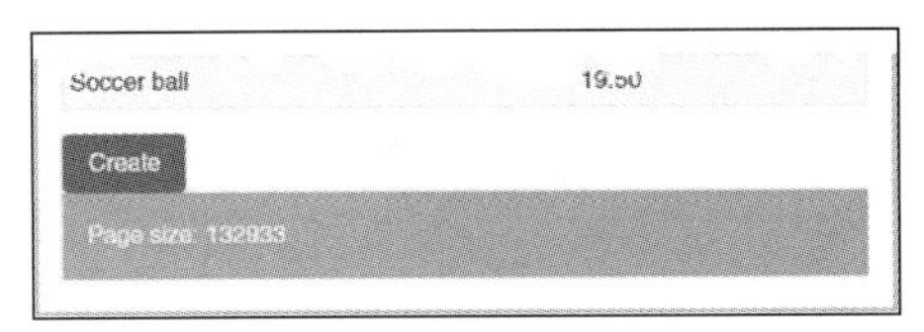

图 22-8 创建异步视图组件

22.4 创建混合的控制器/视图组件类

视图组件通常提供由控制器深入处理的功能的摘要或快照。例如，对于购物车视图组件，通常会存在一个链接，目标是一个控制器，该控制器提供购物车中商品的详细列表，并且可以用于筛选并完成购买。

在这种情况下，可以创建混合的控制器/视图组件类，这样可以将相关功能组合在一起并减少代码重复。为了演示，将一个名为 CityController.cs 的类文件添加到 Controllers 文件夹中，并用它定义代码清单 22-30 所示的控制器。

代码清单 22-30 Controllers 文件夹下的 CityController.cs 文件的内容

```
using System.Collections.Generic;
using Microsoft.AspNetCore.Mvc;
using Microsoft.AspNetCore.Mvc.ViewComponents;
using Microsoft.AspNetCore.Mvc.ViewFeatures;
using UsingViewComponents.Models;

namespace UsingViewComponents.Controllers {

    [ViewComponent(Name = "ComboComponent")]
    public class CityController : Controller {
        private ICityRepository repository;

        public CityController(ICityRepository repo) {
            repository = repo;
        }
        public ViewResult Create() => View();

        [HttpPost]
        public IActionResult Create(City newCity) {
            repository.AddCity(newCity);
            return RedirectToAction("Index", "Home");
        }

        public IViewComponentResult Invoke() => new ViewViewComponentResult() {
            ViewData = new ViewDataDictionary<IEnumerable<City>>(ViewData,
                repository.Cities)
        };
    }
}
```

将 ViewComponent 特性应用于不从 ViewComponent 基类继承且名称不以 ViewComponent 结尾的类，这意味着正常的发现过程通常不会将类看作视图组件。Name 属性用于设置在父视图中使用@Component.Invoke 表达式应用类时可以引用类的名称。在此例中，使用 Name 属性将类的视图组件部分的名称设置成 ComboComponent。该名称将用于调用视图组件并查找视图。

因为混合类不能从 ViewComponent 基类继承，所以它们无法访问 IViewComponentResult 对象提供的便利方法，这意味着必须直接创建 ViewViewComponentResult 对象，就像 POCO 视图组件中所需要的那样。

22.4.1 创建混合视图

混合类需要两组视图——当类用作控制器时渲染的视图以及在将类用作视图组件时呈现的视图。首先，创建Views/City文件夹，并添加一个名为Create.cshtml的视图文件，内容如代码清单22-31所示。

代码清单22-31 Views/City文件夹下的Create.cshtml文件的内容

```
@model City
@{
    ViewData["Title"] = "Create City";
    Layout = "_Layout";
}

<form method="post" asp-action="Create">
    <div class="form-group">
        <label asp-for="Name">Name:</label>
        <input class="form-control" asp-for="Name" />
    </div>
    <div class="form-group">
        <label asp-for="Country">Country:</label>
        <input class="form-control" asp-for="Country" />
    </div>
    <div class="form-group">
        <label asp-for="Population">Population:</label>
        <input class="form-control" asp-for="Population" />
    </div>
    <button type="submit" class="btn btn-primary">Create</button>
    <a class="btn btn-secondary" asp-controller="Home"
       asp-action="Index">
        Cancel
    </a>
</form>
```

Create视图提供了一个简单的用于创建City对象的表单。Create按钮会向City控制器的Create操作方法发送POST请求，而Cancel按钮会向Home控制器的Index操作方法发送GET请求。

接下来，创建Views/Shared/Components/ComboComponent文件夹，并添加一个名为Default.cshtml的视图文件，内容如代码清单22-32所示。将分部视图放置在Views/Shared文件夹中，视图组件的名称将包含在用于定位视图的路径中。

代码清单22-32 Views/Shared/Components/ComboComponent文件夹下的Default.cshtml文件的内容

```
@model IEnumerable<City>

<table class="table table-sm table-bordered">
    <tr>
        <td>Biggest City:</td>
        <td>
            @Model.OrderByDescending(c => c.Population).First().Name
        </td>
    </tr>
</table>
<a class="btn btn-sm btn-info" asp-controller="City" asp-action="Create">
    Create City
</a>
```

这个分部视图接收使用LINQ排序的City对象序列，以选择具有最大Population值的那个。还有一个可设置样式为按钮的锚点元素，按钮将指向City控制器的Create操作方法。

提 示

代码清单22-32明确指定了用于a元素的City控制器。可使用父视图提供的上下文数据生成URL，这意味着默认控制器就是处理请求的控制器，而不是使用视图组件的控制器。如果省略asp-controller属性，链接将在Home控制器上定位Create操作方法。

22.4.2　应用混合类

最后一步是使用 ViewComponent 特性指定的名称将混合类作为视图组件应用于共享布局，如代码清单 22-33 所示。

代码清单 22-33　在 Views/Shared 文件夹下的_Layout.cshtml 文件中应用混合类

```
<!DOCTYPE html>
<html>
<head>
    <meta name="viewport" content="width=device-width" />
    <title>@ViewBag.Title</title>
    <link asp-href-include="lib/bootstrap/dist/css/*.min.css" rel="stylesheet" />
</head>
<body class="m-1 p-1">
    <div class="bg-primary m-1 p-1">
        <div class="row text-white">
            <div class="col-7"><h1>Products</h1></div>
            <div class="col-5">
                @await Component.InvokeAsync("ComboComponent")
            </div>
        </div>
    </div>
    <div class="m-1 p-1">@RenderBody()</div>
    @await Component.InvokeAsync("PageSize")
</body>
</html>
```

结果是由自己的集成控制器（或者如果喜欢，也可以是具有自己的集成视图组件的控制器）备份的视图组件。如果运行应用程序，你将看到伦敦被列为人口最多的城市。单击 Create 按钮，你将看到一个表单，可以向应用程序添加新的城市。填写并提交表单，控制器将收到数据，更新存储库，并将浏览器重定向到应用程序的默认 URL。如果添加人口更多的城市，视图组件的输出结果将会改变，如图 22-9 所示。

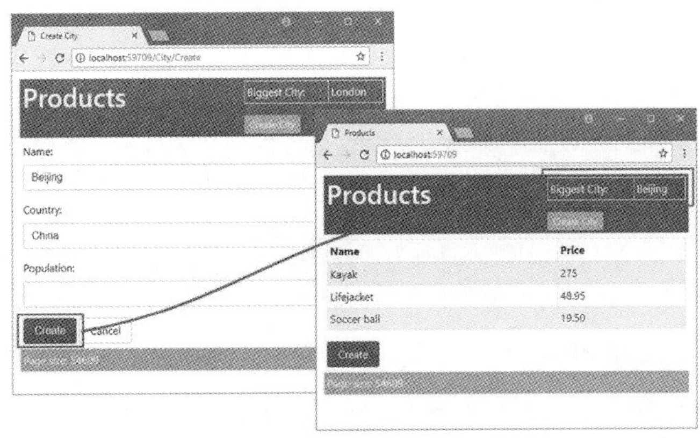

图 22-9　使用混合的控制器/视图组件类

22.5　小结

本章介绍了视图组件，这是 ASP.NET Core MVC 的新增功能，以替代以前的 MVC 版本的子操作功能。本章演示了如何创建 POCO 视图组件以及如何使用 ViewComponent 基类，还展示了组件可以生成的 3 种不同类型的结果，包括在父视图中包含分部视图的选择。本章最后演示了如何将视图组件功能添加到控制器类中，以减少代码重复并简化应用程序。下一章将介绍标签助手，用于在视图中转换 HTML 元素。

第 23 章 标签助手

标签助手是 ASP.NET Core MVC 中引入的新特性，用于在视图中使用 C#类转换 HTML 元素。通常的用法包括使用应用程序的路由配置来为表单生成 URL、确保特定类型的元素样式能够一致地被修饰、将自定义的缩写元素替换为常用的内容片段等。本章将介绍标签助手是如何工作的，以及定制的标签助手是如何创建和应用的。第 24 章将说明内置的支持 HTML 表单的标签助手，第 25 章将说明 MVC 提供的其他内置标签助手。表 23-1 介绍了标签助手的背景。

标签助手组件

ASP.NET Core 2 平台引入了标签助手，从而可以在发送到客户端的响应中修改特定的 HTML 片段，它们也可以通过标签助手来使用。因为它难以使用，对于大多数 MVC 开发人员来说用处不大，所以这里不再介绍。如果要转换发送到客户端的响应中的内容，请使用本章介绍的标签助手特性。

表 23-1 标签助手的背景

问题	答案
标签助手是什么？	标签助手是操作 HTML 元素的类，可使用某种方式改变它们，提供附加的内容，也可使用新的内容整个替换它们
标签助手有何作用？	标签助手允许使用 C#逻辑来呈现或转换视图内容，确保发送到客户端的 HTML 反映应用程序的状态
如何使用标签助手？	可基于元素的类名或者使用 HTMLTargetElement 特性来选择应用标签助手的 HTML 元素。当视图被呈现时，元素通过标签助手被转换并包含在发送到客户端的 HTML 中
标签助手有何缺陷或限制？	标签助手很容易被过度使用，用于生成复杂的 HTML 内容片段，使用视图组件更易于做到
是否有其他的替代方案？	不一定必须使用标签助手，但是它们可以使你在 MVC 应用程序中更加容易地生成复杂的 HTML 内容

表 23-2 列出了本章要介绍的操作。

表 23-2 本章要介绍的操作

操作	方法	代码清单
转换 HTML 元素	创建标签助手，然后使用@addTagHelper 表达式在视图中或在视图导入文件中进行注册	代码清单 23-10～代码清单 23-12
管理标签助手的作用范围	使用 HtmlTargetElement 特性	代码清单 23-13～代码清单 23-17
使用缩写元素	使用 TagHelperOutput 对象生成替换元素	代码清单 23-18 和代码清单 23-19
为目标元素插入内容或环绕内容	使用 TagHelperOutput 提供的 Pre-和 Post-属性	代码清单 23-22 和代码清单 23-23
在标签助手中接收上下文数据	使用 ViewContext 和 HtmlAttributeNotBound 特性装饰属性	代码清单 23-24 和代码清单 23-25
访问视图模型	使用 ModelExpression 属性	代码清单 23-26 和代码清单 23-27
组织标签助手	使用 TagHelperContext.Items 属性	代码清单 23-28 和代码清单 23-29
抑制元素	使用 SuppressOutput 方法	代码清单 23-30 和代码清单 23-31

23.1 准备示例项目

在本章中，使用 ASP.NET Core Web Application（.NET Core）模板新建名为 Cities 的 Empty 项目。

23.1.1 创建模型和存储库

创建 Models 文件夹，添加名为 City.cs 的类文件，并用它定义代码清单 23-1 所示的类。

代码清单 23-1 Models 文件夹下的 City.cs 文件的内容

```
namespace Cities.Models {

    public class City {
        public string Name { get; set; }
        public string Country { get; set; }
        public int? Population { get; set; }
    }
}
```

为了给 City 对象创建存储库，在 Models 文件夹中添加名为 Repository.cs 的类文件，并用它定义代码清单 23-2 所示的接口和实现。

代码清单 23-2 Models 文件夹下的 Repository.cs 文件的内容

```
using System.Collections.Generic;

namespace Cities.Models {

    public interface IRepository {

        IEnumerable<City> Cities { get; }
        void AddCity(City newCity);
    }

    public class MemoryRepository : IRepository {

        private List<City> cities = new List<City> {
            new City { Name = "London", Country = "UK", Population = 8539000},
            new City { Name = "New York", Country = "USA", Population = 8406000 },
            new City { Name = "San Jose", Country = "USA", Population = 998537 },
            new City { Name = "Paris", Country = "France", Population = 2244000 }
        };

        public IEnumerable<City> Cities => cities;

        public void AddCity(City newCity) {
            cities.Add(newCity);
        }
    }
}
```

23.1.2 创建控制器、布局与视图

本章的示例只需要一个控制器。创建 Controllers 文件夹，添加名为 HomeController.cs 的类，然后用它定义代码清单 23-3 所示的控制器。

代码清单 23-3 Controllers 文件夹下的 HomeController.cs 文件的内容

```
using Microsoft.AspNetCore.Mvc;
using Cities.Models;

namespace Cities.Controllers {

    public class HomeController : Controller {
        private IRepository repository;

        public HomeController(IRepository repo) {
            repository = repo;
        }
        public ViewResult Index() => View(repository.Cities);
```

```
        public ViewResult Create() => View();

        [HttpPost]
        public IActionResult Create(City city) {
            repository.AddCity(city);
            return RedirectToAction("Index");
        }
    }
}
```

控制器提供了 Index 操作方法用于列出存储库中的对象，一对 Create 方法将允许用户使用表单来创建新的 City 对象，可延续使用与前面章节中相同的模式。

应用程序中的视图将使用共享的布局。创建 Views/Shared 文件夹，在 Views/Shared 文件夹中添加名为 _Layout.cshtml 的布局文件，加入代码清单 23-4 所示的标记内容。

注 意

本章的目的是演示标签助手如何工作，布局和示例应用程序中的视图仅仅使用标准的 HTML 元素来编写，它们将被介绍的其他标签助手替换。

代码清单 23-4 Views/Shared 文件夹下的_Layout.cshtml 文件的内容

```
<!DOCTYPE html>
<html>
<head>
    <meta name="viewport" content="width=device-width" />
    <title>Cities</title>
    <link href="/lib/bootstrap/dist/css/bootstrap.css" rel="stylesheet" />
</head>
<body class="m-1 p-1">
    <div>@RenderBody()</div>
</body>
</html>
```

然后，创建 Views/Home 文件夹，添加名为 Index.cshtml 的文件，内容如代码清单 23-5 所示。

代码清单 23-5 Views/Home 文件夹下的 Index.cshtml 文件的内容

```
@model IEnumerable<City>

@{  Layout = "_Layout"; }

<table class="table table-sm table-bordered">
    <thead class="bg-primary text-white">
        <tr>
            <th>Name</th>
            <th>Country</th>
            <th class="text-right">Population</th>
        </tr>
    </thead>
    <tbody>
        @foreach (var city in Model) {
            <tr>
                <td>@city.Name</td>
                <td>@city.Country</td>
                <td class="text-right">@city.Population?.ToString("#,###")</td>
            </tr>
        }
    </tbody>
```

```
</table>
<a href="/Home/Create" class="btn btn-primary">Create</a>
```

Index 视图使用 City 对象的序列来填充表格，还包含一个 a 元素用来指向 URL 地址/Home/Create，a 元素已使用 Bootstrap 修饰为按钮样式。对于第二个视图，添加名为 Create.cshtml 的文件到 Views/Home 文件夹中，内容如代码清单 23-6 所示。

代码清单 23-6　Views/Home 文件夹下的 Create.cshtml 文件的内容

```
@model City

@{ Layout = "_Layout"; }

<form method="post" action="/Home/Create">
    <div class="form-group">
        <label for="Name">Name:</label>
        <input class="form-control" name="Name" />
    </div>
    <div class="form-group">
        <label for="Country">Country:</label>
        <input class="form-control" name="Country" />
    </div>
    <div class="form-group">
        <label for="Population">Population:</label>
        <input class="form-control" name="Population" />
    </div>

    <button type="submit" class="btn btn-primary">Add</button>
    <a class="btn btn-primary" href="/Home/Index">Cancel</a>
</form>
```

这里还在 Views 文件夹中创建了名为_ViewImports.cshtml 的视图导入文件，在其中添加代码清单 23-7 所示的表达式。这允许引用 Models 文件夹中的类而不需要使用命名空间。

代码清单 23-7　Views 文件夹下的_ViewImports.cshtml 文件的内容

```
@using Cities.Models
```

示例项目中的视图基于 Bootstrap CSS 包。为了将 BootStrap 添加到示例项目中，使用 Bower Configuration File 模板在项目的根目录中创建名为 bower.json 的文件，并添加代码清单 23-8 所示的包到 dependencies 部分。

代码清单 23-8　在 Cities 文件夹下的 bower.json 文件中添加 Bootstrap

```
{
  "name": "asp.net",
  "private": true,
  "dependencies": {
    "bootstrap": "4.0.0-alpha.6"
  }
}
```

23.1.3 配置应用程序

最终的准备步骤是配置应用程序，如代码清单 23-9 所示。本书在所有示例项目中使用同样的基本配置，另外使用单例生命周期作为服务注册存储库。

代码清单 23-9　Cities 文件夹下的 Startup.cs 文件的内容

```
using System;
using System.Collections.Generic;
using System.Linq;
using System.Threading.Tasks;
```

```
using Microsoft.AspNetCore.Builder;
using Microsoft.AspNetCore.Hosting;
using Microsoft.AspNetCore.Http;
using Microsoft.Extensions.DependencyInjection;
using Cities.Models;

namespace Cities {
    public class Startup {

        public void ConfigureServices(IServiceCollection services) {
            services.AddSingleton<IRepository, MemoryRepository>();
            services.AddMvc();
        }
        public void Configure(IApplicationBuilder app, IHostingEnvironment env) {
            app.UseStatusCodePages();
            app.UseDeveloperExceptionPage();
            app.UseStaticFiles();
            app.UseMvcWithDefaultRoute();
        }
    }
}
```

如果运行应用程序，将会显示来自存储库默认创建的 City 对象的清单。单击 Create 按钮，填写表单，然后单击 Add 按钮；新的对象将会被添加到存储库中，如图 23-1 所示。

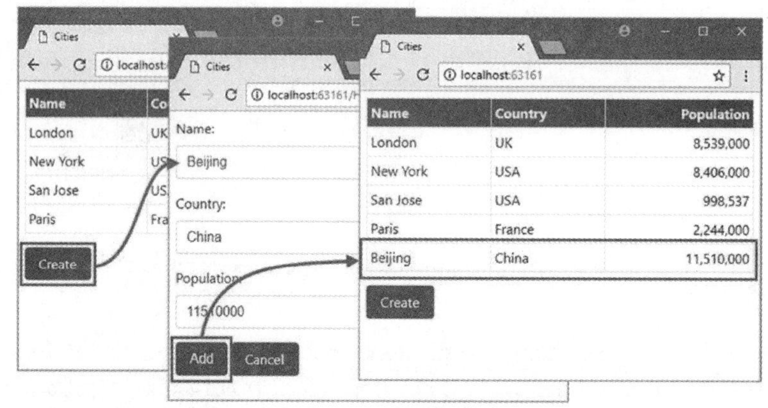

图 23-1 运行应用程序

23.2 创建标签助手

与许多的 MVC 特性一样，理解标签助手的最佳方式就是创建它们，这可以揭示它们是如何运作以及融入应用程序的。后面将介绍创建和应用标签助手的过程，将为 button 元素应用 Bootstrap CSS 样式，以便对如下元素进行转换：

```
...
<button type="submit" bs-button-color="danger">Add</button>
...
```

转换后的形式如下：

```
...
<button type="submit" class="btn btn-danger">Add</button>
...
```

标签助手将会识别 bs-button-color 属性，然后使用该属性的值设置发送到浏览器的元素的 class 属性。

这种转换虽然不是很有用，但可以为说明标签助手如何工作打下基础。

23.2.1 定义标签助手类

标签助手可以在项目的任何位置定义，但将它们保存在一起是有帮助的，与大多数 MVC 组件不同，它们在使用之前需要注册。在 Infrastructure/TagHelpers 文件夹中创建这些标签助手，并将它们添加到项目中。

标签助手类需要派生自 TagHelper 类，且定义在命名空间 Microsoft.AspNetCore.Razor.TagHelpers 中。为了创建标签助手，在 Infrastructure/TagHelpers 文件夹中添加名为 ButtonTagHelper.cs 的文件，然后用它定义代码清单 23-10 所示的类。

代码清单 23-10 Infrastructure/TagHelpers 文件夹下的 ButtonTagHelper.cs 文件的内容

```
using Microsoft.AspNetCore.Razor.TagHelpers;

namespace Cities.Infrastructure.TagHelpers {

    public class ButtonTagHelper : TagHelper {

        public string BsButtonColor { get; set; }

        public override void Process(TagHelperContext context,
                                     TagHelperOutput output) {

            output.Attributes.SetAttribute("class", $"btn btn-{BsButtonColor}");
        }
    }
}
```

TagHelper 类定义了 Process 方法，该方法可由子类重写以实现转换元素的行为。标签助手的名称由被转换元素的名称加上 TagHelper 组成。在这个示例中，类名 ButtonTagHelper 告诉 MVC 这是用于转换 button 元素的标签助手。标签助手的作用域可以使用特性进行拓宽或限制，这里是默认行为。

提 示

异步标签助手可以通过重写 ProcessAsync 方法而不是 Process 方法来创建，但是对大多数标签助手是不必要的，而倾向于创建小的且专注于改变的 HTML 元素。默认的 ProcessAsync 实现会调用 Process 方法。

1. 接收上下文数据

标签助手通过 TagHelperContext 类的实例来接收关于转换元素的信息，实例可作为 Process 方法的参数接收，且定义了表 23-3 定义的属性。

表 23-3 TagHelperContext 类定义的属性

属性	描述
AllAttributes	返回一个只读的待转换元素的特性字典，以属性名称作为索引
Items	返回一个用于协调标签助手的字典
UniqueId	返回待转换元素的唯一标识

尽管可以通过 AllAttributes 字典访问待转换元素的属性详情，但更方便的方式是定义一个属性，属性名对应你感兴趣的元素属性，如下所示：

```
...
public string BsButtonColor { get; set; }
...
```

当标签助手被应用时，MVC 检查标签助手类定义的属性，并使用名称得到完整匹配的 HTML 元素属性的值来设置。作为上述处理过程的一部分，MVC 将试图转换 HTML 属性的值为 C#属性的类型，bool 类型的属性可以接收 HTML 属性的 true 或 false 值，int 类型的属性可以接收类似于 1 和 2 的 HTML 属性值。

HTML Helper 发生了什么？

早期版本的 ASP.NET MVC 使用 HTML helper 来生成表单元素。HTML helper 是一组通过 Razor 表达式访问的以@Html 开始的方法，创建 input 元素并填充属性的方式可能如下所示：

```
...
@Html.TextBoxFor(m => m.Population)
...
```

HTML helper 表达式的问题是它们不适合 HTML 元素的结构，这导致表达式十分笨拙，类似于下面这样：

```
...
@Html.TextBoxFor(m => m.Population, new { @class = "form-control" })
...
```

属性不得不表示为动态对象，并且如果名称为 C#保留字，例如 class，那么不得不使用@前缀。由于 HTML 元素需要变得越来越复杂，HTML helper 表达式也变得更加笨拙。标签助手通过使用 HTML 属性避免了这一点，类似于下面这样：

```
...
<input class="form-control" asp-for="Population" />
...
```

结果就是生成更自然的 HTML，并且处理的视图也更易读和易于理解。MVC 仍然支持 HTML helper（并且标签助手在幕后使用 HTML helper），可以在原来开发的 MVC5 视图中继续使用 HTML Helper 来实现后向兼容性，但是，新的视图应当使用标签助手。

属性的名称会从默认的 HTML 风格执行自动转换，比如从 bs-button-color 转换为 C#风格的 BsButtonColor。可以使用除了 asp-（由微软使用）和 data（由发送到客户端的自定义属性保留）以外的任何前缀。本例的 Process 方法使用 BsButtonColor 属性来接收应用于 button 元素的颜色模式，如下所示：

```
...
output.Attributes.SetAttribute("class", $"btn btn-{BsButtonColor}");
...
```

没有关联 HTML 属性的属性值不会被设置，这意味着应当检查并确认没有在处理 null 或默认值。

提　示

使用 HTML 属性名称用于标签助手属性并不总是导致易读或易于理解的类。可以使用 HtmlAttributeName 特性来打破属性名称与其表示的属性之间的联系，从而用来指定属性将要表示的 HTML 属性。

2. 产生输出

Process 方法通过配置作为参数接收的 TagHelperOutput 对象来转换元素。TagHelperOuput 从自身在 Razor 视图中出现的位置开始输出 HTML 元素，并通过表 23-4 所示的属性和方法来改变元素。

表 23-4　　TagHelperOutput 类定义的属性和方法

名称	描述
TagName	该属性用于获取或设置输出元素的标记名称
Attributes	该属性返回用于输出元素的属性字典

续表

名称	描述
Content	该属性返回 TagHelperContent 对象，用于设置元素内容
PreElement	该属性返回 TagHelperContent 对象，用于在视图中输出元素之前插入内容
PostElement	该属性返回 TagHelperContent 对象，用于在视图中输出元素之后追加内容
PreContent	该属性返回 TagHelperContext 对象，用于在输出元素的内容之前插入内容
PostContent	该属性返回 TagHelperContext 对象，用于在输出元素的内容之后插入内容
TagMode	该属性指定输出元素的输出编写方式，可使用来自 TagMode 枚举的值
SupressOuput()	该方法用于防止元素被包含到视图中

在 ButtonTagHelper 类中，可使用 Attributes 字典添加 class 属性到 HTML 元素中，用于设置 Bootstrap 按钮样式，包含来自 BsButtonColor 属性的值，这意味着可以使用 Bootstrap 名称（比如 primary、info、和 danger）来指定不同的颜色。

23.2.2 注册标签助手

标签助手类只有在使用 Razor 的@addTagHelper 表达式注册之后才能使用。被应用的视图集将基于@addTagHelper 表达式的应用位置。对于单个视图，表达式出现在视图本身之中。对于应用程序中的一组视图，表达式出现在包含这些视图的文件夹或父文件夹中名为_ViewImports.cshtml 的文件中。比如，/Views/Home/_ViewImports.cshtml 文件中使用@addTagHelper 表达式注册的标签助手将作用于 Home 控制器的所有视图。为了让本章创建的标签助手可以被整个应用程序中的视图使用，在 Views/_ViewImports.cshtml 文件添加@addTagHelper 表达式，如代码清单 23-11 所示。

代码清单 23-11　在 Views 文件夹下的_ViewImports.cshtml 文件中注册标签助手

```
@using Cities.Models
@addTagHelper Cities.Infrastructure.TagHelpers.*, Cities
```

参数的第一部分指定了标签助手的类名，支持通配符；第二部分指定了定义所在的程序集名称。以上代码注册了 Cities 程序集中命名空间 Cities.Infrastructure.TagHelpers 里的所有标签助手。

23.2.3 使用标签助手

最终的步骤是使用标签助手转换元素。代码清单 23-12 从 Create.cshtml 视图文件的 button 元素中删除了 class 属性，替换为 ButtonTagHelper 类中定义的属性。

代码清单 23-12　在 Views/Home 文件夹下的 Create.cshtml 视图文件中使用标签助手

```
@model City

@{ Layout = "_Layout"; }

<form method="post" action="/Home/Create">
    <div class="form-group">
        <label for="Name">Name:</label>
        <input class="form-control" name="Name" />
    </div>
    <div class="form-group">
        <label for="Country">Country:</label>
        <input class="form-control" name="Country" />
    </div>
    <div class="form-group">
```

```
        <label for="Population">Population:</label>
        <input class="form-control" name="Population" />
    </div>

    <button type="submit" bs-button-color="danger">Add</button>
    <a class="btn btn-primary" href="/Home/Index">Cancel</a>
</form>
```

如果运行应用程序，单击 Create 按钮，浏览器将会访问/Home/Create，你将会看到 Add 按钮的颜色和样式发生了变化，如图 23-2 所示。

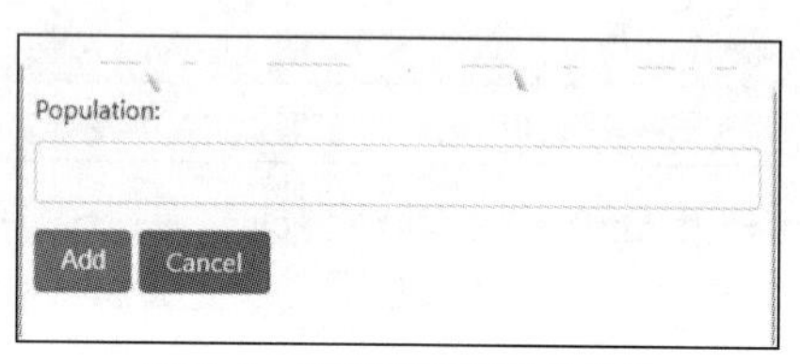

图 23-2 使用标签助手修饰按钮

测试标签助手

用于标签助手的单元测试是一个相对简单的过程。下面是为代码清单 23-12 所示的标签助手提供的示例测试：

```
using System.Collections.Generic;
using System.Linq;
using System.Threading.Tasks;
using Cities.Infrastructure.TagHelpers;
using Microsoft.AspNetCore.Razor.TagHelpers;
using Xunit;

namespace Cities.Tests {

    public class TagHelperTests {

        [Fact]
        public void TestTagHelper() {
            // Arrange
            var context = new TagHelperContext(
                new TagHelperAttributeList(),
                new Dictionary<object, object>(),
                "myuniqueid");

            var output = new TagHelperOutput("button",
                new TagHelperAttributeList(), (cache, encoder) =>
                    Task.FromResult<TagHelperContent>
                        (new DefaultTagHelperContent()));

            // Act
            var tagHelper = new ButtonTagHelper {
                BsButtonColor = "testValue"
            };
            tagHelper.Process(context, output);

            // Assert
            Assert.Equal($"btn btn-{tagHelper.BsButtonColor}",
                output.Attributes["class"].Value);
        }
    }
}
```

> 在以上单元测试中，多数工作是设置 TagHelperContext 和 TagHelperOutput 对象，以便它们可以被传递给标签助手的 Process 方法，以检查并确保 HTML 元素被正确转换。准备用于测试的标签助手所需的工作量自然取决于它们所操作的 HTML 的复杂性以及转换程度。但是，大多数标签助手相对简单，可以按照前面讲述的基本模式进行测试。

23.2.4 管理标签助手的作用域

标签助手将被应用于给定类型的所有元素，这意味着之前创建的 ButtonTagHelper 类的 Process 方法，将由应用程序中每个视图的每个 button 元素调用。这不总是有用的。为了针对这个问题提供示例，在 Create.cshtml 文件中添加另一个 button 元素，如代码清单 23-13 所示。

代码清单 23-13 在 Views/Home 文件夹下的 Create.cshtml 文件中添加另一个 button 元素

```
@model City

@{ Layout = "_Layout"; }

<form method="post" action="/Home/Create">
    <div class="form-group">
        <label for="Name">Name:</label>
        <input class="form-control" name="Name" />
    </div>
    <div class="form-group">
        <label for="Country">Country:</label>
        <input class="form-control" name="Country" />
    </div>
    <div class="form-group">
        <label for="Population">Population:</label>
        <input class="form-control" name="Population" />
    </div>

    <button type="submit" bs-button-color="danger">Add</button>
    <button type="reset" class="btn btn-primary" >Reset</button>
    <a class="btn btn-primary" href="/Home/Index">Cancel</a>
</form>
```

新的 button 元素已经拥有 class 属性，并且不需要通过 ButtonTagHelper 类执行转换。但是，如果运行应用程序并导航到 URL 地址/Home/Create，你将会看到有问题产生，如图 23-3 所示。

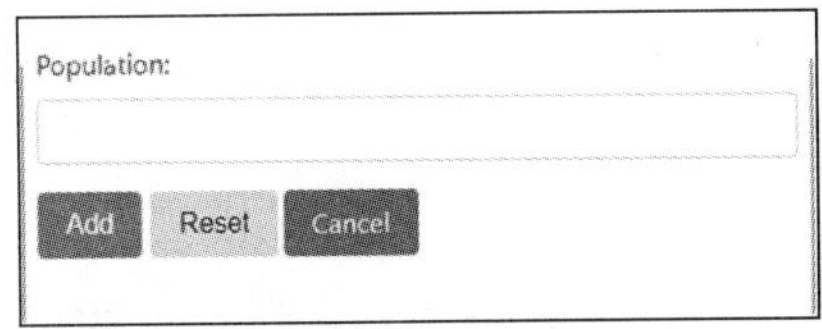

图 23-3 标签助手的默认作用域的效果

可以通过查看发送到浏览器的 HTML 来查看格式缺失的原因，这揭示了 class 属性的问题，如下所示：

```
<button type="reset" class="btn btn-">Reset</button>
```

MVC 应用 ButtonTagHelper 到新的 button 元素，但是由于 HTML 元素没有相应的 bs-button-color 属性，因此没有设置 BsButtonColor 属性的值。结果，标签助手使用不正确的 Bootstrap 样式替换 class 属性，生成格式缺失的元素。

1. 限制标签助手的作用域

解决这个问题有两种方式。一种方式是修改 ButtonTagHelper 类，以使其对可能遇到的 button 元素敏感。对于示例应用程序来说，这需要增加额外的检查，例如确认是否拥有 bs-button-color 属性，并确保在定义了 class 属性之后不会被替换掉。这种方式的问题在于，随着包含 button 元素的视图不断添加到应用程序中，标签助手类会变得越来越复杂，并且基于 ButtonTagHelper 的所有新添加的额外复杂条件描述并不执行转换。

另一种方式是允许标签助手描述对使用方式的限制，缩小作用域。可使用 HtmlTargetElement 特性限制标签助手的作用域，如代码清单 23-14 所示。

代码清单 23-14 限制 Infrastructure/TagHelpers 文件夹下的 ButtonTagHelper.cs 的作用域

```
using Microsoft.AspNetCore.Razor.TagHelpers;

namespace Cities.Infrastructure.TagHelpers {

    [HtmlTargetElement("button", Attributes = "bs-button-color", ParentTag = "form")]
    public class ButtonTagHelper : TagHelper {

        public string BsButtonColor { get; set; }

        public override void Process(TagHelperContext context,
                                     TagHelperOutput output) {

            output.Attributes.SetAttribute("class", $"btn btn-{BsButtonColor}");
        }
    }
}
```

HtmlTargetElement 特性描述了标签助手应用的元素。第一个参数指定了元素类型，并支持表 23-5 所示的额外属性。

表 23-5 HtmlTargetElement 特性支持的属性

属性	描述
Attributes	用于指定标签助手应作用于拥有给定属性集的指定元素，可使用以逗号分隔属性的列表来提供。元素必须拥有全部指定的属性。以星号结束的属性名称将被视为前缀，所以 bs-button-*将匹配 bs-button-color、bs-button-size 等
ParentTag	用于指定标签助手仅应作用于包含指定类型元素的元素
TagStructure	用于指定标签助手仅应作用于元素标记结构对应于 TagStructure 枚举中给定值的元素，其中定义了 Unspecified、NormalOrSelfClosing 和 WithoutEndTag

代码清单 23-14 限制了 ButtonTagHelper 类仅作用于父元素是 form 元素，且拥有 bs-button-color 属性的 button 元素。如果运行应用程序并导航到/Home/Create，你将看到 Reset 按钮因为缺失要求的属性而不再被转换，如图 23-4 所示。

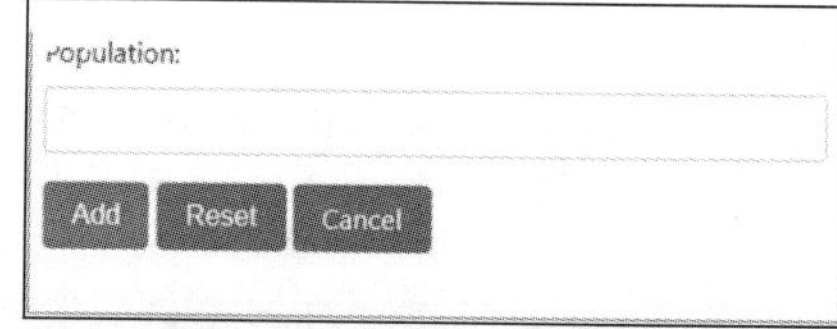

图 23-4 限制标签助手的作用域

2. 拓宽标签助手的作用域

特性 HtmlTargetElement 也可以用来拓宽标签助手的作用域，以便匹配更广泛的元素。当需要对多种不同类型的元素执行相同的转换时，这是很有用的。这与基于标签助手类名匹配元素的前提相违背，如代码清单 23-15 所示。

代码清单 23-15 在 Infrastructure/TagHelpers 文件夹下的 ButtonTagHelper.cs 中拓宽标签助手的作用域

```
using Microsoft.AspNetCore.Razor.TagHelpers;

namespace Cities.Infrastructure.TagHelpers {

    [HtmlTargetElement(Attributes = "bs-button-color", ParentTag = "form")]
    public class ButtonTagHelper : TagHelper {

        public string BsButtonColor { get; set; }

        public override void Process(TagHelperContext context,
                                     TagHelperOutput output) {

            output.Attributes.SetAttribute("class", $"btn btn-{BsButtonColor}");
        }
    }
}
```

以上代码忽略了 HtmlTargetElement 中的元素类型，这意味着标签助手将被应用于任何拥有 bs-button-color 属性的元素，而无论元素的类型如何。代码清单 23-16 修改了表单中的 a 元素，使用与按钮元素相同的 Bootstrap 样式集，以便可以由标签助手进行转换。

代码清单 23-16　在 Views/Home 文件夹下的 Create.cshtml 文件中修改锚元素

```
@model City

@{  Layout = "_Layout"; }

<form method="post" action="/Home/Create">
    <div class="form-group">
        <label for="Name">Name:</label>
        <input class="form-control" name="Name" />
    </div>
    <div class="form-group">
        <label for="Country">Country:</label>
        <input class="form-control" name="Country" />
    </div>
    <div class="form-group">
        <label for="Population">Population:</label>
        <input class="form-control" name="Population" />
    </div>

    <button type="submit" bs-button-color="danger">Add</button>
    <button type="reset" class="btn btn-primary" >Reset</button>
    <a bs-button-color="primary" href="/Home/Index">Cancel</a>
</form>
```

标签助手的作用域更广泛意味着对于不同类型的元素不必重复同样的操作。但是注意，应用程序中的视图内容在演进的时候，很容易创建匹配过于宽泛的元素的标签助手。更平衡的方式是应用多次 HtmlTargetElement 特性，以组合限制定义匹配的方式指定元素的完整集合，如代码清单 23-17 所示。

代码清单 23-17　在 Infrastructure/TagHelpers 文件夹下的 ButtonTagHelper.cs 文件中平衡标签助手的作用域

```
using Microsoft.AspNetCore.Razor.TagHelpers;

namespace Cities.Infrastructure.TagHelpers {

    [HtmlTargetElement("button", Attributes = "bs-button-color", ParentTag = "form")]
    [HtmlTargetElement("a", Attributes = "bs-button-color", ParentTag = "form")]
    public class ButtonTagHelper : TagHelper {

        public string BsButtonColor { get; set; }

        public override void Process(TagHelperContext context,
                                     TagHelperOutput output) {

            output.Attributes.SetAttribute("class", $"btn btn-{BsButtonColor}");
        }
    }
}
```

上述配置对应用程序有着同样的效果，但请确保在未来的开发过程中，出于其他的原因添加 bs-button-color 属性到其他元素时不会导致问题。

排序标签助手的执行

作为一般原则，好的做法是对于特定的 HTML 元素最好仅仅应用一个标签助手，这是因为很容易导致一个标签助手破坏另一个标签助手的情况。如果需要应用多个标签助手，那么通过设置 Order 属性可以控制执行的顺序。管理执行顺序可以帮助你在标签助手之间最小化冲突，虽然仍然容易导致问题。

23.3 高级标签助手特性

前面演示了如何创建基本的标签助手，但是仅仅展示了表面的可能性，后面将展示标签助手提供的更高级的用法和特性。

23.3.1 创建缩写元素

标签助手并不限于转换标准 HTML 元素，也可以用来转换常用内容的自定义元素。这是一个使得创建更简洁、意图更明显的视图的有用特性。为了演示，使用自定义元素替换 Create.cshtml 文件中的 button 元素，如代码清单 23-18 所示。

代码清单 23-18 在 Create.cshtml 文件中添加自定义元素

```
@model City

@{ Layout = "_Layout"; }

<form method="post" action="/Home/Create">
    <div class="form-group">
        <label for="Name">Name:</label>
        <input class="form-control" name="Name" />
    </div>
    <div class="form-group">
        <label for="Country">Country:</label>
        <input class="form-control" name="Country" />
    </div>
    <div class="form-group">
        <label for="Population">Population:</label>
        <input class="form-control" name="Population" />
    </div>
    <formbutton type="submit" bg-color="danger" />
    <formbutton type="reset" />
    <a bs-button-color="primary" href="/Home/Index">Cancel</a>
</form>
```

formbutton 元素不是 HTML 规范的一部分，也不被浏览器所理解。相反，使用这些元素作为生成表单所需的 button 元素的缩写。在 Infrastructure/TagHelper 文件夹中添加一个名为 FormButtonTagHelper.cs 的类文件，并用它定义代码清单 23-19 所示的类。

代码清单 23-19 Infrastructure/TagHelpers 文件夹下的 FormButtonTagHelper.cs 文件的内容

```
using Microsoft.AspNetCore.Razor.TagHelpers;

namespace Cities.Infrastructure.TagHelpers {

    [HtmlTargetElement("formbutton")]
    public class FormButtonTagHelper : TagHelper {

        public string Type { get; set; } = "Submit";

        public string BgColor { get; set; } = "primary";

        public override void Process(TagHelperContext context,
                                     TagHelperOutput output) {

            output.TagName = "button";
            output.TagMode = TagMode.StartTagAndEndTag;
```

```
            output.Attributes.SetAttribute("class", $"btn btn-{BgColor}");
            output.Attributes.SetAttribute("type", Type);
            output.Content.SetContent(Type == "submit" ? "Add" : "Reset");
        }
    }
}
```

提 示

当处理不是 HTML 规范一部分的自定义元素时，必须应用 HtmlTargetElement 特性，并指定元素名称，如代码清单 23-19 所示。基于类型将标签助手应用到元素的便捷方式仅仅适用于标准元素名称。

Process 方法使用 TagHelperOuput 对象的属性来生成复杂的不同元素：TagName 属性用于指定 button 元素，TagMode 属性用于指定元素使用开始和结束标记写入。Attributes.SetAttribute 方法用于定义 Bootstrap 样式的 class 属性，Content 属性用于设置元素内容。

提 示

代码清单 23-19 设置了输出元素的 type 属性。这在输出元素中忽略了标签助手定义的任何属性，这通常是个好主意，因为阻止了用于标签助手的属性出现在发送给浏览器的 HTML 中。但是，在这种情况下需要使用 type 属性来配置标签助手，以便它们出现在输出元素中。

设置 TagName 属性是很重要的，因为对于自定义元素，输出元素默认使用同样的样式。代码清单 23-19 使用了自结束标记：

```
...
<formbutton type="submit" bg-color="danger" />
...
```

为了在输出元素中包含内容，不得不显式指定 TagMode.StartTagAndEndTag 枚举值，以便分离开始和结束标记。

Content 属性会返回一个 TagHelperContent 实例，用来设置元素的内容。表 23-6 列出了 TagHelperContent 定义的主要方法。

表 23-6 TagHelperContent 定义的主要方法

方法	描述
SetContent(text)	设置输出元素的内容，字符串参数将被编码以便被 HTML 元素安全包含
SetHtmlContent(html)	设置输出元素的内容，字符串参数假定已经安全编码，要慎用
Append(text)	安全编码指定的字符串，并添加到输出元素的内容中
AppendHtml(html)	添加指定的字符串到输出元素的内容中，而不执行任何编码，要慎用
Clear()	删除输出元素的内容

在代码清单 23-19 中，标签助手基于 type 特性的值，使用 SetContent 方法来设置输出元素的内容，值由 Type 属性提供。如果运行应用程序，并访问/Home/Create，你将会看到自定义的 formbutton 元素将被替换为标准的 HTML 元素，所以需要转换以下元素：

```
...
<formbutton type="submit" bg-color="danger" />
<formbutton type="reset" />
...
```

转换后的元素如下：

```
<button class="btn btn-danger" type="submit">Add</button>
<button class="btn btn-primary" type="reset">Reset</button>
```

23.3.2 前置和追加内容与元素

TagHelperOutput 类提供了四个属性，使得更容易在视图中注入新的内容，以便环绕元素或内容，如表 23-7 所示。后面将说明如何在目标元素中插入内容。

表 23-7 用于添加内容与元素的 TagHelperOutput 属性

属性	描述
PreElement	用于在目标元素之前插入元素
PostElement	用于在目标元素之后插入元素
PreContent	用于在目标元素的任何现存内容之前插入内容
PostContent	用于在目标元素的任何现存内容之后插入内容

1. 在环绕的输出元素中插入内容

第一组 TagHelperOuput 属性是 PreElement 和 PostElement，它们用来在输出元素之前和之后插入元素到视图中。作为演示，添加名为 ContentWrapperTagHelper.cs 的类文件，并用它创建代码清单 23-20 所示的标签助手类。

代码清单 23-20 Infrastructure/TagHelpers 文件夹下的 ContentWrapperTagHelper.cs 文件的内容

```
using Microsoft.AspNetCore.Mvc.Rendering;
using Microsoft.AspNetCore.Razor.TagHelpers;

namespace Cities.Infrastructure.TagHelpers {

    [HtmlTargetElement("div", Attributes = "title")]
    public class ContentWrapperTagHelper : TagHelper {

        public bool IncludeHeader { get; set; } = true;
        public bool IncludeFooter { get; set; } = true;

        public string Title { get; set; }

        public override void Process(TagHelperContext context,
                                     TagHelperOutput output) {

            output.Attributes.SetAttribute("class", "m-1 p-1");

            TagBuilder title = new TagBuilder("h1");
            title.InnerHtml.Append(Title);

            TagBuilder container = new TagBuilder("div");
            container.Attributes["class"] = "bg-info m-1 p-1";

            container.InnerHtml.AppendHtml(title);

            if (IncludeHeader) {
                output.PreElement.SetHtmlContent(container);
            }

            if (IncludeFooter) {
                output.PostElement.SetHtmlContent(container);
            }
        }
    }
}
```

该标签助手用于转换拥有 title 属性的 div 元素，该 div 元素使用 PreElement 和 PostElement 属性来添加页头和页脚元素以环绕输出元素。

当生成新的 HTML 元素时，可以使用标准的 C#字符串格式来创建输出内容，但除非用于最简单的元素，否则这是一种笨拙且易错的方式。更稳健的方式是使用 TagBuilder 类，它定义在命名空间 Microsoft.AspNetCore.Mvc.Rendering 中，该类支持以更结构化的方式创建元素。TagHelperContent 类定义了接收 TagBuilder 对象的方法，使得在标签助手中更易于创建 HTML 内容。

这个标签助手使用 TagBuilder 类来创建 h1 元素，其包含在使用 Bootstrap 类修饰的 div 元素内。可选的 bool 类型的 include-header 与 include-footer 属性用于指定内容插入何处，默认行为是在输出元素之前和之后添加元素。代码清单 23-21 更新了共享布局，以便包含一个将被标签助手转换的元素。

代码清单 23-21　在 Views/Shared 文件夹下的_Layout.cshtml 文件中使用标签助手

```
<!DOCTYPE html>
<html>
<head>
    <meta name="viewport" content="width=device-width" />
    <title>Cities</title>
    <link href="/lib/bootstrap/dist/css/bootstrap.css" rel="stylesheet" />
</head>
<body class="m-1 p-1">
    <div title="Cities">@RenderBody()</div>
</body>
</html>
```

如果运行应用程序，你将会看标签助手被应用到应用程序中，还在所有页面中添加了页头和页脚，如图 23-5 所示。

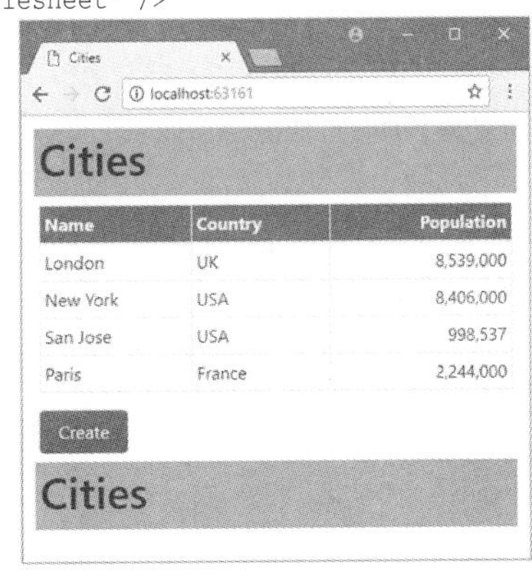

图 23-5　使用标签助手插入 HTML 元素

2. 在输出元素中插入内容

PreContent 和 PostContent 属性用于在输出元素内环绕源内容插入内容。为了演示，在 Infrastructure/TagHelpers 文件夹中添加名为 TableCellTagHelper.cs 的类文件，并用它定义代码清单 23-22 所示的类。

代码清单 23-22　Infrastructure/TagHelpers 文件夹下的 TableCellTagHelper.cs 文件的内容

```
using Microsoft.AspNetCore.Razor.TagHelpers;

namespace Cities.Infrastructure.TagHelpers {

    [HtmlTargetElement("td", Attributes = "wrap")]
    public class TableCellTagHelper : TagHelper {

        public override void Process(TagHelperContext context,
                                     TagHelperOutput output) {

            output.PreContent.SetHtmlContent("<b><i>");
            output.PostContent.SetHtmlContent("</i></b>");
        }
    }
}
```

该标签助手处理拥有 wrap 属性的 td 元素，插入 b 和 i 元素以环绕输出元素的内容。代码清单 23-23 为 Index.cshtml 视图文件中的单元格添加了 wrap 属性。

代码清单 23-23　在 Views/Home 文件夹下的 Index.cshtml 文件中添加 HTML 属性

```
@model IEnumerable<City>

@{ Layout = "_Layout"; }
```

```
<table class="table table-sm table-bordered">
    <thead class="bg-primary text-white">
        <tr>
            <th>Name</th>
            <th>Country</th>
            <th class="text-right">Population</th>
        </tr>
    </thead>
    <tbody>
        @foreach (var city in Model) {
            <tr>
                <td wrap>@city.Name</td>
                <td>@city.Country</td>
                <td class="text-right">@city.Population?.ToString("#,###")</td>
            </tr>
        }
    </tbody>
</table>
<a href="/Home/Create" class="btn btn-primary">Create</a>
```

如果运行应用程序，你会看到表格中第一列的单元格内显示的 City 对象被显示为加粗的斜体文本。检查发送到浏览器的 HTML 内容，你将会看到如何通过 PreContent 与 PostContent 属性在源内容的前后添加内容，如下所示：

```
...
<tr>
    <td wrap><b><i>London</i></b></td>
    <td>UK</td>
    <td class="text-right">8,539,000</td>
</tr>
...
```

提　示

wrap 属性保留在输出元素中。这是因为没有在标签助手类中定义这个属性的相关属性。如果希望阻止属性包含在输出中，可以在标签助手类中为它们定义属性，即使不需要属性的值。

23.3.3　使用依赖注入获取视图上下文数据

标签助手的一种常见用法是转换元素以便包含当前请求或当前视图模型的详细内容。作为示例，在 Infrastructure/TagHelpers 文件夹中添加名为 FormTagHelper.cs 的类文件，并用代码清单 23-24 所示的内容定义此类。

代码清单 23-24　Infrastructure/TagHelpers 文件夹下的 FormTagHelper.cs 文件的内容

```
using Microsoft.AspNetCore.Mvc;
using Microsoft.AspNetCore.Mvc.Rendering;
using Microsoft.AspNetCore.Mvc.Routing;
using Microsoft.AspNetCore.Mvc.ViewFeatures;
using Microsoft.AspNetCore.Razor.TagHelpers;

namespace Cities.Infrastructure.TagHelpers {

    public class FormTagHelper : TagHelper {
        private IUrlHelperFactory urlHelperFactory;

        public FormTagHelper(IUrlHelperFactory factory) {
```

```
            urlHelperFactory = factory;
        }

        [ViewContext]
        [HtmlAttributeNotBound]
        public ViewContext ViewContextData { get; set; }
        public string Controller { get; set; }
        public string Action { get; set; }

        public override void Process(TagHelperContext context,
                                     TagHelperOutput output) {

            IUrlHelper urlHelper = urlHelperFactory.GetUrlHelper(ViewContextData);

            output.Attributes.SetAttribute("action", urlHelper.Action(
                Action ??
                    ViewContextData.RouteData.Values["action"].ToString(),
            Controller ??
                ViewContextData.RouteData.Values["controller"].ToString()));
        }
    }
}
```

顾名思义，FormTagHelper 类处理 form 元素，设置 action 属性以指定将表单数据发送到何方。如果 form 元素拥有 controller 和 action 属性，它们的属性值将用于生成目标 URL；否则，将使用来自路由数据的 controller 和 action 值。

为了获取上下文数据，添加名为 ViewContextData 的属性，并使用两个特性进行修饰，如下所示：

```
...
[ViewContext]
[HtmlAttributeNotBound]
public ViewContext ViewContextData { get; set; }
...
```

ViewContext 特性表示当创建 FormTagHelper 类的实例时，应该为相应的属性分配 ViewContext 对象，如第 18 章所示。ViewContext 类提供了正在渲染的视图的详情、路由数据以及当前的 HTTP 请求，见第 21 章。

如果 HTML 元素 input 拥有 view-context 属性，那么 HtmlAttributeNotBound 特性可防止 MVC 给该属性赋值。这种做法很好，特别是在开发其他开发者使用的标签助手时。

提 示

内置的用于表单的标签助手类可用于设定操作方法，并用于实际项目中。本书中的 HTML helper 仅用于演示上下文数据是如何使用的。请查看第 24 章以了解内置标签助手的详情。

标签助手可以在构造函数中定义作为服务的依赖项，它们将使用依赖注入特性进行解析。在本例中，定义用于 IUrlHelperFactory 服务的依赖，以允许从路由数据创建传出的 URL。在 Process 方法中，标签助手使用 IUrlHelperFactory.GetUrlHelper 方法来获取使用 ViewContext 对象配置的 IUrlHelper 对象，为输出元素的 action 属性创建 URL。代码清单 23-25 展示了准备的视图，这里移除了 action 属性，以便使用标签助手来设置。

代码清单 23-25 在 Create.cshtml 文件中删除表单元素的属性

```
@model City

@{ Layout = "_Layout"; }

<form method="post">
```

```
    <div class="form-group">
        <label for="Name">Name:</label>
        <input class="form-control" name="Name" />
    </div>
    <div class="form-group">
        <label for="Country">Country:</label>
        <input class="form-control" name="Country" />
    </div>
    <div class="form-group">
        <label for="Population">Population:</label>
        <input class="form-control" name="Population" />
    </div>
    <formbutton type="submit" bg-color="danger" />
    <formbutton type="reset" />
    <a bs-button-color="primary" href="/Home/Index">Cancel</a>
</form>
```

如果运行应用程序并访问/Home/Create，然后检查发送到浏览器的 HTML，你将会看到 form 元素拥有 action 属性，其值可使用上下文数据来设置，如下所示：

```
...
<form method="post" action="/Home/Create">
...
```

23.3.4　使用视图模型

标签助手还可以处理视图模型，裁剪执行的转换与创建的输出。为了进行演示，在 Infrastructure/TagHelpers 文件夹中创建名为 LabelAndInputTagHelper.cs 的类文件，并使用代码清单 23-26 所示内容定义此类。

代码清单 23-26　Infrastructure/TagHelpers 文件夹下的 LabelAndInputTagHelper.cs 文件的内容

```
using Microsoft.AspNetCore.Mvc.ViewFeatures;
using Microsoft.AspNetCore.Razor.TagHelpers;

namespace Cities.Infrastructure.TagHelpers {

    [HtmlTargetElement("label", Attributes = "helper-for")]
    [HtmlTargetElement("input", Attributes = "helper-for")]
    public class LabelAndInputTagHelper : TagHelper {

        public ModelExpression HelperFor { get; set; }
        public override void Process(TagHelperContext context,
                                     TagHelperOutput output) {

            if (output.TagName == "label") {
                output.TagMode = TagMode.StartTagAndEndTag;
                output.Content.Append(HelperFor.Name);
                output.Attributes.SetAttribute("for", HelperFor.Name);

            } else if (output.TagName == "input") {
                output.TagMode = TagMode.SelfClosing;
                output.Attributes.SetAttribute("name", HelperFor.Name);
                output.Attributes.SetAttribute("class", "form-control");
                if (HelperFor.Metadata.ModelType == typeof(int?)) {
                    output.Attributes.SetAttribute("type", "number");
                }
            }
        }
    }
}
```

该标签助手转换拥有 helper-for 属性的 label 与 input 元素。该标签助手的重要之处在于 HelperFor 属性的类型，该属性用于接收 helperfor 属性的值。

```
...
public ModelExpression HelperFor { get; set; }
...
```

当希望对视图模型的一部分进行处理时，将使用 ModelExpression 类。最简单的方式是跳过说明，如代码清单 23-27 所示，展示标签助手是如何应用于视图的。

代码清单 23-27 在 Views/Home 文件夹下的 Create.cshtml 文件中应用标签助手处理模型

```
@model Cities.Models.City

@{ Layout = "_Layout"; }

<form method="post">
    <div class="form-group">
        <label helper-for="Name" />
        <input helper-for="Name" />
    </div>
    <div class="form-group">
        <label helper-for="Country" />
        <input helper-for="Country" />
    </div>
    <div class="form-group">
        <label helper-for="Population"/>
        <input helper-for="Population" />
    </div>
    <formbutton type="submit" bg-color="danger" />
    <formbutton type="reset" />
    <a bs-button-color="primary" href="/Home/Index">Cancel</a>
</form>
```

helper-for 属性的值来自 Model 类，该类由 MVC 检测并作为 ModelExpression 对象提供给标签助手。

本节不会深入说明 ModelExpression 类，因为对类型执行的任何检查都会导致无尽的类和属性列表。此外，MVC 附带了一组有用的内置标签助手，可使用视图模型来转换元素，如第 24 章所述，这意味着不必自己创建。

对于标签助手，这里使用了两个值得说明的基本特性。第一个特性是获取模型属性的名称，以便可以包含在输出元素中，如下所示：

```
...
output.Content.Append(HelperFor.Name);
output.Attributes.SetAttribute("for", HelperFor.Name);
...
```

Name 属性返回模型属性的名称。

第二个特性是获取模型属性的类型，以便可以改变 input 元素的 type 属性，如下所示：

```
...
if (HelperFor.Metadata.ModelType == typeof(int?)) {
    output.Attributes.SetAttribute("type", "number");
}
...
```

如果执行应用程序并访问/Home/Create，然后检查发送到浏览器的 HTML 内容，你将会看到如下元素：

```
<div class="form-group">
    <label for="Name">Name</label>
```

```
    <input name="Name" class="form-control" />
</div>
<div class="form-group">
    <label for="Country">Country</label>
    <input name="Country" class="form-control" />
</div>
<div class="form-group">
    <label for="Population">Population</label>
    <input name="Population" class="form-control" type="number" />
</div>
```

以上代码把用于 input 元素 Population 的 type 属性设置为 number，以反映 City.Population 属性在 C#类中为 int 类型。本节展示了标签助手如何反映不同的模型特征来生成 HTML。基于使用的浏览器，input 元素将只允许输入数字。

23.3.5 协调标签助手

TagHelperContext.Items 属性提供了一个字典，用于在操作元素的标签助手与后继操作元素的标签助手之间进行协调。为了演示 Items 集合的用法，在 Infrastructure/TagHelpers 文件夹中添加名为 CoordinatingTagHelpers.cs 的类文件，在其中使用代码清单 23-28 所示的内容定义一对标签助手。

代码清单 23-28 Infrastructure/TagHelpers 文件夹下的 CoordinatingTagHelpers.cs 文件的内容

```
using Microsoft.AspNetCore.Razor.TagHelpers;

namespace Cities.Infrastructure.TagHelpers {

    [HtmlTargetElement("div", Attributes = "theme")]
    public class ButtonGroupThemeTagHelper : TagHelper {

        public string Theme { get; set; }

        public override void Process(TagHelperContext context,
                                     TagHelperOutput output) {
            context.Items["theme"] = Theme;
        }
    }

    [HtmlTargetElement("button", ParentTag = "div")]
    [HtmlTargetElement("a", ParentTag = "div")]
    public class ButtonThemeTagHelper : TagHelper {

        public override void Process(TagHelperContext context,
                                     TagHelperOutput output) {

            if (context.Items.ContainsKey("theme")) {
                output.Attributes.SetAttribute("class",
                    $"btn btn-{context.Items["theme"]}");
            }
        }
    }
}
```

第一个标签助手是 ButtonGroupThemeTagHelper 类，用于处理拥有 theme 属性的 div 元素。协调标签助手可以转换自己的元素，但是这个示例只是简单地添加 theme 属性的值到 Items 字典，以便包含在 div 元素内的标签助手可以使用。

第二个标签助手是 ButtonThemeTagHelper 类，用于处理拥有 a 元素和 button 元素的 div 元素，并使用来自 Items 字典的 theme 属性值来设置输出元素的 Bootstrap 样式。代码清单 23-29 展示了这些标签助手应用的元素集合。

代码清单 23-29　在 Views/Home 文件夹下的 Create.cshtml 文件中应用协调标签助手

```
@model Cities.Models.City

@{  Layout = "_Layout"; }

<form method="post">
    <div class="form-group">
        <label helper-for="Name" />
        <input helper-for="Name" />
    </div>
    <div class="form-group">
        <label helper-for="Country" />
        <input helper-for="Country" />
    </div>
    <div class="form-group">
        <label helper-for="Population" />
        <input helper-for="Population" />
    </div>
    <div theme="primary">
        <button type="submit">Add</button>
        <button type="reset">Reset</button>
        <a href="/Home/Index">Cancel</a>
    </div>
</form>
```

如果运行应用程序并访问/Home/Create，你将会看到按钮组都以同样的方式修饰。如果修改 div 元素的 theme 属性为其他的 Bootstrap 主题设置，例如 info、danger 或 primary，并重新加载页面，你将会看到修改已反映到按钮上，如图 23-6 所示。

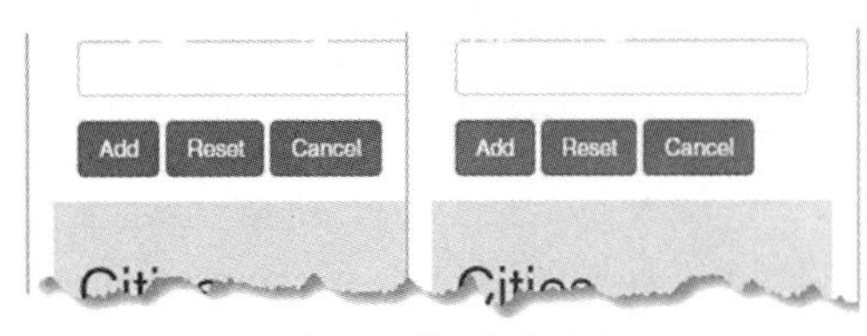

图 23-6　协调标签助手

23.3.6　抑制输出元素

通过在使用 Process 方法的参数接收的 TagHelperOutput 对象上调用 SuppressOuput 方法，标签助手可以用于防止元素被包含在输出到浏览器的 HTML 中。代码清单 23-30 在共享布局中添加了一个元素来显示高亮信息，但是这里仅仅对于特定的请求方式才显示。

代码清单 23-30　在 Views/Shared 文件夹下的_Layout.cshtml 文件中添加可视信息

```
<!DOCTYPE html>
<html>
<head>
    <meta name="viewport" content="width=device-width" />
    <title>Cities</title>
    <link href="/lib/bootstrap/dist/css/bootstrap.css" rel="stylesheet" />
</head>
<body class="m-1 p-1">
    <div show-for-action="Index" class="m-1 p-1 bg-danger">
        <h2>Important Message</h2>
    </div>
    <div title="Cities">@RenderBody()</div>
</body>
</html>
```

show-for-action 属性指定了希望显示警告信息的操作方法的名称。这不是在实际应用程序中控制包含内容的有用方式，但是对于仅有一个控制器和两个操作方法名称的示例应用来说是有效的。代码清单 23-31 展示了 SelectiveTagHelper.cs 类的内容，将它添加到 Infrastructure/TagHelpers 文件夹中。

代码清单 23-31 Infrastructure/TagHelpers 文件夹下的 SelectiveTagHelper.cs 文件的内容

```
using System;
using Microsoft.AspNetCore.Mvc.Rendering;
using Microsoft.AspNetCore.Mvc.ViewFeatures;
using Microsoft.AspNetCore.Razor.TagHelpers;

namespace Cities.Infrastructure.TagHelpers {

    [HtmlTargetElement(Attributes = "show-for-action")]
    public class SelectiveTagHelper : TagHelper {

        public string ShowForAction { get; set; }

        [ViewContext]
        [HtmlAttributeNotBound]
        public ViewContext ViewContext { get; set; }
        public override void Process(TagHelperContext context,
                                    TagHelperOutput output) {

            if (!ViewContext.RouteData.Values["action"].ToString()
                    .Equals(ShowForAction, StringComparison.OrdinalIgnoreCase)) {
                output.SuppressOutput();
            }
        }
    }
}
```

这个标签助手使用 ViewContext 从路由数据中获取 action 值，然后与 HTML 元素的 show-for-action 属性做比较。如果它们不匹配，则 SuppressOutput 方法被调用。为查看效果，启动应用程序并访问/Home/Index 和/Home/Create。如图 23-7 所示，信息仅仅在 Index 这个操作方法被调用时显示。

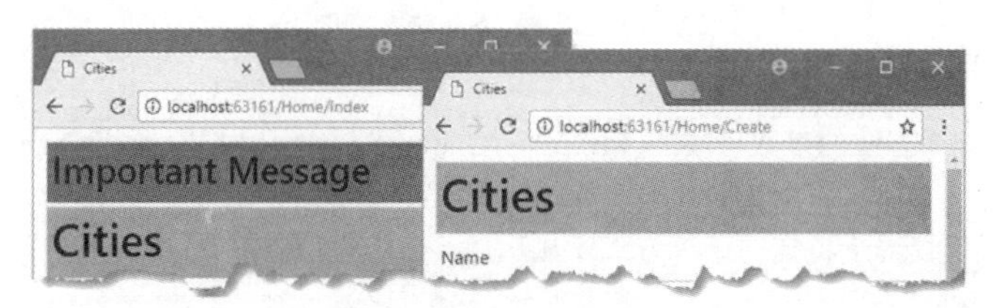

图 23-7 使用标签助手抑制元素

23.4 小结

本章介绍了标签助手的用法，标签助手是 ASP.NET Core MVC 的新增特性。本章还介绍了标签助手在 Razor 视图中的角色，并演示了自定义标签助手是如何创建、注册并应用的。本章接下来展示了如何控制标签助手的作用域，并说明了使用标签助手转换 HTML 元素的各种不同方式。下一章将说明标签助手在 HTML 表单元素中的使用。

第 24 章　使用表单标签助手

MVC 提供了一套内置的标签助手来执行日常必需的 HTML 转换。本章介绍操作 HTML 表单的标签助手，HTML 表单往往包括 form、input、label、select、option 和 textarea 元素。第 25 章将介绍其他内置的标签助手，它们提供非表单元素的特性。表 24-1 介绍了表单标签助手的背景。

表 24-1　　表单标签助手的背景

问题	答案
表单标签助手是什么？	表单标签助手用于转换 HTML 表单元素，从而不必编写自定义的标签助手就可以处理常见的多数问题
表单标签助手有何作用？	表单标签助手确保 HTML 表单元素（包括在表单内的元素，比如 label 和 input 元素）被一致地生成。大部分情况下，标签助手确保重要的属性（如 id、name 和 for）已使用视图模型类正确设置，但是有些标签助手也可以生成内容，例如填充 select 元素中的 option 子元素
如何使用表单标签助手？	使用内置的标签助手查询带有 asp-前缀的属性，比如 asp-for 属性
表单标签助手有何缺陷或限制？	仅有的限制是提供给标签助手的用以生成 select 元素中 option 子元素的模型数据的方式。24.5 节将介绍这个问题，并提供自定义的标签助手来处理
表单标签助手有何替代品？	可以不使用这些标签助手在视图中编写 HTML 表单，也可以使用第 23 章介绍的技术，开发自己的标签助手

表 24-2 列出了本章要介绍的操作。

表 24-2　　本章要介绍的操作

操作	方法	代码清单
为 form 元素设置 action 属性	使用 form 元素标签助手	代码清单 24-5
防止跨站请求伪造（cross-site request forgery）	为操作方法应用 ValidateAntiForgeryToken 特性，在表单元素上设置 asp-antiforgery 属性（可选）	代码清单 24-6 和代码清单 24-7
设置 input 元素的 id、name 和 value 属性	应用 asp-for 属性	代码清单 24-8
格式化 input 元素显示的值	为 input 元素应用 asp-format 属性，或者在模型类上应用 DisplayFormat 特性	代码清单 24-9～代码清单 24-12
设置 label 元素的 for 属性和元素内容	应用 asp-for 属性	代码清单 24-13
修改应用 asp-for 属性的 label 元素的内容	在模型类上应用 Display 特性，且使用 Name 属性指定内容	代码清单 24-14
为 select 元素设置 id 和 name 属性	应用 asp-for 属性	代码清单 24-15
生成 option 元素	应用 asp-items 属性	代码清单 24-16～代码清单 24-21
为 textarea 元素设置 id 和 name 属性	应用 asp-for 属性	代码清单 24-22 和代码清单 24-23

24.1　准备示例项目

本章继续使用第 23 章创建的 Cities 项目。对于本章，需要启用来自 MVC 的内置标签助手，并禁用第 23 章创建的自定义标签助手。代码清单 24-1 展示了在视图导入文件中所做的变更，这里使用 MVC 的标签助手替换了@addTagHelper 在 Cities 程序集中的标签助手，定义在命名空间 Microsoft.AspNetCore.Mvc.TagHelpers 中。

代码清单 24-1　修改 View 文件夹下的_ViewImports.cshtml 文件中的标签助手

```
@using Cities.Models
@addTagHelper *, Microsoft.AspNetCore.Mvc.TagHelpers
```

重置视图和布局

代码清单 24-2 展示了 Index.cshtml 文件的内容，这里移除了使用自定义标签助手类的属性。

代码清单 24-2 Views/Home 文件夹下的 Index.cshtml 文件中的内容

```
@model IEnumerable<City>

@{  Layout = "_Layout"; }

<table class="table table-sm table-bordered">
    <thead class="bg-primary text-white">
        <tr>
            <th>Name</th>
            <th>Country</th>
            <th class="text-right">Population</th>
        </tr>
    </thead>
    <tbody>
        @foreach (var city in Model) {
            <tr>
                <td>@city.Name</td>
                <td>@city.Country</td>
                <td class="text-right">@city.Population?.ToString("#,###")</td>
            </tr>
        }
    </tbody>
</table>
<a href="/Home/Create" class="btn btn-primary">Create</a>
```

代码清单 24-3 展示了对 Create.cshtml 文件所做的相关修改，这里回到了标准的 HTML 元素，而没有使用第 23 章介绍的属性。

代码清单 24-3 Views/Home 文件夹下的 Create.cshtml 文件的内容

```
@model City

@{  Layout = "_Layout"; }

<form method="post" action="/Home/Create">
    <div class="form-group">
        <label for="Name">Name:</label>
        <input class="form-control" name="Name" />
    </div>
    <div class="form-group">
        <label for="Country">Country:</label>
        <input class="form-control" name="Country" />
    </div>
    <div class="form-group">
        <label for="Population">Population:</label>
        <input class="form-control" name="Population" />
    </div>
    <button type="submit" class="btn btn-primary">Add</button>
    <a class="btn btn-primary" href="/Home/Index">Cancel</a>
</form>
```

最后修改共享布局，如代码清单 24-4 所示。

代码清单 24-4 Views/Shared 文件夹下的_Layout.cshtml 文件的内容

```
<!DOCTYPE html>
<html>
```

```
<head>
    <meta name="viewport" content="width=device-width" />
    <title>Cities</title>
    <link href="/lib/bootstrap/dist/css/bootstrap.css" rel="stylesheet" />
</head>
<body class="m-1 p-1">
    <div>@RenderBody()</div>
</body>
</html>
```

如果运行应用程序，你将会看到城市列表，可以单击 Create 按钮，填充表单后提交新的数据给服务器，如图 24-1 所示。

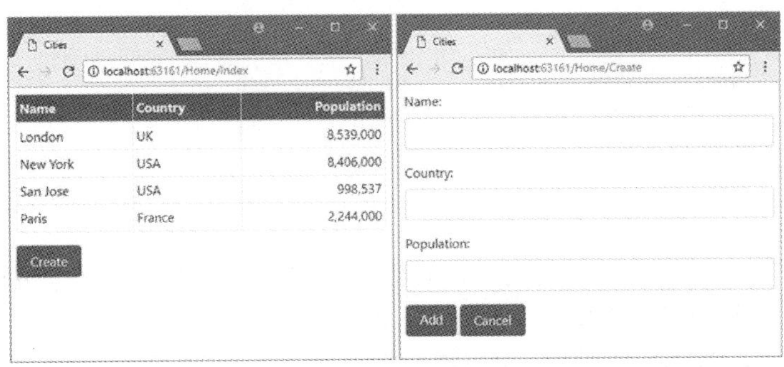

图 24-1 运行示例应用

24.2 使用 form 元素

FormTagHelper 类是 form 元素的内置标签助手，用于管理 HTML 表单的配置以便基于应用程序的路由配置指向正确的操作方法。该标签助手支持表 24-3 所示的属性。

表 24-3 form 元素的内置标签助手类定义的属性

属性	描述
asp-controller	用于指定 action 属性 URL 对应的路由系统的 controller 值。如果忽略，将使用呈现视图的控制器
asp-action	用于指定 action 属性 URL 对应的路由系统的 action 值。如果忽略，呈现视图的 action 将会被使用
asp-route-*	以 asp-route-开头的属性名用于指定 action 属性 URL 的附加值，所以 asp-route-id 属性用来为路由系统提供 id 片段的值
asp-route	用于指定生成 action 属性 URL 的路由名称
asp-area	指定用于生成 action 属性 URL 的 area 名称
asp-antiforgery	控制是否为视图添加 anti-forgery 信息

24.2.1 设置 form 目标

FormTagHelper 类的主要目的是使用应用程序的路由配置来设置 form 元素的 action 属性，确保即使路由架构发生了变化，表单数据也总是被发送到正确的 URL 地址。代码清单 24-5 使用 asp-action 和 asp-controller 属性来指向 Home 控制器的 Create 操作方法。

注 意

标签助手并不设置 method 属性，如果在 form 元素中忽略该属性，那么浏览器将会使用 GET 请求来发送表单数据。如第 17 章所述，如果表单数据用于修改应用程序中的数据，这可能导致问题。最佳实践是设置 method 属性，尤其当希望使用 GET 请求时，显然不能忘了设置 method 属性为 GET。

代码清单 24-5 设置 Views/Home 文件夹下的 Create.cshtml 文件中的 form 目标

```
@model City

@{  Layout = "_Layout"; }

<form method="post" asp-controller="Home" asp-action="Create">
    <div class="form-group">
        <label for="Name">Name:</label>
        <input class="form-control" name="Name" />
    </div>
    <div class="form-group">
        <label for="Country">Country:</label>
        <input class="form-control" name="Country" />
    </div>
    <div class="form-group">
        <label for="Population">Population:</label>
        <input class="form-control" name="Population" />
    </div>
    <button type="submit" class="btn btn-primary">Add</button>
    <a class="btn btn-primary" href="/Home/Index">Cancel</a>
</form>
```

如果运行应用程序，访问/Home/Create 并检查发送到客户端的 HTML，你将会看到标签助手已添加 action 属性到 form 元素中并使用路由系统设置了值，如下所示：

```
<form method="post" action="/Home/Create">
```

24.2.2 使用防伪特性

跨站请求伪造（Cross-Site Request Forgery，CSRF）是一种利用用户请求验证来攻击 Web 应用程序的方式。大多数 Web 应用程序（包括使用 ASP.NET Core 创建的）使用 Cookie 来标识哪些请求与特定会话相关，用户标识通常与之相关。

CSRF（也称为会话劫持）的详细说明参见维基百科，基于用户在使用应用程序，但是没有通过单击 Logout 按钮显式结束会话之后访问恶意站点，应用程序仍然将用户会话视为活动的，存储在浏览器中的 Cookie 仍没有过期。恶意站点包含 JavaScript 代码，它向应用程序发送表单请求，在未经用户同意的情况下执行操作，操作的性质取决于被攻击的应用程序。由于 JavaScript 代码在用户浏览器中执行，因此向应用程序发出的请求包含会话的 Cookie 值，而应用程序会在用户不知情或不同意的情况下执行操作。

如果 form 元素没有包含 action 属性（因为是路由系统使用 asp-controller 和 asp-acton 属性生成的），那么 FormTagHelper 类将自动启用反 CSRF 特性，将安全令牌包含在表单的一个 hidden 类型的 input 元素中，并同 Cookie 一起发送到客户端。应用程序将仅仅处理同时包含 Cookie 和来自表单隐藏域的值的请求，恶意站点不能访问。表单生成的每个请求都有新的唯一安全令牌。

如果运行应用程序，访问/Home/Create 并查看发送到浏览器的 HTML，你将会看到 input 隐藏域，如下所示：

```
<input name="__RequestVerificationToken" type="hidden" value="CfDJ8KuVkH8hFlRApe
    FBxTrhCFTKZe0B9BKwnWDJqLRUDk__PrEwaeCJmiBbGkwW1ZI816c_TrM5XQkJBeqNI5IL8FhuO
    RvjZuYIL-GZvnWZ62OThsZYT02HNX_Lu5LWDNWDdVoS5O5hZtzaoHLeY5lNto" />
```

如果按 F12 键，就可以看到相关的添加到回应中的 Cookie。将安全令牌添加到 HTML 响应中仅仅是处理的一部分，还必须在控制器中进行验证，如代码清单 24-6 所示。

代码清单 24-6 在 Colltrollers 文件夹下的 HomeController.cs 中验证 anti-forgery 令牌

```
using Microsoft.AspNetCore.Mvc;
using Cities.Models;

namespace Cities.Controllers {
```

```
    public class HomeController : Controller {
        private IRepository repository;

        public HomeController(IRepository repo) {
            repository = repo;
        }

        public ViewResult Index() => View(repository.Cities);

        public ViewResult Create() => View();

        [HttpPost]
        [ValidateAntiForgeryToken]
        public IActionResult Create(City city) {
            repository.AddCity(city);
            return RedirectToAction("Index");
        }
    }
}
```

ValidateAntoForgeryToken 特性确保请求中包含有效的反 CSRF 令牌，如果不存在或者没有包含期望的值，将会抛出异常。

FormTagHelper 类提供了 asp-antiforgery 属性来覆写默认的反 CSRF 行为。如果该属性的值为 true，安全令牌将被包含到响应中，即使 form 元素拥有 action 属性。如果该属性的值为 false，安全令牌将被禁用。代码清单 24-7 显式启用了该属性，因为没有在 form 元素中定义 action 属性，所以无论如何都将添加安全令牌。

代码清单 24-7 在 Views/Home 文件夹下的 Create.cshtml 文件中启用反 CSRF 特性

```
@model City

@{  Layout = "_Layout"; }

<form method="post" asp-controller="Home" asp-action="Create"
                    asp-antiforgery="true">
    <div class="form-group">
        <label for="Name">Name:</label>
        <input class="form-control" name="Name" />
    </div>
    <div class="form-group">
        <label for="Country">Country:</label>
        <input class="form-control" name="Country" />
    </div>
    <div class="form-group">
        <label for="Population">Population:</label>
        <input class="form-control" name="Population" />
    </div>
    <button type="submit" class="btn btn-primary">Add</button>
    <a class="btn btn-primary" href="/Home/Index">Cancel</a>
</form>
```

提 示

测试反 CSRF 特性时需要一点技巧。首先让请求包含表单的 URL 地址（对于本例来说是 /Home/Create），然后使用 F12 键来定位并删除表单中的 input 隐藏域。当填充表单后发送回应用程序时，浏览器不会提供请求的数据，请求将会失败并展示错误页面。

24.3 使用 input 元素

input 元素是 HTML 表单的核心，它们提供了用户可以将非结构化数据提供给应用程序的主要方法。InputTagHelper 类用于转换 input 元素，它们反映了在视图模型上获取的数据类型和格式，并使用表 24-4 所示的属性。

表 24-4 用于 input 元素的内置标签助手的属性

属性	描述
asp-for	用于指定 input 元素表示的视图模型属性
asp-format	用于指定 input 元素表示的视图模型属性值的格式

24.3.1 配置 input 元素

asp-for 属性被设置为视图模型属性的名称，从而用于设置 input 元素的 id、type 和 value 属性。代码清单 24-8 为 Create.cshtml 文件中的 input 元素应用了 asp-for 属性。

代码清单 24-8 在 Views/Home 文件夹下的 Create.cshtml 文件中配置 input 元素

```
@model City

@{  Layout = "_Layout"; }

<form method="post" asp-controller="Home" asp-action="Create"
                    asp-antiforgery="true">
    <div class="form-group">
        <label for="Name">Name:</label>
        <input class="form-control" asp-for="Name" />
    </div>
    <div class="form-group">
        <label for="Country">Country:</label>
        <input class="form-control" asp-for="Country" />
    </div>
    <div class="form-group">
        <label for="Population">Population:</label>
        <input class="form-control" asp-for="Population" />
    </div>
    <button type="submit" class="btn btn-primary">Add</button>
    <a class="btn btn-primary" href="/Home/Index">Cancel</a>
</form>
```

如果运行应用程序并访问/Home/Create，你将会看到标签助手使用 asp-for 属性来定制每个 input 元素，如以下代码所示（忽略了反 CSRF 安全令牌）：

```
<form method="post" action="/Home/Create">
    <div class="form-group">
        <label for="Name">Name:</label>
        <input class="form-control" type="text" id="Name" name="Name" value="" />
    </div>
    <div class="form-group">
        <label for="Country">Country:</label>
        <input class="form-control" type="text" id="Country"
               name="Country" value="" />
    </div>
    <div class="form-group">
        <label for="Population">Population:</label>
```

```
            <input class="form-control" type="number" id="Population"
                   name="Population" value="" />
        </div>
        <button type="submit" class="btn btn-primary">Add</button>
        <a class="btn btn-primary" href="/Home/Index">Cancel</a>
    </form>
```

input 元素的 type 属性指定了如何在表单中显示该元素。可以在用于 Population 属性的 input 元素中简单地看到结果，其 type 属性已经设置为 number。因为 C#的 Population 属性类型为 int?，所以标签助手使用 type 属性来指示浏览器只有数字才能接受。

注 意

人们将 type 属性的解释方式留给了浏览器。并非所有的浏览器都响应 HTML5 规范中定义的所有类型，即使处理，它们的实现方式也存在差异。type 属性是对表单中所需数据类型的有效提示，但是你应该使用模型验证特性来确保用户提供有用的数据。

表 24-5 说明了不同的 C#属性类型和它们生成的 input 元素类型。

表 24-5 C#属性类型和它们生成的 input 元素类型

C#属性类型	生成的 input 元素类型
byte、sbyte、int、uint、short、ushort、long、ulong	number
float、double、decimal	text，带有额外的用于模型验证的属性，稍后说明
bool	checkbox
string	text
DateTime	datetime

类型 float、double 和 decimal 能生成 type 属性为 text 的 input 元素，因为不是所有的浏览器都支持可以用于这些类型的全部字符。为辅助用户，标签助手为 input 元素额外添加了用于模型验证的属性，具体将在第 27 章说明。

通过在 input 元素上定义 type 属性可以重写表 24-5 所示的默认映射。标签助手不会重写你定义的值，这允许你利用各种可用的 input 元素类型，例如 password 或 hidden，以及 HTML5 中的新类型，比如 number。

这种方式的不足是，必须在为给定的模型属性生成 input 元素的所有视图中设置 type 属性。如果需要在多个视图中覆写默认映射，可以为 C#模型类的属性应用 UIHint 特性，指定表 24-6 中的值作为特性值。

提 示

如果模型属性不使用表 24-5 中的类型且没有使用 UIHint 特性进行装饰，标签助手将设置 input 元素的 type 属性为 text。

表 24-6 UIHint 特性值及其生成的 input 元素类型

特性值	input 元素类型
HiddenInput	hidden
Password	password
Text	text
PhoneNumber	tel
Url	url
EmailAddress	email
Time	time（用于显示 DateTime 对象的 time 部分）
Date	date（用于显示 DateTime 对象的 date 部分）
DateTime-local	datetime-local（用于显示没有提供时区信息的 DateTime 对象）

24.3.2 格式化数据

当操作方法提供带有视图模型的视图时，标签助手使用 asp-for 属性提供的值来设置 input 元素的 value 属性。asp-format 属性用于指定数据如何格式化。

为了演示，在 Home 控制器中添加一个新的操作方法，如代码清单 24-9 所示。该操作方法从存储中选取第一个 City 对象，并作为 Create 视图的视图模型。

代码清单 24-9 在 Controller 文件夹下的 HomeController.cs 文件中添加操作方法

```
using Microsoft.AspNetCore.Mvc;
using Cities.Models;
using System.Linq;

namespace Cities.Controllers {

    public class HomeController : Controller {
        private IRepository repository;

        public HomeController(IRepository repo) {
            repository = repo;
        }

        public ViewResult Index() => View(repository.Cities);

        public ViewResult Edit() => View("Create", repository.Cities.First());

        public ViewResult Create() => View();

        [HttpPost]
        [ValidateAntiForgeryToken]
        public IActionResult Create(City city) {
            repository.AddCity(city);
            return RedirectToAction("Index");
        }
    }
}
```

如果运行应用程序，访问/Home/Edit 并检查发送给浏览器的 HTML，你将会看到 value 属性已经使用视图模型对象填充了，如下所示：

```
<input class="form-control" type="number" id="Population"
      name="Population" value="8539000" />
```

asp-format 属性接收一个将被传递给标准 C#字符串格式化系统的值，如代码清单 24-10 所示。

代码清单 24-10 在 Views/Home 文件夹下的 Create.cshtml 文件中格式化数据

```
@model City

@{  Layout = "_Layout"; }

<form method="post" asp-controller="Home" asp-action="Create"
                    asp-antiforgery="true">
    <div class="form-group">
        <label for="Name">Name:</label>
        <input class="form-control" asp-for="Name" />
    </div>
    <div class="form-group">
        <label for="Country">Country:</label>
        <input class="form-control" asp-for="Country" />
    </div>
```

```
    <div class="form-group">
        <label for="Population">Population:</label>
        <input class="form-control" asp-for="Population" asp-format="{0:#,###}" />
    </div>
    <button type="submit" class="btn btn-primary">Add</button>
    <a class="btn btn-primary" href="/Home/Index">Cancel</a>
</form>
```

asp-format 属性的值是逐字使用的，这意味着必须包含大括号与“0:引用”以及所需的格式。如果运行应用程序并访问/Home/Edit，你将看到 Population 值被格式化为如下形式：

```
<input class="form-control" type="number" id="Population"
       name="Population" value="8,539,000" />
```

asp-format 属性需要慎用，因为必须确保应用程序的其他部分支持你所使用的格式。在这种情况下，你通过格式化 Population 的值引发了一个问题。标签助手设置 input 元素的 type 属性为 number，并对 Population 使用了表 24-5 中的默认映射，但是由指定的格式化串生成的 value 属性值包含非数字字符。因此，遵守 number 元素类型的浏览器（不是所有，记住）可能不会在元素中显示任何值。

你还必须确保应用程序能够解析你所使用的格式中的值。示例应用程序期望接收可以解析为 int 类型的 Population 值，包含非数字字符将导致验证错误，如第 27 章所述。

通过模型类来指定格式

如果你总是希望对模型属性使用同样的格式，那么可以使用 DisplayFormat 特性修饰 C#类，该特性定义在命名空间 System.ComponentModel.DataAnnotations 中。DisplayFormat 特性需要两个参数来格式化数据的值：DataFormatString 参数指定格式化串，ApplyFormatInEditMode 参数指定当值被编辑时格式化也应被使用。代码清单 24-11 使用 DisplayFormat 特性修饰了 Population 属性，使用的是可以被应用程序和浏览器处理的数字格式。

代码清单 24-11 在 Models 文件夹下的 City.cs 文件中为模型类应用格式化特性

```
using System.ComponentModel.DataAnnotations;

namespace Cities.Models {

    public class City {

        public string Name { get; set; }
        public string Country { get; set; }

        [DisplayFormat(DataFormatString = "{0:F2}", ApplyFormatInEditMode = true)]
        public int? Population { get; set; }
    }
}
```

属性 asp-format 优先于 DisplayFormat 特性，所以需要从视图中删除这个特性，如代码清单 24-12 所示。

代码清单 24-12 从 Views/Home 文件夹下的 Create.cshtml 文件中删除格式化特性

```
@model City

@{ Layout = "_Layout"; }

<form method="post" asp-controller="Home" asp-action="Create"
      asp-antiforgery="true">
    <div class="form-group">
        <label for="Name">Name:</label>
        <input class="form-control" asp-for="Name" />
    </div>
    <div class="form-group">
        <label for="Country">Country:</label>
        <input class="form-control" asp-for="Country" />
```

```
    </div>
    <div class="form-group">
        <label for="Population">Population:</label>
        <input class="form-control" asp-for="Population" />
        </div>
    <button type="submit" class="btn btn-primary">Add</button>
    <a class="btn btn-primary" href="/Home/Index">Cancel</a>
</form>
```

如果运行应用程序并访问/Home/Edit，你将看到 Population 值已经使用两位小数格式化，如下所示：

```
<input class="form-control" type="number" id="Population"
    name="Population" value="8539000.00" />
```

24.4 使用 label 元素

label 元素可使用 LabelTagHelper 类进行转换，这个标签助手使用视图模型类来确保标签的免输入和一致性。支持的属性只有 asp-for，该属性用于指定 label 元素表示的视图模型属性。

Label 标签助手将使用视图模型属性的名称来设置 for 属性的值与 label 元素的内容。代码清单 24-13 在表单中为 label 元素应用了 asp-for 属性，它们将由标签助手进行转换。

代码清单 24-13 在 Views/Home 文件夹下的 Create.cshtml 文件中应用 Label 标签助手

```
@model City

@{  Layout = "_Layout"; }

<form method="post" asp-controller="Home" asp-action="Create"
      asp-antiforgery="true">
    <div class="form-group">
        <label asp-for="Name"></label>
        <input class="form-control" asp-for="Name" />
    </div>
    <div class="form-group">
        <label asp-for="Country"></label>
        <input class="form-control" asp-for="Country" />
    </div>
    <div class="form-group">
        <label asp-for="Population"></label>
        <input class="form-control" asp-for="Population" />
    </div>
    <button type="submit" class="btn btn-primary">Add</button>
    <a class="btn btn-primary" href="/Home/Index">Cancel</a>
</form>
```

因为 label 元素是空的，所以 Label 标签助手将使用模型属性名称作为元素的内容并设置 for 属性，以告诉浏览器每个标签关联的 input 元素。如果运行示例应用程序，访问/Home/Create 或/Homc/Edit，并且检查发送到浏览器的 HTML 内容，你将看到如下输出元素：

```
<form method="post" action="/Home/Create">
    <div class="form-group">
        <label for="Name">Name</label>
        <input class="form-control" type="text" id="Name"
               name="Name" value="London" />
    </div>
    <div class="form-group">
```

```
        <label for="Country">Country</label>
        <input class="form-control" type="text" id="Country"
               name="Country" value="UK" />
    </div>
    <div class="form-group">
        <label for="Population">Population</label>
        <input class="form-control" type="number" id="Population"
               name="Population" value="8539000.00" />
    </div>
    <button type="submit" class="btn btn-primary">Add</button>
    <a class="btn btn-primary" href="/Home/Index">Cancel</a>
</form>
```

可以通过为模型类的属性应用 Display 特性来覆写作为 label 元素的内容，如代码清单 24-14 所示。

代码清单 24-14 在 Models 文件夹下的 City.cs 文件中为模型属性改变说明信息

```
using System.ComponentModel.DataAnnotations;

namespace Cities.Models {

    public class City {

        [Display(Name = "City")]
        public string Name { get; set; }

        public string Country { get; set; }

        [DisplayFormat(DataFormatString = "{0:F2}", ApplyFormatInEditMode = true)]
        public int? Population { get; set; }
    }
}
```

参数 Name 指定用于替代属性名称的值。如果运行示例，访问/Home/Create 并检查发送给浏览器的 HTML 内容，你将会看到 label 元素的内容已经改变，如下所示：

```
<div class="form-group">
    <label for="Name">City</label>
    <input class="form-control" type="text" id="Name" name="Name" value="London" />
</div>
```

注意，for 属性的值没有变化，所以浏览器知道 label 元素已关联到特定的 input 元素，且不受 Display 特性影响。

提 示

可以通过自定义 label 元素的内容来阻止标签助手进行设置。如果希望 label 元素不只包含属性的名称，这将很有用，这就是 Label 标签助手可以提供的全部内容。

24.5 使用 select 和 option 元素

select 和 option 元素为用户提供固定的选择集合，而不像 input 元素那样支持所有可能的开放数据选项。SelectTagHelper 类负责转换 select 元素，支持的属性如表 24-7 所示。

表 24-7 用于 select 元素的内置标签助手类定义的属性

属性	描述
asp-for	用于指定 select 元素表示的视图模型属性
asp-items	用于指定包含在 select 元素中的 option 元素值的来源

属性 asp-for 用于设置 for 和 id 属性以反映接收的模型属性。代码清单 24-15 使用定义了 asp-for 属性的 select 元素来替换用于 Country 属性的 input 元素。

代码清单 24-15 在 Views/Home 文件夹下的 Create.cshtml 文件中使用 select 元素

```
@model City

@{ Layout = "_Layout"; }

<form method="post" asp-controller="Home" asp-action="Create"
      asp-antiforgery="true">
    <div class="form-group">
        <label asp-for="Name"></label>
        <input class="form-control" asp-for="Name" />
    </div>
    <div class="form-group">
        <label asp-for="Country"></label>
        <select class="form-control" asp-for="Country">
            <option disabled selected value="">Select a Country</option>
            <option>UK</option>
            <option>USA</option>
            <option>France</option>
            <option>China</option>
        </select>
    </div>
    <div class="form-group">
        <label asp-for="Population"></label>
        <input class="form-control" asp-for="Population" />
    </div>
    <button type="submit" class="btn btn-primary">Add</button>
    <a class="btn btn-primary" href="/Home/Index">Cancel</a>
</form>
```

这里手动填充了 select 元素的 option 元素，以提供用户选择的国家范围。如果运行应用程序并访问/Home/Create，你将看到发送给浏览器的 HTML 包含如下 select 元素：

```
<select class="form-control" id="Country" name="Country">
    <option disabled selected value="">Select a Country</option>
    <option>UK</option>
    <option>USA</option>
    <option>France</option>
    <option>China</option>
</select>
```

如果访问/Home/Edit 并检查发送到浏览器的 HTML 内容，你将看到视图模型属性 Country 的值已经被修改为选中的 option 元素，如下所示：

```
<select class="form-control" id="Country" name="Country">
    <option disabled selected value="">Select a Country</option>
    <option selected="selected">UK</option>
    <option>USA</option>
    <option>France</option>
    <option>China</option>
</select>
```

选中 option 元素的任务是由 OptionTagHelper 类执行的，这个标签助手通过 TagHelperContext.Items 集合接收来自 SelectTagHelper 的指令。如第 23 章所述，标签助手使用的集合需要协同工作，当创建自定义的标签助手以解决内置的限制时，可利用通过 SelectTagHelper 添加到 Items 集合中的数据。

24.5.1 使用数据源填充 select 元素

显式定义 select 元素的 option 元素，对于选择始终具有同样可能性的情况是有帮助的。但是，当需要提供从数据模型中获取选项，或当需要在多个视图中具有相同的选项集合并且不希望手动维护重复的内容时，就没有什么用了。

24.5.2 从枚举中生成 option 元素

如果有固定的选项集合呈现给用户，并且不希望在整个应用程序的视图中重复它们，那么可以使用枚举。在 Models 文件夹中添加名为 CountryNames.cs 的类文件，并用它定义代码清单 24-16 所示的枚举。

代码清单 24-16 Models 文件夹下的 CountryNames.cs 文件的内容

```
namespace Cities.Models {

    public enum CountryNames {
        UK,
        USA,
        France,
        China
    }
}
```

不能直接在 asp-item 属性中使用枚举，因为标签助手期望处理的是 SelectListItem 对象序列。但是，可以使用辅助方法执行必要的转换，如代码清单 24-17 所示。

代码清单 24-17 在 Views/Home 文件夹下的 Create.cshtml 文件中使用枚举生成 option 元素

```
@model City

@{ Layout = "_Layout"; }

<form method="post" asp-controller="Home" asp-action="Create"
      asp-antiforgery="true">
    <div class="form-group">
        <label asp-for="Name"></label>
        <input class="form-control" asp-for="Name" />
    </div>
    <div class="form-group">
        <label asp-for="Country"></label>
        <select class="form-control" asp-for="Country"
                asp-items="@new SelectList(Enum.GetNames(typeof(CountryNames)))">
            <option disabled selected value="">Select a Country</option>
        </select>
    </div>
    <div class="form-group">
        <label asp-for="Population"></label>
        <input class="form-control" asp-for="Population" />
    </div>
    <button type="submit" class="btn btn-primary">Add</button>
    <a class="btn btn-primary" href="/Home/Index">Cancel</a>
</form>
```

当使用枚举时，生成 option 元素的最佳方式，是为 asp-items 属性提供由枚举值的名称填充的 SelectList 对象。在幕后，SelectTagHelper 类从 IEnumerable<SelectListItem>生成 option 元素，SelectList 类实现了这个接口。

如果运行应用程序并访问/Home/Create 或/Home/Edit，你将看到发送到浏览器的 HTML 中包含与枚举值对应的 option 元素，如下所示：

```
<select class="form-control" id="Country" name="Country">
    <option disabled selected value="">Select a Country</option>
    <option>UK</option>
    <option>USA</option>
```

```
    <option>France</option>
    <option>China</option>
</select>
```

注意，标签助手保留了占位用的 option 元素。任何显式定义的 option 元素都保持不变，这意味着不必在数据中混合占位符。

1. 通过模型生成 option 元素

如果需要生成 option 元素以反映模型中的数据，那么最简单的方式是通过 View Bag 来提供生成元素所需的数据，如代码清单 24-18 所示。

代码清单 24-18 在 Controllers 文件夹下的 HomeController.cs 中通过 View Bag 提供数据

```
using Microsoft.AspNetCore.Mvc;
using Cities.Models;
using System.Linq;
using Microsoft.AspNetCore.Mvc.Rendering;

namespace Cities.Controllers {

    public class HomeController : Controller {
        private IRepository repository;

        public HomeController(IRepository repo) {
            repository = repo;
        }

        public ViewResult Index() => View(repository.Cities);

        public ViewResult Edit() {
            ViewBag.Countries = new SelectList(repository.Cities
                .Select(c => c.Country).Distinct());
            return View("Create", repository.Cities.First());
        }

        public ViewResult Create() {
            ViewBag.Countries = new SelectList(repository.Cities
                .Select(c => c.Country).Distinct());
            return View();
        }

        [HttpPost]
        [ValidateAntiForgeryToken]
        public IActionResult Create(City city) {
            repository.AddCity(city);
            return RedirectToAction("Index");
        }
    }
}
```

操作方法 Edit 和 Create 将 ViewBag.Countries 属性设置为使用存储库中的 City.Country 唯一值填充的 SelectList 对象。代码清单 24-19 使用 asp-items 属性告诉标签助手为 option 元素从 Countries 属性中获取数据。

代码清单 24-19 在 Views/Home 文件夹下的 Create.cshtml 文件中为 option 元素使用 View Bag

```
@model City

@{  Layout = "_Layout"; }

<form method="post" asp-controller="Home" asp-action="Create"
```

```
      asp-antiforgery="true">
    <div class="form-group">
        <label asp-for="Name"></label>
        <input class="form-control" asp-for="Name" />
    </div>
    <div class="form-group">
        <label asp-for="Country"></label>
        <select class="form-control" asp-for="Country" asp-items="ViewBag.Countries">
            <option disabled selected value="">Select a Country</option>
        </select>
    </div>
    <div class="form-group">
        <label asp-for="Population"></label>
        <input class="form-control" asp-for="Population" />
    </div>
    <button type="submit" class="btn btn-primary">Add</button>
    <a class="btn btn-primary" href="/Home/Index">Cancel</a>
</form>
```

如果运行应用程序并访问/Home/Create 或/Home/Edit，你将看到生成的 option 元素，如下所示：

```
<select class="form-control" id="Country" name="Country">
    <option disabled selected value="">Select a Country</option>
    <option selected>UK</option>
    <option>USA</option>
    <option>France</option>
</select>
```

2. 使用自定义的标签助手从模型生成 option 元素

通过 View Bag 为 option 元素传递数据的问题在于，必须在每个渲染视图的操作方法中留意为标签助手生成数据，这将导致代码的重复。你可以在代码清单 24-18 中获取一些感觉，这使得测试和维护控制器变得更加困难。

更好的方式是创建自定义的标签助手来支持内置的 SelectTagHelper 类。在 Infrastructure/TagHelper 文件夹中创建名为 SelectOptionTagHelper.cs 的类文件，定义代码清单 24-20 所示的类。

代码清单 24-20 Infrastructure/TagHelper 文件夹下的 SelectOptionTagHelper.cs 文件的内容

```
using System;
using System.Linq;
using System.Reflection;
using System.Threading.Tasks;
using Cities.Models;
using Microsoft.AspNetCore.Mvc.ViewFeatures;
using Microsoft.AspNetCore.Razor.TagHelpers;

namespace Cities.Infrastructure.TagHelpers {

    [HtmlTargetElement("select", Attributes = "model-for")]
    public class SelectOptionTagHelper : TagHelper {
        private IRepository repository;

        public SelectOptionTagHelper(IRepository repo) {
            repository = repo;
        }

        public ModelExpression ModelFor { get; set; }

        public override async Task ProcessAsync(TagHelperContext context,
                TagHelperOutput output) {
```

```
            output.Content.AppendHtml(
                (await output.GetChildContentAsync(false)).GetContent());

            string selected = ModelFor.Model as string;

            PropertyInfo property = typeof(City)
                .GetTypeInfo().GetDeclaredProperty(ModelFor.Name);
            foreach (string country in repository.Cities
                    .Select(c => property.GetValue(c)).Distinct()) {
                if (selected != null && selected.Equals(country,
                        StringComparison.OrdinalIgnoreCase)) {
                    output.Content
                        .AppendHtml($"<option selected>{country}</option>");
                } else {
                    output.Content.AppendHtml($"<option>{country}</option>");
                }
            }
            output.Attributes.SetAttribute("Name", ModelFor.Name);
            output.Attributes.SetAttribute("Id", ModelFor.Name);
        }
    }
}
```

该标签助手通过 model-for 属性在 select 元素上执行操作，并使用依赖注入接收可以不依赖渲染视图的来自控制器的用于访问模型数据的存储库对象。该标签助手定义了异步的 ProcessAsync 方法，从而简化了获取和保留 select 元素中任何现有内容的过程，这是通过 GetChildContentAsync 方法实现的。

SelectTagHelper 指示应通过 Items 集合中的选项（使用自己的类型为键）选择 option 元素的名称。该标签助手获取已选中选项的列表，并与 LINQ 查询结果结合使用，以便为存储库中的每个唯一值生成 option 元素。代码清单 24-21 更新了 select 元素，以便 asp-items 属性被 model-for 属性替换，还添加了@addTagHelper 表达式，从而仅为 Create 视图启用自定义的标签助手。

代码清单 24-21 在 Views/Home 文件夹下的 Create.cshtml 文件中启用自定义的标签助手

```
@model City
@addTagHelper Cities.Infrastructure.TagHelpers.SelectOptionTagHelper, Cities

@{  Layout = "_Layout"; }

<form method="post" asp-controller="Home" asp-action="Create"
    asp-antiforgery="true">
    <div class="form-group">
        <label asp-for="Name"></label>
        <input class="form-control" asp-for="Name" />
    </div>
    <div class="form-group">
        <label asp-for="Country"></label>
        <select class="form-control" model-for="Country">
            <option disabled selected value="">Select a Country</option>
        </select>
    </div>
    <div class="form-group">
        <label asp-for="Population"></label>
        <input class="form-control" asp-for="Population" />
    </div>
    <button type="submit" class="btn btn-primary">Add</button>
    <a class="btn btn-primary" href="/Home/Index">Cancel</a>
</form>
```

新的标签助手会生成同样的内容，但是不需要内置标签助手要求的 View Bag 数据。这种方式让操作方法可以专注于它们的特定任务，并保持应用程序的整体外观。

24.6 使用 textarea 元素

textarea 元素用于从用户获取更多的文本，通常用于非结构化数据，比如注释或观测结果。TextAreaTagHelper 负责转换 textarea 元素，并支持单个属性 asp-for，该属性用于指定 textarea 元素表示的视图模型属性。

TextAreaTagHelper 类相对简单，为 asp-for 属性提供的值用于为 textarea 元素设置 id 和 name 属性。为了演示这个标签助手，为 City 模型类添加了一个新的属性，如代码清单 24-22 所示。

代码清单 24-22 在 Models 文件夹下的 City.cs 文件中添加属性

```
using System.ComponentModel.DataAnnotations;

namespace Cities.Models {

    public class City {

        [Display(Name = "City")]
        public string Name { get; set; }

        public string Country { get; set; }

        [DisplayFormat(DataFormatString = "{0:F2}", ApplyFormatInEditMode = true)]
        public int? Population { get; set; }

        public string Notes { get; set; }
    }
}
```

代码清单 24-23 在 Create.cshtml 视图文件中添加了一个 textarea 元素，并使用 asp-for 属性关联这个元素到 City 类的 Notes 属性

代码清单 24-23 在 Views/Home 文件夹下的 Create.cshtml 文件中添加文本区域

```
@model City
@addTagHelper Cities.Infrastructure.TagHelpers.SelectOptionTagHelper, Cities

@{ Layout = "_Layout"; }

<form method="post" asp-controller="Home" asp-action="Create"
      asp-antiforgery="true">
    <div class="form-group">
        <label asp-for="Name"></label>
        <input class="form-control" asp-for="Name" />
    </div>
    <div class="form-group">
        <label asp-for="Country"></label>
        <select class="form-control" asp-for="Country" asp-items="ViewBag.Countries">
            <option disabled selected value="">Select a Country</option>
        </select>
    </div>
    <div class="form-group">
        <label asp-for="Population"></label>
        <input class="form-control" asp-for="Population" />
    </div>
    <div class="form-group">
```

```
        <label asp-for="Notes"></label>
        <textarea class="form-control" asp-for="Notes"></textarea>
    </div>
    <button type="submit" class="btn btn-primary">Add</button>
    <a class="btn btn-primary" href="/Home/Index">Cancel</a>
</form>
```

如果运行应用程序并访问/Home/Create 或/Home/Edit。你将看到发送给浏览器的 HTML 包含如下所示的 textarea 元素：

```
<div class="form-group">
    <label for="Notes">Notes</label>
    <textarea id="Notes" name="Notes"></textarea>
</div>
```

TextAreaTagHelper 类相对简单，但它提供了本章介绍的其他表单标签助手的一致性。

24.7 验证表单标签助手

与 HTML 表单相关的其他标签助手有两个（见表 24-8），将在第 27 章详加说明。这两个标签助手用于在用户提供的数据不满足应用程序的期望时向用户提供反馈。

表 24-8 验证表单标签助手

名称	描述
ValidationMessage	用于提供单个表单元素的验证反馈
ValidationSummary	用于提供关于表单所有元素的验证反馈

24.8 小结

本章介绍了用于转换 HTML 表单元素的内置标签助手。这些标签助手确保表单直接从模型类生成，这减少了潜在的错误并提供一致的方式来编写 Razor 视图。下一章将介绍其他内置标签助手，它们用于对一系列各不相同的 HTML 元素进行操作。

第 25 章　使用其他内置标签助手

第 24 章介绍的标签助手专注于生成 HTML 表单，但这不是仅有的 ASP.NET Core MVC 内置标签助手。本章将介绍用于管理 JavaScript 脚本与 CSS 样式表，为超链接元素创建 URL，为图片元素提供缓存清除，以及支持数据缓存的标签助手。本章还会介绍提供应用程序相关的 URL 支持的标签助手，以确保在应用程序被发布到与其他应用程序共享的环境中时，浏览器可以访问静态内容。表 25-1 列出了本章要介绍的操作。

表 25-1　　本章要介绍的操作

操作	方法	代码清单
包含基于宿主环境的内容	使用 environment 元素	代码清单 25-2 和代码清单 25-8
选择 JavaScript 文件	为 script 元素应用 asp-src-include 与 asp-src-exclude 属性	代码清单 25-3～代码清单 25-7
为 JavaScript 使用 CDN	为 script 元素应用 asp-fallback 属性	代码清单 25-9 和代码清单 25-10
选择 CSS 文件	为 link 元素应用 asp-href-include 与 asp-href-exclude 属性	代码清单 25-11
为 CSS 文件使用 CDN	为 link 元素应用 asp-fallback 属性	代码清单 25-12
为 anchor 元素生成 URL	使用 AnchorTagHelper	代码清单 25-13
确保图像的修改被检测	为 img 元素应用 asp-append-version 属性	代码清单 25-14
缓存数据	使用 cache 元素	代码清单 25-15～代码清单 25-23
创建应用程序相关的 URL	使用~字符添加 URL 前缀	代码清单 25-24～代码清单 25-26

25.1　准备示例项目

本章继续使用第 24 章的 Cities 项目，创建 wwwroot/images 文件夹，并在其中添加名为 city.png 的图片文件。这是一张纽约天际线的公用全景照片，如图 25-1 所示。

图 25-1　添加图片到项目中

这幅图片包含在本章的源代码中。如果需要下载示例应用，也可以替换为自己的图片。

本章要做的其他调整是使用 Bower 将 jQuery 加入项目中，如代码清单 25-1 所示。

代码清单 25-1　在 Cities 文件夹下的 bower.json 文件中添加 jQeury

```
{
  "name": "asp.net",
  "private": true,
  "dependencies": {
    "bootstrap": "4.0.0-alpha.6",
    "jquery": "3.2.1"
  }
}
```

如果运行应用程序，你将能够列出存储库中的对象，并创建新的对象，如图 25-2 所示。

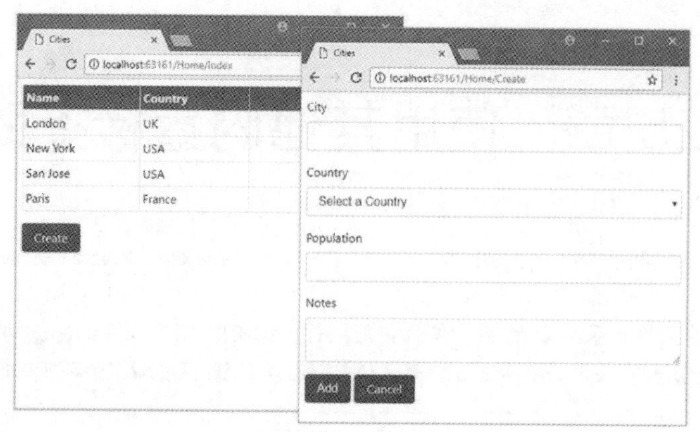

图 25-2 运行示例应用

25.2 使用宿主环境标签助手

通过将 EnvironmentTagHelper 类应用于自定义的 environment 元素，可基于宿主环境决定是否将一块内容发送到浏览器，第 14 章对此曾经介绍过。开始时这可能看起来不太激动人心，但是随后需要借助这个标签助手以充分利用一些相关的功能。environment 元素依赖于 names 属性，该属性用于指定一个以逗号分隔的宿主环境名称列表，包含在 environment 元素中的内容将被包含在发送到客户端的 HTML 中。

代码清单 25-2 在共享的布局文件中添加了 environment 元素，对于开发和生产环境，分别在视图中包含不同的内容。

代码清单 25-2 在 Views/Shared 文件夹下的_Layout.cshtml 中使用 environment 元素

```
<!DOCTYPE html>
<html>
<head>
    <meta name="viewport" content="width=device-width" />
    <title>Cities</title>
    <link href="/lib/bootstrap/dist/css/bootstrap.css" rel="stylesheet" />
</head>
<body class="m-1 p-1">
    <environment names="development">
        <div class="m-1 p-1 bg-info"><h2>This is Development</h2></div>
    </environment>
    <environment names="production">
        <div class="m-1 p-1 bg-danger"><h2>This is Production</h2></div>
    </environment>
    <div>@RenderBody()</div>
</body>
</html>
```

图 25-3 展示了在开发和生产宿主环境下运行应用程序的效果。environment 元素检查当前的宿主环境名称，以及是否包含或忽略内容（environment 元素本身不会发送到客户端的 HTML 中）。

图 25-3 使用宿主环境管理内容

25.3　使用 JavaScript 和 CSS 标签助手

另一类内置的标签助手用来通过 script 和 link 元素管理 JavaScript 和 CSS 样式表。通常它们包含于共享布局中。这些标签助手强大而灵活，但是需要密切注意以避免产生意外的结果。

25.3.1　管理 JavaScript 文件

内置的 ScriptTagHelper 类是用于 script 元素并在视图中管理 JavaScript 文件的标签助手，可使用表 25-2 所示的属性包含 JavaScript 文件，随后将进行说明。

表 25-2　用于 script 元素的内置标签助手类定义的属性

属性	描述
asp-src-include	用于指定 JavaScript 文件将被包含在视图中
asp-src-exclude	用于指定 JavaScript 文件将被排除在视图之外
asp-append-version	用于缓存清除
asp-fallback-src	用于指定在使用 CDN 有问题时，需要使用的回退 JavaScript 文件
asp-fallback-src-include	用于指定在使用 CDN 有问题时，将被使用的 JavaScript 文件
asp-fallback-src-exclude	用于指定当 CDN 有问题时，将被排除的 JavaScript 文件
asp-fallback-test	用于指定一段 JavaScript，从而判断 JavaScript 代码是否已从 CDN 正确下载

1. 选择 JavaScript 文件

属性 asp-src-include 使用通配符在视图中包含 JavaScript 文件。如第 7 章所述，通配符模板支持一系列通配符用于匹配文件。表 25-3 描述了常用的通配符。

表 25-3　常用的通配符

通配符	示例	描述
?	js/src?.js	匹配除/之外的任意单个字符。示例匹配包含于 js 文件夹中，能匹配名称为 src 后跟任意一个字符且以.js 结尾的任何文件，例如 js/src1.js 和 js/srcX.js，但是不匹配 js/src123.js 或 js/mydir/src1.js
*	js/*.js	匹配除/之外任意数量的任意字符。示例匹配包含于 js 文件夹中，能匹配以.js 为扩展名的任意文件，例如 js/src1.js 和 js/src123.js，但是不包括 js/mydir/src1.js
**	js/**/*.js	匹配包含/的任意数量的任意字符。示例匹配包括 js 目录及其子目录中以.js 为扩展名的任何文件，例如/js/src1.js 和/js/mydir/src1.js

通过 asp-src-include 属性使用通配符模式，意味着应用程序中的视图将总是包含这些 JavaScript 文件，即使文件的名称或路径发生变化，以及文件被添加或被删除。代码清单 25-3 选择了 jQuery 包中的文件，可使用 Bower 将 jQuery 安装到 wwwroot/lib/jquery/dist 文件夹中。

代码清单 25-3　在 Views/Shared 文件夹下的_Layout.cshtml 文件中选择 JavaScript 文件

```
<!DOCTYPE html>
<html>
<head>
    <meta name="viewport" content="width=device-width" />
    <title>Cities</title>
    <script asp-src-include="/lib/jquery/dist/**/*.js"></script>
    <link href="/lib/bootstrap/dist/css/bootstrap.css" rel="stylesheet" />
</head>
<body class="m-1 p-1">
    <div>@RenderBody()</div>
</body>
</html>
```

以上示例中使用的通配符很常用。通配符在 wwwroot 文件夹中被计算，其中 jQuery 库可作为单个名为 jquery.js 的文件发布。

以上通配符试图选择文件，以适应 jQuery 在未来发布中的变化，例如改变 JavaScript 文件名称。如果运行示例应用程序并检查发送到浏览器的 HTML 内容，你将会发现里面包含错误，如下所示：

```
<head>
    <meta name="viewport" content="width=device-width" />
    <title>Cities</title>
    <script src="/lib/jquery/dist/core.js"></script>
    <script src="/lib/jquery/dist/jquery.js"></script>
    <script src="/lib/jquery/dist/jquery.min.js"></script>
    <script src="/lib/jquery/dist/jquery.slim.js"></script>
    <script src="/lib/jquery/dist/jquery.slim.min.js"></script>
    <link href="/lib/bootstrap/dist/css/bootstrap.css" rel="stylesheet" />
</head>
```

ScriptTagHelper 类为匹配通过 asp-src-include 属性传递的通配符的每个文件生成一个 script 元素，而不只是选择 jquery.js 文件，jquery.min.js 是紧缩版本的 jquery.js 文件，此处还有正常和紧缩版本的 jQuery 库 core 和 slim。

由于 Visual Studio 默认隐藏了，因此你可能没有意识到 jQuery 发行版包含紧缩文件。为了展现 wwwroot/lib/jquery/dist 文件夹中的全部内容，必须在 Solution Explorer 窗格中展开 jquery.js，然后同样展开 jquery.min.js，如图 25-4 所示。

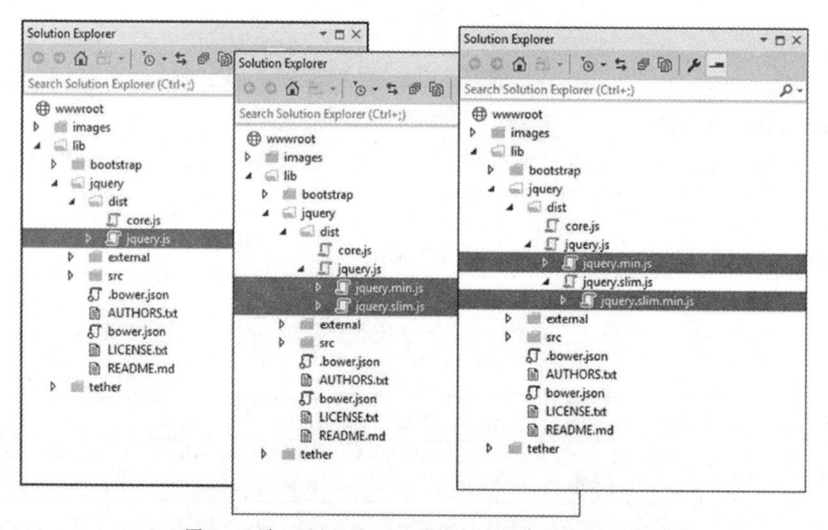

图 25-4　在 Solution Explorer 窗格中展现目录的填充内容

代码清单 25-3 中使用的通配符会将 jQuery 代码加入浏览器多次，这不仅浪费带宽而且拖慢了应用程序的下载速度。对于某些库，可能导致错误或未预期的行为。解决这个问题有 3 种方式，这些方式随后进行说明。

使用源码映射

JavaScript 文件通过紧缩来减少尺寸，这意味着可以更快地发送到客户端，使用更少的带宽。紧缩过程会从源文件中删除所有的空白，重命名函数和变量，比如 myHelpfullyNamedFunction 可使用更少数量的字符来表示，例如 x1。当在紧缩代码中使用浏览器的调试器来追踪错误时，x1 这样的名字使得几乎不可能进行代码跟踪。

jquery.min.map 文件是源码映射文件，有些浏览器可以用它提供紧缩代码与开发者可读的未紧缩源码间的映射。

在写作本书时，源码映射还不是通用的支持特性，但可以在最新版本的 Chrome 和 Edge 浏览器中使用。例如，在 Chrome 浏览器中，如果打开开发者工具窗口，浏览器将自动请求源码映射文件，这意味着在发送紧缩版本的 JavaScript 文件时，仍然可以轻松启用调试。

2. 限制通配符

许多库同时提供正常和紧缩版本的 JavaScript 文件，如果仅仅希望使用紧缩版本，那么可以限制通配符匹配的文件，如代码清单 25-4 所示。如果并不期望使用调试版本的 jQuery 库，因为它们编写良好，问题很少，或者知道浏览器支持源码映射，那么这是一种不错的方式。

代码清单 25-4　在 Views/Shared 文件夹下的_Layout.cshtml 文件中仅仅使用紧缩版本的 JavaScript 文件

```
<!DOCTYPE html>
<html>
<head>
    <meta name="viewport" content="width=device-width" />
    <title>Cities</title>
    <script asp-src-include="/lib/jquery/dist/**/*.min.js"></script>
    <link href="/lib/bootstrap/dist/css/bootstrap.css" rel="stylesheet" />
</head>
<body class="m-1 p-1">
    <div>@RenderBody()</div>
</body>
</html>
```

如果运行应用程序并检查发送到浏览器的 HTML，你将会看到仅包含紧缩版本的 jQuery 文件。

```
<head>
    <meta name="viewport" content="width=device-width" />
    <title>Cities</title>
    <script src="/lib/jquery/dist/jquery.min.js">
    </script><script src="/lib/jquery/dist/jquery.slim.min.js"></script>
    <link href="/lib/bootstrap/dist/css/bootstrap.css" rel="stylesheet" />
</head>
```

虽然限制通配符会起到作用，但是浏览器仍然最终发送了正常和轻量版本的 jQuery 库（轻量版本忽略了一些不常用的功能，详见 jQuery 网站）。为了进一步限制，可以在统配模板中只包含轻量版本，如代码清单 25-5 所示。

代码清单 25-5　在 Views/Shared 文件夹下的_Layout.cshtml 文件中做进一步限制

```
<!DOCTYPE html>
<html>
<head>
    <meta name="viewport" content="width=device-width" />
    <title>Cities</title>
    <script asp-src-include="/lib/jquery/dist/**/*.slim.min.js"></script>
    <link href="/lib/bootstrap/dist/css/bootstrap.css" rel="stylesheet" />
</head>
<body class="m-1 p-1">
    <div>@RenderBody()</div>
</body>
</html>
```

结果是只有一个版本的 jQuery 文件被发送到浏览器中，但仍然保持了文件位置的灵活性。

```
<head>
    <meta name="viewport" content="width=device-width" />
    <title>Cities</title>
    <script src="/lib/jquery/dist/jquery.slim.min.js"></script>
    <link href="/lib/bootstrap/dist/css/bootstrap.css" rel="stylesheet" />
</head>
```

1）排除文件

当希望选择的文件名中包含特定术语时，限制通配符是有帮助的，但是在没有特定术语的时候，就没有什么用了。比如，当希望包含完全版本的紧缩文件时。幸运的是，可以使用 asp-src-exclude 属性，从 asp-src-include 属性匹配的文件中移除文件，如代码清单 25-6 所示。

代码清单 25-6 在 Views/Shared 文件夹下的_Layout.cshtml 文件中排除文件

```
<!DOCTYPE html>
<html>
<head>
    <meta name="viewport" content="width=device-width" />
    <title>Cities</title>
    <script asp-src-include="/lib/jquery/dist/**/*.min.js"
        asp-src-exclude="**.slim.**">
    </script>
    <link href="/lib/bootstrap/dist/css/bootstrap.css" rel="stylesheet" />
</head>
<body class="m-1 p-1">
    <div>@RenderBody()</div>
</body>
</html>
```

运行示例应用程序并检查发送到浏览器的 HTML，你将会看到只包含完全紧缩版本的 JavaScript 文件。

```
<head>
    <meta name="viewport" content="width=device-width" />
    <title>Cities</title>
    <script src="/lib/jquery/dist/jquery.min.js"></script>
    <link href="/lib/bootstrap/dist/css/bootstrap.css" rel="stylesheet" />
</head>
```

同样的技术也可以用于包含非紧缩版本的文件，在开发时这很有用，如代码清单 25-7 所示。

代码清单 25-7 在 Views/Shared 文件夹下的_Layout.cshtml 文件中选取非紧缩文件

```
<!DOCTYPE html>
<html>
<head>
    <meta name="viewport" content="width=device-width" />
    <title>Cities</title>
    <script asp-src-include="/lib/jquery/dist/**/j*.js"
            asp-src-exclude="**.slim.**,**.min.**">
    </script>
    <link href="/lib/bootstrap/dist/css/bootstrap.css" rel="stylesheet" />
</head>
<body class="m-1 p-1">
    <div>@RenderBody()</div>
</body>
</html>
```

注意，可以通过逗号分隔来指定多个术语。如果运行示例应用程序并检查发送到浏览器的 HTML，你将会看到只有非紧缩版本的 JavaScript 文件被包含：

```
<head>
    <meta name="viewport" content="width=device-width" />
    <title>Cities</title>
    <script src="/lib/jquery/dist/jquery.js"></script>
    <link href="/lib/bootstrap/dist/css/bootstrap.css" rel="stylesheet" />
</head>
```

2）使用宿主环境来选择文件

常见的工作方式是在开发过程中使用常规版本的 JavaScript 文件，这使调试变得简单，然后在生产环境中使用紧缩版本的 JavaScript 文件，从而节省了带宽。这可以通过使用 environment 元素，基于宿主环境选择包含的 script 元素来实现，如代码清单 25-8 所示。

代码清单 25-8 在 Views/Shared 文件夹下的_Layout.cshtml 中使用宿主环境来选择文件

```
<!DOCTYPE html>
<html>
```

```
<head>
    <meta name="viewport" content="width=device-width" />
    <title>Cities</title>
    <environment names="development">
        <script asp-src-include="/lib/jquery/dist/**/j*.js"
                asp-src-exclude="**.slim.**,**.min.**">
        </script>
    </environment>
    <environment names="staging, production">
        <script asp-src-include="/lib/jquery/dist/**/*.min.js"
                asp-src-exclude="**.slim.**">
        </script>
    </environment>
    <link href="/lib/bootstrap/dist/css/bootstrap.css" rel="stylesheet" />
</head>
<body class="m-1 p-1">
    <div>@RenderBody()</div>
</body>
</html>
```

这种方式的优势在于可基于宿主环境适配应用，但也意味着不得不编写和维护多个 script 元素。

缓存清除

可经常缓存静态内容（如图片、CSS 样式表以及 JavaScript 文件）以停止从应用服务器请求很少变化的内容。缓存可以通过多种途径完成：浏览器可以被告知缓存来自服务器的内容，应用程序可以使用缓存服务器来支持应用服务器，或者通过内容分发网络来分发内容。不是所有的缓存都在你的控制之下。例如，大的企业经常安装缓存来减少带宽，因为后继请求趋向于访问同样的站点或应用程序。

缓存的问题是，客户端在部署时不会立即收到新版本的静态文件，因为它们的请求仍由以前缓存的内容处理。最终，缓存的内容将过期，并且将使用新内容，但这会留下一段时间，在这段时间，应用程序控制器生成的动态内容与缓存交付的静态内容不同步。根据更新的内容，这可能导致布局问题或应用程序异常。

解决这个问题的方法称为缓存清除。思路是允许缓存处理静态内容，但是当内容在服务器上被修改之后立即做出反应。标签助手通过在用于静态内容的 URL 查询串后包含校验和作为版本来支持。例如，ScriptTagHelper 类通过 asp-append-version 属性来支持缓存清除。

```
...
<script asp-src-include="/lib/jquery/dist/**/j*.js"
        asp-src-exclude="**.slim.**,**.min.**"
        asp-append-version="true">
</script>
...
```

启用缓存清除导致生成的发送到浏览器的 HTML 元素如下所示：

```
...
<script src="/lib/jquery/dist/jquery.min.js?v=3zRSQ1HF-ocUiVcdv9yKTXqM">
</script>
...
```

同样的版本号也可以用于标签助手，直至改变文件的内容，比如更新 JavaScript 库，此刻另一个不同的校验和将被计算出来。附加版本号意味着每次修改文件时，客户端将请求不同的 URL，此前缓存的内容无法达成，缓存视为对新内容的请求并发送到应用服务器。内容将正常缓存直到下一次更新，这将生成另一个具有不同版本的 URL。

3. 使用 CDN

内容分发网络（Content Delivery Network，CDN）用于将用户请求分流到离用户更近的服务器。浏览器不是从应用服务器请求 JavaScript 文件，而是从解析为本地服务器的主机来请求，这减少了加载文件所需的时间，降低了为应用程序提供的带宽。如果拥有庞大的、广域分布的用户群，那么可以使用商业注册的 CDN，但是即使对于最小和最简单的应用程序，也可以通过受益于由主流技术公司管理的免费 CDN 来分发通用的 JavaScript 包，例如 jQuery。

本章将使用微软提供的 CDN，它们对流行的包提供免费访问，可以在 ASP.NET 网站上找到微软提供的 CDN 清单。对于 jQuery 3.2.1，有 6 个包：

- jquery-3.2.1.js；
- jquery-3.2.1.min.js；
- jquery-3.2.1.min.map；
- jquery-3.2.1.slim.js；
- jquery-3.2.1.slim.min.js；
- jquery-3.2.1.slim.min.map。

这些包可以为完全版本和轻量版本提供常规的 JavaScript 文件、紧缩的 JavaScript 文件以及紧缩文件的源码映射文件。代码清单 25-9 通过修改示例应用中的布局文件来使用 CDN 获取的紧缩文件替换本地文件。

代码清单 25-9 在 Views/Shared 文件夹下的_Layout.cshtml 中使用 CDN

```
<!DOCTYPE html>
<html>
<head>
    <meta name="viewport" content="width=device-width" />
    <title>Cities</title>
    <script src="****://ajax.aspnetcdn.***/ajax/jQuery/jquery-3.2.1.min.js"></script>
    <link href="/lib/bootstrap/dist/css/bootstrap.css" rel="stylesheet" />
</head>
<body class="m-1 p-1">
    <div>@RenderBody()</div>
</body>
</html>
```

指定 CDN 意味着对于 jQuery 文件没有请求到达应用服务器。使用 CDN 的问题是 CDN 不在你的控制之下，这意味着 CDN 可能失效，只剩下应用程序在运行，但是不能如预期那样工作，因为 CDN 内容不可用。为了帮助解决这一问题，ScriptTagHelper 类提供了在 CDN 内容不能加载到客户端时回退到本地文件的能力，如代码清单 25-10 所示。

代码清单 25-10 在 Views/Shared 文件夹下的 Layout.cshtml 中使用 CDN 回退

```
<!DOCTYPE html>
<html>
<head>
    <meta name="viewport" content="width=device-width" />
    <title>Cities</title>
    <script src="****://ajax.aspnetcdn.***/ajax/jQuery/jquery-3.2.1.min.js"
            asp-fallback-src-include="/lib/jquery/dist/**/*.min.js"
            asp-fallback-src-exclude="**.slim.**"
            asp-fallback-test="window.jQuery">
    </script>
    <link href="/lib/bootstrap/dist/css/bootstrap.css" rel="stylesheet" />
</head>
<body class="m-1 p-1">
    <div>@RenderBody()</div>
</body>
</html>
```

属性 asp-fallback-src-include 与 asp-fallback-src-exclude 用于在 CDN 不能通过常规的 src 属性分发内容

时选择和排除本地文件。为了判断 CDN 是否正常工作，asp-fallback-test 属性用于定义在浏览器端执行的 JavaScript 片段，如果执行结果为 false，那么回退文件将被请求。

运行应用程序并检查发送到客户端的 HTML，你将看到 ScriptTagHelper 类从 asp-fallback-test 属性取出 JavaScript 片段并用它生成另一个 script 元素，如下所示：

```
<head>
    <meta name="viewport" content="width=device-width" />
    <title>Cities</title>
    <script src="****://ajax.aspnetcdn.***/ajax/jQuery/jquery-3.2.1.min.js">
    </script>
    <script>
        (window.jQuery||document.write("\u003Cscript
           src=\u0022\/lib\/jquery\/dist\/jquery.min.js
           \u0022\u003E\u003C\/script\u003E"));
    </script>
    <link href="/lib/bootstrap/dist/css/bootstrap.css" rel="stylesheet" />
</head>
```

如果文件从 CDN 加载，那么你在 asp-fallback-test 属性中指定的 JavaScript 片段必须返回 true；否则，返回 false。最简单的方式通常是检查 JavaScript 代码提供的功能入口点。jQuery 库在公共的 window 对象上创建了名为 jQuery 的函数，这也是代码清单 25-10 要测试的内容。你需要为从 CDN 加载的每个文件找到等价的测试。

测试回退设置也很重要，因为在 CDN 停止工作导致用户无法访问应用程序之前，无法发现 CDN 是否失效。最简单的检查回退的方式是将 src 属性指定的文件名改为某个不存在的文件名（可在文件名后追加 FAIL），然后使用 F12 键来查看浏览器的网络请求。你应该看到 CDN 文件出错了，后面跟着对回退文件的请求。

警　告

CDN 回退特性依赖于浏览器同步加载和执行 script 元素，并按照定义的顺序依次执行。有很多异步处理技术可用来加速 JavaScript 文件的加载和执行，但是它们将导致在浏览器从 CDN 获取文件之前执行回退测试，结果导致在 CDN 完美工作时，首先淘汰 CDN 的使用。不要和 CDN 回退特性一起混用异步脚本加载。

25.3.2　管理 CSS 样式表

LinkTagHelper 类是内置的作用于 link 元素的标签助手，用来管理视图中包含的 CSS。该标签助手支持的属性如表 25-4 所示。

表 25-4　用于 link 元素的内置标签助手类定义的属性

属性	描述
asp-href-include	用于为输出元素的 href 属性选择文件
asp-href-exclude	用于为输出元素的 href 属性排除文件
asp-append-version	用于启用缓存清除
asp-fallback-href	用于在 CDN 出现问题时指定回退文件
asp-fallback-href-include	用于在 CDN 出现问题时选择将要使用的文件
asp-fallback-href-exclude	用于在 CDN 出现问题时选择将被排除的文件
asp-fallback-href-test-class	用于指定测试 CDN 的 CSS 类名
asp-fallback-href-test-property	用于指定测试 CDN 的 CSS 属性
asp-fallback-href-test-value	用于指定测试 CDN 的 CSS 值

1. 选择样式表

LinkTagHelper 与 ScriptTagHelper 支持许多相同的特性，包括支持使用通配符选择或排除 CSS 文件，所以不必单个指定。能够准确选择 CSS 文件与选择 JavaScript 文件一样重要，因为样式表文件也有常规和紧缩版本，同样支持源码映射。对于流行的 Bootstrap 包，本书一直用它来修饰元素。wwwroot/lib/bootstrap/dist/css 文件夹中包含一些样式表文件，如果在 Solution Explorer 窗格中展开所有的项目，你将看到多个文件，如图 25-5 所示。

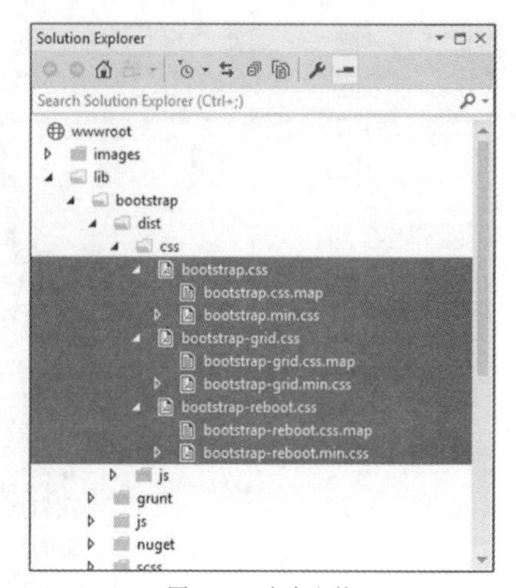

图 25-5 多个文件

文件 boostrap.css 表示常规样式表，文件 boostrap.min.css 表示紧缩版本，文件 bootstrap.css.map 表示源码映射，这里的其他文件用于提供特定功能，但本章并不涉及。代码清单 25-11 使用 link 元素的 asp-href-include 属性来选择紧缩版本的样式表（还删除了加载 jQuery 的 script 元素，因为这里已不再需要）。

代码清单 25-11 在 Views/Shared 文件夹下的_Layout.cshtml 文件中选择样式表

```
<!DOCTYPE html>
<html>
<head>
    <meta name="viewport" content="width=device-width" />
    <title>Cities</title>
    <link rel="stylesheet"
          asp-href-include="/lib/bootstrap/dist/**/*.min.css"
          asp-href-exclude="**/*-reboot*,**/*-grid*"/>
</head>
<body class="m-1 p-1">
    <div>@RenderBody()</div>
</body>
</html>
```

在选择 JavaScript 文件时同样需要注意细节，因为很容易非你所愿地生成用于选择多个同样文件的不同版本的 link 元素。选择 JavaScript 文件有 3 种方式：限制通配符，使用 asp-href-exclude 属性排除文件，使用 environment 元素在重复的元素集合中进行选择。

2. 使用 CDN

LinkTag 类提供了一系列属性用于在 CDN 失效时提供到本地内容的回退控制，但测试样式表是否加载的过程相比 JavaScript 文件有一点复杂。代码清单 25-12 使用 MaxCDN URL 来加载 Bootstrap 库，这仅仅用来展示微软平台之外的另一个选择（MaxCDN 是 Bootstrap 项目推荐的 CDN）。

代码清单 25-12 在 Views/Shared 文件夹下的_Layout.cshtml 文件中使用 CDN 加载 CSS

```
<!DOCTYPE html>
<html>
<head>
    <meta name="viewport" content="width=device-width" />
    <title>Cities</title>
    <link href="*****://maxcdn.bootstrapcdn.***/bootstrap/4.0.0-alpha.6/css/bootstrap.min.css"
          asp-fallback-href-include="/lib/bootstrap/dist/**/*.min.css"
          asp-fallback-href-exclude="**/*-reboot*,**/*-grid*"
          asp-fallback-test-class="btn"
          asp-fallback-test-property="display"
          asp-fallback-test-value="inline-block"
          rel="stylesheet" />
</head>
```

```
<body class="m-1 p-1">
    <div>@RenderBody()</div>
</body>
</html>
```

属性 href 用于指定 CDN 地址，这里还使用 asp-fallback-href-include 属性来选择在 CDN 不可用时将被使用的文件。为了测试 CDN 是否工作，需要使用 3 个不同的属性，并理解所使用的 CSS 样式表定义的 CSS 类。

CSS 回退特性通过为文档添加 meta 元素而起作用，其由 asp-fallback-test-class 属性定义的类添加。以上代码指定了 btn 类，这意味着如下元素将被添加到发送到浏览器的 HTML 中：

```
<meta name="x-stylesheet-fallback-test" class="btn" />
```

指定的 CSS 类必须在通过 CDN 加载的样式表中定义。这里指定的 btn 类为 Bootstrap 按钮元素提供基本的格式化。

属性 asp-fallback-test-property 用于指定由 CSS 类定义的属性，属性 asp-fallback-test-value 用于指定将被设置的值。标签助手会添加 JavaScript 到视图中以测试 meta 元素中的 CSS 属性值，从而确定样式表是否被加载，并且如果没有加载，就为回退文件添加 link 元素。Bootstrap 的 btn 类已设置 display 属性为 inline-block，从而测试并查看浏览器是否能从 CDN 加载 Bootstrap 样式表。

提　示

要测试第三方库，比如 Bootstrap，最简单的方法是按 F12 键。为检验代码清单 25-12 中的测试，为按钮指定 btn 类，然后在浏览器中检查，查看类修改的 CSS 属性，这比试图阅读长且复杂的样式表要容易。

25.4 使用超链接元素

超链接元素 a 是在应用程序中进行导航的基本工具，可通过发送 GET 请求到应用程序来请求不同的内容。AnchorTagHelper 类用于转换 a 元素的 href 属性以便使用路由系统生成指向的 URL 地址。该标签助手支持的属性如表 25-5 所示。

表 25-5　用于超链接元素的内置标签助手类定义的属性

属性	描述
asp-action	指定 URL 指向的操作方法
asp-controller	指定 URL 指向的控制器
asp-area	指定 URL 指向的区域
asp-fragment	用于指定 URL 片段（在#字符之后应用）
asp-host	指定 URL 指向的主机名
asp-protocol	指定 URL 使用的协议
asp-route	指定用于生成 URL 的路由名称
asp-route-*	以 asp-route-开头的属性用于指定 URL 附加值，例如 asp-route-id 属性用于指定路由系统中 id 片段的值

AnchorTagHelper 简单且可预测，使得使用应用程序的路由系统在 a 元素中生成 URL 变得容易。代码清单 25-13 更新了 Index.cshtml 文件中的 a 元素，以便其 href 属性由标签助手生成。

代码清单 25-13　在 Views/Home 文件夹下的 Index.cshtml 文件中转换超链接元素

```
@model IEnumerable<City>

@{  Layout = "_Layout"; }
```

```
<table class="table table-sm table-bordered">
    <thead class="bg-primary text-white">
        <tr>
            <th>Name</th>
            <th>Country</th>
            <th class="text-right">Population</th>
        </tr>
    </thead>
    <tbody>
        @foreach (var city in Model) {
            <tr>
                <td>@city.Name</td>
                <td>@city.Country</td>
                <td class="text-right">@city.Population?.ToString("#,###")</td>
            </tr>
        }
    </tbody>
</table>
<a asp-action="Create" class="btn btn-primary">Create</a>
```

如果运行应用程序并访问/Home/Index，你将看到标签助手将 a 元素转换为如下形式：

```
<a class="btn btn-primary" href="/Home/Create">Create</a>
```

25.5 使用图像元素

ImageTagHelper 类用于通过 img 元素的 src 属性为图片提供缓存清除功能，在确保对图片的修改能够立即得到反馈的同时，允许应用程序获得缓存的好处。可使用 img 元素的 asp-append-version 属性启用缓存清除功能。

代码清单 25-14 在共享布局中添加了一个 img 元素（为了简化，可以重置 style 元素以便使用本地文件）。

代码清单 25-14 在 Views/Shared 文件夹下的_Layout.cshtml 文件中添加图片

```
<!DOCTYPE html>
<html>
<head>
    <meta name="viewport" content="width=device-width" />
    <title>Cities</title>
    <link href="https://maxcdn.bootstrapcdn.com/bootstrap/4.0.0-alpha.6/css/bootstrap.min.css"
          asp-fallback-href-include="/lib/bootstrap/dist/**/*.min.css"
          asp-fallback-href-exclude="**/*-reboot*,**/*-grid*"
          asp-fallback-test-class="btn"
          asp-fallback-test-property="display"
          asp-fallback-test-value="inline-block"
          rel="stylesheet" />
</head>
<body class="m-1 p-1">
    <img src="/images/city.png" asp-append-version="true" />
    <div>@RenderBody()</div>
</body>
</html>
```

如果运行应用程序，你将看到图片显示在每个页面的顶部。如果检查发送到浏览器的 HTML，你将看到用于请求图片文件的 URL 包含了版本校验和，如下所示：

```
<img src="/images/city.png?v=KaMNDSZFbzNpE8Pkb30EXcAJufRcRDpKh0K_IIPNc7E" />
```

与 JavaScript 和 CSS 样式表的缓存清除类似，包含在 URL 中的校验和将保持不变，直到文件被修改为止。

25.6　使用数据缓存

MVC 包含的内存缓存用于缓存内容片段以便加速视图的渲染。被缓存的内容使用视图中的 cache 元素指定，由 CacheTagHelper 类使用表 25-6 所示的属性处理。

注　意

缓存是重用内容段落的有用工具，以便它们不需要为每次请求生成一次。但是，使用缓存事实上需要仔细考虑和规划。缓存可以改进应用程序的性能，也会导致奇怪的效果，比如用户收到过时的内容，或者多个缓存包含不一致的版本，等等，因为应用程序缓存的前一个版本的内容与新版本的内容混杂在一起会导致更新部署被中断。除非清晰定义了要解决的性能问题，并确认理解缓存将导致的影响，否则不要启用缓存。

表 25-6　　用于 cache 元素的内置标签助手类定义的属性

属性	描述
enabled	用于控制 cache 元素的内容是否被缓存，若省略，表示启用缓存
expires-on	用于指定缓存内容的绝对过期时间，表达式为 DateTime 值
expires-after	用于指定缓存内容过期的相对时间，表达式为 TimeSpan 值
expires-sliding	用于指定当缓存内容过期时，从最后使用开始的区间，表达式为 TimeSpan 值
vary-by-header	用于指定请求头的名称，它将用于管理不同版本的缓存内容
vary-by-query	用于指定查询串的键，它将用于管理不同版本的缓存内容
vary-by-route	用于指定路由变量的名称，它将用于管理不同版本的缓存内容
vary-by-cookie	用于指定 Cookie 的名称，它将用于管理不同版本的缓存内容
vary-by-user	用于指定是否验证用户的名称，它将用于管理不同版本的缓存内容
vary-by	用于提供管理不同版本缓存内容的键值
priority	用于指定相对优先级，在内存缓存耗尽并清理未过期缓存内容时，该属性的值可供参考

为了演示 cache 属性的操作方式，创建 Components 文件夹，并添加名为 TimeViewComponent.cs 的类文件，用它定义代码清单 25-15 所示的视图组件。

代码清单 25-15　Components 文件夹下的 TimeViewComponent.cs 文件的内容

```
using System;
using Microsoft.AspNetCore.Mvc;

namespace Cities.Components {

    public class TimeViewComponent : ViewComponent {

        public IViewComponentResult Invoke() {
            return View(DateTime.Now);
        }
    }
}
```

Invoke 方法使用默认视图，并提供 DateTime 对象作为视图模型。为了给视图组件提供视图，创建 Views/Home/Components/Time 文件夹，添加名为 Default.cshtml 的视图文件，内容如代码清单 25-16 所示。

代码清单 25-16　Views/Home/Components/Time 文件夹下的 Default.cshtml 文件的内容

```
@model DateTime

<div class="m-1 p-1 bg-info text-white">
    Rendered at @Model.ToString("HH:mm:ss")
</div>
```

DateTime 模型对象用于显示当前时间，准确到秒。代码清单 25-17 已将 25.5 节的 img 元素替换为@await Component.InvokeAsync 表达式以调用视图组件。

代码清单 25-17　在 Views/Shared 文件夹下的_Layout.cshtml 文件中使用视图组件

```
<!DOCTYPE html>
<html>
<head>
    <meta name="viewport" content="width=device-width" />
    <title>Cities</title>
    <link href="https://maxcdn.bootstrapcdn.com/bootstrap/4.0.0-alpha.6/css/bootstrap.min.css"
          asp-fallback-href-include="/lib/bootstrap/dist/**/*.min.css"
          asp-fallback-href-exclude="**/*-reboot*,**/*-grid*"
          asp-fallback-test-class="btn"
          asp-fallback-test-property="display"
          asp-fallback-test-value="inline-block"
          rel="stylesheet" />
</head>
<body class="m-1 p-1">
    @await Component.InvokeAsync("Time")
    <div>@RenderBody()</div>
</body>
</html>
```

如果运行应用程序，你将看到标题栏显示了内容渲染的时间。等待几秒后重新加载页面，你将看到显示的时间发生了变化，如图 25-6 所示。

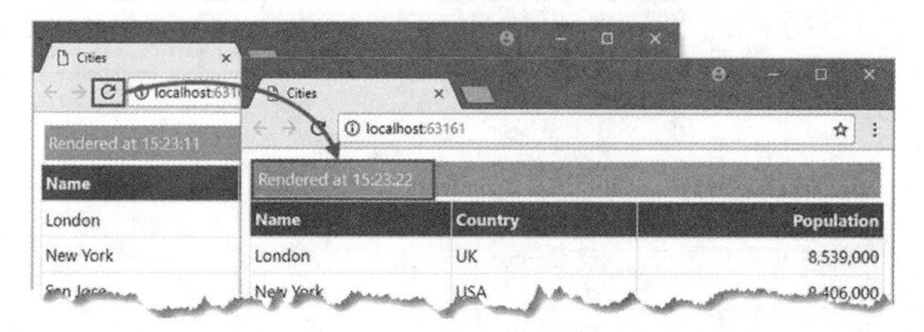

图 25-6　在示例应用中显示时间

元素 cache 用来环绕将被添加到缓存中的内容。代码清单 25-18 使用 cache 元素将视图组件添加到了缓存中。

代码清单 25-18　在 Views/Shared 文件夹下的_Layout.cshtml 文件中缓存内容

```
<!DOCTYPE html>
<html>
<head>
    <meta name="viewport" content="width=device-width" />
    <title>Cities</title>
    < link href="https://maxcdn.bootstrapcdn.com/bootstrap/4.0.0-alpha.6/css/bootstrap.min.css"
          asp-fallback-href-include="/lib/bootstrap/dist/**/*.min.css"
          asp-fallback-href-exclude="**/*-reboot*,**/*-grid*"
          asp-fallback-test-class="btn"
          asp-fallback-test-property="display"
          asp-fallback-test-value="inline-block"
          rel="stylesheet" />
</head>
<body class="m-1 p-1">
    <cache>
        @await Component.InvokeAsync("Time")
```

```
    </cache>
    <div>@RenderBody()</div>
</body>
</html>
```

在没有任何属性的情况下应用 cache 元素会使 MVC 重用这些内容以满足所有未来的请求。如果启动应用程序，则会缓存视图组件生成的内容，所以即使页面重新加载也会显示同样的时间。

提　示

CacheTagHelper 类使用的缓存是基于内存的，这意味着缓存容量受限于可用的内存。当缓存容量短缺的时候，将从缓存中退出内容，在应用程序停止或重启时，整个内容都将丢失。

25.6.1　设置缓存过期时间

属性 expires-*允许指定内容何时过期，使用绝对时间（相对于当前时间而言）或缓存内容不被请求的区间来表示。代码清单 25-19 使用 expires-after 属性指定内容应被缓存 15s。

代码清单 25-19　在 Views/Shared 文件夹下的_Layout.cshtml 文件中设置缓存过期时间

```
<!DOCTYPE html>
<html>
<head>
    <meta name="viewport" content="width=device-width" />
    <title>Cities</title>
    <link href="https://maxcdn.bootstrapcdn.com/bootstrap/4.0.0-alpha.6/css/bootstrap.min.css"
          asp-fallback-href-include="/lib/bootstrap/dist/**/*.min.css"
          asp-fallback-href-exclude="**/*-reboot*,**/*-grid*"
          asp-fallback-test-class="btn"
          asp-fallback-test-property="display"
          asp-fallback-test-value="inline-block"
          rel="stylesheet" />
</head>
<body class="m-1 p-1">
    <cache expires-after="@TimeSpan.FromSeconds(15)">
        @await Component.InvokeAsync("Time")
    </cache>
    <div>@RenderBody()</div>
</body>
</html>
```

如果运行应用程序，你将看到缓存的数据在 15s 后过期，重新加载页面后将调用视图组件，并创建另一新的缓存过期时间为 15s 的条目。

1. 设置固定过期时间

可以使用 expires-on 属性指定固定的缓存内容过期时间，该属性接收一个 DateTime 值，如代码清单 25-20 所示。

代码清单 25-20　在 Views/Shared 文件夹下的_Layout.cshtml 中设置固定过期时间

```
<!DOCTYPE html>
<html>
<head>
    <meta name="viewport" content="width=device-width" />
    <title>Cities</title>
    <link href="https://maxcdn.bootstrapcdn.com/bootstrap/4.0.0-alpha.6/css/bootstrap.min.css"
          asp-fallback-href-include="/lib/bootstrap/dist/**/*.min.css"
          asp-fallback-href-exclude="**/*-reboot*,**/*-grid*"
          asp-fallback-test-class="btn"
          asp-fallback-test-property="display"
          asp-fallback-test-value="inline-block"
          rel="stylesheet" />
</head>
```

```
<body class="m-1 p-1">
    <cache expires-on="@DateTime.Parse("2100-01-01")">
        @await Component.InvokeAsync("Time")
    </cache>
    <div>@RenderBody()</div>
</body>
</html>
```

以上代码指定数据将被缓存到 2100 年。由于应用程序很可能在 22 世纪开始前重新启动，因此这不是什么有用的缓存策略，但它展示了在内容被缓存的时候，如何指定固定的未来时间，而不是相对于当前时刻的过期时间。

2. 设置最后使用缓存期限

属性 expires-sliding 用于指定如果没有被缓存访问过，多长时间之后内容将过期。代码清单 25-21 指定过期时间为 10s。

代码清单 25-21 在 Views/Shared 文件夹下的_Layout.cshtml 文件中指定最后使用缓存期限

```
<!DOCTYPE html>
<html>
<head>
    <meta name="viewport" content="width=device-width" />
    <title>Cities</title>
    <link href="https://maxcdn.bootstrapcdn.com/bootstrap/4.0.0-alpha.6/css/bootstrap.min.css"
          asp-fallback-href-include="/lib/bootstrap/dist/**/*.min.css"
          asp-fallback-href-exclude="**/*-reboot*,**/*-grid*"
          asp-fallback-test-class="btn"
          asp-fallback-test-property="display"
          asp-fallback-test-value="inline-block"
          rel="stylesheet" />
</head>
<body class="m-1 p-1">
    <cache expires-sliding="@TimeSpan.FromSeconds(10)">
        @await Component.InvokeAsync("Time")
    </cache>
    <div>@RenderBody()</div>
</body>
</html>
```

通过运行应用程序并定期重新加载页面，就可以看到 express-sliding 属性的效果。只要在 10s 内重新加载页面，缓存的内容将被使用。如果等待的时间超过 10s 并重新加载页面，那么缓存的内容将被丢弃，视图组件将用于生成新的内容，并且此过程将重新开始。

25.6.2 使用缓存变体

默认情况下，所有请求接收相同的缓存内容。CacheTagHelper 类可以对缓存的内容维护多种不同版本，并使用它们来满足不同种类的 HTTP 请求，使用名称以不同方式开头的 vary-by 属性之一进行设置即可。代码清单 25-22 展示了如何使用 vary-by-route 属性来基于路由系统匹配的 action 名称创建缓存变体。

代码清单 25-22 在 Views/Shared 文件夹下的_Layout.cshtml 文件中创建缓存变体

```
<!DOCTYPE html>
<html>
<head>
    <meta name="viewport" content="width=device-width" />
    <title>Cities</title>
    <link href="https://maxcdn.bootstrapcdn.com/bootstrap/4.0.0-alpha.6/css/bootstrap.min.css"
          asp-fallback-href-include="/lib/bootstrap/dist/**/*.min.css"
          asp-fallback-href-exclude="**/*-reboot*,**/*-grid*"
          asp-fallback-test-class="btn"
          asp-fallback-test-property="display"
          asp-fallback-test-value="inline-block"
          rel="stylesheet" />
</head>
```

```
<body class="m-1 p-1">
    <cache expires-sliding="@TimeSpan.FromSeconds(10)" vary-by-route="action">
        @await Component.InvokeAsync("Time")
    </cache>
    <div>@RenderBody()</div>
</body>
</html>
```

如果运行应用程序并使用两个浏览器窗口来访问/Home/Index 和/Home/Create，你将看到每个窗口接收各自的缓存内容，因为每个请求会处理不同的 action 路由值。CacheTagHelper 类支持不同变体的一系列属性，包括为单个用户缓存内容。

vary-by 请求头允许使用任何数据值定义任意缓存变体。代码清单 25-23 通过指定直接从路由数据中提取的值再现了 vary-by-route 属性的效果。

代码清单 25-23　在 Views/Shared 文件夹下的_Layout.cshtml 文件中指定自定义的缓存变体

```
<!DOCTYPE html>
<html>
<head>
    <meta name="viewport" content="width=device-width" />
    <title>Cities</title>
    <link href="https://maxcdn.bootstrapcdn.com/bootstrap/4.0.0-alpha.6/css/bootstrap.min.css"
          asp-fallback-href-include="/lib/bootstrap/dist/**/*.min.css"
          asp-fallback-href-exclude="**/*-reboot*,**/*-grid*"
          asp-fallback-test-class="btn"
          asp-fallback-test-property="display"
          asp-fallback-test-value="inline-block"
          rel="stylesheet" />
</head>
<body class="m-1 p-1">
    <cache expires-sliding="@TimeSpan.FromSeconds(10)"
           vary-by="@ViewContext.RouteData.Values["action"]">
        @await Component.InvokeAsync("Time")
    </cache>
    <div>@RenderBody()</div>
</body>
</html>
```

属性 vary-by 可以用于创建更复杂的缓存变体，应该尽量慎用，因为它很容易失控，导致创建的变体如此特殊，以至于缓存的内容在过期前永远不会被重用。

使用应用程序相关的 URL

最后介绍的内置标签助手类是 UrlResolutionTagHelper，用于对应用程序相关的 URL 提供支持，它们是以~字符作为前缀的 URL。代码清单 25-24 修改了共享布局中的 link 元素以便使用显式定义的 URL，而不是使用标签助手从路由系统生成 URL。

代码清单 25-24　在 Views/Shared 文件夹下的_Layout.cshtml 文件中使用显式定义的 URL

```
<!DOCTYPE html>
<html>
<head>
    <meta name="viewport" content="width=device-width" />
    <title>Cities</title>
    <link href="/lib/bootstrap/dist/css/bootstrap.min.css" rel="stylesheet" />
</head>
<body class="m-1 p-1">
    <cache expires-sliding="@TimeSpan.FromSeconds(10)"
           vary-by="@ViewContext.RouteData.Values["action"]">
        @await Component.InvokeAsync("Time")
    </cache>
    <div>@RenderBody()</div>
</body>
</html>
```

显式的 URL 是完全可以接受的，但你要明白，如果更新应用程序的 URL 架构，你将不得不更新它们。

但是，某些应用程序将被部署到共享环境中，单个服务器通过向 URL 添加前缀来支持多个应用程序。代码清单 25-25 修改了应用程序的配置，以便将请求管线设置为处理具有 mvcpp 前缀的请求，从而模拟共享环境。

代码清单 25-25　在 Cities 文件夹下的 Startup.cs 文件中添加 URL 前缀

```
using System;
using System.Collections.Generic;
using System.Linq;
using System.Threading.Tasks;
using Microsoft.AspNetCore.Builder;
using Microsoft.AspNetCore.Hosting;
using Microsoft.AspNetCore.Http;
using Microsoft.Extensions.DependencyInjection;
using Cities.Models;

namespace Cities {
    public class Startup {

        public void ConfigureServices(IServiceCollection services) {
            services.AddSingleton<IRepository, MemoryRepository>();
            services.AddMvc();
        }
        public void Configure(IApplicationBuilder app, IHostingEnvironment env) {
            app.Map("/mvcapp", appBuilder => {
                appBuilder.UseStatusCodePages();
                appBuilder.UseDeveloperExceptionPage();
                appBuilder.UseStaticFiles();
                appBuilder.UseMvcWithDefaultRoute();
            });
        }
    }
}
```

Map 方法允许使用不同的前缀设置多个请求管线。在日常 MVC 开发中这通常不是什么有用的功能，因为可以使用路由系统在 MVC 应用程序内创建 URL 前缀。但对于本章，这是一个十分有用的特性，因为它意味着被客户端请求的任何 URL 将包括对静态内容的请求。

通过运行应用程序并访问/mvcapp（该地址现在是应用程序的默认地址，并且指向 Home 控制器的 Index 操作方法），你可以查看出现的问题。现在所有的 URL 都以/mvcapp 开始，对于 link 元素内的样式表，显式的 URL 不起作用，这意味着应用程序的内容不能被修饰，如图 25-7 所示。

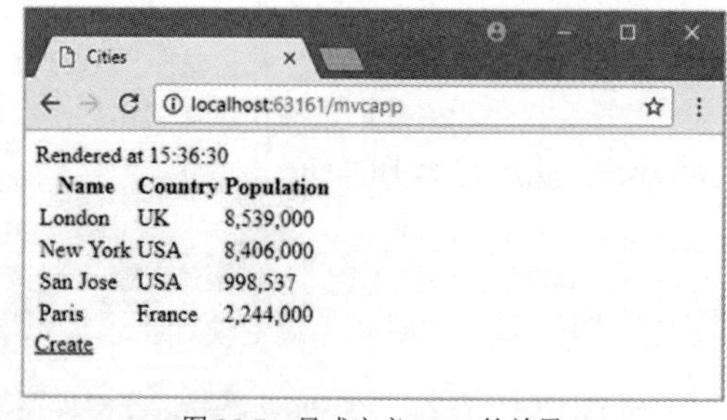

图 25-7　显式定义 URL 的效果

通过在 URL 中包含这个前缀，你可以解决该问题，但是因为前缀可能在部署时更新或在开发时不知道，这种方案并不可行。更好的方案是使用应用程序相关的 URL，这样静态内容的路径就可以表示为相对于已配置的任何前缀，如代码清单 25-26 所示。

代码清单 25-26　在 Views/Shared 文件夹下的_Layout.cshtml 文件中使用应用程序的相对 URL

```
<!DOCTYPE html>
<html>
<head>
    <meta name="viewport" content="width=device-width" />
    <title>Cities</title>
    <link href="~/lib/bootstrap/dist/css/bootstrap.min.css" rel="stylesheet" />
</head>
```

```
<body class="m-1 p-1">
    <cache expires-sliding="@TimeSpan.FromSeconds(10)"
           vary-by="@ViewContext.RouteData.Values["action"]">
        @await Component.InvokeAsync("Time")
    </cache>
    <div>@RenderBody()</div>
</body>
</html>
```

~符号由 UrlResolutionTagHelper 检测，将~符号替换为到达 wwwroot 文件夹内容所需的路径。如果运行应用程序，你将看到内容被修饰了，检查发送到浏览器的 HTML 内容，你将看到 link 元素包含了使用 mvcapp 前缀的 URL：

```
<link href="/mvcapp/lib/bootstrap/dist/css/bootstrap.min.css" rel="stylesheet" />
```

UrlResolutionTaghelper 能在各种元素中查询 URL，如表 25-7 所示。

表 25-7　　可由 UrlResolutionTagHelper 转换的元素及其属性

元素	属性
a	href
applet	archive
area	href
audio	src
base	href
blockquote	cite
button	formaction
del	cite
embed	src
form	action
html	manifest
iframe	src
img	src 和 srcset
input	src 和 formaction
ins	cite
link	href
menuitem	icon
object	archive 和 data
q	cite
script	src
source	src 和 srcset
track	src
video	src 和 poster

提　示

如果使用另一种内置的标签助手从路由系统生成 URL，那么生成的 HTML 将自动包括任何必要的前缀，这是从 HttpRequest.PathBase 上下文属性中获取的，并且值由宿主应用程序的服务器提供。

25.7　小结

本章介绍了除了表单标签助手之外的内置标签助手。这些标签助手用于帮助管理对 JavaScript 文件和 CSS 样式表文件的访问，为超链接元素创建 URL，为图片执行缓存清理，缓存数据，以及转换与应用程序相关的 URL，等等。下一章将介绍模型绑定，用于处理 HTTP 请求中的数据，以便数据可以在 MVC 应用程序内部轻松使用。

第 26 章　模型绑定

模型绑定是使用 HTTP 请求中的数据来创建.NET 对象的过程，以便为操作方法提供需要的参数。本章将描述模型绑定的工作方式，展示模型绑定如何绑定简单类型、复杂类型以及集合，并演示如何控制绑定过程来指定将请求的哪一部分作为操作方法要求的值。表 26-1 介绍了模型绑定的背景。

表 26-1　模型绑定的背景

问题	答案
模型绑定是什么？	模型绑定是使用 HTTP 请求中的数据来创建.NET 对象的过程，以便为操作方法提供需要的参数
模型绑定有何作用？	模型绑定允许操作方法使用 C#类型来定义参数，参数可自动从请求中接收数据，而不需要直接检查、解析和处理请求数据
如何使用模型绑定？	对于最简单的场景，操作方法会定义参数，参数名称用来从 HTTP 请求中抽取数据。可以通过对操作方法的参数应用特性来配置请求的哪一部分用于数据抽取
模型绑定有何陷阱或限制？	主要的陷阱是从请求的错误部分获取数据。26.2 节将阐述用于搜索请求数据的方式，搜索的位置可以使用特性显式指定
模型绑定有何替代品？	操作方法可以完全不需要声明参数，可以使用上下文对象直接从 HTTP 请求对象中获取数据。但是，代码更复杂，且难以理解和维护

表 26-2 列出了本章要介绍的操作。

表 26-2　本章要介绍的操作

操作	方法	代码清单
绑定简单类型或集合	为操作方法添加参数	代码清单 26-1～代码清单 26-10 和代码清单 26-23～代码清单 26-29
绑定复杂类型	确认视图生成的 HTML 是结构良好的	代码清单 26-11～代码清单 26-19
选择绑定属性	使用 Bind 特性指定数据值的名称，或者使用 BindNever 特性从绑定过程中排除模型属性	代码清单 26-20～代码清单 26-22
指定数据绑定值的来源	为操作方法的参数或模型属性应用特性，标识绑定值应该来自何方	代码清单 26-30～代码清单 26-38

26.1　准备示例项目

本章将使用 ASP.NET Core Web Application (.NET Core)模板创建一个新的名为 MvcModels 的 Empty 项目。

26.1.1　创建模型和存储库

这里创建 Models 文件夹，并在其中创建名为 Person.cs 的类文件，用它定义代码清单 26-1 所示的类和枚举。

代码清单 26-1　Models 文件夹下的 Person.cs 文件的内容

```
using System;

namespace MvcModels.Models {

    public class Person {
        public int PersonId { get; set; }
        public string FirstName { get; set; }
```

```
        public string LastName { get; set; }
        public DateTime BirthDate { get; set; }
        public Address HomeAddress { get; set; }
        public bool IsApproved { get; set; }
        public Role Role { get; set; }
    }

    public class Address {
        public string Line1 { get; set; }
        public string Line2 { get; set; }
        public string City { get; set; }
        public string PostalCode { get; set; }
        public string Country { get; set; }
    }
    public enum Role {
        Admin,
        User,
        Guest
    }
}
```

接着，添加名为 Repository.cs 的类文件到 Models 文件夹中以定义接口和实现，如代码清单 26-2 所示。

代码清单 26-2 Models 文件夹下的 Repository.cs 文件的内容

```
using System.Collections.Generic;

namespace MvcModels.Models {

    public interface IRepository {
        IEnumerable<Person> People { get; }

        Person this[int id] { get; set; }
    }

    public class MemoryRepository : IRepository {
        private Dictionary<int, Person> people
                = new Dictionary<int, Person> {
            [1] = new Person {PersonId = 1, FirstName = "Bob",
                LastName = "Smith", Role = Role.Admin},
            [2] = new Person {PersonId = 2, FirstName = "Anne",
                LastName = "Douglas", Role = Role.User},
            [3] = new Person {PersonId = 3, FirstName = "Joe",
                LastName = "Able", Role = Role.User},
            [4] = new Person {PersonId = 4, FirstName = "Mary",
                LastName = "Peters", Role = Role.Guest}
        };

        public IEnumerable<Person> People => people.Values;

        public Person this[int id] {
            get {
                return people.ContainsKey(id) ? people[id] : null;
            }
            set {
                people[id] = value;
            }
        }
    }
}
```

IRepository 接口定义了 People 属性来获取模型中的所有对象，还定义了索引器以允许获取或存储单个对象。MemoryRepository 类使用字典实现这个接口并提供了一些默认内容。这个存储库实现是非持久的，所以在应用停止或重新启动的时候，应用的状态将被重置。

26.1.2 创建控制器和视图

这里创建 Controllers 文件夹，添加名为 HomeController.cs 的类文件，并用它定义代码清单 26-3 所示的控制器。Home 控制器基于依赖注入来接收存储库，在 Index 方法中，通过 PersonId 属性可在存储库中获取单个 Person 对象。

代码清单 26-3 Controllers 文件夹下的 HomeController.cs 文件的内容

```
using Microsoft.AspNetCore.Mvc;
using MvcModels.Models;

namespace MvcModels.Controllers {

    public class HomeController : Controller {
        private IRepository repository;

        public HomeController(IRepository repo) {
            repository = repo;
        }

        public ViewResult Index(int id) => View(repository[id]);
    }
}
```

为了给操作方法提供视图，这里创建 Views/Home 文件夹，使用代码清单 26-4 所示的内容在其中添加名为 Index.cshtml 的 Razor 文件，将模型对象的一些属性展示在表格中。

代码清单 26-4 Views/Home 文件夹下的 Index.cshtml 文件的内容

```
@model Person
@{ Layout = "_Layout"; }

<div class="bg-primary m-1 p-1 text-white"><h2>Person</h2></div>

<table class="table table-sm table-bordered table-striped">
    <tr><th>PersonId:</th><td>@Model.PersonId</td></tr>
    <tr><th>First Name:</th><td>@Model.FirstName</td></tr>
    <tr><th>Last Name:</th><td>@Model.LastName</td></tr>
    <tr><th>Role:</th><td>@Model.Role</td></tr>
</table>
```

Index 视图基于共享布局。创建 Views/Shared 文件夹，在其中添加名为_Layout.cshtml 的布局文件，内容如代码清单 26-5 所示。

代码清单 26-5 Views/Shared 文件夹下的_Layout.cshtml 文件的内容

```
<!DOCTYPE html>
<html>
<head>
    <meta name="viewport" content="width=device-width" />
    <title>@ViewBag.Title</title>
    <link asp-href-include="/lib/bootstrap/dist/**/*.min.css" rel="stylesheet" />
    @RenderSection("scripts", false)
</head>
<body class="m-1 p-1">
    @RenderBody()
</body>
</html>
```

以上布局包含用于 Bootstrap 样式表的 link 元素并且渲染视图内容，还包含可选的 scripts 部分（section），本章的后面将会用到。为了简化本章用到的视图，在 Views 文件夹的_ViewImports.cshtml 文件中添加包含模型类的命名空间，如代码清单 26-6 所示。

代码清单 26-6 在 Views 文件夹下的_ViewImports.cshtml 文件中导入命名空间

```
@using MvcModels.Models
@addTagHelper *, Microsoft.AspNetCore.Mvc.TagHelpers
```

视图基于 Bootstrap CSS 框架。下面使用 Bower Configuration File 模板，在项目的根目录中创建 bower.json 文件，将 Bootstrap 添加到项目中。添加的包如代码清单 26-7 所示。

代码清单 26-7 在 MvcModels 文件夹下的 bower.json 文件中添加 Bootstrap 包

```
{
  "name": "asp.net",
  "private": true,
  "dependencies": {
    "bootstrap": "4.0.0-alpha.6"
  }
}
```

26.1.3 配置应用

为了初始化示例应用，在 Startup 类中启用 MVC 框架和开发中用到的其他中间件，如代码清单 26-8 所示。这里还为存储库创建了服务以便控制器可以访问数据模型。

代码清单 26-8 MvcModels 文件夹下的 Startup.cs 文件的内容

```
using System;
using System.Collections.Generic;
using System.Linq;
using System.Threading.Tasks;
using Microsoft.AspNetCore.Builder;
using Microsoft.AspNetCore.Hosting;
using Microsoft.AspNetCore.Http;
using Microsoft.Extensions.DependencyInjection;
using MvcModels.Models;
namespace MvcModels {

    public class Startup {

        public void ConfigureServices(IServiceCollection services) {
            services.AddSingleton<IRepository, MemoryRepository>();
            services.AddMvc();
        }

        public void Configure(IApplicationBuilder app, IHostingEnvironment env) {
            app.UseStatusCodePages();
            app.UseDeveloperExceptionPage();
            app.UseStaticFiles();
            app.UseMvc(routes => {
                routes.MapRoute(
                    name: "default",
                    template: "{controller=Home}/{action=Index}/{id?}");
            });
        }
    }
}
```

启动应用并访问/Home/Index/1，将生成图 26-1 所示的结果。现在，默认的 URL 地址将导致错误。

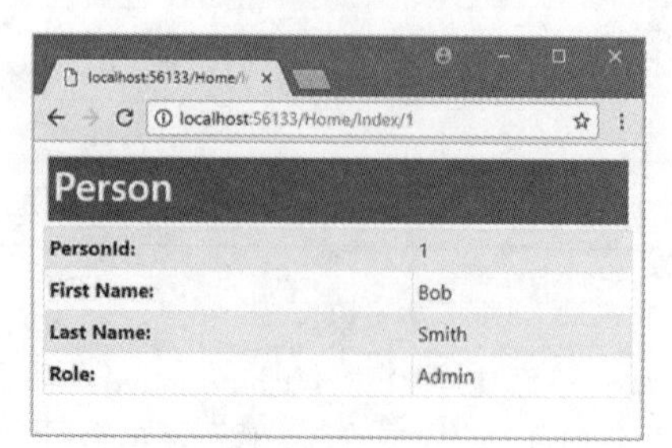

图 26-1 运行示例应用

26.2 理解模型绑定

模型绑定是 HTTP 请求与 C#操作方法之间优雅的桥梁。多数 MVC 应用在某种程度上依赖模型绑定，包括本章的示例应用。前面在测试示例应用的时候，模型绑定就被应用了。请求的 URL 中包含了希望查阅的 Person 对象的 PersonId 属性值，如下所示：

```
/Home/Index/1
```

MVC 对这一部分 URL 进行转换，并且在调用 Home 控制器的 Index 方法以服务请求的时候将之作为参数。

```
...
public ViewResult Index(int id) => View(repository[id]);
...
```

为了调用 Index 方法，MVC 需要一个值作为 id 参数，提供这个值就是模型绑定系统的责任，模型绑定系统负责为调用操作方法提供数据。

模型绑定基于模型绑定器，模型绑定器的职责是从请求的某个部分或应用本身提供数据。默认的模型绑定从 3 个方面提供数据：

- 表单数据；
- 路由变量；
- 查询串。

每个数据源按照顺序被检查，直到参数的值被发现。示例应用中没有表单数据，所以也就不用从表单中进行查找。但是，应用的配置中包含名为 id 的路由片段，因而允许模型绑定系统提供用于调用 Index 方法的值。搜索在发现合适的值之后就会停止，这意味着查询串不会用来搜索数据。

提 示

26.3 节将说明如何使用特性来指定模型绑定的数据源。这将允许从指定的数据源获取数据，例如查询串，即使表单或路由数据中存在合适的数据。

知道数据的搜索顺序很重要，因为请求中可能包含多个值，比如下面这个 URL：

```
/Home/Index/3?id=1
```

路由系统将处理请求并匹配 URL 模板中的 id 片段为 3，查询串中包含 id 为 1 的值。由于路由系统在查询串之前搜索数据，Index 方法将收到值 3，查询串的值将被忽略。

另外，如果请求 URL 中没有 id 片段，查询串将会被处理，这意味着这样的 URL 也允许模型绑定系统为 MVC 提供 id 值，以便调用 Index 方法。

```
/Home/Index?id=1
```

图 26-2 显示了这两种 URL 的结果。

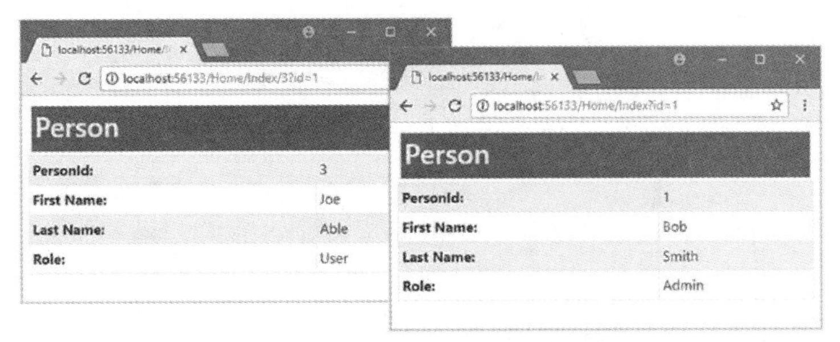

图 26-2　数据源次序对模型绑定的影响

26.2.1　默认绑定值

模型绑定是尽力而为的特性，这意味着 MVC 将使用模型绑定来试着获取调用操作方法所需的值，但是在值不能提供时也仍然调用操作方法，这可能导致一些意想不到的行为。例如，访问/Home/Index 将导致图 26-3 所示的异常。

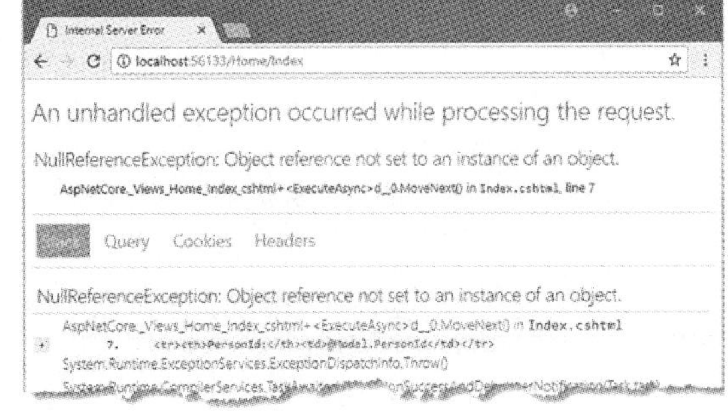

图 26-3　异常

图 26-3 所示的异常不是由模型绑定系统报告的。当处理由 Index 方法选择的 Index 视图时，该异常出现。为了调用 Index 方法，MVC 不得不为 id 参数提供值，所以 MVC 请求每个模型绑定器检查各自的部分来提供值。

示例中没有表单数据，没有 id 路由片段，URL 中也没有查询串，这意味着模型绑定不能提供值，MVC 必须为 id 参数提供某些值来调用 Index 方法，MVC 使用了默认值并期望这是最好的。对于 int 类型参数，默认值是 0，这就是导致异常的原因。在 Index 方法的定义中，使用 id 参数的值从存储库中获取模型对象。

```
...
public ViewResult Index(int id) => View(repository[id]);
...
```

当 MVC 使用默认值时，操作方法试图使用 id 参数值 0 来获取数据对象。没有这样的对象，存储库返回 null，这个结果随后被传递给控制器的 View 方法以为 Index.cshtml 文件提供视图模型对象。当 Index.cshtml 文件中的 Razor 表达式试图访问视图模型对象的属性时，就会导致图 26-3 所示的 NullReferenceException 异常。

这意味着操作方法必须应对模型绑定系统提供的默认值。这可以通过多种方式来完成。可以添加默认值到 URL 路由模板中，为操作方法提供默认参数，或者确保操作方法不会将错误值作为一部分传递给响应。最好的处理方式取决于操作方法做什么：代码清单 26-9 采用了最后一种方式，修改操作方法以确保一个 Person 对象总是能传递给 View 方法，即使 id 参数并不对应数据模型中的对象。

代码清单 26-9　在 Controllers 文件夹下的 HomeController.cs 文件中防范默认的模型绑定值

```
using Microsoft.AspNetCore.Mvc;
using MvcModels.Models;
using System.Linq;

namespace MvcModels.Controllers {

    public class HomeController : Controller {
        private IRepository repository;

        public HomeController(IRepository repo) {
            repository = repo;
```

```
        }

        public ViewResult Index(int id) =>
            View(repository[id] ?? repository.People.First());
    }
}
```

当使用 id 参数的值不能取回对象的时候，操作方法使用 LINQ 和 null 合并操作符来返回存储库中的第一个对象。

26.2.2 绑定简单值

当有合适的值可用时，该值必须被转换为 C#值，以便可以用于调用操作方法。使用请求中的简单类型可以从字符串中解析得到数据项，包括数值、布尔值、日期以及字符串值。

Index 方法的 id 参数为 int 类型，所以模型绑定过程通过将 id 片段解析为整数值提供给 MVC。

如果请求值不能被转换（例如，提供 apple 给需要 int 类型值的参数），模型绑定系统将不能为应用程序提供值，默认值将会被使用。

这会出现问题，因为在两种情况下操作方法都将收到默认值 0。第一种情况是，请求中包含不能被解析为参数类型的值，如 URL 为/Home/Index/Apple。第二种情况是，请求中包含可以解析的值但碰巧是 0，如 URL 为/Home/Index/0。

大多数应用程序需要区分这些状况，最简单的方式是使用可空类型作为操作方法的参数，如代码清单 26-10 所示。

代码清单 26-10 在 Controllers 文件夹下的 HomeController.cs 文件中使用可空类型

```
using Microsoft.AspNetCore.Mvc;
using MvcModels.Models;

namespace MvcModels.Controllers {

    public class HomeController : Controller {
        private IRepository repository;

        public HomeController(IRepository repo) {
            repository = repo;
        }

        public IActionResult Index(int? id) {
            Person person;
            if (id.HasValue && (person = repository[id.Value]) != null) {
                return View(person);
            } else {
                return NotFound();
            }
        }
    }
}
```

默认可空类型的值是 null，这允许区分请求中不包括可以解析为整数值的情况和值碰巧是 0 的情况。如果可空参数没有值或者值没有关联模型中的对象，那么示例中 Index 方法的实现将使用 NotFound 方法来返回 404 错误。这是一种比简单地期望模型中的第一个对象十分适当的更稳健的方式，也是之前采用的方式。

26.2.3 绑定复杂类型

当操作方法的参数是复杂类型的时候（换句话说，任何不能从单个字符串值解析到的类型），模型绑定过程将使用反射来获取目标类型的公共属性集合，然后依次对每个属性执行绑定过程。为了演示这是如何工作的，在 Home 控制器中添加两个操作方法，如代码清单 26-11 所示。

代码清单 26-11 在 Controllers 文件夹下的 HomeController.cs 中添加两个操作方法

```
using Microsoft.AspNetCore.Mvc;
using MvcModels.Models;

namespace MvcModels.Controllers {

    public class HomeController : Controller {
        private IRepository repository;

        public HomeController(IRepository repo) {
            repository = repo;
        }

        public IActionResult Index(int? id) {
            Person person;
            if (id.HasValue && (person = repository[id.Value]) != null) {
                return View(person);
            } else {
                return NotFound();
            }
        }

        public ViewResult Create() => View(new Person());

        [HttpPost]
        public ViewResult Create(Person model) => View("Index", model);
    }
}
```

没有参数的 Create 方法版本会创建一个新的 Person 对象并传递给 View 方法，View 方法将影响选择关联到操作方法的默认视图。在 Views/Home 文件夹中创建名为 Create.cshtml 的视图文件，并添加代码清单 26-12 所示的标记。

代码清单 26-12 Views/Home 文件夹下的 Create.cshtml 文件的内容

```
@model Person
@{
    ViewBag.Title = "Create Person";
    Layout = "_Layout";
}

<form asp-action="Create" method="post">
    <div class="form-group">
        <label asp-for="PersonId"></label>
        <input asp-for="PersonId" class="form-control" />
    </div>
    <div class="form-group">
        <label asp-for="FirstName"></label>
        <input asp-for="FirstName" class="form-control" />
    </div>
    <div class="form-group">
        <label asp-for="LastName"></label>
        <input asp-for="LastName" class="form-control" />
    </div>
    <div class="form-group">
        <label asp-for="Role"></label>
        <select asp-for="Role" class="form-control"
                asp-items="@new SelectList(Enum.GetNames(typeof(Role)))"></select>
    </div>
    <button type="submit" class="btn btn-primary">Submit</button>
</form>
```

Create 视图包含一个将数据回送给 Home 控制器中使用 HttpPost 特性修饰的 Create 方法的表单元素，可通过它为 Person 对象的某些属性提供值。

操作方法接收表单数据，然后使用/Views/Home/Index.cshtml 文件显示它们。启动应用，导航到/Home/Create，填写表单，然后单击 Submit 按钮来看看它们是如何工作的，如图 26-4 所示。

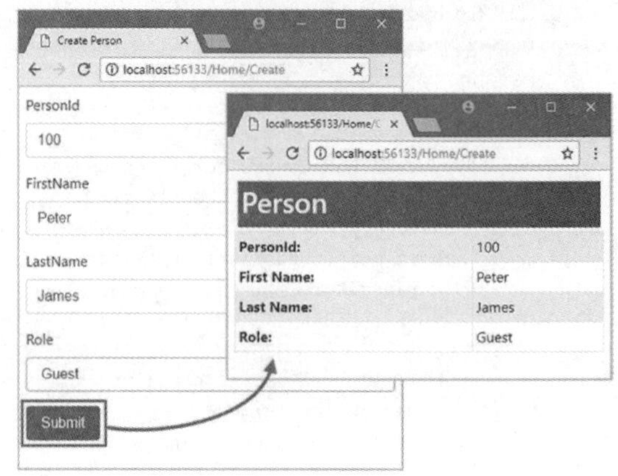

图 26-4 填写表单并单击 Submit 按钮

在把表单数据发送给服务器后，模型绑定系统发现操作方法需要一个 Person 对象。分析 Person 类及其公共属性。对于每个简单类型，模型绑定器试图查找请求值，就像前一个示例那样。

对于这个示例，模型绑定器发现了 PersonId 属性，并寻找 PersonId 值，由于表单中包含合适的值——在 input 元素的设置中使用了 asp-for 属性的标签助手，因此这个值将被使用。

如果属性为另一复杂类型，处理过程将在新的类型上重复进行。获取公共属性集合，然后模型绑定器试图寻找所有属性的值。不同的是，属性名称是嵌套的。例如，Person 类的 HomeAddress 属性为 Address 类型，如下所示：

```
using System;

namespace MvcModels.Models {

    public class Person {
        public int PersonId { get; set; }
        public string FirstName { get; set; }
        public string LastName { get; set; }
        public DateTime BirthDate { get; set; }
        public Address HomeAddress { get; set; }
        public bool IsApproved { get; set; }
        public Role Role { get; set; }
    }

    public class Address {
        public string Line1 { get; set; }
        public string Line2 { get; set; }
        public string City { get; set; }
        public string PostalCode { get; set; }
        public string Country { get; set; }
    }

    public enum Role {
        Admin,
        User,
        Guest
    }
}
```

当寻找 Line1 属性的值时，模型绑定器为 HomeAddress.Line1 查找值——由模型对象属性的名称合并嵌套的模型类型属性名组成。

1. 创建易于绑定的 HTML

前缀意味着视图必须包含模型绑定器查找的信息。使用标签助手可以很容易实现，标签助手能自动将需要的前缀添加到它们想要转换的元素上。代码清单 26-13 扩展了表单以便可以获取地址数据。

代码清单 26-13 在 Views/Home 文件夹下更新 Create.cshtml 文件中的表单

```
@model Person
@{
    ViewBag.Title = "Create Person";
    Layout = "_Layout";
}

<form asp-action="Create" method="post">
    <div class="form-group">
        <label asp-for="PersonId"></label>
        <input asp-for="PersonId" class="form-control" />
    </div>
    <div class="form-group">
        <label asp-for="FirstName"></label>
        <input asp-for="FirstName" class="form-control" />
    </div>
    <div class="form-group">
        <label asp-for="LastName"></label>
        <input asp-for="LastName" class="form-control" />
    </div>
    <div class="form-group">
        <label asp-for="Role"></label>
        <select asp-for="Role" class="form-control"
                asp-items="@new SelectList(Enum.GetNames(typeof(Role)))"></select>
    </div>
    <div class="form-group">
        <label asp-for="HomeAddress.City"></label>
        <input asp-for="HomeAddress.City" class="form-control" />
    </div>
    <div class="form-group">
        <label asp-for="HomeAddress.Country"></label>
        <input asp-for="HomeAddress.Country" class="form-control" />
    </div>
    <button type="submit" class="btn btn-primary">Submit</button>
</form>
```

当使用标签助手的时候，嵌套的属性名称使用 C#约定以便外包和内嵌的属性名称以句点分隔——HomeAddress.Country。如果运行应用程序，访问/Home/Create，然后检查发送给浏览器的 HTML，你将看到对于一些属性使用了不一样的约定：

```
<div class="form-group">
    <label for="HomeAddress_City">City</label>
    <input class="form-control" type="text" id="HomeAddress_City"
        name="HomeAddress.City" value="" />
</div>
<div class="form-group">
    <label for="HomeAddress_Country">Country</label>
    <input class="form-control" type="text" id="HomeAddress_Country"
        name="HomeAddress.Country" value="" />
</div>
```

input 元素的 name 属性遵循 C#风格，但是 label 元素的 for 属性和 input 元素的 id 属性使用下画线分隔属性名称。如果希望不使用标签助手来定义 HTML 元素，那么需要确保使用了同样的命名模式。

作为这一特性的结果，不需要采用任何特殊的行动就可以确保模型绑定器会为 HomeAddress 属性创建 Address 对象。在数据可以通过表单提交之后，可以通过编辑 Index.cshtml 视图文件来显示 HomeAddress 属性，如代码清单 26-14 所示。

代码清单 26-14 在 Views/Home 文件夹下的 Index.cshtml 文件中显示 HomeAddress 属性

```
@model Person
@{ Layout = "_Layout"; }
```

```
<div class="bg-primary m-1 p-1 text-white"><h2>Person</h2></div>

<table class="table table-sm table-bordered table-striped">
    <tr><th>PersonId:</th><td>@Model.PersonId</td></tr>
    <tr><th>First Name:</th><td>@Model.FirstName</td></tr>
    <tr><th>Last Name:</th><td>@Model.LastName</td></tr>
    <tr><th>Role:</th><td>@Model.Role</td></tr>
    <tr><th>City:</th><td>@Model.HomeAddress?.City</td></tr>
    <tr><th>Country:</th><td>@Model.HomeAddress?.Country</td></tr>
</table>
```

如果启动应用并导航到 URL 地址/Home/Create，就可以为 City 和 Country 文本框输入值，通过提交表单可以检查它们是否已被绑定到模型对象，如图 26-5 所示。

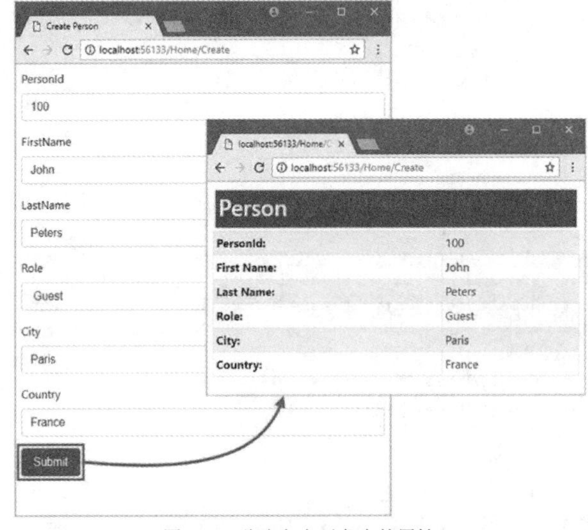

图 26-5　绑定复杂对象中的属性

2. 指定自定义前缀

在有的场合中，生成的 HTML 已关联一种类型，但是希望将 HTML 绑定到另外一种类型。这意味着视图包含的前缀不对应模型绑定期望的结构，所以数据不会被正确处理。为了说明这个问题，添加一个名为 AddressSummary.cs 的类文件到 Models 文件夹中，用它定义代码清单 26-15 所示的类。

代码清单 26-15　Models 文件夹下的 AddressSummary.cs 文件的内容

```
namespace MvcModels.Models {

    public class AddressSummary {
        public string City { get; set; }
        public string Country { get; set; }
    }
}
```

在 Home 控制器中添加一个新的使用 AddressSummary 类的操作方法，如代码清单 26-16 所示。

代码清单 26-16　在 Controllers 文件夹下的 HomeController.cs 文件中添加一个操作方法

```
using Microsoft.AspNetCore.Mvc;
using MvcModels.Models;

namespace MvcModels.Controllers {
```

```
    public class HomeController : Controller {
        private IRepository repository;

        public HomeController(IRepository repo) {
            repository = repo;
        }

        public IActionResult Index(int? id) {
            Person person;
            if (id.HasValue && (person = repository[id.Value]) != null) {
                return View(person);
            } else {
                return NotFound();
            }
        }

        public ViewResult Create() => View(new Person());

        [HttpPost]
        public ViewResult Create(Person model) => View("Index", model);

        public ViewResult DisplaySummary(AddressSummary summary) => View(summary);
    }
}
```

新的操作方法名为 DisplaySummary，它的 AddressSummary 参数被传递给 View 方法以便能被默认视图显示。在/Views/Home 文件夹中创建 DisplaySummary.cshtml 文件，添加代码清单 26-17 所示的标记内容。

代码清单 26-17 Views/Home 文件夹下的 DisplaySummary.cshtml 文件的内容

```
@model AddressSummary
@{
    ViewBag.Title = "DisplaySummary";
    Layout = "_Layout";
}

<div class="bg-primary m-1 p-1 text-white"><h2>Address</h2></div>

<table class="table table-sm table-bordered table-striped">
    <tr><th>City:</th><td>@Model.City</td></tr>
    <tr><th>Country:</th><td>@Model.Country</td></tr>
</table>
```

DisplaySummary 视图显示了 AddressSummary 类中定义的两个属性的值。为了演示使用前缀绑定到不同的模型类型时引发的问题，修改 Create.cshtml 视图文件中的表单元素以发送数据给 DisplaySummary 操作，如代码清单 26-18 所示。

代码清单 26-18 在 Views/Home 文件夹下的 Create.cshtml 文件中修改表单的操作目标

```
@model Person
@{
    ViewBag.Title = "Create Person";
    Layout = "_Layout";
}

<form asp-action="DisplaySummary" method="post">

    <!-- HTML elements omitted for brevity -->

</form>
```

启动应用并导航到 URL 地址/Home/Create，看看发生了什么。提交表单后，你为 City 和 Country 文本框输入的值不会显示在 DisplaySummary 视图生成的 HTML 中。

问题在于 form 元素的 name 属性有 HomeAddress 前缀，在试图绑定到 AddressSummary 类型时，它不是模型绑定器想要寻找的内容。

为解决这个问题，Bind 特性可以被应用于操作方法的参数，从而在模型绑定中指定前缀，如代码清单 26-19 所示。

代码清单 26-19　在 Controllers 文件夹下的 HomeController.cs 中变更模型绑定前缀

```
using Microsoft.AspNetCore.Mvc;
using MvcModels.Models;

namespace MvcModels.Controllers {

    public class HomeController : Controller {
        private IRepository repository;

        public HomeController(IRepository repo) {
            repository = repo;
        }

        public IActionResult Index(int? id) {
            Person person;
            if (id.HasValue && (person = repository[id.Value]) != null) {
                return View(person);
            } else {
                return NotFound();
            }
        }

        public ViewResult Create() => View(new Person());
        [HttpPost]
        public ViewResult Create(Person model) => View("Index", model);

        public ViewResult DisplaySummary(
            [Bind(Prefix = nameof(Person.HomeAddress))] AddressSummary summary)
                => View(summary);
    }
}
```

语法虽然笨拙，但是效果非常好。在填充 AddressSummary 对象的属性时，模型绑定器将在请求中寻找 HomeAddress.City 和 HomeAddress.Country 的值。如果运行应用并重新提交表单，你将看 City 和 Country 文本框中的值现在正确显示，如图 26-6 所示。对于简单问题来说，这看起来有些复杂，但是绑定到不同类型的对象很常见，并且是值得学习的技术。

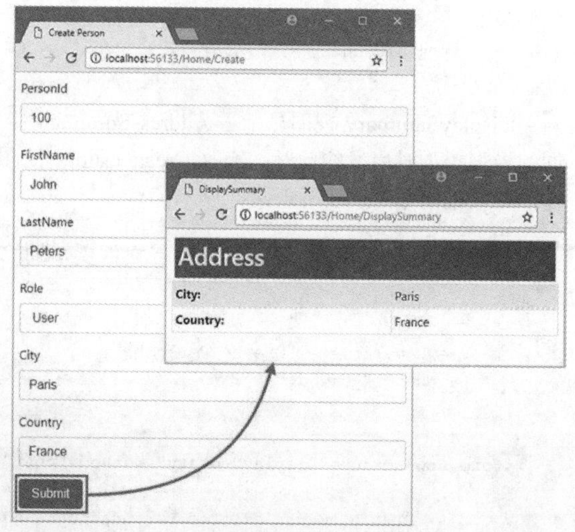

图 26-6　绑定到不同类型对象的属性

3. 选择性绑定属性

想象一下，AddressSummary 类的 Country 属性特别敏感，用户不应该能够为它指定值。你可以做的第一件事，就是通过确保不会在应用的任何涉及这个属性的视图的 HTML 元素中包含它，以防止用户查看或编辑这个属性。

然而，恶意用户可以简单地在提交表单数据的时候，编辑发送给服务器的表单数据并挑选合适的Country属性值。这里需要告诉模型绑定器不要从请求中为Country属性绑定值。这可以通过配置操作方法参数的Bind特性来实现，如代码清单26-20所示。

代码清单 26-20 在 Controllers 文件夹下的 HomeController.cs 文件中指定属性

```
using Microsoft.AspNetCore.Mvc;
using MvcModels.Models;

namespace MvcModels.Controllers {

    public class HomeController : Controller {
        private IRepository repository;

        public HomeController(IRepository repo) {
            repository = repo;
        }

        public IActionResult Index(int? id) {
            Person person;
            if (id.HasValue && (person = repository[id.Value]) != null) {
                return View(person);
            } else {
                return NotFound();
            }
        }

        public ViewResult Create() => View(new Person());

        [HttpPost]
        public ViewResult Create(Person model) => View("Index", model);

        public ViewResult DisplaySummary(
            [Bind(nameof(AddressSummary.City), Prefix = nameof(Person.HomeAddress))]
                AddressSummary summary) => View(summary);
    }
}
```

Bind特性的第一个参数是以逗号分隔的应当被模型绑定过程包含的属性名称列表。在这个列表中，指定了City属性应当包含在过程中，由于Country没有列入，这意味着Country属性将被排除。

如果运行应用，访问/Home/Create，然后填充表单并提交，你将看到Country属性的值没有显示出来，即使已作为浏览器HTTP POST请求的一部分被发送，如图26-7所示。

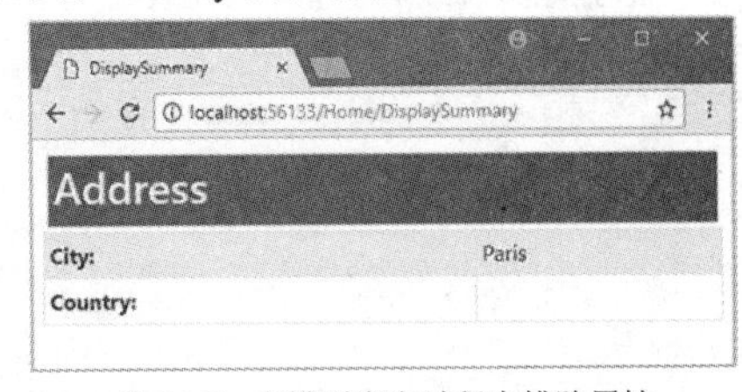

图 26-7 在模型绑定过程中排除属性

当Bind特性被应用于操作方法的参数时，它仅仅影响为这个类的实例绑定的操作方法；其他所有的操作方法将继续试图绑定参数类型定义的所有属性。为了使Bind特性发挥更广泛的影响，可以将Bind特性应用于模型类本身，如代码清单26-21所示。

代码清单 26-21 在 Models 文件夹下的 AddressSummary.cs 文件中应用 Bind 特性

```
using Microsoft.AspNetCore.Mvc;

namespace MvcModels.Models {

    [Bind(nameof(City))]
    public class AddressSummary {
        public string City { get; set; }
```

```
        public string Country { get; set; }
    }
}
```

也可以使用 BindNever 特性来显式地排除属性，如代码清单 26-22 所示，虽然这意味着新添加到模型类的属性将被包含在模型绑定过程中，除非为它们应用这个特性。

代码清单 26-22 在 Models 文件夹下的 AddressSummary.cs 文件中应用 BindNever 特性

```
using Microsoft.AspNetCore.Mvc;
using Microsoft.AspNetCore.Mvc.ModelBinding;

namespace MvcModels.Models {

    public class AddressSummary {
        public string City { get; set; }

        [BindNever]
        public string Country { get; set; }
    }
}
```

提 示

BindRequired 特性用于告诉模型绑定过程，请求中必须包含属性的值。如果请求中没有包含必要的值，将会导致模型验证错误，详见第 27 章。

26.2.4 绑定数组和集合

模型绑定有一些不错的特性用于绑定请求数据到数组和集合，下面进行说明。

1. 绑定到数组

模型绑定默认的优雅特性在于支持数组类型的操作方法参数。为了说明这一点，在 Home 控制器中添加名为 Names 的新方法，如代码清单 26-23 所示。

代码清单 26-23 在 Controllers 文件夹下的 HomeController.cs 文件中添加 Names 方法

```
using Microsoft.AspNetCore.Mvc;
using MvcModels.Models;

namespace MvcModels.Controllers {

    public class HomeController : Controller {
        private IRepository repository;

        public HomeController(IRepository repo) {
            repository = repo;
        }

        public IActionResult Index(int? id) {
            Person person;
            if (id.HasValue && (person = repository[id.Value]) != null) {
                return View(person);
            } else {
                return NotFound();
            }
        }

        public ViewResult Create() => View(new Person());
        [HttpPost]
        public ViewResult Create(Person model) => View("Index", model);
```

```
        public ViewResult DisplaySummary(
            [Bind(nameof(AddressSummary.City), Prefix = nameof(Person.HomeAddress))]
                AddressSummary summary) => View(summary);

        public ViewResult Names(string[] names) => View(names ?? new string[0]);
    }
}
```

Names 方法有一个名为 names 的字符串数组参数。模型绑定器将查询名为 names 且包含这些值的任何数据项来创建数组，为了给这个操作方法提供视图，在 Views/Home 文件夹中创建名为Names.cshtml 的 Razor 文件，并添加代码清单 26-24 所示的标记内容。

代码清单 26-24 Views/Home 文件夹下的 Names.cshtml 文件的内容

```
@model string[]
@{
    ViewBag.Title = "Names";
    Layout = "_Layout";
}

@if (Model.Length == 0) {
    <form asp-action="Names" method="post">
        @for (int i = 0; i < 3; i++) {
            <div class="form-group">
                <label>Name @(i + 1):</label>
                <input id="names" name="names" class="form-control" />
            </div>
        }
        <button type="submit" class="btn btn-primary">Submit</button>
    </form>
} else {
    <table class="table table-sm table-bordered table-striped">
        @foreach (string name in Model) {
            <tr><th>Name:</th><td>@name</td></tr>
        }
    </table>
    <a asp-action="Names" class="btn btn-primary">Back</a>
}
```

Names 视图基于视图模型中包含的数据项的个数展示不同的内容。如果没有数据项，该视图将显示包含 3 个完全相同的 input 元素的表单，如下所示：

```
...
<form method="post" action="/Home/Names">
    <div class="form-group">
        <label>Name 1:</label>
        <input id="names" name="names" class="form-control" />
    </div>
    <div class="form-group">
        <label>Name 2:</label>
        <input id="names" name="names" class="form-control" />
    </div>
    <div class="form-group">
        <label>Name 3:</label>
        <input id="names" name="names" class="form-control" />
    </div>
    <button type="submit" class="btn btn-primary">Submit</button>
</form>
...
```

当表单被提交之后，模型绑定过程查看目标操作方法并得到一个数组，查询与操作方法参数具有相同名称的数据项。对于这个示例来说，这意味着所有 name 属性为 names 的 input 元素的值将被收集在一起以创建数组，并作为参数调用操作方法。为了查看效果，启动应用，导航到/Home/Names，填充表单并提交，你将看到输入的所有值都显示出来，如图 26-8 所示。

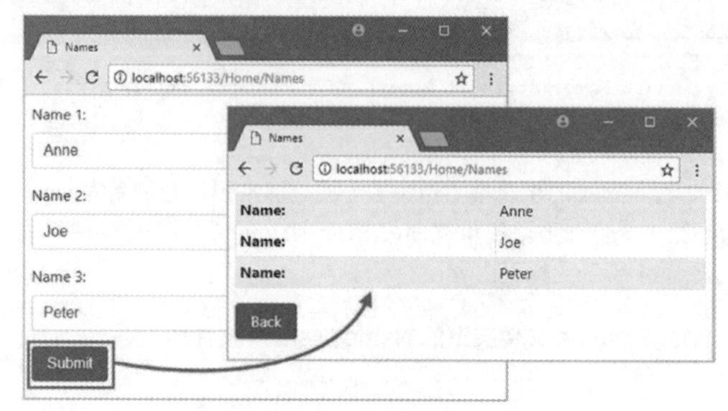

图 26-8 绑定到数组

2. 绑定到集合

模型绑定器不仅可以绑定数组，还可以绑定集合。代码清单 26-25 将 Names 方法的参数类型修改为强类型集合。

代码清单 26-25 在 Controllers 文件夹下的 HomeController.cs 文件中使用强类型集合

```
using Microsoft.AspNetCore.Mvc;
using MvcModels.Models;
using System.Collections.Generic;

namespace MvcModels.Controllers {
    public class HomeController : Controller {
        private IRepository repository;

        public HomeController(IRepository repo) {
            repository = repo;
        }

        // ...other action methods omitted for brevity...

        public ViewResult Names(IList<string> names) =>
            View(names ?? new List<string>());
    }
}
```

这里使用了 IList< T >接口。不需要指定具体的实现类，虽然根据个人意愿也可以这么做。代码清单 26-26 修改了 Names.cshtml 视图文件以使用新的模型类型。

代码清单 26-26 在 Views/Home 文件夹下的 Names.cshtml 文件中使用集合作为模型类型

```
@model IList<string>
@{
    ViewBag.Title = "Names";
    Layout = "_Layout";
}

@if (Model.Count == 0) {
    <form asp-action="Names" method="post">
        @for (int i = 0; i < 3; i++) {
            <div class="form-group">
                <label>Name @(i + 1):</label>
                <input id="names" name="names" class="form-control" />
            </div>
        }
        <button type="submit" class="btn btn-primary">Submit</button>
    </form>
```

```
} else {
    <table class="table table-sm table-bordered table-striped">
        @foreach (string name in Model) {
            <tr><th>Name:</th><td>@name</td></tr>
        }
    </table>
    <a asp-action="Names" class="btn btn-primary">Back</a>
}
```

Names 方法的功能没有变化，但是现在可以使用集合而不是数组进行工作。

3. 绑定到复杂类型的集合

单个数据值可以绑定到复杂类型的集合，这允许在单个请求中收集多个对象（比如示例中的 AddressSummary 模型类）。代码清单 26-27 在 Home 控制器中添加了名为 Address 的操作方法，参数是一个 AddressSummary 对象列表。

代码清单 26-27 在 Controllers 文件夹下的 HomeController.cs 文件中定义操作方法

```
using Microsoft.AspNetCore.Mvc;
using MvcModels.Models;
using System.Collections.Generic;

namespace MvcModels.Controllers {

    public class HomeController : Controller {
        private IRepository repository;

        public HomeController(IRepository repo) {
            repository = repo;
        }

        public IActionResult Index(int? id) {
            Person person;
            if (id.HasValue && (person = repository[id.Value]) != null) {
                return View(person);
            } else {
                return NotFound();
            }
        }

        public ViewResult Create() => View(new Person());

        [HttpPost]
        public ViewResult Create(Person model) => View("Index", model);

        public ViewResult DisplaySummary(
            [Bind(nameof(AddressSummary.City), Prefix = nameof(Person.HomeAddress))]
                AddressSummary summary) => View(summary);

        public ViewResult Names(IList<string> names) =>
            View(names ?? new List<string>());

        public ViewResult Address(IList<AddressSummary> addresses) =>
            View(addresses ?? new List<AddressSummary>());
    }
}
```

为了给这个新的操作方法提供视图，在 Views/Home 文件夹中添加名为 Address.cshtml 的文件，并添加代码清单 26-28 所示的标记内容。

代码清单 26-28 Views/Home 文件夹下的 Address.cshtml 文件的内容

```
@model IList<AddressSummary>
@{
    ViewBag.Title = "Address";
```

```
        Layout = "_Layout";
    }

    @if (Model.Count() == 0) {
        <form asp-action="Address" method="post">
            @for (int i = 0; i < 3; i++) {
                <fieldset class="form-group">
                    <legend>Address @(i + 1)</legend>
                    <div class="form-group">
                        <label>City:</label>
                        <input name="[@i].City" class="form-control" />
                    </div>
                    <div class="form-group">
                        <label>Country:</label>
                        <input name="[@i].Country" class="form-control" />
                    </div>
                </fieldset>
            }
            <button type="submit" class="btn btn-primary">Submit</button>
        </form>
    } else {
        <table class="table table-sm table-bordered table-striped">
            <tr><th>City</th><th>Country</th></tr>
            @foreach (var address in Model) {
                <tr><td>@address.City</td><td>@address.Country</td></tr>
            }
        </table>
        <a asp-action="Address" class="btn btn-primary">Back</a>
    }
```

如果模型集合中没有条目，Address 视图将生成一个 form 元素。这个 form 元素包含 name 属性带有数组索引前缀的 input 元素对，如下所示：

```
...
<form method="post" action="/Home/Address">
    <fieldset class="form-group">
        <legend>Address 1</legend>
        <div class="form-group">
            <label>City:</label>
            <input name="[0].City" class="form-control" />
        </div>
        <div class="form-group">
            <label>Country:</label>
            <input name="[0].Country" class="form-control" />
        </div>
    </fieldset>
    <fieldset class="form-group">
        <legend>Address 2</legend>
        <div class="form-group">
            <label>City:</label>
            <input name="[1].City" class="form-control" />
        </div>
        <div class="form-group">
            <label>Country:</label>
            <input name="[1].Country" class="form-control" />
        </div>
    </fieldset>
    <fieldset class="form-group">
        <legend>Address 3</legend>
        <div class="form-group">
            <label>City:</label>
            <input name="[2].City" class="form-control" />
        </div>
        <div class="form-group">
```

```
            <label>Country:</label>
            <input name="[2].Country" class="form-control" />
        </div>
    </fieldset>
    <button type="submit" class="btn btn-primary">Submit</button>
</form>
...
```

表单被提交后，模型绑定器意识到需要创建 AddressSummary 对象集合，使用 name 属性的数组索引前缀来获取对象的属性值。属性前缀[0]用于第一个 AddressSummary 对象，属性前缀[1]用于第二个 AddressSummary 对象，以此类推。

Address 视图为三个索引的对象定义了 input 元素，在模型集合含有条目的时候显示它们。在加以说明之前，需要从 AddressSummary 模型类中删除 BindNever 特性，如代码清单 26-29 所示；否则，模型绑定器将会忽略 Country 属性。

代码清单 26-29　在 Models 文件夹下的 AddressSummary.cs 文件中删除 BindNever 特性

```
using Microsoft.AspNetCore.Mvc;
using Microsoft.AspNetCore.Mvc.ModelBinding;

namespace MvcModels.Models {

    public class AddressSummary {

        public string City { get; set; }

        //[BindNever]
        public string Country { get; set; }
    }
}
```

通过启动应用并导航到 URL 地址/Home/Address，单击 Submit 按钮发送表单到服务器，就可以看到自定义对象集合的绑定过程是如何工作的。

模型绑定过程将寻找并处理索引的数据值，然后使用它们创建 AddressSummary 对象集合并提供给操作方法，最后使用 View 便捷方法将它们传递回视图以便可以显示出来，如图 26-9 所示。

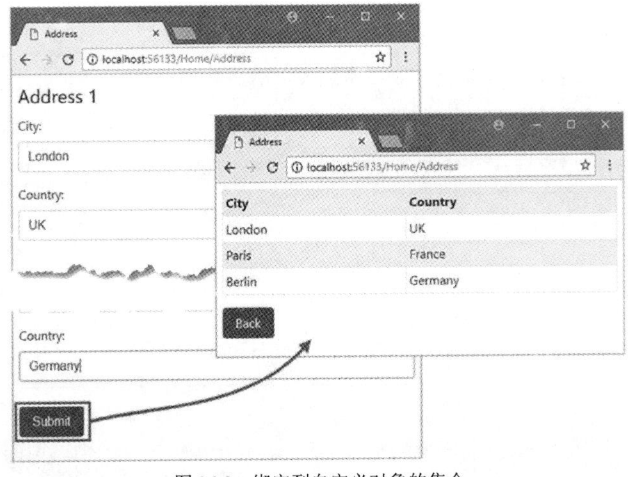

图 26-9　绑定到自定义对象的集合

26.3　指定模型绑定源

模型绑定过程默认从 3 个位置（表单数据、路由变量以及查询串）寻找数据。

默认的搜索顺序并不总是有用，可能是因为希望数据来自请求的某个特定部分，也可能因为希望使用默认不搜索的数据源。模型绑定源包含一系列用来覆盖默认搜索行为的特性，如表 26-3 所示。

表 26-3　　模型绑定源的特性

特性	描述
FromForm	指定表单数据作为数据绑定源。参数名称默认用于定位表单值，但可以使用 Name 属性指定不同的名称
FromRoute	指定路由变量作为数据绑定源。参数名称默认用于定位路由变量值，但可以使用 Name 属性指定不同的名称
FromQuery	指定查询串作为数据绑定源。参数名称默认用于定位查询串值，但可以使用 Name 属性指定不同的名称
FromHeader	指定请求头作为数据绑定源。参数名称默认用于定位请求头的名称，但可以使用 Name 属性指定不同的名称
FromBody	指定请求体作为数据绑定源。如果需要从不是表单编码的请求中接收数据（例如在 API 控制器中，就需要使用这个特性）

26.3.1　选择标准绑定源

FromForm、FromRoute 以及 FromQuery 特性允许指定模型绑定数据将从标准位置获取，但不使用正常的搜索顺序。本章开头使用的是下面这个 URL：

```
/Home/Index/3?id=1
```

这个 URL 包含可以用于 Home 控制器中 Index 方法的 id 参数的两个可能值。路由系统将 URL 的最后部分赋予名为 id 的变量，id 变量定义在 Startup 类中的 URL 模板中，查询串也包含 id 值。使用默认的搜索模式意味着模型绑定数据将从路由变量中获取，查询串将被忽略。

为了改变这个行为，代码清单 26-30 将 FromQuery 特性应用于 Index 方法。为了保持示例简单，这里还将前面定义的其他操作方法删除了。

代码清单 26-30　在 Controllers 文件夹下的 HomeController.cs 中使用查询串

```
using Microsoft.AspNetCore.Mvc;
using MvcModels.Models;

namespace MvcModels.Controllers {

    public class HomeController : Controller {
        private IRepository repository;

        public HomeController(IRepository repo) {
            repository = repo;
        }
        public IActionResult Index([FromQuery] int? id) {
            Person person;
            if (id.HasValue && (person = repository[id.Value]) != null) {
                return View(person);
            } else {
                return NotFound();
            }
        }
    }
}
```

以上代码为 id 参数应用了 FromQuery 特性，这意味着在模型绑定过程中，查询 id 值的时候仅仅查询串被使用。

提　示

在指定模型绑定源（比如查询串）时，仍然可以绑定复杂类型。对于参数类型的每个简单属性，模型绑定过程将使用同样的名字搜索查询串的键。

26.3.2 使用请求头作为绑定源

FromHeader 特性允许将 HTTP 请求头作为数据绑定源。代码清单 26-31 在 Home 控制器中添加了一个简单的操作方法，用于从标准的 HTTP 请求头中接收参数。

代码清单 26-31 在 Controllers 文件夹下的 HomeController.cs 文件中通过请求头进行绑定

```
using Microsoft.AspNetCore.Mvc;
using MvcModels.Models;

namespace MvcModels.Controllers {

    public class HomeController : Controller {
        private IRepository repository;

        public HomeController(IRepository repo) {
            repository = repo;
        }

        public IActionResult Index([FromQuery] int? id) {
            Person person;
            if (id.HasValue && (person = repository[id.Value]) != null) {
                return View(person);
            } else {
                return NotFound();
            }
        }

        public string Header([FromHeader]string accept) => $"Header: {accept}";
    }
}
```

Header 方法定义了一个 accept 参数，当前请求中的 Accept 请求头的值将被接收并作为方法的结果返回。如果运行应用并访问/Home/Header，你将看到类似于下面的结果（虽然基于使用的浏览器不同，确切的执行结果可能也不同）：

```
Header: text/html,application/xhtml+xml,application/xml;q=0.9,image/webp,*/*;q=0.8
```

并不是所有的 HTTP 请求头名称都可以依靠操作方法的参数名称易于选择，因为模型绑定系统不能使用 HTTP 请求头从 C#命名约定进行转换。在这些情况下，必须使用 Name 属性配置 FromHeader 特性以指定请求头的名称，如代码清单 26-32 所示。

代码清单 26-32 在 Controllers 文件夹下的 HomeController.cs 中指定请求头的名称

```
using Microsoft.AspNetCore.Mvc;
using MvcModels.Models;

namespace MvcModels.Controllers {

    public class HomeController : Controller {
        private IRepository repository;

        public HomeController(IRepository repo) {
            repository = repo;
        }

        public IActionResult Index([FromQuery] int? id) {
            Person person;
            if (id.HasValue && (person = repository[id.Value]) != null) {
                return View(person);
```

```
            } else {
                return NotFound();
            }
        }

        public string Header([FromHeader(Name = "Accept-Language")] string accept)
            => $"Header: {accept}";
    }
}
```

不能使用 Accept-Language 作为 C#参数名称，模型绑定器也不能自动将 AcceptLanguage 转换为 Accept-Language 以便匹配请求头。相反，这里使用 Name 属性来配置特性，进而匹配正确的请求头。如果启动应用并访问/Home/Header，你将看到类似下面的响应，这将基于不同的区域设置而有所变化。

```
Header: en-US,en;q=0.8
```

从请求头绑定复杂类型

尽管需求十分罕见，但仍然可以通过应用 FromHeader 特性到模型类的属性来使用请求头中的值绑定复杂类型。作为示例，在 Models 文件夹中添加名为 HeaderModel.cs 的类文件，定义代码清单 26-33 所示的类。

代码清单 26-33 Models 文件夹下的 HeaderModel.cs 文件的内容

```
using Microsoft.AspNetCore.Mvc;

namespace MvcModels.Models {

    public class HeaderModel {

        [FromHeader]
        public string Accept { get; set; }

        [FromHeader(Name = "Accept-Language")]
        public string AcceptLanguage { get; set; }

        [FromHeader(Name = "Accept-Encoding")]
        public string AcceptEncoding { get; set; }
    }
}
```

这个类定义了 3 个属性，其中的每个都使用 FromHeader 特性进行了修饰。可在其中两个 FromHeader 特性中使用 Name 属性来指定不能表示为 C#参数名的请求头名称。代码清单 26-34 更新了 Home 控制器的 Header 方法以接收 HeaderModel 对象。

代码清单 26-34 在 Controllers 文件夹下的 HomeController.cs 文件中使用 Header 模型类

```
using Microsoft.AspNetCore.Mvc;
using MvcModels.Models;

namespace MvcModels.Controllers {

    public class HomeController : Controller {
        private IRepository repository;

        public HomeController(IRepository repo) {
            repository = repo;
        }

        public IActionResult Index([FromQuery] int? id) {
            Person person;
            if (id.HasValue && (person = repository[id.Value]) != null) {
                return View(person);
```

```
            } else {
                return NotFound();
            }
        }

        public ViewResult Header(HeaderModel model) => View(model);
    }
}
```

为了完成示例，在 Views/Home 文件夹中添加名为 Header.cshtml 的视图文件，并添加代码清单 26-35 所示的标记内容。

代码清单 26-35 Views/Home 文件夹下的 Header.cshtml 文件的内容

```
@model HeaderModel
@{
    ViewBag.Title = "Headers";
    Layout = "_Layout";
}

<table class="table table-sm table-bordered table-striped">
    <tr><th>Accept:</th><td>@Model.Accept</td></tr>
    <tr><th>Accept-Encoding:</th><td>@Model.AcceptEncoding</td></tr>
    <tr><th>Accept-Language:</th><td>@Model.AcceptLanguage</td></tr>
</table>
```

模型绑定过程将检查复杂类型的属性以查找表 26-3 中的特性。如果运行应用并访问/Home/Header，你将会看到，这允许使用 FromHeader 特性来定义从请求头绑定复杂类型的属性。产生的结果如图 26-10 所示。

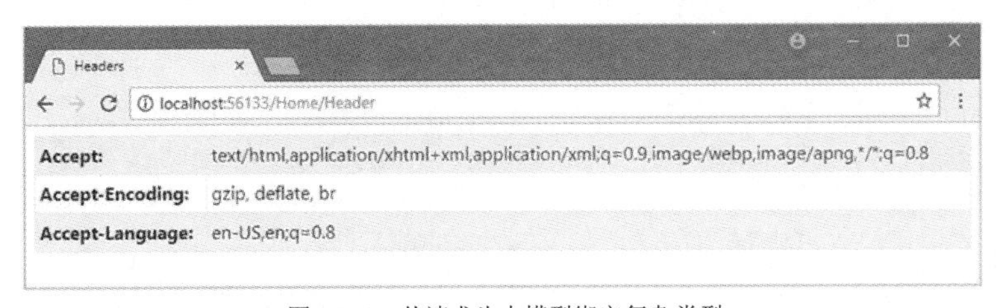

图 26-10 从请求头中模型绑定复杂类型

26.3.3 使用请求体作为绑定源

不是所有的客户端数据都以表单形式发送，比如当 JavaScript 客户端发送 JSON 数据给 API 控制器时。FromBody 特性指定请求体应当被解码并作为模型绑定源。代码清单 26-36 添加了名为 Body 的操作方法来演示具体如何工作。

代码清单 26-36 在 Controllers 文件夹下的 HomeController.cs 文件中添加操作方法

```
using Microsoft.AspNetCore.Mvc;
using MvcModels.Models;

namespace MvcModels.Controllers {

    public class HomeController : Controller {
        private IRepository repository;

        public HomeController(IRepository repo) {
            repository = repo;
        }
```

```
        public IActionResult Index([FromQuery] int? id) {
            Person person;
            if (id.HasValue && (person = repository[id.Value]) != null) {
                return View(person);
            } else {
                return NotFound();
            }
        }

        public ViewResult Header(HeaderModel model) => View(model);

        public ViewResult Body() => View();

        [HttpPost]
        public Person Body([FromBody]Person model) => model;
    }
}
```

这里使用 FromBody 特性修饰了 Body 方法的参数以接收 POST 请求，这意味着请求体的内容将被解码并用于模型绑定。参考第 20 章，MVC 拥有可扩展的系统，但是默认仅处理 JSON 数据。

然后，编辑 bower.json 文件以添加 jQuery 包到应用中，如代码清单 26-37 所示。

代码清单 26-37 在 MvcModels 文件夹下的 bower.json 文件中添加 jQuery 包

```
{
  "name": "asp.net",
  "private": true,
  "dependencies": {
    "bootstrap": "4.0.0-alpha.6",
    "jquery": "3.2.1"
  }
}
```

为了提供操作方法需要的数据，添加一个名为 Body.cshtml 的视图文件到 Views/Home 文件夹中，添加代码清单 26-38 所示的内容。

代码清单 26-38 Views/Home 文件夹下的 Body.cshtml 文件的内容

```
@{
    ViewBag.Title = "Address";
    Layout = "_Layout";
}

@section scripts {
    <script src="/lib/jquery/dist/jquery.min.js"></script>
    <script type="text/javascript">
        $(document).ready(function () {
            $("button").click(function (e) {
                $.ajax("/Home/Body", {
                    method: "post",
                    contentType: "application/json",
                    data: JSON.stringify({
                        firstName: "Bob",
                        lastName: "Smith"
                    }),
                    success: function (data) {
                        $("#firstName").text(data.firstName);
                        $("#lastName").text(data.lastName);
                    }
                });
            });
```

```
        });
    </script>
}

<table class="table table-sm table-bordered table-striped">
    <tr><th>First Name:</th><td id="firstName"></td></tr>
    <tr><th>Last Name:</th><td id="lastName"></td></tr>
</table>
<button class="btn btn-primary">Submit</button>
```

为了简化，Body 视图包含一些内联的 JavaScript 代码，当 button 元素被单击时，使用 jQuery 发送包含 JSON 数据的 HTTP POST 请求到/Home/Body。服务器编码模型绑定创建的对象并编码为 JSON 后发送回客户端，运行应用并访问/Home/Body，然后单击 Submit 按钮，就可以看到效果，如图 26-11 所示。

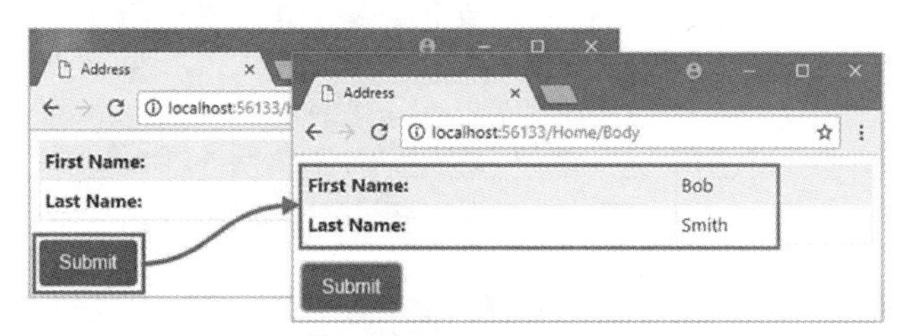

图 26-11 将请求体用于模型绑定

提 示

不是所有的 JavaScript 客户端代码都需要使用 FromBody 特性。这个示例不得不避免使用 jQuery 便捷方法来发送 Ajax POST 请求，因为数据已编码为表单数据。相反，不得不使用其他方法来允许发送 JSON 数据。

FromBody 特性只可以用于一个操作方法参数，如果这个特性被用于一个以上的操作方法参数，将会导致异常。如果需要为一个请求体创建多个模型绑定对象，那么在操作方法中，你将不得不创建一个简单的拥有所有所需属性的数据传输类来创建对象。

26.4 小结

本章描述了模型绑定过程，模型绑定用来从 HTTP 请求中获取数据，以便为操作方法提供需要的参数。本章还解释了如何绑定简单类型和复杂类型，以及如何处理数组和集合，可通过应用特定到操作方法参数或模型类属性来控制模型绑定过程。下一章将描述模型验证功能。

第 27 章　模型验证

上一章展示了 MVC 如何通过模型绑定过程从 HTTP 请求中创建模型对象。本章介绍如何对用户提供的数据进行基本验证。实际情况是，用户经常输入无效或无用的数据，因此本章讨论模型验证。

模型验证是确认应用接收到的数据适合用来绑定到模型的过程，如果情况不是这样，则为用户提供有助于解释问题的有用信息。

要完成模型验证过程，第一步，检查接收到的数据是保持领域模型完整性的关键。拒绝从领域角度看没有意义的数据可以防止应用程序中出现奇怪或不希望的状态。第二步，帮助用户纠正问题也同样重要。如果没有用户需要的与应用程序进行交互的信息和反馈，用户就会变得沮丧和困惑。在面向公众的应用中，这意味着用户会简单地放弃使用这个应用；在企业级应用中，这意味着用户的工作流程会阻滞。以上所有结果都是不可取的。但幸运的是，MVC 为模型验证提供广泛的支持。表 27-1 给出了模型验证的背景。

表 27-1　　模型验证的背景

问题	答案
模型验证是什么？	模型验证是确保请求中提供的数据可以在应用中有效使用的过程
为什么模型验证是有用的？	用户不总是输入有效的数据，在应用中使用无效数据将导致无法预料或不期望的错误
如何使用模型验证？	控制器检查验证过程的结果，标签助手用来在用户显示的视图中包含验证反馈。验证在模型绑定过程中自动执行，通常在控制器类中或者通过使用验证特性来支持自定义验证
模型验证有何缺陷或局限性？	测试验证代码的有效性是非常重要的，确保它们能够阻止应用接收所有范围的值
模型验证有其他替代选择吗？	没有，模型验证它与 ASP.NET Core MVC 紧密集成

表 27-2 列出了本章要介绍的操作。

表 27-2　　本章要介绍的操作

操作	方法	代码清单
显式地验证模型	使用 ModelState 记录验证错误	代码清单 27-9 和代码清单 27-10
汇总验证错误	为 div 元素应用 asp-validation-summary 属性	代码清单 27-11
修改默认的模型绑定信息	在模型绑定信息提供器中重新定义消息函数	代码清单 27-12
生成属性级别的验证错误信息	为 span 元素应用 asp-validation-for 属性	代码清单 27-13
生成模型级别的验证错误信息	使用 ModelState 记录没有关联到特定属性的验证错误，为 div 元素的 asp-validation-summary 属性使用 ModelOnly 值	代码清单 27-14 和代码清单 27-15
定义自验证模型	为模型属性应用数据验证特性	代码清单 27-16 和代码清单 27-17
创建自定义验证特性	实现 IModelValidator 接口	代码清单 27-18 和代码清单 27-19
执行客户端验证	使用 jQuery 验证和 jQuery 无痕验证包	代码清单 27-20 和代码清单 27-21
执行远程验证	定义操作方法以执行验证，为模型属性应用 Remote 特性	代码清单 27-22 和代码清单 27-23

27.1　准备示例项目

在本章中，使用 ASP.NET Core Web Application（.NET Core）模板创建名为 ModelValidation 的 Empty 项目。代码清单 27-1 展示了 Startup 类，这里添加了 MVC 框架并启用了开发中用到的中间件。

代码清单 27-1　ModelValidation 文件夹下的 Startup.cs 文件的内容

```
using System;
using System.Collections.Generic;
```

```
using System.Linq;
using System.Threading.Tasks;
using Microsoft.AspNetCore.Builder;
using Microsoft.AspNetCore.Hosting;
using Microsoft.AspNetCore.Http;
using Microsoft.Extensions.DependencyInjection;

namespace ModelValidation {

    public class Startup {

        public void ConfigureServices(IServiceCollection services) {
            services.AddMvc();
        }

        public void Configure(IApplicationBuilder app, IHostingEnvironment env) {
            app.UseStatusCodePages();
            app.UseDeveloperExceptionPage();
            app.UseStaticFiles();
            app.UseMvcWithDefaultRoute();
        }
    }
}
```

27.1.1 创建模型

首先创建 Models 文件夹，添加一个名为 Appointment.cs 的类文件，并用它定义代码清单 27-2 所示的类。

代码清单 27-2 Models 文件夹下的 Appointment.cs 文件的内容

```
using System;
using System.ComponentModel.DataAnnotations;

namespace ModelValidation.Models {
    public class Appointment {

        public string ClientName { get; set; }

        [UIHint("Date")]
        public DateTime Date { get; set; }

        public bool TermsAccepted { get; set; }
    }
}
```

Appointment 模型类定义了 3 个属性，还使用了 UIHint 特性来指示 Date 属性应该展示为没有时间部分的日期。

27.1.2 创建控制器

然后创建 Controllers 文件夹，添加一个名为 HomeController.cs 的类文件，并用它定义代码清单 27-3 所示的控制器，用以处理 Appointment 模型类。

代码清单 27-3 Controllers 文件夹下的 HomeController.cs 文件的内容

```
using System;
using Microsoft.AspNetCore.Mvc;
using ModelValidation.Models;

namespace ModelValidation.Controllers {
```

```
    public class HomeController : Controller {
        public IActionResult Index() =>
            View("MakeBooking", new Appointment { Date = DateTime.Now });

        [HttpPost]
        public ViewResult MakeBooking(Appointment appt) =>
            View("Completed", appt);
    }
}
```

Index 方法使用新的 Appointment 对象作为视图模型来渲染 MakeBooking 视图。在本章中，MakeBooking 方法更有意义，因为模型验证将在这个操作方法中处理。

注 意

这个应用十分简单，既没有定义存储库，也没有添加任何代码来存储模型绑定过程中产生的 Appointment 对象。这就是说，需要铭记在心的是，验证模型的主要目的是防止坏的或无意义的数据被放置到存储库中并导致问题（无论是在试图存储数据时还是以后试图处理数据时）。

27.1.3 创建布局和视图

在本章中，对于某些示例使用需要简单的布局。创建 Views/Shared 文件夹，在其中添加名为_Layout.cshtml 的视图文件，内容如代码清单 27-4 所示。

代码清单 27-4 Views/Shared 文件夹下的_Layout.cshtml 文件的内容

```
<!DOCTYPE html>
<html>
<head>
    <meta charset="utf-8" />
    <meta name="viewport" content="width=device-width" />
    <title>Model Validation</title>
    <link asp-href-include="/lib/bootstrap/dist/css/bootstrap.min.css" rel="stylesheet" />
    @RenderSection("scripts", false)
</head>
<body class="m-1 p-1">
    @RenderBody()
</body>
</html>
```

下面为操作方法提供视图。创建 Views/Home 文件夹，并添加名为 MakeBooking.cshtml 的视图文件，内容如代码清单 27-5 所示。

代码清单 27-5 Views/Home 文件夹下的 MakeBooking.cshtml 文件的内容

```
@model Appointment

@{ Layout = "_Layout"; }

<div class="bg-primary m-1 p-1 text-white"><h2>Book an Appointment</h2></div>

<form class="m-1 p-1" asp-action="MakeBooking" method="post">
    <div class="form-group">
        <label asp-for="ClientName">Your name:</label>
        <input asp-for="ClientName" class="form-control" />
    </div>
    <div class="form-group">
```

```
        <label asp-for="Date">Appointment Date:</label>
        <input asp-for="Date" type="text" asp-format="{0:d}" class="form-control" />
    </div>
    <div class="radio form-group">
        <input asp-for="TermsAccepted" />
        <label asp-for="TermsAccepted" class="form-check-label">
            I accept the terms & conditions
        </label>
    </div>
    <button type="submit" class="btn btn-primary">Make Booking</button>
</form>
```

当 Index.cshtml 文件中的表单被回发到应用程序时，MakeBooking 方法使用 Views/Home 文件夹中的 Completed 视图显示用户创建的预约详情，如代码清单 27-6 所示。

代码清单 27-6　Views/Home 文件夹下的 Completed.cshtml 文件的内容

```
@model Appointment
@{ Layout = "_Layout"; }

<div class="bg-success m-1 p-1 text-white"><h2>Your Appointment</h2></div>

<table class="table table-bordered">
    <tr>
        <th>Your name is:</th>
        <td>@Model.ClientName</td>
    </tr>
    <tr>
        <th>Your appontment date is:</th>
        <td>@Model.Date.ToString("d")</td>
    </tr>
</table>
<a class="btn btn-success" asp-action="Index">Make Another Appointment</a>
```

Completed 视图基于 Bootstrap CSS 包来修饰 HTML 元素。下面为项目添加 Bootstrap 包，使用 Bower Configuration File 模板创建 bower.json 文件，在 dependencies 部分添加 Bootstrap 包，如代码清单 27-7 所示，这里还添加了 jQuery 包到项目中，本章后面会用到 jQuery。

代码清单 27-7　添加 Bootstrap 包到 bower.json 文件中

```
{
  "name": "asp.net",
  "private": true,
  "dependencies": {
    "bootstrap": "4.0.0-alpha.6",
    "jquery": "3.2.1"
  }
}
```

最后的准备工作是在 Views 文件夹中创建_ViewImports.cshtml 文件，在其中设置内置的标签助手用于 Razor 视图，并导入模型命名空间，如代码清单 27-8 所示。

代码清单 27-8　Views 文件夹下的_ViewImports.cshtml 文件的内容

```
@using ModelValidation.Models
@addTagHelper *, Microsoft.AspNetCore.Mvc.TagHelpers
```

如你所知，本章的示例围绕着如何创建预约。可以通过启动应用并且访问默认 URL 地址来看看具体是如何工作的。在表单中输入预约详情，单击 Make Booking 按钮将数据发送回服务器，执行模型绑定过程以创建 Appointment 对象，预约详情将使用 Completed 视图渲染出来，如图 27-1 所示。

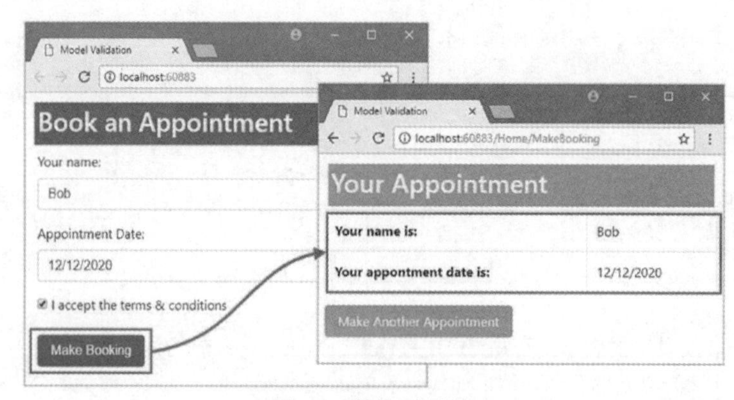

图 27-1 使用示例应用

27.2 理解模型验证的需求

模型验证是确保应用程序从客户端接收的数据满足要求的过程。没有验证，应用程序将会试图处理接收的任何数据，这将导致异常或意外的行为，还将面临随着存储中填充不良数据，不完整或恶意的数据逐渐出现的长期问题。

目前，应用会接收用户提交的任何数据。为保护应用和领域模型的完整性，在知道用户提交可接收的预约对象之前，需要做以下 3 件事：

- 提供名字。
- 提供未来的日期。
- 选中复选框以接受条款。

下面将演示如何使用模型验证，通过检查应用接收的数据，以及在应用不能使用提交的数据时为用户提供反馈来强制满足这些要求。

27.3 显式地验证模型

多数直接验证模型的途径定义在操作方法中，在代码清单 27-9 所示的 MakeBooking 方法中，已加入对 Appointment 类的每个属性的显式检查。

代码清单 27-9 在 Controllers 文件夹下的 HomeController.cs 文件中显式地验证模型

```
using System;
using Microsoft.AspNetCore.Mvc;
using ModelValidation.Models;
using Microsoft.AspNetCore.Mvc.ModelBinding;

namespace ModelValidation.Controllers {

    public class HomeController : Controller {

        public IActionResult Index() =>
            View("MakeBooking", new Appointment { Date = DateTime.Now });

        [HttpPost]
        public ViewResult MakeBooking(Appointment appt) {
            if (string.IsNullOrEmpty(appt.ClientName)) {
```

```
                ModelState.AddModelError(nameof(appt.ClientName),
                    "Please enter your name");
            }

            if (ModelState.GetValidationState("Date")
                    == ModelValidationState.Valid && DateTime.Now > appt.Date) {
                ModelState.AddModelError(nameof(appt.Date),
                    "Please enter a date in the future");
            }
            if (!appt.TermsAccepted) {
                ModelState.AddModelError(nameof(appt.TermsAccepted),
                    "You must accept the terms");
            }

            if (ModelState.IsValid) {
                return View("Completed", appt);
            } else {
                return View();
            }
        }
    }
}
```

检查模型绑定器已经赋予参数对象的属性值，并使用从控制器基类继承的 ModelState 属性返回的 ModelStateDictionary 类型对象来注册任何错误。

顾名思义，ModelStateDictionary 类是用来跟踪模型对象状态详情的字典，主要用于验证错误，表 27-3 描述了 ModelStateDictionary 类的重要成员。

表 27-3　ModelStateDictionary 类的重要成员

成员	描述
AddModelError(property,message)	这个方法用于对特定的属性记录模型验证错误
GetValidationState(property)	这个方法用来检测特定的属性是否存在模型验证错误，表示为 ModelValidationState 枚举中的值
IsValid	如果所有的模型属性都是有效的，那么这个属性返回 true；否则，返回 false

作为使用 ModelStateDictionary 的示例，考虑 ClientName 属性是如何被验证的：

```
...
if (string.IsNullOrEmpty(appt.ClientName)) {
    ModelState.AddModelError(nameof(appt.ClientName), "Please enter your name");
}
...
```

以上示例中，验证的目标之一是确认用户为这个属性提供了有效值，所以可使用 string.IsNullOrEmpty 静态方法来测试模型绑定已经从请求中提取的属性值。如果 ClientName 是 null 或空串，就会得出验证目的没有达成的结论，然后使用 ModelState.AddModelError 方法注册验证错误信息，指定属性的名字（ClientName）以及将会用来显示给用户以说明问题原因的信息（Please enter your name）。

模型绑定系统还使用 ModelStateDictionary 记录查找和分配值给模型属性的任何问题。GetValidationState 方法用于查看模型属性是否存在任何错误，无论是从模型绑定过程还是由于在操作方法中进行显式验证期间调用了 AddModelError 方法。GetValidationState 方法会返回 ModelValidationState 枚举中的一个值，如表 27-4 所示。

表 27-4　ModelValidationState 枚举中的值

值	描述
Unvalidated	这个值意味着没有在模型属性上执行验证，通常是因为在请求中没有值关联到属性名称
Valid	这个值意味着关联到属性的请求值是有效的
Invalid	这个值意味着关联到属性的请求值是无效的，并且不能使用
Skipped	这个值意味着模型属性还没有处理，通常意味着出现了太多的验证错误，以至于无法继续执行验证检查

对于 Date 属性，检查模型绑定过程是否报告了关于将浏览器发送的值解析为 DateTime 对象的错误，如下所示：

```
...
if (ModelState.GetValidationState("Date") == ModelValidationState.Valid
        && DateTime.Now > appt.Date) {
    ModelState.AddModelError(nameof(appt.Date), "Please enter a date in the future");
}
...
```

对于 Date 属性，验证目标是确认用户提供了正确的未来日期。使用 GetValidationState 方法检查 ModelValidationState.Valid 的值以判定模型绑定过程能够解析请求的值到 DateTime 对象。如果存在有效的日期，再确认是正确的未来日期，如果不是，就使用 AddModelError 方法记录存在验证问题。

在验证模型对象的所有属性之后，检查 ModelState.IsValid 属性以查看是否存在错误。如果在检查过程中 ModelState.AddModelError 方法被调用了，或者模型绑定器在创建 Appointment 对象时出现问题，这个方法将返回 true。

```
...
if (ModelState.IsValid) {
    return View("Completed", appt);
} else {
    return View();
}
...
```

如果 IsValid 属性返回 true，那么 Appointment 对象是有效的，在这种情况下，操作方法渲染 Completed 视图。如果 IsValid 属性返回 false，就表示出现验证问题，可通过调用 View 方法来渲染默认视图。

27.3.1 为用户显示验证错误消息

通过调用 View 方法来处理验证错误消息看起来很奇怪，MVC 提供给视图的上下文数据中包含模型验证错误的详细信息，它们被自动检测并由用于转换 input 元素的标签助手使用。

为了查看具体是如何工作的，启动应用并在不填充任何表单内容的情况下单击 Make Booking 按钮。看起来没有任何浏览器窗口中的可视内容发生变化。但是，如果查看 MVC 为这个 POST 请求返回的 HTML 内容，你将看到表单元素的 class 属性发生了变化。以下是 ClientName 元素在提交表单之前的内容：

```
<input class="form-control" type="text" id="ClientName" name="ClientName" value="">
```

以下是提交空的表单之后的 input 元素：

```
<input class="form-control input-validation-error" type="text" id="ClientName"
    name="ClientName" value="">
```

标签助手为验证失败的元素添加了 input-validation-error 样式类，然后设置为向用户突出显示问题。这可以通过在样式表中定制 CSS 样式来实现，但是如果希望使用内置的样式表，比如 Bootstrap 提供的样式表，那么还有一点额外的工作要做。添加到 form 元素的样式类名不能修改，这意味着某些 JavaScript 代码需要在 MVC 使用的名称和 Bootstrap 提供的 CSS 错误类名之间进行映射。

提 示

使用这样的 JavaScript 代码可能很尴尬，即使使用 Bootstrap 这样的样式库，也可以使用自定义样式。然而，Bootstrap 中用于验证类的颜色可以使用主题或者通过定制包以及定义自己的样式来覆盖。这意味着你将不得不确认对于主题的任何修改都与自定义样式的相应更改相匹配。理想情况下，微软将使未来发行版本的 ASP.NET Core MVC 中验证类的名称可配置，但在此之前，使用 JavaScript 来应用 Bootstrap 样式是比创建自定义样式更加强大的方法。

代码清单 27-10 添加了 jQuery 代码到 MakeBooking 视图以查找拥有 inputvalidation-error 样式类的元素，

定位最临近的被赋予 form-group 样式类的父元素，然后为这个元素添加 has-error 样式类（Bootstrap 用它为 form 元素设置出错颜色）。

代码清单 27-10 在 Views/Home 文件夹下的 MakeBooking.cshtml 文件中赋予元素验证类

```
@model Appointment

@{ Layout = "_Layout"; }

@section scripts {
    <script src="/lib/jquery/dist/jquery.min.js"></script>
    <script type="text/javascript">
        $(document).ready(function () {
            $("input.input-validation-error")
                .closest(".form-group").addClass("has-danger");
        });
    </script>
}

<div class="bg-primary m-1 p-1 text-white"><h2>Book an Appointment</h2></div>

<form class="m-1 p-1" asp-action="MakeBooking" method="post">
    <div class="form-group">
        <label asp-for="ClientName">Your name:</label>
        <input asp-for="ClientName" class="form-control" />
    </div>
    <div class="form-group">
        <label asp-for="Date">Appointment Date:</label>
        <input asp-for="Date" type="text" asp-format="{0:d}" class="form-control" />
    </div>
    <div class="radio form-group">
        <input asp-for="TermsAccepted" />
        <label asp-for="TermsAccepted" class="form-check-label">
            I accept the terms & conditions
        </label>
    </div>
    <button type="submit" class="btn btn-primary">Make Booking</button>
</form>
```

在浏览器完成解析 HTML 文档中的所有元素之后，jQuery 代码被执行，效果是突出显示被赋予 input-validation-error 样式类的 input 元素。可以通过运行应用并在没有填写任何文本框的情况下提交表单来查看效果，处理的结果如图 27-2 所示。

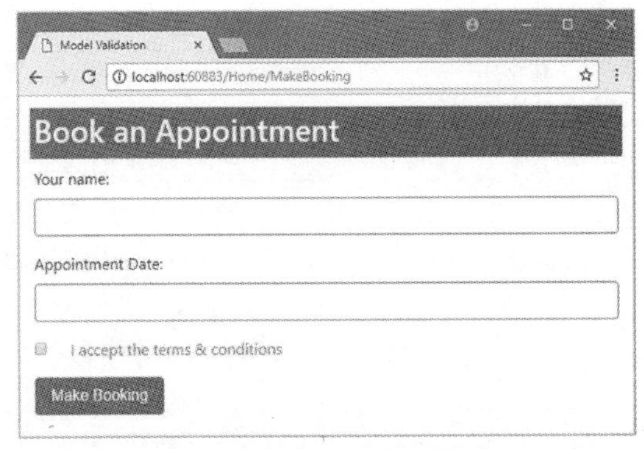

图 27-2 高亮显示验证错误信息

可在没有输入任何数据的情况下提交表单，对于 3 个文本框，错误消息都将突出显示。除非提交的表单可以被模型绑定器解析，并且通过了 MakeBooking 方法的显式验证，否则用户将看不到 Completed 视图。在此之前，提交表单将会导致使用当前验证错误消息的 MakeBooking 视图被渲染。

27.3.2 显示验证消息

一些标签助手应用到 input 元素的 CSS 样式类标志着表单字段存在问题，但是它们没有告诉用户是什么问题。要为用户提供更加详尽的消息，可使用其他标签助手添加问题的摘要到视图中，如代码清单 27-11 所示。

代码清单 27-11　在 Views/Home 文件夹下的 MakeBooking.cshtml 文件中显示验证摘要

```
@model Appointment

@{ Layout = "_Layout"; }

@section scripts {
<script src="/lib/jquery/dist/jquery.min.js"></script>
<script type="text/javascript">
    $(document).ready(function () {
        $("input.input-validation-error")
            .closest(".form-group").addClass("has-danger");
    });
</script>
}
<div class="bg-primary m-1 p-1 text-white"><h2>Book an Appointment</h2></div>

<form class="m-1 p-1" asp-action="MakeBooking" method="post">
    <div asp-validation-summary="All" class="text-danger"></div>
    <div class="form-group">
        <label asp-for="ClientName">Your name:</label>
        <input asp-for="ClientName" class="form-control" />
    </div>
    <div class="form-group">
        <label asp-for="Date">Appointment Date:</label>
        <input asp-for="Date" type="text" asp-format="{0:d}" class="form-control" />
    </div>
    <div class="radio form-group">
        <input asp-for="TermsAccepted" />
        <label asp-for="TermsAccepted" class="form-check-label">
            I accept the terms & conditions
        </label>
    </div>
    <button type="submit" class="btn btn-primary">Make Booking</button>
</form>
```

ValidationSummaryTagHelper 类监测 div 元素的 asp-validation-summary 属性，通过添加描述由操作方法检测的任何验证错误消息进行响应。asp-validation-summary 属性的值来自 ValidationSummary 枚举，如表 27-5 所示。

表 27-5　ValidationSummary 枚举中的值

值	描述
All	用来显示所有被记录的验证错误
ModelOnly	用来显示仅限于整个模型的验证错误，不包括被记录的属于独立属性的验证错误
None	用来禁用标签助手以便不转换 HTML 元素

如果运行应用，并在没有做任何修改的情况下提交表单，你将会看到标签助手生成的摘要信息。本示例中的文本颜色由 text-dangerBootstrap 类定义，该类确保文本匹配文本框中突出显示的颜色，如图 27-3 所示。

如果查看浏览器接收到的 HTML，你将会看到验证信息被作为列表发送，像下面这样：

```
<div class="text-danger validation-summary-errors" data-valmsg-summary="true">
  <ul>
    <li>Please enter your name</li>
    <li>Please enter a date in the future</li>
    <li>You must accept the terms</li>
  </ul>
</div>
```

配置默认验证错误消息

当第 26 章描述的模型绑定过程尝试提供调用操作方法所需的数据时，会执行自己的验证。要了解具体是如何工作的，可启动应用，清除 Appointment Date 文本框中的内容，然后提交表单。你将看到显示的错误消息之一已经变化，并且与操作方法中传递给 AddModelError 方法的任何字符串都不匹配。

```
The value '' is invalid
```

当模型绑定过程找不到属性的值或者虽然找到值但是不能解析时，该消息将被添加到 ModelStateDictionary 中。在这个示例中，出现错误是因为表单中提供的空字符串不能解析为 Appointment 对象的 Date 属性要求的 DateTime 对象。

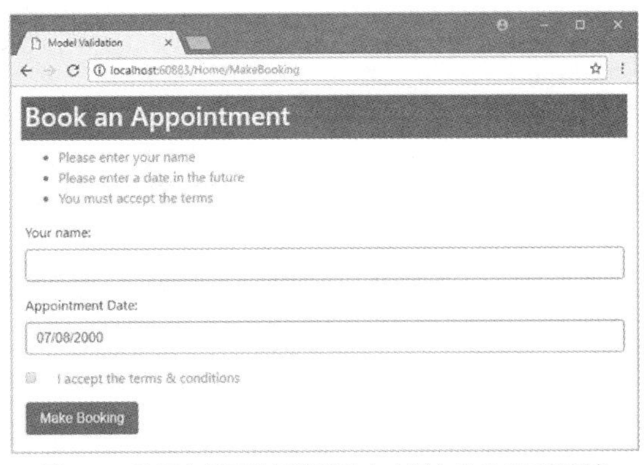

图 27-3 为用户显示验证摘要时文本框中突出显示的颜色

模型绑定器有一组用于验证错误的预定义信息。通过将函数赋予由 DefaultModelBindingMessageProvider 类定义的方法，可以将它们替换为自定义的消息，如表 27-6 所示。

表 27-6 DefaultModelBindingMessageProvider 类定义的方法

方法	描述
SetValueMustNotBeNullAccessor	当非空的模型属性值为空时，生成验证错误消息
SetMissingBindRequiredValueAccessor	当请求中没有包含必需属性的值时，生成验证错误消息
SetMissingKeyOrValueAccessor	当字典模型对象要求的数据包含 null 键或值时，生成验证错误消息
SetAttemptedValueIsInvalidAccessor	当尝试提供给模型绑定系统的值无效时，生成验证错误消息
SetUnknownValueIsInvalidAccessor	当模型绑定系统不能将数据值转换为要求的 C#类型时，生成验证错误消息
SetValueMustBeANumberAccessor	当数据值不能被解析为 C#数值类型时，生成验证错误消息
SetValueIsInvalidAccessor	生成备用的验证错误消息，用作最后的手段

表 27-6 中描述的每个方法接收一个函数，它将被调用以获取显示给用户的验证消息。这些方法用于在 Startup 类中配置应用，代码清单 27-12 替换了当值为 null 时的默认消息。

代码清单 27-12 在 ModelValidation 文件夹下的 Startup.cs 文件中替换绑定的消息

```
using System;
using System.Collections.Generic;
using System.Linq;
using System.Threading.Tasks;
using Microsoft.AspNetCore.Builder;
using Microsoft.AspNetCore.Hosting;
using Microsoft.AspNetCore.Http;
using Microsoft.Extensions.DependencyInjection;
namespace ModelValidation {

    public class Startup {

        public void ConfigureServices(IServiceCollection services) {
            services.AddMvc().AddMvcOptions(opts =>
                opts.ModelBindingMessageProvider
                    .SetValueMustNotBeNullAccessor(value => "Please enter a value")
            );
        }

        public void Configure(IApplicationBuilder app, IHostingEnvironment env) {
            app.UseStatusCodePages();
```

```
            app.UseDeveloperExceptionPage();
            app.UseStaticFiles();
            app.UseMvcWithDefaultRoute();
        }
    }
}
```

可指定用来接收用户提供的值的函数，尽管这对于 null 值并不是特别有帮助。为了改变模型绑定的错误消息，可重新启动应用并在清空 Appointment Date 文本框之后提交表单，如图 27-4 所示。

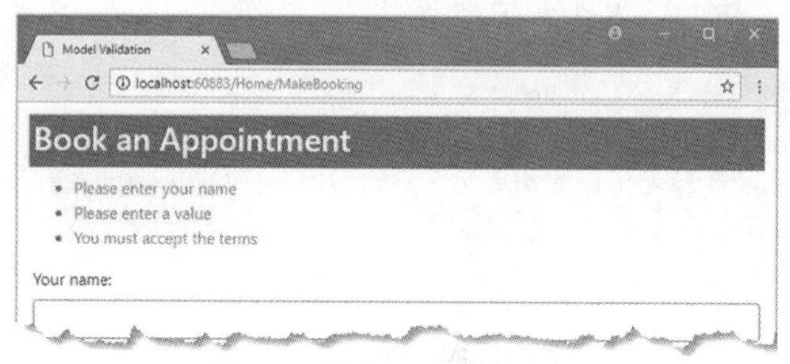

图 27-4　改变模型绑定的错误消息

27.3.3　显示属性级验证消息

尽管自定义的错误消息比默认的更有意义，但它仍然没有帮助，因为它没有清楚地为用户标识问题。对于这种错误，在包含错误数据的 HTML 元素旁显示错误消息更有帮助，这可以使用 ValidationMessageTag 标签助手来完成，该标签助手将查找拥有 asp-validation-for 属性的 span 元素，该元素用于指定应显示错误消息的模型属性。

代码清单 27-13 为表单中的每个 input 元素添加了属性级的验证消息元素。这里还删除了 scripts 部分，因为独立的验证消息将使验证错误的元素足够醒目。

代码清单 27-13　在 Views/Home 文件夹下的 MakeBooking.cshtml 文件中添加属性级验证消息

```
@model Appointment

@{ Layout = "_Layout"; }

@section scripts {
<script src="/lib/jquery/dist/jquery.min.js"></script>
<script type="text/javascript">
    $(document).ready(function () {
        $("input.input-validation-error")
            .closest(".form-group").addClass("has-danger");
    });
</script>
}

<div class="bg-primary m-1 p-1 text-white"><h2>Book an Appointment</h2></div>

<form class="m-1 p-1" asp-action="MakeBooking" method="post">
    <div asp-validation-summary="All" class="text-danger"></div>
    <div class="form-group">
        <label asp-for="ClientName">Your name:</label>
        <div><span asp-validation-for="ClientName" class="text-danger"></span></div>
        <input asp-for="ClientName" class="form-control" />
    </div>
    <div class="form-group">
        <label asp-for="Date">Appointment Date:</label>
        <div><span asp-validation-for="Date" class="text-danger"></span></div>
        <input asp-for="Date" type="text" asp-format="{0:d}" class="form-control" />
```

```
    </div>
    <span asp-validation-for="TermsAccepted" class="text-danger"></span>
    <div class="radio form-group">
        <input asp-for="TermsAccepted" />
        <label asp-for="TermsAccepted" class="form-check-label">
            I accept the terms & conditions
        </label>
    </div>
    <button type="submit" class="btn btn-primary">Make Booking</button>
</form>
```

由于 span 元素作为行内元素显示，因此必须注意提供的验证消息与哪个元素相关。通过运行应用程序并在不输入任何数据的情况下提交表单，可以看到新的验证消息的影响，如图 27-5 所示。

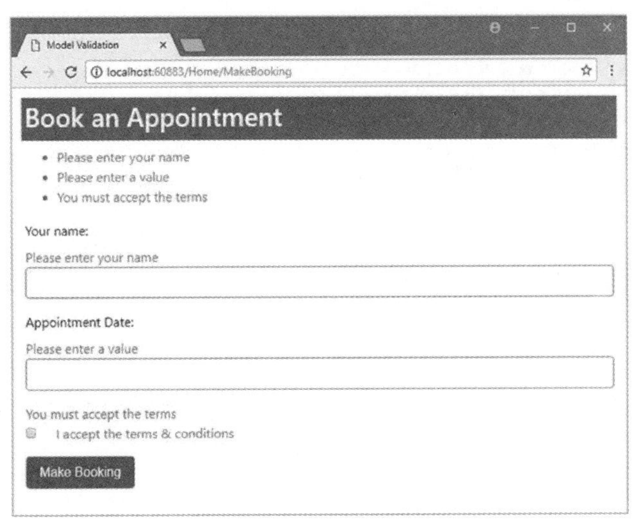

图 27-5 使用属性级验证消息

27.3.4 显示模型级验证消息

验证摘要信息似乎是多余的，因为它们仅仅重复了属性级验证信息。属性级验证信息由于出现在需要解决问题的表单元素旁边，通常对用户更有帮助。但是验证摘要信息有显示用于整个模型的信息的能力，这意味着可以报告由组合的单个属性引起的错误，例如，在给定的日期仅仅对特定名称的组合才有效的情况下。

代码清单 27-14 添加了验证检查，用来阻止名为 Joe 的客户于星期一的预约。

代码清单 27-14 在 Controllers 文件夹下的 HomeController.cs 文件中执行模型级验证

```
using System;
using Microsoft.AspNetCore.Mvc;
using ModelValidation.Models;
using Microsoft.AspNetCore.Mvc.ModelBinding;

namespace ModelValidation.Controllers {

    public class HomeController : Controller {

        public IActionResult Index() =>
            View("MakeBooking", new Appointment() { Date = DateTime.Now });
        [HttpPost]
        public ViewResult MakeBooking(Appointment appt) {
            if (string.IsNullOrEmpty(appt.ClientName)) {
                ModelState.AddModelError(nameof(appt.ClientName),
```

```
                "Please enter your name");
            }

            if (ModelState.GetValidationState("Date")
                    == ModelValidationState.Valid && DateTime.Now > appt.Date) {
                ModelState.AddModelError(nameof(appt.Date),
                    "Please enter a date in the future");
            }

            if (!appt.TermsAccepted) {
                ModelState.AddModelError(nameof(appt.TermsAccepted),
                    "You must accept the terms");
            }

            if (ModelState.GetValidationState(nameof(appt.Date))
                    == ModelValidationState.Valid
                && ModelState.GetValidationState(nameof(appt.ClientName))
                    == ModelValidationState.Valid
                && appt.ClientName.Equals("Joe", StringComparison.OrdinalIgnoreCase)
                && appt.Date.DayOfWeek == DayOfWeek.Monday) {
                    ModelState.AddModelError("",
                        "Joe cannot book appointments on Mondays");
            }

            if (ModelState.IsValid) {
                return View("Completed", appt);
            } else {
                return View();
            }
        }
    }
}
```

这段代码看起来比实际上要复杂，这是数据验证的特点。通过检查模型状态，确认已经收到有效的 ClientName 和 Date 值，然后检查特定的日期是否为星期一，以及 ClientName 属性的值是否为 Joe。如果 Joe 试图于星期一预约，那么使用空字符串（""）作为第一个参数来调用 AddModelError 方法，这表示错误将应用于整个模型而不是单个属性。

代码清单 27-15 通过将 asp-validation-summary 属性的值改为 ModelOnly，排除了属性级别的错误消息，这意味着摘要将仅仅显示应用于整个模型的错误消息。

代码清单 27-15 在 Views/Home 文件夹下的 MakeBooking.cshtml 文件中显示模型级验证错误

```
@model Appointment

@{ Layout = "_Layout"; }

@section scripts {
<script src="/lib/jquery/dist/jquery.min.js"></script>
<script type="text/javascript">
    $(document).ready(function () {
        $("input.input-validation-error")
            .closest(".form-group").addClass("has-danger");
    });
</script>
}

<div class="bg-primary m-1 p-1 text-white"><h2>Book an Appointment</h2></div>

<form class="m-1 p-1" asp-action="MakeBooking" method="post">
    <div asp-validation-summary="ModelOnly" class="text-danger"></div>
    <div class="form-group">
        <label asp-for="ClientName">Your name:</label>
        <div><span asp-validation-for="ClientName" class="text-danger"></span></div>
```

```
            <input asp-for="ClientName" class="form-control" />
        </div>
        <div class="form-group">
            <label asp-for="Date">Appointment Date:</label>
            <div><span asp-validation-for="Date" class="text-danger"></span></div>
            <input asp-for="Date" type="text" asp-format="{0:d}" class="form-control" />
        </div>
        <span asp-validation-for="TermsAccepted" class="text-danger"></span>
        <div class="radio form-group">
            <input asp-for="TermsAccepted" />
            <label asp-for="TermsAccepted" class="form-check-label">
                I accept the terms & conditions
            </label>
        </div>
        <button type="submit" class="btn btn-primary">Make Booking</button>
    </form>
```

通过运行应用可以查看效果，输入 Joe 到 Your name 文本框中，选择预约日期为 01/18/2027，提交表单，你将看到图 27-6 所示的响应。

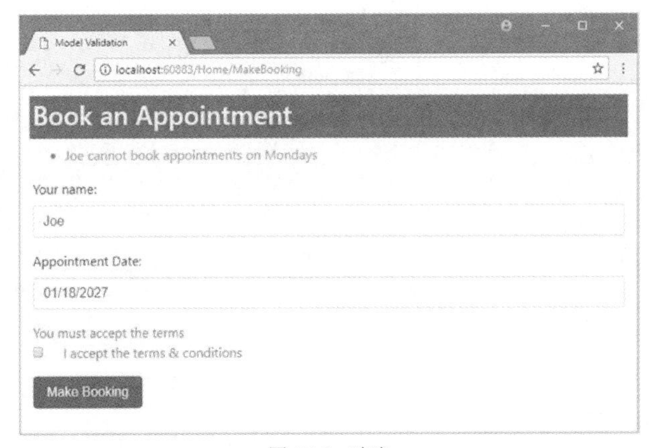

图 27-6 响应

27.4 使用元数据指定验证规则

将验证逻辑放入操作方法的问题是会在每个接收数据的操作方法中导致重复。为了减少重复，验证过程支持使用特性直接在模型类中表达验证规则，确保无论哪个操作方法用来处理请求，都使用相同的一组验证规则，

代码清单 27-16 已在 Appointment 类中应用验证特性以实施属性级验证规则。

代码清单 27-16 在 Models 文件夹下的 Appointment.cs 文件中应用验证特性

```
using System;
using System.ComponentModel.DataAnnotations;

namespace ModelValidation.Models {

    public class Appointment {

        [Required]
        [Display(Name = "name")]
        public string ClientName { get; set; }
```

```
        [UIHint("Date")]
        [Required(ErrorMessage = "Please enter a date")]
        public DateTime Date { get; set; }
        [Range(typeof(bool), "true", "true",
         ErrorMessage = "You must accept the terms")]
        public bool TermsAccepted { get; set; }
    }
}
```

以上代码使用了两个验证特性——Required 和 Range。Required 特性指定了如果用户没有为属性提供值，就是验证错误。Range 特性指定了可接收的子集。表 27-7 展示了 MVC 应用中内置的验证特性。

表 27-7 内置的验证特性

特性	示例	描述
Compare	[Compare("OtherProperty")]	这个特性确保属性拥有相同的值，当要求用户提供两次相同的信息时很有用，比如电子邮件地址或口令
Range	[Range(10,20)]	这个特性确保数值（或者任何实现了 IComparable 的属性类型）不超出指定的最小值和最大值。为了仅指定单侧边界，可使用 MinValue 或 MaxValue 常量，例如[Range(int.MinValue,50)]
RegularExpression	[RegularExpression ("pattern")]	这个特性确保字符串值与指定的正则表达式匹配。注意，模式必须匹配整个用户提供的值，而不仅仅是其中的子串。默认情况下，匹配时区分大小写，但可以通过应用（?i）修饰符（如[RegularExpression("(?i)mypattern")]）来不区分大小写
Required	[Required]	这个特性确保值不为空或是仅仅包含空格的字符串。如果要将空格视为有效，请使用[Required（AllowEmptyStrings = true）]
StringLength	[StringLength(10)]	这个特性确保字符串的值不长于指定的最大长度，还可以指定最小长度，如[StringLength(10,MinimumLength=2)]

所有的验证特性都支持通过为 ErrorMessage 属性设置值来定制错误消息，类似于下面这样：

```
...
[UIHint("Date")]
[Required(ErrorMessage = "Please enter a date")]
public DateTime Date { get; set; }
...
```

如果这里没有定制错误消息，那么默认的信息将被应用，但是它们倾向于显示对用户没有意义的模型类的详细信息，除非还使用 Display 特性，并组合应用于 ClientName 属性。

```
...
[Required]
[Display(Name = "name")]
public string ClientName { get; set; }
...
```

通过 Required 特性生成的默认消息反映了使用 Display 特性指定的名称，因此不会将属性的名称显示给用户。

需要注意使验证保持一致，例如，考虑应用于 TermAccepted 属性的特性：

```
...
[Range(typeof(bool), "true", "true", ErrorMessage="You must accept the terms")]
public bool TermsAccepted { get; set; }
...
```

虽然十分希望用户选中复选框并接受条款，但不能使用 Required 特性，因为如果用户没有选中复选框，浏览器将为这个属性发送 false 值。为了绕过这个问题，可使用 Range 特性的一项功能：提供 Type，并将上限和下限指定为字符串值。通过设置两个界限都为 true，为基于复选框的 bool 属性创建 Required 特性的等价物。这可能需要做一些实验，以确保浏览器发送的验证特性和数据一起工作。

在模型类上使用验证特性意味着控制器中的操作方法可以简化，如代码清单 27-17 所示。

代码清单 27-17 在 Controllers 文件夹下的 HomeController.cs 文件中移除属性级验证

```
using System;
using Microsoft.AspNetCore.Mvc;
using ModelValidation.Models;
using Microsoft.AspNetCore.Mvc.ModelBinding;

namespace ModelValidation.Controllers {

    public class HomeController : Controller {

        public IActionResult Index() =>
            View("MakeBooking", new Appointment() { Date = DateTime.Now });

        [HttpPost]
        public ViewResult MakeBooking(Appointment appt) {

            if (ModelState.GetValidationState(nameof(appt.Date))
                    == ModelValidationState.Valid
                && ModelState.GetValidationState(nameof(appt.ClientName))
                    == ModelValidationState.Valid
                && appt.ClientName.Equals("Joe", StringComparison.OrdinalIgnoreCase)
                && appt.Date.DayOfWeek == DayOfWeek.Monday) {
                ModelState.AddModelError("",
                    "Joe cannot book appointments on Mondays");
            }

            if (ModelState.IsValid) {
                return View("Completed", appt);
            } else {
                return View();
            }
        }
    }
}
```

验证特性在调用操作方法之前被应用，这意味着可以仍然基于模型来确定在执行模型验证时单个属性是否有效。下面在操作方法中查看验证特性，启动应用，然后不输入任何数据就提交表单，如图 27-7 所示。

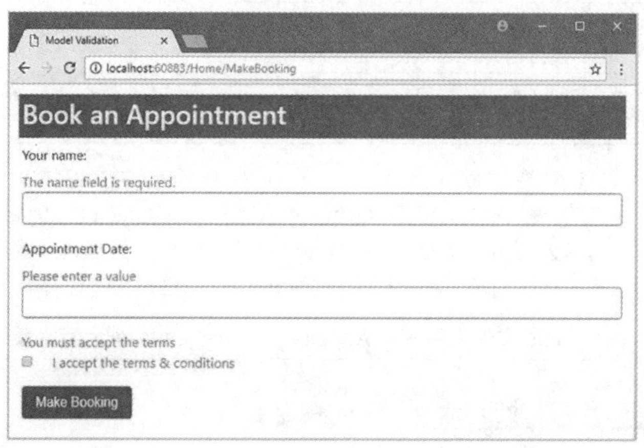

图 27-7 使用验证特性

创建自定义的属性验证特性

验证过程可以通过创建实现了 IModelValidator 接口的特性来扩展。为了展示，创建 Infrastructure 文件夹，然后添加一个名为 MustBeTrueAttribute.cs 的类文件，定义代码清单 27-18 所示的类。

代码清单 27-18 Infrastructure 文件夹下的 MustBeTrueAttribute.cs 文件的内容

```
using System;
using System.Collections.Generic;
using System.Linq;
using Microsoft.AspNetCore.Mvc.ModelBinding.Validation;

namespace ModelValidation.Infrastructure {
    public class MustBeTrueAttribute : Attribute, IModelValidator {

        public bool IsRequired => true;

        public string ErrorMessage { get; set; } = "This value must be true";

        public IEnumerable<ModelValidationResult> Validate(
                ModelValidationContext context) {
            bool? value = context.Model as bool?;
            if (!value.HasValue || value.Value == false) {
                return new List<ModelValidationResult> {
                    new ModelValidationResult("", ErrorMessage)
                };
            } else {
                return Enumerable.Empty<ModelValidationResult>();
            }
        }
    }
}
```

IModelValidator 接口定义了名为 IsRequired 的属性，用于指示是否需要验证该类（这里有一点误导，因为这个属性返回的值仅仅被用于排序验证特性，以便需要的先执行）。Validate 方法用来执行验证并通过 ModelValidationContext 类的实例接收信息，该类定义的主要属性如表 27-8 所示。

表 27-8 ModelValidationContext 类定义的属性

属性	描述
Model	返回被验证的属性值，在示例中是 TermsAccepted 属性的值
Container	返回包含属性的对象，在示例中是 Appointment 对象
ActionContext	返回 ActionContext 对象，该对象提供上下文数据，并描述将处理请求的操作方法
ModelMetadata	返回 ModelMetadata 对象，该对象描述用于进行详细验证的模型类

Validate 方法返回一个 ModelValidationResult 对象序列，其中每一个对象描述了一个验证错误。在示例特性中，如果模型值不是真的，就创建 ModelValidationResult。ModelValidationResult 构造函数的第一个参数是属性的名称，该属性与错误是相关联的。当验证每个错误的时候，指定该属性是空字符串。错误消息的第二个参数是显示给用户的错误消息。在代码清单 27-19 中，用自定义特性替换了 Range 特性。

代码清单 27-19 在 Models 文件夹中应用 Appointment.cs 文件中的自定义验证特性

```
using System;
using System.ComponentModel.DataAnnotations;
using ModelValidation.Infrastructure;

namespace ModelValidation.Models {

    public class Appointment {

        [Required]
```

```
        [Display(Name = "name")]
        public string ClientName { get; set; }

        [UIHint("Date")]
        [Required(ErrorMessage = "Please enter a date")]
        public DateTime Date { get; set; }

        [MustBeTrue(ErrorMessage = "You must accept the terms")]
        public bool TermsAccepted { get; set; }
    }
}
```

使用自定义的验证特性的结果与使用 Range 特性的结果相同，但是在阅读代码时，自定义特性更加容易理解。

27.5 执行客户端验证

到目前为止，已展示的验证技术都是服务器端验证方面的，这意味着用户提交数据到服务器，然后服务器验证数据并返回验证结果（不管是成功处理数据还是需要更正的错误列表）。

在 Web 应用中，用户通常需要立即得到验证回馈，而无须向服务器提交任何内容，这称为客户端验证，并使用 JavaScript 来实现。用户输入的数据在发送到服务器之前被验证，为用户提供即时反馈并有机会更正任何错误。

MVC 支持无痕的客户端验证。无痕意味着使用视图生成的、添加到 HTML 元素的属性来表示验证规则。这些属性由作为 MVC 一部分的 JavaScript 库进行解释，反过来，配置执行实际验证工作的 jQuery 验证库。后面将展示内置的验证支持如何工作，并展示如何扩展它的功能以支持自定义的客户端验证。

提 示

客户端验证的重点是验证单个属性。事实上，使用 MVC 的内置支持很难设置模型级客户端验证。为此，大多数 MVC 应用程序使用客户端验证来处理属性级别的问题，并依赖服务器端验证处理整体模型。

第一步是使用 Bower 将新的 JavaScript 包加入应用，如代码清单 27-20 所示。

代码清单 27-20 在 ModelValidation 文件夹下的 bower.json 文件中添加包

```
{
  "name": "asp.net",
  "private": true,
  "dependencies": {
    "bootstrap": "4.0.0-alpha.6",
    "jquery": "3.2.1",
    "jquery-validation": "1.17.0",
    "jquery-validation-unobtrusive": "3.2.6"
  }
}
```

使用客户端验证意味着将 3 个 JavaScript 文件——jQuery 库、jQuery 验证库和微软无痕验证库添加到视图中，所有这些如代码清单 27-21 所示。

提 示

Bower 工具不会始终正确地安装验证包。如果你发现 wwwroot/lib 文件夹不包含需要的文件，那么可以删除 wwwroot/lib 及其内容。打开新的 PowerShell 窗口，找到项目文件夹，运行 bower cache clean 和 bower install，下载验证包的新副本。

代码清单 27-21 在 Views/Home 文件夹下的 MakeBooking.cshtml 文件中添加验证脚本元素

```
@model Appointment

@{ Layout = "_Layout"; }

@section scripts {
    <script src="/lib/jquery/dist/jquery.min.js"></script>
    <script src="/lib/jquery-validation/dist/jquery.validate.min.js"></script>
    <script
        src="/lib/jquery-validation-unobtrusive/jquery.validate.unobtrusive.min.js">
    </script>
}

<div class="bg-primary m-1 p-1 text-white"><h2>Book an Appointment</h2></div>

<form class="m-1 p-1" asp-action="MakeBooking" method="post">
    <div asp-validation-summary="ModelOnly" class="text-danger"></div>
    <div class="form-group">
        <label asp-for="ClientName">Your name:</label>
        <div><span asp-validation-for="ClientName" class="text-danger"></span></div>
        <input asp-for="ClientName" class="form-control" />
    </div>
    <div class="form-group">
        <label asp-for="Date">Appointment Date:</label>
        <div><span asp-validation-for="Date" class="text-danger"></span></div>
        <input asp-for="Date" type="text" asp-format="{0:d}" class="form-control" />
    </div>
    <span asp-validation-for="TermsAccepted" class="text-danger"></span>
    <div class="radio form-group">
        <input asp-for="TermsAccepted" />
        <label asp-for="TermsAccepted" class="form-check-label">
            I accept the terms & conditions
        </label>
    </div>
    <button type="submit" class="btn btn-primary">Make Booking</button>
</form>
```

这些文件必须按照顺序添加。当使用标签助手转换 input 元素的时候，它们会检查应用与模型类的验证特性并向输出元素添加属性。如果运行应用并检查发送到浏览器的 HTML，你将会看到类似于下面这样的元素：

```
<input class="form-control" type="text" data-val="true"
    data-val-required="The name field is required." id="ClientName"
    name="ClientName" value="" />
```

JavaScript 代码使用 data-val 属性来查找元素，并在用户提交表单时在浏览器中执行本地验证，而不向服务器发送 HTTP 请求。在使用 F12 键时，可以通过运行应用和提交表单来查看效果。注意，即使把 HTTP 请求发送到服务器，也会显示验证错误消息。

避免与浏览器验证发生冲突

当前的一些 HTML5 浏览器支持基于 input 元素属性的简单客户端验证。例如，如果为 input 元素应用了 required 属性，那么当用户试图在不提供值的情况下提交表单时，会导致浏览器显示验证错误。

如果从模型中生成表单元素，那么浏览器验证将不会有任何问题，因为 MVC 将生成和使用 data-属性来表示验证规则(例如，为必须具有值的 input 元素使用 data-val-required 属性，浏览器将不会识别该元素)。

但是，如果无法完全控制应用程序中的标记，那么你可能会遇到问题。当你处理以其他途径生成的内容时，通常会发生这种错误。结果是 jQuery 验证和浏览器验证同时作用于表单，对用户造成困扰。为避免此类问题，可以将 novalidate 属性添加到 form 元素中。

MVC 客户端验证的一个很好特性是用来指定验证规则的属性可同时应用于客户端和服务器端，这意味着来自不支持 JavaScript 的浏览器的数据将受同样的验证，而不需要做任何额外的工作。但是，这也意味着客户端验证不支持自定义的验证特性，因为 JavaScript 无法在客户端实现自定义验证逻辑。换句话说，如果要使用客户端验证，则需要遵守表 27-7 中描述的内置特性。

比较 MVC 客户端验证与 jQuery 验证

MVC 客户端验证建立在 jQuery 验证库之上。如果愿意，你可以直接使用验证库而忽略 MVC 特性。验证库不仅灵活而且功能丰富。如果仅仅期望理解如何自定义 MVC 特性以最大限度利用可用的验证选项，这是值得的。*Pro jQuery 2.0* 一书深入介绍了 jQuery 验证库。

27.6 执行远程验证

本章将要描述的最后一个验证特性是远程验证。这是一种客户端验证技术，可以通过调用服务器端的操作方法来执行验证。

远程验证的常见示例是在应用中检查用户名是否可用，这些名称必须是唯一的，用户提交数据，然后执行客户端验证。作为处理的一部分，向服务器发出 Ajax 请求以验证请求的用户名。如果用户名已经被占用，则显示验证错误以便用户输入另外的名称。

这可能看起来类似于常规的服务器端验证，但这种方法有一些好处。首先，仅有少量属性被远程验证；客户端验证的好处仍然适用于用户输入的所有其他属性值。其次，请求是轻量级的并专注于验证，而不是处理整个模型对象。

远程验证在后台执行，用户不需要单击提交按钮并等待新的视图被渲染和返回。这使得用户体验更好，尤其是在浏览器和服务器之间网速较慢的时候。

也就是说，远程验证做了妥协，它在客户端验证和服务器端验证之间取得平衡，但它确实需要向应用服务器发出请求，因而不像通常的客户端验证那样快速。

使用远程验证的第一步是创建一个可以验证模型属性的操作方法，下面将验证 Appointment 模型的 Date 属性来确保请求的预约发生在未来（这是本章开头使用的原有验证规则之一，但是不可能使用标准的客户端验证特性来实现）。代码清单 27-22 展示了如何将操作方法 ValidateDate 添加到 Home 控制器中。

代码清单 27-22 在 Controllers 文件夹下的 HomeController.cs 文件中加入验证用的操作方法

```
using System;
using Microsoft.AspNetCore.Mvc;
using ModelValidation.Models;
using Microsoft.AspNetCore.Mvc.ModelBinding;
namespace ModelValidation.Controllers {

    public class HomeController : Controller {

        public IActionResult Index() =>
            View("MakeBooking", new Appointment() { Date = DateTime.Now });

        [HttpPost]
        public ViewResult MakeBooking(Appointment appt) {

            if (ModelState.GetValidationState(nameof(appt.Date))
                    == ModelValidationState.Valid
                && ModelState.GetValidationState(nameof(appt.ClientName))
                    == ModelValidationState.Valid
                && appt.ClientName.Equals("Joe", StringComparison.OrdinalIgnoreCase)
                && appt.Date.DayOfWeek == DayOfWeek.Monday) {
```

```
                ModelState.AddModelError("",
                    "Joe cannot book appointments on Mondays");
            }

            if (ModelState.IsValid) {
                return View("Completed", appt);
            } else {
                return View();
            }
        }

        public JsonResult ValidateDate(string Date) {
            DateTime parsedDate;

            if (!DateTime.TryParse(Date, out parsedDate)) {
                return Json("Please enter a valid date (mm/dd/yyyy)");
            } else if (DateTime.Now > parsedDate) {
                return Json("Please enter a date in the future");
            } else {
                return Json(true);
            }
        }
    }
}
```

支持远程验证的操作方法必须返回 JsonResult 类型，以告诉 MVC 正在使用 JSON 数据。如第 20 章所述，除结果之外，验证用的操作方法必须定义与要验证的数据字段同名的参数。对于这个示例来说，在操作方法内部，通过将值解析为 DateTime 对象并检查是否为未来的时间来执行验证。

提 示

你可以利用模型绑定带来的好处将操作方法的参数定义为 DateTime 对象，但是这样做意味着如果用户输入没有意义的值，比如 apple，验证方法将无法处理。这是因为模型绑定器无法从 apple 创建 DateTime 对象，并在试图处理时抛出异常。远程验证没有办法表达这个异常，因此被无提示地丢弃。这导致不能突出显示数据字段的未预期结果，产生用户输入的数据是有效的印象。作为一般规则，远程验证的最佳方法是在操作方法中接收字符串类型的参数，并执行显式的类型转换、解析或模型绑定。

使用 Json 方法可以表达验证结果，Json 方法创建了客户端远程验证脚本可以解析和处理的 JSON 格式的结果。如果值是有效的，就将 true 作为参数传递给 Json 方法，如下所示：

```
...
return Json(true);
...
```

如果有问题，就将用户应该看到的验证错误信息作为参数传递给 Json 方法，如下所示：

```
...
return Json("Please enter a date in the future");
...
```

为了使用远程验证方法，可在模型类的属性上应用 Remote 特性，如代码清单 27-23 所示。

代码清单 27-23 在 Models 文件夹下的 Appointment.cs 文件中使用 Remote 特性

```
using System;
using System.ComponentModel.DataAnnotations;
using ModelValidation.Infrastructure;
using Microsoft.AspNetCore.Mvc;
```

```
namespace ModelValidation.Models {

    public class Appointment {

        [Required]
        [Display(Name = "name")]
        public string ClientName { get; set; }

        [UIHint("Date")]
        [Required(ErrorMessage = "Please enter a date")]
        [Remote("ValidateDate", "Home")]
        public DateTime Date { get; set; }
        [MustBeTrue(ErrorMessage = "You must accept the terms")]
        public bool TermsAccepted { get; set; }
    }
}
```

Remote 特性的参数是用于生成 JavaScript 验证库将要调用的 URL 以便执行验证的操作方法名称和控制器名称，在本例中是 Home 控制器的 ValidateDate 方法。

启动应用，导航到/Home 并输入过去的日期来查看远程验证如何工作。当你选择一个值并将焦点转移到另外一个元素时，将会显示验证消息，如图 27-8 所示。

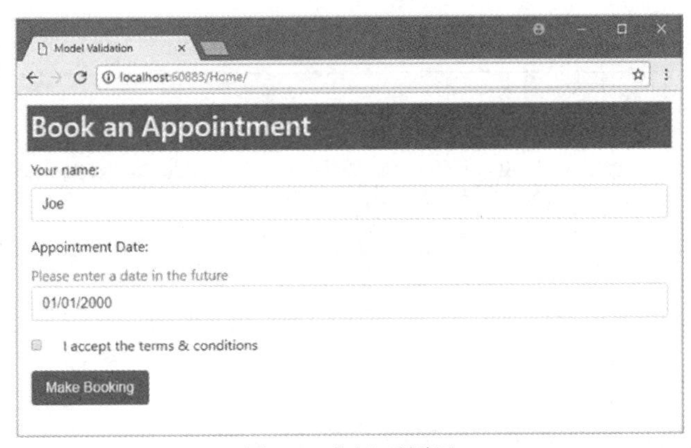

图 27-8　执行远程验证

警　告

在用户首次提交表单以及以后每次编辑数据时，都会调用验证用的操作方法。对于文本类型的 input 元素，每次单击都将导致调用到服务器。对于某些应用，这可能是大量请求，在生产中指定服务器容量和带宽时必须考虑这些请求。此外，可以选择不对验证成本高昂的属性使用远程验证（例如，当必须查询较慢的服务器以确定用户名是否唯一时）。

27.7　小结

本章说明了可用于执行模型验证的各种技术，确保用户提供的数据与对数据模型的约束一致。模型验证是一个重要的话题，为应用得到正确的验证对于确保用户具有良好和无挫败感的体验至关重要。下一章将说明如何使用 ASP.NET Core Identity 来保护 MVC 应用程序。

第 28 章　ASP.NET Core Identity 入门

ASP.NET Core Identity 是来自微软的用于在 ASP.NET 应用程序中管理用户的 API。本章将演示它的设置过程，并创建一个简单的用户管理工具来管理存储在数据库中的独立用户账户。

ASP.NET Core Identity 也支持其他类型的用户账户，比如存储在活动目录中的用户账户，但不会描述它们，因为它们不经常用于企业之外的环境（活动目录的实现往往很复杂，很难提供有用的通用示例）。

注　意

本章要求为 Visual Studio 安装 SQL Server LocalDB 功能。可以通过运行 Visual Studio 安装器和安装 Microsoft SQL Server Data Tools 来添加 SQL Server LocalDB。

第 29 章将展示如何使用这些用户账户执行身份验证和授权，第 30 章将展示如何在这些基础之上应用一些高级技术。表 28-1 提供了关于 ASP.NET Core Identity 的问题。

表 28-1　关于 ASP.NET Core Identity 的问题

问题	答案
是什么？	ASP.NET Core Identity 是管理用户并通过 Entity Framework Core 将用户数据存储到诸如数据库的存储库中的 API
有何作用？	对于大多数应用程序来说，用户管理是重要的功能。ASP.NET Core Identity 提供了现成的和经过良好测试的平台，对于常见功能甚至不需要创建自定义的版本
如何使用？	可通过添加到 Startup 类的服务和中间件，以及作为应用和 ASP.NET Core Identity 之间桥梁的类来使用
有何缺陷或限制？	ASP.NET Core Identity 更加灵活且可配置，能够弥补早期的 ASP.NET 用户管理 API 的不足
有何替代方案？	可以实现自己的 API，但是需要做大量的工作，而且除非非常小心，否则可能导致安全漏洞

表 28-2 列出了本章要介绍的操作。

表 28-2　本章要介绍的操作

操作	方法	代码清单
为项目添加 ASP.NET Core Identity	为 ASP.NET Core Identity 和 Entity Framework Core 添加包和中间件，创建用户类和数据库上下文类，创建数据库迁移	代码清单 28-1～代码清单 28-13
读取用户数据	使用上下文类查询 ASP.NET Core Identity 数据库	代码清单 28-14 和代码清单 28-15
创建用户账户	调用 UserManager.CreateAsync 方法	代码清单 28-16～代码清单 28-18
创建默认密码策略	在 Startup 类中设置密码配置	代码清单 28-19
实现自定义密码验证	实现 IPasswordValidator 接口或者从 PasswordValidator 派生子类	代码清单 28-20～代码清单 28-22
改变账户验证策略	在 Startup 类中设置用户配置	代码清单 28-23
实现自定义账户验证	实现 IUserValidator 接口或者从 UserValidator 派生子类	代码清单 28-24～代码清单 28-26
删除用户账户	调用 UserManager.DeleteAsync 方法	代码清单 28-27 和代码清单 28-28
编辑用户账户	调用 UserManager.UpdateAsync 方法	代码清单 28-29～代码清单 28-31

28.1　准备示例项目

在本章中，使用 ASP.NET Core Web Application (.NET Core)模板创建新的名为 Users 的 Empty 项目。这个示

例应用需要 Entity Framework Core 命令行工具，但该工具必须通过手动编辑.csproj 文件来加入项目。在 Solution Explorer 窗格中右击 Users 项目，从弹出的菜单中选择 Edit Users.csproj 文件，并添加代码清单 28-1 所示的元素。

代码清单 28-1 在 Users 文件夹下的 Users.csproj 文件中添加包

```
<Project Sdk="Microsoft.NET.Sdk.Web">

  <PropertyGroup>
    <TargetFramework>netcoreapp2.0</TargetFramework>
  </PropertyGroup>

  <ItemGroup>
    <Folder Include="wwwroot\" />
  </ItemGroup>

  <ItemGroup>
    <PackageReference Include="Microsoft.AspNetCore.All" Version="2.0.0" />
    <DotNetCliToolReference Include="Microsoft.EntityFrameworkCore.Tools.DotNet"
        Version="2.0.0" />
  </ItemGroup>

</Project>
```

代码清单 28-2 展示了 Startup 类，在其中配置 MVC 框架并启用开发中要用到的中间件。

代码清单 28-2 Users 文件夹下的 Startup.cs 文件的内容

```
using Microsoft.AspNetCore.Builder;
using Microsoft.Extensions.DependencyInjection;

namespace Users {

    public class Startup {

        public void ConfigureServices(IServiceCollection services) {
            services.AddMvc();
        }

        public void Configure(IApplicationBuilder app) {
            app.UseStatusCodePages();
            app.UseDeveloperExceptionPage();
            app.UseStaticFiles();
            app.UseMvcWithDefaultRoute();
        }
    }
}
```

创建控制器和视图

创建 Controllers 文件夹，添加名为 HomeController.cs 的类文件，定义代码清单 28-3 所示的控制器。后面将使用 Home 控制器来详尽描述用户账户和数据，操作方法 Index 通过 View 方法传递了一个字典给默认视图。

代码清单 28-3 Controllers 文件夹下的 HomeController.cs 文件的内容

```
using System.Collections.Generic;
using Microsoft.AspNetCore.Mvc;

namespace Users.Controllers {

    public class HomeController : Controller {
```

```
        public ViewResult Index() =>
            View(new Dictionary<string, object>
                {["Placeholder"] = "Placeholder" });
    }
}
```

为了给控制器提供视图，创建 Views/Home 文件夹，添加名为 Index.cshtml 的视图文件，内容如代码清单 28-4 所示。

代码清单 28-4 Views/Home 文件夹下的 Index.cshtml 文件的内容

```
@model Dictionary<string, object>

<div class="bg-primary m-1 p-1 text-white"><h4>User Details</h4></div>

<table class="table table-sm table-bordered m-1 p-1">
    @foreach (var kvp in Model) {
        <tr><th>@kvp.Key</th><td>@kvp.Value</td></tr>
    }
</table>
```

Index 视图使用表格来显示模型字典的内容。为支持视图，创建 Views/Shared 文件夹，添加名为_Layout.cshtml 的视图文件，内容如代码清单 28-5 所示。

代码清单 28-5 Views/Shared 文件夹下的_Layout.cshtml 文件的内容

```
<!DOCTYPE html>
<html>
<head>
    <title>Users</title>
    <meta name="viewport" content="width=device-width" />
    <link href="/lib/bootstrap/dist/css/bootstrap.css" rel="stylesheet" />
</head>
<body class="m-1 p-1">
    @RenderBody()
</body>
</html>
```

这个视图基于 Bootstrap CSS 包来修饰 HTML 元素。为了将 Bootstrap 包添加到项目中，可使用 Bower Configuration File 模板创建 bower.json 文件，然后在 dependencies 部分添加 Bootstrap 包，如代码清单 28-6 所示。

代码清单 28-6 在 Users 文件夹下的 bower.json 文件中添加 Bootstrap 包

```
{
  "name": "asp.net",
  "private": true,
  "dependencies": {
    "bootstrap": "4.0.0-alpha.6"
  }
}
```

要做的最后的准备工作是在 Views 文件夹中创建_ViewImports.cshtml 文件，以设置视图中使用的内置标签助手，如代码清单 28-7 所示。

代码清单 28-7 Views 文件夹下的_ViewImports.cshtml 文件的内容

```
@addTagHelper *, Microsoft.AspNetCore.Mvc.TagHelpers
```

在 Views 文件夹中添加名为_ViewStart.cshtml 的视图文件，内容如代码清单 28-8 所示，用以确保你在代码清单 28-5 中创建的布局将被用于应用的所有视图。

代码清单 28-8 Views 文件夹下的_ViewStart.cshtml 文件的内容

```
@{
    Layout = "_Layout";
}
```

执行示例应用，你将看到图 28-1 所示的输出。

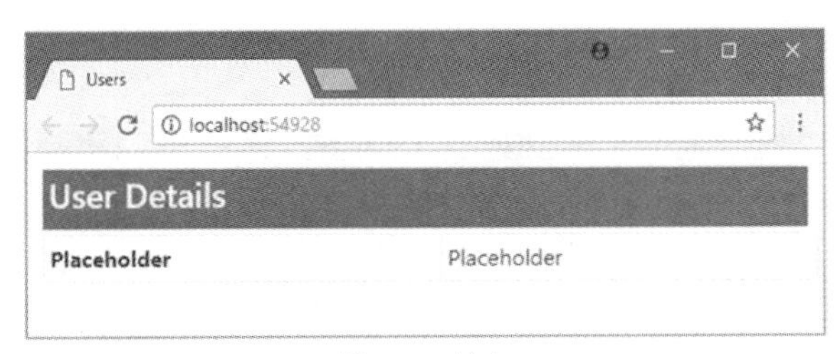

图 28-1 输出

28.2 设置 ASP.NET Core Identity

设置 ASP.NET Core Identity 的过程几乎涉及应用的各个部分，并且需要新的模型类、配置更新、控制器和操作方法来支持验证和授权操作。下面将介绍 ASP.NET Core Identity 的基本配置以展示涉及的不同步骤。有多种不同的方式在应用中使用 ASP.NET Core Identity，本章遵循最简单、最常用的配置。

28.2.1 创建用户类

第一步是定义用来表示应用中用户的类，称为用户类。

用户类派生自 IdentityUser 类，该类定义在 Microsoft.AspNetCore.Identity.EntityFrameworkCore 命名空间中。IdentityUser 类提供基本的用户表示，可以通过在派生类中添加属性来扩展，详见第 30 章。表 28-3 展示了 IdentityUser 类定义的属性。

表 28-3 IdentityUser 类定义的属性

属性	描述
Id	包含用户的唯一 ID
UserName	返回用户的名字
Claims	返回用户的凭据集合
Email	包含用户的电子邮件地址
Logins	返回用户的登录集合，用于第三方验证
PasswordHash	返回用户密码的散列值
Roles	返回用户所属的角色集合
PhoneNumber	返回用户的电话号码
SecurityStamp	当用户的身份发生变化之后，返回新的值，例如当密码发生变更之后

单独的属性此刻并不重要，重要的是，IdentityUser 类可以访问用户的基本信息——用户名、电子邮件、电话号码、密码的散列值、成员角色以及其他信息。如果期望保存用户的额外信息，就必须在从 IdentityUser 类派生的类中添加属性，从而在应用中用来表示用户。

为了在应用中创建用户，创建 Models 文件夹，并添加名为 AppUser.cs 的类文件来创建 AppUser 类，如代码清单 28-9 所示。

代码清单 28-9 Models 文件夹下的 AppUser.cs 文件的内容

```
using Microsoft.AspNetCore.Identity;

namespace Users.Models {

    public class AppUser : IdentityUser {
```

```
        // no additional members are required
        // for basic Identity installation
    }
}
```

这就是此刻必须要做的事性，第 30 章在展示如何添加特定于应用的用户数据属性时，还会讨论这个类。

配置视图导入

虽然与设置 ASP.NET Core Identity 没有直接关系，但仍需要在视图中使用 AppUser 对象。为了简化视图的创建，可在视图导入文件中添加 Users.Models 命名空间，如代码清单 28-10 所示。

代码清单 28-10　在 Views 文件夹下的_ViewImports.cshtml 文件中添加命名空间

```
@using Users.Models
@addTagHelper *, Microsoft.AspNetCore.Mvc.TagHelpers
```

28.2.2　创建数据库上下文类

下一步是创建操作 AppUser 类的 Entity Framework Core 数据库上下文类。数据库上下文类派生自 IdentityDbContext<T>，这里的 T 就是用户类（示例项目中的 AppUser 类）。添加名为 AppIdentityDbContext.cs 的类文件到 Models 文件夹中，并定义代码清单 28-11 所示的类。

代码清单 28-11　Models 文件夹下的 AppIdentityDbContext.cs 文件的内容

```
using Microsoft.AspNetCore.Identity.EntityFrameworkCore;
using Microsoft.EntityFrameworkCore;

namespace Users.Models {

    public class AppIdentityDbContext : IdentityDbContext<AppUser> {

        public AppIdentityDbContext(DbContextOptions<AppIdentityDbContext> options)
            : base(options) { }
    }
}
```

数据库上下文类可以扩展为自定义数据库的设置和使用方式，但是对于基本的 ASP.NET Core Identity 应用来说，仅仅定义这个类即可开始使用，并为将来的任何定制提供扩展空间。

注　意

不必担心这些类是否有意义。如果不熟悉 Entity Framework Core，建议将其视为黑盒。一旦基本的构件块就绪，就可以将它们复制到项目中并使用，几乎不需要修改它们。

28.2.3　配置数据库连接串

ASP.NET Core Identity 的第一个配置步骤是定义用于数据库的数据库连接串。默认会将数据库连接串放在 appsettings.json 文件中，在应用启动时，该文件将由 Startup 类加载并由 Startup 类访问。使用 ASP.NET Configuration File 模板在项目的根目录中创建 appsettings.json 文件，并添加代码清单 28-12 所示的配置。

代码清单 28-12　Users 文件夹下的 appsettings.json 文件的内容

```
{
  "Data": {
    "SportStoreIdentity": {
      "ConnectionString":
  "Server=(localdb)\\MSSQLLocalDB;Database=IdentityUsers;Trusted_Connection=True;
  MultipleActiveResultSets=true"
    }
```

```
    }
}
```

在数据库连接串中指定 localdb 选项，从而为开发人员提供方便的数据库支持。对于数据库本身，可指定名称 IdentityUsers。

注　意

数据库连接串必须在一行中，这在 Visual Studio 中效果很好，但是对于印刷而言，受限于纸张大小，不得不分多行显示。

使用数据库连接串，可以更新 Startup 类来读取配置文件并使配置可用，如代码清单 28-13 所示。

代码清单 28-13　在 Users 文件夹下的 Startup.cs 文件中读取应用配置

```
using Microsoft.AspNetCore.Builder;
using Microsoft.Extensions.DependencyInjection;
using Microsoft.Extensions.Configuration;
using Microsoft.AspNetCore.Identity;
using Microsoft.EntityFrameworkCore;
using Users.Models;

namespace Users {

    public class Startup {

        public Startup(IConfiguration configuration) =>
            Configuration = configuration;

        public IConfiguration Configuration { get; }

        public void ConfigureServices(IServiceCollection services) {
            services.AddDbContext<AppIdentityDbContext>(options =>
                options.UseSqlServer(
                    Configuration["Data:SportStoreIdentity:ConnectionString"]));

            services.AddIdentity<AppUser, IdentityRole>()
                .AddEntityFrameworkStores<AppIdentityDbContext>()
                .AddDefaultTokenProviders();

            services.AddMvc();
        }
        public void Configure(IApplicationBuilder app) {
            app.UseStatusCodePages();
            app.UseDeveloperExceptionPage();
            app.UseStaticFiles();
            app.UseAuthentication();
            app.UseMvcWithDefaultRoute();
        }
    }
}
```

创建基本的 ASP.NET Core Identity 安装需要 3 组更新。第一组是设置 Entity Framework (EF) Core，从而为 MVC 应用提供数据访问服务。

```
...
services.AddDbContext<AppIdentityDbContext>(options =>
    options.UseSqlServer(Configuration["Data:SportStoreIdentity:ConnectionString"]));
...
```

AddDbContext 方法用于添加 Entity Framework Core 所需的服务，UseSqlServer 方法用于设置使用

Microsoft SQL Server 存储数据所需的支持。AddDbContext 方法允许你应用之前创建的数据库上下文类，使用从应用配置中获取的数据库连接串中的 SQL Server 数据库进行备份（对于示例应用，这意味着读取 appsettings.json 文件）。

你还需要设置 ASP.NET Core Identity 服务，如下所示：

```
...
services.AddIdentity<AppUser, IdentityRole>()
    .AddEntityFrameworkStores<AppIdentityDbContext>()
    .AddDefaultTokenProviders();
...
```

AddIdentity 方法的类型参数用来指定表示用户和角色的类。以上代码为用户指定了 AppUser 类，为角色指定了内置的 IdentityRole 类。使用之前创建的数据库上下文类，AddEntityFrameworkStores 方法指定了应当使用 Entity Framework Core 来存储和检索数据。AddDefaultTokenProviders 方法用于使用默认配置来支持获取 token，例如修改口令。

对 Startup 类所做的最终修改是将 ASP.NET Core Identity 添加到请求处理管线中，这意味着允许用户凭据与基于 Cookie 或 URL 重写的请求相关联，还意味着用户账户的详细内容不会直接包含在发送给应用的 HTTP 请求或生成的响应中。

```
...
app.UseAuthentication();
...
```

28.2.4 创建 ASP.NET Core Identity 数据库

一切几乎就绪，仅剩的步骤是实际创建用来存储 ASP.NET Core Identity 数据的数据库。打开命令行窗口或 PowerShell 窗口，导航到 Users 文件夹（包含 Startup.cs 类文件的那个），然后执行如下命令：

```
dotnet ef migrations add Initial
```

正如为 SportsStore 应用设置数据库时说明的那样，Entity Framework Core 通过名为迁移的特性管理对数据库模式所做的更新。当修改用于生成架构的模型类时，可以生成包含 SQL 命令的迁移文件来更新数据库。该命令将创建为 ASP.NET Core Identity 设置数据库的迁移文件。

当执行完这个命令之后，你将在 Solution Explorer 窗格中看到 Migrations 文件夹。检查其中的内容，你就可以看到用来创建初始数据库的 SQL 命令。为了使用这些迁移文件来创建数据库，可运行如下命令：

```
dotnet ef database update
```

完成这个过程可能需要一点时间，一旦命令完成，数据库将被创建并备用。

28.3 使用 ASP.NET Core Identity

现在，基本的设置已经完成，可以开始使用 ASP.NET Core Identity 为示例应用增加用户管理支持了。下一步将展示 ASP.NET Core Identity API 如何被用来创建集中管理用户的管理工具。

集中化的用户管理工具在所有应用中都很有用，甚至可以允许用户创建和管理自己的账户。总有一些客户需要批量创建账户，比如，需要检查和调整用户数据的支持问题。管理工具非常有用，因为它们将很多基本的用户管理功能整合到了少量的类中，使之成为演示 ASP.NET Core Identity 基本功能的有用示例。

28.3.1 列举用户账户

本节的起点是列举数据库中所有的用户账户，这可以使你看到稍后添加到应用中的代码的效果。Controllers 文件夹中添加名为 AdminController.cs 的类文件，并定义代码清单 28-14 所示的控制器，进而定义用户管理功能。

代码清单 28-14 Controllers 文件夹下的 AdminController.cs 文件的内容

```
using Microsoft.AspNetCore.Identity;
using Microsoft.AspNetCore.Mvc;
using Users.Models;

namespace Users.Controllers {
    public class AdminController : Controller {
        private UserManager<AppUser> userManager;

        public AdminController(UserManager<AppUser> usrMgr) {
            userManager = usrMgr;
        }

        public ViewResult Index() => View(userManager.Users);
    }
}
```

操作方法 Index 将枚举 ASP.NET Core Identity 系统管理的用户。当然，现在没有任何用户，但很快就会有。可通过控制器类的构造函数接收的 UserManager<AppUser>对象来访问用户数据，该对象由依赖注入系统提供。

有了 UserManager<AppUser>对象，便可以开始查询数据存储。Users 属性将返回用户对象的枚举值——在应用中就是 AppUser 实例——可以使用 LINQ 进行查询和操控。在操作方法中，将 Users 属性传递给 View 方法，以便我们可以显示账户的详情。为了给操作方法提供视图，创建 Views/Admin 文件夹，并添加名为 Index.cshtml 的视图文件，应用代码清单 28-15 所示的标记内容。

代码清单 28-15 Views/Admin 文件夹下的 Index.cshtml 文件的内容

```
@model IEnumerable<AppUser>

<div class="bg-primary m-1 p-1 text-white"><h4>User Accounts</h4></div>

<table class="table table-sm table-bordered">
    <tr><th>ID</th><th>Name</th><th>Email</th></tr>
    @if (Model.Count() == 0) {
        <tr><td colspan="3" class="text-center">No User Accounts</td></tr>
    } else {
        foreach (AppUser user in Model) {
            <tr>
                <td>@user.Id</td>
                <td>@user.UserName</td>
                <td>@user.Email</td>
            </tr>
        }
    }
</table>

<a class="btn btn-primary" asp-action="Create">Create</a>
```

Index 视图包含一个表格，其中的每一行对应一个用户，表格的列则包括唯一 ID、用户名和电子邮件地址。如果数据库中没有用户，那么会显示一条消息，如图 28-2 所示，启动应用并访问/Admin 就可以看到。

Index 视图中包含了按钮样式的 Create 锚点元素，目标为 Admin 控制器的 Create 方法，用来支持添加用户的操作。

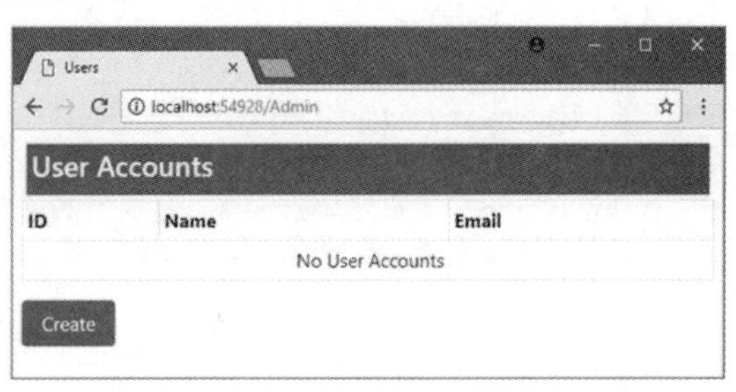

图 28-2 显示用户列表（目前是空的）

重置数据库

可以通过打开 Visual Studio 的 SQL Server 对象资源管理器来查看创建的数据库。如果这是你第一次使用 SQL Server 对象资源管理器，那么需要从 Tools 菜单中选择 Connect to Database 来告诉 Visual Studio 要使用的数据库。对于数据源，选择 Microsoft SQL Server，使用(localdb)\mssqllocaldb 作为服务名，选择使用 Windows 身份验证，单击 Select Or Enter a Database Name 字段的下拉箭头。几秒后，你将看到可用的 LocalDB 数据库列表，你应该可以选择 IdentityUsers，这是示例应用的数据库。单击 OK 按钮，SQL Server Object Explorer 窗口中会显示一个新的条目。Visual Studio 将记住该数据库，因此只需要执行这个过程一次。

在 SQL Server Object Explorer 窗口中，选择(localdb)\mssqllocaldb→Databases→IdentityUsers，浏览数据库。你将看到由迁移文件创建的表，名称类似于 AspNetUsers 和 AspNetRoles。一旦将用户添加到数据库中，就可以查询数据库以便查看表的内容。

要删除数据库，可右击 SQL Server Object Explorer 窗口中的 IdentityUsers 项目，然后从弹出的菜单中选择 Delete，在数据库删除对话框中选择所有的选项，然后单击 OK 按钮删除。

要重新创建数据库，可打开 Package Manager Console 窗口，并运行如下命令：

```
dotnet ef database update
```

数据库将被重建，并在下次启动应用时可用。

28.3.2 创建用户

要为应用接收的输入应用 MVC 模型验证，最简单的方式是为控制器支持的每个操作创建简单的视图模型。添加名为 UserViewModels.cs 的类文件到 Models 文件夹中，并用它定义代码清单 28-16 所示的类。

代码清单 28-16 Models 文件夹下的 UserViewModels.cs 文件的内容

```
using System.ComponentModel.DataAnnotations;

namespace Users.Models {

    public class CreateModel {
        [Required]
        public string Name { get; set; }
        [Required]
        public string Email { get; set; }
        [Required]
        public string Password { get; set; }
    }
}
```

定义的初始模型名为 CreateModel，它定义了创建用户账户所需的基本属性——用户名、电子邮件地址和密码。可使用来自 System.ComponentModel.DataAnnotations 命名空间的 Required 特性来表示定义在模型中的 3 个属性是必需的。

代码清单 28-17 为 Admin 控制器添加了一对 Create 操作方法，它们可由 Index 视图中的链接定位，并使用标准的控制器模式对 GET 请求向用户呈现视图，使用 POST 请求处理表单数据。

代码清单 28-17 在 Controllers 文件夹下的 AdminController.cs 文件中定义 Create 操作方法

```
using Microsoft.AspNetCore.Identity;
using Microsoft.AspNetCore.Mvc;
using Users.Models;
using System.Threading.Tasks;
```

```
namespace Users.Controllers {

    public class AdminController : Controller {
        private UserManager<AppUser> userManager;

        public AdminController(UserManager<AppUser> usrMgr) {
            userManager = usrMgr;
        }

        public ViewResult Index() => View(userManager.Users);

        public ViewResult Create() => View();

        [HttpPost]
        public async Task<IActionResult> Create(CreateModel model) {
            if (ModelState.IsValid) {
                AppUser user = new AppUser {
                    UserName = model.Name,
                    Email = model.Email
                };
                IdentityResult result
                    = await userManager.CreateAsync(user, model.Password);

                if (result.Succeeded) {
                    return RedirectToAction("Index");
                } else {
                    foreach (IdentityError error in result.Errors) {
                        ModelState.AddModelError("", error.Description);
                    }
                }
            }
            return View(model);
        }
    }
}
```

以上代码中的 Create 操作方法接收一个 CreateModel 参数，它将在管理员提交表单数据时被调用。ModelState.IsValid 属性用于检查数据是否包含所需的值，如果包含，则创建 AppUser 类的新实例并传递给异步的 UserManager.CreateAsync 方法，如下所示：

```
...
AppUser user = new AppUser { UserName = model.Name, Email = model.Email };
IdentityResult result = await userManager.CreateAsync(user, model.Password);
...
```

CreateAsync 方法的返回结果是一个 IdentityResult 对象，可通过表 28-4 所示的属性来描述操作结果。

表 28-4 IdentityResult 类定义的属性

属性	描述
Succeeded	如果操作成功，返回 true
Errors	返回用于描述在尝试操作时遇到的错误的 IdentityError 对象序列。IdentityError 类提供了 Description 属性来汇总问题

在 Create 操作方法中检查 Succeeded 属性，以确定是否在数据库中创建了新的用户。如果 Succeeded 属性为 true，客户将被重定向到 Index 操作以便用户列表显示出来。

```
...
if (result.Succeeded) {
    return RedirectToAction("Index");
```

```
} else {
    foreach (IdentityError error in result.Errors) {
        ModelState.AddModelError("", error.Description);
    }
}
...
```

如果 Succeeded 属性为 false，那么 Errors 属性提供的 IdentityError 对象序列将被枚举，Description 属性将被 ModelState.AddModelError 方法用来创建模型级验证错误，如第 27 章所述。

为了给新的操作方法提供视图，在 Views/Admin 文件夹中创建名为 Create.cshtml 的视图文件，并添加代码清单 28-18 所示的标记内容。

代码清单 28-18 Views/Admin 文件夹下的 Create.cshtml 文件的内容

```
@model CreateModel

<div class="bg-primary m-1 p-1 text-white"><h4>Create User</h4></div>
<div asp-validation-summary=" All" class="text-danger"></div>
<form asp-action="Create" method="post">
    <div class="form-group">
        <label asp-for="Name"></label>
        <input asp-for="Name" class="form-control" />
    </div>
    <div class="form-group">
        <label asp-for="Email"></label>
        <input asp-for="Email" class="form-control" />
    </div>
    <div class="form-group">
        <label asp-for="Password"></label>
        <input asp-for="Password" class="form-control" />
    </div>
    <button type="submit" class="btn btn-primary">Create</button>
    <a asp-action="Index" class="btn btn-secondary">Cancel</a>
</form>
```

Create 视图没什么特别之处，只是一个简单的表单，用于收集 MVC 用来绑定到模型类的属性的值，模型类的属性被传递给 Create 操作方法，并包含有验证错误时的摘要。

测试创建功能

为了测试创建新用户账户的能力，启动应用并导航到 URL 地址/Admin，然后单击 Create 按钮，使用表 28-5 提供的值填写表单。

提　示

人们为测试保留了一些域名，如 example.com。你可以在 IETF 网站上看到完整列表。

表 28-5 创建示例用户的值（一）

名字	值
Name	Joe
Email	joe@example.com
Password	Secret123$

输入值之后，单击 Create 按钮，ASP.NET Core Identity 将会创建用户账户。当浏览器被重定向到 Index 操作方法之后，将显示创建的用户账户，如图 28-3 所示（你将会看到不一样的 ID，因为每个账户会随机生成 ID）。

再次单击 Create 按钮并在表单中输入表 28-5 中同样的值。提交表单后，你将看到由模型验证摘要报告的错误，如图 28-4 所示。

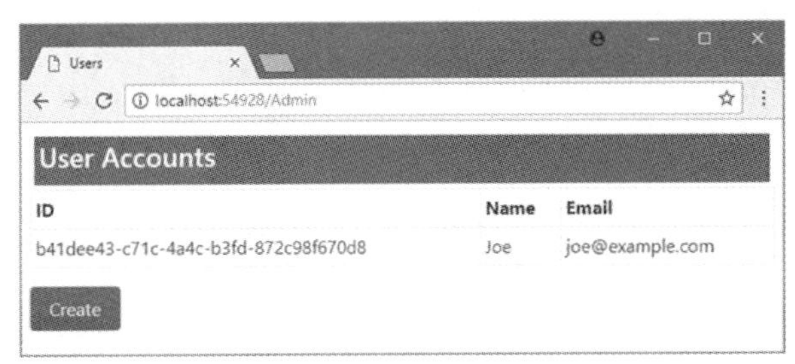

图 28-3 添加新的用户账户

图 28-4 因试图创建相同用户账户产生的错误

28.3.3 验证密码

对于企业级应用，最常见的需求之一是强制密码策略。可以通过运行应用来查看默认的策略，访问/Admin/Create，然后使用表 28-6 所示的数据填充表单，与之前的重要区别在于 Password 字段。

表 28-6 创建示例用户的值（二）

名称	值
Name	Alice
Email	alice@example.com
Password	secret

当提交表单到服务器之后，ASP.NET Core Identity 系统检查备选的密码，如果不符合要求，将生成错误消息，生成的错误消息如图 28-5 所示。

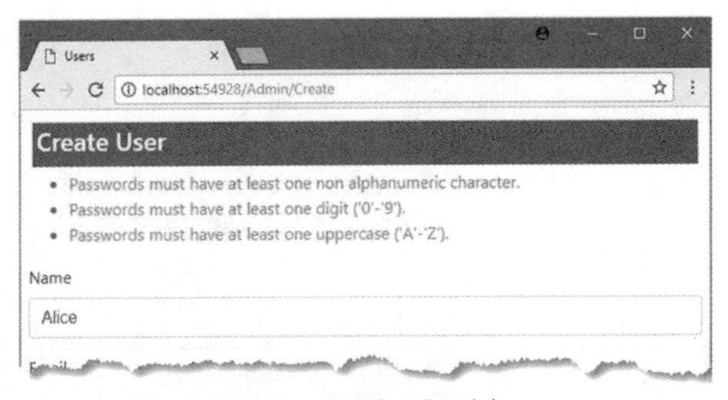

图 28-5 密码验证错误消息

可以在 Startup 类中配置密码验证规则，如代码清单 28-19 所示。

代码清单 28-19 在 Users 文件夹下的 Startup.cs 文件中配置密码验证规则

```
using Microsoft.AspNetCore.Builder;
using Microsoft.Extensions.DependencyInjection;
```

```
using Microsoft.Extensions.Configuration;
using Microsoft.AspNetCore.Identity;
using Microsoft.EntityFrameworkCore;
using Users.Models;
namespace Users {

    public class Startup {

        public Startup(IConfiguration configuration) =>
            Configuration = configuration;

        public IConfiguration Configuration { get; }

        public void ConfigureServices(IServiceCollection services) {

            services.AddDbContext<AppIdentityDbContext>(options =>
                options.UseSqlServer(
                    Configuration["Data:SportStoreIdentity:ConnectionString"]));

            services.AddIdentity<AppUser, IdentityRole>(opts => {
                opts.Password.RequiredLength = 6;
                opts.Password.RequireNonAlphanumeric = false;
                opts.Password.RequireLowercase = false;
                opts.Password.RequireUppercase = false;
                opts.Password.RequireDigit = false;
            }).AddEntityFrameworkStores<AppIdentityDbContext>()
                .AddDefaultTokenProviders();

            services.AddMvc();
        }

        public void Configure(IApplicationBuilder app) {
            app.UseStatusCodePages();
            app.UseDeveloperExceptionPage();
            app.UseStaticFiles();
            app.UseAuthentication();
            app.UseMvcWithDefaultRoute();
        }
    }
}
```

AddIdentity 方法可以与一个接收 IdentityOptions 对象的函数一起使用，该函数的 Password 属性将返回一个 PasswordOptions 实例，PasswordOptions 类提供了如表 28-7 所示的用于管理密码策略的属性。

表 28-7　　PasswordOptions 类定义的属性

属性	描述
RequiredLength	用于指定密码的最小长度
RequireNonAlphanumeric	为 true 时要求密码至少包含一个非字符或数字的字符
RequireLowercase	为 true 时要求密码至少包含一个小写字母
RequireUppercase	为 true 时要求密码至少包含一个大写字母
RequireDigit	为 true 时要求密码至少包含一个数字

启动应用，导航到/Admin/Create，然后重新提交表单，你将看到密码 secret 被接受，并为 Alice 创建了一个新的账户，如图 28-6 所示。

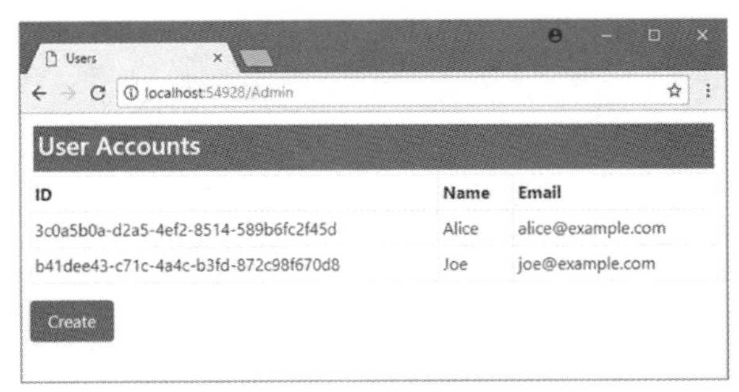

图 28-6 改变密码验证策略

实现自定义的密码验证策略

内置的密码验证策略对于大多数应用是足够的，但是你可能需要实现自定义的密码验证策略，特别是当你正在为企业级在线业务应用实施常见的复杂密码验证策略时。密码验证功能由 Microsoft.AspNetCore.Identity 命名空间的 IPasswordValidator<T>接口定义，这里的 T 是应用特定的用户类（在示例中是 AppUser）。

```
using System.Threading.Tasks;

namespace Microsoft.AspNetCore.Identity {

    public interface IPasswordValidator<TUser> where TUser : class {

        Task<IdentityResult> ValidateAsync(UserManager<TUser> manager,
            TUser user, string password);
    }
}
```

ValidateAsync 方法被调用以验证密码，UserManager 对象提供了数据上下文（允许查询 ASP.NET Core Identity 数据库），该对象表示用户和候选密码。返回的结果是一个 IdentityResult 对象，如果验证没有问题，则使用静态的 Success 属性创建这个对象；否则，使用静态的 Failed 方法，该方法会传递一个 IdentityError 数组对象，其中的每个对象描述一个验证问题。

为了演示自定义的密码验证策略，创建 Infrastructure 文件夹，添加名为 CustomPasswordValidator.cs 的类文件，并定义代码清单 28-20 所示的类。

代码清单 28-20 Infrastructure 文件夹下的 CustomPasswordValidator.cs 文件的内容

```
using System.Collections.Generic;
using System.Threading.Tasks;
using Microsoft.AspNetCore.Identity;
using Users.Models;

namespace Users.Infrastructure {

    public class CustomPasswordValidator : IPasswordValidator<AppUser> {

        public Task<IdentityResult> ValidateAsync(UserManager<AppUser> manager,
                AppUser user, string password) {

            List<IdentityError> errors = new List<IdentityError>();

            if (password.ToLower().Contains(user.UserName.ToLower())) {
                errors.Add(new IdentityError {
                    Code = "PasswordContainsUserName",
                    Description = "Password cannot contain username"
                });
            }
            if (password.Contains("12345")) {
```

```
                errors.Add(new IdentityError {
                    Code = "PasswordContainsSequence",
                    Description = "Password cannot contain numeric sequence"
                });
            }

            return Task.FromResult(errors.Count == 0 ?
                IdentityResult.Success : IdentityResult.Failed(errors.ToArray()));
        }
    }
}
```

验证器类检查密码并确保不包含用户名和序列 12345。代码清单 28-21 为 AppUser 对象注册了 CustomPasswordValidator 类作为密码验证器。

代码清单 28-21 在 Startup.cs 文件中注册自定义的密码验证器

```
using Microsoft.AspNetCore.Builder;
using Microsoft.Extensions.DependencyInjection;
using Microsoft.Extensions.Configuration;
using Microsoft.AspNetCore.Identity;
using Microsoft.EntityFrameworkCore;
using Users.Models;
using Users.Infrastructure;

namespace Users {

    public class Startup {

        public Startup(IConfiguration configuration) =>
            Configuration = configuration;

        public IConfiguration Configuration { get; }

        public void ConfigureServices(IServiceCollection services) {

            services.AddTransient<IPasswordValidator<AppUser>,
                CustomPasswordValidator>();

            services.AddDbContext<AppIdentityDbContext>(options =>
                options.UseSqlServer(
                    Configuration["Data:SportStoreIdentity:ConnectionString"]));

            services.AddIdentity<AppUser, IdentityRole>(opts => {
                opts.Password.RequiredLength = 6;
                opts.Password.RequireNonAlphanumeric = false;
                opts.Password.RequireLowercase = false;
                opts.Password.RequireUppercase = false;
                opts.Password.RequireDigit = false;
            }).AddEntityFrameworkStores<AppIdentityDbContext>()
                .AddDefaultTokenProviders();

            services.AddMvc();
        }

        public void Configure(IApplicationBuilder app) {
            app.UseStatusCodePages();
            app.UseDeveloperExceptionPage();
            app.UseStaticFiles();
            app.UseAuthentication();
            app.UseMvcWithDefaultRoute();
```

```
            }
        }
    }
```

为了测试自定义的密码验证策略，启动应用并访问/Admin/Create，使用表 28-8 所示的数据填充表单。

表 28-8　　创建示例用户的值（三）

名称	值
Name	Bob
Email	bob@example.com
Password	bob12345

表 28-8 中的密码违反了两个验证规则，因而返回图 28-7 所示的错误消息。

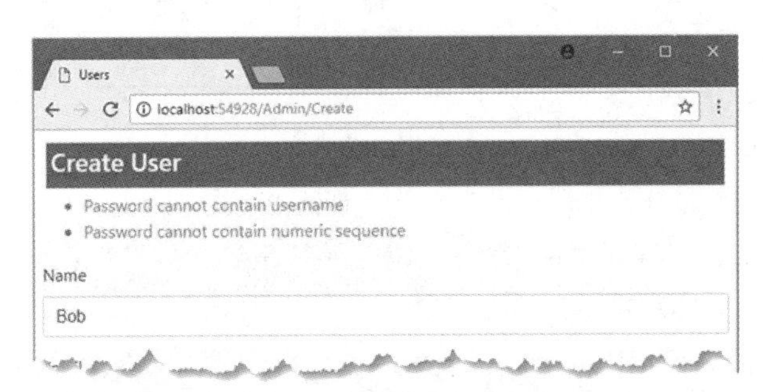

图 28-7　使用自定义的密码验证器

也可以基于默认的内置验证器来实施自定义的验证规则。默认的验证器为定义在 Microsoft.AspNetCore.Identity 命名空间中的 PasswordValidator。代码清单 28-22 将自定义的验证器修改为派生自 PasswordValidator 类，并基于提供的基本检查来实现验证。

代码清单 28-22　在 CustomPasswordValidator.cs 文件中从内置的验证器派生自定义的验证器

```
using System.Collections.Generic;
using System.Threading.Tasks;
using Microsoft.AspNetCore.Identity;
using Users.Models;
using System.Linq;

namespace Users.Infrastructure {

    public class CustomPasswordValidator : PasswordValidator<AppUser> {
        public override async Task<IdentityResult> ValidateAsync(
                UserManager<AppUser> manager, AppUser user, string password) {

            IdentityResult result = await base.ValidateAsync(manager,
                user, password);

            List<IdentityError> errors = result.Succeeded ?
                new List<IdentityError>() : result.Errors.ToList();

            if (password.ToLower().Contains(user.UserName.ToLower())) {
                errors.Add(new IdentityError {
                    Code = "PasswordContainsUserName",
                    Description = "Password cannot contain username"
                });
            }
            if (password.Contains("12345")) {
                errors.Add(new IdentityError {
```

```
                Code = "PasswordContainsSequence",
                Description = "Password cannot contain numeric sequence"
            });
        }

        return errors.Count == 0 ? IdentityResult.Success
            : IdentityResult.Failed(errors.ToArray());
    }
  }
}
```

为了测试组合验证，可运行应用并在/Admin/Create 返回的表单中使用表 28-9 所示的数据填充表单。

表 28-9　　创建示例用户的值（四）

名称	值
Name	Bob
Email	bob@example.com
Password	12345

提交表单后，你将看到组合了自定义验证和内置验证的错误消息，如图 28-8 所示。

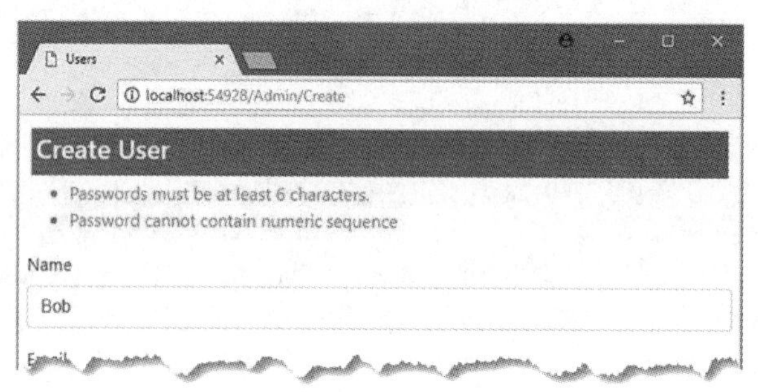

图 28-8　组合自定义验证和内置验证

28.3.4　验证用户详情

在创建账户的时候，验证也同样作用于用户名和电子邮件地址。为了查看内置验证，启动应用并使用表 28-10 所示的数据填充地址/Admin/Create 返回的表单。

表 28-10　　创建示例用户的值（五）

名称	值
Name	Bob!
Email	alice@example.com
Password	secret

提交表单后，你将看到图 28-9 所示的验证错误消息。

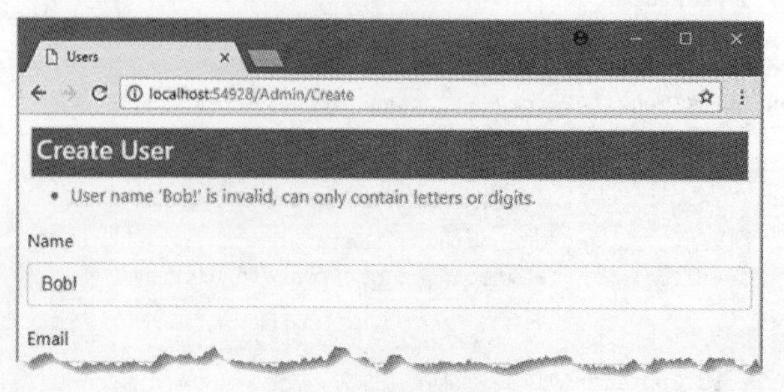

图 28-9　验证错误消息

验证可以在 Startup 类中使用 IdentityOptions.User 属性进行配置，该属性将返回 UserOptions 类的实例，表 28-11 描述了 UserOptions 类定义的属性。

表 28-11　UserOptions 类定义的属性

属性	描述
AllowedUserNameCharacters	包含可以用于用户名的所有字符，默认值为 a～z、A～Z 和 0～9 以及连字符（-）、句点（.）、下画线（_）和@。这个属性不是正则表达式，并且任何合法字符必须显式包含在这个字符串中
RequireUniqueEmail	为 true 表示要求新账户的电子邮件地址尚未使用

代码清单 28-23 修改了应用的配置以便要求唯一的电子邮件地址，以及仅有小写字母可用于用户名。

代码清单 28-23　在 Startup.cs 文件中修改用户账户的设置

```
using Microsoft.AspNetCore.Builder;
using Microsoft.Extensions.DependencyInjection;
using Microsoft.Extensions.Configuration;
using Microsoft.AspNetCore.Identity;
using Microsoft.EntityFrameworkCore;
using Users.Models;
using Users.Infrastructure;

namespace Users {

    public class Startup {

        public Startup(IConfiguration configuration) =>
            Configuration = configuration;

        public IConfiguration Configuration { get; }

        public void ConfigureServices(IServiceCollection services) {

            services.AddTransient<IPasswordValidator<AppUser>,
                CustomPasswordValidator>();

            services.AddDbContext<AppIdentityDbContext>(options =>
                options.UseSqlServer(
                    Configuration["Data:SportStoreIdentity:ConnectionString"]));

            services.AddIdentity<AppUser, IdentityRole>(opts => {
              opts.User.RequireUniqueEmail = true;
              opts.User.AllowedUserNameCharacters = "abcdefghijklmnopqrstuvwxyz";
                opts.Password.RequiredLength = 6;
                opts.Password.RequireNonAlphanumeric = false;
                opts.Password.RequireLowercase = false;
                opts.Password.RequireUppercase = false;
                opts.Password.RequireDigit = false;
            }).AddEntityFrameworkStores<AppIdentityDbContext>()
                .AddDefaultTokenProviders();

            services.AddMvc();
        }

        public void Configure(IApplicationBuilder app) {
            app.UseStatusCodePages();
            app.UseDeveloperExceptionPage();
            app.UseStaticFiles();
            app.UseAuthentication();
```

```
            app.UseMvcWithDefaultRoute();
        }
    }
}
```

重新提交测试的上一个表单，你将会看到电子邮件地址导致新的错误，而且用于用户名的字符仍然被拒绝，如图 28-10 所示。

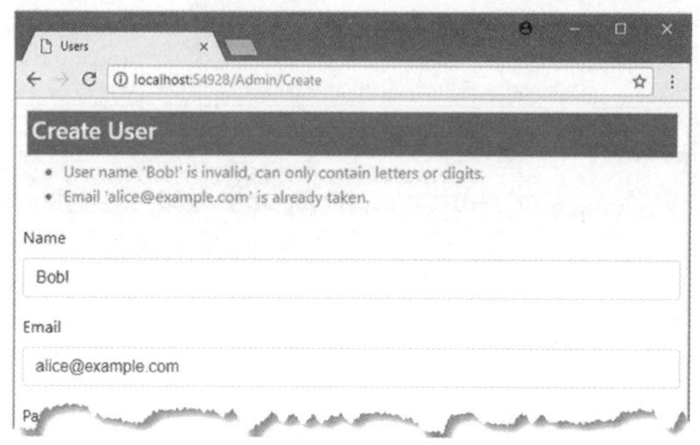

图 28-10 修改账户验证设置

实现自定义用户验证

验证功能由接口 IUserValidator<T>定义，它定义在命名空间 Microsoft.AspNetCore.Identity 中。

```
using System.Threading.Tasks;

namespace Microsoft.AspNetCore.Identity {

    public interface IUserValidator<TUser> where TUser : class {
        Task<IdentityResult> ValidateAsync(UserManager<TUser> manager, TUser user);
    }
}
```

ValidateAsync 方法被调用以执行验证。返回结果是一个 IdentityResult 对象，与验证密码的类相同。为了演示自定义验证，在 Infrastructure 文件夹中添加名为 CustomUserValidator.cs 的类文件，并用它定义代码清单 28-24 所示的类。

代码清单 28-24 Infrastructure 文件夹下的 CustomUserValidator.cs 文件的内容

```
using System.Threading.Tasks;
using Microsoft.AspNetCore.Identity;
using Users.Models;

namespace Users.Infrastructure {

    public class CustomUserValidator : IUserValidator<AppUser> {

        public Task<IdentityResult> ValidateAsync(UserManager<AppUser> manager,
                AppUser user) {

            if (user.Email.ToLower().EndsWith("@example.com")) {
                return Task.FromResult(IdentityResult.Success);
            } else {
                return Task.FromResult(IdentityResult.Failed(new IdentityError {
                    Code = "EmailDomainError",
                    Description = "Only example.com email addresses are allowed"
                }));
```

```
                }
            }
        }
    }
```

上述验证器检查电子邮件的域以确认是 example.com 域的一部分，代码清单 28-25 对这个自定义类（作为 AppUser 对象的验证器）进行了注册。

代码清单 28-25　在 Startup.cs 文件中注册自定义的用户验证器

```
...
public void ConfigureServices(IServiceCollection services) {

    services.AddTransient<IPasswordValidator<AppUser>,
        CustomPasswordValidator>();
    services.AddTransient<IUserValidator<AppUser>,
        CustomUserValidator>();

    services.AddDbContext<AppIdentityDbContext>(options =>
        options.UseSqlServer(
            Configuration["Data:SportStoreIdentity:ConnectionString"]));

    services.AddIdentity<AppUser, IdentityRole>(opts => {
        opts.User.RequireUniqueEmail = true;
        opts.User.AllowedUserNameCharacters = "abcdefghijklmnopqrstuvwxyz";
        opts.Password.RequiredLength = 6;
        opts.Password.RequireNonAlphanumeric = false;
        opts.Password.RequireLowercase = false;
        opts.Password.RequireUppercase = false;
        opts.Password.RequireDigit = false;
    }).AddEntityFrameworkStores<AppIdentityDbContext>()
        .AddDefaultTokenProviders();

    services.AddMvc();
}
...
```

为了测试自定义验证器，运行应用并使用表 28-12 所示的数据填充地址/Admin/Create 返回的表单。

表 28-12　创建示例用户的值（六）

名字	值
Name	bob
Email	bob@invalid.com
Password	secret

用户名和密码已通过验证，但电子邮件地址不是正确的域。提交表单后，你将看到图 28-11 所示的错误消息。

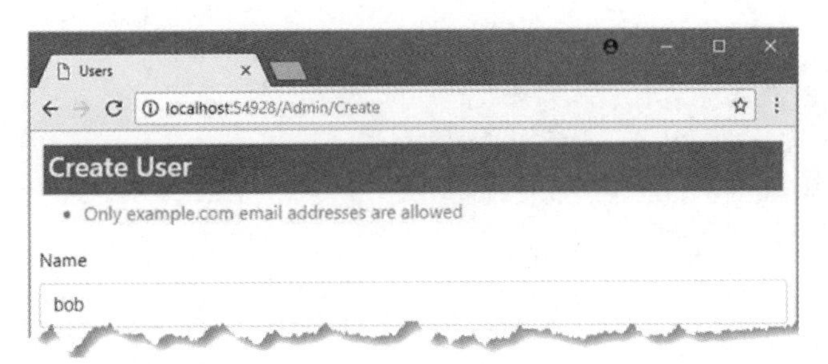

图 28-11　错误消息

将 UserValidator<T>类提供的内置验证与自定义验证相结合的处理过程与验证密码遵循相同的模式，如代码清单 28-26 所示。

代码清单 28-26 在 CustomUserValidator.cs 文件中扩展内置的用户验证

```
using System.Collections.Generic;
using System.Linq;
using System.Threading.Tasks;
using Microsoft.AspNetCore.Identity;
using Users.Models;

namespace Users.Infrastructure {

    public class CustomUserValidator : UserValidator<AppUser> {

        public override async Task<IdentityResult> ValidateAsync(
                UserManager<AppUser> manager,
                AppUser user) {

            IdentityResult result = await base.ValidateAsync(manager, user);

            List<IdentityError> errors = result.Succeeded ?
                new List<IdentityError>() : result.Errors.ToList();

            if (!user.Email.ToLower().EndsWith("@example.com")) {
                errors.Add(new IdentityError {
                    Code = "EmailDomainError",
                    Description = "Only example.com email addresses are allowed"
                });
            }

            return errors.Count == 0 ? IdentityResult.Success
                : IdentityResult.Failed(errors.ToArray());
        }
    }
}
```

28.4 完成管理功能

只需要实现编辑和删除用户的功能即可完成管理工具。在代码清单 28-27 中，可以看到对 Views/Admin/Index.cshtml 文件所做的修改，以便在管理控制台中处理 Edit 和 Delete 操作。

代码清单 28-27 在 Views/Admin 文件夹下的 Index.cshtml 文件中添加 Edit 和 Delete 按钮

```
@model IEnumerable<AppUser>

<div class="bg-primary m-1 p-1 text-white"><h4>User Accounts</h4></div>

<div class="text-danger" asp-validation-summary="ModelOnly"></div>

<table class="table table-sm table-bordered">
    <tr><th>ID</th><th>Name</th><th>Email</th></tr>
    @if (Model.Count() == 0) {
        <tr><td colspan="3" class="text-center">No User Accounts</td></tr>
    } else {
        foreach (AppUser user in Model) {
            <tr>
                <td>@user.Id</td><td>@user.UserName</td><td>@user.Email</td>
                <td>
                    <form asp-action="Delete" asp-route-id="@user.Id" method="post">
```

```
                    <a class="btn btn-sm btn-primary" asp-action="Edit"
                        asp-route-id="@user.Id">Edit</a>
                    <button type="submit"
                        class="btn btn-sm btn-danger">Delete</button>
                </form>
            </td>
        </tr>
    }
}
</table>
<a class="btn btn-primary" asp-action="Create">Create</a>
```

Delete 按钮将表单发送到 Admin 控制器的 Delete 操作方法，这很重要，因为修改应用状态需要 POST 请求。Edit 按钮是锚点元素，用于发送 GET 请求，因为编辑过程的第一步是显示当前的数据。Edit 按钮包含在表单元素内以便 Bootstrap CSS 样式不会将其垂直堆叠起来。加入模型验证的摘要信息到视图中，以便可以轻松地显示由其他管理功能导致的错误消息。

28.4.1 实现删除功能

UserManager<T>类定义了 DeleteAsync 方法，该方法接收 AppUser 类的实例，并将其从数据库中删除。在代码清单 28-28 中，可以看到如何使用 DeleteAsync 方法在 Admin 控制器中实现删除功能。

代码清单 28-28 在 Controllers 文件夹下的 AdminController.cs 文件中删除用户

```
using Microsoft.AspNetCore.Identity;
using Microsoft.AspNetCore.Mvc;
using Users.Models;
using System.Threading.Tasks;

namespace Users.Controllers {

    public class AdminController : Controller {
        private UserManager<AppUser> userManager;

        public AdminController(UserManager<AppUser> usrMgr) {
            userManager = usrMgr;
        }

        // ...other actions omitted for brevity...

        [HttpPost]
        public async Task<IActionResult> Delete(string id) {
            AppUser user = await userManager.FindByIdAsync(id);
            if (user != null) {
                IdentityResult result = await userManager.DeleteAsync(user);
                if (result.Succeeded) {
                    return RedirectToAction("Index");
                } else {
                    AddErrorsFromResult(result);
                }
            } else {
                ModelState.AddModelError("", "User Not Found");
            }
            return View("Index", userManager.Users);
        }

        private void AddErrorsFromResult(IdentityResult result) {
            foreach (IdentityError error in result.Errors) {
                ModelState.AddModelError("", error.Description);
            }
        }
    }
}
```

操作方法接收用户的唯一 ID 作为参数，并使用 FindByIdAsync 方法找到相应的 user 对象，以便可以传递给 DeleteAsync 方法。DeleteAsync 方法的返回结果是一个 IdentityResult 对象。使用前面同样的方式进行处理，以确保向用户显示任何错误。可以通过创建用户，然后单击 Index 视图中的 Delete 按钮来测试删除功能，如图 28-12 所示。

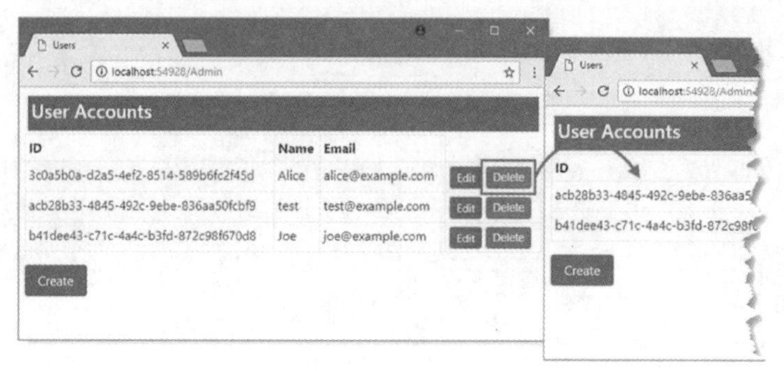

图 28-12　删除用户账户

28.4.2　实现编辑功能

要完成管理工具，需要加入对用户账户的电子邮件地址和密码的编辑支持。此刻，它们是用户定义的仅有的两个属性，第 30 章将介绍如何使用自定义属性来扩展架构。代码清单 28-29 展示了添加到 Admin 控制器的 Edit 操作方法。

代码清单 28-29　为 Controllers 文件夹下的 AdminController.cs 文件中实现编辑功能

```
using Microsoft.AspNetCore.Identity;
using Microsoft.AspNetCore.Mvc;
using Users.Models;
using System.Threading.Tasks;

namespace Users.Controllers {

    public class AdminController : Controller {
        private UserManager<AppUser> userManager;
        private IUserValidator<AppUser> userValidator;
        private IPasswordValidator<AppUser> passwordValidator;
        private IPasswordHasher<AppUser> passwordHasher;

        public AdminController(UserManager<AppUser> usrMgr,
                IUserValidator<AppUser> userValid,
                IPasswordValidator<AppUser> passValid,
                IPasswordHasher<AppUser> passwordHash) {
            userManager = usrMgr;
            userValidator = userValid;
            passwordValidator = passValid;
            passwordHasher = passwordHash;
        }
        // ...other action methods omitted for brevity...

        public async Task<IActionResult> Edit(string id) {
            AppUser user = await userManager.FindByIdAsync(id);
            if (user != null) {
                return View(user);
            } else {
                return RedirectToAction("Index");
            }
        }
```

```
        [HttpPost]
        public async Task<IActionResult> Edit(string id, string email,
                string password) {
            AppUser user = await userManager.FindByIdAsync(id);
            if (user != null) {
                user.Email = email;
                IdentityResult validEmail
                    = await userValidator.ValidateAsync(userManager, user);
                if (!validEmail.Succeeded) {
                    AddErrorsFromResult(validEmail);
                }
                IdentityResult validPass = null;
                if (!string.IsNullOrEmpty(password)) {
                    validPass = await passwordValidator.ValidateAsync(userManager,
                        user, password);
                    if (validPass.Succeeded) {
                        user.PasswordHash = passwordHasher.HashPassword(user,
                            password);
                    } else {
                        AddErrorsFromResult(validPass);
                    }
                }
                if ((validEmail.Succeeded && validPass == null)
                        || (validEmail.Succeeded
                        && password != string.Empty && validPass.Succeeded)) {
                    IdentityResult result = await userManager.UpdateAsync(user);
                    if (result.Succeeded) {
                        return RedirectToAction("Index");
                    } else {
                        AddErrorsFromResult(result);
                    }
                }
            } else {
                ModelState.AddModelError("", "User Not Found");
            }
            return View(user);
        }
        private void AddErrorsFromResult(IdentityResult result) {
            foreach (IdentityError error in result.Errors) {
                ModelState.AddModelError("", error.Description);
            }
        }
    }
}
```

Edit 操作通过在 Index 视图中嵌入 ID 字符串的 GET 请求来调用 FindByIdAsync 方法以获取表示用户的 AppUser 对象。

更复杂的实现则接收 POST 请求，其中带有用户 ID、新的电子邮件地址和密码。必须执行几个任务才能完成 Edit 操作。

第一个任务是验证接收的值。此刻，正在使用简单的 user 对象，尽管第 30 章将展示如何自定义为用户存储的数据。即便如此，也仍然需要验证数据以确保不会违反 28.3.3 节和 28.3.4 节中的自定义验证策略。要验证电子邮件地址，可以这样做：

```
...
user.Email = email;
IdentityResult validEmail = await userValidator.ValidateAsync(userManager, user);
if (!validEmail.Succeeded) {
    AddErrorsFromResult(validEmail);
}
...
```

为 IUserValidator<AppUser>在控制器的构造函数中添加依赖，以便可以验证新的电子邮件地址。注意，在验证之前，必须修改 Email 属性，因为 ValidateAsync 方法仅仅接收 user 对象。

下一步是更改密码（如果已经提供）。ASP.NET Core Identity 存储密码的哈希值而不是密码本身，这是为了防止密码被盗。接下来获取通过验证的密码并生成将存储到数据库中的哈希值，这将在第 29 章中演示。密码通过实现了 IPasswordHasher<AppUser>接口的类被转换为哈希值，可通过声明依赖注入的构造函数参数来获取。IPasswordHasher 接口定义了 HashPassword 方法，该方法接收一个字符串参数并返回对应的哈希值，如下所示：

```
...
if (!string.IsNullOrEmpty(password)) {
    validPass = await passwordValidator.ValidateAsync(userManager, user, password);
    if (validPass.Succeeded) {
        user.PasswordHash = passwordHasher.HashPassword(user, password);
    } else {
        AddErrorsFromResult(validPass);
    }
}
...
```

对 AppUser 类的所做修改不会被存储，直到 UpdateAsync 方法被调用，如下所示：

```
...
if ((validEmail.Succeeded && validPass == null) || (validEmail.Succeeded
        && password != string.Empty && validPass.Succeeded)) {
    IdentityResult result = await userManager.UpdateAsync(user);
    if (result.Succeeded) {
        return RedirectToAction("Index");
    } else {
        AddErrorsFromResult(result);
    }
}
...
```

创建视图

最后的组件是用来向用户显示当前值并允许将新值提交给控制器的视图，代码清单 28-30 展示了创建在 Views/Admin 文件夹下的 Edit.cshtml 文件中的内容。

代码清单 28-30　Views/Admin 文件夹下的 Edit.cshtml 文件的内容

```
@model AppUser

<div class="bg-primary m-1 p-1"><h4>Edit User</h4></div>

<div asp-validation-summary="All" class="text-danger"></div>

<form asp-action="Edit" method="post">
    <div class="form-group">
        <label asp-for="Id"></label>
        <input asp-for="Id" class="form-control" disabled />
    </div>
    <div class="form-group">
        <label asp-for="Email"></label>
        <input asp-for="Email" class="form-control" />
    </div>
    <div class="form-group">
        <label for="password">Password</label>
        <input name="password" class="form-control" />
    </div>
    <button type="submit" class="btn btn-primary">Save</button>
    <a asp-action="Index" class="btn btn-secondary">Cancel</a>
</form>
```

Edit 视图将显示以静态文本显示时无法修改的用户 ID，并提供编辑电子邮件地址和密码的表单，如图 28-13

所示。注意，没有对密码使用标签助手，因为 AppUser 类不包含密码信息，只有哈希值保存在数据库中。

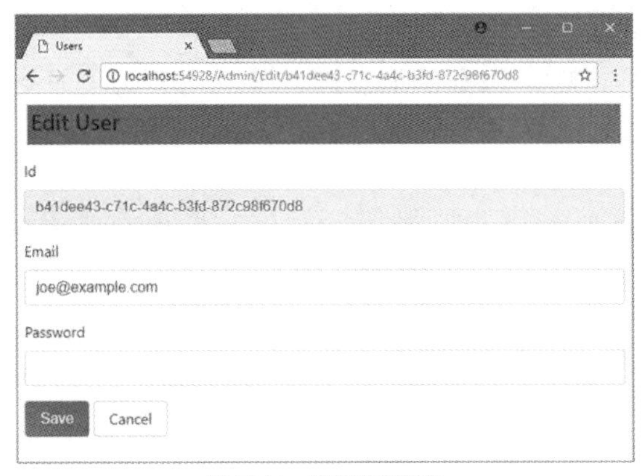

图 28-13 编辑用户账户

最后的修改是从 Startup 类中注释掉用户验证设置，以便用于用户名的默认字符可用，如代码清单 28-31 所示。由于数据库中的某些账户是在更改验证设置之前创建的，因此无法编辑它们，这是因为用户名不会通过验证。另外，当电子邮件地址被验证时，验证将作用于整个 user 对象，结果就是用户账户不能修改。

代码清单 28-31 在 Startup.cs 文件中禁用自定义的验证设置

```
...
public void ConfigureServices(IServiceCollection services) {

    services.AddTransient<IPasswordValidator<AppUser>,
        CustomPasswordValidator>();
    services.AddTransient<IUserValidator<AppUser>,
        CustomUserValidator>();

    services.AddDbContext<AppIdentityDbContext>(options =>
        options.UseSqlServer(
            Configuration["Data:SportStoreIdentity:ConnectionString"]));

    services.AddIdentity<AppUser, IdentityRole>(opts => {
        opts.User.RequireUniqueEmail = true;
        //opts.User.AllowedUserNameCharacters = "abcdefghijklmnopqrstuvwxyz";
        opts.Password.RequiredLength = 6;
        opts.Password.RequireNonAlphanumeric = false;
        opts.Password.RequireLowercase = false;
        opts.Password.RequireUppercase = false;
        opts.Password.RequireDigit = false;
    }).AddEntityFrameworkStores<AppIdentityDbContext>()
        .AddDefaultTokenProviders();

    services.AddMvc();
}
...
```

要测试编辑功能，可运行应用，然后访问/Admin 并单击 Edit 按钮。修改电子邮件地址或者输入新密码，单击 Save 按钮以更新数据库并返回/Admin。

28.5 小结

本章展示了如何创建使用 ASP.NET Core Identity 所需的配置和类，并演示了如何应用它们来创建用户管理工具。下一章将展示如何使用 ASP.NET Core Identity 执行身份验证和授权。

第 29 章　应用 ASP.NET Core Identity

本章将展示如何为上一章创建的用户账号使用 ASP.NET Core Identity 进行验证和授权。表 29-1 列出本章要介绍的操作。

表 29-1　本章要介绍的操作

操作	方法	代码清单
限制访问操作方法	应用 Authorize 特性	代码清单 29-1
验证用户	创建 Account 控制器以接收用户凭据，使用 UserManager 类检查凭据	代码清单 29-2～代码清单 29-5
创建和管理角色	使用 UserManager 类	代码清单 29-6～代码清单 29-10
使用角色授权访问操作方法	将用户账户加入角色，使用 Authorize 特性指定可以访问操作方法的角色	代码清单 29-11～代码清单 29-18
确认管理账户	使用种子数据库自动创建账户	代码清单 29-19～代码清单 29-24

29.1　准备示例项目

本章将继续使用你在第 28 章中创建的项目。请运行应用程序并导航到 URL 地址/Admin，使用 Create 按钮确认表 29-2 中的用户账户已存入数据库中。

表 29-2　本章需要的用户账户

用户名	电子邮件	密码
Joe	joe@example.com	secret123
Alice	alice@example.com	secret123
Bob	bob@example.com	secret123

完成后，访问/Admin，将会展示一个用户列表，其中包含表 29-2 中的用户（不必介意创建的其他用户，只要表 29-2 中的用户存在即可），如图 29-1 所示。

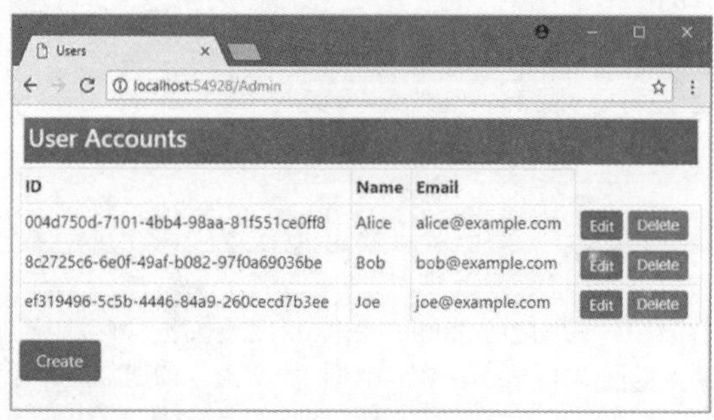

图 29-1　执行示例应用

29.2　验证用户

ASP.NET Core Identity 最基本的用途就是验证用户。限制访问操作方法的关键工具是 Authorize 特性，它告诉 MVC 只有来自已验证用户的请求才应该被处理。代码清单 29-1 为 Home 控制器的 Index 操作方法应用了 Authorize 特性。

代码清单 29-1　在 Controllers 文件夹下的 HomeController.cs 文件中限制访问

```
using System.Collections.Generic;
using Microsoft.AspNetCore.Mvc;
using Microsoft.AspNetCore.Authorization;

namespace Users.Controllers {

    public class HomeController : Controller {

        [Authorize]
        public ViewResult Index() =>
            View(new Dictionary<string, object> { ["Placeholder"] = "Placeholder" });
    }
}
```

如果启动应用，浏览器将发送请求到默认的 URL 地址，该 URL 将针对使用 Authorize 修饰的操作方法。目前用户无法自行验证，结果如图 29-2 所示。

图 29-2　访问受保护的 action 方法的结果

Authorize 特性没有指定如何验证用户，也没有直接链接到 ASP.NET Core Identity。验证服务和中间件将跨越整个 ASP.NET Core 平台，通过修改描述 HTTP 请求的上下文对象，可以简单流畅地与 MVC 应用集成，将验证处理的详细结果提供给 MVC 应用而不用涉及内部细节。

ASP.NET Core 平台通过 HttpContext 对象来提供关于用户的信息，该对象被 Authorize 特性用来检查当前请求的用户状态和查看用户是否已被验证。HttpContext.User 属性返回一个实现了 IPrincipla 接口的实例，这个接口定义在命名空间 System.Security.Principal 中。IPrincipal 接口中的重要成员包括 Identity 属性和 IsInRole(role)方法。

Identity 属性返回接口 Iidentity 的实现，IIdentity 接口描述了请求关联的用户。

如果用户是特定角色的成员，IsInRole(role)方法则返回 true。

你可通过 IPrincipal.Identity 属性返回 IIdentity 接口的实现，并通过表 29-3 中的属性返回一些关于当前用户的信息。

表 29-3　定义在 IIdentity 接口中的重要属性

属性	描述
AuthenticationType	返回用于用户验证机制的描述字符串
IsAuthenticated	如果用户已经被验证，则返回 true
Name	返回当前用户的名字

提　示

第 30 章将介绍 ASP.NET Core Identity 中关于 IIdentity 接口的实现类。

ASP.NET Core Identity 中间件使用浏览器返回的 Cookie 来检查用户是否已经通过验证。如果用户已经通过验证，就把 IIdentity.IsAuthenticated 属性设置为 true。由于当前的示例应用还没有提供验证机制，因此 IsAuthenticated 属性总是返回 false。这将导致验证错误，使客户端被重定向到地址/Account/Login，这是提供验证凭据的默认地址。

浏览器被重定向到地址/Account/Login，但是示例应用没有任何相关的控制器或操作存在，服务器返回

404-Not Found 响应，导致的错误消息参见图 29-2。

改变登录地址

虽然/Account/Login 是需要验证时客户端默认的重定向地址，但是在设置 ASP.NET Core Identity 服务的时候，可以在 Startup 类的 ConfigureServices 方法中修改配置选项来指定自己的地址，如下所示：

```
...
services.ConfigureApplicationCookie(opts => opts.LoginPath = "/Users/Login");
...
```

由于不能依赖路由系统来生成 URL 地址，因此必须使用字面值来设定重定向目标。如果为应用修改了路由模式，就必须确保同时修改了 ASP.NET Core Identity 设置，以便这个地址还可以到达目标控制器。

29.2.1 准备实现验证

请求虽然以错误结束，但还是展示了 ASP.NET Core Identity 如何适配标准的 ASP.NET 请求生命周期。下一步，我们将实现一个接收请求并验证用户的控制器。如代码清单 29-2 所示，添加一个新的模型类到 UserViewModels.cs 文件中。

代码清单 29-2 在 Models 文件夹下的 UserViewModels.cs 文件中添加新的模型类

```
using System.ComponentModel.DataAnnotations;

namespace Users.Models {

    public class CreateModel {
        [Required]
        public string Name { get; set; }
        [Required]
        public string Email { get; set; }
        [Required]
        public string Password { get; set; }
    }
    public class LoginModel {
        [Required]
        [UIHint("email")]
        public string Email { get; set; }

        [Required]
        [UIHint("password")]
        public string Password { get; set; }
    }
}
```

这个新的模型拥有 Email 和 Password 属性，它们都使用 Required 特性进行修饰，所以可以使用模型验证来检查用户是否提供了这些值。还可以使用 UIHint 特性来修饰这些属性，这将使标签助手在视图中渲染的 input 元素拥有合适的 type 属性。

提 示

在真实项目中，客户端验证可以在提交表单到服务器之前检查用户提供的用户名和密码，详见第 27 章。

添加名为 AccountController.cs 的类文件到 Controllers 文件夹中，定义代码清单 29-3 所示的控制器。

代码清单 29-3 Controllers 文件夹下的 AccountController.cs 文件的内容

```
using System.Threading.Tasks;
using Microsoft.AspNetCore.Authorization;
using Microsoft.AspNetCore.Mvc;
```

```
using Users.Models;

namespace Users.Controllers {

    [Authorize]
    public class AccountController : Controller {

        [AllowAnonymous]
        public IActionResult Login(string returnUrl) {
            ViewBag.returnUrl = returnUrl;
            return View();
        }

        [HttpPost]
        [AllowAnonymous]
        [ValidateAntiForgeryToken]
        public async Task<IActionResult> Login(LoginModel details, string returnUrl) {
            return View(details);
        }
    }
}
```

这里还没有实现验证逻辑，下面首先定义视图，然后演示检验用户凭据的过程以及用户登录应用的过程。

虽然还没有验证用户，但 Account 控制器包含了一些有用的基础架构，可从 ASP.NET Core Identity 代码中分别说明这些不久就要添加到 Login 操作方法的内容。

注意，两个版本的 Login 操作方法都有一个名为 returnUrl 的参数。当用户访问受限的地址时，他们将被重定向到带有指向受限地址的查询串的地址/Account/Login，一旦用户被验证，就可以返回这个地址。如果启动应用并访问地址/Home/Index，就可以看到这些。你的浏览器将会被重定向，如下所示：

```
/Account/Login?ReturnUrl=%2FHome%2FIndex
```

ReturnUrl 参数的字符串值使得打开页面和安全处理之间的页面重定向处理变得平滑和无缝。

接下来，注意应用于 Account 控制器的特性。管理用户账号的控制器包含只有验证用户才能使用的功能，例如重置口令。我们最终为控制器应用了 Authorize 特性，然后为个别操作方法应用了 AllowAnonymous 特性。这样，默认的操作方法需要验证用户，但是允许未验证用户登录应用。这里还应用了 ValidateAntiForgeryToken 特性，第 24 章介绍过，从而与表单元素的标签助手协作以防止跨站仿冒攻击。

最后的准备步骤是创建用户获取凭据的视图。创建 Views/Account 文件夹，添加名为 Login.cshtml 的视图文件，内容如代码清单 29-4 所示。

代码清单 29-4 Views/Account 文件夹下的 Login.cshtml 文件的内容

```
@model LoginModel

<div class="bg-primary m-1 p-1 text-white"><h4>Log In</h4></div>

<div class="text-danger" asp-validation-summary="All"></div>

<form asp-action="Login" method="post">
    <input type="hidden" name="returnUrl" value="@ViewBag.returnUrl" />
    <div class="form-group">
        <label asp-for="Email"></label>
        <input asp-for="Email" class="form-control" />
    </div>
    <div class="form-group">
        <label asp-for="Password"></label>
        <input asp-for="Password" class="form-control" />
    </div>
    <button class="btn btn-primary" type="submit">Log In</button>
</form>
```

Login 视图中唯一需要注意的就是 hidden 输入元素，这里用来保持 returnUrl 参数。从任何其他方面看，Login 都是标准的 Razor 视图，并且完成了验证的准备工作，演示了未经验证的请求被拦截和重定向的方式。测试这个新的控制器，启动应用，当浏览器请求应用程序的默认地址时，重定向到地址/Account/Login，生成的内容如图 29-3 所示。

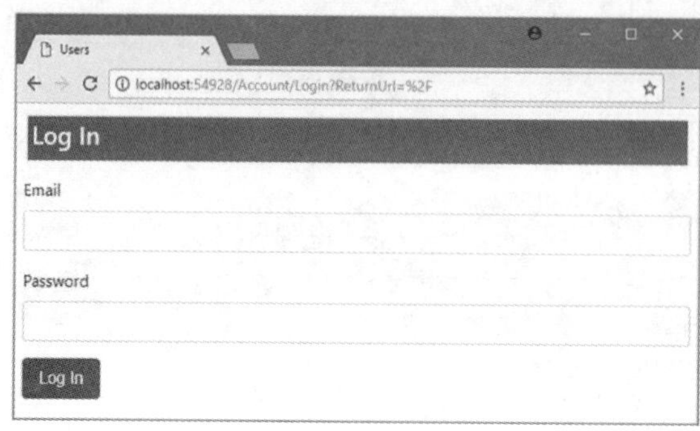

图 29-3 生成的内容

29.2.2 添加用户验证

请求保护的操作方法已被正确重定向到 Account 控制器，但是用户提供的凭据还没有被验证。代码清单 29-5 完整实现了 Login 操作方法，并使用 ASP.NET Core Identity 服务根据存储在数据库中的详细内容来验证用户。

代码清单 29-5 在 Controllers 文件夹下的 AccountController.cs 文件中添加用户验证

```
using System.Threading.Tasks;
using Microsoft.AspNetCore.Authorization;
using Microsoft.AspNetCore.Mvc;
using Users.Models;
using Microsoft.AspNetCore.Identity;

namespace Users.Controllers {

    [Authorize]
    public class AccountController : Controller {
        private UserManager<AppUser> userManager;
        private SignInManager<AppUser> signInManager;

        public AccountController(UserManager<AppUser> userMgr,
                SignInManager<AppUser> signinMgr) {
            userManager = userMgr;
            signInManager = signinMgr;
        }

        [AllowAnonymous]
        public IActionResult Login(string returnUrl) {
            ViewBag.returnUrl = returnUrl;
            return View();
        }

        [HttpPost]
        [AllowAnonymous]
        [ValidateAntiForgeryToken]
        public async Task<IActionResult> Login(LoginModel details,
                string returnUrl) {
            if (ModelState.IsValid) {
                AppUser user = await userManager.FindByEmailAsync(details.Email);
                if (user != null) {
                    await signInManager.SignOutAsync();
                    Microsoft.AspNetCore.Identity.SignInResult result =
```

```
                    await signInManager.PasswordSignInAsync(
                        user, details.Password, false, false);
                if (result.Succeeded) {
                    return Redirect(returnUrl ?? "/");
                }
            }
            ModelState.AddModelError(nameof(LoginModel.Email),
                "Invalid user or password");
        }
        return View(details);
    }
  }
}
```

以上代码中最简单的部分是获取用于表示用户的 AppUser 对象，可使用 UserManager<AppUser>类的 FindByEmailAsync 方法来完成：

```
...
AppUser user = await userManager.FindByEmailAsync(details.Email);
...
```

这个方法使用已经创建的电子邮件地址来定位用户。另外一些候选的用于定位用户的方式有使用 ID、用户名以及登录等。这里使用电子邮件地址进行登录，因为这种方式可由大多数面向互联网的 Web 应用采用，而且在企业级应用中变得越来越流行。

如果存在包含用户指定的电子邮件地址的账户，下一步就是执行验证，由于需要使用 SignInManager<AppUser> 类，因此添加通过依赖注入获取的构造函数参数。可使用 SignInManager 类执行两个验证步骤：

```
...
await signInManager.SignOutAsync();
Microsoft.AspNetCore.Identity.SignInResult result =
    await signInManager.PasswordSignInAsync(user, details.Password, false, false);
...
```

SignOutAsync 方法取消用户拥有的任何现有会话，PasswordSignIn 方法执行验证。PasswordSignInAsync 方法的参数包括用户对象、用户提供的密码以及用于控制身份验证的 Cookie 是否应该持久的 bool 参数（已被禁用）。这个 bool 参数还决定了如果密码正确，账号是否应该被锁定（也已禁用）。

PasswordSignInAsync 方法的返回结果是一个 SignInResult 对象，其中定义了名为 Succeeded 的 bool 属性来标识验证是否已经成功。

在这个示例中，检查 Succeeded 属性。如果值为 true，就将用户重定向到 returnUrl 地址；否则，添加验证错误消息，然后重新显示 Login 视图以便用户可以重新尝试登录。

作为验证过程的一部分，ASP.NET Core Identity 添加了一个 Cookie 到响应中，以便浏览器在随后的请求中携带这个 Cookie，它将用来标识用户的会话和关联的用户账号。不必直接创建或管理这个 Cookie，它将被 ASP.NET Core Identity 中间件自动处理。

考虑双因子验证

执行单因子验证，以使用户可以使用他们已经提前知道的信息——密码进行验证。

ASP.NET Core Identity 同样支持双因子验证，此时用户需要一些额外信息，通常是在验证用户身份时给予用户的信息。最常见的是来自 SecureID 令牌的值以及来自邮件或短信的验证码（严格来说，双因子可以是任何因素，包括指纹、虹膜扫描、声音识别，虽然这些是大多数 Web 应用很少需要的选项）。

由于攻击者需要知道用户的密码并且需要第二个因子，因此电子邮件账户或手机的安全性被加强了。

有两个原因导致本书不会展示双因子验证。首先，因为需要做大量的准备工作，比如设置分发作为第二个因子的电子邮件地址和短信的基础架构，并且实现验证逻辑，所有这些都超出了本书的讨论范围。

其次，因为双因子验证迫使用户记住要跳到额外的步骤进行验证，比如记住自己的电话号码或者保持 SecureID 令牌在手边，有些方式不便于在 Web 应用中使用。

> 如果对双因子验证感兴趣，建议基于第三方系统（比如 Google）进行验证，这将允许用户选择自己希望的双因子验证额外的安全信息。第 30 章将演示第三方验证。

29.2.3　测试验证

为了测试验证，可启动应用并请求地址/Home/Index。当被重定向到地址/Account/Login 后，输入本章开头列出的用户信息（例如，电子邮件地址 joe@example.com 和密码 secret123）。单击 Log In 按钮，浏览器将被重定向到地址/Home/Index。但是，此时将提交验证 Cookie，使得被授权访问这个操作方法，如图 29-4 所示。

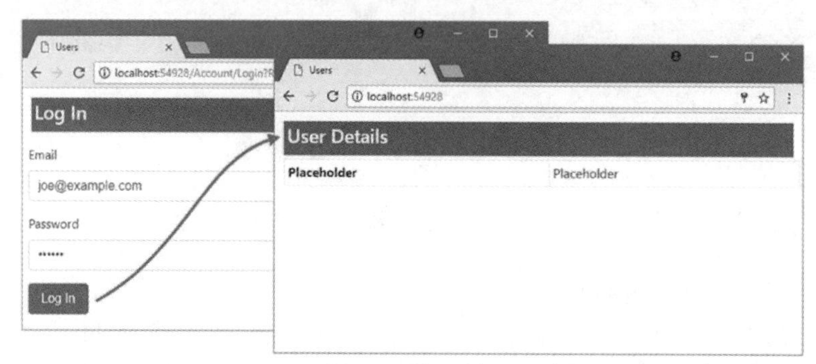

图 29-4　验证用户

提　示

使用浏览器的开发人员工具可以查看用于标识验证请求的 Cookie。

29.3　使用角色授权用户

之前，特性 Authorize 以最基本的方式使用，这将允许任何经过验证的用户执行操作方法。基于用户的成员角色，还可以优化授权，对根据哪些用户执行哪些操作进行精细控制。

角色仅仅是在应用中自定义的表示一组活动权限的定制标签。几乎所有的应用都区分可以执行管理功能的用户和没有管理功能的用户。在角色的世界中，可通过创建 Administrators 角色并赋予用户来实现授权。用户可以隶属于多个角色，角色的权限可以是粗放的，也可以是细粒度的。因此，可以使用单独的角色来区分可以执行基本任务的管理员（比如创建新的账号），以及可以执行更敏感操作的用户（比如访问付款数据）。

ASP.NET Core Identity 负责管理定义在应用中的角色，并跟踪隶属于每个角色的用户成员，但是它不知道每个角色的意义，这些信息包含在应用的 MVC 部分，并基于角色成员来限制对操作方法的访问。

ASP.NET Core Identity 提供了基于强类型的名为 RoleManager<T>的基类来访问和管理角色，这里的 T 就是存储机制中表示角色的类。Entity Framework Core 使用名为 IdentityRole 的类，其中定义的属性如表 29-4 所示。

表 29-4　IdentityRole 类定义的属性

属性	描述
Id	定义角色的唯一标识
Name	定义角色名称
Users	返回表示角色成员的 IdentityUserRole 对象集合

如果希望扩展内置的功能，可以创建应用特定的角色类，第 30 章将说明用户对象，但是这里将继续使用 IdentityRole 类，这是因为该类可以完成大多数应用需要的所有工作。第 28 章在配置应用的时候已

经告诉 ASP.NET Core Identity 使用 IdentityRole 来表示角色，详见 Startup 类的 ConfigureService 方法中的代码：

```
...
services.AddIdentity<AppUser, IdentityRole>(opts => {
    opts.User.RequireUniqueEmail = true;
    //opts.User.AllowedUserNameCharacters = "abcdefghijklmnopqrstuvwxyz";
    opts.Password.RequiredLength = 6;
    opts.Password.RequireNonAlphanumeric = false;
    opts.Password.RequireLowercase = false;
    opts.Password.RequireUppercase = false;
    opts.Password.RequireDigit = false;
}).AddEntityFrameworkStores<AppIdentityDbContext>()
    .AddDefaultTokenProviders();
...
```

AddIdentity 方法的类型参数指定了将用来表示用户和角色的类。在示例应用中，AppUser 类用来表示用户，内置的 IdentityRole 类用来表示角色。

29.3.1 创建与删除角色

为了演示如何使用角色，下面创建一个管理工具来管理它们，首先创建可以用来创建和删除角色的操作方法。在 Controllers 文件夹中创建名为 RoleAdminController.cs 的类文件，用它定义代码清单 29-6 所示的控制器。

代码清单 29-6 Controllers 文件夹下的 RoleAdminController.cs 文件的内容

```
using System.ComponentModel.DataAnnotations;
using System.Threading.Tasks;
using Microsoft.AspNetCore.Identity;
using Microsoft.AspNetCore.Identity.EntityFrameworkCore;
using Microsoft.AspNetCore.Mvc;

namespace Users.Controllers {

    public class RoleAdminController : Controller {
        private RoleManager<IdentityRole> roleManager;
        public RoleAdminController(RoleManager<IdentityRole> roleMgr) {
            roleManager = roleMgr;
        }

        public ViewResult Index() => View(roleManager.Roles);

        public IActionResult Create() => View();

        [HttpPost]
        public async Task<IActionResult> Create([Required]string name) {
            if (ModelState.IsValid) {
                IdentityResult result
                    = await roleManager.CreateAsync(new IdentityRole(name));
                if (result.Succeeded) {
                    return RedirectToAction("Index");
                } else {
                    AddErrorsFromResult(result);
                }
            }
            return View(name);
        }

        [HttpPost]
        public async Task<IActionResult> Delete(string id) {
            IdentityRole role = await roleManager.FindByIdAsync(id);
```

```
            if (role != null) {
                IdentityResult result = await roleManager.DeleteAsync(role);
                if (result.Succeeded) {
                    return RedirectToAction("Index");
                } else {
                    AddErrorsFromResult(result);
                }
            } else {
                ModelState.AddModelError("", "No role found");
            }
            return View("Index", roleManager.Roles);
        }

        private void AddErrorsFromResult(IdentityResult result) {
            foreach (IdentityError error in result.Errors) {
                ModelState.AddModelError("", error.Description);
            }
        }
    }
}
```

角色使用 RoleManager<T>类进行管理，这里的 T 是用来表示角色的类（示例应用使用了内置的 IdentityRole 类）。RoleAdminController 构造函数定义了基于 RoleManager<IdentityRole>的构造函数依赖，当控制器被创建时，可通过依赖注入获取。

RoleManager<T>类定义了表 29-5 所示的方法和属性，它们允许创建和管理角色。

表 29-5 RoleManager<T>类定义的成员

成员	描述
CreateAsync(role)	创建新角色
DeleteAsync(role)	删除特定角色
FindByIdAsync(id)	使用 ID 发现角色
FindByNameAsync(name)	使用角色名称发现角色
RoleExistsAsync(name)	如果特定名称的角色存在，则返回 true
UpdateAsync(role)	为特定角色存储更新
Roles	返回定义角色的枚举

新控制器的 Index 操作方法将显示应用中的所有角色。Create 操作方法被用来显示和接收表单，来自表单的信息将被 CreateAsync 方法用来创建新的角色。Delete 操作方法接收 POST 类型的请求和唯一的角色标识，使用 FindByIdAsync 方法来定位 ID 表示的角色对象，并使用 DeleteAsync 方法从应用中删除角色。

1. 创建视图

为了在应用中显示角色的详情，创建 Views/RoleAdmin 文件夹，在其中添加名为 Index.cshtml 的视图文件，内容如代码清单 29-7 所示。

代码清单 29-7 Views/RoleAdmin 文件夹下的 Index.cshtml 文件的内容

```
@model IEnumerable<IdentityRole>

<div class="bg-primary m-1 p-1"><h4>Roles</h4></div>

<div class="text-danger" asp-validation-summary="ModelOnly"></div>

<table class="table table-sm table-bordered table-bordered">
    <tr><th>ID</th><th>Name</th><th>Users</th><th></th></tr>
    @if (Model.Count() == 0) {
        <tr><td colspan="4" class="text-center">No Roles</td></tr>
    } else {
```

```
        foreach (var role in Model) {
            <tr>
                <td>@role.Id</td>
                <td>@role.Name</td>
                <td identity-role="@role.Id"></td>
                <td>
                    <form asp-action="Delete" asp-route-id="@role.Id" method="post">
                        <a class="btn btn-sm btn-primary" asp-action="Edit"
                           asp-route-id="@role.Id">Edit</a>
                        <button type="submit"
                                class="btn btn-sm btn-danger">
                            Delete
                        </button>
                    </form>
                </td>
            </tr>
        }
    }
</table>
<a class="btn btn-primary" asp-action="Create">Create</a>
```

Index 视图使用表格来显示应用角色的详细信息，其中的第三列使用了定制的元素属性，如下所示：

```
...
<td identity-role="@role.Id"></td>
...
```

为了显示每个角色相关的成员列表，需要在视图中包含大量的代码。为了保持视图简单，添加一个名为 RoleUsersTagHelper.cs 的类文件到 Ingrastructure 文件夹中，用它定义代码清单 29-8 所示的标签助手。

代码清单 29-8　Infrastructure 文件夹下的 RoleUsersTagHelper.cs 文件的内容

```
using System.Collections.Generic;
using System.Threading.Tasks;
using Microsoft.AspNetCore.Identity;
using Microsoft.AspNetCore.Identity.EntityFrameworkCore;
using Microsoft.AspNetCore.Razor.TagHelpers;
using Users.Models;

namespace Users.Infrastructure {

    [HtmlTargetElement("td", Attributes = "identity-role")]
    public class RoleUsersTagHelper : TagHelper {
        private UserManager<AppUser> userManager;
        private RoleManager<IdentityRole> roleManager;

        public RoleUsersTagHelper(UserManager<AppUser> usermgr,
                                  RoleManager<IdentityRole> rolemgr) {
            userManager = usermgr;
            roleManager = rolemgr;
        }

        [HtmlAttributeName("identity-role")]
        public string Role { get; set; }

        public override async Task ProcessAsync(TagHelperContext context,
                TagHelperOutput output) {

            List<string> names = new List<string>();
            IdentityRole role = await roleManager.FindByIdAsync(Role);
            if (role != null) {
                foreach (var user in userManager.Users) {
```

```
                    if (user != null
                        && await userManager.IsInRoleAsync(user, role.Name)) {
                        names.Add(user.UserName);
                    }
                }
            }

            output.Content.SetContent(names.Count == 0 ?
                "No Users" : string.Join(", ", names));
        }
    }
}
```

这个标签助手处理 td 元素的 identity-role 属性，用它接收被处理角色的 ID。RoleManager<IdentityRole>和 UserNamager<AppUser>对象允许查询 Identity 数据库来构建角色中的用户名列表。代码清单 29-9 将这个标签助手添加到视图的导入文件中，并且添加@using expression，以便在视图中不通过命名空间使用 Entity Framework Core 中的类型。

代码清单 29-9　在 Views 文件夹下的 ViewImports.cshtml 文件中添加标签助手

```
@using Users.Models
@using Microsoft.AspNetCore.Identity
@addTagHelper *, Microsoft.AspNetCore.Mvc.TagHelpers
@addTagHelper Users.Infrastructure.*, Users
```

然后，添加名为 Create.cshtml 的视图文件到 Views/RoleAdmin 文件夹中，并且添加代码清单 29-10 所示的标记来支持添加角色。

代码清单 29-10　Views/RoleAdmin 文件夹下的 Create.cshtml 文件的内容

```
@model string

<div class="bg-primary m-1 p-1"><h4>Create Role</h4></div>

<div asp-validation-summary="All" class="text-danger"></div>

<form asp-action="Create" method="post">
    <div class="form-group">
        <label for="name"></label>
        <input name="name" class="form-control" />
    </div>
    <button type="submit" class="btn btn-primary">Create</button>
    <a asp-action="Index" class="btn btn-secondary">Cancel</a>
</form>
```

创建角色时唯一需要的表单数据是角色名称，这就是使用 string 类型作为 Create 视图的视图模型类的原因。借用模型验证的优势可以确认在提交表单的时候用户提供了值，但是不值得为了这么简单的任务创建专门的模型类。相反，观察代码清单 29-6 中接收 POST 请求的 Create 操作方法，你会看到直接在参数中使用了 Required 验证特性。这与在模型类中使用这个特性有同样的效果，而且允许获得内置的模型验证带来的好处。

2. 测试、创建与删除角色

为了测试新的控制器，启动应用后导航到 URL 地址/RoleAdmin，单击 Create 按钮，在 input 元素中输入名字，然后单击另一个 Create 按钮。新的角色将保存到数据库中，在浏览器被重定向到 Index 操作后将显示出来，如图 29-5 所示。单击 Delete 按

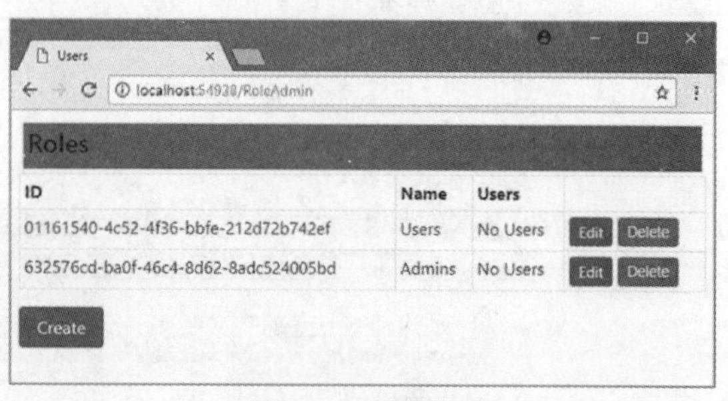

图 29-5　创建新角色

钮，从应用中删除这个角色。

29.3.2 管理角色成员

下一步是在角色中添加和删除用户。这个过程并不复杂，但是它调用了来自 RoleManager 类的角色数据，并将它们与单个用户的详情相关联。

首先定义一些视图模型类，它们用来表示角色的成员资格，并从用户那里接收一组新的成员资格，代码清单 29-11 展示了在 Models 文件夹中创建的 UserViewModels.cs 文件。

代码清单 29-11 添加视图模型到 UserViewModels.cs 文件中

```
using System.ComponentModel.DataAnnotations;
using System.Collections.Generic;
using Microsoft.AspNetCore.Identity;

namespace Users.Models {
    public class CreateModel {
        [Required]
        public string Name { get; set; }
        [Required]
        public string Email { get; set; }
        [Required]
        public string Password { get; set; }
    }

    public class LoginModel {
        [Required]
        [UIHint("email")]
        public string Email { get; set; }
        [Required]
        [UIHint("password")]
        public string Password { get; set; }
    }

    public class RoleEditModel {
        public IdentityRole Role { get; set; }
        public IEnumerable<AppUser> Members { get; set; }
        public IEnumerable<AppUser> NonMembers { get; set; }
    }

    public class RoleModificationModel {
        [Required]
        public string RoleName { get; set; }
        public string RoleId { get; set; }
        public string[] IdsToAdd { get; set; }
        public string[] IdsToDelete { get; set; }
    }
}
```

RoleEditModel 类表示系统中角色及用户的详细信息，根据用户是否是角色的成员进行分类。RoleModificationModel 类表示针对角色的一组更新。

代码清单 29-12 展示了 RoleAdmin 控制器中新添加的操作方法，它们使用代码清单 29-11 所示的视图模型来管理角色成员。

代码清单 29-12 在 Controllers 文件夹下的 RoleAdminController.cs 文件中添加操作方法

```
using System.ComponentModel.DataAnnotations;
using System.Threading.Tasks;
using Microsoft.AspNetCore.Identity;
using Microsoft.AspNetCore.Mvc;
```

```
using Users.Models;
using System.Collections.Generic;

namespace Users.Controllers {

    public class RoleAdminController : Controller {
        private RoleManager<IdentityRole> roleManager;
        private UserManager<AppUser> userManager;
        public RoleAdminController(RoleManager<IdentityRole> roleMgr,
                                   UserManager<AppUser> userMrg) {
            roleManager = roleMgr;
            userManager = userMrg;
        }

        // ...other action methods omitted for brevity...

        public async Task<IActionResult> Edit(string id) {

            IdentityRole role = await roleManager.FindByIdAsync(id);
            List<AppUser> members = new List<AppUser>();
            List<AppUser> nonMembers = new List<AppUser>();
            foreach (AppUser user in userManager.Users) {
                var list = await userManager.IsInRoleAsync(user, role.Name)
                    ? members : nonMembers;
                list.Add(user);
            }
            return View(new RoleEditModel {
                Role = role,
                Members = members,
                NonMembers = nonMembers
            });
        }

        [HttpPost]
        public async Task<IActionResult> Edit(RoleModificationModel model) {
            IdentityResult result;
            if (ModelState.IsValid) {
                foreach (string userId in model.IdsToAdd ?? new string[] { }) {
                    AppUser user = await userManager.FindByIdAsync(userId);
                    if (user != null) {
                        result = await userManager.AddToRoleAsync(user,
                            model.RoleName);
                        if (!result.Succeeded) {
                            AddErrorsFromResult(result);
                        }
                    }
                }
                foreach (string userId in model.IdsToDelete ?? new string[] { }) {
                    AppUser user = await userManager.FindByIdAsync(userId);
                    if (user != null) {
                        result = await userManager.RemoveFromRoleAsync(user,
                            model.RoleName);
                        if (!result.Succeeded) {
                            AddErrorsFromResult(result);
                        }
                    }
                }
            }
            if (ModelState.IsValid) {
                return RedirectToAction(nameof(Index));
```

```
            } else {
                return await Edit(model.RoleId);
            }
        }

        private void AddErrorsFromResult(IdentityResult result) {
            foreach (IdentityError error in result.Errors) {
                ModelState.AddModelError("", error.Description);
            }
        }
    }
}
```

在 GET 版本的 Edit 操作方法中，大部分代码用来生成针对所选角色的成员集合和非成员集合。一旦所有用户分类完成，一个新的 RoleEditModel 实例就被传递给 View 方法，以便这些数据可以使用默认视图显示出来。

POST 版本的 Edit 操作方法负责在角色中添加或删除用户。UserManager<T>类提供了一些处理角色的方法，如表 29-6 所示。

表 29-6　UserManager<T>类定义的角色相关方法

方法	描述
AddToRoleAsync(user,name)	将用户 ID 添加到具有指定名称的角色中
GetRoleAsync(user)	返回用户的角色成员名称列表
IsInRoleAsync(user,name)	如果用户是指定角色的成员，返回 true
RemoveFromRoleAsync(user, name)	从指定名称的角色中删除用户

这些方法的奇怪之处在于角色相关的操作方法基于角色名称来执行操作，即便角色拥有唯一的标识。正因为如此，RoleModificationModel 视图模型类有一个 RoleName 属性。

创建 Edit.cshtml 文件并把它添加到 Views/RoleAdmin 文件夹中，用它定义代码清单 29-13 所示的标记以允许用户编辑角色成员。

代码清单 29-13　Views/RoleAdmin 文件夹下的 Edit.cshtml 文件的内容

```
@model RoleEditModel

<div class="bg-primary m-1 p-1 text-white"><h4>Edit Role</h4></div>

<div asp-validation-summary="All" class="text-danger"></div>

<form asp-action="Edit" method="post">
    <input type="hidden" name="roleName" value="@Model.Role.Name" />
    <input type="hidden" name="roleId" value="@Model.Role.Id" />

    <h6 class="bg-info p-1 text-white">Add To @Model.Role.Name</h6>
    <table class="table table-bordered table-sm">
        @if (Model.NonMembers.Count() == 0) {
            <tr><td colspan="2">All Users Are Members</td></tr>
        } else {
            @foreach (AppUser user in Model.NonMembers) {
                <tr>
                    <td>@user.UserName</td>
                    <td>
                        <input type="checkbox" name="IdsToAdd" value="@user.Id">
                    </td>
                </tr>
```

```
            }
        }
    </table>

    <h6 class="bg-info p-1 text-white">Remove From @Model.Role.Name</h6>
    <table class="table table-bordered table-sm">
        @if (Model.Members.Count() == 0) {
            <tr><td colspan="2">No Users Are Members</td></tr>
        } else {
            @foreach (AppUser user in Model.Members) {
                <tr>
                    <td>@user.UserName</td>
                    <td>
                        <input type="checkbox" name="IdsToDelete" value="@user.Id">
                    </td>
                </tr>
            }
        }
    </table>
    <button type="submit" class="btn btn-primary">Save</button>
    <a asp-action="Index" class="btn btn-secondary">Cancel</a>
</form>
```

Edit 视图中包含两个表格：一个用于不是所选角色成员的用户，另一个用于是所选角色成员的用户。每个用户名后跟着允许修改成员状态的复选框。两个表格被包含在将发送到 Edit 操作方法的表单中，数据将被绑定到 RoleModificationModel 类，从而可以轻松地访问角色成员的变更列表。

测试和编辑角色成员

为了测试角色成员，启动应用并导航到 URL 地址/RoleAdmin。如果需要，创建名为 Users 的新角色。单击 Edit 按钮，你将看到应用中的用户显示为角色的非成员用户，如图 29-6 所示。

选中复选框以添加 Alice 和 Joe（在本章开始的时候，这两个账号已添加到 ASP.NET Core Identity 系统中），然后单击 Save 按钮。在角色列表中，你将看到 Alice 和 Joe 现在位于角色成员列表中，如图 29-7 所示。

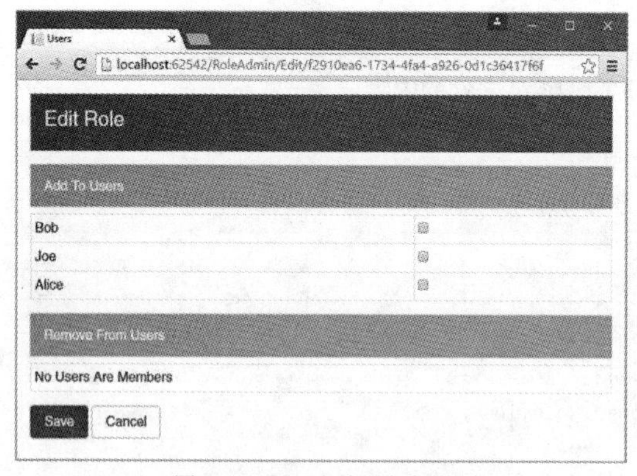

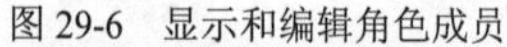
图 29-6 显示和编辑角色成员

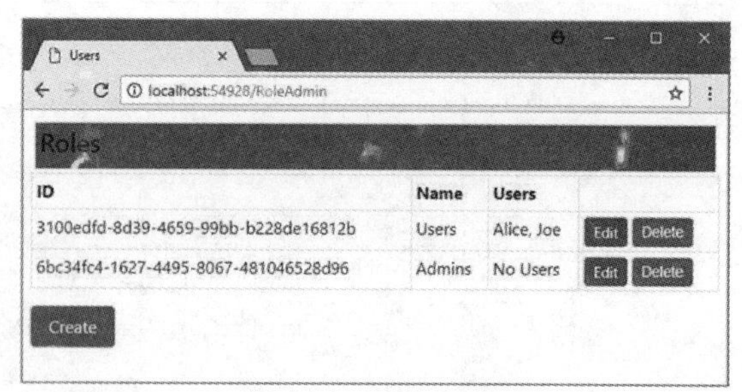

图 29-7 管理角色成员

29.3.3 使用角色进行授权

示例应用现在有了角色，它们可以通过 Authorize 特性作为授权的基础。为了易于测试基于角色的授权，在 Account 控制器中添加 Logout 操作方法，如代码清单 29-14 所示，这将允许用户退出后以另一个用户的身份重新登录并查看角色成员的效果。

代码清单 29-14 在 Controllers 文件夹下的 AccountController.cs 文件中添加 Logout 操作方法

```
using System.Threading.Tasks;
using Microsoft.AspNetCore.Authorization;
```

```
using Microsoft.AspNetCore.Mvc;
using Users.Models;
using Microsoft.AspNetCore.Identity;

namespace Users.Controllers {

    [Authorize]
    public class AccountController : Controller {
        private UserManager<AppUser> userManager;
        private SignInManager<AppUser> signInManager;

        // ...other action methods omitted for brevity...

        [Authorize]
        public async Task<IActionResult> Logout() {
            await signInManager.SignOutAsync();
            return RedirectToAction("Index", "Home");
        }
    }
}
```

下一步是更新Home控制器以加入新的操作方法，并将认证用户的信息传递给视图，如代码清单29-15所示。

代码清单 29-15　扩展 Controllers 文件夹下的 HomeController.cs 文件

```
using System.Collections.Generic;
using Microsoft.AspNetCore.Mvc;
using Microsoft.AspNetCore.Authorization;

namespace Users.Controllers {

    public class HomeController : Controller {

        [Authorize]
        public IActionResult Index() => View(GetData(nameof(Index)));

        [Authorize(Roles = "Users")]
        public IActionResult OtherAction() => View("Index",
            GetData(nameof(OtherAction)));
        private Dictionary<string, object> GetData(string actionName) =>
            new Dictionary<string, object> {
                ["Action"] = actionName,
                ["User"] = HttpContext.User.Identity.Name,
                ["Authenticated"] = HttpContext.User.Identity.IsAuthenticated,
                ["Auth Type"] = HttpContext.User.Identity.AuthenticationType,
                ["In Users Role"] = HttpContext.User.IsInRole("Users")
            };
    }
}
```

Index操作方法的Authorize特性没有变化，但是在为OtherAction操作方法应用这个特性时设置了Roles属性，指定只有Users角色的成员可以访问。这里还定义了GetData方法，以返回关于用户标识的一些基本信息，可使用HttpContext对象的属性获取这些信息。

提　示

Authorize特性也可以基于独立用户名的列表授权访问。对于小项目来说，这是一个很有吸引力的特性，但是这意味着每次要授权的用户发生变化时，都不得不修改控制器中的代码，也这意味着不得不再次完成测试和发布周期。使用角色来授权可以将应用从独立账户的变化中隔离出来，并且允许通过存储在ASP.NET Core Identity中的成员来控制应用的访问。

最后，修改 Views/Home 文件夹下的 Index.cshtml 文件，该文件由 Home 控制器的两个操作方法使用。添加目标为 Account 控制器的 Logout 操作的链接，如代码清单 29-16 所示。

代码清单 29-16 在 Views/Home 文件夹下的 Index.cshtml 文件中添加退出链接

```
@model Dictionary<string, object>

<div class="bg-primary m-1 p-1 text-white"><h4>User Details</h4></div>

<table class="table table-sm table-bordered m-1 p-1">
    @foreach (var kvp in Model) {
        <tr><th>@kvp.Key</th><td>@kvp.Value</td></tr>
    }
</table>

@if (User?.Identity?.IsAuthenticated ?? false) {
    <a asp-controller="Account" asp-action="Logout"
       class="btn btn-danger">Logout</a>
}
```

为了测试验证，启动应用并导航到/Home/Index，浏览器将被重定向以便输入用户凭据。由于应用于 Index 操作的 Authorize 特性允许任何通过验证的用户访问，因此从表 29-2 中选择哪个用户进行验证显得并不重要。

但是，如果访问 URL 地址/Home/OtherAction，你在表 29-2 中所做的选择就会导致不同，因为只有 Alice 和 Joe 是 Users 角色的成员，这对访问 OtherAction 操作是必需的。如果以 Bob 用户身份登录，浏览器将被重定向到/Account/AccessDenied，这适用于当用户不能访问操作方法时。为了处理这种情况，这里在 Account 控制器中添加 AccessDenied 操作，以便有一个操作可以处理这种请求，如代码清单 29-17 所示。

提 示

设置 IdentityOptions.Cookies.ApplicationCookie.AccessDeniedPath 属性，改变/Account/AccessDenied 地址。

代码清单 29-17 在 AccountController.cs 文件中添加操作方法

```
using System.Threading.Tasks;
using Microsoft.AspNetCore.Authorization;
using Microsoft.AspNetCore.Mvc;
using Users.Models;
using Microsoft.AspNetCore.Identity;

namespace Users.Controllers {

    [Authorize]
    public class AccountController : Controller {
        private UserManager<AppUser> userManager;
        private SignInManager<AppUser> signInManager;

        public AccountController(UserManager<AppUser> userMgr,
                SignInManager<AppUser> signinMgr) {
            userManager = userMgr;
            signInManager = signinMgr;
        }

        // ...other action methods omitted for brevity...

        [AllowAnonymous]
        public IActionResult AccessDenied() {
            return View();
```

```
        }
    }
}
```

要为 AccessDenied 操作提供视图，可在 Views/Account 文件夹中创建名为 AccessDenied.cshtml 的文件，然后添加代码清单 29-18 所示的内容。

代码清单 29-18 Views/Account 文件夹下的 AccessDenied.cshtml 文件的内容

```
<div class="bg-danger mb-1 p-2 text-white"><h4>Access Denied</h4></div>
<a asp-action="Index" asp-controller="Home" class="btn btn-primary">OK</a>
```

启动应用并访问/Account/Login，以 bob@example.com 身份登录。当验证过程完成后，浏览器将被重定向到/Home/Index，这将显示账户的详请，如图 29-8 左边的截图所示，这表明 Bob 不是 Users 角色的成员。现在访问/Home/OtherAction，Bob 不是所要求角色的成员，浏览器被重定向到 URL 地址/Account/AccessDenied，如图 29-8 右边的截图所示。

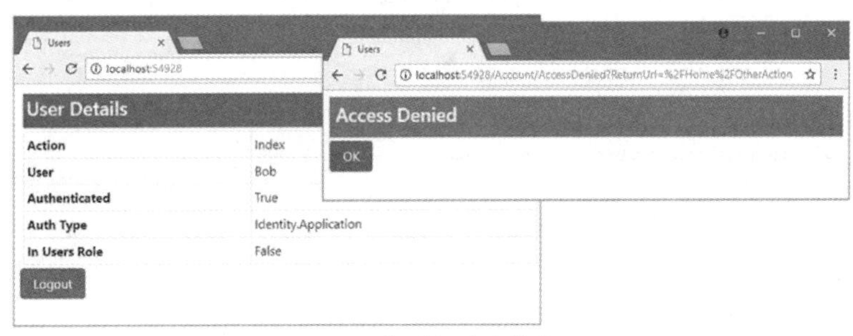

图 29-8 使用基于角色的验证

提 示

角色在用户登录的时候加载，这意味着如果改变当前已登录用户的角色，那么在注销和重新登录之前，更改不会生效。

29.4 播种数据库

在示例项目中，访问 Admin 和 RoleAdmin 控制器是不受限制的。需要创建用户和角色，但 Admin 和 RoleAdmin 控制器是用户管理工具，如果使用 Authorize 特性保护了它们，就不能授权任何凭据来访问它们了，特别是在第一次部署应用之前。

解决方案是当应用启动的时候，使用一些初始数据来播种数据库。如代码清单 29-19 所示，添加一些新的配置数据到 appsettings.json 文件中以指定将要创建的账户的详情。

代码清单 29-19 添加配置数据到 appsettings.json 文件中

```
{
  "Data": {
    "AdminUser": {
      "Name": "Admin",
      "Email": "admin@example.com",
      "Password": "secret",
      "Role": "Admins"
    },
```

```
    "SportStoreIdentity": {
      "ConnectionString": "Server=(localdb)\\MSSQLLocalDB;Database=IdentityUsers;Trusted_Con
nection=True;MultipleActiveResultSets=true"
    }
  }
}
```

Data: AdminUser 类别提供了创建账户需要的 4 个值，并赋予了用于管理工具的角色。

警 告

将口令放在明文配置文件中意味着当部署应用并第一次初始化数据时，必须将修改默认口令作为部署过程的一部分。

然后，在 AppIdentityDbContext 类中添加一个静态方法，如代码清单 29-20 所示。创建默认账户的代码不需要定义在这个类中，但是作者把它用在了这里的项目中。

代码清单 29-20 在 Models 文件夹下的 AppIdentityDbContext.cs 文件中添加静态方法

```
using Microsoft.AspNetCore.Identity;
using Microsoft.AspNetCore.Identity.EntityFrameworkCore;
using Microsoft.EntityFrameworkCore;
using Microsoft.Extensions.Configuration;
using Microsoft.Extensions.DependencyInjection;
using System;
using System.Threading.Tasks;

namespace Users.Models {

    public class AppIdentityDbContext : IdentityDbContext<AppUser> {

        public AppIdentityDbContext(DbContextOptions<AppIdentityDbContext> options)
            : base(options) { }

        public static async Task CreateAdminAccount(IServiceProvider serviceProvider,
            IConfiguration configuration) {

            UserManager<AppUser> userManager =
                serviceProvider.GetRequiredService<UserManager<AppUser>>();
            RoleManager<IdentityRole> roleManager =
                serviceProvider.GetRequiredService<RoleManager<IdentityRole>>();

            string username = configuration["Data:AdminUser:Name"];
            string email = configuration["Data:AdminUser:Email"];
            string password = configuration["Data:AdminUser:Password"];
            string role = configuration["Data:AdminUser:Role"];
            if (await userManager.FindByNameAsync(username) == null) {
                if (await roleManager.FindByNameAsync(role) == null) {
                    await roleManager.CreateAsync(new IdentityRole(role));
                }

                AppUser user = new AppUser {
                    UserName = username,
                    Email = email
                };

                IdentityResult result = await userManager
                    .CreateAsync(user, password);
                if (result.Succeeded) {
                    await userManager.AddToRoleAsync(user, role);
                }
```

```
            }
        }
    }
}
```

CreateAdminAccount 方法接收一个 IServiceProvider 对象，它用来获取 UserManager、RoleManager 和 IConfiguration 对象，还可以用来从 appsetting.json 文件中获取数据。CreateAdminAccount 方法中的代码会检查用户是否已经存在，如果不存在，则创建用户并赋予指定的角色。如果角色也不存在，就创建角色。代码清单 29-21 在 Startup 类中添加了一些代码，可在设置和配置应用之后调用 CreateAdminAccount 方法。

代码清单 29-21　在 Users 文件夹下的 Startup.cs 文件中调用数据库方法

```
...
public void Configure(IApplicationBuilder app) {
    app.UseStatusCodePages();
    app.UseDeveloperExceptionPage();
    app.UseStaticFiles();
    app.UseAuthentication();
    app.UseMvcWithDefaultRoute();
    AppIdentityDbContext.CreateAdminAccount(app.ApplicationServices,
        Configuration).Wait();
}
...
```

由于正在通过 IApplicationBuilder.ApplicationServices 提供程序来访问受限的服务，因此还必须在 Program 类中禁用依赖注入范围验证特性，如代码清单 29-22 所示。

代码清单 29-22　在 Users 文件夹下的 Program.cs 文件中禁用依赖注入范围验证特性

```
using System;
using System.Collections.Generic;
using System.IO;
using System.Linq;
using System.Threading.Tasks;
using Microsoft.AspNetCore;
using Microsoft.AspNetCore.Hosting;
using Microsoft.Extensions.Configuration;
using Microsoft.Extensions.Logging;

namespace Users {
    public class Program {
        public static void Main(string[] args) {
            BuildWebHost(args).Run();
        }

        public static IWebHost BuildWebHost(string[] args) =>
            WebHost.CreateDefaultBuilder(args)
                .UseStartup<Startup>()
                .UseDefaultServiceProvider(options =>
                    options.ValidateScopes = false)
                .Build();
    }
}
```

现在，Identity 数据库中有了可信的默认账号，可以使用 Authorize 特性来保护 Admin 和 RoleAdmin 控制器，代码清单 29-23 展示了对 Admin 控制器所做的修改。

代码清单 29-23　在 Users 文件夹下的 AdminController.cs 文件中限制访问

```
using Microsoft.AspNetCore.Identity;
using Microsoft.AspNetCore.Mvc;
using Users.Models;
```

```
using System.Threading.Tasks;
using Microsoft.AspNetCore.Authorization;

namespace Users.Controllers {

    [Authorize(Roles = "Admins")]
    public class AdminController : Controller {

        // ...statements omitted for brevity...
    }
}
```

代码清单 29-24 展示了在 RoleAdmin 控制器中所做的修改。

代码清单 29-24 在 Controllers 文件夹下的 RoleAdminController.cs 文件中限制访问

```
using System.ComponentModel.DataAnnotations;
using System.Threading.Tasks;
using Microsoft.AspNetCore.Identity;
using Microsoft.AspNetCore.Mvc;
using Users.Models;
using System.Collections.Generic;
using Microsoft.AspNetCore.Authorization;

namespace Users.Controllers {
    [Authorize(Roles = "Admins")]
    public class RoleAdminController : Controller {
        // ...statements omitted for brevity...
    }
}
```

启动应用并访问 URL 地址/Admin 或/RoleAdmin。如果已经以其他用户身份登录，则必须先注销。否则，系统将提醒你输入凭据，可以使用账号 admin@example.com 和密码 secret 进行验证以访问管理功能。

29.5 小结

本章展示了如何使用 ASP.NET Core Identity 进行验证和授权，说明了如何收集和验证用户凭据，以及如何基于用户所属的角色来限制对操作方法的访问。下一章将演示 ASP.NET Core Identity 提供的高级特性。

第 30 章　ASP.NET Core Identity 进阶

本章将通过展示 ASP.NET Core Identity 提供的一些高级特性来完成对 ASP.NET Core Identity 的介绍，演示如何通过为 AppUser 类上定义自定义属性来扩展数据库架构，以及如何在不删除 ASP.NET Core Identity 数据库中数据的情况下使用数据库迁移来应用这些属性。此外，本章还将说明 ASP.NET Core Identity 如何支持声明（claim）的概念，并展示如何通过策略灵活地授权对操作方法的访问。最后，本章通过展示 ASP.NET Core Identity 如何使用第三方对用户进行身份验证来结束本章。可使用 Google 账户来演示身份验证，但 ASP.NET Core Identity 还内置对微软、Facebook 和 Twitter 账户的支持。表 30-1 列出了本章要介绍的操作。

表 30-1　本章要介绍的操作

操作	方法	代码清单
存储用户自定义数据	添加属性到 AppUser 类，并更新 Identity 数据库	代码清单 30-1～代码清单 30-3
执行精细授权	使用声明	代码清单 30-4～代码清单 30-7
创建自定义声明	使用声明转换	代码清单 30-8 和代码清单 30-9
使用声明评估用户访问	创建策略	代码清单 30-10～代码清单 30-14
使用策略访问资源	评估操作方法中的策略	代码清单 30-15～代码清单 30-20
允许使用第三方执行验证	接收来自验证提供器的声明，比如微软、Google 和 Facebook	代码清单 30-21～代码清单 30-24

30.1　准备示例项目

本章将继续使用你在第 28 章创建且在第 29 章扩展的 Users 项目。启动应用并确认数据库中已经有一些用户。图 30-1 展示了数据库的状态，其中包含来自上一章的用户 Admin、Alice、Bob 和 Joe。为了检查这些用户，启动应用并访问/Admin，以 Admin 用户身份进行验证，使用电子邮件地址 admin@example.com 和密码 secret。

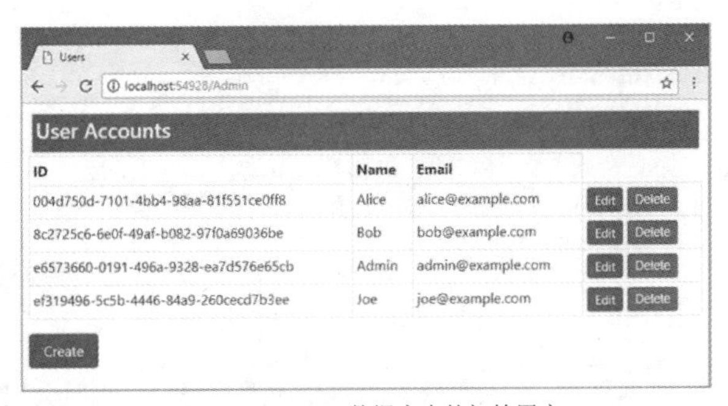

图 30-1　Identity 数据库中的初始用户

本章还需要一些角色。导航到/RoleAdmin，然后创建名为 Users 和 Employees 的角色并为这些角色分配用户，如表 30-2 所示。

表 30-2　示例应用要求的角色和成员

角色	成员
Users	Alice、Joe
Employees	Alice、Bob

图 30-2 展示了 RoleAdmin 控制器的必要角色配置。

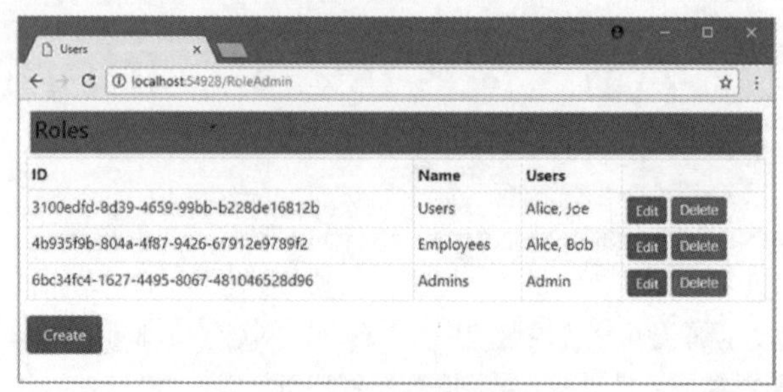

图 30-2　配置本章要求的角色

30.2　添加自定义用户属性

在第 28 章，当创建 AppUser 类来表示用户的时候，基类定义了用户的基本属性，比如电子邮件地址和电话号码。

大多数应用程序需要保存更多的用户信息，包括持久化的应用程序首选项和详细信息，比如地址。简而言之，任何运行应用程序有用且在会话之间保留的数据。由于 ASP.NET Core Identity 系统默认使用 Entity Framework Core 来保存数据，因此定义额外的用户信息意味着为 AppUser 类添加属性并使用 Entity Framework Core 创建存储它们所需的数据库模式。

代码清单 30-1 展示了如何为 AppUser 类添加两个简单属性以表示用户生活的城市和资格等级。

代码清单 30-1　在 Models 文件夹下的 AppUser.cs 文件中添加属性

```
using Microsoft.AspNetCore.Identity;

namespace Users.Models {

    public enum Cities {
        None, London, Paris, Chicago
    }

    public enum QualificationLevels {
        None, Basic, Advanced
    }

    public class AppUser : IdentityUser {
        public Cities City { get; set; }
        public QualificationLevels Qualifications { get; set; }
    }
}
```

枚举 Cities 和 QualificationLevels 定义了用于城市的值以及资格水平的不同级别。这些枚举由添加到 AppUser 类的 City 和 Qualification 属性使用。

在代码清单 30-2 中，添加到 Home 控制器的操作方法允许用户查看和编辑它们的 City 和 Qualification 属性。

代码清单 30-2　在 Controllers 文件夹下的 HomeController.cs 文件中添加对自定义用户属性的支持

```
using System.Collections.Generic;
using Microsoft.AspNetCore.Mvc;
using Microsoft.AspNetCore.Authorization;
using Users.Models;
using Microsoft.AspNetCore.Identity;
```

```
using System.Threading.Tasks;
using System.ComponentModel.DataAnnotations;

namespace Users.Controllers {
    public class HomeController : Controller {
        private UserManager<AppUser> userManager;

        public HomeController(UserManager<AppUser> userMgr) {
            userManager = userMgr;
        }

        [Authorize]
        public IActionResult Index() => View(GetData(nameof(Index)));

        [Authorize(Roles = "Users")]
        public IActionResult OtherAction() => View("Index",
            GetData(nameof(OtherAction)));

        private Dictionary<string, object> GetData(string actionName) =>
            new Dictionary<string, object> {
                ["Action"] = actionName,
                ["User"] = HttpContext.User.Identity.Name,
                ["Authenticated"] = HttpContext.User.Identity.IsAuthenticated,
                ["Auth Type"] = HttpContext.User.Identity.AuthenticationType,
                ["In Users Role"] = HttpContext.User.IsInRole("Users"),
                ["City"] = CurrentUser.Result.City,
                ["Qualification"] = CurrentUser.Result.Qualifications
            };

        [Authorize]
        public async Task<IActionResult> UserProps() {
            return View(await CurrentUser);
        }

        [Authorize]
        [HttpPost]
        public async Task<IActionResult> UserProps(
                [Required]Cities city,
                [Required]QualificationLevels qualifications) {
            if (ModelState.IsValid) {
                AppUser user = await CurrentUser;
                user.City = city;
                user.Qualifications = qualifications;
                await userManager.UpdateAsync(user);
                return RedirectToAction("Index");
            }
            return View(await CurrentUser);
        }

        private Task<AppUser> CurrentUser =>
            userManager.FindByNameAsync(HttpContext.User.Identity.Name);
    }
}
```

新的 CurrentUser 属性使用 UserManager<AppUser>类来获取表示当前用户的 AppUser 实例。AppUser 实例可作为 GET 版本的 UserProps 操作方法的视图模型，POST 方法则使用它来更新 City 和 QualificationLevel 属性的值。

已经更新的 GetData 方法则返回包含当前用户自定义属性的值的字典，这意味着这些属性的值将在 Index 和 OtherAdction 操作方法的视图中展示出来。

为了给 UserProps 操作方法提供视图，在 Views/Home 文件夹中添加名为 UserProps.cshtml 的文件，并添加代码清单 30-3 所示的标记。

代码清单 30-3　Views/Home 文件夹下的 UserProps.cshtml 文件的内容

```
@model AppUser

<div class="bg-primary m-1 p-1 text-white"><h4>@Model.UserName</h4></div>

<div asp-validation-summary="All" class="text-danger"></div>

<form asp-action="UserProps" method="post">
    <div class="form-group">
        <label asp-for="City"></label>
        <select asp-for="City" class="form-control"
                asp-items="@new SelectList(Enum.GetNames(typeof(Cities)))">
            <option disabled selected value="">Select a City</option>
        </select>
    </div>
    <div class="form-group">
        <label asp-for="Qualifications"></label>
        <select asp-for="Qualifications" class="form-control"
            asp-items="@new SelectList(Enum.GetNames(typeof(QualificationLevels)))">
                <option disabled selected value="">Select a City</option>
        </select>
    </div>

    <button type="submit" class="btn btn-primary">Submit</button>
    <a asp-action="Index" class="btn btn-secondary">Cancel</a>
</form>
```

UserProps 视图包含一个具有 select 元素的表单，其中填充了代码清单 30-1 中的枚举值。当表单被提交时，从 ASP.NET Core Identity 中获取表示当前用户的 AppUser 实例，用户的自定义属性的值被更新为当前选中的值，如下所示：

```
...
AppUser user = await CurrentUser;
user.City = city;
user.Qualifications = qualifications;
await userManager.UpdateAsync(user);
return RedirectToAction("Index");
...
```

注意，必须通过调用 UpdateAsync 方法显式地告诉用户管理器为用户更新数据库记录以反映更新。以前不必这样做是因为在更新 ASP.NET Core Identity 的方法中已经调用了 UpdateAsync 方法，但是当直接修改属性时，你有责任通知用户管理器执行更新。

30.2.1　准备数据库迁移

用以支持新增属性的所有应用内部支持已经就绪，剩下的就是更新数据库，以便其中的数据表可以存储自定义属性的值。

Entity Framework Core 没有对处理种子数据提供集成支持，在创建迁移以禁用种子的时候必须小心，如代码清单 30-4 所示；否则，代码清单 30-1 中新添加到模型类的属性将会导致错误。种子语句可以在创建和应用数据库迁移之后再次启用。

代码清单 30-4　在 Users 文件夹下的 Startup.cs 文件禁用数据库播种

```
...
public void Configure(IApplicationBuilder app) {
    app.UseStatusCodePages();
    app.UseDeveloperExceptionPage();
```

```
    app.UseStaticFiles();
    app.UseAuthentication();
    app.UseMvcWithDefaultRoute();
    //    AppIdentityDbContext.CreateAdminAccount(app.ApplicationServices,
    //        Configuration).Wait();
}
...
```

在禁用种子语句的情况下，下一步是创建新的数据库迁移文件，其中包含用来更新数据库架构的 SQL 命令。使用命令行窗口或 PowerShell 窗口在 Users 项目文件夹中执行如下命令：

```
dotnet ef migrations add CustomProperties
```

当命令完成的时候，你将会在 Migrations 文件夹中发现一个名称中含有 CustomProperties 的新文件。确切的名称中还包含数字标识，但是如果打开这个文件，你将会发现一个包含名为 Up 的方法的 C#类，Up 方法执行为支持向数据库添加自定义属性所需的 SQL 命令。另外一个名为 Down 的方法用于执行将数据库降级到上一个模式的命令。

下一步是迁移数据库到新的模式，可通过执行以下命令完成：

```
dotnet ef database update
```

当命令执行完之后，数据库中存储用户数据的表将包含新的表示自定义属性的列。

警　告

在包含真实用户数据的产品数据库上执行数据库迁移时，务必小心。创建包含删除列或整个表的迁移非常容易，这可能导致毁灭性的影响。确保彻底测试数据库迁移的影响，确保备份关键数据以防万一。

30.2.2 测试自定义属性

要测试数据库迁移的影响，可启动应用程序，并使用已标识的用户之一进行验证（例如，使用电子邮件地址 alice@example.com 和密码 secret123）。一旦验证通过，你将会看到 City 和 QualificationLevel 属性的默认值。这两个属性可以通过访问/Home/UserProps 进行修改，选择新的值，单击 Submit 按钮，这将会更新数据库并重定向到/Home，这里将显示新的值，如图 30-3 所示。

图 30-3　使用自定义用户属性

30.3 使用声明和策略

较早的用户管理系统（比如 ASP.NET Membership）是 ASP.NET Core Identity 的前身，应用程序本身假定自己是所有用户信息的权威来源，基本上把应用看作封闭的世界，只信任应用自身的数据。

这是软件开发中很久远的处理方式。在第 29 章中，当基于保存在数据库中的凭据验证用户，然后基于这些凭据授权访问时，你已经看到了一些示例。在本章中，当添加属性到 AppUser 类的时候，可以做同样的事情。管理用户的验证和授权所需的任何信息都来自应用。对于许多 Web 应用来说，这是一种完美的方式，这也是本章如此深入地演示这些技术的原因。

ASP.NET Core Identity 还支持另外一种管理用户的替代方式——使用声明，当 MVC 应用程序并不是用户信息的唯一来源时同样工作良好，相对于传统方式，以更加灵活、流畅的方式授权用户。

提　示

并一定需要使用声明，如第 29 章所述，ASP.NET Core Identity 完美提供了验证和授权服务，完全不需要理解声明。

30.3.1　声明

声明是关于用户的一段信息，它还包括关于信息来源的信息。理解声明的最简单方式是进行一些实际的演示，否则任何讨论都太过抽象而无用。为了开始演示，在 Controllers 文件夹中添加名为 ClaimsController.cs 的类文件，并用它定义代码清单 30-5 所示的控制器。

提　示

你可能对这个示例中的代码和类感到一点迷惑。现在不必担心细节，只要坚持下去，直至看到操作方法和定义的视图的输出。重要的是，这将有助于理解声明。

代码清单 30-5　Controllers 文件夹下的 ClaimsController.cs 文件的内容

```
using Microsoft.AspNetCore.Authorization;
using Microsoft.AspNetCore.Mvc;

namespace Users.Controllers {

    public class ClaimsController : Controller {

        [Authorize]
        public ViewResult Index() => View(User?.Claims);
    }
}
```

可以通过多种途径来获取用户相关的声明。User 属性（也可以通过 HttpContext.User 属性）返回一个 ClaimsPrincipal 对象，这也是这个示例中使用的方式。关联到用户的声明集合可通过 ClaimsPrincipal 类访问，该类的重要成员如表 30-3 所示。

表 30-3　ClaimsPrincipal 类的重要成员

成员	描述
Identity	获取关联到当前用户的 IIdentity
FindAll(type) FindAll(<predicate>)	返回特定类型或匹配条件谓词的声明
FindFirst(type) FindFirst(<predicate>)	返回第一个特定类型或匹配条件谓词的声明
HasClaim(type,value) HasClaim(<predicate>)	如果用户拥有特定声明类型的特定值，或者拥有匹配条件谓词的声明，就返回 true
IsInRole(name)	如果用户是特定角色的成员，就返回 true

如第 28 章所述，HttpContext.User.Identity 属性将返回一个实现了 IIdentity 接口的对象，在 ASP.NET Core

Identity 中是 ClaimsIdentity 对象。表 30-4 展示了定义在 ClaimsIdentity 类中的重要成员。

表 30-4　　定义在 ClaimsIdentity 类中的重要成员

成员	描述
Claims	返回表示用户声明的 Claim 对象的枚举
AddClaim(claim)	为用户标识添加声明
AddClaims(claims)	为用户标识添加一组声明
HasClaim(predicate)	如果用户标识包含匹配特定谓词的声明，就返回 true
RemoveClaim(claim)	从用户标识中删除声明

还有其他的方法和属性可用，但是表 30-4 中的这些经常用于 Web 应用，原因显而易见，它们演示了声明（Claim）如何适用于广泛的 ASP.NET Core 平台。

代码清单 30-5 使用 Controller.User 属性来获取 ClaimsPrincipal 对象并传递 Claims 属性值作为视图模型给默认视图。Cliam 对象表示关于用户的单片数据，Claim 类定义的属性如表 30-5 所示。

表 30-5　　Claim 类定义的属性

属性	描述
Issuer	返回提供声明的系统的名称
Subject	返回声明所表示用户的 ClaimsIdentity 对象
Type	返回声明所表示的信息类型
Value	返回声明所表示的信息片段

为了显示用户所关联声明的详情，创建 Views/Claims 文件夹，在其中创建名为 Index.cshtml 的文件，并添加代码清单 30-6 所示的标记内容。

代码清单 30-6　Views/Claims 文件夹下的 Index.cshtml 文件的内容

```
@model IEnumerable<System.Security.Claims.Claim>

<div class="bg-primary m-1 p-1 text-white"><h4>Claims</h4></div>

<table class="table table-sm table-bordered">
    <tr>
        <th>Subject</th><th>Issuer</th><th>Type</th><th>Value</th>
    </tr>
    @if (Model == null || Model.Count() == 0) {
        <tr><td colspan="4" class="text-center">No Claims</td></tr>
    } else {
        @foreach (var claim in Model.OrderBy(x => x.Type)) {
            <tr>
                <td>@claim.Subject.Name</td>
                <td>@claim.Issuer</td>
                <td identity-claim-type="@claim.Type"></td>
                <td>@claim.Value</td>
            </tr>
        }
    }
</table>
```

Index 视图使用表格来显示视图模型中提供的每个声明。Claim.Type 属性是用于微软架构的 URI，不是特别有用。常用的架构可作为 System.Security.Claims.ClaimTypes 的字段值，所以为了使 Index 视图的输出更易读，可添加自定义属性到 td 元素以显示 Type 属性的值：

```
...
<td identity-claim-type="@claim.Type"></td>
...
```

添加名为 ClaimTypeTagHelper.cs 的类文件到 Infrastructure 文件夹中，并用它创建标记助手以转换属性的值到更易读的字符串中，如代码清单 30-7 所示。

代码清单 30-7 Infrastructure 文件夹下的 ClaimTypeTagHelper.cs 文件的内容

```
using System.Linq;
using System.Reflection;
using System.Security.Claims;
using Microsoft.AspNetCore.Razor.TagHelpers;

namespace Users.Infrastructure {

    [HtmlTargetElement("td", Attributes = "identity-claim-type")]
    public class ClaimTypeTagHelper : TagHelper {
        [HtmlAttributeName("identity-claim-type")]
        public string ClaimType { get; set; }

        public override void Process(TagHelperContext context,
                                    TagHelperOutput output) {
            bool foundType = false;
            FieldInfo[] fields = typeof(ClaimTypes).GetFields();
            foreach (FieldInfo field in fields) {
                if (field.GetValue(null).ToString() == ClaimType) {
                    output.Content.SetContent(field.Name);
                    foundType = true;
                }
            }
            if (!foundType) {
                output.Content.SetContent(ClaimType.Split('/', '.').Last());
            }
        }
    }
}
```

要了解为什么只创建使用声明的控制器而不解释声明是什么，可启动应用程序，以用户 Alice 的身份进行验证（使用电子邮件地址 alice@example.com 以及密码 secret123）。一旦通过验证，就访问/Claims 来查看用户关联的声明，如图 30-4 所示。

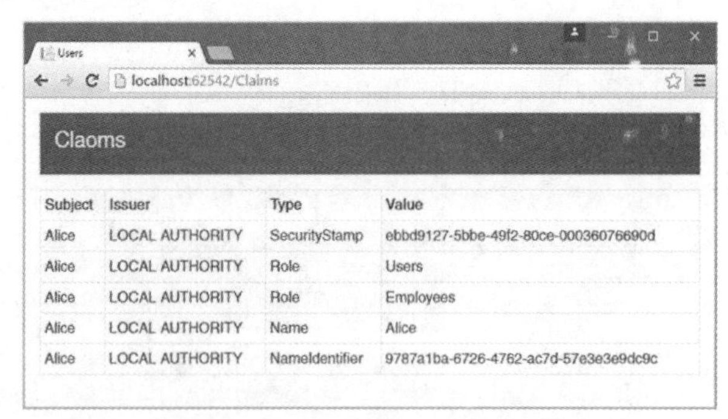

图 30-4 Claims 控制器的 Index 操作的输出

通过图 30-4 你很难弄清楚细节，所以这里重写了表 30-6 所示的内容。

表 30-6 图 30-4 中展示的内容

Subject	Issuer	Type	Value
Alice	LOCAL AUTHORITY	SecurityStamp	唯一 ID
Alice	LOCAL AUTHORITY	Role	Users
Alice	LOCAL AUTHORITY	Role	Employees
Alice	LOCAL AUTHORITY	Name	Alice
Alice	LOCAL AUTHORITY	NameIdentifier	Alice 的用户 ID

表 30-6 展示了声明的一些重要方面，第 29 章在实现传统的身份验证与授权功能时，已经使用过它们。可以看到某些声明涉及用户标识（Name 声明是 Alice，NameIdentifier 声明是 Alice 在 ASP.NET Core Identity 数据库中的唯一用户标识）。其他声明展示了角色成员，表 30-6 中有两个角色声明，反映了 Alice 被赋予 Users 和 Employees 角色。

当这些信息被表示为一组声明时，区别在于可以确定数据的来源。表 30-6 中所有声明的 Issuer 属性都被设置为 LOCAL AUTHORITY，这表示用户的身份由应用程序建立。

现在你已经看到一些示例声明，可以更容易地描述声明了：声明是任何关于用户的可用于应用程序的信息，包括用户的标识和角色成员资格，并且如你所见，在早期章节中定义的关于用户的信息由 ASP.NET Core Identity 自动提供。虽然一开始声明令人困惑，但是如同 MVC 应用程序的其他方面一样，一旦看清它们是如何工作的，它们就变得不那么困难了。

30.3.2 创建声明

有意思的是，可以从多种来源获取声明，而不仅仅依赖本地数据库中的用户信息。30.4 节会演示实际的示例，但是此刻将通过添加一个类到示例项目中来模拟提供声明信息的系统。代码清单 30-8 展示了添加到 Infrastructure 文件夹的 LocationClaimsProvider.cs 文件的内容。

代码清单 30-8 Infrastructure 文件夹下的 LocationClaimsProvider.cs 文件的内容

```
using System.Security.Claims;
using System.Threading.Tasks;
using Microsoft.AspNetCore.Authentication;

namespace Users.Infrastructure {

    public class LocationClaimsProvider : IClaimsTransformation {

        public Task<ClaimsPrincipal> TransformAsync(ClaimsPrincipal principal) {
            if (principal != null && !principal.HasClaim(c =>
                    c.Type == ClaimTypes.PostalCode)) {
                ClaimsIdentity identity = principal.Identity as ClaimsIdentity;
                if (identity != null && identity.IsAuthenticated
                        && identity.Name != null) {
                    if (identity.Name.ToLower() == "alice") {
                        identity.AddClaims(new Claim[] {
                            CreateClaim(ClaimTypes.PostalCode, "DC 20500"),
                            CreateClaim(ClaimTypes.StateOrProvince, "DC")
                        });
                    } else {
                        identity.AddClaims(new Claim[] {
                            CreateClaim(ClaimTypes.PostalCode, "NY 10036"),
                            CreateClaim(ClaimTypes.StateOrProvince, "NY")
                        });
                    }
                }
            }
            return Task.FromResult(principal);
        }

        private static Claim CreateClaim(string type, string value) =>
            new Claim(type, value, ClaimValueTypes.String, "RemoteClaims");
    }
}
```

由接口 IClaimsTransformation 定义的 TransformAsync 方法接收 ClaimsPrincipal 对象并检查有效性，然后将 Identity 属性的值转型为 ClaimsIdentity 对象。Name 属性的值用于创建关于用户邮编和州的声明。

这个类能够模拟中心化的 HR 数据库系统，作为员工位置信息的权威来源。例如，为注册声明的来源，可在 Startup 类的 ConfigureServices 方法中定义来源，如代码清单 30-9 所示。

代码清单 30-9 在 Users 文件夹下的 Startup.cs 文件中启用声明转换

```
using Microsoft.AspNetCore.Builder;
using Microsoft.Extensions.DependencyInjection;
using Microsoft.Extensions.Configuration;
using Microsoft.AspNetCore.Identity;
using Microsoft.EntityFrameworkCore;
using Users.Models;
using Users.Infrastructure;
using Microsoft.AspNetCore.Authentication;

namespace Users {

    public class Startup {

        public Startup(IConfiguration configuration) =>
            Configuration = configuration;

        public IConfiguration Configuration { get; }

        public void ConfigureServices(IServiceCollection services) {

            services.AddTransient<IPasswordValidator<AppUser>,
                CustomPasswordValidator>();
            services.AddTransient<IUserValidator<AppUser>,
                CustomUserValidator>();
            services.AddSingleton<IClaimsTransformation,
                LocationClaimsProvider>();

            services.AddDbContext<AppIdentityDbContext>(options =>
                options.UseSqlServer(
                    Configuration["Data:SportStoreIdentity:ConnectionString"]));

            services.AddIdentity<AppUser, IdentityRole>(opts => {
                opts.User.RequireUniqueEmail = true;
                //opts.User.AllowedUserNameCharacters = "abcdefghijklmnopqrstuvwxyz";
                opts.Password.RequiredLength = 6;
                opts.Password.RequireNonAlphanumeric = false;
                opts.Password.RequireLowercase = false;
                opts.Password.RequireUppercase = false;
                opts.Password.RequireDigit = false;
            }).AddEntityFrameworkStores<AppIdentityDbContext>()
                .AddDefaultTokenProviders();

            services.AddMvc();
        }

        public void Configure(IApplicationBuilder app) {
            app.UseStatusCodePages();
            app.UseDeveloperExceptionPage();
            app.UseStaticFiles();
            app.UseAuthentication();
            app.UseMvcWithDefaultRoute();
        }
    }
}
```

每当一个请求被接收时，声明转换中间件就调用 LocationClaimsProvider.AddClaims 方法，模拟 HR 数据源并创建自定义声明。可以通过启动应用程序，以用户身份验证，然后访问/Claim 来查看自定义声明的效果。图 30-5 展示了 Alice 的声明。你可能需要先注销再重新登录才能查看变化。

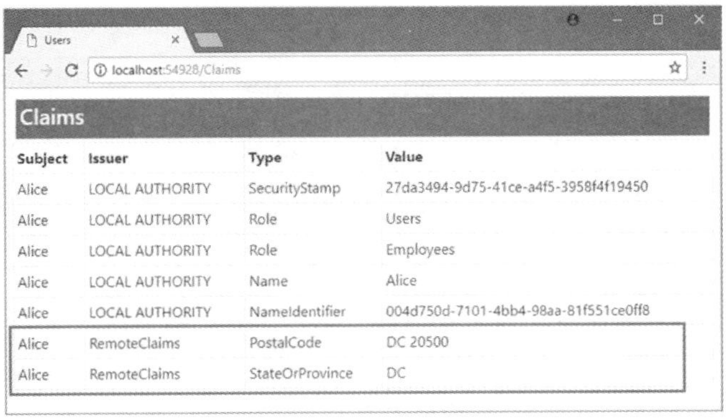

Claims

Subject	Issuer	Type	Value
Alice	LOCAL AUTHORITY	SecurityStamp	27da3494-9d75-41ce-a4f5-3958f4f19450
Alice	LOCAL AUTHORITY	Role	Users
Alice	LOCAL AUTHORITY	Role	Employees
Alice	LOCAL AUTHORITY	Name	Alice
Alice	LOCAL AUTHORITY	NameIdentifier	004d750d-7101-4bb4-98aa-81f551ce0ff8
Alice	RemoteClaims	PostalCode	DC 20500
Alice	RemoteClaims	StateOrProvince	DC

图 30-5 为用户定义额外的声明

从多个位置获取声明意味着应用程序不必重复在其他地方已经持有的数据，并允许集成外部数据。Claim.Issuer 属性指出声明来自何方，这有助于判断数据的准确程度，以及在应用程序中赋予的权重。从中心化的 HR 数据库中获取的位置信息可能比从外部邮件列表供应商获取的数据更准确和可靠。

创建自定义的标识声明

如果希望向应用程序添加自定义的本地声明，可以在创建新用户时执行此操作。UserManager<T>类提供了可用于定义本地声明的 AddClaimAsync 和 AddClaimsAsync 方法，随后声明会存储在数据库中，并在用户被验证时自动提取（这意味着不需要依赖与声明转换特性）。但是，在使用这些方法之前，请考虑如何将数据存储在最新状态，以及如何从数据源中动态检索数据以更好地服务于应用程序。声明用于授权检查，而陈旧的声明数据允许用户访问本应该被禁止访问的应用程序区域，并阻止已经被授权访问的应用程序区域。

30.3.3 使用策略

一旦一些声明可用，就可以用相比标准角色更为灵活的方式来管理用户对应用程序的访问。角色的问题在于它们是静态的，一旦用户被分配到角色，用户就保持成员直到角色被显式移除。例如，大公司中的长期雇员最终获得内部系统难以置信的访问权限的原因，就在于他们被赋予每个新职位所需的角色，但是旧的角色很少被删除。

声明用于构建授权策略，它们是应用程序配置的一部分，并将 Authorize 特性应用于操作方法或控制器。代码清单 30-10 展示的简单策略只允许拥有特定声明类型和值的用户访问。

代码清单 30-10 在 Users 文件夹下的 Startup.cs 文件中创建声明策略

```
using Microsoft.AspNetCore.Builder;
using Microsoft.Extensions.DependencyInjection;
using Microsoft.Extensions.Configuration;
using Microsoft.AspNetCore.Identity;
using Microsoft.EntityFrameworkCore;
using Users.Models;
using Users.Infrastructure;
using Microsoft.AspNetCore.Authentication;
using System.Security.Claims;

namespace Users {

    public class Startup {
```

```
        public Startup(IConfiguration configuration) =>
            Configuration = configuration;

        public IConfiguration Configuration { get; }

        public void ConfigureServices(IServiceCollection services) {

            services.AddTransient<IPasswordValidator<AppUser>,
                CustomPasswordValidator>();
            services.AddTransient<IUserValidator<AppUser>,
                CustomUserValidator>();
            services.AddSingleton<IClaimsTransformation, LocationClaimsProvider>();

            services.AddAuthorization(opts => {
                opts.AddPolicy("DCUsers", policy => {
                    policy.RequireRole("Users");
                    policy.RequireClaim(ClaimTypes.StateOrProvince, "DC");
                });
            });

            services.AddDbContext<AppIdentityDbContext>(options =>
                options.UseSqlServer(
                    Configuration["Data:SportStoreIdentity:ConnectionString"]));
            services.AddIdentity<AppUser, IdentityRole>(opts => {
                opts.User.RequireUniqueEmail = true;
                //opts.User.AllowedUserNameCharacters = "abcdefghijklmnopqrstuvwxyz";
                opts.Password.RequiredLength = 6;
                opts.Password.RequireNonAlphanumeric = false;
                opts.Password.RequireLowercase = false;
                opts.Password.RequireUppercase = false;
                opts.Password.RequireDigit = false;
            }).AddEntityFrameworkStores<AppIdentityDbContext>()
                .AddDefaultTokenProviders();

            services.AddMvc();
        }

        public void Configure(IApplicationBuilder app) {
            app.UseStatusCodePages();
            app.UseDeveloperExceptionPage();
            app.UseStaticFiles();
            app.UseAuthentication();
            app.UseMvcWithDefaultRoute();
        }
    }
}
```

AddAuthorization 方法用于设置授权策略并提供 AuthorizationOptions 对象，AuthorizationOption 类定义的成员如表 30-7 所示。

表 30-7 AuthorizationOptions 类定义的成员

成员	描述
DefaultPolicy	返回默认的授权策略，用于在没有使用任何参数的情况下使用 Authorize 特性时。默认情况下，检查用户是否已经被验证
AddPolicy(name,expression)	用于定义新策略

策略是使用 AddPolicy 方法定义的，可在 AuthorizationPolicyBuilder 对象上使用表 30-8 描述的方法逐步构建策略。

表 30-8 AuthorizationPolicyBuilder 类定义的成员方法

方法	描述
RequireAuthenticatedUser()	要求请求关联于已验证的用户
RequireUserName(name)	要求请求关联于特定用户
RequireClaim(type)	要求用户拥有特定类型的声明。只要求特定类型的声明存在，声明的任何值都可以接收
RequireClaim(type,values)	要求用户拥有特定类型的声明，且拥有特定范围内的某个值。值可以表示为使用逗号分隔的参数，或是 IEnumerable<string >类型值
RequireRole(roles)	要求用户是角色的成员。多个角色可以使用逗号分隔的参数或 IEnumerable<string >表示，任何角色的成员都可以满足要求
AddRequirements(requirement)	用于添加自定义的需求到策略中

代码清单 30-10 中的策略要求用户拥有 Users 角色以及带 DC 值的 StateOrProvince 声明。当有多个要求时，必须满足所有的要求才能授权。

AddPolicy 方法的第一个参数是应用策略时引用的策略名称。在代码清单 30-10 中，策略的名称是 DCUsers。在代码清单 30-11 中，这个名称由 Home 控制器中的 Authorize 特性使用以应用策略。

代码清单 30-11 在 Controllers 文件夹下的 HomeController.cs 文件中应用授权策略

```
using System.Collections.Generic;
using Microsoft.AspNetCore.Mvc;
using Microsoft.AspNetCore.Authorization;
using Users.Models;
using Microsoft.AspNetCore.Identity;
using System.Threading.Tasks;
using System.ComponentModel.DataAnnotations;

namespace Users.Controllers {

    public class HomeController : Controller {
        private UserManager<AppUser> userManager;

        public HomeController(UserManager<AppUser> userMgr) {
            userManager = userMgr;
        }

        [Authorize]
        public IActionResult Index() => View(GetData(nameof(Index)));
        //[Authorize(Roles = "Users")]
        [Authorize(Policy = "DCUsers")]
        public IActionResult OtherAction() => View("Index",
            GetData(nameof(OtherAction)));

        // ...other methods omitted for brevity...

        private Task<AppUser> CurrentUser =>
            userManager.FindByNameAsync(HttpContext.User.Identity.Name);
    }
}
```

Policy 属性用于指定将用于保护操作方法的策略名称。结果就是当请求针对 OtherAction 方法时，将对用户拥有的角色和声明进行组合检查。只有在 Alice 拥有角色成员和声明的正确组合时，才可以通过运行应用程序进行检查，以不同的身份进行验证，并访问/Home/OtherAction。

创建自定义的策略需求

内置的策略需求会检查特定的值，这是很好的起点，但是并不能处理所有的授权场景。如果某个声明值的访问应当被禁止，那么事情就会变得棘手，内置的策略需求不是为应对这种检查而建立的。

幸运的是，可以通过自定义需求来扩展策略系统，它们是实现了接口 IAuthorizationRequirement 并自定义授权处理的类，也是评估给定请求的需求的 AuthorizationHandler 类的子类。为了演示，在 Infrastructure 文件夹中添加名为 BlockUsersRequirement.cs 的类文件，如代码清单 30-12 所示，用它定义自定义需求和处理程序。

代码清单 30-12 Infrastructure 文件夹下的 BlockUsersRequirement.cs 文件的内容

```
using System;
using System.Linq;
using System.Threading.Tasks;
using Microsoft.AspNetCore.Authorization;

namespace Users.Infrastructure {

    public class BlockUsersRequirement : IAuthorizationRequirement {

        public BlockUsersRequirement(params string[] users) {
            BlockedUsers = users;
        }

        public string[] BlockedUsers { get; set; }
    }

    public class BlockUsersHandler : AuthorizationHandler<BlockUsersRequirement> {

        protected override Task HandleRequirementAsync(
                AuthorizationHandlerContext context,
                BlockUsersRequirement requirement) {
            if (context.User.Identity != null && context.User.Identity.Name != null
                && !requirement.BlockedUsers
                    .Any(user => user.Equals(context.User.Identity.Name,
                        StringComparison.OrdinalIgnoreCase))) {
                context.Succeed(requirement);
            } else {
                context.Fail();
            }
            return Task.CompletedTask;
        }
    }
}
```

BlockUserRequirement 类是需求指定用来创建策略的数据，在这个示例中是不会被授权访问的用户列表。BlockUsersHandler 类的职责是使用需求数据评估授权中的请求，该类派生自 AuthorizationHandler<T>类，T 是需求的类型。

当授权系统需要检查对资源的访问时，处理程序中的 Handle 方法被调用。这个方法的参数是 AuthorizationHandlerContext 对象，其中的重要成员如表 30-9 所示。

表 30-9 AuthorizationHandlerContext 对象中的重要成员

成员	描述
User	返回请求关联的 ClaimsPrincipal 对象
Succeed(requirement)	如果请求符合需求，这个方法将被调用，参数对象 IAuthorizationRequirement 由 Handle 方法接收
Fail()	如果请求不符合需求，这个方法将被调用
Resource	返回用于授权访问单个应用程序资源的对象

代码清单 30-12 中的需求处理程序检查用户的名称是否在 BlockUsersRequirement 对象提供的被禁止的名单

中，并分别调用 Succeed 或 Fail 方法。应用自定义的授权需求需要执行两处配置变更，如代码清单 30-13 所示。

代码清单 30-13 在 Users 文件夹下的 Startup.cs 文件中应用自定义的授权需求

```
using Microsoft.AspNetCore.Builder;
using Microsoft.Extensions.DependencyInjection;
using Microsoft.Extensions.Configuration;
using Microsoft.AspNetCore.Identity;
using Microsoft.EntityFrameworkCore;
using Users.Models;
using Users.Infrastructure;
using Microsoft.AspNetCore.Authentication;
using System.Security.Claims;
using Microsoft.AspNetCore.Authorization;
namespace Users {

    public class Startup {

        public Startup(IConfiguration configuration) =>
            Configuration = configuration;

        public IConfiguration Configuration { get; }

        public void ConfigureServices(IServiceCollection services) {

            services.AddTransient<IPasswordValidator<AppUser>,
                CustomPasswordValidator>();
            services.AddTransient<IUserValidator<AppUser>,
                CustomUserValidator>();
            services.AddSingleton<IClaimsTransformation, LocationClaimsProvider>();
            services.AddTransient<IAuthorizationHandler, BlockUsersHandler>();

            services.AddAuthorization(opts => {
                opts.AddPolicy("DCUsers", policy => {
                    policy.RequireRole("Users");
                    policy.RequireClaim(ClaimTypes.StateOrProvince, "DC");
                });
                opts.AddPolicy("NotBob", policy => {
                    policy.RequireAuthenticatedUser();
                    policy.AddRequirements(new BlockUsersRequirement("Bob"));
                });
            });

            services.AddDbContext<AppIdentityDbContext>(options =>
                options.UseSqlServer(
                    Configuration["Data:SportStoreIdentity:ConnectionString"]));

            services.AddIdentity<AppUser, IdentityRole>(opts => {
                opts.User.RequireUniqueEmail = true;
                //opts.User.AllowedUserNameCharacters = "abcdefghijklmnopqrstuvwxyz";
                opts.Password.RequiredLength = 6;
                opts.Password.RequireNonAlphanumeric = false;
                opts.Password.RequireLowercase = false;
                opts.Password.RequireUppercase = false;
                opts.Password.RequireDigit = false;
            }).AddEntityFrameworkStores<AppIdentityDbContext>()
                .AddDefaultTokenProviders();

            services.AddMvc();
        }
```

```
        public void Configure(IApplicationBuilder app) {
            app.UseStatusCodePages();
            app.UseDeveloperExceptionPage();
            app.UseStaticFiles();
            app.UseAuthentication();
            app.UseMvcWithDefaultRoute();
        }
    }
}
```

第一步是将处理程序类与服务提供者注册为 IAuthorizationHandler 接口的实现。第二步是使用 AddRequirements 方法将自定义的需求添加到策略中，如下所示：

```
...
opts.AddPolicy("NotBob", policy => {
    policy.RequireAuthenticatedUser();
    policy.AddRequirements(new BlockUsersRequirement("Bob"));
});
...
```

得到的结果是一个策略，它要求用户不是 Bob，并且可以使用 Authorize 特性通过指定策略名称来应用，如代码清单 30-14 所示。

代码清单 30-14 在 HomeController.cs 文件中应用自定义策略

```
...
//[Authorize(Roles = "Users")]
[Authorize(Policy = "DCUsers")]
public IActionResult OtherAction() => View("Index", GetData(nameof(OtherAction)));

[Authorize(Policy = "NotBob")]
public IActionResult NotBob() => View("Index", GetData(nameof(NotBob)));
...
```

如果以 Bob 身份进行验证，将不能访问 URL 地址/Home/NotBob，但是其他的账户将被授权访问。

30.3.4 使用策略对资源授权访问

策略也可以对单个资源进行授权访问，资源是应用程序中任意数据项的广义术语，并且相比操作方法级别需要更精细的管理。作为演示，将一个名为 ProtectedDocument.cs 的类文件添加到 Models 文件夹中，用它定义一个类，以表示具有所有权属性的文档，如代码清单 30-15 所示。

代码清单 30-15 Models 文件夹下的 ProtectedDocument.cs 文件的内容

```
namespace Users.Models {

    public class ProtectedDocument {
        public string Title { get; set; }
        public string Author { get; set; }
        public string Editor { get; set; }
    }
}
```

对于真实文档来说，这只是一个占位符，关键是每个文档只能由两种人修改，分别是作者和编辑。真实的文档需要内容以及变更跟踪和许多其他的特性，但是对于这个示例已经足够了。在 Controllers 文件夹中添加名为 DocumentController.cs 的类文件，并用它定义代码清单 30-16 所示的控制器。

代码清单 30-16 Controllers 文件夹下的 DocumentController.cs 文件的内容

```
using Microsoft.AspNetCore.Authorization;
using Microsoft.AspNetCore.Mvc;
using System.Linq;
```

```
using Users.Models;

namespace Users.Controllers {

    [Authorize]
    public class DocumentController : Controller {
        private ProtectedDocument[] docs = new ProtectedDocument[] {
            new ProtectedDocument { Title = "Q3 Budget", Author = "Alice",
                Editor = "Joe"},
            new ProtectedDocument { Title = "Project Plan", Author = "Bob",
                Editor = "Alice"}
        };

        public ViewResult Index() => View(docs);

        public ViewResult Edit(string title) {
            return View("Index", docs.FirstOrDefault(d => d.Title == title));
        }
    }
}
```

Document 控制器维护一个固定的 ProtectedDocument 对象集合，ProtectedDocument 对象用在 Index 操作方法中，并传递所有的文档给 View 方法。Edit 操作方法基于 title 参数选择文档。这两个操作方法都使用名为 Index.cshtml 的视图文件，将它添加到新创建的名为 Views/Document 的文件夹中，内容如代码清单 30-17 所示。

代码清单 30-17　Views/Document 文件夹下的 Index.cshtml 文件的内容

```
@if (Model is IEnumerable<ProtectedDocument>) {
    <div class="bg-primary m-1 p-1 text-white">
        <h4>Documents (@User?.Identity?.Name)</h4>
    </div>
    <table class="table table-sm table-bordered">
        <tr><th>Title</th><th>Author</th><th>Editor</th><th></th></tr>
        @foreach (var doc in Model) {
            <tr>
                <td>@doc.Title</td>
                <td>@doc.Author</td>
                <td>@doc.Editor</td>
                <td>
                    <a class="btn btn-sm btn-primary" asp-action="Edit"
                       asp-route-title="@doc.Title">
                        Edit
                    </a>
                </td>
            </tr>
        }
    </table>
} else {
    <div class="bg-primary m-1 p-1">
        <h4>Editing @Model.Title (@User?.Identity?.Name)</h4>
    </div>
    <div class="m-1 p-1">
        Document editing feature would go here...
    </div>
    <a asp-action="Index" class="btn btn-primary">Done</a>
}
<a asp-action="Logout" asp-controller="Account" class="btn btn-danger">Logout</a>
```

如果视图模型是 ProtectedDocument 对象序列，那么 Index 视图将以表格的形式，在每行中显示一个文档、作者和编辑的姓名以及用于链接到 Edit 操作的链接。如果视图模型是单个 ProtectedDocument 对象，那么 Index 视图将显示一些占位符，用于为应用程序提供编辑功能。

此刻，只有应用于 DocumentController 类的 Authorize 特性的授权限制，这意味着任何用户都可以编辑任意文档，而不仅仅是作者和编辑。可以通过执行应用程序来查看，访问/Document，以任意应用程序用户进行验证，然后单击文档的 Edit 按钮。例如，图 30-6 表明用户 Joe 可以编辑 Project Plan 文档。

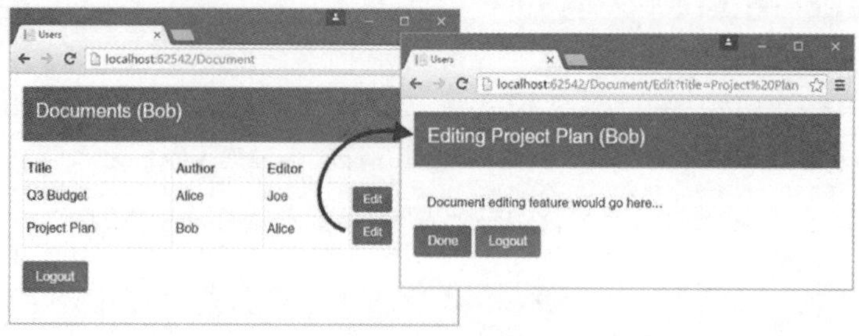

图 30-6　编辑文档

创建资源授权策略和处理程序

很难在操作方法级别限制对单个文档的访问，因为 Authorize 特性是在操作方法执行前评估的。这意味着在提取和检查 ProtectedDocument 对象之前，哪些用户应该被允许访问文档的决定已经做完。

这个问题的解决方案是创建授权策略和处理程序，它们知道如何处理 ProtectedDocument 对象，并在用户的详情被展现之后在操作方法中使用它们。为了演示，在 Infrastructure 文件夹中添加名为 DocumentAuthorization.cs 的类文件，并用它定义代码清单 30-18 所示的类。

代码清单 30-18　Infrastructure 文件夹夹下的 DocumentAuthorization.cs 文件的内容

```
using System;
using System.Threading.Tasks;
using Microsoft.AspNetCore.Authorization;
using Users.Models;

namespace Users.Infrastructure {

    public class DocumentAuthorizationRequirement : IAuthorizationRequirement {
        public bool AllowAuthors { get; set; }
        public bool AllowEditors { get; set; }
    }

    public class DocumentAuthorizationHandler
        : AuthorizationHandler<DocumentAuthorizationRequirement> {

        protected override Task HandleRequirementAsync(
                AuthorizationHandlerContext context,
                DocumentAuthorizationRequirement requirement) {
            ProtectedDocument doc = context.Resource as ProtectedDocument;
            string user = context.User.Identity.Name;
            StringComparison compare = StringComparison.OrdinalIgnoreCase;
            if (doc != null && user != null &&
                (requirement.AllowAuthors && doc.Author.Equals(user, compare))
                || (requirement.AllowEditors && doc.Editor.Equals(user, compare))) {
                context.Succeed(requirement);
            } else {
                context.Fail();
            }
            return Task.CompletedTask;
        }
    }
}
```

AuthorizationHandlerContext 对象提供了 Resource 属性，从而提供对可以检查授权的对象的访问。DocumentAuthorizationHandler 类检查 Resource 属性是否为 ProtectedDocument 对象，如果是，就检查当前用户是否为作者和编辑，以及 DocumentAuthorizationRequirement 对象是否允许编辑或作者访问文档。

代码清单 30-19 已经注册 DocumentAuthorizationHandler 类作为 DocumentAuthorizationRequirement 需求的处理类，并定义了具有这一需求的策略。

代码清单 30-19 在 Users 文件夹下的 Startup.cs 文件中注册处理程序和定义策略

```
...
public void ConfigureServices(IServiceCollection services) {

    services.AddTransient<IPasswordValidator<AppUser>,
        CustomPasswordValidator>();
    services.AddTransient<IUserValidator<AppUser>,
        CustomUserValidator>();
    services.AddSingleton<IClaimsTransformation, LocationClaimsProvider>();
    services.AddTransient<IAuthorizationHandler, BlockUsersHandler>();
    services.AddTransient<IAuthorizationHandler, DocumentAuthorizationHandler>();

    services.AddAuthorization(opts => {
        opts.AddPolicy("DCUsers", policy => {
            policy.RequireRole("Users");
            policy.RequireClaim(ClaimTypes.StateOrProvince, "DC");
        });
        opts.AddPolicy("NotBob", policy => {
            policy.RequireAuthenticatedUser();
            policy.AddRequirements(new BlockUsersRequirement("Bob"));
        });
        opts.AddPolicy("AuthorsAndEditors", policy => {
            policy.AddRequirements(new DocumentAuthorizationRequirement {
                AllowAuthors = true,
                AllowEditors = true
            });
        });
    });

    services.AddDbContext<AppIdentityDbContext>(options =>
        options.UseSqlServer(
            Configuration["Data:SportStoreIdentity:ConnectionString"]));

    services.AddIdentity<AppUser, IdentityRole>(opts => {
        opts.User.RequireUniqueEmail = true;
        //opts.User.AllowedUserNameCharacters = "abcdefghijklmnopqrstuvwxyz";
        opts.Password.RequiredLength = 6;
        opts.Password.RequireNonAlphanumeric = false;
        opts.Password.RequireLowercase = false;
        opts.Password.RequireUppercase = false;
        opts.Password.RequireDigit = false;
    }).AddEntityFrameworkStores<AppIdentityDbContext>()
        .AddDefaultTokenProviders();

    services.AddMvc();
}
...
```

最后的步骤是为操作方法应用授权策略，如代码清单 30-20 所示。

代码清单 30-20 在 Controllers 文件夹下的 DocumentController.cs 文件中应用授权策略

```
using Microsoft.AspNetCore.Authorization;
using Microsoft.AspNetCore.Mvc;
using System.Linq;
using Users.Models;
using System.Threading.Tasks;

namespace Users.Controllers {

    [Authorize]
```

```
    public class DocumentController : Controller {
        private ProtectedDocument[] docs = new ProtectedDocument[] {
            new ProtectedDocument { Title = "Q3 Budget", Author = "Alice",
              Editor = "Joe"},
            new ProtectedDocument { Title = "Project Plan", Author = "Bob",
                Editor = "Alice"}
        };
        private IAuthorizationService authService;

        public DocumentController(IAuthorizationService auth) {
            authService = auth;
        }

        public ViewResult Index() => View(docs);

        public async Task<IActionResult> Edit(string title) {
            ProtectedDocument doc = docs.FirstOrDefault(d => d.Title == title);
            AuthorizationResult authorized = await authService.AuthorizeAsync(User,
                doc, "AuthorsAndEditors");
            if (authorized.Succeeded) {
                return View("Index", doc);
            } else {
                return new ChallengeResult();
            }
        }
    }
}
```

DocumentController 控制器类的构造函数定义了 IAuthorizationService 参数，从而提供了用于评估授权策略的方法，并通过依赖注入提供。在 Edit 操作方法中，调用 AuthorizeAsync 方法，传入当前的用户、ProtectedDocument 对象以及应用的策略名称。如果 AuthorizeAsync 方法返回 true，授权被批准，然后 View 方法被调用；如果返回 false，就表示存在授权问题，返回一个 ChallengeResult 对象，如第 17 章所述，以告诉 MVC 验证失败。

可以通过执行应用程序并访问 URL 地址/Document 来查看效果，以不同的用户身份进行验证，例如以 Joe 身份进行验证，你将能够编辑预算文档，但是不能编辑项目计划文档。

30.4 使用第三方验证

基于声明的系统（如 ASP.NET Core Identity）的优势之一是，任何声明都可能来自外部系统，甚至来自识别用户的应用程序。这意味着其他系统可以代表应用程序对用户进行身份验证，ASP.NET Core Identity 就建立在这种想法之上，可以简单且容易地添加通过 Microsoft、Google、Facebook 和 Twitter 等第三方对用户进行身份验证的功能。

使用第三方验证有很多好处：许多用户已经拥有账户，用户可以选择使用双因子验证，而不必在应用中管理用户凭据。随后的章节将展示如何为 Google 用户设置和使用第三方身份验证。

30.4.1 注册 Google 应用

在可以验证用户身份之前，第三方身份验证服务通常需要注册应用程序。注册结果是包含在针对第三方服务的认证请求中的凭据。注册过程可在 Google 开发人员网站上进行，请按照其中的说明进行操作。必须指定回调地址，默认配置为/signin-google。如果正在开发，可设置回调地址为 http://localhost:port/signin-google。对于生产型应用，可创建包含公共主机名和端口的 URL 地址。

完成注册过程后，你将收到客户端 ID，用于将应用程序标识给 Google，还将收到客户密钥，以防止其他应用程序伪装成你的应用程序。

注 意

你必须注册自己的应用程序，并使用注册过程生成的客户端标识（客户端 ID）和密钥。除非使用应用程序的唯一值更改凭据，否则代码无法工作。微软提供了名为用户密钥（user secret）的功能，允许你将安全信息存储到外部配置文件中，但为了简单起见，将假设你的配置文件不是共享的，并且可以安全地包含 Google 身份验证凭据。

30.4.2 启用 Google 验证

ASP.NET Core Identity 内置支持通过 Microsoft、Google、Facebook 和 Twitter 账号对用户身份进行验证，还提供对支持 OAuth 的任意身份验证服务的普遍支持。每种服务都有自己的在 Startup 类中注册到应用的扩展方法，代码清单 30-21 展示了 Google 服务是如何设置的。为了简洁起见，这里删除了上一示例中的配置语句。

代码清单 30-21 在 Users 文件夹下的 Startup.cs 文件中启用 Google 验证

```
...
public void ConfigureServices(IServiceCollection services) {

    services.AddTransient<IPasswordValidator<AppUser>,
        CustomPasswordValidator>();
    services.AddTransient<IUserValidator<AppUser>,
        CustomUserValidator>();
    services.AddSingleton<IClaimsTransformation, LocationClaimsProvider>();
    services.AddTransient<IAuthorizationHandler, BlockUsersHandler>();
    services.AddTransient<IAuthorizationHandler, DocumentAuthorizationHandler>();
    services.AddAuthorization(opts => {
        opts.AddPolicy("DCUsers", policy => {
            policy.RequireRole("Users");
            policy.RequireClaim(ClaimTypes.StateOrProvince, "DC");
        });
        opts.AddPolicy("NotBob", policy => {
            policy.RequireAuthenticatedUser();
            policy.AddRequirements(new BlockUsersRequirement("Bob"));
        });
        opts.AddPolicy("AuthorsAndEditors", policy => {
            policy.AddRequirements(new DocumentAuthorizationRequirement {
                AllowAuthors = true,
                AllowEditors = true
            });
        });
    });

    services.AddAuthentication().AddGoogle(opts => {
        opts.ClientId = "<enter client id here>";
        opts.ClientSecret = "<enter client secret here>";
    });

    services.AddDbContext<AppIdentityDbContext>(options =>
        options.UseSqlServer(
            Configuration["Data:SportStoreIdentity:ConnectionString"]));

    services.AddIdentity<AppUser, IdentityRole>(opts => {
        opts.User.RequireUniqueEmail = true;
        opts.Password.RequiredLength = 6;
        opts.Password.RequireNonAlphanumeric = false;
        opts.Password.RequireLowercase = false;
        opts.Password.RequireUppercase = false;
        opts.Password.RequireDigit = false;
    }).AddEntityFrameworkStores<AppIdentityDbContext>()
        .AddDefaultTokenProviders();
```

```
        services.AddMvc();
    }
    ...
```

AddAuthentication.AddGoogle 方法设置了使用 Google 验证用户时必要的服务，以及在注册过程中创建的客户端 ID 和用户密钥。

当使用第三方验证用户时，可以选择在 Identity 数据库中创建用户，然后用于管理角色和声明，就像普通用户一样。在第 28 章中，添加了用户验证类，如果电子邮件地址不在 example.com 域中，则可以阻止创建用户。由于要处理来自任意域的所有用户，因此必须在该例中禁用电子邮件地址检查，如代码清单 30-22 所示。

代码清单 30-22 在 Infrastructure 文件夹下的 CustomUserValidator.cs 文件中禁用验证

```
using System.Collections.Generic;
using System.Linq;
using System.Threading.Tasks;
using Microsoft.AspNetCore.Identity;
using Users.Models;

namespace Users.Infrastructure {

    public class CustomUserValidator : UserValidator<AppUser> {

        public override async Task<IdentityResult> ValidateAsync(
                UserManager<AppUser> manager,
                AppUser user) {

            IdentityResult result = await base.ValidateAsync(manager, user);

            List<IdentityError> errors = result.Succeeded ?
                new List<IdentityError>() : result.Errors.ToList();

            //if (!user.Email.ToLower().EndsWith("@example.com")) {
            //    errors.Add(new IdentityError {
            //        Code = "EmailDomainError",
            //        Description = "Only example.com email addresses are allowed"
            //    });
            //}

            return errors.Count == 0 ? IdentityResult.Success
                : IdentityResult.Failed(errors.ToArray());
        }
    }
}
```

然后在 Views/Account/Login.cshtml 文件中添加一个按钮，以允许用户通过 Google 登录，如代码清单 30-23 所示。Google 为按钮提供了图像，使它们与支持 Google 账户的其他应用程序保持一致。但为了简单起见，这里仅创建一个标准按钮。

代码清单 30-23 在 Views/Account 文件夹下的 Login.cshtml 文件中添加按钮

```
@model LoginModel

<div class="bg-primary m-1 p-1 text-white"><h4>Log In</h4></div>

<div class="text-danger" asp-validation-summary="All"></div>

<form asp-action="Login" method="post">
    <input type="hidden" name="returnUrl" value="@ViewBag.returnUrl" />
    <div class="form-group">
        <label asp-for="Email"></label>
```

```
            <input asp-for="Email" class="form-control" />
        </div>
        <div class="form-group">
            <label asp-for="Password"></label>
            <input asp-for="Password" class="form-control" />
        </div>
        <button class="btn btn-primary" type="submit">Log In</button>
        <a class="btn btn-info" asp-action="GoogleLogin"
           asp-route-returnUrl="@ViewBag.returnUrl">
            Log In With Google
        </a>
    </form>
```

新按钮指向 Account 控制器的 GoogleLogin 操作方法。可以看看这个操作方法，以及对控制器所做的其他更改，如代码清单 30-24 所示。

代码清单 30-24 在 Controllers 文件夹下的 AccountController.cs 文件中添加对 Google 验证的支持

```
using System.Threading.Tasks;
using Microsoft.AspNetCore.Authorization;
using Microsoft.AspNetCore.Mvc;
using Users.Models;
using Microsoft.AspNetCore.Identity;
using System.Security.Claims;
using Microsoft.AspNetCore.Http.Authentication;

namespace Users.Controllers {

    [Authorize]
    public class AccountController : Controller {
        private UserManager<AppUser> userManager;
        private SignInManager<AppUser> signInManager;

        // ...methods omitted for brevity...

        [AllowAnonymous]
        public IActionResult GoogleLogin(string returnUrl) {
            string redirectUrl = Url.Action("GoogleResponse", "Account",
                new { ReturnUrl = returnUrl });
            var properties = signInManager
                .ConfigureExternalAuthenticationProperties("Google", redirectUrl);
            return new ChallengeResult("Google", properties);
        }

        [AllowAnonymous]
        public async Task<IActionResult> GoogleResponse(string returnUrl = "/") {
            ExternalLoginInfo info = await signInManager.GetExternalLoginInfoAsync();
            if (info == null) {
                return RedirectToAction(nameof(Login));
            }
            var result = await signInManager.ExternalLoginSignInAsync(
                info.LoginProvider, info.ProviderKey, false);
            if (result.Succeeded) {
                return Redirect(returnUrl);
            } else {
                AppUser user = new AppUser {
                    Email = info.Principal.FindFirst(ClaimTypes.Email).Value,
                    UserName =
                        info.Principal.FindFirst(ClaimTypes.Email).Value
                };
                IdentityResult identResult = await userManager.CreateAsync(user);
                if (identResult.Succeeded) {
                    identResult = await userManager.AddLoginAsync(user, info);
                    if (identResult.Succeeded) {
                        await signInManager.SignInAsync(user, false);
```

```
                    return Redirect(returnUrl);
                }
            }
            return AccessDenied();
        }
    }
  }
}
```

GoogleLogin 操作方法创建了一个 AuthenticationProperties 实例，并将 RedirectUri 属性设置为同一控制器中的 GoogleResponse 操作方法的 URL。以下代码导致 ASP.NET Core Identity 通过将用户重定向到 Google 身份验证页面而不是应用程序定义的内容来响应未授权错误：

```
...
return new ChallengeResult("Google", properties);
...
```

这意味着当用户通过单击 Log In 按钮登录时，他们的浏览器将被重定向到 Google 验证服务，然后一旦被验证，就被重定向到 GoogleResponse 操作方法。在 GoogleResponse 操作方法中，可通过调用 SigninManager 的 GetExternalLoginInfoAsync 方法来获取外部登录的详细信息，如下所示：

```
...
ExternalLoginInfo info = await signInManager.GetExternalLoginInfoAsync();
...
```

ExternalLoginInfo 类定义了 ExternalPrincipal 属性，该属性返回一个 ClaimsPrincipal 对象，该对象包含由 Google 为用户生成的声明。可使用 ExternalLoginSignInAsync 方法让用户登录应用程序，如下所示：

```
...
var result = await signInManager.ExternalLoginSignInAsync(
                info.LoginProvider, info.ProviderKey, false);
...
```

登录失败，这是因为数据库中没有代表 Google 用户的账户，可创建新用户并将 Google 账户与之关联，方法是使用如下两条语句：

```
...
IdentityResult identResult = await userManager.CreateAsync(user);
...
identResult = await userManager.AddLoginAsync(user, info);
...
```

注 意

当创建 Identity 用户时，可使用 Google 提供的电子邮件声明并用于 AppUser 对象的 Email 和 UserName 属性，以便不与数据库中现存的任何用户发生命名冲突。

为了测试验证，启动应用程序，单击 Log In 按钮进行登录，并提供有效的 Google 账户凭据。完成验证之后，浏览器将被重定向到应用程序。

30.5 小结

本章展示了 ASP.NET Core Identity 支持的一些高级功能，演示了使用自定义属性以及如何使用数据库迁移来更新数据库模式以执行它们，解释了声明如何工作以及如何使用它们通过策略来创建更灵活的授权用户，还解释了如何使用策略来控制对应用程序管理单个资源的访问，以及如何使用 Google 验证用户。下一章将介绍如何实现 MVC 应用程序中使用的一些最重要的约定，以及如何在自己的应用程序中自定义它们。

第 31 章　模型约定与操作约束

本章描述两个用于自定义 MVC 工作方式的有用特性：模型约定允许替换用于创建控制器和操作的默认约定；操作约束允许指定操作可用于何种类型的请求，当 MVC 选择操作以处理请求时，可提供指导。

表 31-1 列出了本章要介绍的操作。

表 31-1　　本章要介绍的操作

操作	方法	代码清单
自定义应用模型	使用内置的特性之一或者创建自定义的模型约定	代码清单 31-1～代码清单 31-14
在整个应用程序中应用自定义约定	定义全局模型约定	代码清单 31-15 和代码清单 31-16
区分可以处理请求的两种操作方法	使用 action 约定	代码清单 31-17～代码清单 31-25

31.1　准备示例项目

在本章中，使用 ASP.NET Core Web Application（.NET Core）模板创建新的名为 ConventionsAndConstraints 的 Empty 项目。代码清单 31-1 展示的 Startup 类用来设置 MVC 框架和用于开发的中间件。

代码清单 31-1　ConventionsAndConstraints 文件夹下的 Startup.cs 文件的内容

```
using System;
using System.Collections.Generic;
using System.Linq;
using System.Threading.Tasks;
using Microsoft.AspNetCore.Builder;
using Microsoft.AspNetCore.Hosting;
using Microsoft.AspNetCore.Http;
using Microsoft.Extensions.DependencyInjection;

namespace ConventionsAndConstraints {

    public class Startup {

        public void ConfigureServices(IServiceCollection services) {
            services.AddMvc();
        }

        public void Configure(IApplicationBuilder app, IHostingEnvironment env) {
            app.UseStatusCodePages();
            app.UseDeveloperExceptionPage();
            app.UseStaticFiles();
            app.UseMvcWithDefaultRoute();
        }
    }
}
```

创建视图模型、控制器和视图

对于本章的大多数示例，知道哪个方法用于响应请求是有帮助的。为此，创建 Models 文件夹并在其中添加名为 Result.cs 的类文件，用它定义代码清单 31-2 所示的类。这个类将允许本章中的控制器将请求是如何处理的信息传递给视图。

代码清单 31-2 Models 文件夹下的 Result.cs 文件的内容

```
using System.Collections.Generic;

namespace ConventionsAndConstraints.Models {

    public class Result {
        public string Controller { get; set; }
        public string Action { get; set; }
    }
}
```

本章只需要一个控制器和视图。创建 Controllers 文件夹，在其中添加名为 HomeController.cs 的类文件，并用它定义代码清单 31-3 所示的类。

代码清单 31-3 Controllers 文件夹下的 HomeController.cs 文件的内容

```
using ConventionsAndConstraints.Models;
using Microsoft.AspNetCore.Mvc;

namespace ConventionsAndConstraints.Controllers {

    public class HomeController : Controller {

        public IActionResult Index() => View("Result", new Result {
            Controller = nameof(HomeController),
            Action = nameof(Index)
        });

        public IActionResult List() => View("Result", new Result {
            Controller = nameof(HomeController),
            Action = nameof(List)
        });
    }
}
```

Home 控制器中的两个操作方法都渲染名为 Result 的视图，创建 Views/Home 文件夹，并使用代码清单 31-4 所示的标记内容在其中创建 Result.cshtml 视图文件。

代码清单 31-4 Views/Home 文件夹下的 Result.cshtml 文件的内容

```
@model Result
@{ Layout = null; }

<!DOCTYPE html>
<html>
<head>
    <meta name="viewport" content="width=device-width" />
    <link href="/lib/bootstrap/dist/css/bootstrap.min.css" rel="stylesheet" />
    <title>Result</title>
</head>
<body class="m-1 p-1">
    <table class="table table-sm table-bordered">
        <tr><th>Controller:</th><td>@Model.Controller</td></tr>
        <tr><th>Action:</th><td>@Model.Action</td></tr>
    </table>
</body>
</html>
```

Result 视图基于 Bootstrap CSS 包为 HTML 元素设置样式。要将 Bootstrap 包添加到项目中，可以使用 Bower Configuration File 模板创建 bower.json 文件，并将 Bootstrap 包添加到 dependencies 部分，如代码清单 31-5 所示。

代码清单 31-5 添加 Bootstrap 包到 ConventionsAndConstraints 文件夹下的 bower.json 文件中

```
{
  "name": "asp.net",
  "private": true,
  "dependencies": {
    "bootstrap": "4.0.0-alpha.6"
  }
}
```

最后的准备步骤是在 Views 文件夹中创建_ViewImports.cshtml 文件，从而设置内置的标签助手，将之用于 Razor 视图并导入 Models 命名空间，如代码清单 31-6 所示。

代码清单 31-6 Views 文件夹下的_ViewImports.cshtml 文件的内容

```
@using ConventionsAndConstraints.Models
@addTagHelper *, Microsoft.AspNetCore.Mvc.TagHelpers
```

如果运行应用程序，你将看到图 31-1 所示的结果。

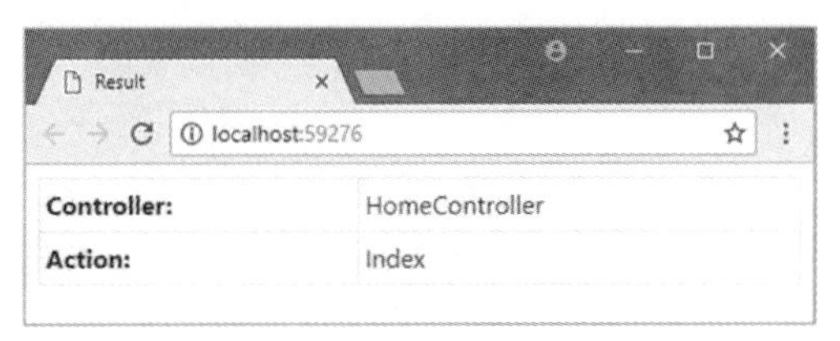

图 31-1 运行示例应用程序的结果

31.2 使用应用程序模型和模型约定

MVC 偏爱约定胜于配置，因此，可以简单地创建名称以 Controller 结尾的类，并开始定义操作方法。在运行时，MVC 使用发现过程来定位应用中所有的控制器和操作方法，并检查它们是否使用了过滤器等特性。

发现过程的最终结果就是应用程序模型，由描述每个找到的控制器类、操作方法和参数的对象组成。MVC 依赖的约定在构建时被应用于应用程序模型。例如，当发现控制器类时，类的名称用于在模型中作为表示控制器的基础；换句话说，HomeController 类用来创建名为 Home 的控制器。当路由系统识别必须由 Home 控制器处理的请求时，应用程序模型提供到 HomeController 类的映射。

应用程序模型可以通过模型约定进行定制，模型约定是用于检查应用程序模型内容并进行调整的类，例如合成新的操作方法或者用于创建控制器的方式等。本节将说明应用程序模型的结构，介绍不同类型的模型约定，并演示模型约定的使用方式。表 31-2 提供了应用程序模型和模型约定的上下文。

表 31-2 应用程序模型和模型约定的上下文

问题	答案
它们是什么？	应用程序模型是对应用程序中发现的控制器和操作方法的完整描述。模型约定允许将自定义的更改应用于应用程序模型
它们为什么有用？	模型约定有用是因为它们允许更改将类和方法映射到控制器和操作方法的方式。可以执行其他的定制化操作，例如限制操作可以接收的 HTTP 方法，或者应用操作约束等（本章稍后介绍）

续表

问题	答案
如何使用它们?	模型约定使用一系列接口来定义，后面将进行说明，可作为特性使用或在 Startup 类中进行配置
是否有任何陷阱或限制?	在模型约定的应用方式上有一些怪异
有其他备选方案吗?	没有，如果默认的应用程序模型不满足需要，可以通过引入自定义组件来创建自定义的应用程序模型

31.2.1 理解应用程序模型

在发现过程中，MVC 创建 ApplicationModel 类的实例，并使用找到的控制器和操作方法详情进行填充。当发现过程完成之后，应用模型约定以执行指定的任何自定义更新。理解应用程序模型的起点是查看由 Microsoft.AspNetCore.Mvc.ApplicationModels.ApplicationModel 类定义的属性，如表 31-3 所示。

表 31-3 ApplicationModel 类定义的属性

属性	描述
Controllers	返回应用程序中包含的所有控制器的 IList<ControllerModel>
Filters	返回应用程序中包含的全局过滤器的 IList<IFilterMetadata>

注 意

刚开始看起来有些枯燥，特别是当想深入细节时，本节介绍的类完全描述了 MVC 应用程序的核心部分，值得花点时间来体会。理解应用程序模型的工作原理将有助于你理解更多的高级功能是如何在幕后工作的，当在自己的项目中遇到未预期的问题时，这更有助于诊断问题。

Controllers 属性会返回一个包含应用程序中发现的对应每个控制器的 ControllerModel 对象列表。表 31-4 介绍了 ControllerModel 类定义的属性。

表 31-4 ControllerModel 类定义的属性

属性	描述
ControllerName	这个 string 类型的属性定义了控制器的名称，以匹配 controller 路由片段中的名称
ControllerType	这个 TypeInfo 类型的属性定义了控制器的类型
ControllerProperties	返回由控制器定义的描述所有属性的 IList<PropertyModel>列表
Actions	返回由控制器定义的描述所有操作方法的 IList<ActionModel>列表
Filters	返回控制器中应用于所有操作方法的过滤器的 IList<IFilterMetadata>列表
RouteConstraints	返回由控制器定义的如何路由目标操作方法的路由约束的 IList<IRouteConstraintProvider>列表
Selectors	返回包含操作方法约束详情的 IList<SelectorModel>列表，以及通过特性应用于控制器的路由信息

由此可以看到 MVC 的一些核心功能是如何被应用程序模型捕获的。例如，ControllerName 属性用于设置路由系统用来匹配 URL 的名称，而 ControllerType 属性用于设置控制器名称关联的类。

ControllerProperties 属性返回一个 PropertyModel 对象列表，其中的每个 PropertyModel 对象描述控制器定义的一个属性。表 31-5 介绍了 PropertyModel 类定义的属性。

表 31-5 PropertyModel 类定义的属性

属性	描述
PropertyName	返回属性的名称
Attributes	返回应用于属性的特性列表

Actions 属性返回一个 ActionModel 对象列表，其中的每个 ActionModel 对象描述由单个控制器类定义的一个操作方法。表 31-6 描述了 ActionModel 类定义的属性。

表 31-6 ActionModel 类定义的属性

属性	描述
ActionName	这个 string 类型的属性定义了操作的名称，用于匹配 action 路由片段
ActionMethod	这个 MethodInfo 类型的属性用来指定实现操作的方法
Controller	返回用于描述操作所属控制器的 ControllerModel 对象
Filters	返回应用于操作的过滤器列表 IList<IFilterMetadata>
Parameters	返回操作的方法参数描述列表 IList<PropertyModel>
RouteConstraints	返回用来限制如何路由到操作的列表 IList<IRouteConstraintProvider>
Selectors	返回描述操作约束的列表 IList<SelectorModel>，以及通过特性作用于控制器的路由信息

终极的详细信息可通过 Parameters 属性访问，该属性返回描述操作方法定义的每个参数的 ParameterModel 对象列表。表 31-7 描述了 ParameterModel 类定义的属性。

表 31-7 ParameterModel 类定义的属性

属性	描述
ParameterName	这个 string 类型的属性表示参数名称
ParameterInfo	表示指定的参数信息
BindingInfo	用于配置模型绑定过程

ApplicationModel、ControllerModel、PropertyModel、ActionModel 和 ParameterModel 用于描述应用程序中控制器类的各个方面，比如方法、属性以及每个方法定义的参数。

自定义应用程序模型

MVC 有一些内置的约定，可使用 ControllerModel、PropertyModel、ActionModel 和 ParameterModel 对象来填充 ApplicationModel，以描述发现的控制器。

一些约定是显式的，例如将 Controller 从控制器的类名中移除，用来设置 ControllerModel 对象的 ControllerName 属性。这个约定意味着定义名为 HomeController 的类，但在 URL 片段中使用 Home 来定位。

其他约定是隐式的，例如类用来创建控制器，方法用来创建操作。大多数 MVC 开发人员将这些约定视为理所当然，但是实际上，应用程序模型的各个方面都可以改变。

前面的章节描述了用来改变 MVC 工作方式的特性，它们实际上是模型约定。表 31-8 介绍了这些特性。

表 31-8 用来改变默认应用程序约定的一些基本特性

特性	描述
ActionName	允许显式指定 ActionModel 的 ActionName 属性，而不是从方法名中派生
NonController	防止类被用于创建 ControllerModel 对象
NonAction	防止方法被用于创建 ActionModel 对象

代码清单 31-7 使用 ActionName 特性修改了创建于 HomeController 类中的 List 方法的操作名称。

代码清单 31-7 在 Controllers 文件夹下的 HomeController.cs 中定制应用程序模型

```
using ConventionsAndConstraints.Models;
using Microsoft.AspNetCore.Mvc;

namespace ConventionsAndConstraints.Controllers {
```

```
    public class HomeController : Controller {

        public IActionResult Index() => View("Result", new Result {
            Controller = nameof(HomeController),
            Action = nameof(Index)
        });

        [ActionName("Details")]
        public IActionResult List() => View("Result", new Result {
            Controller = nameof(HomeController),
            Action = nameof(List)
        });
    }
}
```

名称 Details 将被用于创建操作，并替换默认的名称 List。可以通过启动应用程序并访问/Home/Details 来查看效果，如图 31-2 所示，请求被 List 方法处理。

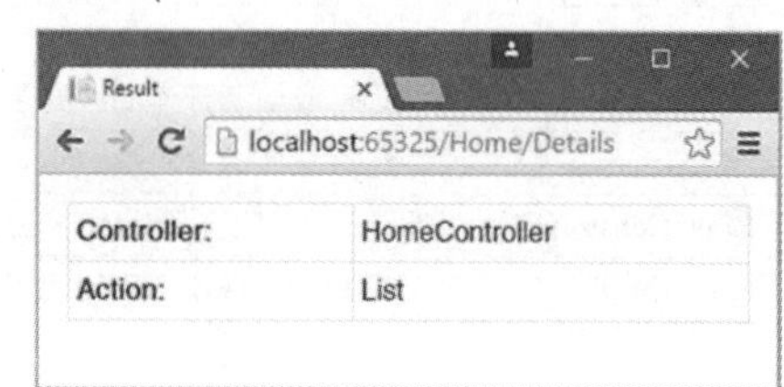

图 31-2 定制应用程序模型

31.2.2 理解模型约定角色

表 31-8 描述的特性允许对应用程序模型进行基本的变更，但是范围有限。对于更实质的定制，需要使用模型约定。

来自表 31-8 的特性允许在应用程序模型对象创建之前指定变更，例如覆盖操作的名称。相比之下，模型约定允许在模型对象创建之后更新应用程序模型，从而允许进行更广泛的更新。有 4 种类型的模型约定，每种通过不同的接口进行定义，如表 31-9 所示。

表 31-9 应用程序模型约定接口

接口	描述
IApplicationModelConvention	应用模型约定到 ApplicationModel 对象
IControllerModelConvention	应用模型约定到应用程序模型的 ControllerModel 对象
IActionModelConvention	应用模型约定到应用程序模型的 ActionModel 对象
IParameterModelConvention	应用模型约定到应用程序模型的 ParameterModel 对象

这 4 种接口都以相同的方式工作，只是它们修改的应用程序模型层面不同。例如，下面是 IControllerModel-Convention 接口的定义：

```
namespace Microsoft.AspNetCore.Mvc.ApplicationModels {

    public interface IControllerModelConvention {

        void Apply(ControllerModel controller);
    }
}
```

通过调用 Apply 方法可提供对应用程序模型约定的目标修改 ControllerModel 的机会，作为方法的参数接收。其他接口也定义了 Apply 方法，并且每个都接收对应类型的模型对象，比如 IActionModelConvention 接口接收 ActionModel 对象，而 IParameterModelConvention 接口接收 ParameterModel 对象。

31.2.3 创建模型约定

控制器、操作和参数模型约定可以作为特性应用，这样可以很容易设置应用更改的范围。作为演示，创建 Infrastructure 文件夹并添加名为 ActionNamePrefixAttribute.cs 的类文件，用它定义代码清单 31-8 所示的类。

代码清单 31-8 Infrastructure 文件夹下的 ActionNamePrefixAttribute.cs 文件的内容

```
using System;
using Microsoft.AspNetCore.Mvc.ApplicationModels;
```

```
namespace ConventionsAndConstraints.Infrastructure {

    [AttributeUsage(AttributeTargets.Method, AllowMultiple = false)]
    public class ActionNamePrefixAttribute : Attribute, IActionModelConvention {
        private string namePrefix;

        public ActionNamePrefixAttribute(string prefix) {
            namePrefix = prefix;
        }

        public void Apply(ActionModel action) {
            action.ActionName = namePrefix + action.ActionName;
        }
    }
}
```

ActionNamePrefixAttribute 类派生自 Attribute 类并实现了 IActionModelConvention 接口。ActionNamePrefix-Attribute 类的构造函数接受用作前缀的字符串，该字符串用于修改 Apply 方法接收的 ActionModel 对象的 ActionName 属性。

提 示

这里限制了 ActionNamePrefix 特性的使用，它只能应用于方法。当以特性的方式应用模型约定时，控制器约定只有当作用于类时才能生效，操作约定只有当作用于方法时才能生效，参数约定只有当作用于参数才能生效。在错误的级别应用约定会被简单地忽略而不产生任何错误。为避免混淆，可使用 AttributeUsage 来限制创建的特性。

代码清单 31-9 将模型约定应用到了 Home 控制器的操作方法上。

代码清单 31-9 在 Controllers 文件夹下的 HomeController.cs 文件中应用模型约定

```
using ConventionsAndConstraints.Models;
using Microsoft.AspNetCore.Mvc;
using ConventionsAndConstraints.Infrastructure;

namespace ConventionsAndConstraints.Controllers {

    public class HomeController : Controller {

        public IActionResult Index() => View("Result", new Result {
            Controller = nameof(HomeController),
            Action = nameof(Index)
        });

        [ActionNamePrefix("Do")]
        public IActionResult List() => View("Result", new Result {
            Controller = nameof(HomeController),
            Action = nameof(List)
        });
    }
}
```

当 MVC 执行发现过程的时候，将创建用于描述 List 方法的 ActionModel 对象，检测 ActionNamePrefix 并调用 Apply 方法。可以通过运行应用程序并访问/Home/DoList 来查看效果，在默认约定下，定位到 List 方法的 URL 已经被替换了，如图 31-3 所示。

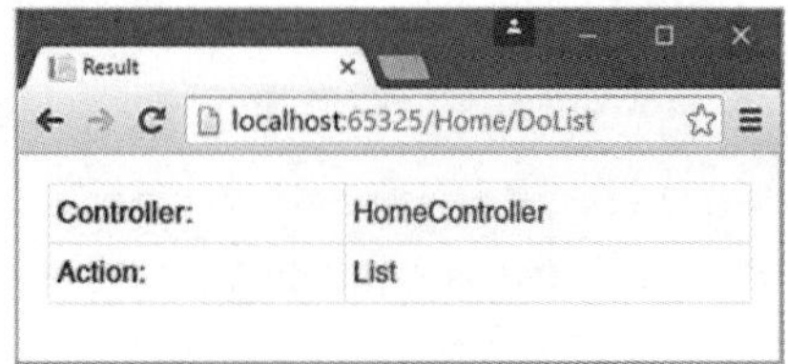

图 31-3 应用模型约定

使用约定添加或删除模型

使用模型约定的一种奇怪方式是阻止在应用程序模型中添加或删除对象。例如，假设要创建一个约定，可以通过两种不同的操作来访问某些方法。为了演示，在 Infrastructure 文件夹中创建名为 AddActionAttribute.cs 的类文件，并用它定义代码清单 31-10 所示的类。

代码清单 31-10 Infrastructure 文件夹下的 AddActionAttribute.cs 文件的内容

```
using System;
using Microsoft.AspNetCore.Mvc.ApplicationModels;

namespace ConventionsAndConstraints.Infrastructure {

    [AttributeUsage(AttributeTargets.Method, AllowMultiple = true)]
    public class AddActionAttribute : Attribute, IActionModelConvention {
        private string additionalName;

        public AddActionAttribute(string name) {
            additionalName = name;
        }

        public void Apply(ActionModel action) {
            action.Controller.Actions.Add(new ActionModel(action) {
                ActionName = additionalName
            });
        }
    }
}
```

这个模型约定使用 ActionModel 构造函数来复制现有对象并修改新对象的 ActionName 属性。可通过 ActionModel.Controller 属性将新的 ActionModel 添加到控制器的操作集合中。在代码清单 31-11 中，可以看到如何将模型约定应用于 Home 控制器。

代码清单 31-11 在 Controllers 文件夹下的 HomeController.cs 文件中应用模型约定

```
using ConventionsAndConstraints.Models;
using Microsoft.AspNetCore.Mvc;
using ConventionsAndConstraints.Infrastructure;

namespace ConventionsAndConstraints.Controllers {

    public class HomeController : Controller {

        public IActionResult Index() => View("Result", new Result {
            Controller = nameof(HomeController),
            Action = nameof(Index)
        });

        [AddAction("Details")]
        public IActionResult List() => View("Result", new Result {
            Controller = nameof(HomeController),
            Action = nameof(List)
        });
    }
}
```

启动应用程序，MVC 将开始发现过程并报告如下错误：

```
InvalidOperationException: Collection was modified; enumeration operation may not execute.
```

当操作集合被发现过程枚举时，模型约定试图修改 action 模型对象的集合，从而导致异常。为避免错误，可采用不同的方法，如代码清单 31-12 所示。

代码清单 31-12 在 Infrastructure 文件夹下的 AddActionAttribute.cs 文件中创建安全的模型约定

```
using System;
using Microsoft.AspNetCore.Mvc.ApplicationModels;
```

```
using System.Linq;

namespace ConventionsAndConstraints.Infrastructure {

    [AttributeUsage(AttributeTargets.Method, AllowMultiple = true)]
    public class AddActionAttribute : Attribute {

        public string AdditionalName { get; }

        public AddActionAttribute(string name) {
            AdditionalName = name;
        }
    }

    [AttributeUsage(AttributeTargets.Class, AllowMultiple = false)]
    public class AdditionalActionsAttribute : Attribute,
            IControllerModelConvention {

        public void Apply(ControllerModel controller) {
            var actions = controller.Actions
                .Select(a => new {
                    Action = a,
                    Names = a.Attributes.Select(attr =>
                        (attr as AddActionAttribute)?.AdditionalName)
                });

            foreach (var item in actions.ToList()) {
                foreach (string name in item.Names) {
                    controller.Actions.Add(new ActionModel(item.Action) {
                        ActionName = name
                    });
                }
            }
        }
    }
}
```

在 action 模型约定中，不能修改控制器相关的操作方法集合。但是，仍然需要以某种方式表达要做的修改。因此，将 AddActionAttribute 作为特性而不是模型约定使用。

你可以在控制器的模型约定中修改操作方法集合，这就是为什么创建 AdditionalActionsAttribute 类。Apply 方法使用 LINQ 来定位使用了 AddActionAttribute 特性的方法并使用指定的名称创建新的 ActionModel 对象。

对于 AdditionalActionsAttribute 类，最重要的部分是在 LINQ 查询结果上调用 ToList 方法：

```
...
foreach (var item in actions.ToList()) {
...
```

ToList 方法强制对 LINQ 查询结果求值并用结果生成一个新的集合。这意味着 foreach 循环将在另外的对象集上枚举。若没有 ToList 调用，就会收到与代码清单 31-12 相同的错误；若使用 ToList 调用，就能创建新的 action 模型对象。代码清单 31-13 展示了如何将修改后的特性应用于 Home 控制器。

代码清单 31-13 在 Controllers 文件夹下的 HomeController.cs 文件中应用修改后的模型约定

```
using ConventionsAndConstraints.Models;
using Microsoft.AspNetCore.Mvc;
using ConventionsAndConstraints.Infrastructure;

namespace ConventionsAndConstraints.Controllers {
```

```
    [AdditionalActions]
    public class HomeController : Controller {

        public IActionResult Index() => View("Result", new Result {
            Controller = nameof(HomeController),
            Action = nameof(Index)
        });

        [AddAction("Details")]
        public IActionResult List() => View("Result", new Result {
            Controller = nameof(HomeController),
            Action = nameof(List)
        });
    }
}
```

可以通过启动应用程序并访问/Home/Details 和/Home/List 来查看修改后的模型约定效果。如图 31-4 所示，模型约定添加了由 List 方法处理的新操作，补充了默认创建的操作模型。

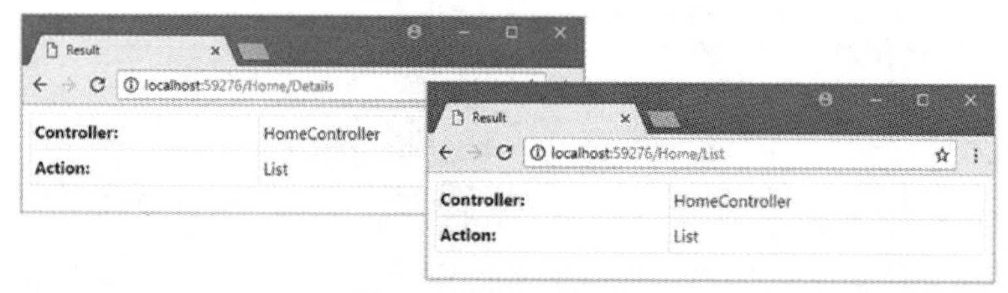

图 31-4 创建新的操作模型

31.2.4 理解模型约定的执行顺序

模型约定将以特定的次序应用，从最大范围开始：首先应用控制器模型约定，然后应用操作模型约定，最后应用参数模型约定。为了演示，将前面几个示例中创建的自定义模型约定应用于 HomeController 类的 List 方法，如代码清单 31-14 所示。

代码清单 31-14 在 Controllers 文件夹下的 HomeController.cs 文件中应用多个模型约定

```
using ConventionsAndConstraints.Models;
using Microsoft.AspNetCore.Mvc;
using ConventionsAndConstraints.Infrastructure;

namespace ConventionsAndConstraints.Controllers {

    [AdditionalActions]
    public class HomeController : Controller {

        public IActionResult Index() => View("Result", new Result {
            Controller = nameof(HomeController),
            Action = nameof(Index)
        });

        [ActionNamePrefix("Do")]
        [AddAction("Details")]
        public IActionResult List() => View("Result", new Result {
            Controller = nameof(HomeController),
            Action = nameof(List)
        });
    }
}
```

首先应用作为控制器模型约定的 AdditionalActions 特性，并创建名为 Details 的操作。然后，应用作为操作模型约定的 ActionNamePrefix，将 Do 前缀应用于操作方法关联的操作。结果导致 List 方法实现了两个操作——DoList 和 DoDetails，它们可以使用 URL 地址/Home/DoList 和/Home/DoDetails 访问，如图 31-5 所示。

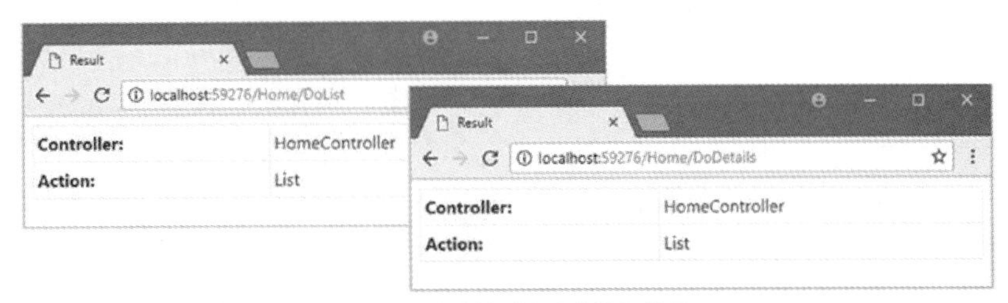

图 31-5 多个模型约定的执行效果

31.2.5 创建全局模型约定

如果需要更改默认模型约定，那么可能必须为应用程序中的每个控制器、操作或参数执行更改操作。如果是这种情况，那么可以创建全局模型约定，而不必将特性一一应用于每个控制器类。全局模型约定在 Startup 类中配置，如代码清单 31-15 所示。

代码清单 31-15 在 ConventionsAndConstraints 文件夹下的 Startup.cs 文件中创建全局过滤器

```
using System;
using System.Collections.Generic;
using System.Linq;
using System.Threading.Tasks;
using Microsoft.AspNetCore.Builder;
using Microsoft.AspNetCore.Hosting;
using Microsoft.AspNetCore.Http;
using Microsoft.Extensions.DependencyInjection;
using ConventionsAndConstraints.Infrastructure;

namespace ConventionsAndConstraints {

    public class Startup {

        public void ConfigureServices(IServiceCollection services) {
            services.AddMvc().AddMvcOptions(options => {
                options.Conventions.Add(new ActionNamePrefixAttribute("Do"));
                options.Conventions.Add(new AdditionalActionsAttribute());
            });
        }

        public void Configure(IApplicationBuilder app, IHostingEnvironment env) {
            app.UseStatusCodePages();
            app.UseDeveloperExceptionPage();
            app.UseStaticFiles();
            app.UseMvcWithDefaultRoute();
        }
    }
}
```

由 AddMvcOptions 扩展方法接收的 MvcOptions 对象定义了 Conventions 属性。这个属性将返回用于添加模型约定对象的列表集合，其中的列表可全局应用两个自定义模型约定。这意味着所有的操作名称将以 Do 为前缀，并将检查所有的 AddAction 特性。由于这些模型约定将被全局应用，因此从 HomeController 类中将它们删除，如代码清单 31-16 所示。

代码清单 31-16　从 Controllers 文件夹下的 HomeController.cs 文件中删除模型约定

```
using ConventionsAndConstraints.Models;
using Microsoft.AspNetCore.Mvc;
using ConventionsAndConstraints.Infrastructure;

namespace ConventionsAndConstraints.Controllers {

    //[AdditionalActions]
    public class HomeController : Controller {

        public IActionResult Index() => View("Result", new Result {
            Controller = nameof(HomeController),
            Action = nameof(Index)
        });

        //[ActionNamePrefix("Do")]
        [AddAction("Details")]
        public IActionResult List() => View("Result", new Result {
            Controller = nameof(HomeController),
            Action = nameof(List)
        });
    }
}
```

在模型约定作用于类之前，应用全局模型约定。如果有多个全局模型约定，那么它们将按照注册的顺序应用，而不考虑类型。在控制器模型约定之前注册操作模型约定，这意味着在将 ActionNamePrefixAttribute 约定应用于所有操作之后，通过 AddAction 特性指定的 Details 操作才被创建。因此，List 方法实现了两个操作——DoList 和 Details，它们可以通过 URL 地址/Home/DoList 和/Home/Details 进行访问，如图 31-6 所示。

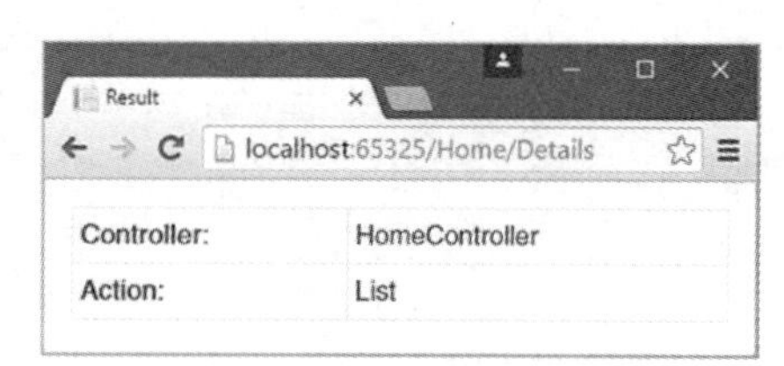

图 31-6　全局模型约定的执行结果

31.3　使用操作约束

操作约束决定了操作方法是否适合用来处理特定请求，这可能导致你认为操作约束类似于第 19 章描述的授权过滤器。

实际上，操作约束更受限。当 MVC 接收到 HTTP 请求时，将通过一个选择过程来识别用于处理 HTTP 请求的操作方法。如果有多个操作方法可以处理请求，那么 MVC 需要使用一些手段来决定使用哪个操作方法，这就是使用操作约束的原因。表 31-10 提供操作约束的背景。

表 31-10　操作约束的背景

问题	答案
它们是什么？	操作约束是 MVC 用来确定 HTTP 请求是否可以由特定操作方法处理的类
它们为什么有用？	如果有两个或更多个操作方法可以处理某个请求，那么 MVC 需要使用一些手段来决定使用哪个操作方法，操作约束用来提供这些信息
如何使用？	操作约束通过特性来使用，这允许它们在整个应用程序中重用，且意味着确定操作是否应该处理请求的逻辑不必在操作方法中定义
有什么陷阱或限制吗？	操作约束可以被广泛应用，并阻止请求被任何适当的操作方法处理，导致向客户端发送无用的 404 - Not Found 响应
有备选方案吗？	如果需要在特定情况下限制对操作的访问，那么过滤器更为有用，可以重定向客户端以便显示有用的错误页面

31.3.1 准备示例项目

操作约束的目的是帮助 MVC 在两个或多个可以用于处理请求的操作方法之间进行选择。代码清单 31-17 在 Home 控制器中添加了另一个新的操作方法。

代码清单 31-17 在 Controllers 文件夹下的 HomeController.cs 文件中创建两个合适的操作方法

```
using ConventionsAndConstraints.Models;
using Microsoft.AspNetCore.Mvc;
using ConventionsAndConstraints.Infrastructure;

namespace ConventionsAndConstraints.Controllers {

    //[AdditionalActions]
    public class HomeController : Controller {

        public IActionResult Index() => View("Result", new Result {
            Controller = nameof(HomeController),
            Action = nameof(Index)
        });

        [ActionName("Index")]
        public IActionResult Other() => View("Result", new Result {
            Controller = nameof(HomeController),
            Action = nameof(Other)
        });

        //[ActionNamePrefix("Do")]
        [AddAction("Details")]
        public IActionResult List() => View("Result", new Result {
            Controller = nameof(HomeController),
            Action = nameof(List)
        });
    }
}
```

以上代码添加了一个新的名为 Other 的操作方法并应用 ActionName 特性以便处理为名为 Index 的操作。更新 Startup 类以删除本章前面的全局模型约定，如代码清单 31-18 所示。

代码清单 31-18 在 ConventionsAndConstraints 文件夹下的 Startup.cs 文件中删除全局模型约定

```
using System;
using System.Collections.Generic;
using System.Linq;
using System.Threading.Tasks;
using Microsoft.AspNetCore.Builder;
using Microsoft.AspNetCore.Hosting;
using Microsoft.AspNetCore.Http;
using Microsoft.Extensions.DependencyInjection;
using ConventionsAndConstraints.Infrastructure;

namespace ConventionsAndConstraints {

    public class Startup {
        public void ConfigureServices(IServiceCollection services) {
            services.AddMvc().AddMvcOptions(options => {
                //options.Conventions.Add(new ActionNamePrefixAttribute("Do"));
                //options.Conventions.Add(new AdditionalActionsAttribute());
            });
        }
```

```
        public void Configure(IApplicationBuilder app, IHostingEnvironment env) {
            app.UseStatusCodePages();
            app.UseDeveloperExceptionPage();
            app.UseStaticFiles();
            app.UseMvcWithDefaultRoute();
        }
    }
}
```

这意味着 Home 控制器中存在两个名为 Index 的操作，如果运行应用，将看到图 31-7 所示的错误消息，指出 MVC 不知道应用使用哪个操作方法。

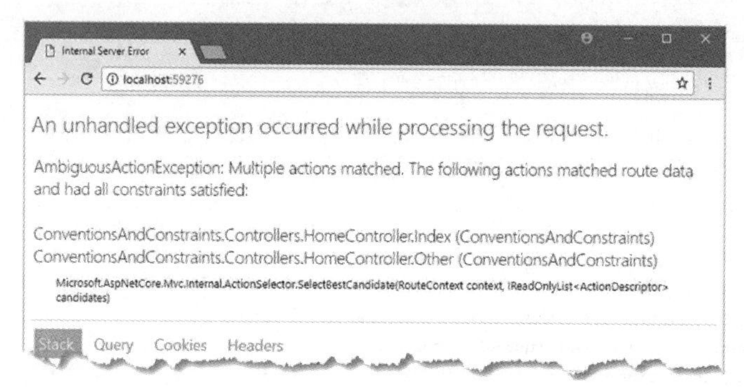

图 31-7 错误消息

以上是错误信息的相关内容：

```
AmbiguousActionException: Multiple actions matched. The following actions matched route
data and had all constraints satisfied:
ConventionsAndConstraints.Controllers.HomeController.Index
ConventionsAndConstraints.Controllers.HomeController.Other
```

31.3.2 操作约束的作用

操作约束用来告诉 MVC 是否可以使用操作方法处理请求和实现 IActionConstraint 接口，这个接口的定义如下：

```
namespace Microsoft.AspNetCore.Mvc.ActionConstraints {

    public interface IActionConstraint : IActionConstraintMetadata {

        int Order { get; }

        bool Accept(ActionConstraintContext context);
    }
}
```

MVC 在经过选择操作方法来处理请求的过程后，将检查是否存在与之关联的约束。如果有，它们将按照 Order 属性的值依次组织，并且依次调用每个约束的 Accept 方法。只要任何约束的 Accept 方法返回 false，MVC 就知道操作方法不能用于处理当前请求。

提 示

IActionConstraint 接口派生自 IActionConstraintMetadata，后者是没有定义成员的接口，并不直接使用。当创建自定义约束时，应当使用 IActionConstraint 接口；当期望创建有依赖要处理的约束时，应当使用 IActionConstraintFactory 接口。

为了有助于操作约束做出决定，MVC 提供了 ActionConstraintContext 实例作为上下文数据，ActionConstraintContext 类定义的属性如表 31-11 所示。

表 31-11 ActionConstraintContext 类定义的属性

属性	描述
Candidates	返回 ActionSelectorCandidate 对象列表，描述 MVC 用于处理当前请求的候选操作方法列表
CurrentCandidate	返回 ActionSelectorCandidate 对象，描述约束当前正在请求评估的操作方法
RouteContext	返回 RouteContext 对象，提供路由数据（通过 RouteData 属性）以及 HTTP 请求（通过 HttpContext 属性）

31.3.3 创建操作约束

最常见的操作约束是确认请求符合某些策略，比如提供特定的 HTTP 请求头。为了演示此种类型的操作约束，在示例项目的 Infrastructure 文件夹中添加名为 UserAgentAttribute.cs 的类文件，用它定义代码清单 31-19 所示的类。

代码清单 31-19 Infrastructure 文件夹夹下的 UserAgentAttribute.cs 文件的内容

```
using System;
using System.Linq;
using Microsoft.AspNetCore.Mvc.ActionConstraints;

namespace ConventionsAndConstraints.Infrastructure {

    public class UserAgentAttribute : Attribute, IActionConstraint {
        private string substring;

        public UserAgentAttribute(string sub) {
            substring = sub.ToLower();
        }

        public int Order { get; set; } = 0;

        public bool Accept(ActionConstraintContext context) {
            return context.RouteContext.HttpContext
                .Request.Headers["User-Agent"]
                .Any(h => h.ToLower().Contains(substring));
        }
    }
}
```

以上操作约束用于在 UserAgent 请求头不包含特定字符串的时候，阻止匹配的操作方法处理请求。在 Accept 方法中，从 HttpContext 对象中获取请求头，然后使用 LINQ 来检查其中之一是否包含通过构造函数接收的子串。

注 意

在实际的应用程序中不要基于 User-Agent 头来识别浏览器，因为请求头的值经常会误导人。例如，在编写本书时，Microsoft Edge 浏览器发送的 User-Agent 请求头中包含了 Android、Apple、Chrome 和 Safari，使得很容易误认为其他的浏览器。更稳健的方式是在应用中使用 JavaScript 库（比如 Modernizr）来检测浏览器。

代码清单 31-20 为 HomeController 类中的一个方法应用了操作约束。

代码清单 31-20 在 Controllers 文件夹下的 HomeController.cs 文件中应用操作约束

```
using ConventionsAndConstraints.Models;
using Microsoft.AspNetCore.Mvc;
```

```
using ConventionsAndConstraints.Infrastructure;

namespace ConventionsAndConstraints.Controllers {

    //[AdditionalActions]
    public class HomeController : Controller {

        public IActionResult Index() => View("Result", new Result {
            Controller = nameof(HomeController),
            Action = nameof(Index)
        });

        [ActionName("Index")]
        [UserAgent("Edge")]
        public IActionResult Other() => View("Result", new Result {
            Controller = nameof(HomeController),
            Action = nameof(Other)
        });

        //[ActionNamePrefix("Do")]
        [AddAction("Details")]
        public IActionResult List() => View("Result", new Result {
            Controller = nameof(HomeController),
            Action = nameof(List)
        });
    }
}
```

为Other操作方法应用这个特性，并指定如果请求的User-Agent请求头中没有包含Edge，就不接收请求。

启动应用程序，分别使用Google Chrome和Microsoft Edge浏览器访问/Home/Index，你将看到不同的处理结果，如图31-8所示。

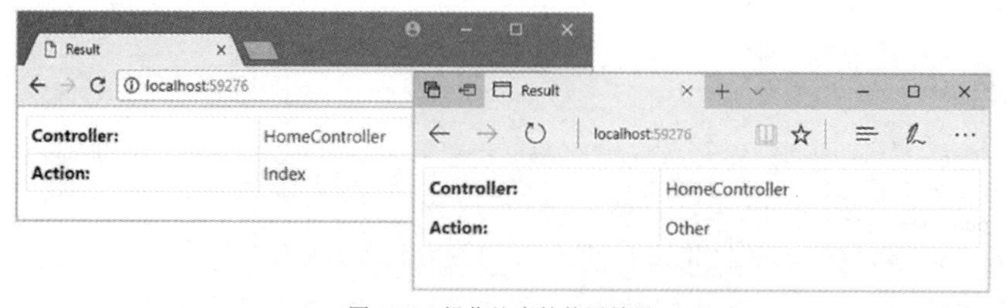

图31-8 操作约束的使用效果

> **理解操作方法的约束效果**
>
> 前面的示例揭示了使用操作约束的如下方面：对于给定的请求，带有Accept方法且返回true的应用了操作约束的操作方法优先于没有应用操作约束的操作方法。
>
> Home控制器中有两个名为Index的操作，分别由Index和Other操作方法创建，它们都可以处理带有请求头User-Agent且值为Edge的请求。Other操作方法用来处理来自Edge浏览器的请求这是由于为之应用了操作约束，并且Accept方法返回true。理念就是，带接收请求的操作约束的操作方法总是优先于没有操作约束的操作方法。

创建比较用的操作约束

通过ActionConstraintContext对象的Candidates和CurrentCandidate属性，操作约束提供了处理请求的其他候选操作方法的详情。每个潜在的匹配可使用ActionSelectorCandidate实例来描述，用到的属性如表31-12所示。

表 31-12 ActionSelectorCandidate 类定义的属性

属性	描述
Action	返回描述候选操作方法的 ActionDescriptor 对象
Constraints	返回包含应用到候选操作方法的 IActionConstraint 约束列表

ActionDescriptor 类用于通过表 31-13 所示的属性描述操作方法，与其他上下文对象提供的内容类似。

表 31-13 ActionDescriptor 类定义的属性

属性	描述
Name	返回操作方法的名称
RouteConstraints	返回用来限制如何路由到操作方法的 IList<IRouteConstraintProvider>列表
Parameters	返回操作方法参数描述的 IList<PropertyModel>列表
ActionConstraints	返回操作方法的约束列表 IList<IActionConstraintMetadata>

操作约束可以检查候选的操作方法并洞察如何以及在哪里应用，还可以微调它们如何工作。作为示例，考虑代码清单 31-21 中应用到 Home 控制器的操作约束。

代码清单 31-21 在 Controllers 文件夹下的 HomeController.cs 文件中为控制器应用操作约束

```
using ConventionsAndConstraints.Models;
using Microsoft.AspNetCore.Mvc;
using ConventionsAndConstraints.Infrastructure;

namespace ConventionsAndConstraints.Controllers {

    public class HomeController : Controller {

        public IActionResult Index() => View("Result", new Result {
            Controller = nameof(HomeController),
            Action = nameof(Index)
        });

        [ActionName("Index")]
        [UserAgent("Edge")]
        public IActionResult Other() => View("Result", new Result {
            Controller = nameof(HomeController),
            Action = nameof(Other)
        });

        [UserAgent("Edge")]
        public IActionResult List() => View("Result", new Result {
            Controller = nameof(HomeController),
            Action = nameof(List)
        });
    }
}
```

这里仅为名为 List 的操作方法应用了操作约束，这意味着仅在请求头 User-Agent 中包含 Edge 的请求可使用这个操作方法处理。例如，如果使用 Google Chrome 发出请求，将收到 404 - Not Found 响应。

这没什么用，因为用户不知道为什么收到错误，也没有说明性的文本来建议使用其他的浏览器替代 Google Chrome。在控制处理请求的操作方法的选择时，操作约束很有用；使用过滤器将允许把客户重定向到有错误说明的页面，这才是更实用的响应。

为了处理这个问题，更新 UserAgentAttribute 类，以便在只有一个候选操作方法的时候不拒绝请求，如代码清单 31-22 所示。

代码清单 31-22 在 Infrastructure 文件夹下的 UserAgentAttribute.cs 文件中检查其他候选操作方法

```
using System;
using System.Linq;
using Microsoft.AspNetCore.Mvc.ActionConstraints;
namespace ConventionsAndConstraints.Infrastructure {

    public class UserAgentAttribute : Attribute, IActionConstraint {
        private string substring;

        public UserAgentAttribute(string sub) {
            substring = sub.ToLower();
        }

        public int Order { get; set; } = 0;

        public bool Accept(ActionConstraintContext context) {
            return context.RouteContext.HttpContext
                    .Request.Headers["User-Agent"]
                    .Any(h => h.ToLower().Contains(substring))
                || context.Candidates.Count() == 1;
        }
    }
}
```

附加的 LINQ 查询将检查 CurrentCandidate 返回的候选操作方法是否为 Candidates 属性返回的集合中的唯一元素。如果是，那么操作约束就知道 MVC 没有其他的候选操作方法而允许请求。使用 Google Chrome 浏览器运行应用程序并请求 URL 地址/Home/List 来查看效果。即使通过 Google Chrome 发送请求的 User-Agent 请求头中没有包含 Edge，约束类也将因为检测到没有其他的候选操作方法而允许请求被处理。

31.3.4 在操作约束中处理依赖

当需要通过 service provider 在操作约束中处理依赖的时候，使用 IActionConstraintFactory 接口。如第 18 章所示，下面是这个接口的定义：

```
using System;

namespace Microsoft.AspNetCore.Mvc.ActionConstraints {

    public interface IActionConstraintFactory : IActionConstraintMetadata {

        IActionConstraint CreateInstance(IServiceProvider services);

        bool IsReusable { get; }
    }
}
```

CreateInstance 方法将被调用以创建操作约束，IsReusable 属性用来指示通过 CreateInstance 方法创建的对象是否可以用于多个请求。

为了演示这个接口的使用，要用到依赖关系。为此，在 Infrastructure 文件夹中添加名为 UserAgentComparer.cs 的类文件，并用它定义代码清单 31-23 所示的类。

代码清单 31-23 Infrastructure 文件夹下的 UserAgentComparer.cs 文件的内容

```
using System.Linq;
using Microsoft.AspNetCore.Http;

namespace ConventionsAndConstraints.Infrastructure {

    public class UserAgentComparer {
```

```
        public bool ContainsString(HttpRequest request, string agent) {
            string searchTerm = agent.ToLower();
            return request.Headers["User-Agent"]
                .Any(h => h.ToLower().Contains(searchTerm));
        }
    }
}
```

UserAgentComparer 类定义了一个用于在 User-Agent 请求头中查询字符串的方法。功能与前面的示例相同，但是打包成单独的类，以便可以使用 service provider 来管理生命周期，如代码清单 31-24 所示。

代码清单 31-24 在 ConventionsAndConstraints 文件夹下的 Startup.cs 文件中为 service provider 注册类型

```
using System;
using System.Collections.Generic;
using System.Linq;
using System.Threading.Tasks;
using Microsoft.AspNetCore.Builder;
using Microsoft.AspNetCore.Hosting;
using Microsoft.AspNetCore.Http;
using Microsoft.Extensions.DependencyInjection;
using ConventionsAndConstraints.Infrastructure;

namespace ConventionsAndConstraints {

    public class Startup {

        public void ConfigureServices(IServiceCollection services) {
            services.AddSingleton<UserAgentComparer>();
            services.AddMvc().AddMvcOptions(options => {
                //options.Conventions.Add(new ActionNamePrefixAttribute("Do"));
                //options.Conventions.Add(new AdditionalActionsAttribute());
            });
        }

        public void Configure(IApplicationBuilder app, IHostingEnvironment env) {
            app.UseStatusCodePages();
            app.UseDeveloperExceptionPage();
            app.UseStaticFiles();
            app.UseMvcWithDefaultRoute();
        }
    }
}
```

以上代码选择了单例生命周期，这意味着 UserAgentComparer 类只有单个实例。代码清单 31-25 更新了 UserAgent 约束以便将检查请求头的任务委托给通过 service provider 获取的 UserAgentComparer 对象。

代码清单 31-25 在 Infrastructure 文件夹下的 UserAgentAttribute.cs 文件中解决依赖问题

```
using System;
using System.Linq;
using Microsoft.AspNetCore.Mvc.ActionConstraints;
using Microsoft.Extensions.DependencyInjection;

namespace ConventionsAndConstraints.Infrastructure {

    public class UserAgentAttribute : Attribute, IActionConstraintFactory {
        private string substring;

        public UserAgentAttribute(string sub) {
            substring = sub;
```

```
        }

        public IActionConstraint CreateInstance(IServiceProvider services) {
            return new UserAgentConstraint(services.GetService<UserAgentComparer>(),
                substring);
        }

        public bool IsReusable => false;

        private class UserAgentConstraint : IActionConstraint {
            private UserAgentComparer comparer;
            private string substring;

            public UserAgentConstraint(UserAgentComparer comp, string sub) {
                comparer = comp;
                substring = sub.ToLower();
            }

            public int Order { get; set; } = 0;

            public bool Accept(ActionConstraintContext context) {
                return comparer.ContainsString(context.RouteContext
                        .HttpContext.Request, substring)
                    || context.Candidates.Count() == 1;
            }
        }
    }
}
```

在以上模型中，应用于操作方法的特性负责在调用 CreateInstance 方法时创建约束类的实例。CreateInstance 方法的参数是一个 IserviceProvider 对象，本例使用了 UserAgentComparer，以便可以创建私有约束类的实例，然后在选择过程中使用。

避免范围陷阱

与其他基于特性的功能一样，将约束特性应用于控制器类等同于将约束特性应用于每个独立的方法。然而，这通常会导致未预期的结果，因为操作约束的目的是帮助 MVC 选择操作方法，而不是对控制器中所有的操作方法应用相同的约束。

例如，如果将 UserAgent 特性应用于 HomeController 类，那么任何浏览器都将无法访问 Index 这个操作方法。两个 Index 操作方法都同样适用于处理 User-Agent 字符串中包含 Edge 的浏览器，这将导致异常。对于所有其他的浏览器，这两个 Index 操作方法都不合适，这将导致 404-Not Found 响应。

使用约束中的上下文对象可以查找其他的约束，并查看它们是否可能拒绝请求，但这会导致每个约束的 Accept 方法被调用多次。当有多个操作方法可以处理相同的请求时，可将约束应用于这些操作方法，此时最有效。

31.4 小结

本章描述了两个用于定制 MVC 运行方式的特性，解释了如何使用模型约束来改变将类和方法映射到控制器和操作的方式，还讲述了如何使用操作约束来限制操作方法可能处理的请求的范围，以及如何使用它们在请求到达时从识别的候选列表中选择一个操作方法。

本书从创建一个简单的应用程序开始，带你全面了解了 ASP.NET Core MVC 框架中不同的组件，理解如何配置、定制或完全替换它们。

最后，祝你在 MVC 项目中取得圆满成功！